Geomorphological Techniques

Geomorphological Techniques

SECOND EDITION

Edited for the
British Geomorphological Research Group by

Andrew Goudie
University of Oxford

with the assistance of

John Lewin
Keith Richards
Malcolm Anderson
Tim Burt
Brian Whalley
Peter Worsley

London and New York

First published 1990
by Unwin Hyman

Reprinted 1994 by Routledge
11 New Fetter Lane, London EC4P 4EE
29 West 35th Street, New York, NY 10001

Typeset in Times by Computape (Pickering) N. Yorkshire
Printed and bound in Great Britain by Cambridge University Press

British Library Cataloguing in Publication Data
A catalogue record for this book is available from the British Library

Library of Congress Cataloging in Publication Data
A catalog record for this book is available from the Library of Congress

ISBN 0-415-11939-1

Preface to the first edition

This book is an attempt to provide geomorphologists and others in the Earth and environmental sciences with a manual of useful techniques. The constraints of length have necessitated the omission from its contents of such matters as statistical analysis, but fortunately we are already well served in this field. Likewise, we have not been able to include all the material that we should have liked on the theme of laboratory simulation and hardware modelling. Throughout the development of this volume we have been forced to recognise that geomorphology grades imperceptibility into related disciplines such as engineering, hydrology, geology, geography and soil science, all of which have an array of techniques with which geomorphologists may have to be familiar. Indeed, even relatively limited topics (such as grain size analysis or scanning electron microscopy) have their own handbooks of techniques. We have thus had to be extremely selective and extremely concise. There are, however, large numbers of techniques employed regularly by geomorphologists which are not covered adequately in other handbooks produced by other disciplines, and this volume aims to concentrate on them. In particular, we have concentrated on methods of obtaining data in the field and on methods of obtaining, through laboratory analysis, data from materials obtained in the field. We have also attempted to present some critical evaluation of the relative merits of the large number of techniques that are often available to deal with a particular problem. In many cases, we have not been able to provide a complete recipe to enable a particular technique to be employed, but at worst we hope that we will have pointed the reader to a source where such a recipe can be obtained. This book should also be used in conjunction with its companion volume (*Geomorphological field manual*, by Gardiner & Dackombe, George Allen & Unwin, 1983).

Geomorphologists are essentially concerned with form (hence Part Two), with the materials upon which landforms occur (hence Part Three), with the processes that mould the materials (hence Part Four), and with the evolution and history of the landscape (hence Part Five). This is the logic behind the structure of the volume and it arose at a meeting of various members of the British Geomorphological Research Group in July 1976, attended by R.U. Cooke, K.J. Gregory, A.S. Goudie, D.Q. Bowen, D. Brunsden, J.C. Doornkamp, D. Walling and M.J. Kirkby.

My main debt lies with my co-editors, without whose efficiency and expertise a volume of such complexity and breadth could not have been produced.

In addition, I must express my gratitude to the draughtsmen of Oxford Illustrators, to all the contributors who have met their stiff deadlines with some fortitude, and to the staff of the School of Geography at Oxford University, who have helped me in a multitude of ways.

I am grateful to the following for making constructive comments on some of the material included in this volume, or for offering assistance of various types:

Professor R. U. Cooke
Professor V. R. Baker
Professor J. O. Mattsson
Dr M. J. Selby
Dr H. J. Walker
Dr K. Hewitt
Dr D. Q. Bowen
The Director, Centre de Géomorphologie, Caen
Dr P. H. T. Beckett
Professor A. Young
Dr D. Walling
Dr G. Butterfield
Mr A. Watson
Dr J. Gerrard
Miss R. Fraser

For section 4.8 Dr A. C. Imeson made stimulating remarks and Dr R. A. van Zuidam made helpful suggestions which were much appreciated by Dr van Zon.

A. S. Goudie
OXFORD

Preface to the second edition

In this second edition I have once again been greatly in the debt of my fellow editors. We have kept to the basic framework of the first edition, but have made substantial changes to portions of the text, and have tried to incorporate some of the new literature which has continued to pour out during the 1980s.

A. S. Goudie
OXFORD

Contents

Acknowledgements

The following individuals and organisations are thanked for permission to reproduce illustrative material (figure numbers in parentheses):

PART ONE
Edward Arnold (1.1): Figure 1.3 reprinted with permission from *The fluvial system* (S. A. Schumm), © 1977 John Wiley & Sons Ltd; Figure 1.9 reprinted with permission from *Timescales in geomorphology* (R. A. Cullingford *et al*., eds), © 1979 John Wiley & Sons Ltd; Geological Society of London (1.11); the Editor, *Journal of Materials* (1.12); Figure 1.14 reprinted with permission from *Hillslope form and process* (M. Carson & M. J. Kirkby), Cambridge University Press; Figure 1.16 reprinted with permission from *American Journal of Science*

PART TWO
British Geomorphological Research Group (2.21a); Figure 2.21b reprinted with permission from *Journal of Geology* (1968) vol. 76, p. 718, © 1968 University of Chicago Press; the Editor, *Australian Geographical Studies* and R. J. Blong (2.21c); British Geomorphological Research Group (2.21d); British Geomorphological Research Group (2.25)

PART THREE
Figure 3.2 reprinted with permission from *Sand and sandstone* (F. J. Pettijohn *et al.*), Springer Verlag

PART FOUR
R. U. Cooke (4.5, 4.6); B. I. Smith (4.7); Butterworth and the Centre de Géomorphologie de Caen 4.8) F. D. Miotke (4.11); Cambridge University Press (4.12); Geologists' Association of London (4.13); The Royal Society (4.14); J. N. Hutchinson (4.15); Institution of Mining and Metallurgy (4.21); Figures 4.51 and 4.53 reproduced from the *Journal of Glaciology* by permission of the International Glaciological Society, W. D. Harrison (4.51) and G. S. Boulton (4.53); P. W. West and Academic Press Inc. (4.62); Pergamon Press (4.63); E. R. Hendrickson and Academic Press Inc. (4.64, 4.65c); Pergamon Press (4.65a); Figure 4.66 reproduced by permission of B. Steen and the publishers from *Saharan dust* (C. Morales, ed.), copyright 1979 John Wiley & Sons Ltd; Figures 4.68–4.70 reprinted from papers by Ph. H. Kuenen in *Journal of Geology* by permission of the University of Chicago Press; C. S. Breed and Academic Press Inc. 4.73).

PART FIVE
The Royal Society (5.1); the Editor, *Geologiska Föreningens i Stockholm Förhandlingar* (5.2); the editor, *Norsk Geografisk Tidsskrift* (5.11); the Editor, *Quaestiones Geographicae* (5.27); Figure 5.28 reproduced from the *Scottish Journal of Geology* with the permission of J. B. Sissons and Scottish Academic Press Ltd; the Editor, *Geologiska*

Föreningens i Stockholm Förhandlingar (5.30); Figure 5.32 reproduced from *Quaternary Research* (1974), vol. 4, p. 203 with the permission of Academic Press Inc. and A. L. Bloom; the Editor *Norges Geologiske Undersökelse* and J. B. Sissons (5.34); the Regents of the University of Colorado (5.35); The Royal Meteorological Society (5.36); Figure 5.37 reproduced from *Palaeogeography, Palaeoclimatology, Palaeoecology* (1971), vol. 9, p. 372 with the permission of Elsevier Scientific Publishing Co.; The Royal Society of Edinburgh (5.38); Institute of British Geographers (5.39); the Regents of the University of Colorado (5.41); Figure 5.42 reprinted by permission from *Nature* vol. 202, pp. 165–6, © 1963 Macmillan Journals Ltd.

Contributors

M. G. Anderson, *Department of Geography, University of Bristol, Bristol, England.*
P. Bull, *School of Geography, University of Oxford, Oxford, England.*
T. P. Burt, *School of Geography, University of Oxford, Oxford, England.*
N. J. Clifford, *Department of Geography, University of Cambridge, Cambridge, England.*
D. Collins, *Department of Geography, University of Manchester, Manchester, England.*
N. J. Cox, *Department of Geography, University of Durham, Durham, England.*
R. Cryer, *Department of Geography, University of Sheffield, Sheffield, England.*
R. E. Dugdale, *Department of Geography, University of Nottingham, Nottingham, England.*
I. S. Evans, *Department of Geography, University of Durham, Durham, England.*
I. Fenwick, *Department of Geography, University of Reading, Reading, England.*
D. C. Ford, *Department of Geography, McMaster University, Hamilton, Ontario, Canada.*
V. Gardiner, *Department of Geography, University of Leicester, Leicester, England.*
A. S. Goudie, *School of Geography, University of Oxford, Oxford, England.*
R. Jarvis, *Department of Geography, State University of New York at Buffalo, New York, USA.*
P. Knott, *formerly Department of Geography, University College, London, England.*
J. Lewin, *Department of Geography, University College of Wales, Aberystwyth, Wales.*
J. McAllister, *Department of Geography, Queen's University, Belfast, Northern Ireland.*
W. S. McGreal, *School of Surveying, Ulster Polytechnic, Newtownabbey, Northern Ireland.*
G. Miller, *Institute of Arctic and Alpine Research, University of Colorado, Boulder, Colorado.*
F. Oldfield, *Department of Geography, University of Liverpool, Liverpool, England.*
J. D. Orford, *Department of Geography, Queen's University, Belfast, Northern Ireland.*
D. Petley, *Department of Civil Engineering, University of Warwick, Coventry, England.*
K. S. Richards, *Department of Geography, University of Cambridge, Cambridge, England.*
N. Richardson, *Department of Geography, University of Liverpool, Liverpool, England.*
H. Roberts, *Coastal Studies, Louisiana State University, Baton Rouge, Louisiana, USA.*
J. Rose, *Department of Geography, Royal Holloway Bedford New College, University of London, Egham, Surrey, England.*
W. C. Rouse, *Department of Geography, University College, Wales, Swansea, Wales.*
H. P. Schwarcz, *Department of Geology, McMaster University, Hamilton, Ontario, Canada.*
S. Shimoda, *Institute of Geoscience, University of Tsukuba, Ibarati, Japan.*
I. Statham, *Ove Arup and Partners, Cardiff, Wales.*
H. van Zon, *Fysisch Geografisch en Bodenkundig Laboratorium, University of Amsterdam, Amsterdam, Netherlands.*
A. Warren, *Department of Geography, University College, London, England.*
W. B. Whalley, *Department of Geography, Queen's University, Belfast, Northern Ireland.*
P. Worsley, *Sedimentology Research Institute, University of Reading, Reading, England.*
E. Yatsu, *Institute of Geoscience, University of Tsukuba, Ibarati, Japan.*

PART ONE
INTRODUCTION

1.1 Methods of geomorphological investigation

1.1.1 Changing scientific method within geomorphology

The main concept of William Morris Davis was to arrange landforms into a cycle of development. His legacy was a concern with historical theory, which lasted until the 1950s (Chorley 1978). Despite claims by some commentators that more recent changes have been essentially incremental (Stoddart 1986), it seems quite clear that there has been a 'revolution' within the science of geomorphology over the last three decades. Following Kuhn (1962), Cohen (1985, p. 41) suggests several criteria by which a revolution may be seen to have occurred: 'these include conceptual changes of a fundamental kind, new postulates or axioms, new forms of acceptable knowledge, and new theories that embrace some or all of these features and others'. Following the research of Horton (1945) and Strahler (1950), geomorphologists have developed a radically different framework for their science into which knowledge about landforms can be put. This recent work centres on the study of contemporary processes, a topic neglected by Davis. Geomorphology has become overtly 'scientific' in that the deductive route to explanation is followed (see below). This has entailed the adoption of quantitative techniques and mathematics within the subject, and has established experiment and observation as the basis of our knowledge. Thus, it is perhaps only natural that there has been much effort devoted to the invention, development and use of measurement techniques over the last 30 years.

The abandonment by most geomorphologists of the Davisian model, and its attendant concern with denudation chronology, in favour of a process-based approach, has carried with it an implicit change in the mode of scientific explanation employed. The aim of scientific explanation is to establish a general statement that covers the behaviour of the objects or events with which the science in question is concerned, thereby enabling connection to be made between separate known events and reliable predictions to be made of events as yet unknown. During the development of modern science, two very different methods have been followed in order to arrive at a satisfactory scientific explanation (Fig. 1.1); a clear distinction exists between these two routes of induction and deduction. In the inductive or 'Baconian' approach, generalisations are obtained through experiment and observation. The final explanation is greatly dependent on the data that become available to the investigator, since these facts cannot be divorced from the theory (i.e. explanation) that is eventually produced. The process of classification is a central mechanism within this approach, since the explanation depends entirely upon the grouping procedure employed. The inductive approach thus rests heavily both on the data available and on the conceptual basis used to classify those data, and, even though a general model may exist within which reality may be viewed, many individual explanations remain a classification of the unique data set rather than a verification of the model itself.

The domination of Anglo-American geomorphology in the earlier part of this century by the concept of the Davisian cycle ensured that the classification procedure was

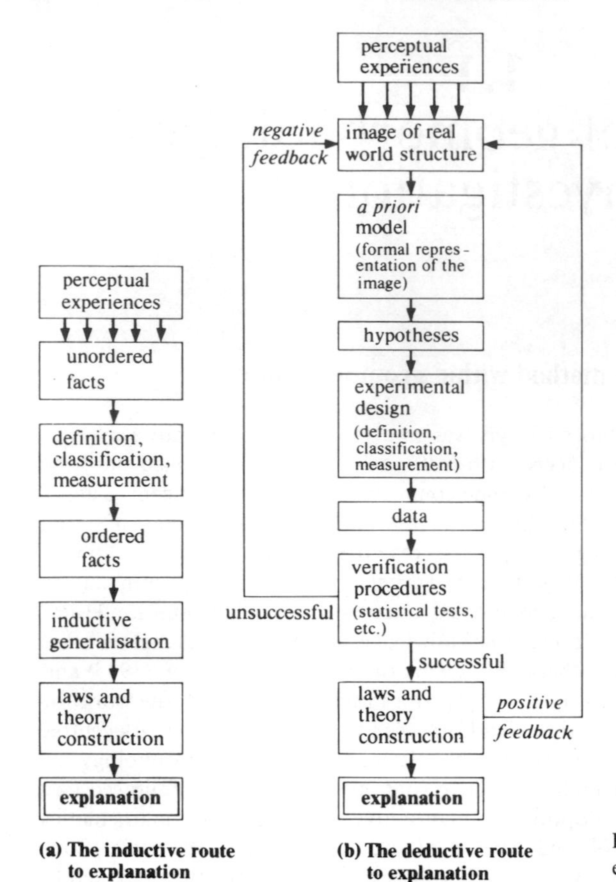

Figure 1.1 The routes to scientific explanation. (After Harvey 1969b).

dominated by considerations of time and the evolutionary stages of landform development. Thus, the division of landscapes into 'youthful', 'mature' and 'old' immediately prejudiced the explanation of the landscape since it assumed that a continuous process of erosion was involved, with each part of the landscape passing through each stage of development. In essence, time itself is seen as a process, indeed as the explanatory variable. Harvey (1969b, p. 424) notes that the assumption of time as a process is reasonable only if the process producing change is known, but frequently within the Davisian cycle the processes operating were not understood in detail. Moreover, the mechanism producing change must act continuously to allow the whole landscape to pass through the necessary sequence of stages. However, it often proved difficult to locate landforms into an 'exact' stage, especially where interruptions in the cycle had occurred. Indeed, the explanations produced by denudation chronologists illustrate only too well that interruptions in the cycle can occur so often that the explanation of a particular suite of landforms becomes very much concerned with a unique sequence of events, rather than with the application of the universal (classificatory) model. The temporal mode of explanation is not invalid methodologically, but it does require a process to act clearly and continuously through time and also demands an exact definition of the individual stages so that any particular situation may be located within them. However, as noted above, these requirements were generally not met by the Davisian approach (although perhaps ironically, as noted below, computer simulation models adopt such assumptions in order to allow retrodiction or prediction of landform evolution). It was gradually realised that what was important was not that change was occurring, but rather the following questions. What were the processes inducing change? At what rate did these processes act? And what effect did this activity have on the form and evolution of the landforms being considered? Thus, the temporal mode of explanation in geomorphology proved a poor mechanism with which to classify the unordered data being used in the inductive scientific model, and, because of the weaknesses inherent in the Davisian cycle, there was no proven unifying theoretical structure within which the individual explanations could be embraced. The trend towards a more detailed consideration of process–response systems was accompanied therefore by the abandonment of both the temporal mode of explanation and the inductive scientific approach with which it was associated. Geomorphology thus moved away from being essentially a classification procedure and adopted a radically different method of investigation based on the use of the deductive scientific method.

The adoption of the deductive scientific method can be seen as both a cause and an effect of recent trends in geomorphology. By the early 1950s, the Davisian cycle had become more than simply a model of reality, and formed a paradigm within which geomorphological problems could be organised. However, as Kuhn (1962) points out, any paradigm that leaves unanswered more than it successfully explains leaves itself open for replacement by a completely new approach. The advent of the new paradigm was heralded by the introduction of quantification into geomorphology and by the adoption of statistical analysis as a means of testing hypotheses (Strahler 1950, Melton 1957, Chorley 1966). Implicit in the quantitative approach is the use of the deductive scientific method (Fig. 1.1b) whereby a theoretical approach to landscape form and development has replaced the elucidation of unique landform sequences which was dominant during the denudation chronology era. This theoretical approach is dependent on the formulation of an idealised view or model of reality. Such models may then be tested, either to confirm that they are indeed an acceptable (if ideal) reflection of the real world, or, if this is not the case, so that they may be revised and improved so as to become one. The testing of a model involves the independent collection of data. Thus, explanation of individual objects or events becomes, under the deductive approach, a more efficient process since general statements are produced to cover all such events, rather than producing a unique explanation based on the objects or events themselves.

The deductive route depends on a clear distinction between the origin and testing of theories: only the latter is based upon observation and logic. Popper (1972) has argued that the purpose of scientific experiments is to attempt to falsify theories: the best established are those which over a long period have withstood a gruelling procedure of testing. By ruling out what is false, a theory approaches the truth: though supported, it is not conclusively 'proven' since it remains possible that an alternative is a better explanation. Such an approach comes close to the deductive route proposed by Harvey (Fig. 1.1), notwithstanding semantic arguments over the difference between 'falsification' and 'verification'. Even so, as Burt and Walling (1984) point out, the deductive route to explanation is more complex than Harvey (1969b) implies. Initial structuring of the model will depend in part upon the operational constraints known to apply – what experiments are required, what instruments or techniques are available, what field or laboratory facilities exist. The ideal separation of theory and fact may not initially be possible, therefore. Failure of an experiment may not condemn the theory; rather the experiment itself may be judged deficient in some way. In this case, the 'unsuccessful' feedback loop (Fig. 1.1) should stop at the experimental design stage; a new measurement technique may be required before the theory can be adequately tested. Strict formulation of an 'experiment' may not always be possible either. As noted below, many geomorphological studies do not constitute experiments *sensu stricto*, particularly in the early stage of theory development (Church 1984). It is clear, therefore, that neither the theory nor the experimental design is likely to remain static.

The new quantitative geomorphology has been very much concerned with elucidation of the relevant geomorphological processes and with considerations of the rates at which such processes operate. From knowledge gained about process activity, it has been possible to theorise on the landforms that should be produced by such processes. Certainly one major result of process study has been the relegation of time to the position of a parameter to be measured rather than as a process in its own right. Another major result of the change in geomorphological emphasis has been a reduction in the spatial and temporal scales within which landforms are now considered. The scale of landforms being examined has been reduced, partly because process and form are best related at small scales and partly because, over the large scales with which the denudational chronologists were concerned, process operation is not invariate as Davis envisaged but different processes have been seen to operate at different locations in a catchment (Dunne & Black 1970). In the early 1930s, it was not unusual for whole regions to be considered in one paper (e.g. Mackin 1936), whereas more recently, in fluvial geomorphology for example, landforms are considered at no more than the scale of a drainage basin (Chorley 1969), and frequently at a far smaller scale: for instance, differences between the processes operating on individual hillslope segments have been identified (Anderson & Burt 1977).

Schumm (1977a) has, however, augmented the Davisian model of landscape evolution by relating field and laboratory investigations to the mechanisms of long-term landscape change. It is argued that thresholds exist in the drainage basin system; channel pattern change associated with changing flume channel slopes is one such example, with meandering thalwegs introduced only at a threshold slope (Fig. 1.2). Evidence of this form has led Schumm to propose the replacement of Davis's progressive change notions of landscape change by the inclusion of episodes of adjustment occasioned by thresholds – a model stressing episodic erosion resulting from the exceeding of thresholds. Figure 1.3 illustrates the nature of the Davisian model modification, as proposed by Schumm, with the inclusion of the concept of 'dynamic' metastable equilibrium, or thresholds.

Of course, evidence of long-term landform development stems not only from contemporary laboratory investigations but from other more qualitative areas, too. Geomorphological systems are not only complex, but our knowledge of them becomes less exact as we move away from the present into the recent past. Environmental change, especially since the Pleistocene, represents a second major line of enquiry into the processes of long-term landscape change. Knowledge of such geomorphological indicators as those given in Table 1.1 can provide a most significant base from which to assess and explain aspects of change. Chandler and Pook (1971), for example, have shown that knowledge of general Pleistocene water table conditions combined with hillslope stability analysis is capable of providing a mean soil creep rate since the Pleistocene at a site proved to have suffered a landslide at that time. Therefore, the more qualitative aspects of environmental change can be seen to provide sound evidence contributing to more analytical approaches. While denudation chronology as a means of investigation is unlikely to find such a significant role, qualitative indicators of landscape environments and changes will continue to do so, and, together with improved methods of dating (see Part Five), their significance in the interpretation and modelling of long-term landform development should not be belittled.

A significant feature of the geomorphological 'revolution' has involved the introduction of systems theory as an overall explanatory structure within the subject. There has been a move away from the preoccupation with change and development towards the view that landforms may be balanced systems, with process and form closely interrelated,

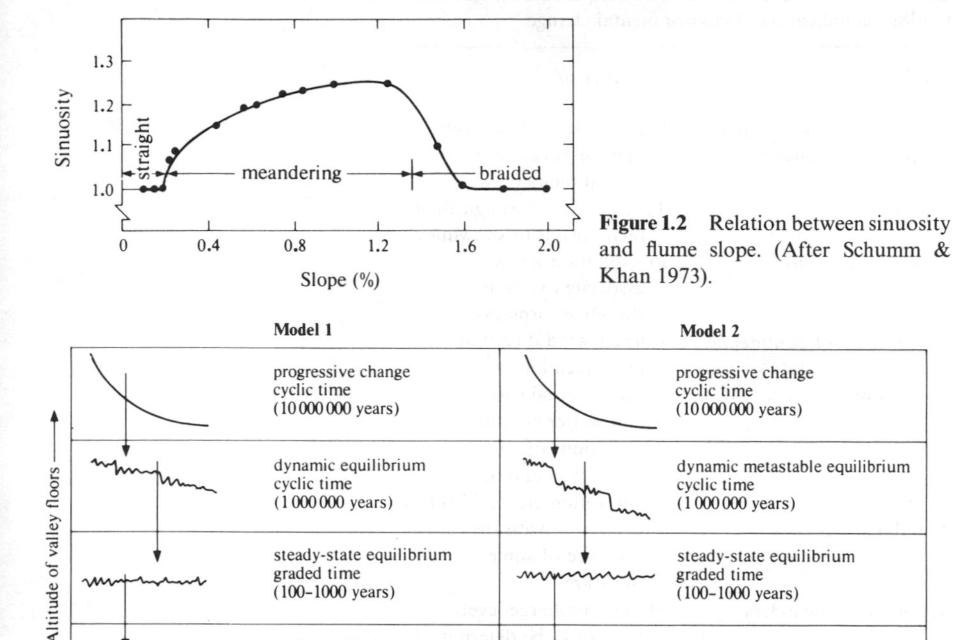

Figure 1.2 Relation between sinuosity and flume slope. (After Schumm & Khan 1973).

Figure 1.3 Models of landscape evolution. *Model 1*: equilibrium components of Davis's model of progressive denudation. *Model 2*: equilibrium components of model based upon episodic erosion as proposed by Schumm. (After Schumm 1977a).

yet with the system activity aimed at maintaining a stable form rather than producing progressive change. Landscape development is thus seen as the end-result of system operation and not as a major product of system activity. In this sense, the new geomorphological paradigm has subsumed the old according to the Kuhnian model: the successful use of mathematical process–response models to predict landform evolution is an example of this (e.g. Kirkby 1984). The use of a systems approach as the organising paradigm in geomorphology has accompanied the general development of model building within the subject. Stoddart (1967) has proposed four reasons why the system is such a useful and fundamental organisational concept. First, systems are monistic, bringing together all relevant components of the physical and human environment. Secondly, systems are structured in an orderly and rational way so that the form of the system can be easily investigated. Thirdly, systems are functional entities, incorporating the throughput of matter and energy, so that the system is not just a framework but a dynamic operational unit. Finally, the majority of physical systems are governed by negative-feedback mechanisms, which limit change and maintain system stability.

Systems can be disaggregated in terms of equations wherein the output Y is a function of the input X and operators whose parameters may be estimated act upon the input. Bennett and Chorley (1978) observe that the form of the equations is based upon personal knowledge that the investigator has of the system and secondly upon mathematical and physical principles. Figure 1.4 shows the development of (a) an *a priori* postulate of the variables involved (structural knowledge) and (b) *a posteriori* measurement knowledge from sampling variables in nature. This leads to a robust theoretical

Table 1.1 Some quantitative and semiquantitative geomorphological indicators of environmental change.

Landform	*Indicative of*
Pingos, palsen, ice-wedge casts, giant polygons	Permafrost, and through that, of negative mean annual temperatures
Cirques	Temperatures through their relationship to snowlines
Closed lake basins	Precipitation levels associated with ancient shoreline formation
Fossil dunes of continental interiors	Former wind directions and precipitation levels
Tufa mounds	Higher groundwater levels and wetter conditions
Caves	Alternations of solution (humidity) and aeolian deposition etc. (aridity)
Angular screes	Frost action with the presence of some moisture
Misfit valley meanders	Higher discharge levels which can be determined from meander geometry
Aeolian-fluted bedrock	Aridity and wind direction
Oriented and shaped deflation basins	Deflation under a limited vegetation cover

After Goudie (1977).

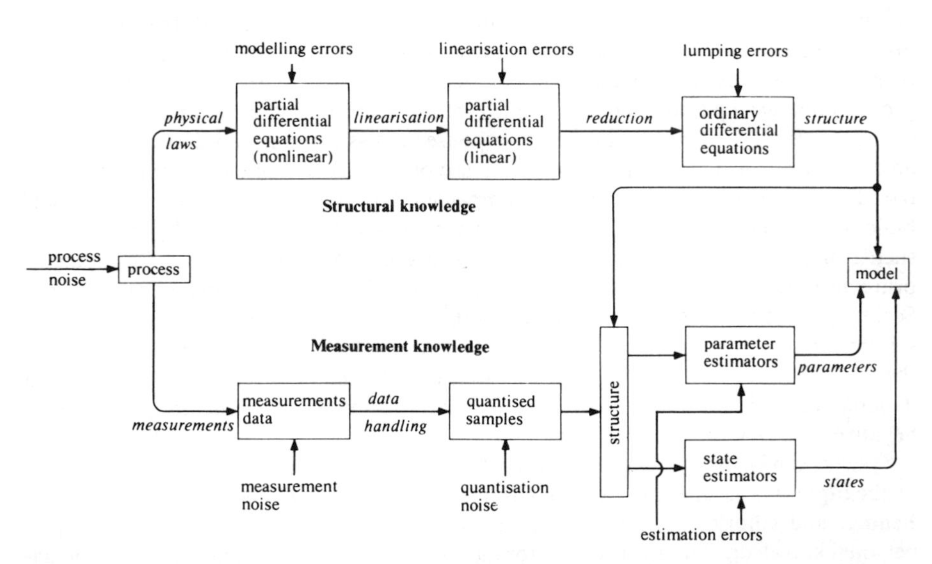

Figure 1.4 A systems representation of model construction. (From Bennett & Chorley 1978, after Eykhoff 1974).

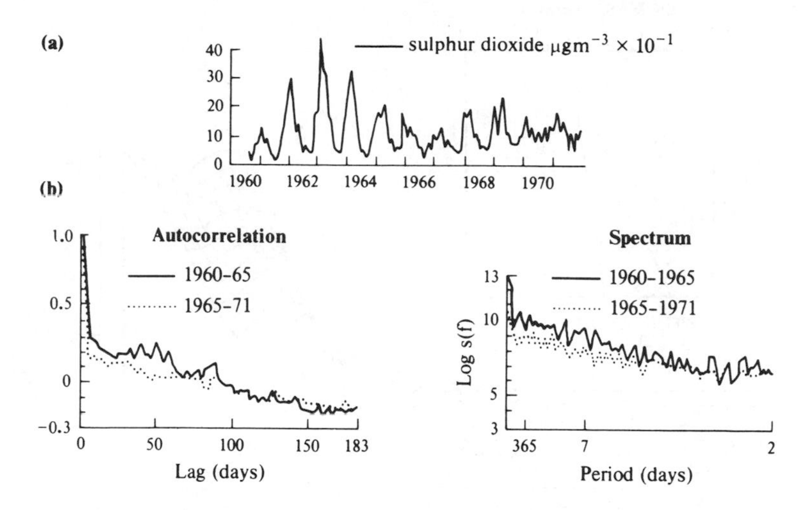

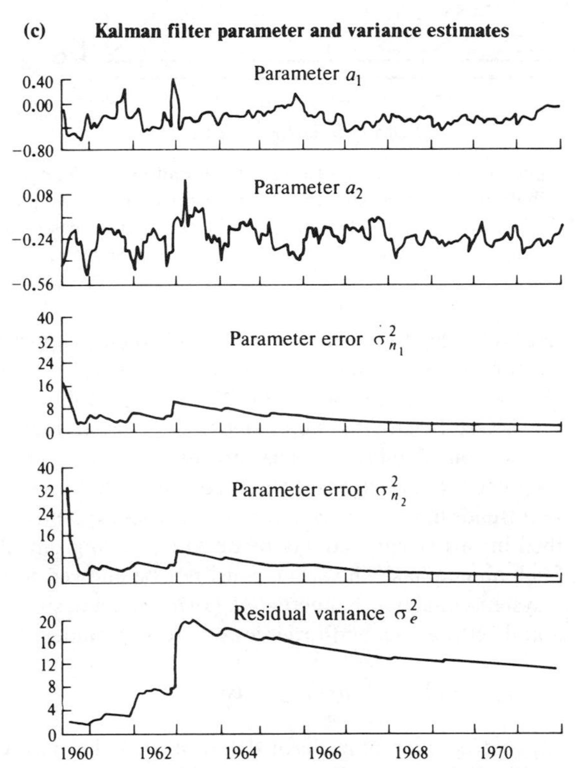

Figure 1.5 Analysis of daily SO_2 concentrations at Kew for the period 1960–71 using Kalman filter estimates. Estimates of autoregressive model are derived from least-squares regression; estimates of the non-linear parameter evolution model derive from the use of a Sage–Husa filter, which gives the value of each parameter a_i^t at time t as

$$a_1^t = a_1^{t-1} + n_t^1 \qquad a_2^t = a_2^{t-1} + n_t^2$$

(From Bennett & Chorley 1978, after Bennett *et al.* 1976).

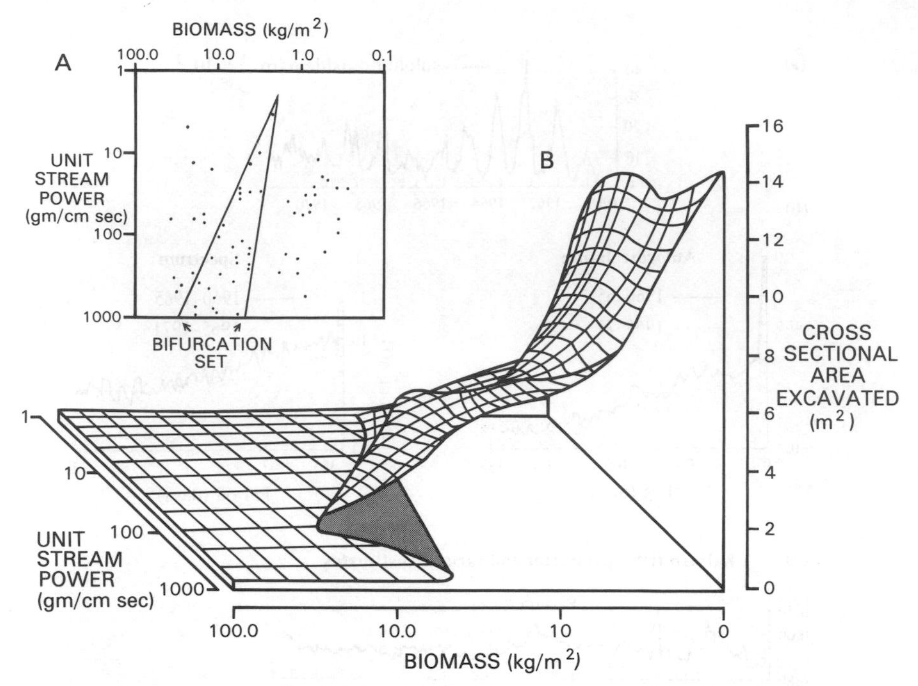

Figure 1.6 The cusp catastrophe of arroyo development in the Colorado Front Range as defined by empirical data: (A) the distribution of sampled sites; (B) the catastrophe surface. Major trenching occurs in and to the right of the bifurcation set. Note that the ordinate of (A) increases downwards so that the two-dimensional representation is oriented in the same way as the perspective view of (B). (After Graf 1979, Fig. 8).

model of the system (Bennett & Chorley 1978). However desirable such an approach may be, rarely have these two components been interdigitated successfully in physical geography modelling. Rather, as Amorocho and Hart (1964) have postulated, complex systems can more easily be analysed by either systems analysis (the use of mathematical functions to relate X and Y without detailed physical process knowledge) or systems synthesis (the use of detailed process knowledge and observed systems linkages). We have observed already that current trends in physical geography emphasise systems synthesis, and the techniques described in this volume largely relate to improving our ability to synthesise such systems. Such an emphasis, however, must not be allowed to bias our judgement on the utility of systems analysis. Bennett *et al.* (1976), in analysing SO_2 concentrations as a time series and getting a second-order autoregressive model of the form

$$Y_t = a_1 Y_{t-1} + a_2 Y_{t-2} + \varepsilon_t \tag{1.1}$$

and tracking changes in a_1 and a_2 over time, have demonstrated the nature of the non-linearities in the system (Fig. 1.5). Non-linear responses can be modelled in terms of a systems synthesis framework, too. Graf (1979, 1983) has illustrated threshold change in the formation of arroyos using catastrophe theory. Figure 1.6 shows that major trenching occurs with low biomass and high unit stream power, but that, at intermediate values, a steady-state equilibrium may exist for eroded *or* uneroded floodplains. With low stream power and good vegetation cover, floodplains will gradually aggrade. Landform change may be complex: at a threshold level of unit stream power or biomass, a jump can occur from a depositing to an eroding state; reversals can also happen, though usually such changes are hysteretic. Within each steady state, small fluctuations in the form of the system may take place. Thus, catastrophe theory accommodates the opposing situations of equilibrium and disequilibrium, gradual versus abrupt changes, depositional as well as erosional phases of arroyo development. Of course, it may be some time before quantitative analysis using catastrophe theory becomes widespread; Graf's work shows that such an approach may be feasible. Non-linearity in system responses has been evidenced by both systems analysis and systems synthesis, and techniques for identification of this important characteristic from qualitative environmental change through to parameter estimation methods for autoregressive schemes must be used to augment the knowledge we have acquired of steady and progressive process operation.

The adoption of a systems approach has provided an overall structure within which geomorphological models can be formulated. More importantly perhaps, as will be stressed in Section 1.1.4, the use of rigorously tested models should provide geomorphologists with a powerful means of participating in problem-solving exercises, particularly where human manipulation of physical systems is involved.

1.1.2 The main branches of geomorphological enquiry

Chorley (1966) has described the main branches of geomorphological activity (Fig. 1.7). Examination of these branches, together with the steps involved in the deductive scientific method (Fig. 1.1), indicates that several lines of enquiry are available to the geomorphologist and, indeed, all these approaches may be necessary if models are to be properly evaluated. Chorley's scheme correctly isolated the theoretical work from the data-collecting stage: this division is necessary if the deductive approach is being followed correctly, although, as noted above, such a strict division may not be possible during the period of 'exploratory' experimentation.

The development of the deductive scientific method within geomorphology has not proceeded in a smooth or unified manner. The initial application of quantitative techniques preceded the period of model building by a decade. More recently, much fieldwork has been conducted, aimed at providing data with which to elucidate the processes operating. However, such fieldwork has not always succeeded in adequately providing the link between form and process. Moreover, the data collection has often remained in isolation from the model builders, so that the functions of model verification and improvement have not been carried out efficiently. Indeed, it cannot be doubted that in some branches of the subject, model building has occurred largely without the model verification stage. In contrast, much of the data collected, though aimed at individual hypotheses, has avoided the wider implications contained within the results. Dunne (1981) has noted the need for 'field measurement programmes which will generate the critical data required for modelling rather than just the data that are easy to obtain. Such planning requires that from the outset the study should be designed by someone skilled in both theory and fieldwork or by a partnership of such interests'.

The branches of data acquisition identified by Chorley do not represent equal portions of geomorphological enquiry. During the predominance of denudation chronology, data acquisition was dominated largely by qualitative field observations and by subjective map analysis. The application of statistical methods in geomorphology has demanded quantitative data, which has led in particular to the gathering of a vast amount of field data. Much work has also been devoted to the numerical analysis of field samples in the laboratory, particularly methods investigating the strength of materials, the grain size of

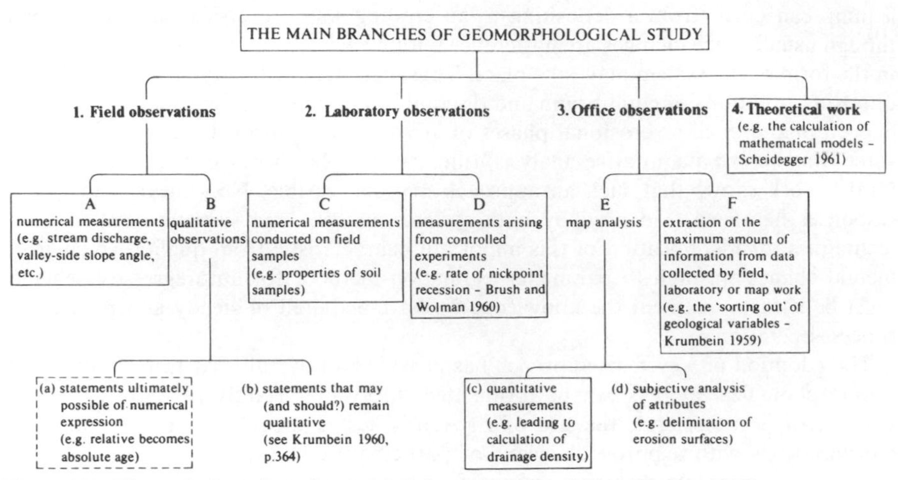

Figure 1.7 The main branches of geomorphological studies. (After Chorley 1966).

sediments, and the chemical analysis of sediments and water samples. Perhaps one of the most neglected avenues has involved the controlled laboratory simulation of field conditions. Such experiments allow cause-and-effect relationships to be examined in detail since many independent variables, which in the field could vary in unknown or uncontrollable fashion (e.g. climate, bedrock, soil, topography), can be effectively eliminated in the laboratory by being held constant. Though some laboratory experiments have been conducted (see Table 1.2), this branch of geomorphology remains relatively neglected, even though it represents the closest line of geomorphological study to the deductive laboratory studies of chemists and physicists. Church (1984) has, however, pointed out the difficulty of scaling inherent in such experiments. Especially where detailed process studies are required as a prerequisite to field analysis, there seems ample opportunity for the development of laboratory studies, and some examples of this appear in this volume.

It must be admitted that thus far the major line of data collection in geomorphology has involved quantitative field observations. However, Soulliere and Toy (1986) remark

Table 1.2 Selected examples of laboratory simulations in geomorphology.

Phenomenon	*Authors*	
Weathering		
Chemical weathering	Daubrée	(1879)
Frost weathering	Martini	(1967)
Salt weathering	Goudie	(1986)
Salt weathering	Sperling and Cooke	(1985)
Insolation weathering	Blackwelder	(1933)
Rillensteine	Whitney and Brewer	(1967)
Dune reddening	Williams and Yaalon	(1977)
Podsolisation	Paton *et al*	(1976)
Vadose and phreatic cementation	Thorstenson *et al.*	(1972)
Rillenkarren	Glew	(1977)
Gypsum crusts	Cody	(1976)
Wetting and drying	Pissart and Lautridou	(1984)
Freeze–thaw structures	Van Vliet-Lanoe *et al.*	(1984)

Table 1.2 (*cont.*

Phenomenon	*Authors*	
Fluvial		
River sediment movement	Gilbert	(1914)
Knickpoints	Lewis	(1944)
Meanders	Friedkin	(1945)
Networks	Schumm	(1977a)
Channel patterns	Schumm and Khan	(1971)
Overland flow erosion	Moss and Walker	(1978)
Confluences	Best	(1986)
Rainsplash	Noble and Morgan	(1983)
Slopes		
Angle of repose	van Burkalow	(1945)
Angle of sliding friction	van Burkalow	(1945)
Frost heave	Taber	(1930)
Drainage	Anderson and Burt	(1977)
Dye tracing and permeability	Anderson and Burt	(1978)
Leaching	Trudgill *et al.*	(1984)
Screes	Statham	(1973)
Liquefaction	De Ploey	(1971)
Greze-litée	Van Steijn	(1984)
Glacial		
Esker fans	Jagger	(1912)
Kaolin glaciers	Lewis and Miller	(1955)
Kettle holes	Maizels	(1977)
Glacial crushing	Whalley	(1978b)
Glacial milk	Keller and Reesman	(1963)
Glacial folding	Ramberg	(1964)
Coastal		
Cusp formation	Flemming	(1964)
Offshore bars	McKee	(1960)
Deltas	Nevin and Trainer	(1927)
Atolls	Purdy	(1974)
Straits	Thompson	(1977b)
Pebble abrasion	Bigelow	(1984)
Miscellaneous		
Aeolian abrasion	Kuenen	(1960b)
Frost sorting	Corte	(1963)
Wind ripples	Bagnold	(1941)
Mountain building	Kenn	(1965)
Stone pavements	Cooke	(1970)
Salt domes	Nettleton	(1934)
Faulting	Horsfield	(1977)
Ventifacts	Whitney and Dietrich	(1973)
Dust abrasion	Dietrich	(1977)
Mud clast durability	Smith	(1972)
Lunar features	Schumm	(1970)

Source: A. S. Goudie, personal communication.

on the relative lack of field observations in certain areas of applied geomorphology. They cite the paucity of erosion data for lands disturbed by surface mining, for example, although more information is now becoming available; see table 1 in Soulliere and Toy (1986) for a summary of the main erosion studies, and also de Ploey and Gabriels (1980).

1.1.3 The role of 'techniques' in data collection

The ability to measure successfully is central to any scientific explanation. The role of a 'technique' (i.e. a measurement procedure) is to allow the variables involved to be quantified in a consistent, rigorous and precise manner. The need to quantify is essential, since general understanding of a problem is inherently interconnected with ability to define the elements contained in that problem. To say that something cannot be measured is tantamount to admitting that one's knowledge is limited and that any ideas on the subject must remain vague and confused. The ability to measure helps to fix one's general conception of a problem, so that quantity is not regarded as opposite to quality but rather an improvement on it.

Having established the need to measure, it is important to employ valid measurements. Harvey (1969b) defines a valid measure as one that measures what it is supposed to measure. The structure of a theory will determine the definition of the object or its attributes that are required to be measured. The measurement procedure must aim to reflect this definition as closely as possible; without successful measurement, the results of statistical tests on hypotheses must inevitably remain inadequate. The measurement procedure must therefore follow a well defined set of rules in order to produce consistent data. In many cases, the rules of measurement are already well established; for example, simple measures of length or weight. However, for more complex procedures, the measurement method may be ill defined: the technique employed may not be wholly valid in that it may not measure exactly the attributes contained in the definition of the variable; even if the technique is valid, the errors involved in the act of measurement may mean that the data are not suitably rigorous for use in sophisticated statistical tests. Thus, even in a physical science like geomorphology, there is still a need to develop a conventional methodology so that the measurement of specific defined variables becomes consistently established within the subject. Thus, an elucidation of a particular 'technique' should describe the exact steps to be taken to use, and if necessary to construct, the equipment and should provide information on the precision of the data obtained. Implicit within the development of any technique that might become established convention is that it will be seen to measure unambiguously and exactly the properties of the variable involved.

To measure effectively requires a deep analytical understanding and considerable thought about the problems and errors involved in making the measurements. Despite the implicit separation of theory, definition, measurement, classification and verification within the deductive approach (Fig. 1.1), it is impossible for these to be totally separated in reality. Definition of the variable to be measured in a given study should logically precede the measurement process, but any definition is likely to be coloured by prior knowledge of what measurement procedures are available. Indeed, the initial structuring of the theory or model may partly depend on the operational constraints that are already known to apply – what operations are required, what instruments and measures are to be used and indeed are available to the researcher, and under what conditions the observations are to be made. Thus, an operational approach must accompany any scientific investigation, producing an inevitable interconnection between theory, measurement and statistical tests, and ultimately controlling the results and conclusions obtained (Fig. 1.4). What is particularly important to the geomorphologist is that the ability to measure a given variable is often the limiting factor in any study. Once a new technique becomes available, then the whole scientific process can circle up to a higher level of investigation and explanation.

Closely related to the need for effective measurement is the requirement of a sound experimental design: few geomorphologists have considered this despite paying close attention to the detailed measurements (e.g. Slaymaker 1980). Church (1984) showed that a properly structured experiment exhibits several distinctive characteristics, which follow from the deductive route shown in Figure 1.1. These are:

(a) a conceptual model of the processes or relations of interest, which will be supported or refuted by the experiment;
(b) specific hypotheses that will be confirmed or controverted by the experimental results;
(c) definition of explicit properties of interest and operational statements of the measurements to be made;
(d) a formal schedule of measurements to be made under conditions controlled to ensure that remaining variability be predictable under the research hypotheses;
(e) a specified scheme for analysis of the measurements that will discriminate the possibilities in (b); and
(f) a data management system designed for this purpose.

Church argues that on this basis very little geomorphological work qualifies as being properly experimental. Most field studies are no more than 'case studies', although it must be said that such empirical work may have great value as a basis for the formulation of theory or for the probing of conceptual models. There is, however, the need to test such concepts. This may be done initially using 'exploratory' experiments that incorporate simple elements of control. If successful, these should be followed by 'confirmatory' experiments, which necessarily involve a formal experimental design, a narrow focus of interest and preferably a distinct location. Church (1984) provides a full account of the types of experimental control that may be employed.

1.1.4 Application of geomorphological models

The crucial test demanded by use of the deductive scientific method is the successful application of models that have been produced. In geomorphology, such application will involve not only the purely academic pursuit of explanation of landform development, but also, increasingly, a concern with 'applied' aspects of the subject. The use of simulation models in geomorphology is long established (Table 1.1). The speed and capacity of modern computers has allowed the development of a new generation of mathematical models that are complex and physically realistic. Once written and calibrated (see Sec. 1.2.2.4), such models may be used in two ways. First, they can show how existing landforms may be expected to evolve, given assumptions about the rates of processes operating (Kirkby 1986). Closely allied to this is the reconstruction of past landform sequences; Kirkby (1984) has simulated the evolutionary sequence of slope development behind an advancing coastal spit in south Wales as originally described by Savigear (1952). Secondly, simulations can be performed in order to explore and experiment with the theoretical structure of the model. Given the complexity of most geomorphological systems, the use of a simulation model may actually uncover new aspects about the way in which the system functions and may pose new questions to be addressed (Burt & Butcher 1985).

Geomorphologists are beginning to demonstrate the ability to produce successful solutions to problems using well constructed and tested models. Further development and strengthening of this model base should allow even greater participation in such exercises. The majority of funding for geomorphological research may well become increasingly linked to particular problems so that the mere ability of the subject to survive and develop may involve a greater participation in 'applied' matters. However, there seems little doubt that the investigation of landforms and processes should remain paramount within the subject. Models can be of general applicability, whereas a unique solution to an individual problem may be of little use in a completely fresh situation. The strength of pure research also lies partly in its ability to deal with subjects that, though not seeming immediately relevant, may become so as a result of new developments. A prime example of this type of application is the sudden need for information on periglacial slope and soil processes to allow successful construction of pipelines in northern Canada. Much pure research into solifluction and frost heave has, perhaps unexpectedly, become central to the accomplishment of such projects (see, for example, Williams 1986). This is not to imply, however, that problem-solving exercises must play a subordinate role: not only must such exercises provide a lucrative outlet for applying many geomorphological models, but the advent of new problems may well stimulate research into new areas, as was shown, for example, by the growth of interest in coastal geomorphology in the UK following the 1953 floods, by the work of G. K. Gilbert in the surveys of the western USA, and by the work of R. E. Horton. The roles and scope of applied geomorphology have been discussed in, for example, Cooke and Doornkamp (1974).

The setting for an applied geomorphology may be viewed within the concept of the control system (Chorley & Kennedy 1971). In such a system, intervention in the physical process–response system is seen as a single element within the larger socio-economic decision-making system. Most human systems are dominated by positive-feedback mechanisms that make change ongoing. Intervention in physical systems in the past has often taken this form, with change accelerating eventually to destroy the system, as with the example of the US Dust Bowl in the 1930s. The natural physical system is generally dominated by negative feedback, which regulates the system, maintaining balance and preventing self-destructive change from occurring, though as King (1970b) has shown at certain scales, notably in glacial systems, positive feedback may predominate. Geomorphologists should therefore ensure that any intervention in landform systems is thoughtfully regulated so as to exploit the system successfully, rather than cause its degradation. Such intervention must therefore be based on proven geomorphological models that can accurately predict the likely impact of any planned intervention in the system.

1.2
Process determination in time and space

1.2.1 Scale and explanation

The majority of geomorphological techniques seek either directly or indirectly to provide a rate of process operation in time or space. Thus, techniques of material properties (Part Three) allow us to predict failure on hillslopes, while drainage basin morphometry (Part Two) frequently facilitates a significant postdiction potential when combined with knowedge of contemporary processes. Thus, techniques capable of both postdiction (extrapolation from the present to the past of contemporary process–form interrelationships) and prediction are to be reviewed in the following sections. The advent in recent time of an ever-improving ability to be more precise over measurement of nearly every type has broadened the base from which process statements either forwards or backwards in time can be made. The geomorphologist now has data available from a spectrum of techniques, providing high resolution, from the micrometre scale to thousands of square kilometres of the Earth's surface.

We have already seen in the preceding section that there has been a change in scientific method, while geomorphology continues to concern itself with the origin and evolution of landforms. New techniques are beginning to provide us with a *choice* of scales at which the geomorphologist can not only operate but also seek explanation if he or she so chooses. Before we comment further on this point, however, we must attempt to extrapolate this progression. At the large scale, any significant increase in scale of imagery seems unlikely, while this is probably not so at the small scale, as X-ray and other scanning techniques develop (Anderson 1979) – see Section 3.2 of this text. A desire to understand more about a process has nearly always been taken to be synonymous with investigating that process with techniques at different scales. In many cases, these techniques have seemed so successful that they have outstripped our capacity to model the entirety of the original process. It may well be the case, therefore, that we are moving towards a new set of objectives in geomorphological investigations defined increasingly by the techniques of laboratory and field, while awaiting the development of analytical solutions capable of aggregating our detailed findings.

Superimposed upon this picture of increasing resolution provided by geomorphological techniques, we have the complication of catastrophic events, which engender non-stationary behaviour into the system. For example, the response of sediment yield to pre-afforestation open ditching in upland Britain provides for a marked change over time (Fig. 1.8), while a high-magnitude large-recurrence-interval storm event can have profound spatial significance in terms of channel change (Fig 1.9). It is generally recognised that the occurrence of such non-linearities, identified by techniques at all scales, is a major constraint upon the establishment of accurate models and represents one restriction in the aggregation of results.

Nevertheless, in terms of certain temporal investigations, this inferential problem has been circumvented to some degree by the use of statistical techniques and by an implied

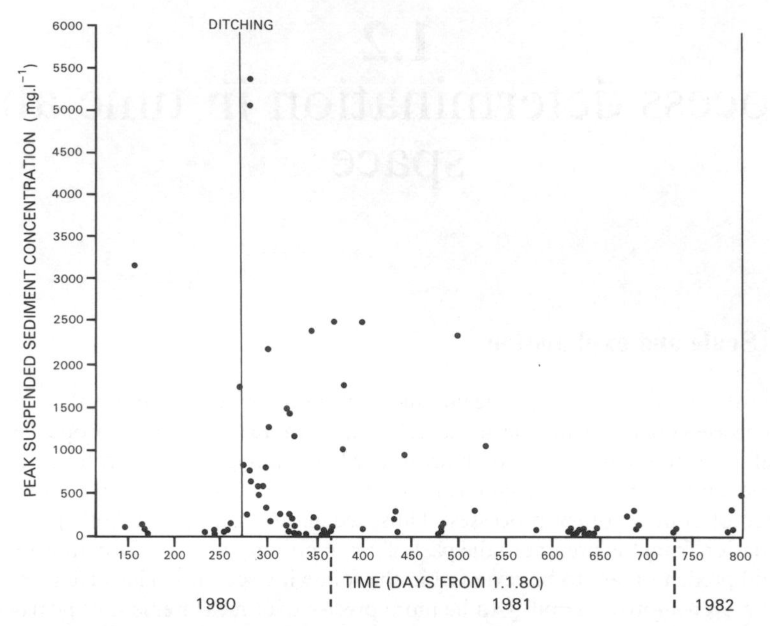

Figure 1.8 Peak suspended sediment concentrations for the period April 1980 to March 1982 at the Hades Clough catchment, Yorkshire. (From Burt *et al.* 1983). Open ditching (deep ribbon ploughing) was carried out between 26 and 30 September 1980.

change in what is accepted as explanation. Rescher (1962) has urged that probabilistic statements, such as Markov chain processes, should become accepted as 'explanation' of the particular phenomenon that they describe. While this represents perhaps an extreme view, Zelenhasic (1970) was able to show that flood peaks in Colorado rivers above a baseflow could be modelled by a mixed event model of Poisson plus harmonic – the Poisson distribution having mean X fitted event distributions in X periods throughout the year and change in X itself being modelled by a harmonic component. Traditionally, we would regard such an approach as perhaps no more than providing a statistical explanation of rare events, and it would certainly not be regarded as providing a scientific explanation. While there is a choice of techniques available for the analysis of recent and contemporary catastrophic events ranging from statistical to sedimentological, abrupt changes in the geological past are of necessity approached in a less technical manner. If one can assume that the Quaternary period has similar vegetation to the present (Krausel 1961), then Schumm (1965) has shown that a family of curves can be constructed depicting sediment concentration as a result of mean annual precipitation and mean annual temperature when climatic change is gradual. However, Knox (1972) argues that in the north central USA climatic changes could be abrupt, leading to a period of landscape instability (Fig. 1.10) and to fluctuations of erosion channel adjustment, which may be difficult to distinguish from high-magnitude, large-recurrence-interval floods (Schumm 1977a), exemplifying the multiple process–single form relation. Such suggestions by Knox and Schumm involve postdiction. Thus postdiction is considering to what extent the present observations are typical of past conditions (Pitty 1971); in the geomorphological context it must interdigitate with similar postdictions in the fields of botany and climatology. Such an expansive, indeed explosive, inclusion of evidence as one looks backwards in time is seen to contrast heavily with more contemporary process investigations, which are frequently more introverted, for reasons of theory, technique and data

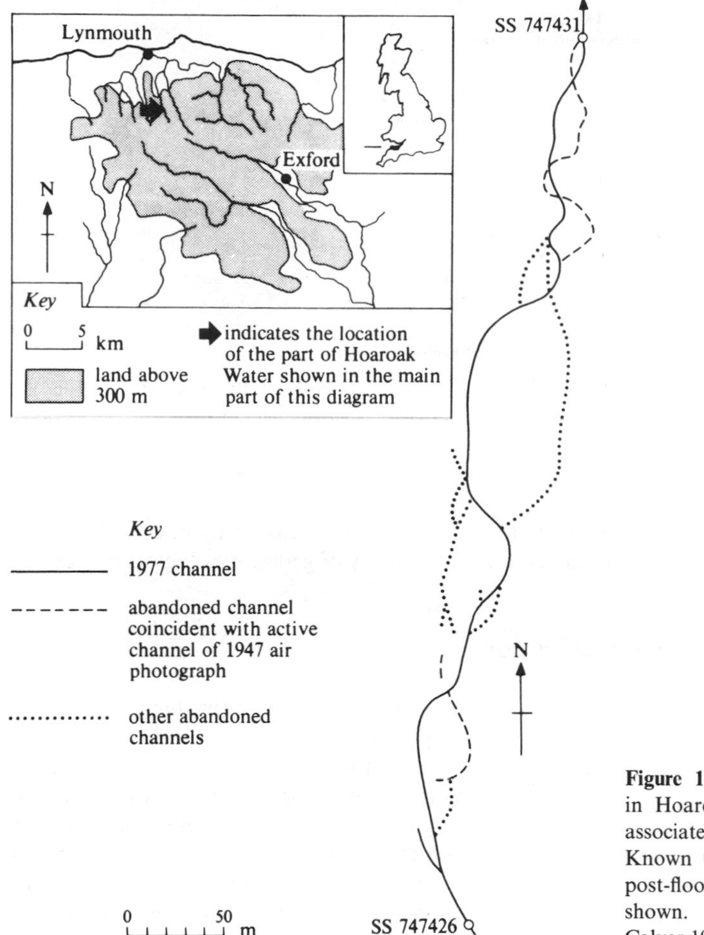

Figure 1.9 Channel changes in Hoaroak Water, Exmoor, associated with the 1952 flood. Known pre-flood (1947) and post-flood (1977) channels are shown. (After Anderson & Calver 1980).

availability, although, as we will discuss, more complete in scientific understanding. The final message here might be that postdiction requires support from other areas, while prediction in geomorphology at levels of explanation we now accept suffers from an inability to generalise at a similar medium scale – inputs from other sources (say botanical) complicate the model to an unmanageable degree. The question must be asked as to whether this is a position reached by rational thought and logical development, or whether we have bypassed the most difficult questions and have spent too much effort on matters of less importance (Leopold 1978).

The increase in techniques now employed in geomorphology must be regarded as akin to a revolution of which quantification can, with hindsight, be considered a subset. A rapid broadening of scales at which measurement can be made presents new choices of explanation to the investigator and, as we have attempted to indicate, it places a new onus upon modelling micro-scale processes in relation to medium-scale landforms; for example, our ability to measure may well currently exceed our ability to explain. Such balance can, as we will discuss below, be redressed to some degree by an examination of technique types used in the verification of different classes of geomorphological investigations. Knowledge of non-stationary, abrupt and catastrophic behaviour, as we

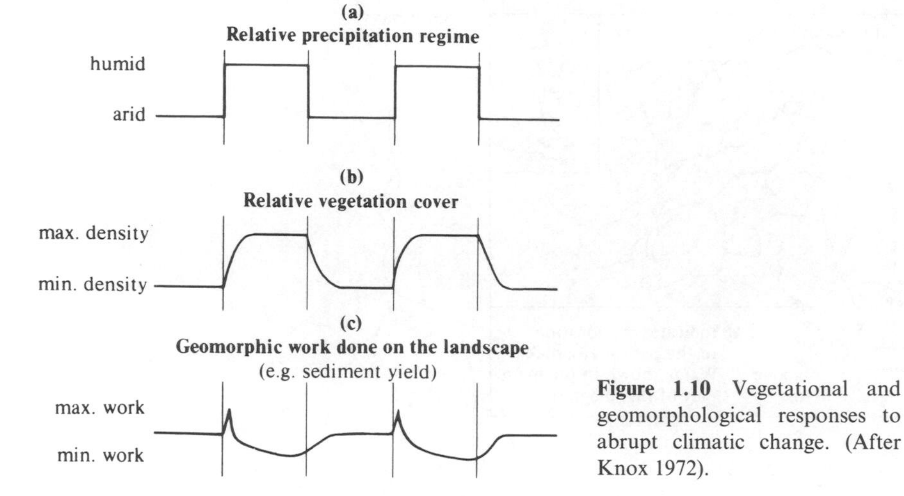

Figure 1.10 Vegetational and geomorphological responses to abrupt climatic change. (After Knox 1972).

have illustrated, is a special class to which we can expect increasing attention to be paid in view of the importance of thresholds in a wide variety of geomorphological processes.

1.2.2 Techniques and verification

We must choose what we measure – we cannot measure everything. Moreover, measurement has a self-regulating, somewhat paradoxical, element, since as Amorocho and Hart (1964) relate: 'to measure precipitation input totally accurately into a basin with the desire to establish precipitation–runoff relationships, then the entire basin must be covered with raingauges, thereby eliminating the runoff and in consequence the ability to establish the relationship'. Experimental design in the context of technique application is therefore an important element in any field science. Techniques are used in geomorphology to aid the verification or rejection of particular hypotheses associated with three main types of investigation: description, direct measurement of processes and finally model calibration. Parts Two to Five of this text reflect the details of techniques that can be associated with these three fields of enquiry. General comment on each of these areas is of interest here since, in practice, techniques are rarely employed singly.

1.2.2.1 Description of form

Simple description of landscape form has provided the basis of geomorphology. Whether such description takes the form of geomorphological mapping (Fig. 1.11) or statistical description analysis (Speight 1971) matters comparatively little, since description of this type cannot relate to or verify the detailed mechanics of process operation. However, such an approach has the value of providing general information on the *spatial* variability of processes, which, if it can be combined with contemporary process rates, provides the basis for improved postdiction and prediction. Such mapping, however, can have superimposed on it a severe sampling constraint, in that the investigator cannot choose freely historical evidence of the times of air photograph cover. Thus, enforced external sampling of this type is made without reference to the rate of process operation, and so caution must be exercised when such data are used for purposes other than description.

While description of landscape form can be accompanied by the use of indices of slope angle, slope length and the like (Part Two) and can be related by similar linear statistical

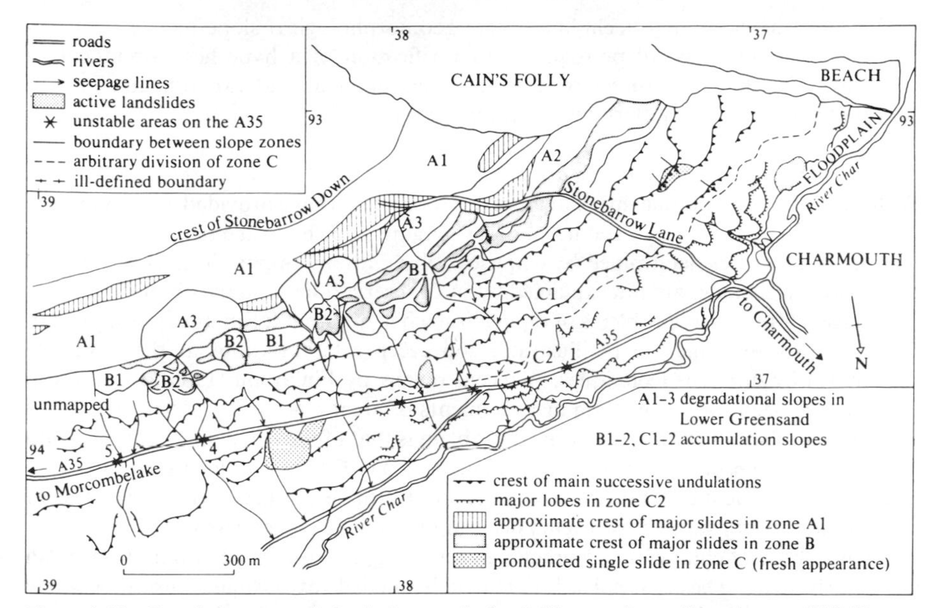

Figure 1.11 Detailed geomorphological map of a landslide complex at Stonebarrow Hill, Dorset. (After Brunsden & Jones 1972).

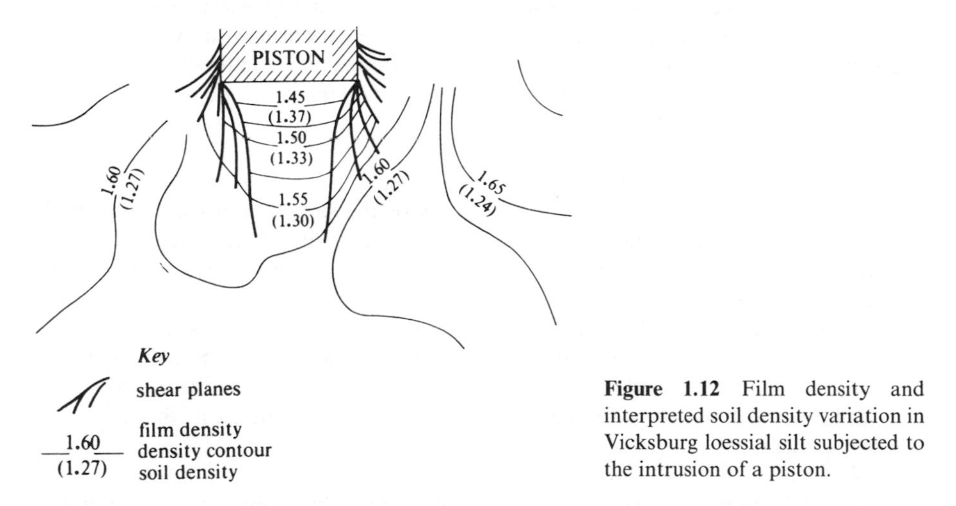

Figure 1.12 Film density and interpreted soil density variation in Vicksburg loessial silt subjected to the intrusion of a piston.

models to other parameters, description in other areas of geomorphology is less readily achieved. Schumm (1960) has noted that in the study of sediment types the representation of the physical property of the sediment in a single parameter is a particularly difficult problem to resolve. In the remote-sensing field, as well as with more conventional scanning techniques like X-ray, image description is now feasible with the use of the further techniques of densitometry and isodensitracers. Figure 1.12 shows soil density variations obtained from an X-ray of a piston intrusion into a sample of loess. The X-ray image density identified by the isodensitracer when used in conjunction with a calibration curve of film density–soil density facilitates accurate delineation of density changes with time as additional external pressure is applied (Krinitzsky 1972). Such a description can, of course, be averaged spatially or temporally as desired.

The range of descriptive techniques from geomorphological slope indices to densitometry, although a useful prerequisite to verification of a hypothesis or model, is generally subordinate to direct process rate measurement and experimental investigations where these two latter lines of enquiry are possible.

1.2.2.2 Direct process rate measurements

Techniques for the empirical measurement of process rates have provided a second major component in contemporary geomorphology. Figure 1.13 shows the rates for a selected number of processes determined by empirical observations. Again, it must be stressed that process mechanics are not explained by such observations, and the observations can themselves suffer severe problems of replication when different techniques are applied to the same process. Figure 1.14 illustrates soil creep rate variations at the same site determined by different techniques. A recent review by Finlayson (1985) of soil creep confirms the need for longer-term measurements of greater accuracy; see Harris (1987) for a review of solifluction rates. A third problem is that of the frequency of observation in relation to the process rate. Thornes and Brunsden (1977) have noted the irony that to record the infrequent event near-continuous monitoring is required, while less frequent regular monitoring suffices for events occuring with greater frequency. The latter two problems, those of equipment accuracy and sampling, are of particular importance – in few cases have these been formalised. Fortunately, there do exist isolated instances where previous conclusions have been questioned and shown to be invalid when proper account of accuracy of the original data is considered (Speight 1971, for example).

There is a real need to assess measurement errors particularly where predictions may be based on such results, since the accuracy of such predictions is entirely dependent on the accuracy of the model providing the prediction. For example, in the estimation of the hydraulic conductivity of a frozen soil sample (Burt 1974) there are two principal sources of error:

(a) Errors involved in the laboratory measurements associated with the use of the permeameter – these include measurements of the sample dimensions, hydraulic head and outflow discharge. The cumulative effect of all possible errors amounted to an absolute error of $\pm 15\%$ on the calculated value. In itself, this represents a wide confidence limit around the true value, but in the context of the range of values encountered (10^{-3} to 10^{-11} cm s^{-1}) the error range is perhaps less significant.
(b) Errors between the measured laboratory value and the actual field value for identical hydraulic conditions – the magnitude of such errors is difficult to estimate and relies entirely on a realistic simulation of field conditions in the laboratory experiment.

It should be noted that these errors are not necessarily important in themselves if simple comparison between results is the sole aim of the study (e.g. differences between two soil types). However, the magnitude of such errors may be much more important when the results are used for field prediction (e.g. to provide estimates of frost heave around a cooled gas pipeline), since accurate estimates are required for economic, and perhaps also ecological, reasons.

In the estimation of river discharge there are three principal sources of error:

(a) changes in the stage–discharge relation with time;
(b) error (E) in the stage estimation, given by Herschy (1971) as

$$E = (S^2 + B^2h^2)^{1/2} \tag{1.2}$$

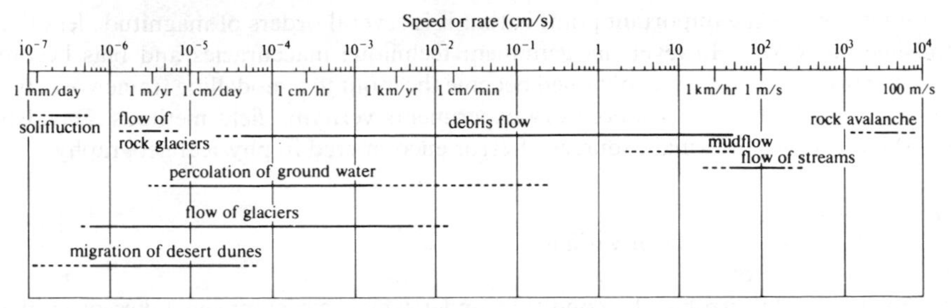

Figure 1.13 Ranges of measured rates of movement of selected processes. (After Longwell *et al.* 1969).

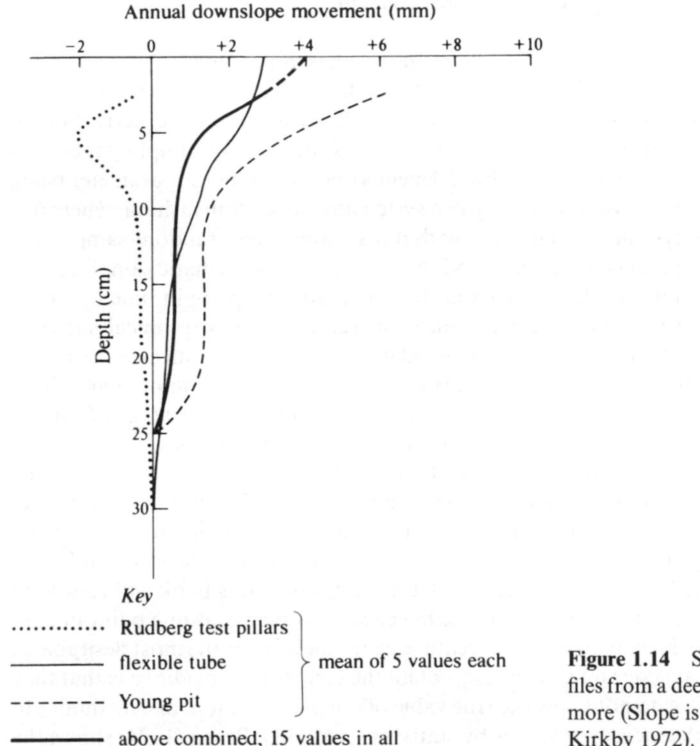

Figure 1.14 Soil creep velocity profiles from a deep stony soil near Baltimore (Slope is 17°). (After Carson & Kirkby 1972).

where S is the standard deviation in the mean relation, B is the exponent in the stage–discharge relation, and h is the error in measurement of the head; and

(c) sampling interval.

Herbotson *et al.* (1971) have used percentage errors based upon these three sources, and such calculations reveal that the most diagnostic flows in many respects, the maximum and minimum, are the least accurately gauged (Bergqvist 1971).

Despite recent advances in measurement techniques (particularly transducers, facilitating very small incremental displacements of material to be recorded – see Section 3.5 of this text), there has been a lack of associated error analysis. Of course, in many applications in geomorphology one is dealing only with the idea of a dominant process, and perhaps in terms of a process–form relationship it matters little if the soil creep rate is

2 or 4 mm yr^{-1} – the important point is that it is several orders of magnitude less than landslide processes. However, measurement technique inaccuracies and bias become more important when data so obtained become the input to a modelling framework, be it statistical, mathematical or laboratory experiments verifying field methods. Davidson (1978) has reviewed the main sources of error encountered in physical geography:

(a) sampling error;
(b) observer error – operator variance;
(c) instrument error;
(d) environmental error – the condition under which measurement takes place may influence the results; and
(e) nature of the observed phenomenon – error due to the method of measurement influencing the results, e.g. soil creep measurement.

The use of statistical methods in physical geography has not resulted in close scrutiny of sampling errors. Carson (1967) has observed that taking a larger sample than is necessary is a waste of resources, while too small a sample will inhibit the derivation of meaningful conclusions from the data. Figure 1.5 shows the types of sampling most frequently used. The choice of sampling method depends upon whether the parameter being sampled is itself known or assumed to vary in a systematic or random fashion: where it is necessary to exclude systematic bias from within a stratum, then random sampling is required, but where a process varies in a random fashion, then systematic sampling may suffice. Walling (1978) has outlined errors in the systematic sampling in time of a continuous variable. However, the recent advances in techniques and technical sophistication may well demand the retraction by physical geographers of statistical modelling in favour of deterministic models in most areas other than time-series approaches. This might be expected since many investigators cannot always adopt the strategy of sample size demanded by statistical sampling theory owing to equipment costs, time and the impossibility in many cases of operating more than one field site. Under such a restricted sampling environment, sampling error can only be removed by making the results site-specific (removing the spatial sampling error by implication) and eliminating the temporal sampling error by knowledge of the continuous behaviour of the phenomenon under study. It is, of course, far easier technically, as later sections of this book will testify, to accommodate the latter than it is to replicate in space. To operate at one point in continuous time is the easiest way to meet this requirement, but it is not the most desirable in the context of physical geography. The result of all the errors (a) – (e) above is that they cumulate, and, since we do not know the true value of the phenomenon, one method is to assess variability of the measured values by statistical methods. Bragg (1974) shows the effect of increasing sample size and reducing class width of a hypothetical attribute. If no biased errors exist, the final curve will be symmetrical about x and the shape of the curve will reflect measurement variability (Davidson 1978) – see also Figure 1.5.

1.2.2.3 Statistical techniques and process inference

We have seen that statistics became a commonly used means of seeking to establish loose cause–effect relationships in the mid-1950s. While such applications 'explained' relationships in terms of variance minimisation, the actual process mechanics in geomorphological terms is rarely explained, as Statham (1977) observes, although useful linkages between variables such as those of a stratigraphic sequence exhibiting a Markovian property with respect to strata and depth (Harbaugh & Bonham-Carter 1970) can be isolated. While it is thus relatively easy to establish links between variables, the nature of these links is less easily ascertained.

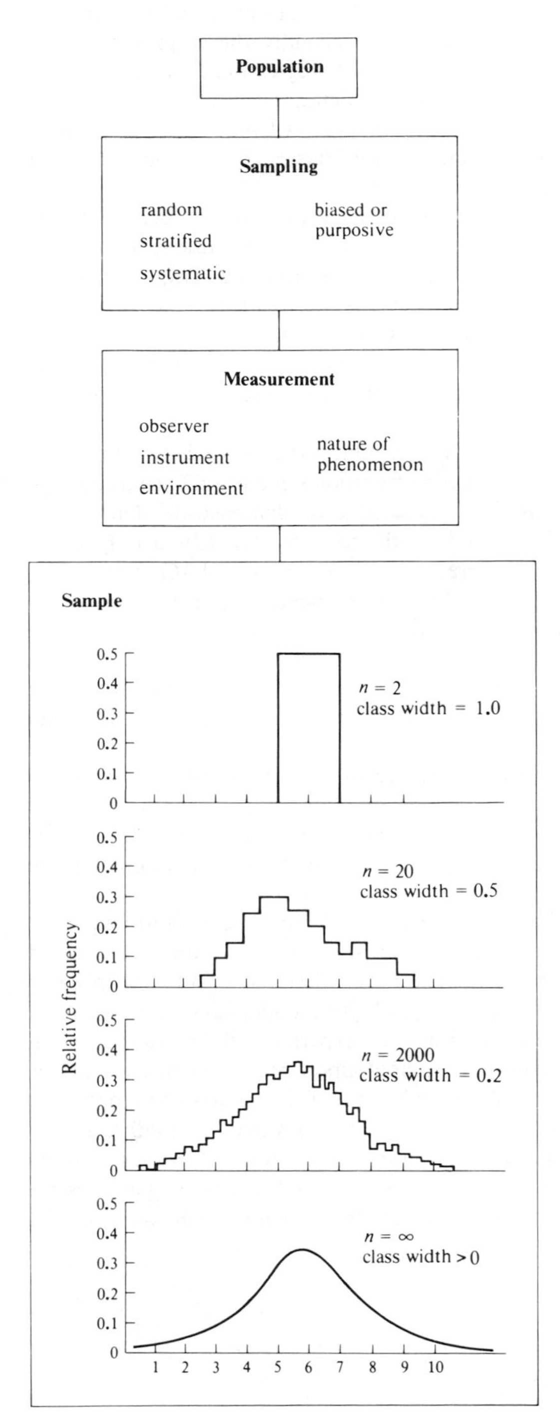

Figure 1.15 Sampling strategies and the combined effect of sources of error with sample size on the values of a hypothetical variable.

The use of statistical techniques in association with geomorphological ones necessitates the following of the assumptions regarding *both* techniques and of making the correct inference from the application of the combination of these techniques to the geomorphological problem. As regards the geomorphological techniques, bias and error of the data are important (Poole & O'Farrell 1971) and, secondly, the assumptions of the statistical method (for example, spatial autocorrelation or regression residuals) must be complied with. It is beyond the scope of this review to detail the spectrum of statistical techniques that geomorphologists employ, but two examples from regression analysis will serve to illustrate the geomorphological importance of testing technique assumptions.

A study of causal factors of flood variability in New England by Benson (1960) provides a rare example of testing a multiple regression model at each stage of inclusion of a new variable for spatial autocorrelation among the residuals. A consequence of such is that, statistically at least, the final model was satisfactory with the exclusion of an orographic factor being recognised 'statistically' in the penultimate regression formulation. While adherence to the strict statistical assumptions of this type is rare in geomorphological applications, it does highlight the importance of ensuring that variables have not been omitted.

A somewhat more complex and more fundamental variable omission has been revealed by Richards (1973) in the hydraulic geometry studies of natural stream channels reported by Leopold *et al.* (1964). Richards was able to demonstrate that certain relationships between discharge, velocity and depth were best fitted by a polynomial regression rather than the linear function previously assumed (Fig. 1.16). The critical omission here was that of the roughness effect, consequential upon an increase in discharge, on the dependent variables. Richards concludes that, as a result of the application of the incorrect (linear) statistical model, oversimplifications resulted in the geomorphological relationships that are inferred, and that this situation probably arose from the tendency of the earlier workers to concentrate on the empirically available data and infer roughness from this.

This latter example shows the importance, therefore, of adhering to the assumptions of the modelling technique, and the consequences in terms of process inference if this is not done. Similar consequences are reported by Dowling (1974) in the context of the analysis of eight water quality variables over time, and the problem of correctly estimating the phase lag unambiguously between each of them.

Interrelationships – the 'mainstay' of geomorphology – cannot be a continuing basis for the organising technique framework *if* such relationships are of the form of statistical probability functions, however elaborate; we must have a desire at least to enhance our modelling techniques to maximise our geomorphological understanding. In certain fields, recent work suggests that mathematical and experimental investigations, in necessitating an initial theoretical hypothesis (based upon presumed mechanical or chemical controls), may provide this enhancement; in other fields, it may be inapplicable for the foreseeable future. Nevertheless, Rescher (1962) argues that strict adherence to the deductive concept of explanation 'may be buttressed by fond memories of what explanation used to be like in *nineteenth*-century physics'. We continue to get plausible and intellectually helpful answers in statistical terms to the question of why did X occur.

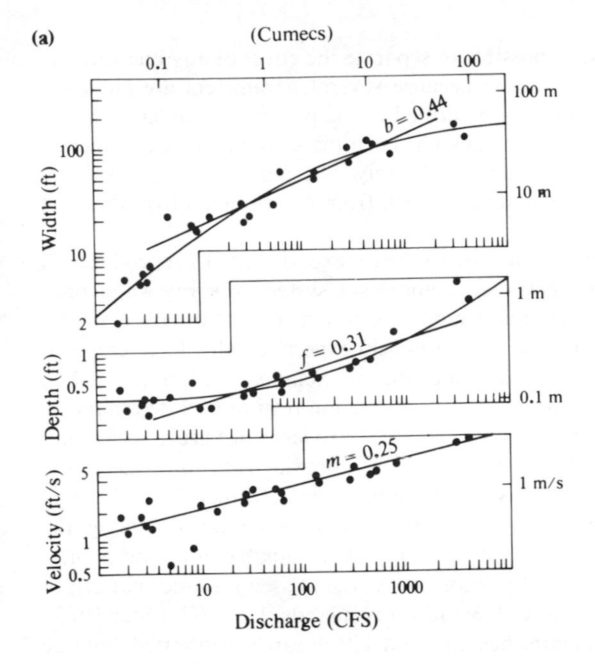

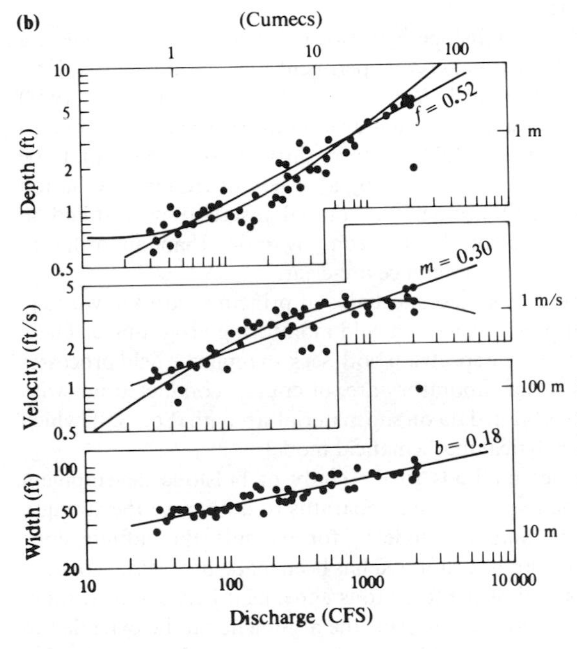

Figure 1.16 (a) Hydraulic geometry relationships on the Rio Galisteo Domingo, New Mexico, gauging station. (b) Hydraulic geometry relationships on the Senerca Creek, Dawsonville. All parts have the original straight-line relationships indicated, together with the best-fit curves. (After Richards 1973).

1.2.2.4 Model calibration

In many field studies, it is not always possible to separate the effect of any one variable controlling the principal one under study; because several parameters are changing simultaneously, such a condition has been described as the principle of indeterminacy. Such temporal as well as spatial indeterminacy has been the stimulus for experimental investigations in which variables can be controlled singly. The analogue to this practical experimental technique is that of attempting to work from a theoretical hypothesis of process operation.

Jagger (1908) was probably among the first to devise experiments to reproduce rill patterns, relying on the use of an electric air compressor system to spray water onto inclined glass plates covered with 'the finest kind of crushed rock'. Moreover, in this early attempt to consider drainage network development experimentally, he attempts to monitor the effects of ground water on pattern evolution. Schumm (1977b, pp. 230–2) observes that the reader of Jagger's work is left with the impression of a deterministic drainage system growth mechanism, in contrast to the approaches of Shreve (1967) and Smart (1972). Schumm (1973) has, however, been able to provide suggestions as to the processes involved in river incision in terms of explaining ancient, former channels. Such models are rarely capable of maintaining the appropriate scale ratios among the hydraulic and morphological variables involved, but they are capable of providing new information on such large-scale natural phenomena as the incised meander pattern of the Colorado River across the structural dome in Canyonlands, Utah (Gardner 1975). Through the control that the investigator has, the model itself can be calibrated, but due to the scale constraint the model cannot be calibrated completely with the natural phenomena it is desired to replicate.

By contrast, experimental studies of hillslope hydrological behaviour can achieve a much closer one-to-one correspondence between experiment and nature, particularly since time and spatial scales can be identical. Accordingly, Anderson and Burt (1978) were able to optimise a design of throughflow trough by laboratory experimentation. In the laboratory, known hydraulic conditions allowed exact throughflow prediction to be made, which could be compared with that received by a variety of trough installation designs at different depths in the soil profile. Suboptimal installations resulted in throughflow estimates that were in error, by horizon, by more than an order of magnitude; the attendant errors in process inference are clear.

Experimental techniques, where the process mechanistic principles are known (e.g. Darcy's law), can be an important prerequisite to a field monitoring programme. These two examples at opposing ends of the scale spectrum both seek to replicate field processes or conditions. Many experiments in the laboratory are, of course, conducted not with this aim in view, but are designed to obtain data on site material strength (Sec. 3.4), which can then be used to calibrate an analytical mathematical model.

The application of soil mechanics methods to the study of hillslope development provides just such an example. The use of triaxial apparatus to determine the strength parameters of material in order that stability models – for example, the 'infinite slope model' (Skempton & Delory 1957) – can be calibrated has been made by several workers. Chandler (1977) has, however, warned of the fortuitous agreement between field slides and laboratory-calibrated stability models. However, the argument can be extended to embrace a situation in which an apparently stable slope has a computed factor of safety less than unity (see Ching *et al.* 1984, Anderson and Howes 1985). There is thus plenty of evidence to suggest the need for particularly accurate measurements of pore-water pressures at the shear surface as well as the need for careful and appropriate field and laboratory strength determinations.

Calibration of a model to the natural phenomenon under investigation is among the most difficult of tasks: it combines several techniques with attendant errors, it can require scale alterations (either qualitative or quantitative) and it can involve field–laboratory iteration, in the context of equipment installation, for example. There can be no escape from the rigour of measurement or from the close definition of the assumptions contained in a mechanistic model of process. As the techniques described in the following sections are in due course refined, the need for such rigour in the science of geomorphological investigations will increase. The consequences of failing to maintain consistency from technique to model or vice versa, as we have seen from examples of statistical analysis, is particularly harmful in terms of inference.

We have argued in general terms that increasingly individual geomorphological studies will be tackled in terms of process mechanics, aided by the techniques described in the following sections; and that, moreover, this is compatible with geomorphological explanation at the meso-scale, *providing* aggregation or microprocess studies prove possible. Chorley (1978) has stressed the need to coarsen physical truth to a degree that 'geomorphologically-viable landform objects' are studied. While many would undoubtedly regard this as an important prerequisite for the maintenance of geomorphology's contribution to science, in practical terms, with new techniques allowing us to undertake more detailed investigations to a higher standard than pre-existing techniques did at a larger scale, it may prove difficult to set or adopt an enquiry level boundary, *however* desirable. Supportive comment to Chorley's views comes from West (1978), who argues that it is easier in geomorphology to quantify and apply techniques to *modifying* variables than to the true controlling ones, and it is possible that Leopold's (1978) observation, that in the case of arroyo investigations the easy questions were asked while difficult, more important, ones were bypassed, may have a much more general relevance, which all geomorphologists should perhaps consider if they are to avoid being 'technique-led' in their investigations.

PART TWO
FORM

2.1
Introduction to morphometry

Rigorous, scientific geomorphology demands quantification in order that its theories can be tested. Qualitative description of surface form must therefore frequently be replaced by morphometry: the *measurement* of landforms. Such measurement generates nominal- and ordinal-scale information (for example, qualitative landform classifications and stream ordering methods), but interval- and ratio-scale data are commonly employed. The nominal and ordinal scales are in fact often used to define 'treatments', or strata, for classification of samples of data measured on the higher scales. Some form of classification system often precedes quantitative measurement of landform attributes. For example, different indices of form may be deemed relevant for different types of sand dune, so that initial classification of such individual features as barchan, parabolic, linguoid, or sief dune forms is necessary. However, it is often preferable to define sets of indices that are generally applicable, and attempt classification after measurement, by using objective geomorphological classification and regionalisation schemes based on principal components analysis and cluster analysis (Mather & Doornkamp 1970).

The objectivity imparted by numerical data can, however, be illusory, since quantitative indices must be operationally defined, reproducible in practice and valid representations of theoretical concepts (Wilks 1961). Furthermore, morphometry often generates relationships whose geomorphological significance is limited, such as the Hortonian stream number–order relationship, which equally satisfactorily fits the veins in bovine livers. The most geomorphologically productive measurements of Earth surface form are those which relate form to process in some way. If equilibrium between form and process can be assumed, this relationship can be exploited in both directions. The correlation between landform characteristics and process variables can be used to interpret the physical adjustment of the landscape to prevailing process regimes, while at the same time process variables can be predicted from landform characteristics. However, in each case, a high level of measurement of landform indices is required, in terms of both the measurement scale adopted and the precision of operational definition of variables.

Two general themes are apparent in the following sections dealing with aspects of this measurement process: those of scale and data source. General geomorphometry (Sec. 2.3) considers large-scale analysis of altitude data and their derivatives, and this is supplemented by a section concerned with altitude measurements for geochronological purposes (Sec. 2.4). Specific morphometry deals with the approaches to the measurement of plan and profile characteristics of specific features (Sec. 2.5). Although the methods outlined in Section 2.3 can be used to investigate specific landforms (e.g. cirques or drumlins; Evans 1987a), the type of index considered in Section 2.5 is less demanding of data and easier to derive rapidly. Landforms of fluvial environments are considered at progressively smaller scales with the analysis of drainage basin (Sec. 2.6), river channel (Sec. 2.7) and slope (Sec. 2.8) forms. The mapping techniques of Section 2.9, while apparently representing a more generalised approach to geomorphology, consider the landscape as the sum of its parts, defining general characteristics from the assemblage of specific landforms and processes, and hence integrating the two scales of study. Parallel with the scale and generalisation theme is that relating to data sources: the earlier sections most often employ secondary map and air-photograph sources, while Sections 2.7 and 2.8, and some methods considered in Section 2.9, require primary field data.

First, however, some important methodological and practical problems common to many areas of morphometry are briefly outlined.

2.2 General problems in morphometry

Several critical decisions precede quantification of landform properties (Evans 1987a). A conceptual image of the landform must be translated into measurable attributes that represent the concept satisfactorily, and can be quantified with accuracy, precision and reproducibility. Thus a rigorous operational definition is essential, and this must consider:

(a) delimitation of the landform boundary;
(b) the quantitative index itself (see, for example, Sec. 2.5);
(c) the sampling scheme required to define a representative subset of the relevant population;
(d) the appropriate and available data sources and methods of measurement (see Sec. 2.2.2); and
(e) the measurement procedures and practices.

2.2.1 Scale problems

Scale considerations clearly influence decisions concerning measurement technique and data source. A regional relief survey may derive sufficient information about gradient from a map-based area–altitude analysis (Secs. 2.3 & 2.4), whereas a study of slope processes necessitates field slope profile survey (Sec. 2.8). This reflects the varying level of generalisation acceptable at these different scales.

Scale and generalisation even control detailed decisions concerning measurement, particularly because the scale of sampling unit influences variability in the index measured (Reynolds 1974). Slope profile survey over fixed distances is preferable if subsequent statistical analysis is to be attempted, but the slope angle *variance* is inversely related to measurement length (Gerrard & Robinson 1971). If Young's (1971) technique for identifying slope segments and elements is then used, the critical coefficients of variation that control this classification are therefore themselves dependent on the field procedure used to measure slope angles.

Awareness of the effects of sampling scale and aggregation of sampling units is thus essential. If the effect of scale is known, prediction between scales is possible. Thus, relief increases systematically with the size of circular sampling units. If this property is understood, it can be used for prediction; relief in small areal units can be estimated from measurements of large areal units on small-scale maps too generalised for direct measurement of the smaller areas (Wood & Snell 1957). Relationships are also scale-dependent. Evans (1972), for example, showed that the correlation between quadrat relief and the standard deviation of altitude can decrease from $r = 0.97$ to $r = 0$ over a range of grid square sizes from 475 to 6000 m. Only in special cases, where phenomena have true fractal patterns (Mandelbrot 1977a, b), is relative variation constant at all scales.

The *nature* of relationships identified between variables depends on the scale of an investigation. If slope angle and drainage density are related when sampling is confined to an area of uniform relative relief, the association is different from that observed if relief is variable over a wide sampling region (Chorley & Kennedy 1971, p. 62). Generally, the larger the spatial scale of a study, the greater the number of potentially interrelated variables that need to be measured or to be controlled by the sampling scheme. In a local investigation, many variables can be assumed constant.

Thus, the scale and purpose of investigation, the initial assessment of the variability of phenomena being measured, and the time available for data collection, all dictate the appropriate data source and measurement technique, and the results obtained are invariably specific to the scale, source and technique adopted.

2.2.2 Data sources

Map, photographic and field data rarely exist independently of one another since maps are derived from remotely sensed imagery and field survey, and ground truth for imagery is provided in the field. The validation of map data may use air photographs (Eyles 1966), but photogrammetric data need field checks. In many cases geomorphological studies exploit all three sources, and the dominant one reflects the availability of maps and imagery at scales appropriate for the investigation. Historical changes in the landscape in periods prior to the development of remote sensing can be assessed using early cartographic sources such as tithe and estate maps, and Hooke and Kain (1982) discuss such sources and their special problems.

2.2.2.1 Topographic maps

The appropriate map scale to employ in studies of Earth surface form depends on the type of study; land systems mapping may employ 1:50 000 base maps, or even smaller scales, whereas detailed morphological mapping (Sec. 2.9) and morphometric measurement require maps of 1:25 000 or larger with contour intervals of 10 m or less. The volume of data required by morphometric studies often demands secondary, published map sources, with attendant problems concerning the realism of landform portrayal. Surveys of map accuracy in drainage network representation (Gardiner 1975, Sec. 2.6) suggest that the 1:25 000 map provides the most acceptable compromise between accuracy and convenience in Britain, in spite of inconsistencies between the two series of this map; and Werritty's (1972) conclusion is that only 1:2500 maps adequately define exterior stream link lengths. However, new 1:10 000 maps with photogrammetric contours are a considerable improvement. The recommended scale in the USA is the 1:24 000 series (Smart 1978), although its reliability in network portrayal is questioned (Morisawa 1957, Giusti & Schneider 1962, Mark 1983).

For maximum comparability of results, morphometry should employ maps near these suggested scales, although cartographic conventions vary sufficiently between national surveys that calibration using air photography is desirable. Unfortunately, little of the Earth is mapped at a scale suitable for detailed morphometry. Topographic maps at scales of 1:20 000 to 1:25 000 exist for most of Europe, Scandinavia, Japan, the USA, New Zealand and some North African and South American countries. Maps at a scale suitable for morphometry do not exist for much of Asia and are not available for the Eastern bloc, but Middle Eastern and South American surveys are increasing their 1:50 000 coverage. Many African countries are also mapped at this scale by the Institut Géographique National in Paris or the Directorate of Overseas Survey in the UK, and since the official surveys of Canada, South Africa, Australia and Iberia are all largely at 1:50 000, it may often prove necessary to resort to these smaller-scale maps or use photographs. Up-to-date information on map coverage can be obtained from national survey organisations, the *Bibliographie Cartographique Internationale*, or *GEO Katalog* (published by GeoCentre Internationales Landkartenhaus, Postfach 800830, D-7000 Stuttgart, FRG).

2.2.2.2 Remote sensing

Remote sensing includes all representations of the Earth's surface generated without direct human measurement, regardless of the type of imagery (visible spectrum, thermal infrared or radar wavebands). Even ground photography may provide remote sensing of phenomena such as glacier motion, gully extension or landslide displacement, either by repeat photography from a fixed platform or by automatic exposure. Photogrammetry from ground photographs usually requires secondary ground survey to provide ground control. A novel application of ground photography involves simultaneously viewing with one eye a scene and with the other a colour transparency taken from the viewpoint and mounted in a viewer with a focusable lens (Kaulagher 1984). This results in 'direct mixed fusion' analogous to stereovision, and permits calculation of displacement of moving objects (e.g. a landslide toe) by measurement of the angle of rotation needed to fuse their images. Ground-level ciné photography can also be used to monitor geomorphological processes, particularly if displacement velocities can be estimated from fixed points providing distance control; Wright (1976) provides an example in the measurement of wave properties.

Conventional air photographs at scales of 1:20 000 or greater yield accurate height data for mapping small-scale features such as beach ridges and floodplain topography, but require good ground control and experience in the use of sophisticated stereoplotting equipment. Relatively inexpensive stereoscopes can, however, provide slope angle data from air photographs (Turner 1977). Photo scales of 1:50 000 are more appropriate for land systems mapping (Sec. 2.9). Sources of air photograph coverage are listed in Table 2.1, and by Curran (1985) and Carter (1986). Detailed explanations of geological and geomorphological applications of photo interpretation and photogrammetry are provided by Colwell (1983), Verstappen (1977, 1983) and van Zuidam (1985/6). The latter authors derive their experience from the Institute for Aerospace Survey and Earth Sciences at Enschede in the Netherlands. They review photo interpretation of landforms in different morphoclimate zones, associated with a range of processes, and in applied contexts.

Air photography often can only supplement map data (Woodruff & Evenden 1962, Eyles 1966), and cannot entirely supplant field investigation of dynamic phenomena because special-purpose, repeat photography is expensive.

Alternative remote-sensing developments involving space-borne platforms (Table 2.2) may be cheaper to the user of the imagery, but have yet to realise their full potential in geomorphology, notwithstanding the early reviews by Cooke and Harris (1970) and Allan (1978), and more recent assessment by Curran (1985). These developments include:

(a) a range of satellite platforms both manned and unmanned (Table 2.2);
(b) sensors employing wavebands within and beyond the visible spectrum; and
(c) image processing methods.

Although spatial resolution is now high (10–40 m ground resolution from the Landsat Thematic Mapper and SPOT; see Table 2.2), systematic temporal sampling at appropriate intervals cannot be guaranteed, and, for strongly process-oriented small-scale geomorphological investigations, satellite imagery is unlikely to prove useful. However, it can provide data for mapping remote and inhospitable terrain, for identifying

Table 2.1 Sources of remotely sensed imagery.

Air photography (vertical)	
England	Air Photo Cover Group, Ordnance Survey, Romsey Road, Maybush, Southampton SO9 4DH, UK
Wales	Central Register of Air Photographs, Welsh Office, Cathays Park, Cardiff CF1 3NQ, UK
Scotland	Central Register of Air Photographs, Scottish Development Department, St James Centre, Edinburgh EH1 3SZ, UK
Northern Ireland	Ordnance Survey of Northern Ireland, 83 Ladas Drive, Belfast BT6 9FT, UK
USA	User Services, EROS Data Centre, Sioux Falls, South Dakota 57198, USA
Canada	National Air Photo Library, 615 Booth Street, Ottawa K1A OE9, Canada
France	Institut Géographique National, Centre de Documentation de Photographies Aériennes, 2 Avenue Pasteur, 94150 St Mande, France
Developing countries	UN Remote Sensing Unit, FAO, Via delle Terme di Caracalla, 00100 Rome, Italy
Satellite imagery (general)	
UK	National Remote Sensing Centre, Space Department, Royal Aircraft Establishment, Farnborough GU14 6TD, UK
USA	EROS (Earth Resources Observations System) Data Centre; as above
Europe	National Points of Contact for EARTHNET, an organisation of the European Space Agency (addresses are given by Carter (1986, pp. 47–8))
Satellite imagery (specific)	
Skylab	Skylab Earth Resources Data Catalog, US Government Printing Office, Washington DC 20402, USA
Meteosat	Meteosat Data Services, European Space Operations Centre, Robert-Bosch-Strasse 5, D-6100 Darmstadt, West Germany
GOES	National Oceanic and Atmospheric Administration, Satellite Data Services Division, Room 100, World Weather Building, Washington DC 20233, USA
Landsat	EARTHNET Points of Contact; Landsat data reception and processing stations (addresses are given by Carter (1986, pp. 227–30))

Source: after Carter (1986), Curran (1985).

large-scale landform patterns such as desert dune trends (active and fossil) or features associated with structural control, and for studying large-scale processes such as aeolian dust transport or major flood inundation. Imagery of value in geomorphological studies (see Table 2.2) can be obtained from sources listed in Table 2.1, or from other agencies detailed by Carter (1986).

Much of the research effort in remote sensing is devoted to calibration and image processing. Modern multispectral sensors/scanners (MSS) provide images of surface reflectance in narrow wavebands within the visible spectrum, as well as thermal infrared wavebands; microwave radar wavelengths are also employed (Table 2.2). Different surface phenomena are recognisable from the 'density' of their reflectance signatures in specific wavebands, and combinations thereof (e.g. Curtis 1974). For example, suspended sediment in water provides a strong reflection of 'red' light (0.6–0.7 μm, which penetrates the water column, so band 5 of the Landsat MSS or band 2 of the Thematic

Table 2.2 Space-borne platforms for remotely sensed imagery.

Manned flights		
Skylab	1973–4	Resolution 30–80 m; Earth Terrain Camera (V)†; 13-channel Multi Spectral Scanner (V, I, M)
Shuttle*	1981 to date	Resolution *c.* 40 m; large-format metric camera (V); shuttle imaging radar (M)
Unmanned satellites		
Landsat*	1972 to date	Five series; latest has six-channel MSS, Thematic Mapper (V) with 30 m resolution; thermal scanner (I) with 120 m resolution; 16-day repeat imagery
Seasat	1978	Resolution 25 m; Synthetic Aperture radar (M; + V, I)
Meteosat	1977 to date	Geostationary; repeat imagery every 30–40 min (V, I)
GOES	1978 to date	Polar orbit; 15 orbits per 24 h (V, I)
SPOT*	1985 to date	Resolution 10 m (panchromatic), 25 m (multispectral) (V); 26-day repeat imagery; off-nadir viewing gives potential for 1–3 day repeat and stereo imagery

* Imagery of greatest value geomorphologically.
† V = visible and near-infrared (0.4–1.1 μm); I = infrared and thermal (3–14 μm); M = microwave radar (1–500 mm).

Mapper display suspensions clearly. However, extensive ground calibration is required so that reflectance characteristics of specific phenomena can be defined and subsequently recognised and quantified from imagery. After calibration, image processing methods permit automated classification of individual pixels within an image, and mapping the distribution of a phenomenon is then readily possible. However, it is likely that many new applications will require calibration and processing *ab initio*, which involves extensive ground control. This may mean that remote sensing cannot be used to circumvent the costs, in money and time, of direct field monitoring. Furthermore, as the calibration is empirical, it cannot be guaranteed that relationships – for example, between the density of 'red' light reflectance and sediment concentration – can be used on other images exposed at different times and places. The benefit of satellite remote sensing is thus largely that it can reveal new phenomena, as well as large-scale patterns. Side-looking radar, for example, can 'see through' cloud cover (Martin-Kaye 1974), dense vegetation (McCoy 1969) and even superficial sedimentary deposits, and reveal hidden features simply as a result of visual appraisal of the image. In addition, some image processing methods that involve spatial and directional filtering can yield geomorphologically useful products such as automatically generated slope maps (Gardiner & Rhind 1974, Statham 1976).

Thus, remote sensing has its greatest geomorphological potential in analysis of relatively static phenomena; unresolved problems exist in the relationship between acceptable images obtained at variable intervals and the dynamism of geomorphological process systems. However, for small-scale, detailed studies requiring accurate height data, and for advance study before field survey, large-scale air photographs are indispensable.

2.2.2.3 Field survey

A detailed explanation of surveying techniques cannot be provided here; the reader is referred instead to the text by Bannister and Raymond (1984). The choice of technique clearly reflects the nature of the problem: Abney levels are sufficiently accurate for slope profile survey; levelling and tacheometric distance survey are necessary for lower-gradient landforms such as stream beds or floodplains; and electro-optical distance measurement (EDM) is preferable where ground disturbance is a problem as on saltmarshes (Carr 1983). 'Total-station' instruments, consisting of a theodolite with built-in EDM, are fast becoming favoured for rapid and accurate survey, since distance and angle measurements can be stored on a data logger, which can download to a microcomputer with automatic mapping software. Nevertheless, plane table, alidade and chain/tape survey remain invaluable because the surveyor operates interactively with the map developing before him.

2.2.3 Sampling problems

The size of most geomorphological populations demands sample extraction before measurement, the sampling strategy being dependent on the purpose of data collection and the nature of the population. The concept of a population is often idealised, and it may be acceptable to treat the complete set of measurable landforms in a particular area as a sample realisation of a random process that could have produced an infinite number of arrangements of such features (Cliff 1973). Thus, *inference* about population characteristics is possible even if all individuals have been measured; but see Gould (1970) and Meyer (1972).

The range of sampling schemes available to the geomorphologist includes random, systematic and probabilistic sampling methods, and mixtures of these, such as systematic random sampling and nested sampling designs, in which random sampling is undertaken within systematically defined areas (Chorley 1966). For subsequent inferential statistical analysis, random sampling is essential. Stratification of the sample is common and is a natural requirement for any statistical comparison of samples. However, some forms of analysis (series analysis, for instance) require systematic sampling. A population of discrete landforms (cirques, drumlins, dunes) can be sampled at random because the observed 'population' can be identified, numbered and formally sampled using random numbers. A continuous phenomenon is not easy to sample; for example, Young *et al.* (1974) demonstrate the problems encountered in selecting measurable slope profiles (Sec. 2.8). In this case, it is impossible to define the size of a proportional sample. In spite of these problems, random sampling is necessary to avoid bias in subsequent analysis, and feasibility or convenience should not dictate sample selection.

It is frequently necessary to define the sample size required to provide with the minimum labour an appropriate level of confidence in estimating population parameters (Carson 1967). For normally distributed variates, this is possible after prior estimation of the population variance from a trial sample (Gardiner & Dackombe 1983, pp. 216–18). Figure 2.1 can be used to identify the sample size necessary to establish a population mean within a chosen confidence limit, at the 95% probability level. An initial small sample (5–30 observations) is collected, and its standard deviation is calculated. This is then divided by the required confidence interval, and this ratio is used to identify the necessary sample size for the chosen level of confidence. For example, if an initial sample of five has a standard deviation twice the required confidence interval, a final sample of 32 is necessary to achieve such a confidence interval.

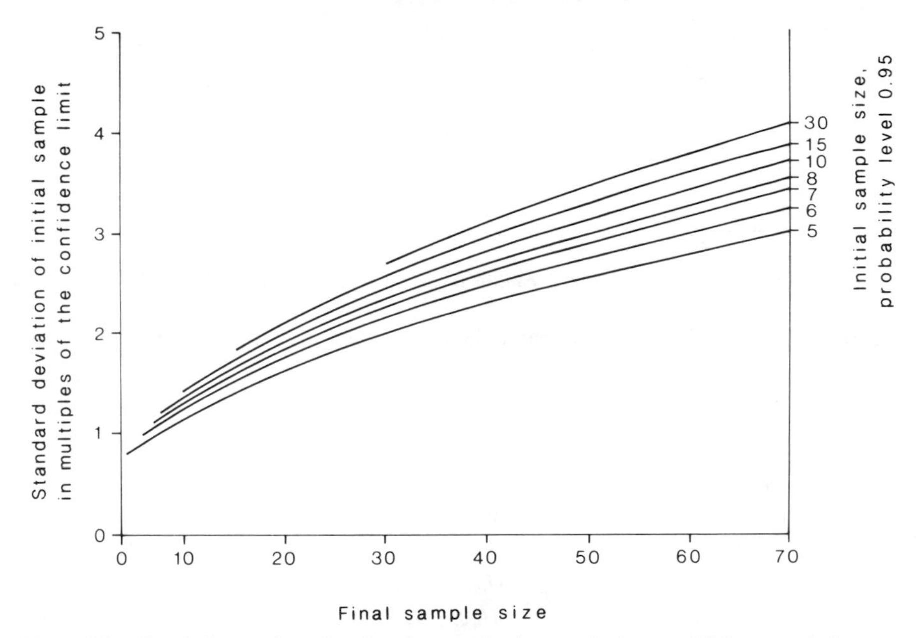

Figure 2.1 Graph for use in estimating the sample size required to establish a population mean within a given confidence limit. (Based on Gardiner & Dackombe 1983, Fig. 12.1, p. 218).

Figure 2.1, and similar charts (Reynolds 1975), assumes random, normal variates. Unfortunately, geomorphological populations are often non-normal (Gardiner 1973) and these rules do not apply. An initial priority is therefore to compare the sample with a theoretical distribution (which needs a minimum sample size of 30–50), and to transform the variable where appropriate. Furthermore, the *effective* sample size is commonly reduced because of spatial autocorrelation (i.e. non-randomness), and larger samples are then needed to specify population parameters. The structure and extent of spatial autocorrelation in geomorphological data are demonstrated by the semi-variogram (Oliver & Webster 1986), which can be used to reveal quite complex patterns of spatial dependence in, for example, variables defining physical and chemical attributes of soils and sediments.

2.2.4 Definition of variables

Quantification is impossible until ideas and concepts are translated into variables that can be measured. For example, landform shape (Sec. 2.5) may be discussed qualitatively, but a clear operational definition of a measurable, reproducible shape index is required before quantification of shape can be achieved. It is necessary to delimit the extent of some features (whether cirque, drumlin, sand dune or drainage basin), and, if this is not satisfactorily and consistently achieved, comparability between measurements may be lost. The operational definition, as well as enabling reproducible measurement, must generate numerical indices that relate directly to the geometric or physical quality that is the subject of analysis.

Frequently, variables have been defined that reproduce information in existing variables; again, landform shape is a clear illustration. Horton's (1932) arbitrarily defined form factor (area/length^2) is essentially an equivalent measure to the lemniscate k

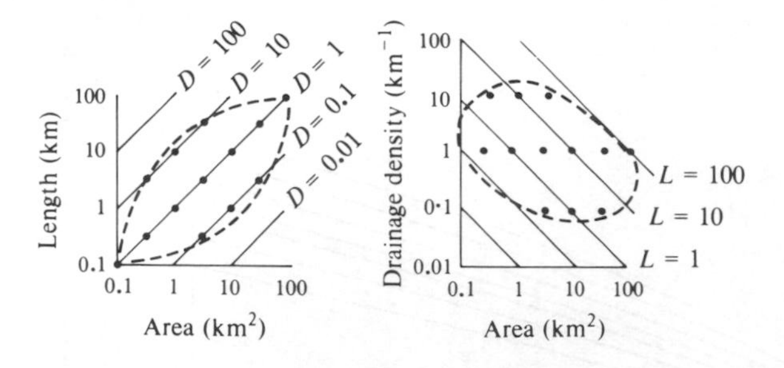

Figure 2.2 Ratio correlation effects; since $D = L/A$, an apparent relationship between D and A arises from a sample in which L is correlated to A.

of Chorley *et al.* (1957), which compares the drainage basin with an assumed teardrop shape and is

$$k = \pi L^2/4A$$

Even less immediately apparent is the mathematical equivalence between the hypsometric integral (Strahler 1952) and the elevation–relief ratio (*ER*) (Pike & Wilson 1971). The latter is simple

$$ER = \frac{H_{max} - H_{mean}}{H_{max} - H_{min}}$$

and is much simpler to estimate given a sample of height measurements (H) to calculate mean height. Thus, when alternative definitions exist of a single variable, the choice rests with the one with least operational error and tedium.

Many variables are derived from two or more primary measurements, which are combined to form ratio variables, and this creates difficulties in subsequent analysis. Ratio correlation (Benson 1965, Chayes 1971) is illustrated in Figure 2.2, where the derived variable, drainage density (D), is obtained from channel length (L) and basin area (A). A log–log plot of length against area is contoured by lines of equal density and an arbitrary ellipse of points is located to represent a sample from a population in which L is positively correlated to A. The same data plotted on a D–A diagram, contoured with values of L, illustrate a slight apparent inverse relation, which simply reflects the derivation of D. Ratio correlation may be a particular problem if dimensional analysis is used as a technique for identifying the nature of interrelationships between variables (Haynes 1983). This creates dimensionless groups of primary variables, some of which may include the *same* variables. Nevertheless, when used carefully, dimensional analysis can reveal the form of relationship to be expected, and helps to identify limitations in a sample of data if this relationship does not materialise; Wong (1963, 1979) provides a salutary example.

2.2.5 Operator variance

However well specified the operational definition, different 'operators' inevitably obtain different values for particular morphometric measures of the same landform. This variation reflects errors associated with specification of the measurable feature, and with measurement itself. The former is critical when secondary data sources are employed. For example, drainage patterns on topographic maps are arbitrarily selected models of reality (Board 1967) and the network to be studied may be defined by the blue lines, or extended by the contour crenulation (Strahler 1957) or slope criterion (Shreve 1974) methods. The subjectivity of the second method may create specification problems.

Nevertheless, Chorley (1958) demonstrated non-significant differences between groups of operators even where first-order streams were defined by the crenulation method.

However, analysis of variance between *groups* is an inappropriate guide to the operator inaccuracy arising in morphometric studies; there is no expectation of differences between randomly selected *groups*, although systematically defined groups may show differences. More marked deviations occur between *individuals*, and comparisons between studies are therefore difficult to achieve. Variations in measurement technique are particularly critical. Mandelbrot (1967) showed that lengths of irregular lines depend on chord approximations made during measurement. Thus, a researcher measuring stream lengths by dividers or chartometer would be ill advised to predict streamflow using a baseflow–stream length regression based on data involving length measurements from a coordinate string generated by a digitiser, without an assessment of the comparability of the different length estimates.

2.3
General geomorphometry

Faced with the complexity of the Earth's surface, the geomorphologist seeking to quantify its form is immediately forced to analyse it into more manageable components. He has two major strategies available. By defining specific landforms, he can abstract these from their surroundings to measure their size, shape and relation to each other. This approach, called *specific* geomorphometry, requires precise operational definitions (Sec. 2.5) and involves problems of landform delimitation and operator variance.

Alternatively, the geomorphologist may attempt to analyse the land surface as a continuous, rough surface, described by attributes at a sample of points, or of arbitrary areas. Such *general* geomorphometry completely avoids problems of landform definition and delimitation, and subjectivity and operator variance are reduced drastically, though not eliminated. Its major problem is the dependence of results on data resolution and sampling pattern.

In both types of geomorphometry, attributes should be defined to facilitate their relationship to Earth surface processes. In this way, a description and analysis provided through general geomorphometry is valuable for:

(a) comparing areas and suggesting differences in form that require explanation;
(b) demonstrating whether sites selected for detailed study (e.g. process measurements) are representative of a study area or of a certain subset of sites comprising the target population;
(c) demonstrating how representative a study area is of a broader region to which extension of conclusions is desired;
(d) expressing the areal importance of a target population (e.g. 'straight slopes') selected for detailed study;
(e) predicting process rates and discharges from their relationships to form attributes; and
(f) showing the presence and importance of various combinations of surface attributes, which may be relevant to predictions from theoretical models.

Land surface form has important effects on construction, 'trafficability', erosion hazard, farming, telecommunications, radar, terrain-following aircraft and missiles, irrigation and drainage. Other influences are via topoclimate, vegetation and soil. In most of these applications, it is not necessary to predefine specific landforms; general geomorphometry is the relevant approach.

The land surface is defined by deviation from the equipotential surface of sea level; for small areas, the latter is treated as a flat plane on which position is defined by the cartesian coordinates (x, y). Deviations (z) are measured at right angles to this equipotential surface, giving land surface altitude. Tractability and conceptual simplicity are greatly improved by excluding caves, overhangs and vertical cliffs, and treating the land surface as *single-valued*, with only one value of z for each position (x, y). Hence, *general* geomorphometry as treated here deals with the measurement and analysis of attributes of a one-sided, single-valued, continuous rough surface.

2.3.1 Point attributes

A useful description of landform at any point is given by altitude and the surface derivatives, i.e. slope and convexity (curvature). Both derivatives are vector quantities and are best resolved into vertical and horizontal components. 'Slope' is defined by a plane tangent to the surface at the given point, and is completely specified by the two components, gradient and aspect. 'Gradient' is the maximum rate of change of altitude (z) with horizontal displacement, while 'aspect' specifies the direction of this maximum in the (x, y) plane. Aspect is orthogonal to contour trend, the rate of change of x with y for constant z. Hence gradient is essentially the first vertical derivative of the altitude surface, while aspect is the first horizontal derivative.

Gradient affects the operation of gravity and aspect does not, so it is essential in relation to process to make this conceptual separation, rather than treating slope as a vector in three dimensions as does Hobson (1967, 1972). Aspect should not, however, be viewed independently of gradient, since aspect is indeterminate if gradient is zero, and the importance of aspect in controlling topoclimate (and hence process) increases with gradient. Analyses such as those of Chapman (1952) usefully combine representation of gradient and aspect, while maintaining their distinction, by plotting them in polar coordinates.

The influence of altitude, especially upon climate, is universally recognised, and gradient appears to be a fundamental geomorphological attribute. There has been less recognition of the importance of a further land surface property, curvature. This may be defined as rate of change of slope, and hence is the second derivative of the altitude surface. Since it can be positive or negative, the term 'convexity' is a reminder of the convention that positive curvature is convexity, while negative curvature is concavity. A point on a straight, planar slope has zero convexity. (This sign convention follows that of Young (1972) and many others, but is contrary to that of Troeh (1965) and Speight (1968).)

The two components of the convexity vector are *profile* convexity, the rate of change of gradient along a slope line (a line of maximum gradient), and *plan* convexity, the rate of change of aspect along a contour. These properties control the acceleration/deceleration and convergence/divergence of near-surface water flows (Heerdegen & Beran 1982, Burt & Butcher 1986) and influence the stress field in near-surface rocks. Zevenbergen and Thorne (1987) use curvature in a vertical section transverse to the slope line, in place of horizontal or contour curvature. The two measures of convexity may be combined to provide a measure of topographic exposure/shelter if average altitude at a fixed distance is subtracted from altitude at the point in question to give *local* convexity (Evans 1972, p. 67). Papo and Gelbman (1984) map a combined radius of curvature, calculating this and slope from the fast Fourier transform.

Where altitude is expressed in metres and gradient and aspect are in degrees, profile and plan convexity are expressed in degrees/100 m. Gradient may vary from 0 to 90°. Aspect, measured clockwise from north, varies from 0 to 360°. Altitude and both convexities may be positive or negative, and have no limits.

These five attributes (altitude, gradient, aspect, profile convexity and plan convexity) are fundamental to general geomorphometry. In theory, defined as derivatives, they are attributes of any *point* on the land surface. In practice, except for altitude itself, they can be calculated from altitude data only by reference to a small *neighbourhood* around each point. Geomorphometric systems emphasising all five have been proposed independently by Evans (1972) and by Krcho (1973). The importance of all five attributes was previously recognised, for example, by Curtis *et al.* (1965), by Speight (1968) and by Ahnert (1970). Young (1972) has used all five in the context of slope analysis, while all figure extensively in discussions of slope development (Carson & Kirkby 1972).

Further applications of these attributes have been in the prediction of landslides (Carrara *et al.* 1978) and gullying (Zevenbergen & Thorne 1987), and in the classification of land cover in remote sensing. The varying illumination of steep slopes often prevents useful classification on the basis of reflectance alone, even with several spectral bands. Justice *et al.* (1981) found that a non-Lambertian model of scattering in preferred directions was necessary to correct the 'topographic effect' for woodland on Appalachian ridges with slopes up to 15–20°: variability between aspects was reduced by 86% when this model was applied to USGS orthophoto-based altitude data. The more complex problem of several land-cover types in a mountainous area can be tackled by including geomorphometric variables along with spectral responses in a discriminant analysis. In this way Franklin (1987) improved accuracy from 58% (or 46%) for four Landsat MSS bands, to 87% (75%) for these plus five geomorphometric variables, with altitude contributing most. This is not just correcting the topographic effect, it is supplementing reflectance data with further predictive variables.

2.3.2 Data

In attempting to analyse the form of a three-dimensional surface, a considerable quantity of data must be processed. For example, if altitudes for a 10 × 10 km area are recorded on a relatively coarse grid of 100 m mesh, there are 10 000 data points. It would be pointless to tackle analysis of such data except by computer. How, then, should (x, y, z) data describing the land surface be generated and formatted for input to a computer?

Hormann (1968, 1971) showed that by defining *surface-specific* points along ridges, valleys and breaks of slope, and relating each to neighbouring points, a continuous mesh of triangles (variable in shape and size) could be built up. Mark (1975b) suggested that this permits considerable economies in data capture and data storage, for purposes such as calculating mean gradient and range in altitude. Unfortunately, it does not permit calculation of meaningful surface convexities. Also, the variable size of the triangles is disturbing, since gradient and other roughness attributes are very sensitive to the size of area over which they are calculated.

Alternatively, a surface may be sampled by digitised contours (strings of x, y coordinates for fixed increments of z) or altitude matrices (z coordinates for fixed increments of x and y) (Evans 1972). The difficulties of processing digitised contours (for anything but aspect and plan convexity) are such that they are often converted, by interpolation, to altitude matrices. High-resolution altitude data (digital elevation/terrain/ground models (DEM, DTM or DGM)) are rapidly becoming available for industrialised countries such as Sweden, France, the UK and the USA, either from digitisation of contours at 10 m interval or better, or directly from air photographs as a byproduct of analytical stereoplotters (Konecny 1977) producing orthophotomaps. Data resolution is limited only by photographic scale: it would be possible, for example, to produce microrelief data by photographing from a frame. In assessing error, it should be remembered that the second derivative is much more sensitive to data error than the first, while altitude statistics are very robust to error.

Even if the geomorphologist needs to analyse a particular area for which digital data are unobtainable, he or she is well advised to generate a matrix from the best available map. From a matrix, it is easy to calculate all four derivatives for each non-peripheral data point, by reference to the eight neighbouring points. This can be done in terms of the coefficients of a quadratic fitted to each 3 × 3 neighbourhood, though least-squares fitting is unnecessary because of the gridded format of the data (Evans 1978). Derivatives and other general geomorphometric attributes can be calculated together from a single data set and for a consistent set of locations. Alternative approaches such as those of Wood and Snell (1960) and Parry and Beswick (1973) involve a series of separate data capture operations and produce attributes that cannot be compared location by location.

2.3.3 Summarisation

The use of complex indices to summarise an area is unnecessary, since all the concepts invoked can be quantified by statistics widely used to describe frequency distributions. The mean and standard deviation describe magnitude and variability in the same units as the original measurements; skewness and kurtosis describe the shape of a frequency distribution and are dimensionless statistics. Each of these four 'moment statistics' has an alternative based on quantiles (percentiles) of the frequency distribution, providing summaries that are more resistant (insensitive) to outlying values. However, the order statistics defined in sedimentology (Tanner 1959, 1960) should be used with caution (if at all), since they often invert the customary sign conventions. Once gross data errors have been corrected, extreme values may be of geomorphometric interest; this suggests use of moment measures. The best check for outliers and other peculiarities is inspection of histograms (Fig. 2.3) and scatter plots.

The mean and standard deviation of gradient, and the standard deviations of altitude and of convexity in profile and in plan, measure components of surface *roughness*. Roughness is a vague and complex concept, for which there is no single satisfactory measurement scale; it is best to analyse it into these components. Ohmori (1978) found close relationships between denudation rates and the standard deviation of altitude. Standard deviations of gradient that are high relative to mean gradient reflect contrasted topography such as glaciated mountains, dissected plateaux, or inselberg landscapes. Variability of plan convexity measures contour curvature and substitutes for drainage density as a measure of texture, which is also a component of surface roughness.

The skewness of altitude shows whether there is a 'tail' of high values (positive skew) or of low values (negative skew). The former may indicate inselbergs or other types of relatively infrequent topographic highs, or broad valleys, while the latter suggests a dissected plateau. Hence skewness of altitude is a more broadly based substitute for the hypsometric integral (Sec. 2.2.4). The altitude frequency distribution has commonly been viewed in cumulated form as the hypsometric curve, and used together with more specialised forms of altimetric analysis in the search for erosion surfaces: Clarke (1966) provided a critical review of this field, which is discussed in the following section.

Gradient frequency distributions vary from gaussian (Strahler 1950) for steep, fluvially dissected areas such as the Ferro basin to very short-tailed and broad-peaked (negative kurtosis, cf. the gaussian value) for glaciated mountains such as Thvera (Fig. 2.3). Most, however, show positive skew, which increases as proximity of the mean to the lower limiting gradient (zero) increases. Hence, although the log tangent transformation is often useful (Speight 1971), no single transformation is universally valid (Strahler 1956, O'Neill & Mark 1987). The square root of sine transformation is a useful compromise, but tends to produce negative skew for high-relief areas while not removing positive skew for lowlands. Negative kurtosis is indicative of either truncation or mixing of frequency distributions.

Convexity distributions, especially for profile convexity, show marked positive skew for glaciated mountains; as expected, convexities are sharper than concavities. (This is shown even more clearly by comparing the number of profile convexities stronger than 20° or 40°/100 m with the number of profile concavities; for plan convexities, take 500°/100 m). Kurtosis of both types of convexity is strongly positive for all areas, because

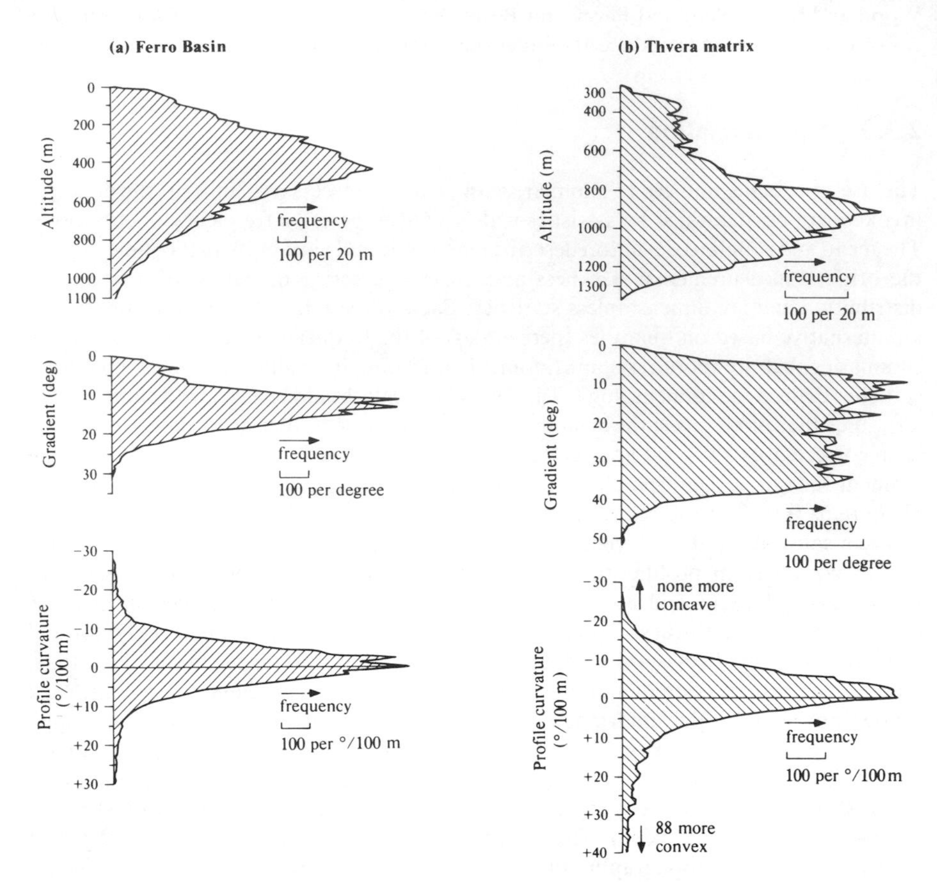

Figure 2.3 Frequency distributions of altitude, gradient and profile convexity for (a) the Ferro basin (11 582 points) and (b) Thvera matrix (9993 points). The deeply dissected Ferro basin is in northern Calabria, Italy, while Thvera forms part of the glaciated mountain range in north-central Iceland. Both have high relief, but Thvera has steeper gradients and a positively skewed profile convexity distribution, together with greater dispersion in both these attributes. Data were interpolated manually every 100 m, to the nearest 5 m vertically for Thvera and 10 m for Ferro.

of long tails of extreme positive and negative values. Since convexity is usually balanced by concavity, mean or median convexities are usually close to zero.

Aspect is measured on a circular scale, with its two 'extremes' (0° and 360°) coincident; hence it is essential to use appropriate statistics (Mardia 1972, 1975). The direction and strength of the resultant vectors should be calculated, with and without weighting by gradient.

2.3.4 Maps

Once altitude data are in a computer, the five point variables can be calculated, summarised and interrelated in a single run. The more complex operations of flow-tracing and scale-related analyses can be performed as discussed below. Statistical analyses must be complemented by maps of located variables, both to give a spatial element to the analysis and to check for errors. Most computer centres provide programs for the production of shaded maps, contoured maps and block diagrams; the tedium of

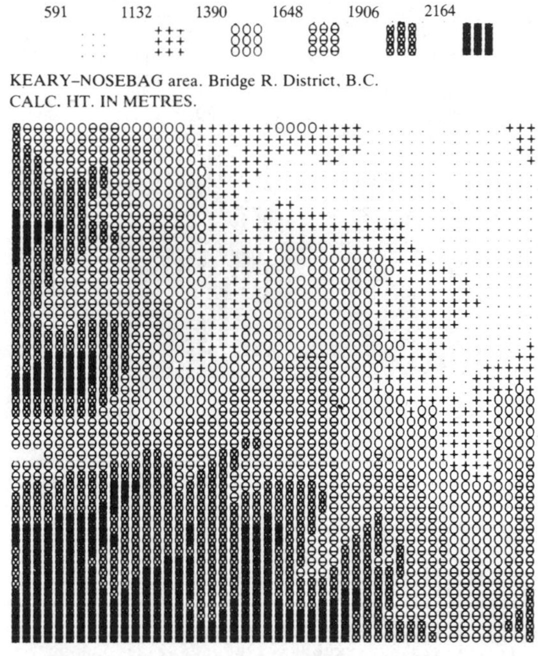

Figure 2.4 Derivative maps of a 4.8 × 4.8 km area of the British Columbia Coast Mountains. Each symbol represents 100 × 100 m.

Fig. 2.4 (*cont.*)

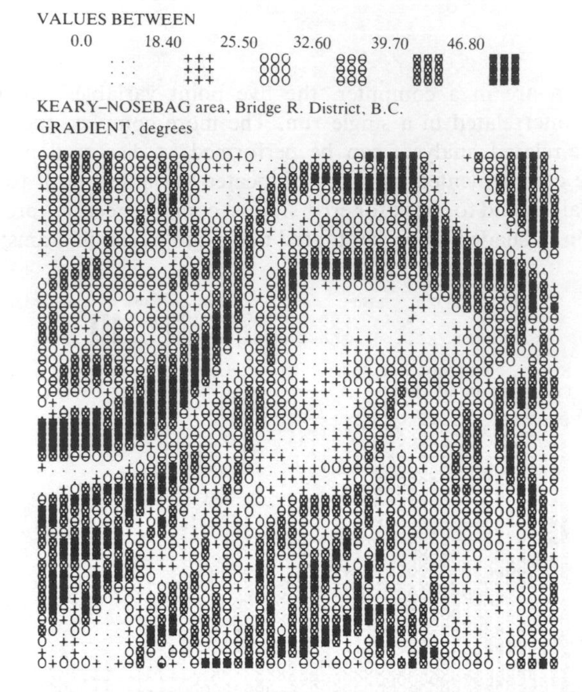
VALUES BETWEEN
0.0
18.40
25.50
32.60
39.70
46.80
KEARY–NOSEBAG area, Bridge R. District, B.C.
GRADIENT, degrees

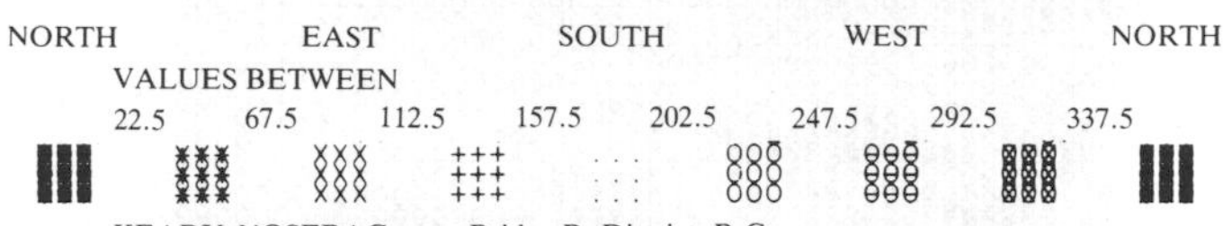
NORTH
EAST
SOUTH
WEST
NORTH
VALUES BETWEEN
22.5
67.5
112.5
157.5
202.5
247.5
292.5
337.5

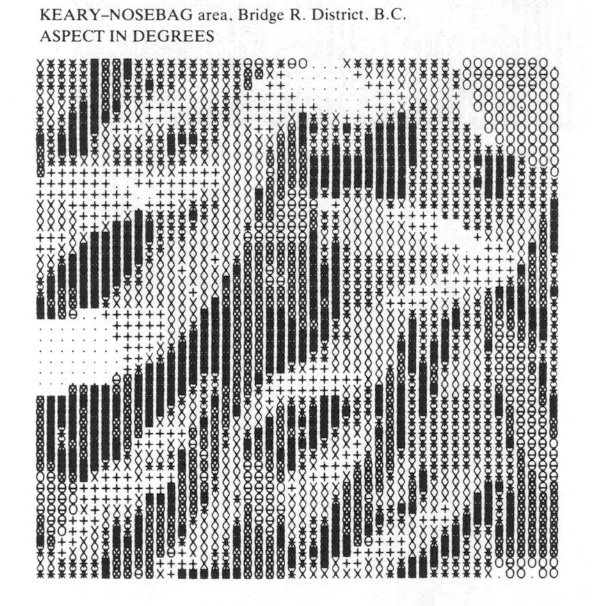
KEARY–NOSEBAG area, Bridge R. District, B.C.
ASPECT IN DEGREES

Fig. 2.4 (*cont.*)

VALUES BETWEEN
−70.00 −6.00 −3.00 0.0 3.00 6.00
KEARY–NOSEBAG area, Bridge R. District, B.C.
profile convexity, degrees per 100 m

VALUES BETWEEN
******* −40.00 −20.00 0.0 20.00 40.00
KEARY–NOSEBAG area, Bridge R. District, B.C.
PLAN CONVEXITY, degrees per 100 m

manual methods (Krcho 1973, Gardiner 1978a) can now be avoided. Guth *et al.* (1987) have made available programs for use on MS-DOS microcomputers, for data in USGS DEM (Digital Elevation Models) format.

Figure 2.4 shows maps of the five point variables for the Keary–Nosebag area of the British Columbia Coast Mountains (as featured in Evans 1972). These were produced on a laser printer, providing clearer output than the line printer for which the programs were written. The aspect map has eight 45° classes, symbolised so that northward slopes are dark, southward light, eastward have crosses and westward have circles: thus it combines a quantitative gradation with a qualitative dichotomy. The other maps have the same, six-class graded shading.

Fixed intervals used for both convexity maps are symmetrical about zero (straight slopes) and open-ended at the extremes (long 'tails' are typical); class width would be adjusted for different regions or grid meshes. The use of black for the most convex values emphasises ridges and crest lines; for some purposes it would be appropriate to invert the shading scale. Note that contouring would not be appropriate for surfaces as rough as these second derivatives.

To make full use of the six shading classes, the altitude and gradient maps are calibrated around their mean value, which gives the central limit, with classes 0.6 standard deviations wide except that the extreme classes are open-ended. This works well unless strong skewness is present (Evans 1977b). But one advantage of automation is that maps can be produced with any class limits, even if a new requirement makes, for example, a different threshold gradient appropriate. Manually or optically, the adoption of new class limits would require repetition of the measurement process.

2.3.5 Relationships

Strong correlations between general geomorphometric variables occur mainly where there is considerable overlap in their definitions, as in Tobler (1969b). Since altitude and its derivatives are defined so as to be distinct surface descriptions, strong correlations between these point variables are not expected. Such relationships as do occur are often complex and the linear correlation coefficient is not an adequate summary. It is essential to inspect scatter plots of each pair of attributes.

The one universal relationship is triangular; extreme (positive and negative) plan convexities are found only on gentle slopes. On steep slopes, contours are relatively straight. Secondly, there is a very weak tendency for plan and profile convexity to increase together. Both (especially profile) have low positive correlations with altitude. Very often, gradient increases with altitude; though moderately strong, this relationship is usually non-linear, with gentler gradients (rounded ridge tops) at the highest altitudes. Figure 2.5 illustrates the diversity and complexity of the altitude–gradient relationship. Because of its variability, this relationship provides a particularly useful additional descriptor of terrain. Traditionally, geomorphologists have plotted average gradients against altitude, ignoring dispersion within each altitudinal class; Clarke (1966) has reviewed the defects in many attempts at defining such 'clinographic curves'.

Multivariate relationships may also be investigated by regression. It seems appropriate to treat altitude and aspect as controlling (predictor) variables, both convexities as dependent variables and gradient as dependent on altitude and aspect, but controlling for the convexities. Regressions involving aspect are necessarily periodic, based on the sine and cosine of aspect angle. Linear regressions would be meaningless unless only a limited range of aspects were included and there was a gap around 0° ≡ 360°.

To compare different areas, and to establish relations between geomorphometric

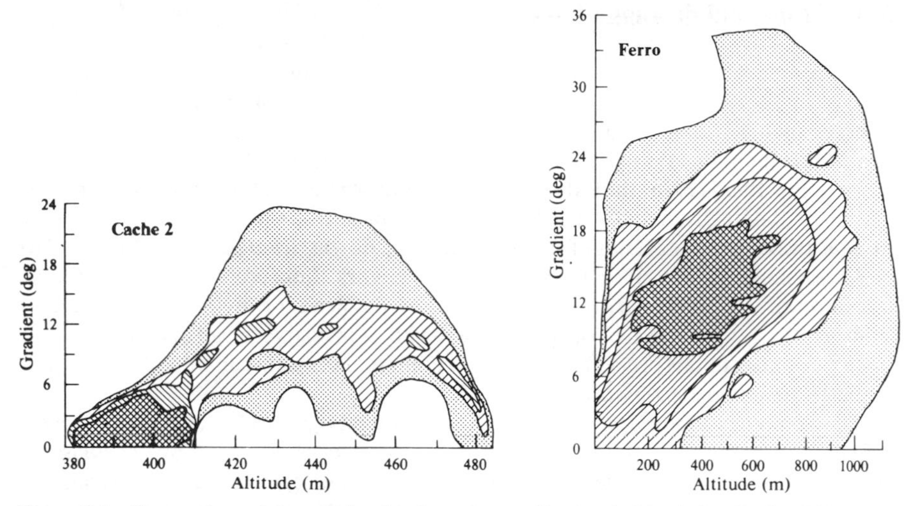

Figure 2.5 Scatterplots of the relationship between gradient and altitude for Cache (8985 points) and Ferro (11 582 points) generalised from computer output; shading is related to the density of points. The Cache area is on the fringe of some hills in south-west Oklahoma, and was digitised in a UNAMACE analytical plotter every 25 m horizontally, and to the nearest 1 m vertically.

statistics over a range of areas, it is essential that each area should be of comparable extent and sampled at the same resolution, with the same data capture technique. Both Evans (1984) and Depraetere (1987) have analysed sets of 10 km × 10 km blocks of 50 m mesh grids. For 54 contiguous blocks in Wessex, Evans found that the variation of 23 summary statistics (moment measures, correlations and vector strengths) could be represented by 5–7 compound dimensions, one of which represented data bias due to the interpolation routine. These represent mean and variability of gradient, with variability of profile convexity; skewness of altitude and gradient; mean and skewness of profile convexity; plan convexity variability; orientation; direction; and altitude–convexity correlations.

Depraetere (1987) analysed 90 variables (including many quantiles of frequency distributions) for 72 blocks providing an even coverage of France. 'Relief', the first principal component, embraced mean and variability of both altitude and gradient, and the variability of horizontal convexity. Skewness of altitude (and hypsometric integral) formed a component separate from skewness of gradient (and kurtosis of altitude). Most convexity variables (vertical and horizontal) contributed to two further components, one of which included col and depression frequency, the other, crest and slope frequency. Clearly the degree of intercorrelation between geomorphometric variables is not sufficient to permit compression to the two dimensions represented by the spectral and fractal models discussed below.

Over large areas or arbitrary map sheets, we expect mean profile and plan convexity to be zero. For specific landforms such as cirques, drumlins (Evans 1987a, b), landslides or dunes, expectations are different: mean convexity, and relationships such as between convexity and altitude, may take on greater interest.

2.3.6 Slope and drainage lines

General geomorphometry is being linked to drainage network studies (Sec. 2.6) by improved methods of detecting slope and drainage lines from altitude matrices. Local properties such as convexity are not sufficient to define drainage networks, but in well dissected terrain points that are never the highest in any block of four provide a suitable starting point. Band (1986) thins these to give a set of segments of points connected to at least one of their eight neighbours. Consideration of altitude permits identification of upstream and downstream ends of each segment, and downstream points are 'drained' to their lowest neighbour to give a more connected network. Intervening areas can then be 'thinned' to give a divide network, constrained by the need for divides to terminate at stream junctions; the pattern of ridge lines, defined by convexity instead of concavity, can be incorporated into this process. Finally each point in the matrix can be coded as part of a drainage area associated with a stream link.

In more complex topography, problems caused by pits (points lower than their eight neighbours) are more difficult to overcome and approaches discussed by O'Callaghan and Mark (1984) may be needed. Pits are common in matrices produced by automated photogrammetry, even in fluvial topography, which should be completely drained. They may be small data errors, but even if all altitudes in a grid are exactly true, pits will arise where drainage is between two of the neighbouring grid points and is thus lost at that grid resolution. Smoothing removes most pits where they are numerous, but it may enlarge some. To 'drain' all pits by seeking a low point among successively broader bands of neighbours may take dozens of iterations (O'Callaghan & Mark 1984).

O'Callaghan and Mark (1984) approximate the whole slope network by allocating one unit of runoff to each grid point and draining it to the lowest of its eight neighbours. Channels are identified where more than a threshold number of units have accumulated: with 30 m mesh data, their figure 4 suggests that 10–20 units is the best runoff threshold. The results are worst on uniform slopes and in flat areas, suggesting a need for combination with a concavity criterion and for separate treatment of floodplains and marshes.

Zevenbergen and Thorne (1987) also use this eight-directional runoff cumulation approach to provide a useful approximation to drainage area for each matrix point. No studies have been made of the distortions produced (and cumulated) by constraining runoff to directions 45° apart, but the cost of overcoming this constraint is a large increase in computer time. Bell (1983) has fitted smooth surfaces (of weighted, overlapping local quadratics) to altitude matrices. This permits 'true' curving slope profiles to be traced up and down from any sample of points. Profile termination criteria are sharpness of direction change and overall direction change. Douglas (1986, p. 43) has constructed slope lines directly from a grid and its diagonals, in order to establish ridge and channel lines and drainage basins.

Hence there has been considerable progress in relating drainage basin analysis (Sec. 2.6) to grid-based general geomorphometry. The deficiencies of 1970s attempts to trace slope and drainage lines through grids seem to have been overcome: catchment area per unit width of slope (Speight 1968), a vital component in hydrological modelling, can now be calculated. This has not satisfied Moor *et al.* (1988), who regard the results of Band (1986) and O'Callaghan and Mark (1984) as too coarse. Instead, they use digitised contours, searching upwards to the nearest points on the next contour (or in plan concavities, searching at right angles to the initial contour) to approximate slope lines. This provides a digital terrain model of polygons bounded by two contour segments and two slope line segments, for which gradient, aspect, width and upslope contributing area can be calculated.

2.3.7 Spectra and fractals

When engineers come to analyse land surface data they often turn to spectral analysis, which has been so successful in signal processing. A profile of equally spaced values is viewed as the result of superimposing sine and cosine curves with various wavelengths, each with its own amplitude and phase. Plots of log amplitude (or power) against log wavelength often show a linear decline from high variability at long wavelengths (low frequencies) to low variability at short wavelengths. The slope of this plot supplements the total variability in the spectrum to give two basic summary properties: marked periodicities show as deviations from a linear trend, and are characterised by their wavelength and amplitude (Pike & Rozema 1975).

Unfortunately, necessary preprocessing operations such as windowing, filtering and tapering involve subjective rule-of-thumb decisions, and the spectrum normally loses a considerable proportion of the variability of the altitude data. Extension to two dimensions (Rayner 1971, 1972) compounds such technical problems; interpretation of variation in spectra with wavelength is not straightforward. Ridges and valleys curve and converge, and it is difficult to define regions within which assumptions of stationarity hold. Spectral analysis is undirected; it does not deal with asymmetry. One-dimensional spectral analysis is useful in specific geomorphometry for profiles along and across fields of bedforms (Sec. 2.7) and the grain of structural relief. Oceanographers make considerable use of spectral analysis, and Fox and Hayes (1985) have modelled the anisotropy of the sea bed in this way, though their examples (pp. 37–44) show the difficulty of recovering even artificial input models from spectral statistics.

Visually attractive surfaces have been simulated by Mandelbrot (1982) using the fractional Brownian model, in which surface statistical properties are invariant with scale (Goodchild & Mark 1987). Assertions that such simulations are realistic pose a challenge to geomorphologists, who have found scale-dependent processes and recognised landforms at specific scales; if this or any other fractal model were fully accepted, there would be little point in specific geomorphometry. Fractal surfaces are characterised by two statistics: the altitude variance at some reference resolution, and the fractal dimension D, which specifies the rate at which variance changes with resolution (separation between points) on a log–log plot.

A perfectly smooth surface has $D = 2$ while one consisting of independent points has $D = 3$, i.e. it fills a volume. These extremes correspond to local autocorrelations of $+1$ and 0 respectively. Simulated surfaces with D values between 2.1 and 2.4 (Goodchild 1980) are closest to land surfaces in appearance, but real surfaces rarely exhibit fractal behaviour over a broad range of scales (Burrough 1985). Mark and Aronson (1984) analysed 17 circular areas some 12 km in diameter, for which USGS Digital Elevation Models were available, giving altitudes on a 30 m square grid mesh. They located 32 000 pairs of points randomly, and grouped these into 100 equal classes of horizontal separation to plot log(variance) against log(separation); classes with fewer than 64 pairs (mainly short separations) were excluded. Only one area out of the 17 exhibited fractal behaviour with a linear plot over the whole range (0.3–11.5 km): Shadow Mountain, Colorado, gave a fractal dimension of $D = 2.28$. Others showed breaks, with low 'fractal dimensions' below separations of 0.4–1.5 km (interpreted as the average length of slopes) and surprisingly high values of D around 2.75 in Pennsylvania and 2.5 in Oregon for separations up to 4–9 km.

Such results are obtained by others (e.g. Culling & Datko 1987); linear plots are found only for small, spatially limited data sets. Land surfaces are *not* self-similar: the limitations of fractal analyses are comparable to those of autocorrelation, spectral and regionalised variable analysis. Indeed, the results of multivariate analysis (Sec. 2.3.5) imply that geomorphologists are never likely to be fully satisfied with the results of two-parameter models of the land surface, except as 'null models', deviations from which are of interest.

2.4
Geomorphometry and geochronology

Present landscapes have evolved through the Tertiary and Quaternary, and often retain the imprint of former sub-aerial and marine processes. These operated under a variety of environmental regimes different from those presently observable, and related to different base-level conditions controlled by eustatic, tectonic and isostatic changes in the relative elevation of land and sea. Geomorphometry has developed one branch whose objective is the identification and correlation of remnants of formerly more extensive erosional and depositional surfaces associated with these processes, in order to provide a quantitative basis for reconstructing the geomorphological history of an area.

Measurement and analysis of the heights of surfaces considered discordant with geological structure first developed as an essential element of denudation chronology, which was concerned with high-level (> 50 m) erosional surfaces of marine or sub-aerial origin. However, in relation to methods of data collection and analysis, and strength of interpretation, a major distinction must be made between these studies of the older, higher, degraded and discontinuous erosional features associated with mid- and early Quaternary and Tertiary erosion, and the younger, lower, more continuous, often depositional and less eroded (but sometimes buried) features associated with the later Quaternary. The latter include marine benches existing as raised beaches at heights below 50 m above sea level, which are often correlated with fluvioglacial terraces and moraines, and are identified by the presence of undercut cliffs, caves, arches and stacks beyond the reach of present marine erosion (Gray 1974). They also include river terraces formed in Quaternary valley-fill sediments, which provide fragmentary evidence of former floodplain surfaces and often preserve on their surface the morphology of palaeochannels that can be used to reconstruct the former discharge regime of the valley (Sec. 2.7). Remnants of *depositional* surfaces clearly provide more potential for geochronological interpretation through the sedimentary structures, palaeoecological evidence and dateable materials that they contain. This section concentrates on the morphological analysis of dissected surfaces; Quaternary shoreline analysis is treated in detail in Section 5.8, and sections relevant to the analysis of Quaternary deposits include Sections 3.2, 4.1, 5.1, 5.4 and 5.9.

2.4.1 Data sources

The traditional field instrument for ascertaining the heights of ancient, scattered erosional remnants of surfaces at elevations above about 50 m is the surveying aneroid. Dissection and erosion of these features make accurate measurement irrelevant, and correlation can only be achieved in height ranges of 10–20 m, and over short distances given the likelihood of post-formation warping. Sparks (1953) and Hamilton *et al.* (1957) discuss the use of the surveying aneroid, or altimeter, emphasising the importance of surveying in suitable meteorological conditions. Standard errors increase to 5 m in strong winds and to 8 m where tangential acceleration of airflow occurs over topographic obstacles.

However, the greatest difficulties have often been associated with interpretation of the data in relation to the sampling method. Benches or surfaces may be initially defined from topographic maps before field survey, and are most evident as a widening of the space between contours. Hodgson *et al.* (1974) demonstrate that 261 of 283 erosional remnants defined on the English South Downs dip slope by Sparks (1949) fail to cross a contour, and that the frequency distribution of these bench heights can be modelled by a single harmonic function with a 50 ft wavelength, 50 ft being the contour interval of the maps used to locate remnants. Thus, they argue that sampling bias has created a systematic sequence of benches, subsequently interpreted as the result of falling sea level, from a random distribution of remnants representing various exhumed relics or a sub-Eocene surface.

Elevation data from topographic maps may be used to define erosion surfaces directly. Clarke (1966) notes considerable variety in the objectivity of sampling methods, with preference being for methods based on sampling spot heights, summit heights, or maximum heights in grid squares rather than those involving measurement of summit and bench areas and shoulder and col lengths. Sampling bias is again a problem: Hollingworth (1938) recognised that preferential omission of spot heights below *c.* 150 ft on 1::63 360 sheet 186 used to define Figure 2.6b indicates significant clustering. The shoulder–summit–col method (Geyl 1961) involves greater subjectivity and potential for error, as well as being time-consuming; and the grid square method (Baulig 1935, p. 41) exemplified by Evans's discussion of general geomorphometry (Sec. 2.3) is objective and readily automated to generate the large sample size necessary before confident interpretation is possible. In a major analysis of the hypsometry of the continents, Cogley (1986) has therefore adopted a 1° grid-based sampling scheme amenable to computer mapping and analysis.

Studies at smaller spatial scales of geologically more recent features such as Quaternary raised beach surfaces or valley-fill terraces clearly require large samples of accurate elevation measurements, because of the altitudinal proximity of successive surfaces, the possibilities for divergence and convergence of adjacent features, and the local factors (e.g. beach exposure) that cause deviations from average elevation. Thus levelling from bench marks, with measurement accuracy of ± 0.01 m and traverse closure errors of ± 0.15 m distributed among the traverse elevations, is the necessary field technique (Sec. 5.8, Gray 1983). Although published topographic maps cannot generate data of sufficient accuracy for these investigations, photogrammetric altitude data derived with a stereoplotter from air photographs at scales of about 1:10 000 provide a valuable supplement to field survey.

2.4.2 Data presentation and analysis

Traditional methods of geomorphometry applicable to the identification of erosion surfaces from map data are reviewed by Clarke (1966) and Pannekoek (1967). The former discusses hypsometric, clinographic and altimetric analyses, and concludes that the last of these, based on point samples of altitude, is the most appropriate for defining erosion levels. Pannekoek (1967) discusses generalised contour maps, summit-level maps and streamline surfaces. Other classic methods include composite and superimposed profiles. Figure 2.6a illustrates superimposed profiles across the Fowey–Looe watershed in Cornwall, taken originally from Ordnance Survey 1:63 360 map sheet 186. The interpretation of these diagrams is hazardous and inclined to be influenced by predispo-

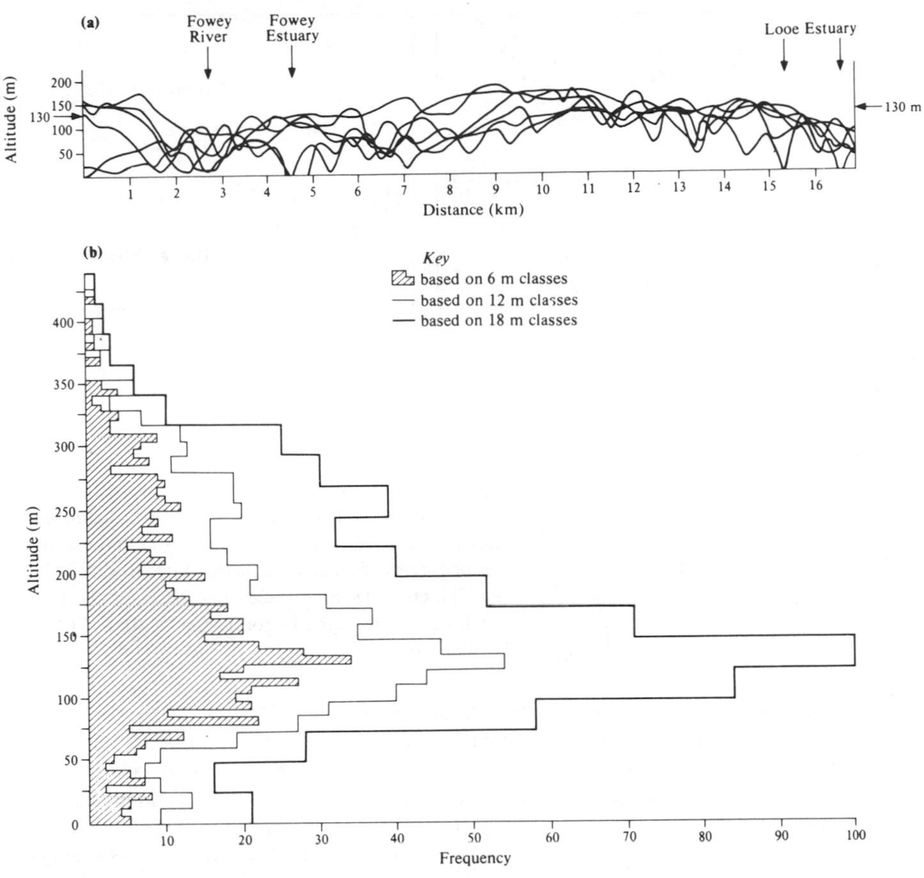

Figure 2.6 (a) Superimposed profiles across the watershed between the rivers Fowey and Looe in south Cornwall; six east–west profiles along northings 54, 55, 60, 61, 62 and 63. Vertical exaggeration is approximately × 11. (b) Altimetric frequency analysis with three different class intervals, based on spot height data from Ordnance Survey 1:63 360 map, sheet 186.

sition; the significance of accordance of height in the range 130–140 m is untestable and the importance of the 130 m (430 ft) erosion surface usually prominent in the denudation chronology of Cornwall is not firmly established by qualitative assessment of such a method of data presentation. One alternative, an altimetric frequency analysis (Fig. 2.6b) based on spot height data from the same map, is a quantitative summary, but the significance of individual modes in a generally log-normal height distribution is difficult to test, particularly since they depend partly on the choice of class interval. It cannot be argued, therefore, that these methods placed the analysis of high-level (> 50 m) erosion surface remnants on a firm quantitative basis.

Svensson (1956) first suggested the idea of using trend surface analysis to generalise the form of erosion surfaces, fitting a linear surface to summit heights in the Lofoten Islands. However, King's (1969a) application of the method to central Pennine erosion surfaces generated considerable criticism. Tarrant (1970) and Unwin and Lewin (1971) especially emphasised that a least-squares surface requires positive and negative residuals, so that subjective data screening is necessary if a true representation by the

A

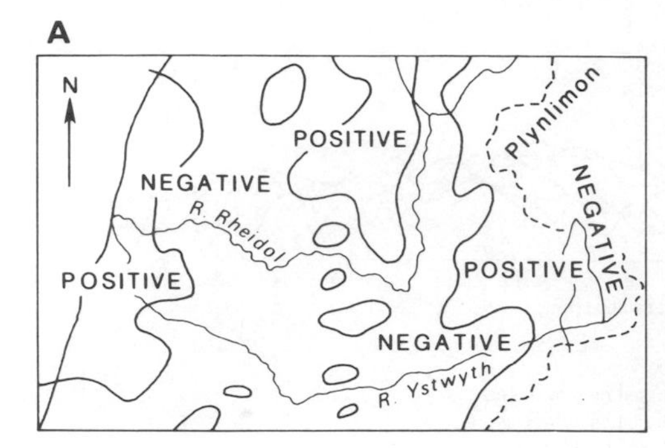

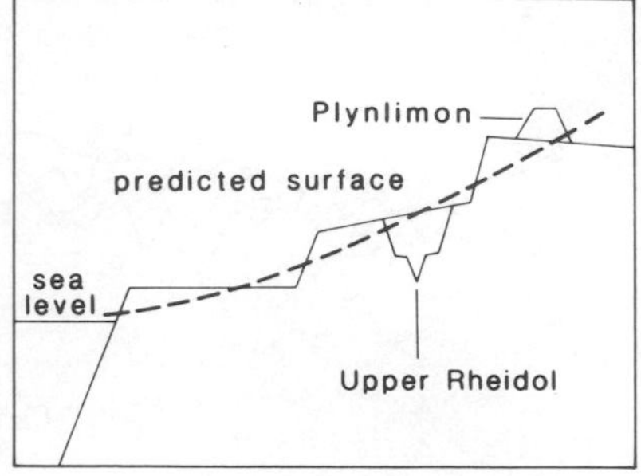

B

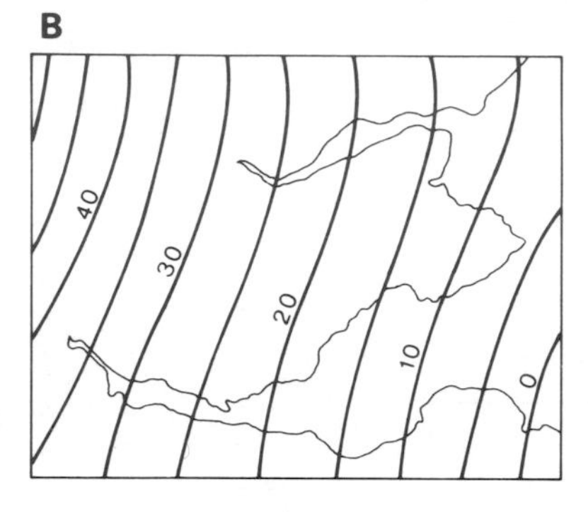

Figure 2.7 (A) Residuals from the quadratic trend surface fitted to altitude data in north Cardiganshire, showing positive spatial autocorrelation, indicating a stepped series of erosion surfaces. (After Rodda 1970). (B) The cubic trend surface of isobases of the main Perth raised shoreline in south-east Scotland. (After Smith *et al.* 1969).

trend surface of erosion surface gradient and direction of slope is required. This is because, otherwise, altitudes of monadnocks and dissected valleys will distort the trend. Rodda (1970) adopted an alternative approach in studying erosion surfaces in north Cardiganshire and successfully demonstrated that Brown's (1950) earlier identification of three main surfaces was appropriate by fitting a single trend and defining erosion surfaces as parallel bands of positive and negative residuals. In this case, stepped surfaces are identified by examining patterns of spatial autocorrelation in the residuals from the general pattern of altitude variation (Fig. 2.7a).

A telling comment on applications of trend surface analysis to erosion levels is that the alternative method of spatial analysis based on three-dimensional spectral analysis (Rayner 1972) starts with the assumption that relief amplitude and wavelength vary over a narrow range such that harmonic functions can model land surface form. Thus, accordance of ridge summits is no longer necessarily referred to a model of past planation and subsequent dissection, but to a model of uniform watershed elevation where stream spacing is roughly constant, as argued by Hack (1960).

The altitudes of late Quaternary raised beaches were initially analysed by projecting measurement locations perpendicularly onto a straight line radiating from the assumed centre of isostatic uplift (Cullingford & Smith 1966). However, following McCann (1966) and Smith *et al.* (1969), trend surface analysis has again been used (Fig. 2.7b). These applications have defined the pattern of isostatic uplift in Scotland, suggested occasional evidence of local tectonic displacement, and distinguished late-glacial from postglacial beaches by their differential tilting and gradient. Some debate has centred on statistical problems related to spatial autocorrelation in data sets consisting of clusters of heights on each beach fragment. The number of degrees of freedom is effectively dependent on the number of fragments rather than on the number of heights (Gray 1972), so that significance testing requires care unless mean fragment heights are used. Such averaging, however, itself smooths the data and inflates the percentage explanation of variance by the fitted surfaces, although having little effect on their gradient and direction of slope. The effect of spatial clustering is less important when the trends are strong (Unwin 1973b), and this is usually the case. However, there may still be problems in selecting the appropriate trend surface model by testing the significance of extra explanation associated with progressively higher-order terms.

In the case of alluvial-fill river terraces, height–range diagrams have commonly been used to effect correlation between fragments and separation between terraces (Fig. 2.8). These involve projection of fragment locations onto the valley centre-line and construction of a plot of fragment height against downvalley distance. Correlation between fragments is based on assumed continuity of the former valley-fill surface, and is usually strongly influenced by some *a priori* model of longitudinal morphology (e.g. a down-valley decrease of gradient).

2.4.3 Interpretation

The dangers of interpretation based solely on morphological information are now well known; morphology and morphological relationships alone are poor guides to relative age, and weak bases for correlation or genetic inference. For example, McArthur's (1977) re-evaluation of the ages of erosion surfaces in the southern Pennines considerably compresses the timescale over which surface formation and erosion are likely to have occurred. This involved estimates of the volume of incision into planation surfaces viewed in the context of contemporary process measurements of erosion rates. Even more fundamentally, the mechanism of planation is rarely identifiable from morphology alone. It is unclear how surfaces created by pediplanation differ from those formed by peneplanation in terms of both preservation and identification, especially in mid-latitude landscapes subjected to Quarternary glaciation. Dury (1972) has attempted to use curve-fitting methods on slope profiles, but it is likely that remnant pockets of deep

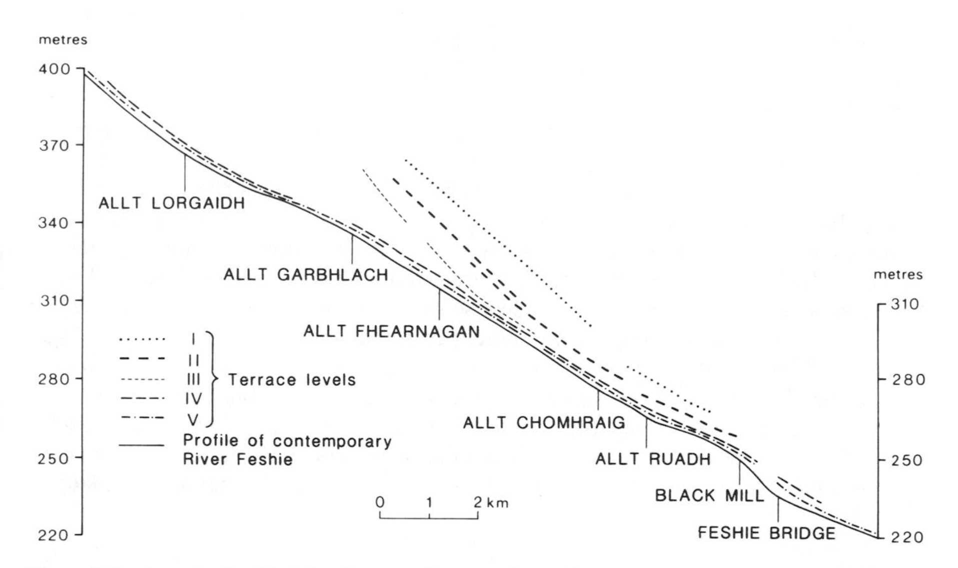

Figure 2.8 A generalised height–distance diagram of correlated terrace fragments in Glen Feshie, Cairngorms, Scotland. Note large vertical exaggeration to facilitate separation of terraces. (After Young 1976).

weathering products are more conclusive evidence of subtropical denudation systems; cf. Battiau-Queney (1984) in a study of Wales. The debate over the efficacy of marine and sub-aerial planation is also unresolved by morphology alone. King (1963) argues against extensive marine planation except in the special circumstance of a slowly rising sea level, and features separated by Sparks (1949) into marine and sub-aerial 'flats' at the same base level have been shown by Small and Fisher (1970) to have virtually indistinguishable heights, indicating that apparent morphological differences in the field may not warrant this discrimination when altitudes are compared statistically.

The additional sedimentological and stratigraphic evidence available in studies of river terraces highlights the limitations of height and gradient data, and points up the problems of denudation chronology based on erosion surfaces. The gradient of a valley-fill surface varies with the local channel pattern and the calibre of sediment introduced by tributaries, and correlating fragments by assuming simple morphological continuity is therefore hazardous. The chronological significance of the terrace surface itself may be obscured by the composite nature of the aggradational sediments beneath the surface, as Dawson's (1985) interpretation of Severn terrace gravels demonstrates. Here, tributary fan gravels and main valley deposits combine to create the terrace form, and the importance of detailed lithofacies analysis of terrace sediment exposures is emphasised. Correlation of fragments, separation of the terrace surfaces they represent, and relative dating of aggradation and incision phases, all rely heavily on additional information such as weathering of sediments (Mills & Wagner 1985) or degree of surface pedogenesis (Robertson-Rintoul 1986). The latter author has used soil properties to demonstrate that the Feshie terraces illustrated in Figure 2.8 are not all late-glacial, as had formerly been assumed, and that several of the morphological correlations between fragments identified by the height–range diagram were unreliable.

2.4.4 Conclusion

Although problems clearly exist in the morphological interpretation of river terraces, the contrast between late Quaternary shoreline studies and investigation of earlier assumed marine and sub-aerial surfaces is clear-cut. The former are better preserved, more continuous features, which can be accurately surveyed in the field and correlated by trend surface analysis. Furthermore, they often connect inland with fluvioglacial terraces or moraines and may be dated by the geochronological methods discussed in Part Five. The latter, however, are more fragmentary, and summit accordance in particular is susceptible to widely differing interpretation. A denudation chronology for these erosional remnants cannot be more than a hypothetical model without supportive geochronological data, since morphological correlation alone is open to debate. There is little justification for the view that widely separated remnants at similar altitude are isochronous, because warping and differential erosional lowering will have increasingly destroyed this correlation for progressively older surfaces. Unfortunately, the potential for finding deposits that can place the interpretation on a firmer foundation also diminishes as the time available for their erosion increases. Nevertheless, useful conclusions can be derived from analysis of fragmentary deposits, as evaluation of supposed marine sediments in the Chiltern Hills demonstrates (Moffat & Catt 1986).

2.5 Specific geomorphometry

Specific geomorphometry is the measurement and analysis of specific surface features defined by one or more processes, and separated from adjacent parts of the land surface according to clear criteria of delimitation (Evans 1974, p. 386). This section is concerned not with description of individual landform features, but with landforms that are distributed regionally as a set of replicates. It is assumed that all individuals are geocoded by coordinate pairs and that the sample replicates required for statistical analysis are compatible in data quality. Most measurements used for analysis of shape and pattern in specific geomorphometry are ultimately drawn from a map source. The map may be compiled from an original field survey or from secondary sources such as air photographs, published topographic maps, or satellite imagery (Ford 1984). The optimal way to analyse map data is with an automatic digitiser, and numerous computer programs are now available to process digitised data (Marble *et al.* 1978). The range of phenomena to which specific geomorphometry has been applied is considerable, and now encompasses investigation of an interplanetary nature (Table 2.3).

2.5.1 Landform pattern

2.5.1.1 Point pattern analysis

Studies of spatial pattern in specific geomorphometry employ the techniques of two-dimensional point pattern analysis derived from work in plant ecology (Greig-Smith 1964, Kershaw 1964). Replicates of a specific landform within a bounded region are mapped as a distribution of points in space. Such patterns have two attributes, which Pielou (1969) has termed intensity and grain (Fig. 2.9). Intensity is the extent to which density varies from place to place, and the distribution may be described as regular (or uniform), random or clustered. Grain is independent of intensity, and concerns the area and spacing of patches of high and low density. Normally, point pattern analysis measures intensity, using the techniques of quadrat analysis, block size analysis of variance and nearest-neighbour analysis. A rigorous statistical treatment is given by Pielou (1969) and some geographical examples are presented in Silk (1979). Only the principal uses and limitations of each technique, as applied in specific geomorphometry, are given here.

Quadrat analysis compares the frequencies of points in small areal subdivisions of a study region (quadrats) with those obtained using hypothesised probability models. A single random process free from locational constraints would generate a Poisson model, while clustered or contiguous models include the negative binomial, double Poisson or Polya–Aeppli (Harvey 1966). Application of these models is discussed by Oeppen and Ongley (1975) for river channel networks and by McConnell and Horn (1972) for karst depressions. Various sampling designs may be used. Oeppen and Ongley use square quadrats in a contiguous grid, while McConnell and Horn use circular units in a stratified systematic unaligned design (Berry 1964). Quadrat analysis has also been applied to areal volcanism (Tinkler 1971) and drumlin distribution (King 1974, Gravenor 1974,

Table 2.3 Selected examples of the application of specific geomorphology.

Landform type	*Morphometric index**	*Data source†*	*Authors*
Glacial			
Cirques, hollows	S,FE	P, 1 : 60	Aniya and Welch (1981)
	S,F	P (digitised)	Sauchyn and Gardner (1983)
	S,F	M, 1 : 100, 1 : 50, P	Vilborg (1984)
Drumlins	Mosaic	M, 1 : 25, P, 1 :15.8, FS	Boots and Burns (1984)
	S	Seasat imagery	Ford (1984)
	S	M	Mills (1980)
Moraines	F	FS	Matthews *et al.* (1979)
Karst			
Depressions	F	P, 1 : 50, FS	Marker and Sweeting (1983)
	S, F	M, 1 : 2.4	Mills and Starnes (1983)
Limestone pavements	F	FS	Goldie (1981)
	F	FS	Lewis (1983)
Trittkarren	S,F	FS	Vincent (1983)
Planetary			
Streamlined islands	S	Viking imagery	Baker (1979), Komar (1983, 1984)
Valley networks	S, F	Viking imagery	Pieri (1980)
Other			
Alluvial fans	FE	FS	Hooke and Rohrer (1979)
	FE	M, 1 : 62.5	Troeh (1965)
Coastlines	S	M, FS	Yasso (1965)
Coral atolls	S	M, FS	Stoddart (1965)
Desert dunes	S, F	P, 1 : 50, 1 : 60	Lancaster (1981)
Lakes	S,F	M, FS	Hakanson (1982)
Landslides	S,F	FS	Crozier (1973)

*S = two-dimensional (planar shape) index; F = three-dimensional shape index; FE = three-dimensional form equation.
†M = map source; P = photogrammetric source; FS = survey; scales are $\times 10^{-3}$ (i.e. 1 : 60 is an abbreviation for 1:60 000).

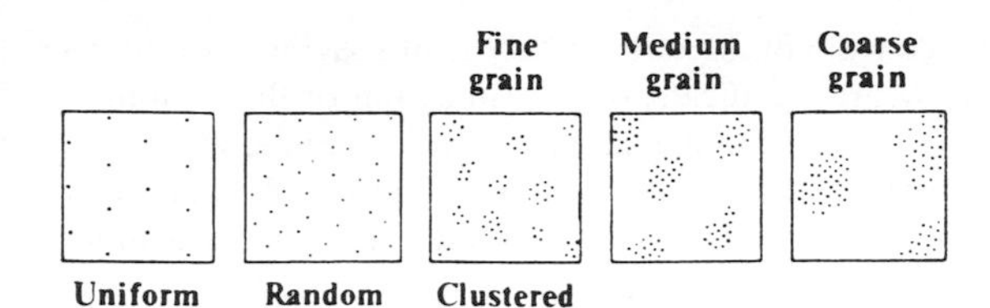

Figure 2.9 Point patterns.

Trenhaile 1971). Quadrat analysis is appropriate only when a deductive hypothesis suggests a probability model that generates a point pattern similar to that observed; it cannot be used to infer a generating function. Since adherence to a given probability model may vary with scale, correct specification of quadrat size is vital (Harvey 1968) and it is important to conduct a series of tests with different-sized quadrats.

Block size analysis of variance investigates scales of non-randomness within a pattern (Usher 1969), and again the only measurement required is a count. The technique examines changes in the variance of a distribution of points per quadrat as the cells in a contiguous grid are successively agglomerated into larger blocks. Peak values of variance occur at block sizes corresponding to the mean area of patches of high and low density. The method has been adapted to linear transects and applied to the distribution within a drumlin field (Hill 1973).

Nearest-neighbour analysis includes a variety of methods for comparing distances between points in a pattern with values that would be generated by a random process. Length measurement is required and the advantages of a data file of digitised point coordinates are apparent. The simplest application measures linear distances between a sample of points and their nearest neighbours. A computed nearest-neighbour statistic may then be compared with a theoretical value for a random distribution. This method has been applied to drumlins (Smalley & Unwin 1968, Jauhiainen 1975, Rose & Letzer 1975), dolines (Williams 1972b, Day 1976), earth hummocks (Bolster 1985) and river basin outlets (Morgan 1970). It has also been used in studies of cirque distributions where fitting trend surfaces to cirque altitudes requires a demonstration of the random distribution of data points (Unwin 1973a, Robinson *et al.* 1971). Order–rank analysis utilises neighbours other than the first nearest and permits a more sensitive study of departures from randomness. These methods have been applied to drumlins (Trenhaile 1971, 1975a) and areal volcanism (Tinkler 1971). A major problem with nearest-neighbour methods is the boundary effect, and an elegant solution by Dacey and Krumbein (1971) computes toroidal distances for patterns of river channel network nodes. Although more demanding in data extraction, nearest-neighbour analyses do yield a statistic that can be used for quantitative comparison of distributions in different regions (Williams 1972b).

A common problem in point pattern analysis is the need to represent three-dimensional landforms as dimensionless dots on a map. Information such as size, shape and orientation is lost, operator variance is introduced and the outcome of statistical analysis may be biased. Few features can be unequivocally defined as points. Dolines may be assigned the point location of their sink or swallet (Williams 1972b, Day 1976) but if, during growth, the sink migrates towards the divides (Williams 1972b), spacing between points will be affected by size variation within the doline field. Other forms lack any internal point of geomorphic relevance and require geometrical resolution to a point location. Cirques have been located by the centre of a semicircle fitted to their outline (Sugden 1969) or the midpoint of the median axis from threshold to headwall crest (Evans 1974). Tinkler (1971) fixed the point location of elliptical volcanic craters at the intersection of the two principal axes. Drumlin studies have variously utilised the midpoint of the long axis, stoss end of long axis, intersection of long and short axes and drumlin summit, although this choice will not affect the analysis of spacing if drumlins are similarly shaped (Muller 1974). More significant is the truncation of the minimum interpoint distance that occurs when cellular features are resolved to points. Minimum distance cannot be less than the average width of the features concerned, and when spacing is tested against a random model, the bias introduced will indicate a more dispersed than random empirical pattern (Boots & Burns 1984). Many of these difficulties may be overcome using mosaic methods.

2.5.1.2 Mosaic methods

Mapped distributions may be examined as *n*-phase mosaics in which every point in an area is assigned to a particular phase, and all the phases are present as patches of finite extent (Pielou 1969). Boots and Burns (1984) present the simplest case when drumlin distribution is analysed as a two-phase mosaic of drumlinised patches and non-drumlinised gaps, sampled parallel and perpendicular to the assumed iceflow direction. An obvious potential exists to extend the technique to many phases by exploiting the rasters assembled automatically from densitometric analysis of aerial photographs or the digital image data from multispectral scanning, which are natural mosaics. Mosaics may be evaluated using theoretical reference structures, but no unambiguous counterpart to the Poisson point pattern is available for this purpose. A random mosaic cannot be defined without specifying exactly the sort of randomness meant (Pielou 1969) and an element of subjectivity in the analysis is thus inevitable. Boots and Burns (1984) adopt Pielou's (1969) suggestion of a random line mosaic, but offer the alternative of a random placement model (Smalley & Unwin 1968).

2.5.2 Landform planar shape

Many attempts to quantify planar shape in specific geomorphometry exploit methods from sedimentary particle shape analysis. This has led to a desire for a univariate shape index, constructed from the basic dimensional attributes of a closed planar loop. A battery of indices exist using ratio combinations of area, perimeter length and axis lengths. Another consequence of the analogy with particle shapes is the development of indices scaled such that a critical value approximates to that of a standard shape. Comparison with the circle has been the traditional approach, with indices variously defined as the form factor (Horton 1932), the circularity ratio (Miller 1953) and the elongation ratio (Schumm 1956). Similar measures are quoted for a range of landforms in more recent literature. Shape indices based on the ellipse (Stoddart 1965), lemniscate loop (Chorley *et al.* 1957) and rose curves (Doornkamp & King 1971) have also been proposed.

The terminology of shape indices has been confused by geometrical relationships between the standard shapes. A simple ratio of long- to short-axis lengths has been variously termed an elongation ratio, used to indicate approximation to circularity (Sperling *et al.* 1977), and an ellipticity index (Tinkler 1971, Stoddart 1965). Given an elliptical model, a ratio of the principal axes can be derived in terms of long-axis length L and area A. The resultant ellipticity index in the form $\pi L^2/4A$ is identical to the lemniscate index k (Chorley *et al.* 1957), algebraically the reciprocal of Horton's form factor as modified by Haggett (1965), and the squared reciprocal of Schumm's elongation ratio. Relationships between length, width, axis ratio and k are examined as power-function regressions for streamlined island shapes on Earth and Mars by Komar (1984).

Standard shape indices have also shown little consideration for the implications of the model shape (Vilborg 1984). Sugden (1969) observes that many cirque headwalls approximate to the arc of a semicircle, and cirque length : width ratios have been computed recently by Aniya and Welch (1981) and Sauchyn and Gardner (1983). Drainage basins have been widely abused by indices of circularity (Morisawa 1958, Krumbein & Shreve 1970) and lemniscate curves (Chorley *et al.* 1957). The basin outlet provides a unique perimeter point and leads to a variety of ways of determining basin length. Failure to detect shape variations in the index measures and absence of correlations with other morphometric and environmental variables probably stem from thc irrclcvancc of thc standard shapc modcl.

The most successful applications of standard shape indices have been where there is some process rationale for generation of the expected model. The streamlined shape of the lemniscate loop has given it a natural application in this respect. Thus the lemniscate k, widely used to describe drumlin shape, may also be related to drumlin formation (Smalley & Unwin 1968, Trenhaile 1971, 1975a). High values of k may reflect greater ice pressure (Chorley 1959), greater constancy of ice direction (Doornkamp & King 1971) and absence of coarse material in the local till (Gravenor 1974), although Shaw (1983), arguing on the basis of a diversity in drumlin form, rejects the ice pressure and constant direction hypotheses. Attempts to relate length : width ratios to landform genesis have met with varying success. Jauhiainen (1975) and Muller (1974) propose nominal classifications of drumlin shape based on this measure, which can be related to the lemniscate index using a power-function regression (Komar 1983). An increase in the ratio as the variability of long-axis orientation declines has been used by Mills (1980) to support consistency of iceflow as the major factor in drumlin formation. Tinkler (1971) terms the ratio an ellipticity index to describe volcanic craters, where the short axis represents the basic explosion width of the crater and the long axis the tendency of the crater to elongate in the direction of some controlling factor. Lavalle (1967) attempts to explain variations in the doline shape index using a multiple regression model that indicates the importance of structural adjustments and relief. Hakanson (1982), however, was unable to find any significant relationship between area of erosion and deposition within lakes and a lake shape index based on a length : width measure.

A second group of shape measures utilise more information from a digitised outline of the planar form. Sauchyn and Gardner (1983) adopt the compactness index of Blair and Biss (1967) in their study of open rock basins. Compactness is the ratio of basin area to the sum of the moments of inertia of the infinitesimal area elements constituting basin shape. Since this is based on the distribution of basin area about the centroid, it is a more complete index of planimetric shape than the conventional length : width measure. This and other indices are reviewed by Whittington *et al.* (1972). Drainage basin shapes have been described by Fourier analysis of unrolled shapes (Jarvis 1976a, McArthur & Ehrlich 1977), also taken from shape studies of sedimentary particles and fossils (Gervitz 1976). The shape is transformed from a closed loop into a bivariate series plot (Benson 1967). Fourier analysis is then a straightforward curve-fitting exercise (Rayner 1971). Drainage basins can be 'unrolled' about an axis from their centroid to the basin outlet (Fig. 2.10), and the centroid determined algebraically from the digitised perimeter points. The shape variables, given by the Fourier coefficients, permit graphical reconstruction of the model shape and statistical expression of the goodness of fit between model and observed shapes. Fourier analysis is less successful when the macro-form of the shape is highly irregular. In such cases, application of the concept of fractal dimensions may offer an alternative (Mandelbrot 1977 a, b). Fractal dimensions arise from the observation that the measured length of a fixed irregular perimeter varies with the length of measurement unit (or step length) used. The dimension is obtained from the slope of the logarithmic plot of perimeter length against step length, and different elements of the plot may represent discrete textural or structural components of shape (Flooks 1978, Kaye

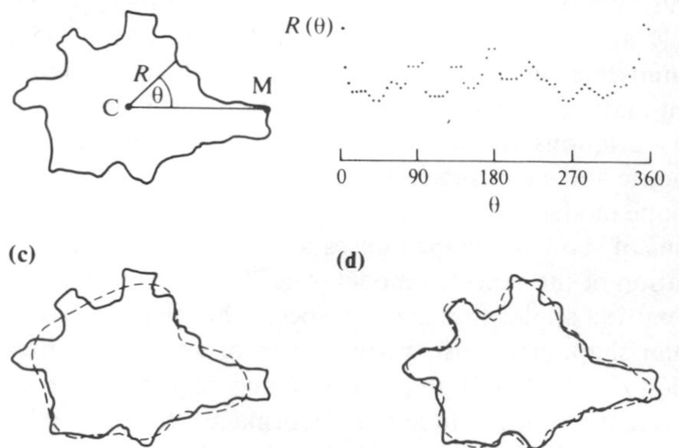

Figure 2.10 Shape of the River Ottery basin, south-west England, by the Fourier method. (a) Perimeter radials of length R and rotation angle Θ, taken from the axis joining centroid C to basin outlet M. (b) Unrolled shape perimeter radials, R, as a function of Θ. (c) Model shape (broken line) generated by harmonics 1–6 giving 63.8% variance explained. (d) Model shape generated by harmonics 1–10 giving 90.5% variance explained.

1978a, b). As yet, no substantive investigation of this has been undertaken (Orford & Whalley 1983). Both Fourier and fractal analyses offer a potential discriminatory power that would be unrealistic using conventional single indices. However, curve fitting may yield satisfactory univariate indices describing simple landforms; the headland–bay beach form modelled by a single-parameter logarithmic spiral is an example (Yasso 1965).

A final category of shape measures applies to landforms whose outline planar shape may be related to geometric properties of features within the perimeter. The sink in a doline, the main channel in a drainage basin and the summit of a drumlin have all been used as foci of indices of perimeter shape asymmetry. Williams (1971, 1972b) devised a symmetry measure for karst depressions, given by the product of (i) the ratio of long-axis segment lengths on either side of the sink, and (ii) the ratio of short-axis segment lengths on either side of the sink. This 'product of symmetry' increases with asymmetry. Within polygonal karst in New Guinea, Williams (1972b) found no change in length : width elongation with depression growth but an increase in internal asymmetry. Although planar shape remained constant as the divides became fixed by competition with neighbouring depressions, the sink migrated towards the divides. Morgan (1970) applied a measure of drainage basin asymmetry devised by Schumm and Hadley (1961): the ratio of maximum lateral distance from the main channel to the southern or western perimeter divided by maximum lateral distance from the main channel to the northern or eastern perimeter. Rose and Letzer (1975) developed a drumlin stoss point ratio given by the ratio of the distances from the highest point to the nearest end divided by the drumlin long-axis length. This measure may aid inference of iceflow direction from drumlin form.

2.5.3 Landform surface shape

Problems of quantifying planar shape are inevitably increased by incorporation of the third dimension. The three-dimensional surface has often been reduced to a transect profile typically along the longest axis of the planar form, and thus reduced to a two-dimensional or planar shape problem. For example, Strahler's (1952) hypsometric integral has been applied to megadunes (Warren 1976) and lake basins (Hakanson 1974, 1977). Simple measurements of profile slope elements convey little about three-dimensional shape without gross assumptions of symmetry. Comparisons of shape to standard reference forms such as the cone have been attempted for lakes (Hakanson 1974) and sinkholes (Mills & Starnes 1983) but are similarly crude. The regression algorithm developed for alluvial fan elevation data used by Hooke and Rohrer (1979) provides a true three-dimensional landform equation of considerably greater flexibility, where the polar coordinate system adopted is suggested by the geometry of the landform. Inevitably, given the variety of forms, solutions vary widely between different landforms and there is little consensus about a general mode of analysis.

Surface form of karst depressions has been measured by the ratio of mean depth and diameter or length (Sperling *et al.* 1977, Marker & Sweeting 1983), although this is merely a height : length relief ratio and must overgeneralise to an intolerable degree. There are operational difficulties in defining the depth of a doline whose rim varies in elevation. Day (1976) computed mean depth from six randomly located profiles, while Sperling *et al.* (1977) take the mean depth of long- and short-axis profiles. Mills and Starnes (1983) utilise four measures of depression form, but these are poorly correlated with drainage basin form by comparison with conventional two-dimensional measures of shape. Attempts to relate form indices to the solution of collapse origin of dolines (Jennings 1971, Coleman & Balchin 1960) are not supported by the findings of Sperling *et al.* (1977) and Day (1976). Width : depth ratios of clints and grikes have been correlated with calcite content (Goldie 1981) and grain size (Lewis 1983). Matthews *et al.* (1979) combined a height : width ratio with the tangent of the proximal slope angle to generate an asymmetry index for saw-tooth moraines, and found statistically different values for teeth and notch components.

Drumlim surface form has been compared to the standard shape of an ellipsoid (Reed *et al.* 1962), although in Trenhaile's (1971) sample of 100 drumlins from southern Ontario, only 33% conform to the ellipsoid model, and profile curve fitting using quartic polynomials failed to reveal any distinctive shape categories. Boulton (1982) uses three basic shapes specified by differing aspect ratios (amplitude/wavelength) and symmetry in this theoretical analysis of bedform stability in subglacial conditions.

Cirque form has been quantified by various height : length ratios (Andrews & Dugdale 1971, Vilborg 1984) and by profile curve fitting. The flattest : steepest ratio (King 1974) and the form flatness index (Derbyshire & Evans 1976) represent more elaborate attempts to specify the degree of cirque development based on height : length ratios. Evans' (1974) index of concavity, given as the plan closure plus four times the profile closure, performs the same task (see Sec. 2.4). Gordon (1977) applied several of these indices to a spatial sample of cirques, and inferred possible temporal development from their intercorrelations, suggesting that a change of shape results as headwall retreat outstrips downcutting. Aniya and Welch (1981) have modelled the three-dimensional form of cirques using a polynomial whose exponents and coefficients are determined from least-squares fits of cross profile, long profile and terrain coordinates, and Wheeler (1984) argues that a quadratic equation is a less restricted and more realistic approximation to the form of glaciated valleys than the more popular power-function curve.

Drainage basin shape measures have made little attempt to incorporate relief elements. The indices developed by Hobba and Robinson (1972) have been criticised by Wilcock (1973) for lacking physical interpretation, and they have been little utilised. Anderson (1973) used a second-order polynomial describing height variation around the basin perimeter to introduce relief information successfully, and in combination with a lemniscate loop perimeter and the mainstream gradient this permits simulation of three-dimensional basin form and correlates to hydrological parameters.

In the analysis of aeolian landforms, linear dunes have been shown to exhibit stable height : width and width : spacing relationships (Lancaster 1981), and the consistent geometry of barchans (Hastenrath 1967) enables inverse correlation of height and displacement rate (Finkel 1959, Long & Sharp 1964), and estimation of bulk transport from the dune volume × celerity product (Lettau & Lettau 1969). Volume is obtained from plan measurement between successive contours (Warren 1976), or from triangular prismoids bounded by cross sections (Norris 1966).

Consistent morphometric relationships also apply to the surface form of fresh and degraded cinder cones (Porter 1972, Wood 1980a). Rates of degradation are sensitive to rainfall and temperature parameters, which, given initially similar geometries, makes morphometric analysis of degraded cones a possible way to gauge long-term climatic change (Wood 1980b).

2.5.4 Conclusion

The greatest potential of specific geomorphometry lies in its ability to indicate processes and controls that operate on landforms in situations where the process cannot easily be measured and controlling variables are not immediately apparent. Perhaps the clearest illustration of this is seen where morphometric analysis of terrestrial features is used to provide possible explanations for the surface forms on other planets. Examples include the study of Martian drainage networks (Pieri 1980), river islands (Baker 1979, Komar 1983, 1984) and curvilinear landforms (Rossbacher 1985).

Traditionally, specific geomorphometry has been hampered by the complexity of three-dimensional surface description, a preference for simpler univariate shape indices, and the frequent inadequacy of data sources. None of these factors need represent the obstacles they once were. Computer analysis makes the handling of multivariate indices comparatively simple, and new sources of high-quality data are emerging to complement this increased analytical capacity. Progress in specific geomorphometry is, therefore, largely dependent on the willingness of investigators to exploit new developments, and on the insights gained from so doing.

2.6
Drainage basin morphometry

Six stages may be recognised in drainage basin morphometry, according broadly with nine stages recognised by Evans (1987a) for morphometry of all specific landforms. For drainage basins the stages are: definition and conceptualisation; network delimitation; sampling; measurement; variable definition; and analysis. These are used here as a framework for outlining techniques, although the first and last are omitted as being beyond the scope of the present section.

Although drainage basin morphometry is emphasised, some techniques are equally applicable to other levels of fluvial investigation (Sec. 2.7) or to non-fluvial landforms (Sec. 2.5), and fluvial morphometry must encompass all aspects of the fluvial system (Gardiner & Park 1978, Gregory 1977b, Gardiner 1983a) and their relationships with other landforms, particularly slopes (Sec. 2.8).

2.6.1 Network delimitation

Defining a drainage network for analysis requires choice of an information source and method for extraction of the network. Problems involved in these have been reviewed by Chorley and Dale (1972) and Gardiner (1975), and some case studies are summarised in Table 2.4. Networks may be categorised according to the source employed and the basis used for delimitation of the network (Table 2.5, Fig. 2.11). Note that it should not be inferred from Table 2.5 that any degree of correspondence necessarily exists between map- and field-derived networks employing the same defining characteristics. Data source and basis for network delimitation depend upon the purpose of the investigation, as morphologically defined networks may include elements significant to palaeohydrological or extreme-event studies, but not to those of contemporary processes of 'normal' magnitude. Hydrologically defined networks are easier to delimit than others, which may be subject to considerable variability of interpretation (e.g. Fig. 2.11), although Smart and Werner (1976) reported that operator variation in the number of exterior stream links identified by the contour crenulation method was always less than about 5%, and differences between two operators were within about 10% for mean exterior link lengths and 2–3% for mean interior link lengths (Smart 1978). It is always necessary to employ a consistent rule for determining the lengths of exterior links according to the contour crenulation method (Table 2.6).

Map data are subject to variance because of varying stream network depiction on different scales and editions, whereas field data are subject to problems concerning temporal variability in the extension (see review in Gardiner & Gregory 1982) and visible expression of stream channels, and difficulties in covering large areas. Aerial photographs offer a partial answer but problems occur because of their non-synchroneity and the difficulties of network identification in vegetated areas (e.g. Jennings 1967). Thus network delimitation is beset with uncertainty, although some recommendations have been made concerning cartographic data sources in particular cases (Table 2.4).

Table 2.4 Some conclusions concerning maps as a data source

Source	*Map*	*Area*	*Conclusions*
United Kingdom			
Gregory (1966)	OS 1:25 000	SW England	Functioning net greater than blue line net. Crenulation net includes palaeohydrological element
Werrity (1972)	OS 1:25 000 and 1:2500	SW England	Exterior link lengths reliably shown by 1:2500 but not 1:25 000
Gardiner (1975)	OS 1:25 000 and 1:10 560	NE and SW England, S Scotland	1:25 000 2nd Series or 1:10 560 Regular Edition best. 1:25 000 Provisional Edition has less blue line net and may differ nationally
Jarvis (1976b)	OS 1:25 000	SW England, S Scotland	2nd Series best available at this scale; comparable in two areas
Stagg (1978)	OS 1:25 000	Wales	Drainage density of gully networks from air photographs 5.34 × that of Provisional Edition map
Richards (1978)	OS 1:25 000	SW England	On Provisional Edition, mean field exterior link lengths less, but mean numbers more, than on crenulation or blue line nets
Gardiner (1982)	OS maps	Severn basin	'Laws' of cartographic generalisation of Topfer and Pillewizer (1966) apply to river networks from maps of various scales
United States			
Melton (1957)	USGS 1:24 000	Western States	High percentage of exterior links faithfully portrayed by crenulations
Coates (1958)	USGS 1:2400	Illinois	Exterior links cannot be inferred from contours. Field lengths 3 – 5 × 4 crenulation lengths
Morisawa (1957, 1959)	USGS 1:24 000	Appalachian Plateau	Crenulation net similar to field net
Maxwell (1960)	USGS 1:24 000	California	USGS maps prepared by plane table survey inadequate for network analysis
Schneider (1961)	(Reanalysed	Morisawa's data)	Contour crenulation inadequate for network delimitation

Giusti and Schneider (1962)	AMS/SCS 1:24 000; USGS 1:24 000, 1:62 500 and 1:125 000; AMS 1:62 500 and 1:250 000	Piedmont	Give factors to convert stream frequency from one map to that from another
Leopold *et al.* (1964)	USGS 1:31 680	New Mexico	First-order streams on map are in reality fifth-order
Coffman (1969)	USGS 1:24 000	Indiana	Blue lines show *c.* 4% exterior links. Crenulations are improvement, but still omit links, and vary according to physiographic province. High-relief areas better mapped.
Yang and Stall (1971)	USGS 1:24 000, 1:625 000 and 1:250 000	Oregon	Strahler order depends upon map scale; ratio measures independent of map scale
Patton and Baker (1976)	USGS 1:25 000	Texas	Crenulation net accurately portrays channels but omits ephemeral gullies
Kheoruenromme *et al.* (1976)	USGS 1:24 000 1:62 500, 1:250 000 and 1:500 000	South Carolina	Mean ratio of link magnitude on small-scale map to that from large-scale is 2 × ratio of map representative fractions
Mark (1983)	USGS 1:24 000	Kentucky	Mapped first-order basins contain 0–5 stream sources, with an average of 2.15 per basin
Elsewhere			
Eyles (1966)	1:63 360	Malaysia	Map underestimates stream numbers, and lengths correspond to next 'true' order. Omission factor greater in high-slope regions
Selby (1968)	1:15 840	New Zealand	Maps useless for detailed analysis, depict only 52% of total length
Eyles (1973)	1:25 000 and 1:63 360	New Zealand	1:63 360 completely unsuitable 1:25 000 depicts 86% of network; quality of representation conditioned by age of map, slope of area
Yoxall (1969)	1:250 000	Ghana	The map edition is non-uniform
Schick (1964)	1:20 000 and 1:2500	Israel	1:20 000 depicts about 10% less stream than 1:2500

Table 2.5 Bases for stream network delimitation on maps and in the field.

	Source	
Basis	*Map**	*Field*
Hydrological	a Blue lines. Governed by cartographic conventions	Flow during survey. Temporally variable.
Morphological	b Contour crenulation network. Consistent rules necessary (Table 2.6)	All channels. Consistent identification rules necessary (see Hupp 1986)
Maximum	c Mesh length extension. Rarely used outside USA.	Channel extensions into pipes and hollows. Difficult to implement.

*Letters refer to Figure 2.11.

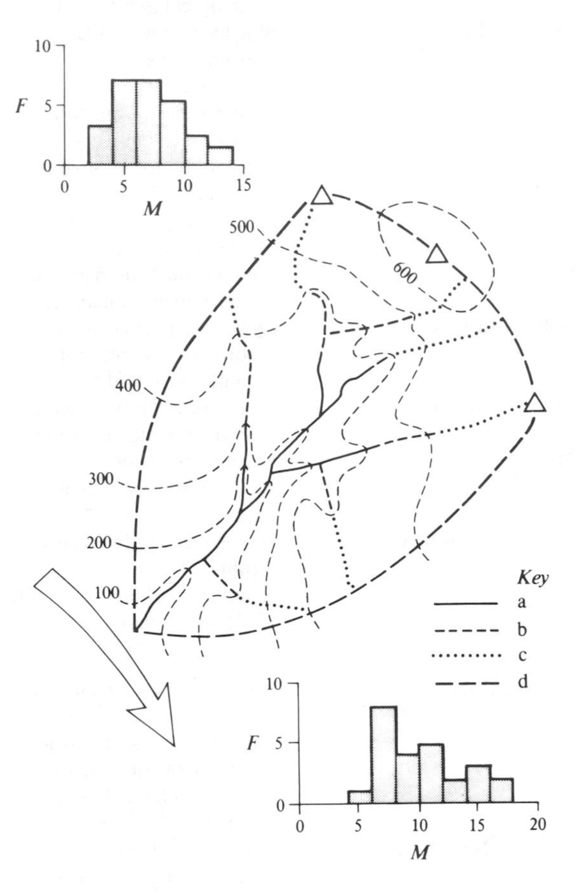

Figure 2.11 Types of map-derived drainage networks (see Table 2.5): (a) blue lines; (b) contour crenulation; (c) mesh length extension; (d) basin perimeter drawn orthogonal to the contours. The insets show histograms of resulting magnitudes of two basins delimited from contour crenulations by a group of 25 operators *without* a common set of strictly defined conventions, using OS 1:25 000 Provisional Edition maps.

2.6.2 Sampling

Most studies have used basins selected according to some ordering method, although computer data acquisition and storage now permit more flexible data structures (e.g. Smart 1970, Jarvis 1972, Coffman *et al.* 1971, Van Asch & Van Steijn 1973) by which data may be collected for individual links and aggregated as required. Many ordering methods have been devised (Table 2.7, Fig. 2.12), but those of Strahler (1964) and Shreve (1967) are most widely employed for selection of basins for morphological examination *per se*, while research and debate continue concerning process-oreinted ordering methods (Gregory & Walling 1973 pp. 46–7, Graf 1975, 1977, Dunkerley 1977, Yoon 1975). No

Table 2.6 Criteria for contour crenulation network delimitation.

Criterion	*Procedure*	*Comments*
Overall form	Morisawa (1957)	V-shaped contours where definite notch occurs.
	Jarvis (1976b)	Wherever cusp in contour indicates significant indentation of the topographic surface.
	Sakaguchi (1969)	Ratio length : width greater than 1.0.
Minimum length	Morgan (1971)	At least 100 m on ground.
	Sakaguchi (1969)	At least 500 m on ground.
Highest crenulation	Morgan (1971)	Extended to highest crenulated contour.
	Gardiner (1975)	As Morgan (1971), providing that at least 1 in 4 contours crenulated.
Slope	Shreve (1974)	Channel source is point where channel slope declines to predefined value.

Table 2.7 Ordering methods: stream links identified in Figure 2.12 are ordered

	*Ordering methods**						
Stream links	*(1)*	*(2)*	*(3)*	*(4)*	*(5)*	*(6)*	*(7)*
a	1	1	1	1	1	1	1
b	3	1	1	1	1	1	1
c	1	1	1	1	1	1	1
d	2	1	1	1	1	1	1
e	1	1	1	1	1	1	1
f	1	1	1	1	1	1	1
g	2	1	1	1	1	1	1
h	1	1	1	1	1	1	1
j	3	2	2	3	2	1.67	2
k	2	2	2	3	2	1.67	2
l	2	2	3	5	2.59	2.06	2
m	2	2	2	3	2	1.67	2
n	3	3	5	9	3.32	2.55	2
p	3	3	7	13	3.81	2.87	3
q	3	3	8	15	4	3	3
Symbol	–	–	*M*	*G*	*S*	*W*	*T*
Derivation	–	–	No. of exterior links	$G = 2M - 1$	$S = \dfrac{\log 2M}{\log 2}$	$W = \left(\dfrac{\log M}{\log R}\right) + 1$	Each order has range of 4^{T-1} magnitude
Properties†	*(1)*	*(2)*	*(3)*	*(4)*	*(5)*	*(6)*	*(7)*
A		●					
B		●	●	●	●	●	●
C	●	●	●	●			●
D			●	●	●	●	●
E			●	●	●	●	
F		●	●	●	●		●
Suggested purposes‡	None	S	T/S	P	T	T	T/S

**Ordering methods*: (1), Horton (1945); (2), Strahler (1952); (3), Shreve (1967); (4), Gregory and Walling (1973), see also Graf (1975); (5), Scheidegger (1965); (6), Woldenberg (1967); R is bifurcation ratio; (7), Smart (1978). Other methods (not illustrated) include those of Lewin (1970), Walsh (1972) and Yoon (1975).

† *Properties of the ordering methods*: A, convenient for sampling; B, completely objective; C, integer orders only; D, does not violate distributive law; E, each junction should change order, F, a network should have only one defined ordering.

(●) denotes that the method satisfies the property.

† *Purposes*: S, sampling for morphological studies; T, topological investigations; P, process investigations.

method can be suitable for all purposes but suggestions may be made concerning those suitable in particular contexts (Table 2.7).

It has been suggested that ordering methods are somewhat unsatisfactory because, although they afford a method by which a network may be subdivided into basins of approximately equal size, they fail to consider geographical position within the basin. Thus, Jarvis (1977a) proposed a distinction between order-excess basins, which are tributary to streams of greater order and have no effect on the order of the receiving stream, and order-formative streams, two of which form a stream of the next higher order. Order-formative streams characterise the headwaters of river systems and may therefore have differing morphometric properties from order-excess streams, which characterise the lower parts of the system and major valley sides, and develop a geometry that is controlled by different processes of spatial competition. Similar ideas have been developed by Abrahams (1980, 1984).

Regular sampling units have also been employed in fluvial morphometry. For example grid square maps of drainage density have been prepared (Gregory & Gardiner 1975), and Dacey and Krumbein (1976) attempted to develop relationships between topological characteristics of networks in sampling quadrats.

2.6.3 Measurement

Measurement of lengths and areas is basic to virtually all morphometric studies and is crucial in establishing the statistical variance of the data (Mark & Church 1977). Available techniques, some of which are illustrated in Figure 2.13, have been reviewed by Gardiner (1975), who concluded that the opisometer (map measurer) offered the most effective method for linear measurement and the polar planimeter for measurement of area. However, their accuracy may differ widely according to the model employed. For example, the four opisometers shown in Figure 2.14 gave values differing by up to 10% for rivers of low sinuosity. The significance of the difference between opisometers diminishes with increasing line sinuosity (Fig. 2.13), because the inherent inaccuracy of the opisometer and the inability to follow tortuous curves lead to increased variance of measurement. Thus, for more sinuous rivers at least, choice of opisometer may not be as critical as *a priori* considerations suggest, although the dependence of measurement variance upon line sinuoisty may render statistical interpretation difficult.

Digitisation offers an attractive alternative to manual methods, combining measurement of length and area in one operation (Fig. 2.13). The limited availability of the necessary equipment has restricted its application, although cheaper microcomputer-

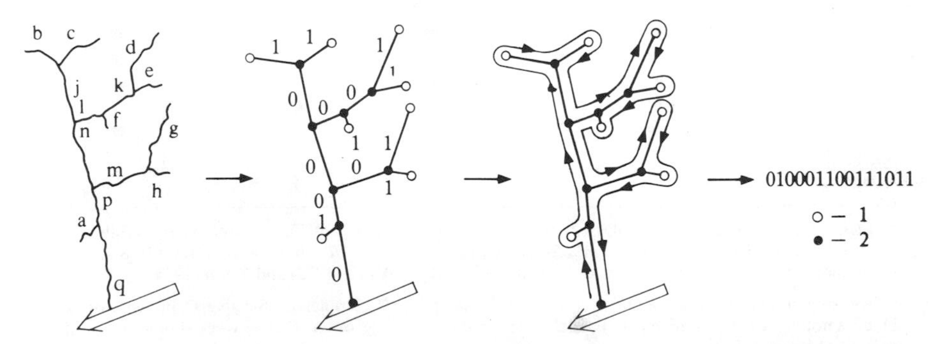

Figure 2.12 Representation of a network as a topological vector. Links are identified and ordered in Table 2.7. Symbol '1' represents stream sources, symbol '2' stream junctions.

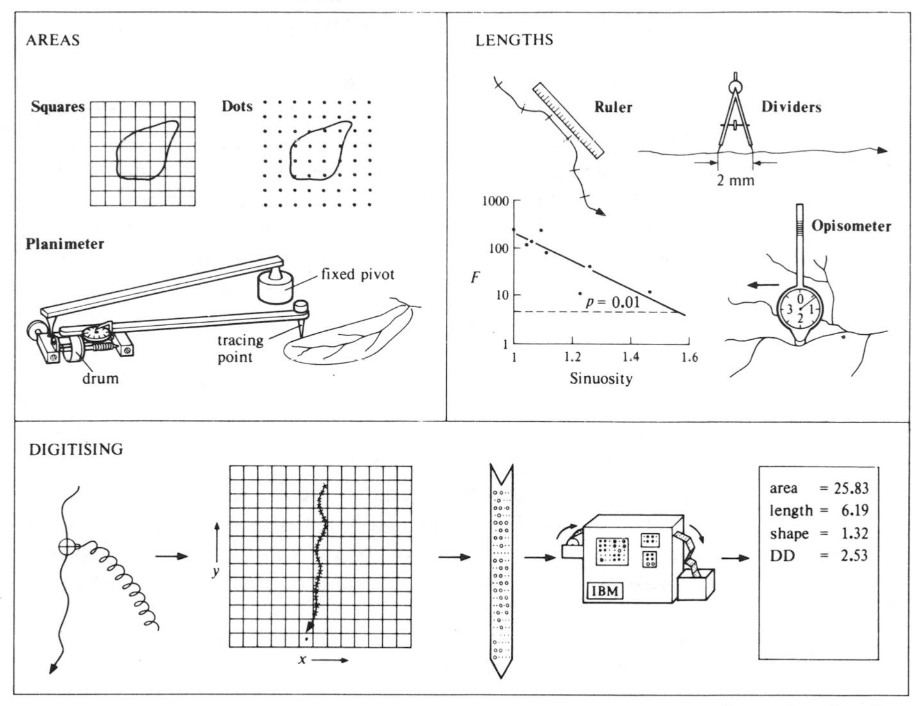

Figure 2.13 Methods of areal and linear measurement. The graph, in the section on lengths, plots the F values from analyses of variance of six repeated measurements of the length of a line employing the four opisometers shown in Figure 2.14 against the sinuosity of each line.

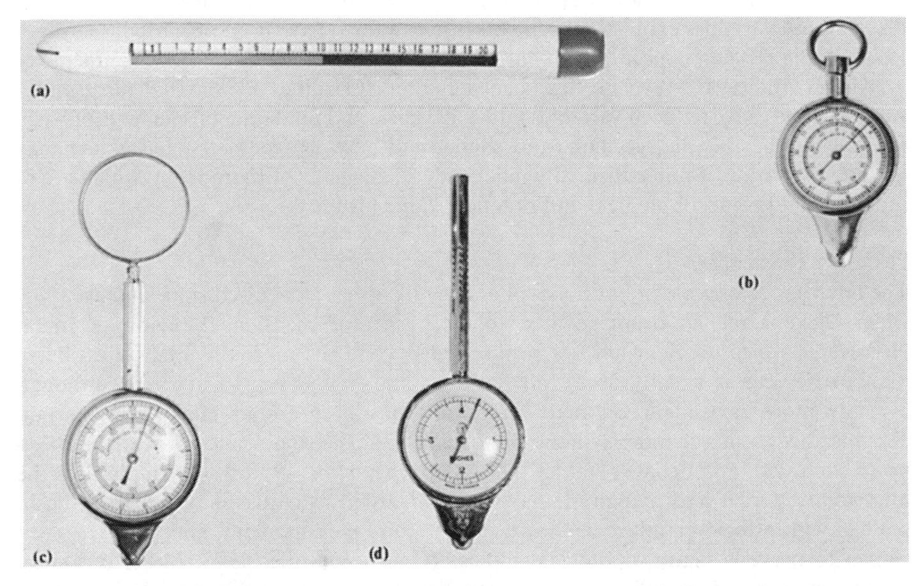

Figure 2.14 Four commonly used opisometers: (a) the Milograph (see Rhind 1976); (b) and (c) Chartometers commonly available from drawing office suppliers; (d) the HB Curvimetre.

based systems are now available (e.g. Davidson & Jones 1985). Algorithmic aspects of area and distance calculation from digitised data are outlined by Gardiner and Unwin (1989). Procedures for digitisation of networks to facilitate examination of network topology and to provide an efficient data structure for morphometric analysis are considered below.

Attempts have also been made to estimate morphometric measurements. Donahue (1974) suggested a method by which basin areas and stream lengths may be estimated in one operation. This requires superposition of the basin and stream network onto a regular grid of points, and counts of the points enclosed within the basin and cut by the network are used as estimates of area and length. This method, an extension of the dot planimeter (Wood 1954), could also be employed for estimation of basin length, perimeter and other measurements. Mark (1974) examined the line intersection method of Carlston and Langbein (1960), and suggested that the conversion from number of stream intersections per unit length of line to drainage density required the coefficient $\pi/2$ and not the reciprocal of sin $(\pi/4)$ as employed by McCoy (1971). He also concluded that an average error of about 10% using this method was outweighed by a 50% saving in time. Gardiner (1979) extended this approach by using additional topological characteristics to estimate grid square drainage densities, and similar approaches have been developed by Karcz (1978) and Hakanson (1978). A combination of quadrat sampling and counts of stream segments has also been used to predict basin drainage density (Richards 1979), although this demands initial establishment of an empirical relationship between basin drainage density and the number of stream segments in a 1 km^2 quadrat within the basin, which will only remain valid within an area of constant mean link length.

2.6.4 Morphometric variable definition

Once basic basin dimensions have been determined, it is necessary to consider how these may be most usefully combined into variables expressing significant and meaningful elements of basin form; indeed entitation may itself influence the measurements to be made. Many variables expressing elements of basin form have been proposed (Gardiner 1975, Strahler 1964), conditioned by a variety of purposes, scales and environments of individual studies. However, it has been suggested (Gardiner 1978b) that most multivariate studies have reduced variables into a few basic and fundamental elements of basin form. Thus, basin scale, gradient and shape, and network geometry and extent may represent separate, if not statistically independent, elements of basin form, which are of fundamental geomorphological and practical value (Table 2.8).

2.6.4.1 Basin scale

The importance of scale in basin morphometry has been recognised since at least 1865, when Dickens related drainage area to discharge for the Bengal region of India. However, its measurement has not progressed beyond the more obvious states. Basin planimetric area is usually used, but basin perimeter is easier to measure, although possibly more dependent on map scale, and has been found effective in certain circumstances as a size index (Chang & Boyer 1977). The significance of scale has been hotly contested, and debate (Pethick 1975, Gardiner *et al.* 1977, 1978) emphasises the uncertainty. Order and magnitude have been regarded as convenient indices of basin scale in within-region studies, as they do not require measurement, and scale has been shown to pervade many topological indices (Onesti & Miller 1978, Miller & Onesti 1975). Order-based sampling is one means by which a scale-free sample is at least notionally achieved, but only within a uniform geomorphological environment for

Table 2.8 Some variables commonly used in drainage basin morphometry.

Basic measurement	*Dimensions*	*Symbol*
Area	length^2	A
Perimeter	length	P
Total channel length	length	C_t
Basin length	length	L
Basin relief	length	H_b
Nos. of 1st- to *n*th-order streams	—	$N_1, \ldots, N_n$

Variables	*Derivation*	*Comments*
Basin scale		
Area	A	Strictly, is only planimetric equivalent of true area; difference negligible except for very steep basins
Perimeter	P	Often difficult to delimit precisely
Order/magnitude	See Table 2.7	Topological analogue of scale
Gradient and relief form		
Relief ratio	$R = H_b/L$	May be unduly influenced by one point. Length variously defined.
Hypsometric integral	See Strahler (1952)	Best determined by approximation (Pike & Wilson 1971). May be area-dependent (see Aronovici 1966)
Basin shape		
Elongation	$S_e = A^{1/2}/L$	Simple ratios – others also possible – see Gardiner (1975)
Relative perimeter crenulation	$S_c = P^2/A$	
Others See Section 2.3 and Gardiner (1975)		
Network extent		
Drainage density	$D_d = C_t/A$	Widely used: estimation methods available – see text
Network geometry		
Stream frequency	(*a*) $F = (\Sigma N_i)/A$ or (*b*) $F = (2N_1 - 1)/A$	Related to D_d – see Melton (1958). Alternative definitions are: (*a*) Strahler segment frequency and (*b*) Shreve link frequency
Magnitude	See Table 2.7	
Network diameter	Greatest no. of links from mouth to source	Topological analogue of mainstream length
For others, see Smart (1978) and Jarvis (1977b)		

which a basin of a given order (or magnitude) will be of a certain size, within reasonable limits.

Basin studies have been slow to adopt alternative indices of basin scale in terms of the functional basin rather than that contained within its potential perimeter, which may not all be hydrologically active in streamflow generation at a given moment. Thus, the contributing area concept as developed by Gregory and Walling (1968) and Kirkby and Chorley (1967) may provide a more meaningful basin area, by identification of those areas satisfying topographic criteria concerning valley bottom location and contour concavity.

2.6.4.2 Gradient aspects of the basin

The simplest relatively effective index of basin gradient is the relief ratio (Schumm 1956). This, however, combines two rather uncertain measurements. Basin length is subject to a variety of alternative definitions; thus Gardiner (1975, p. 18) lists nine alternative definitions based on the basin outline, relief distribution or stream net, and Cannon (1976) offers further definitions and compares their effect upon the values of morphometric variables. Basin relief may be greatly influenced by one isolated peak within the basin and may therefore be unrepresentative. The hypsometric integral (Strahler 1952) offers a conceptually attractive alternative for expressing overall relief aspects of the basin, despite criticisms made of the index (Harlin 1978). Paradoxically, in view of its application to other landforms such as lakes (Hakanson 1977), sand dunes (Warren

1976) and cirques (Gardiner 1983c), it has not been widely applied in fluvial morphometry, because of the laborious data collection required for its determination; the effective approximation method of Pike and Wilson (1971) demands more widespread consideration.

2.6.4.3 Basin shape

In measurement of basin shape, a basic distinction exists between studies of basin form *per se* and studies employing form for process prediction. In the former, many ratios of basic measurements have been proposed in attempts to derive dimensionless indices of basin form, as extensively reviewed by Gardiner (1975). These are effective for morphological description although correlation analysis of such ratio indices should be treated with caution as the null correlation is seldom zero (Sec. 2.2, Chayes 1971). The application to basin form of Fourier analysis, which models the basin outline by a sum of harmonics, provides a useful multivariate description of basin form (Jarvis 1976a). McArthur and Ehrlich (1977) have examined this technique in relation to conventional ratio measurements, and Jarvis and Clifford (Sec. 2.5) have described the procedure.

For process estimation, shape measurement based upon the distribution of area within the basin is probably more meaningful, and the method of Blair and Biss (1967) offers considerable potential. Again, however, greater potential may be realised if it becomes possible to recognise the hydrologically active contributing area for a particular magnitude of discharge event, rather than employing the basin as defined by its perimeter.

2.6.4.4 Network extent and type

The study of network extent has been a central theme of enquiry in drainage basin morphometry as it is of considerable conceptual and practical interest. Drainage density is the most widely employed index, although Singh (1976) has examined several alternatives. Stream frequency may be included as an index of network extent, although it incorporates both topological and areal elements in its calculation; it is related to drainage density in a geomorphologically significant manner (Melton 1958). Alternative measures of network extent are the texture ratio (Smith 1950), drainage texture (Morgan 1976) and drainage intensity (Faniran 1969, Donahue 1972), although none of these have been widely used.

Gregory (1977a, b) has emphasised that fluvial studies should encompass all relevant levels of spatial organisation within the system, and has developed a technique by which network volume may be derived from consideration of channel capacity and stream length. This demands knowledge of both basin morphometry from map observations and channel geometry from field measurements. Reservations have been expressed concerning the derivation of the index (Ferguson 1979), although successful applications have been reported by Gregory (1977 a, b) and Gardiner and Gregory (1982). A further developement is that of Gregory (1979), who incorporated the relief dimension into an index of network power. Attempts have also been made to incorporate channel type in functional terms into morphometric analyses, as reviewed by Gardiner (1983a). These ideas emphasise the need to extend morphometry to all levels of investigation within fluvial systems, as attempted by Onesti and Miller (1974).

Network extent is not a static quantity. Ample observations exist, as collated by Gardiner and Gregory (1982), to show that, during the course of individual storm events, networks expand and contract; a dynamic drainage net must be recognised in a manner analogous to the partial contributing area concept. Drainage networks also vary in extent on much longer timescales, and the morphometry of valley systems may therefore be used as a means of palaeohydrological retrodiction (Gardiner 1983b, 1987).

2.6.4.5 Network geometry

Examination of the topology of stream networks has developed into a major thread of enquiry (Gardiner & Park 1978) and a full account of the indices used is not feasible here; recently reviews have been presented by Jarvis (1977b) and Smart (1978). Analysis of network geometry is facilitated by computer-aided methods of data capture and storage, and these also afford effective storage of non-topological data. Shreve (1967) and Scheidegger (1967) recognised the possibility of binary representation of networks, and Smart (1970) proposed a method for producing a topological vector (see Fig. 2.12). The order of progression through the network was modified (Jarvis 1972) to allow greater flexibility, and a greater range of information was incorporated in the WATER system of drainage network analysis (Coffman *et al.* 1971). Algorithms for extracting information from topological vectors have been suggested by Liao and Scheidegger (1968), Smart (1970, 1978), Jarvis (see Smart 1978) and Ferguson (1977b). Their flexibility has allowed development of a wide range of link-associated variables (see Table 2.8), which have been employed in investigation of the random model of drainage network composition as reviewed by Shreve (1975) and Smart (1978). Topological variables such as mean exterior path length, diameter and network width (Werner & Smart 1973, Kirkby 1976, Kennedy 1978) have also been shown to be viable indices of network structure for prediction of elements of the catchment hydrological process.

One aspect of network geometry to have received particular attention recently is the morphometry of river junctions. Early empirical observations of junction angles (e.g. Zernitz 1932, Lubowe 1964) have been supplemented by studies of changes in particular channel properties at junctions (Richards 1980) and relationships between morphometric characteristics of the network and junction angles (Abrahams 1980, Roy 1983).

2.6.5 Conclusion

It is appropriate to conclude this outline with a consideration of desirable and likely future developments, as a basis for innovations; four such themes may be identified.

Recent studies have stressed the need for larger data sets, implying greater use of automated data collection, storage and analysis, and further development of estimation techniques. With the increasing availability of computer facilities, this need is being realised, allowing wide-ranging comparative studies extending the preliminary investigations of, for example, Gregory (1976a) and Gregory and Gardiner (1975).

The need for a greater awareness and understanding of the analytical techniques employed is also being recognised. Thus, the statistical properties of the data (Gardiner 1973, Mark & Church 1977) and the meaning of correlations between variables (Carson 1966, Gardiner 1978b) are beginning to receive attention.

Application of morphometric techniques to discharge prediction and retrodiction is increasingly being based upon a greater understanding of the processes involved (Gardiner & Gregory 1982), with models such as the geomorphic unit hydrograph (Rodriguez-Iturbe & Valdes 1979, Rodriguez-Iturbe 1982) being developed for contemporary application, and initial attempts being made at palaeohydrological retrodiction (Gardiner 1987).

The final direction in which fluvial morphometry should develop is towards incorporation of a greater range of fluvial characteristics, encompassing cross section, channel and network attributes, and attempts to evolve suitable procedures represent what is probably the most significant and desirable advance since Horton's (1945) original formulation of basin morphometry.

2.7
River channel form

Channel morphometry is concerned with the definition, measurement and analysis of quantitative indices describing the cross section, the bedform and long profile, and the plan geometry of rivers. Field measurement is usually necessary, although the widths of large rivers, long-profile gradients and planform properties can all be derived from topographic maps and air photographs. Reaches are samples for measurement in various ways. Since downstream changes in channel geometry are step functions at tributary junctions (Roy & Woldenberg 1986), a random sample of junctions may be adopted as long as measured sections are beyond the immediate hydraulic effects of the mixing of flow at the confluence. Alternatively, a stratified random sample within the drainage network of link midpoints (reaches) having predefined stream magnitudes or orders (see Sec. 2.6) will give a representative range of channel sizes. Traditionally, however, pseudo-random sampling based largely on convenience of access has favoured identification of channel variations as continuous functions of discharge (Leopold & Maddock 1953) or contributing basin area (Hack 1957). Within a reach, riffle–pool variation introduces the need for sample stratification at another scale, first because downstream (between-reach) analysis requires data from comparable sections (Brush 1961), secondly because gravel riffles provide more stable gauging sections for at-a-station hydraulic geometry studies (Wolman 1955), and thirdly because systematic sampling of riffle and pool sections aids assessment of within-reach variation of channel form in relation to locally operating hydraulic processes (Richards 1976).

The purpose behind channel morphometry is often to relate form characteristics to hydrological or process variables so that these can in turn be inferred from more easily measured and relatively static morphometric properties. This may be to obtain reconnaissance hydrological data in ungauged basins, as when bankfull discharge is estimated from channel widths or cross-section areas (Brown 1971); or it may be to reconstruct palaeohydrology from the cross sections or plan geometry of palaeochannels, as in Williams' (1984b) use of the relationship between meander bend radius of curvature and bankfull discharge, to assess former discharges of palaeochannels.

2.7 Cross-section geometry

Measurements of flow and channel geometry refer, respectively, to the water prism, and to an *assumed* water prism beneath the morphologically defined bankfull stage or an arbitrary stage defined by alternative criteria. Primary variables are the water surface width (w) and bed width, and the mean and maximum depths (d and d_{max}); the cross-section area ($A = wd$) and form ratio ($F = w/d$) are derived from these. The wetted perimeter (p) and hydraulic radius ($R = A/p$) are sometimes preferred to width and depth, as they are more relevant to the mechanics of open channel flow. Measures of cross-section asymmetry are based on the displacement of either area or maximum depth relative to the channel centre-line, and range from a simple ratio of left- and right-hand sub-areas (Milne 1982) to more complex measures tested by Knighton (1981). In studies of flow geometry the mean flow velocity (v) and water surface slope (s) are also important, and are interrelated through resistance equations such as the Manning and Darcy–Weisbach equations. These are

$$v = R^{2/3}s^{1/2}n^{-1} \quad \text{and} \quad v = (8gRs/f)^{1/2} \tag{2.1}$$

in which n is Manning's roughness coefficient and f is the Darcy–Weisbach friction factor. In these equations depth is often used as an approximation for the hydraulic radius in wide, shallow channels ($F > 10$). The coefficients n and f have to be judged from tables or standard photographs (Barnes 1967), although frequently in hydraulic geometry studies these equations have been inverted and the friction measures have been estimated from measured values of velocity, depth and slope. Both resistance equations apply to uniform flow conditions, so measurement reaches must be selected appropriately. The Darcy–Weisbach equation is preferred (Chow 1959), as it is a dimensionless expression with a sound theoretical basis.

2.7.1.1 Flow geometry

Velocity–area discharge estimation (British Standards Institution 1964) necessitates survey of the water prism geometry, which is used in both at-a-station hydraulic geometry studies and analysis of downstream trends in flow characteristics at specific flow frequencies (Richards 1977, 1982 pp. 155–60). Figure 2.15 illustrates the mean section gauging method, a standard procedure outlined and programmed by Midgley *et al.* (1986) in a set of computer routines for streamflow estimation from field data, and for assessment of accuracy in relation to the number of sampled verticals. In this method, each segment area is

$$a_i = w_i\left(\frac{d_{i1} + d_{i2}}{2}\right) \tag{2.2}$$

and discharge is

$$q_i = \left(\frac{v_{i1} + v_{i2}}{2}\right)a_i \tag{2.3}$$

where d_{i1} and d_{i2} are marginal depths, and marginal velocities are means in the verticals measured by current meter at 0.6 × depth (or the average of observations at $0.2d$ and $0.8d$). Total discharge (Q) and area (A) are obtained by summing the segment contributions, which should each be less than 15% of the total:

$$Q = \sum_{i=1}^{n} q_i \quad \text{and} \quad A = \sum_{i=1}^{n} a_i \tag{2.4}$$

At the bank, velocity is assumed to be zero, as is the depth unless the bank is vertical.

Water surface slope is spatially variable at low flow because of large-scale bedforms, so accurate surveying is necessary over short distances, preferably by levelling (Bannister & Raymond 1984). However, where cross-channel flow vectors occur, the line of maximum slope may be difficult to define, and a linear trend surface (Unwin 1975) fitted to a grid of water surface elevation data may provide the best estimate of mean gradient and direction. Figure 2.16 illustrates an alternative method for measuring water surface slopes over short distances, using two vertical graduated staffs and transparent tubing filled with coloured liquid.

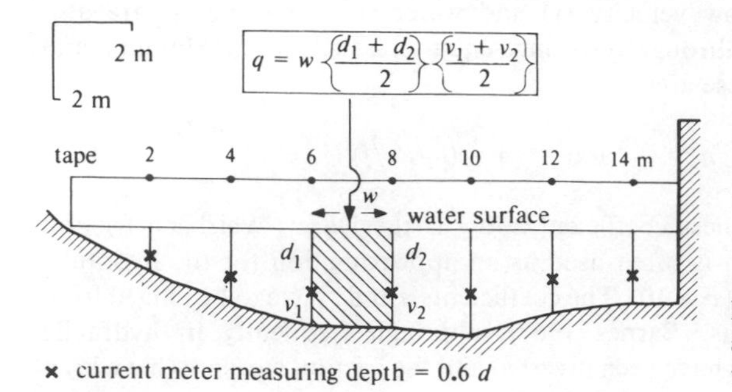

Figure 2.15 An example of velocity–area stream gauging by current meter, using the mean section method.

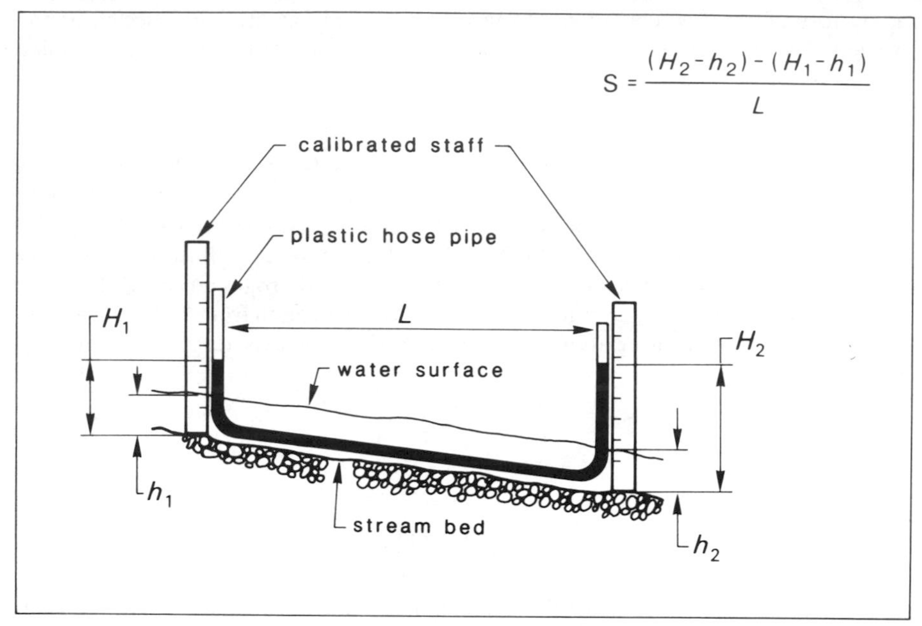

Figure 2.16 A simple device to permit measurement of water-surface slope over a short (*c*.1–5 m) reach in a small stream.

2.7.1.2 Channel geometry

For incised and bedrock channels, non-morphological criteria are necessary to delimit the level attained by flows of a given frequency. However, inundation of the floodplain of an alluvial channel modifies the shear stress distribution (Barashnikov 1967), and this process discontinuity focuses attention on the bankfull channel morphology. Bankfull flow frequency in alluvial channels varies with flow regime and flood duration (Harvey 1968), channel gradient (Kilpatrick & Barnes 1964) and bank material shear resistance (Pickup & Warner 1976), so that a constant return period of 1–2 years on the annual series (Wolman & Leopold 1957, Dury 1961) cannot be assumed. For example, Williams (1978) has compiled data which suggest that the return period of inundation of active floodplain surfaces ranges from 1.1 to 10 years. This complicates attempts to use bankfull channel capacity (cross-section area) as an index of a specific flow frequency in ungauged catchments. Further, Lewin and Manton (1975) demonstrate that bank elevation and floodplain topography vary considerably within a reach, so that it may be necessary to average channel properties from several adjacent comparable sections.

Indirect measurements Channel width may be inferred from vegetation or sediment patterns, obviating the need for field survey, and introducing air photographs as a potential data source. Colour imagery at 1:5000 (Lewin & Manton 1975) provides clear tonal variation identifying vegetation patterns, and permits height estimation to ± 0.10 m, so that channel margins can be defined and elevations measured. Although Riley (1972) argues that vegetation obscures channel margins, this depends on its size relative to the channel, and on the timing of photography. Speight (1965) and Nunally (1967) both used the channelward limit of trees or tall grasses to delimit bankfull widths, and Nunally finds that this corresponds with the transition between channel and overbank deposits on point-bar channel banks, and suggests the upper limit of continuous sand deposition as a convenient sedimentological index of bankfull stage. However, sedimentary evidence is invariably too complex or transitional for definitive boundaries to be drawn. Sigafoos (1964) demonstrated that overbank flow may damage trees or inhibit germination, but vegetational and sedimentological evidence of flood-plain inundation does not relate consistently to a specific flow frequency, and is therefore only useful for reconnaissance purposes, or as a supplement to morphological surveys. Thus, Hupp's (1986) recognition of distinctive floral assemblages between terrace, floodplain and 'channel self' sites provides a basis for identifying the headward limits of first-order streams, and the point of initiation of alluviation, but does not necessarily define lateral channel limits at a specific flow frequency. Only Gregory's (1976b) analysis of the truncated distribution of lichen thalli on *bedrock* channel walls provides an unambiguous correlation of a vegetational limit to a particular flow, the discharge at the lichen limit having a return period of 1.14–1.37 years.

Morphological measurements Field survey is essential to define the cross-section shape in detail. Instruments such as the A-frame (Riley 1969) and pantometer (Pitty 1968) are usable if bank angles are below 60°, but prevent measurement at each identifiable break of slope. Consequently, survey by recording depths to the floodplain surface or channel bed from an arbitrary horizontal datum is preferable. For channels less than 2–3 m wide, a horizontal tape may suffice, but levelling and tacheometry are necessary in larger streams. Width measurement should be accurate to ± 1% and depth measurement to ± 5%; in large perennial rivers with soft sand beds, echo sounding may be appropriate.

The surveyed cross section is plotted (Fig. 2.17a), and bankfull level is defined by reference to floodplain level (Dury 1961) or major benches (Kilpatrick & Barnes 1964). The frequency of exceeding bench elevations is critical, since some are sub-bankfull benches within the contemporary channel (Woodyer 1968), whereas others reflect adjustment of channel geometry to altered hydrological regime (Gregory & Park 1974). Objective definition of bankfull stage may be achieved by noting the depth or stage at which width:depth ratio increases rapidly, reflecting overbank flow (Wolman 1955). A minimum percentage increase in width:depth ratio may be required to avoid defining minor benches as floodplains. Riley's (1972) bench index

$$BI = \frac{w_i - w_{i+1}}{d_i - d_{i+1}} \tag{2.5}$$

peaks strongly at bankfull, but also shows minor peaks at bench levels (the *i* subscript refers to successively higher stages). Both methods work best in straight reaches where left and right banks are approximately equal in height (see Fig. 2.17b); in bend sections with point bars, supplementary information may be required. Field survey data may be input directly to a computer program, which defines the channel as a set of coordinates, with arbitrary, equally spaced 'water' levels located to calculate successive widths and depths. If several sections are surveyed in a reach, bankfull water surface slope can be calculated by comparing bank elevations to define points of overbank spillage. A measure of bed width as the total channel width between banks at the bed (even if part is exposed at low flow) may be useful, because it relates directly to sediment transport process and avoids the frequent problem of banktop curvature, which increases bankfull width, but may be unrelated to fluvial processes.

2.7.2 Long profile

Long-profile measurements are scale-dependent, since gradients obtained over shorter reaches refer to bedforms rather than to general bed elevation trends. Nevertheless, in detailed hydraulic studies, water surface and bed gradients are required for short reaches, and field survey is therefore essential. Floodplain gradients may be measured from map or air photograph sources, but should only be used to estimate changes in bed or water surface elevation if the degree of channel incision into the floodplain is known to be roughly constant.

2.7.2.1 Channel gradient and long-profile shape

Park (1976) measured mean channel gradient in the field using an Abney level, sighting over a reach between three riffle summits, the middle of which provided a measured cross section. This estimate was found to correlate with gradients derived from successive 25 ft (7.62 m) contour intersections with the river, measured from 1:25 000 maps. Thus, for general correlation of basin area, channel slope and bed material size (Hack 1957, Wilcock 1967), cartographic bedslope is considered acceptable. The method assumes that channel incision does not vary in the reach concerned, so that bed and banktop

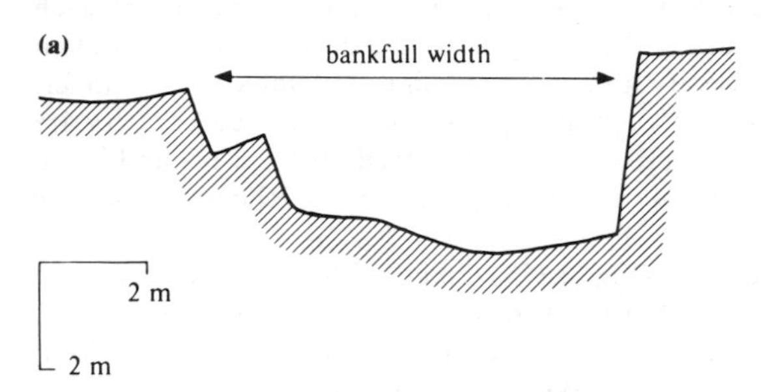

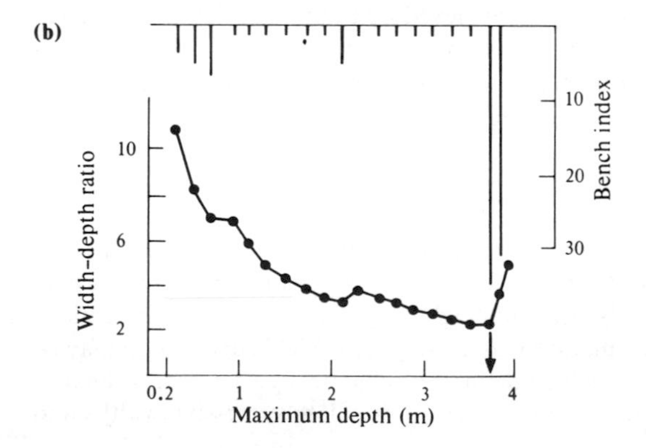

Figure 2.17 (a) A surveyed channel cross section and (b) plots of width: depth ratio and bench index for various arbitrary water levels, and their associated maximum depth measurement, for the section shown in (a).

gradients are comparable. Scatter in the relationship between map and field slopes (Park 1976) reflects error in both measurement techniques. Field measurement should be by levelling and, since riffle amplitudes vary, it is preferable to obtain equally spaced relative elevations through two or three complete riffle–pool cycles, and generalise the trend by a regression line. The error of estimation will be much less than that based on only two height observations. Map-derived valley slope estimates are more accurate if based on spot height data, and may be obtained by projecting spot height locations perpendicularly onto a downvalley trend line and calculating a height–distance regression (Richards 1972, Cullingford & Smith 1966). Channel gradient can be obtained by dividing the regression coefficient (valley gradient) by channel sinuosity.

At a between-reach scale, variations of gradient are summarised by a range of exponential long-profile models, which relate bed elevation (H) to downstream distance (L). On being differentiated to give $\mathrm{d}H/\mathrm{d}L$ (= slope s), these define the downstream variation of slope, which usually decreases in perennial rivers with generally concave profiles (Richards 1982, p. 224). Indices of concavity describe this shape and commonly define a straight line from *source* to mouth, and then compare the total profile relief with either the maximum vertical distance between the straight line and a parallel line just touching the profile (Wheeler 1979).

2.7.2.2 *Bedforms*

Bedforms range from riffle–pool features in gravel-bed streams to dune forms in sand-bed streams. Survey techniques must reflect the type of feature, but analysis of bed elevation data and quantification of morphology involve similar techniques and indices. Typical bedform indices for dunes are described by Nordin and Richardson (1968): average length $\bar{L}$, height $\bar{H}$ and the angles of upstream and downstream faces, α and β, define dune shape (Fig. 2.18a). However, the statistical variation of these indices makes their estimation hazardous and it is usually preferable to use alternative, but closely related, variables derived from series analysis of bed elevation data. Similarly, riffle–pool wavelengths may be estimated from air photographs taken at low flow, although errors may arise where strong reflection occurs from the water surface and where small riffles do not show as white water. However, the inherent variability of wavelength and amplitude of all bedforms has encouraged the use of 'series analysis' as an objective basis for the estimation of these parameters. These techniques require equally spaced data, so that levelling of bed profiles is desirable with successive staff positions separated by equal distances. Sonic depth sounding is a useful method in sand-bed streams because it does not disturb the bedforms and associated sediment transport process (Richard *et al.* 1961); however, some levelling is also necessary to provide a horizontal datum.

Given a set of equally spaced relative elevations, objective methods exist for identifying bedform properties. These often involve generalising the elevation trend using regression analysis. Riffles and pools (for example) may then be defined as positive or negative residuals, and the wavelengths and amplitudes of bedforms can be identified between successive upward crossings of the bed profile through the regression line. O'Neill and Abrahams (1984) note that regression residuals may give misleading information, and propose a semi-objective method for defining successive bedforms, based on the cumulated elevation change since the last extreme maximum or minimum, and the selection of a tolerance level designed to prevent *local* maxima or minima being defined as riffles or pools.

The elevation series may also be analysed by serial correlation methods. Ideally in this case, series should consist of at least $N = 50$ observations, with 10–12 per bedform

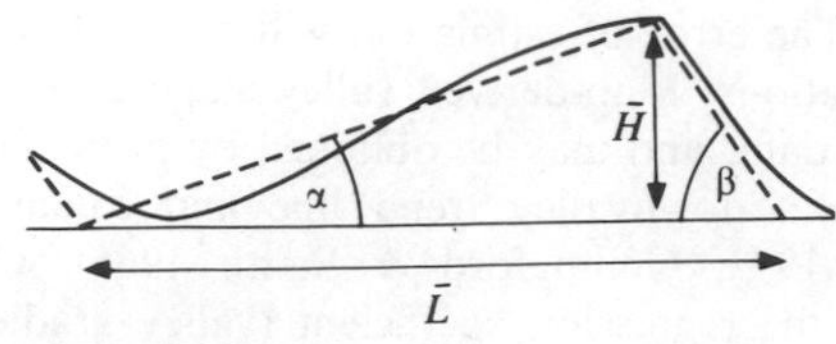

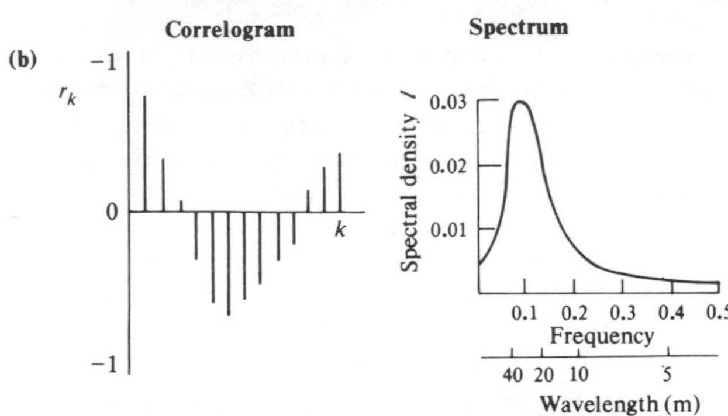

Figure 2.18 (a) Variables describing average sand-dune form; (b) examples of a correlogram and spectral density function for a series of bed elevations of a riffle–pool stream.

wavelength (Nordin & Algert 1966). A plot of the serial correlation r against lag k is the correlogram, which for a stationary series is given by

$$r_k = \sum_{i=1}^{N-k} x_i x_{i+k} \Big/ \sum_{i=1}^{N} x_i^2 \tag{2.6}$$

where the x_i are deviations from the downstream regression trend, so that the series has zero mean. This is a simplified estimator of the serial correlation coefficient, and results should always quote the estimator used (Cox 1983). Correlograms for bed profile series with well developed bedforms often take the form expected for a second-order autoregressive stochastic process (Richards 1976), which is pseudo-oscillatory (Box & Jenkins 1970, p. 59). This has the equation

$$x_i = \phi_1 x_{i-1} + \phi_2 x_{i-2} + e_i \tag{2.7}$$

where the e_i are random-error terms and ϕ_1 and ϕ_2 are coefficients estimated from the serial correlations, or by conventional multiple regression of the x_i on x_{i-1} and x_{i-2}. The mean wavelength of the pseudo-cyclic bed profile oscillation is given by the reciprocal of the frequency f, which is obtained from

$$\cos(2\pi f) = |\phi_1|/[2\sqrt{(-\phi_2)}] \tag{2.8}$$

(Box & Jenkins 1970, p. 60). Alternatively, wavelength may be estimated from the peak of the spectral density function, which is a Fourier transform of the autocovariance function (itself the numerator in Eqn 2.6). Examples of the correlogram and spectrum for a riffle–pool bed profile are shown in Figure 2.18b.

These statistical estimation procedures are complex, but the objective method is preferable to subjective field estimation and little extra field survey is required in addition to that recommended for overall gradient measurement. Series analysis treats the bed profile as two-dimensional, when the sand bedforms in particular are three-dimensional. However, they shift across the channel at random, and a single two-dimensional realisation of the three-dimensional stochastic process may be considered representative and capable of generating satisfactory bedform statistics. However, isolated consideration of two-dimensional topographic form can be misleading in sedimentological and process contexts; for example, riffles and pools are now generally regarded as components of a single lobate bar unit, which 'wraps' itself around meander bends and on which is superimposed the point bar.

2.7.3 Plan geometry

A primary morphological distinction exists between single-thread and multithread (braided or anastomosing) channels, a detailed qualitative classification of channel patterns being provided by Dury (1969). Meander bend statistics are applied particularly to single-thread channels, and traditional indices measured for individual bends have been superseded by statistics derived from series analysis, particularly as it has been recognised that bends are frequently irregular and asymmetric in plan shape.

2.7.3.1 Single-thread channels

Traditional bend statistics illustrated in Figure 2.19a include axial wavelength (λ), path wavelength (L), radius of curvature (r_c) and amplitude (A). Since bends are not, in general, circular arcs (Ferguson 1973), difficulties arise in defining and measuring radius of curvature. A sine-generated curve (Langbein & Leopold 1966) increases in curvature towards the apex, so it requires description by a minimum (apex) radius and mean radius. The latter is best defined by replacing the continuous curve of the bend by equally spaced points (about one channel width apart), connecting these by straight segments, and finding the centroid of the polygon defined by their perpendicular bisectors (Fig. 2.19b). This represents the mean centre of curvature, which can subsequently be used to initiate radii in order to find the mean radius.

The simplest overall reach statistic is the sinuosity index P, the ratio of channel length to straight-line distance. For an individual bend, $P = L/\lambda$. It is difficult to define the 'straight' line, which should start and finish at the channel midpoint, but which may connect successive bend inflections, or even be a smooth midvalley curve. Mueller (1968) defines two statistics, which divide sinuosity into the additive components of topographic and hydraulic sinuosity, the former being valley length divided by straight-line distance and the latter being the additional excess channel sinuosity. Ferguson (1975) suggests that sinuous rivers are characterised by scale (wavelength), sinuosity and irregularity parameters. Successive connected perfect meander bends of equal wavelength are extremely rare, so series analysis provides the basis for estimating these aspects of plan geometry. Quantification of river plan for this purpose demands discrete one-dimensional series of channel direction relative to an arbitrary baseline (θ) or channel direction change (curvature $\Delta\theta$), which can be obtained by digitising the channel path using a digitiser in line mode, then interpolating equally spaced points along the channel at any required spacing Δs (usually 1–2 channel widths; Fig. 2.19c). Serial correlation or spectrum analysis of the series permit wavelength and irregularity estimation, with different characteristic correlograms and spectra depending on the relative irregularity of the meander pattern (Ferguson 1976). Furthermore, the variance of the direction series may be used to estimate sinuosity, since a close theoretical relationship can be proved between these measures (Ferguson 1977a); this can be used to reconstruct palaeochannel sinuosity from fragmentary exposures.

An alternative approach based on the curvature series is suggested by O'Neill and Abrahams (1986); this is particularly useful for identifying the end-points of individual curves within compound or 'double-headed' bends. This is a 'planform differencing technique' in which a half-meander is defined by the constant *sign* of its curvature series, as long as the cumulative summation of the curvature values exceeds a threshold level designed to distinguish true curved reaches from wandering 'straight' reaches.

2.7.3.2 *Multithread channels*

Brice (1960) suggested the braiding index

$$B_I = \frac{2 \times \text{total length of bars in reach}}{\text{reach length at midchannel}} \tag{2.9}$$

which measures the extra bank length created by braiding. However, braiding morphometry has not been investigated in detail except for the application of topological indices for planar circuit networks (Kansky 1963) by Howard *et al.* (1970). For a selected reach, some stream segments (edges) and islands are bisected by the limits of the reach. Defining n as the number of nodes, i and p as the numbers of unbisected and bisected islands, and $t = e + c$ (total number of segments t being the sum of bisected and complete segments), network connectivity indices α and γ may be calculated, where

$$\alpha = \frac{t - (n + e) + 1}{2(n + e) - 5} \quad \text{and} \quad \gamma = \frac{t}{3(n + e - 2)} \tag{2.10}$$

The α index is the ratio of observed number of islands to maximum number possible with the available nodes, and the γ index is the equivalent measure for segments. An example is shown in Figure 2.20. In practice, these indices may have too narrow a range to discriminate between reaches, although they may illustrate the changing complexity of flow patterns within a reach at different discharges. Howard *et al.* (1970) found that simpler topological indices, such as the number of excess segments in a transect of the channel, correlated with gradient and discharge, and hence demonstrated the greater complexity and connectivity of braided patterns in high-energy environments. An alternative approach involves extending the sinuosity index to multithread streams. This demands an operational definition in which are mapped all channels that flow between macro-bars. The *total* channel length is then divided by reach length to give a 'total sinuosity' index, which also correlates with stream power (Richards 1982, pp. 181–2); for the example in Figure 2.20, total sinuosity is 3.54.

In multithread channels, there are commonly complex topographic variations in a cross section, reflecting the bed relief associated with the bar and channel morphology.

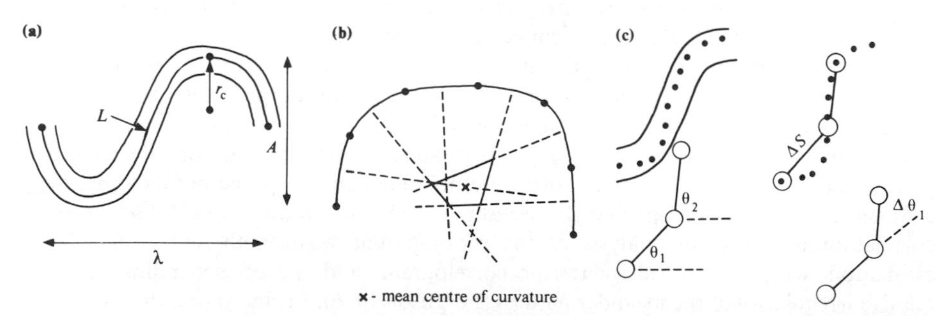

Figure 2.19 (a) Standard meander bend characteristics; (b) a technique for defining mean centre of curvature; and (c) definition of a series of equally spaced points along a channel, for measurement of direction angle Θ and curvature $\Delta\Theta$.

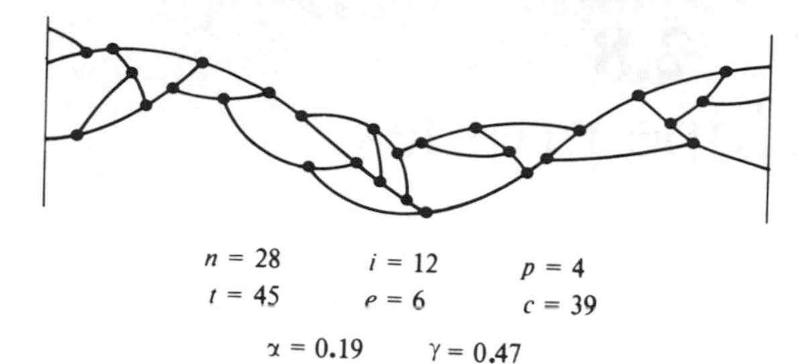

Figure 2.20 A braided reach as a planar circuit network, with topological measurements and connectivity indices.

These variations may be described by Smith's (1970) 'bed relief index', which measures the difference between the sums of height maxima and minima across a braided river section. Downstream in the Platte River particle size diminishes, longitudinal bars give way to transverse bars and the bed relief declines. This example illustrates the interdependent nature of the three-dimensional geometry of river channels (transverse, longitudinal and plan forms), and its relationship to sedimentological processes.

2.8 Hillslope profiles

2.8.1 Introduction

A hillslope profile is a line on a land surface linking a crest on a drainage divide and a base on a drainage line, following a maximum gradient path that runs perpendicular to the contours. The idea of hillslope profiles is most easily applicable in well integrated fluvial topography, but has some relevance in most kinds of landscape.

Hillslope profile determination and analysis can readily be justified as providing basic descriptive information, of academic interest and of background value in various practical applications. Despite a great investment of thought and effort, however, there is little consensus about appropriate methods. Agreement and standardisation are clearly advisable in order to build up a body of adequate and comparable data sets, although the variety of purposes behind hillslope survey and the variety of hillslope profiles must be recognised (Young *et al.* 1974).

2.8.2 Data collection

Hillslope profiles may be recorded directly in the field or indirectly from maps or air photographs. Since profile measurement requires angle measurements to ± 1° and height measurements to ± 0.2 m, most topographic maps are insufficiently accurate for profile determination; even where photogrammetric contours give improved accuracy, the number of contours is often too small to yield a detailed picture of a profile, with information on breaks and changes of slope or microrelief (Pitty 1969, Young *et al.* 1974). Although the best methods of angle measurement from large-scale air photographs give accuracy similar to an Abney level used in the field (Turner 1977), vegetation frequently obscures the ground surface, making these methods difficult to implement (Young *et al.* 1974). Hence profile determination must generally be carried out in the field.

Profile selection in a given area is clearly a sampling problem. The target population is usually *either* all profiles *or* all profiles straight in plan (perhaps with further simple restrictions, e.g. to a specified lithology). Profiles with straight contours are often thought simplest to understand, although it is best if an operational definition of approximate straightness is given (Young *et al.* 1974, Abrahams & Parsons 1977).

All methods of sampling may experience practical difficulties: access to land may be forbidden or dangerous (e.g. precipitous cliffs, military activities, unsympathetic landowners or natives, wild animals) or vegetation or land use may render survey impracticable. Hence it might be necessary to modify an initially chosen sample of profiles.

Attempts to choose representative samples of hillslope profiles should recognise that selection of a set of paths on a surface is a special kind of sampling problem, which is not well understood. Any kind of point set selection, whether of profile end-points or of points within profiles, ideally should be accompanied by a demonstration that the point set is associated with a representative profile set; it is not sufficient that the point set be chosen representatively.

Methods based on the stream or valley network are widely used (e.g. Arnett 1971, Summerfield 1976, Abrahams & Parsons 1977, Parsons 1979, 1982). In particular, Young *et al.* (1974) outline a procedure for profile selection based on the construction of a 'profile sampling baseline', which joins the midpoints of profiles within a drainage basin. Despite yielding a desirably ordered spatial sequence of profiles, this procedure requires several subjective choices, while to produce coverage of an entire basin, the baseline must penetrate deeply into spurs and hollows, thereby oversampling first-order valleys (Bell 1983). Furthermore, all sampling methods employing the stream or valley network as a framework depend on the identification of that network, which may often be difficult in detail (e.g. stream sources may be hard to locate, permanent watercourses may lack geomorphological significance). Alternatively, grid-based samples generally work well, although oversampling of long slopes may require compensatory weighting (Bell 1983).

Hillslope profiles can be generated from an altitude matrix by a FORTRAN program written by Bell (1983), which uses local quadratic interpolation and extends profiles both upslope and downslope, subject to prespecified constraints on changes in direction. At best, map-derived profiles resemble field profiles in altitude, gradient and aspect, but not profile curvature: central tendencies are naturally more reproducible than extreme values. Bell's program allows evaluation and modification of tentative field designs, and thus aids the choice of sample size and location.

Slope profiles may be measured by recording a series of measured lengths along the ground surface and a corresponding series of measured angles. On most kinds of relief, it is sufficient to record angles to the nearest 0.5° and lengths to the nearest 0.1 m (Young *et al.* 1974). In many early surveys, at least some measured lengths were bounded by perceived breaks or changes of slope, and thus lengths were generally unequal, but such practice is unduly subjective and tends to prejudice analysis and interpretation of the data. Consequently, it has been argued that measured lengths should be equal (Pitty 1967) or within a restricted range (Young *et al.* 1974). Current practice is to use relatively short measured lengths, often 5 m or less; lengths of 1–2 m are natural for instruments like pantometers (Pitty 1968), and only on very long regular slopes should lengths much more than 5 m be used. A sensible preference (Leopold & Dunne 1971) is to have at least 50 measurements, whatever the length.

Gerrard and Robinson (1971) discussed the choice of measured length in some detail, warning that the results of profile survey may vary considerably with the measured length adopted and that microrelief will influence readings made at very short measured lengths. These warnings may be slightly exaggerated. Their results are based on repeated surveys and hence mix scale variation with measurement error (Gilg 1973), while the homily against microrelief rests on a dichotomy between microrelief and 'the true nature of the slope', which seems dubious. There do not appear to be any firm physical grounds for distinguishing sharply between these two components of topography, while results from spectral analysis (e.g. Sayles & Thomas 1978) support the contrary idea that variations can be observed at all spatial frequencies of geomorphological interest. Sensitivity to measured length must, however, be borne in mind, and, at a minimum, measured lengths must be stated in research reports.

Cross profiles at selected point downslope may be used to indicate surface roughness or plan curvature (Leopold & Dunne 1971, Parsons 1979, 1982).

The most convenient and widely used instruments are an Abney level, linen tape and ranging poles; these give the required accuracy, are rapid to use and need only one observer (Young 1972). A Suunto inclinometer can be used instead of an Abney level, while in open country one can dispense with ranging poles by sighting from eye level to

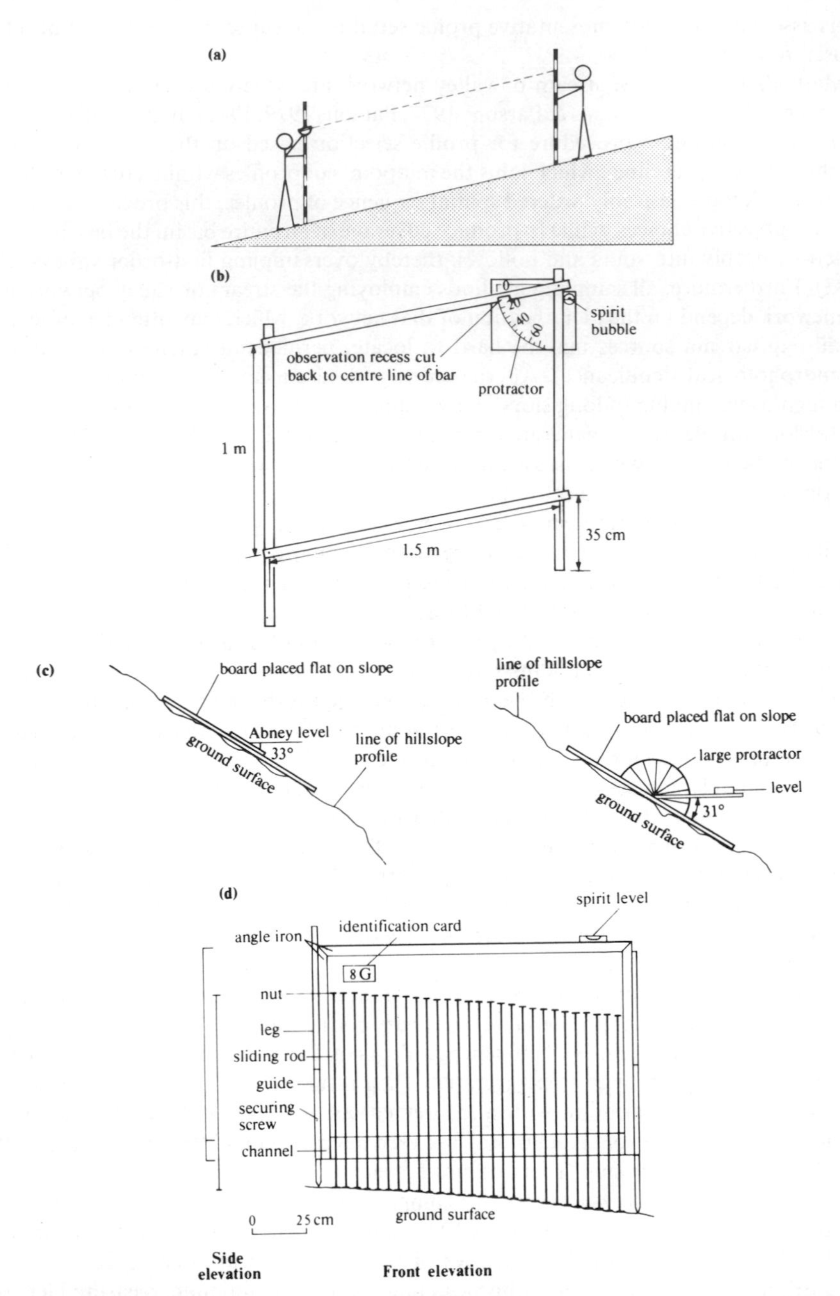

Figure 2.21 Some methods for hillslope profile analysis. (a) Abney level and ranging poles. (After Gardiner & Dackombe 1977, Fig. 4A, p. 10). (b) Pantometer. (After Pitty 1968, Fig. 1, p. 7.8). (c) Inclined board. (After Blong 1972, Fig. 1, p. 185). (d) Slope profiler in side and front elevations. (After Mosley 1975, Fig. A/1, p. 5).

ground level (Young *et al.* 1974, Gardiner & Dackombe 1977) (Fig. 2.21 a). A slope pantometer, suitable for obtaining large numbers of short (1–2 m) and equal measured lengths, has been devised by Pitty (1968, 1969) (Fig. 2.21 b), and other special-purpose instruments have been constructed (e.g. Blong 1972), although none has been used as widely. For small-scale profile recording, it is possible to measure the angle of an inclined board (e.g. Strahler 1950) (Fig. 2.21 c) or to employ an instrument like the slope profiler described by Mosley (1975) (Fig. 2.21 d). A measurement frame for recording the surface of arcas of 1 m^2 was used by Campbell (1970) is a study of badland erosion. A variety of broadly similar relief meters or contour gauges have been devised, especially in agricultural engineering (Kuipers 1957, Haigh 1981; cf. Podmore & Huggins 1981). Conventional surveying instruments are also used frequently, such as a combination of theodolite and Ewing stadi-altimeter (Young *et al.* 1974). Precipitous slopes can be measured with a surveyor's transit or theodolite and an optical rangefinder (Churchill 1979).

In practice, it may be difficult to decide on the path of a hillslope profile near the end-points. One conservative rule is that end-points should be where downslope angles equal those along divide or drainage line (Pitty 1966), although it may be preferable to continue measurements beyond these points, even if the angles recorded are not local maxima (Young *et al.* 1974). The base is defined by Blong (1972) as the point where downslope processes are replaced by downvalley processes (cf. Dalrymple *et al.* 1968), although this practice mixes genetic inferences with morphometric description.

Detailed recommendations and discussions of problems in profile survey are given by Young *et al.* (1974). This reference should be consulted by all undertaking work of this kind.

2.8.3 Data analysis

The first stage in data analysis is to calculate profile coordinates and to plot profile curves. By an arbitrary majority convention, profiles are drawn descending to the right. Other kinds of diagram, such as a plot of measured angles in sequence or a histogram of the angle frequency distribution, are also useful (Pitty 1969). A neglected but useful exploratory technique is to smooth angle series in some way, perhaps by running means (Pitty 1969) or running medians (Velleman & Hoaglin 1981).

Angles may be ordered and presented as frequency distributions. Most workers use descriptive summaries based on moment measures (mean, standard deviation, skewness, kurtosis), but many data series include outliers (especially a few very high angles), which make quantile-based measures more appropriate (Pitty 1970). Much of the work on angle frequency distributions has centred on the identification and interpretation of characteristic (modal) and limiting (threshold) angles (Young 1961, 1972), but workers need to take into account observed sensitivity of results to measured length, the imprecision of angle measurement, the inadequacies of many data sets, and the difficulties of genetic and process interpretation (Gerrard & Robinson 1971, Speight 1971, Statham 1975, Anderson *et al.* 1980). The majority of angle distributions are right-skewed (Young 1972). Speight (1971) suggested that logarithms of tangents tend to be approximately gaussian (normal) in distribution.

Angle series allow the calculation of curvature, properly the rate of change of angle θ with arc length s, that is $d\theta/ds$. Formulae for calculation are given by Young (1972), while Ahnert (1970) and Demirmen (1975) discuss related measures. If measured lengths are constant, it is easy to work with first differences of angles

$$\Delta\theta_i = \theta_{i+1} - \theta_i \qquad (2.11)$$

cf. Ferguson (1975) on river meanders. Higher derivatives have been calculated by some workers (e.g. Parsons 1977).

In many cases, profile forms, or components of profiles, allow curve fitting as a means of descriptive summary. The most popular functional forms have been power-series polynomials, power functions, and exponential and logarithmic curves. Care should be taken in choice of error terms, and in considering the problems of errors in variables, of autocorrelated errors and of significance testing (Cox 1977).

The autocorrelation properties of angle and curvature series are of interest if only to determine the degree of mutual dependence between nearby values, which affects the validity of standard statistical tests (Nieuwenhuis & van den Berg 1971). However, autocorrelation analysis runs into the problem of empirical non-stationarity (i.e. variation of statistical properties within profiles) (Cox 1983) and calculated autocorrelation functions are often difficult to interpret.

Profile analysis may be defined as the partition of a hillslope profile into components that are approximately homogeneous in some sense. If components each have approximately constant angle, then they may be called 'segments'; if they each have approximately constant curvature, then they may be called 'elements'. Segments are a subset of elements by this definition (Cox 1978). Other kinds of component may be identified using different homogeneity criteria (e.g. Parsons 1977). Terminology here is confused: all these terms are used in different senses elsewhere, while references to facets, sections, sectors and units are quite common.

The existence and bounds of distinct components are fairly obvious in some field situations, while graphical display of profile data may yield an adequate partition into components. In other circumstances, a numerical method of profile analysis would be valuable. In all cases, the 'atomistic' hypothesis that a hillslope is essentially a combination of discrete components should be compared with the 'continuous' hypothesis that it is essentially a smoothly changing curve (Cox 1978).

Young (1971) proposed three techniques for profile analysis: best-segments analysis, best-elements analysis and best-units analysis (Cox 1978) and it is not further discussed here. Young measures the variability of a subseries of either angles or curvatures by its coefficient of variation (standard deviation as a percentage of mean), and a subseries is accepted as a segment (element) if its coefficient of variation of angle (curvature) is less than a maximum prespecified by the user. Overlaps are avoided by assigning measured lengths whose status is in doubt to the 'longest acceptable' component, defined in a complicated and rather arbitrary manner.

Unfortunately, the coefficient of variation is not a stable measure and, as mean angles (curvatures) approach zero on nearly flat (straight) slopes, calculated coefficients tend to explode. Young (1971) tackled this difficulty by replacing means below 2 by a value of 2, but this is only a crude solution. Moreover, the coefficient of variation works well only if the standard deviation is proportional to the mean (Lewontin 1966), an assumption that is less realistic for shape variables in geomorphology than for size variables in biology, for which the coefficient is often used. In its present form, Young's algorithm for walking the algorithm along the slope can produce different results if reversed, although this problem has been reduced by a modified version of the program (Young pers. comm. 1980).

Young's methods include *ad hoc* and arbitrary procedures. Using his methods, it is not easy to vary the number of components, since the parameters controlled are related to the number of components in a complex and unknown manner, or to compare between-component and within-component variation. Investigators are therefore recommended to base partitions into components on graphical displays rather than trust complicated numerical methods known to have unsatisfactory properties.

2.9 Cartographic techniques in geomorphology

British and North American geomorphologists might be surprised to read that 'geomorphological mapping is at present the main research method of geomorphology in numerous countries' (Demek & Embleton 1978, p. 313). The excitement of work on process mechanisms, and the decline of the spatial tradition in geography, have pushed mapping to the periphery of academic concern, though it is important everywhere in applied work. The production of comprehensive geomorphological maps has a very low priority in Britain and North America, but many field-based publications contain maps of selected features relevant to their themes; these features may be the dominant landforms within the limited areas involved. Here, content and graphic design will be reviewed, followed by scale and applications.

2.9.1 Content

Geomorphology is the science of Earth surface form, explained in terms of the interaction between processes and materials through space and time. Any or all of these topics can be the subject of maps in geomorphology. Coinciding with the structure of this volume, forms, materials, processes and ages may be mapped; this 'pure' information may be evaluated and regrouped for a specific application. 'Landform genesis' usually involves attributing a form to a process, past or continuing.

2.9.1.1 Form

In Britain, emphasis has been on mapping form, and on assigning genesis for particular groups of landforms. Purely morphometric maps, as in Section 2.3, have been rare, but information on slope gradient has often been combined with morphographic information as in Cooke and Doornkamp (1974, p. 358).

Slope gradient is the most important morphometric variable for many processes and applications, and it is important to select relevant class limits. Table 2.9 suggests some critical gradients (in percentage, as tangents of the angle), and Olofin (1974) compiles both 'theoretical' and 'practical' limits. If new limits become important because of technological change, using manual methods (Monkhouse & Wilkinson 1971 p. 132, Denness & Grainger 1976) it may be necessary to go back to the contour map and take further measurements. But if gridded altitude data are held in a computer (Sec. 2.3), it is a simple matter to produce a gradient map with any class limits; new maps can be produced for particular applications. This may soon become the main method for producing slope and curvature maps, since photomechanical methods do not seem sufficiently robust to process existing contours, and photogrammetric and field measurements are too time-consuming for large sets of measurements.

As used by Tricart (1965, p. 187), the term 'morphographic' covers breaks and changes of slope, convexities and concavities, gullies and terraces, types of ridge and types of

Table 2.9 Critical gradients for selected economic activities.

Activity	*Critical slope gradients (%)*																	
Transport																		
roads	2	4.5					12			19						36		
railways	2				9													
airports	2																	
military vehicles						10							25				50	100
Cultivation	2	5		8–9									25	30				
Agricultural machinery	2	5				10			15		20	23	25	30		36		
Forestry						10					20				33		50	
Industry	2	5				10												
Residential	2	5	8			10					20			30				
Recreation	2		8						15				25					
General development restrictions	2	5	8			10		14										

valley; in other words, morphometric information reduced to an ordinal rather than an interval or ratio scale, and usually produced by subjective interpretation rather than measurement. The stream network is included as a relevant base map. Morphographic information is especially prominent on Polish geomorphological maps (Gilewska 1967). High-quality topographic maps also convey much morphographic information: Imhof's classic text on relief portrayal is now available (revised) in English (Imhof 1982).

In Britain, maps produced by field mapping of morphography have been known as morphological maps following Waters (1958), though Tricart (1965, 1970) insisted that without genetic interpretation these do not constitute geomorphological maps. The basis of the Waters system is a double dichotomy into convex or concave breaks or inflections of slope profile. Breaks are shown as continuous lines, inflections (more gradual changes of gradient) as dashed lines; rows of 'v's pointing downslope are always placed on the steeper part of the slope, and thus distinguish convexity from concavity (Cooke & Doornkamp 1974, Gardiner & Dackombe 1983). (Note that Savigear (1965) and some other authors subvert this simplicity by having upslope-pointing 'v's on convexities.) Further symbols are added for gullies, bluffs and cliffs, and slope arrows between the breaks give gradients in degrees. Unlike the Polish and other systems, there are no signs for valley types, ridge types and summit types.

This British system of morphographic mapping developed at a time when the best-contoured map available as a mapping base was the provisional edition of the Ordnance Survey 1:25 000. On this, master contours had been surveyed in detail but most contours had been sketched in; geomorphologists were dissatisfied with the inevitable smoothing of breaks of slope. Publication during the 1970s and 1980s of many 1:10 000 maps with photogrammetric contours at 10 m or 5 m intervals provides a much better basis for field mapping. The need for morphographic mapping is thus reduced at this scale, but remains for very detailed work. In Sweden, Elvhage (1983) has illustrated the contrast between smoothed contours produced by a standard photogrammetrist and those produced by one with geomorphological training. In a gullied terrace, the latter contours are sharper and show more gradient contrast.

British morphographic maps were pioneered in the Peak District, an area with many structural breaks of slope. They work less well in areas of smoother slopes, such as chalk downlands, and in very broken, complex relief. In the absence of curvature measurements, the distinction between 'break' and 'inflection' is inevitably subjective. Nevertheless, production of such maps is useful in developing 'an eye for landscape', and is educationally valuable (Soons 1975, Gray 1981). By concentrating attention on detailed irregularities of the land surface, they are useful in landslide areas (see Fig. 2.24, Brunsden *et al.* 1975) and may highlight areas of former landsliding, which are easily overlooked.

2.9.1.2 Genesis and materials

When further symbols are added for 'alluvial fans' or 'areas of slumping' (Soons 1975) a genetic element is involved in maps that are mainly morphographic. Likewise, the identification of 'river terraces', often numbered by relative age, is strictly speaking genetic. Maps of terraces are mainly morphographic, but the close spacing of convexities and concavities requires extensive use of the bluff symbol rather than two separate symbols (Fig. 2.22). Proper genetic interpretation requires identification of surface and often subsurface materials. Stratigraphic study often shows river terraces to have a history more complex than required by surface morphology, and areas that have suffered strong gelifluction can be especially misleading, with different materials underlying what appears to be a single surface (Jones & Derbyshire 1983).

Maps of surface materials such as 'till', 'glacial sands and gravels', 'loess' and 'alluvium' clearly have implications for the genesis of landforms; they are produced in Britain mainly by the Geological and Soil Surveys. More relevant to current processes and hydrology are maps of grain size, since gravel, sand, silt and clay differ greatly in both permeability and mobility. In erosional topography, degree of jointing is often the most relevant property of bedrock, but little information is available of this from geological maps.

Morphogenetic maps are common in the field studies of glacial features by J. B. Sissons and his students. Slope breaks are emphasised in the mapping of meltwater

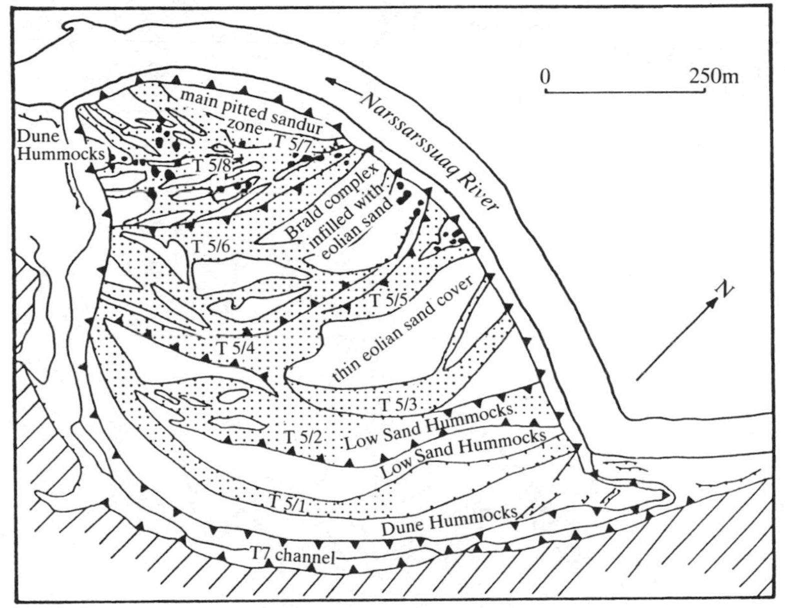

N.B. North-east section of map based on field sketch maps only.
South-west section of map based on aerial photography (7908 45/352,377, Geodetisk Inst., Copenhagen)

Major terrace bluff (>2m high)
Minor terrace bluff (<2m high)
Mid channel bar or island
Bedrock slope
Paleo channel course
T5 Paleo channel course number
Kettle hole

Figure 2.22 Palaeochannels and bluffs on a terrace in south-west Greenland. (From Maizels 1983).

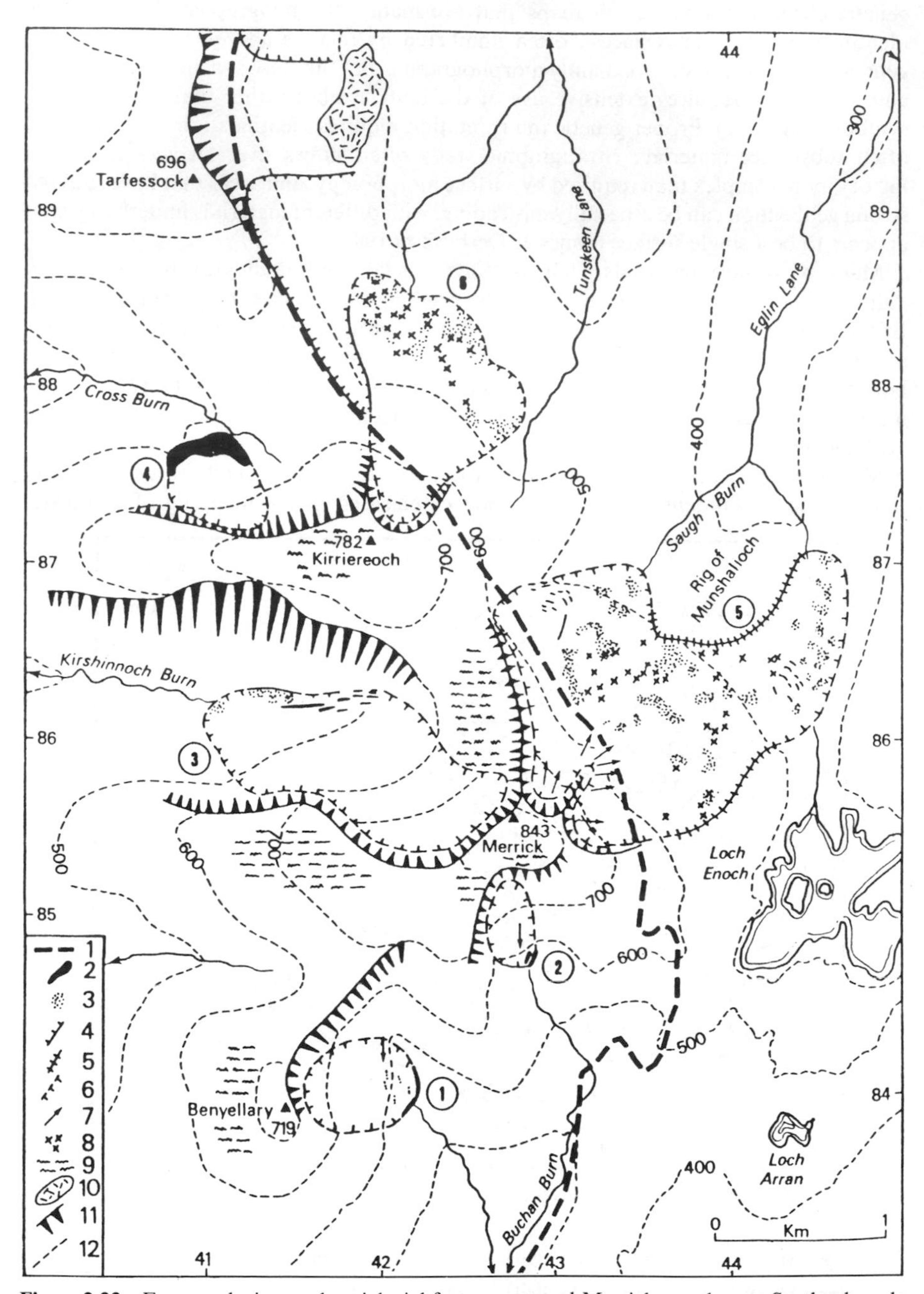

Figure 2.23 Former glaciers and periglacial features around Merrick, south-west Scotland; scale 1:33 000. (From Cornish 1981). *Key*: 1, western edge of granite; 2, linear moraine ridges; 3, hummock moraine; 4, boulder limit; 5, drift limit; 6, inferred ice limit; 7, striae; 8, metamorphosed greywacke erratics; 9, inactive periglacial features; 10, landslide; 11, very steep slopes; 12, contours (100 m interval) on land and hypothesised glacier surfaces.

channels (Sissons 1967, 1974) and glaciofluvial deposits. Where moraines of the late-glacial Loch Lomond stade are mapped, the more pronounced periglacial features are found to fall outside the reconstructed glaciers (Fig. 2.23, Cornish 1981). Mapping of present-day processes is largely confined to hazards such as landslides, avalanches and gullying (Sec. 2.9.4); slower processes such as wash, creep and solution are usually inferred from form, measurements of their rates being too localised to permit mapping. Several papers in de Boodt and Gabriels (1980) map erosion risk or erosivity, mainly at small scales.

2.9.2 Graphic design

For the monochrome British maps reproduced here, graphic design involves a fairly straightforward choice from a limited number of point, line and area symbols. Morphographic maps use mainly line symbols and gradient maps use area shading. Base maps consist of hydrography and usually a few contours, with national grid ticks in the margins to facilitate precise location without cluttering the map. Maps such as Figure 2.23 identify symbols by numbers, which are defined in the caption, a two-stage procedure often insisted on by editors, but clumsy for the map reader and unnecessary with modern technology. Symbols should be defined directly, and with the simplicity of many British maps this legend can often be placed in a corner of the map.

The more diverse the area covered, the more extended the legend needs to be and the greater the need for colour. Hence legends that aim for international standardisation run to many pages; the IGU (International Geographical Union) Commission's proposals include 350 symbols, while Polish and Russian legends exceed 500 symbols, e.g. 36 for marine platforms of different ages (why not use numbers or letters?). Two undesirable consequences stem from legends that cover more space than the map sheet they serve: the map will be comprehensible only after extended study or even a long apprenticeship using such maps, and will therefore be a closed book to non-specialists; and it will be so complex that even such specialists will not obtain a synoptic overview of the distribution of any element. Information should now be stored on computers rathers than maps; each map should be designed for a specific purpose.

The rationale for comprehensive geomorphological maps (in colour) is that it should be possible to interrelate all the elements – morphometry, morphography, materials, landform genesis, current processes, chronology and base map – at any location. This worthy aim has proved difficult to achieve without the high costs mentioned above. By omitting morphogenesis, Leser (1975) produced a 1:25 000 colour map of morphography, morphometry and current processes that is aesthetically pleasing and clearer than the comprehensive maps. The construction of a Geographic Information System (GIS) on a computer permits retrieval either of tabular information for one or several sites, or of maps providing synoptic views of single elements or selected combinations. These are among the aims of a new German program for digital geoscientific maps (Vinken 1986).

The legends of comprehensive detailed geomorphological maps have been discussed by Tricart (1965, 1972), Saint-Onge (1968) and Demek (1972); samples of the symbols involved are presented by Cooke and Doornkamp (1974) and Gardiner and Dackombe (1983). Useful comparisons of different legends have been made, for given areas, by Gilewska (1967) – French, Hungarian, Soviet and Polish legends applied to a Silesian scarpland – and by Salomé and van Dorsser (1982) – six legends applied in the Ardennes fringe (Table 2.10). For the latter area, the ITC and Unified legends seem, to British eyes, to give the best results. They might be improved by leaving some areas blank (as does the

Table 2.10 Interpretation of the use of graphic symbols in six legends for 1:50 000 comprehensive maps demonstrated in Salomé and van Dorsser (1982)*

Design source	Number of colours	Colour hue	Grey pattern intensity	Line or line shading	Point symbols	Letter symbols	symbols
ITC 1968	3	Denudation vs fluvial	Surface vs slope	Lithology	Morphography	–	Chronology
Unified 1968	4	Genesis	Gradient (S&vD)	Gradient (Unified)	–	–	Chronology
Swiss 1958	3	Erosion vs accumulation	–	Surface height (red)	Concavity	–	–
Belgian 1970	3	Morphography (and lithology, on inset)	Valley sharpness	–	–	Genesis	Chronology
Poland 1956	4	Denudation vs fluvial: chronology	Gradient	Surfaces	–	–	–
France 1970	6	Geological resistance	Chronology (omitted by S&vD)	Surfaces, and loam	Concavity	Genesis (colour)	–
Recommendation (ISE)		Genesis	Gradient	Lithology	Morphography	Minor or superimposed forms	Chronology

* All use line symbols to show base material (roads and settlement names) in grey, contours in grey or brown, and streams in blue or red.

Swiss) rather than saturating the whole map with colour; by using point symbols and further line symbols; and by showing gradient (as does the Unified). The Swiss version is clear, but has little information on slopes or morphogenesis, and none on age. The French version was the most expensive to produce, but the most difficult to read. The Belgian symbolisation plays down genesis, and the multiple use of colour is confusing.

Since colour is the most prominent graphic variable, it should be used for the most important, reliably determined variable. In most cases this will be genesis. Where important, relative age can be shown by intensity of shading; otherwise, by numbers or letters. Lithology or surface materials are best shown by grey patterns, as on the modern German maps. Morphography lends itself to line and point symbols, which could be better used in most of the examples given by Salomé and van Dorsser (1982). Table 2.10 was derived independently of the similar tables in Salomé and van Dorsser (1985).

2.9.3 Scale

Except for hurried reconnaissances, field observations are plotted on the largest-scale topographic map available. For upland Britain, this is 1:10 000; for lowland rural areas, 1:2500, though often with fewer contours than the 1:10 000. Often enlargement even of these large-scale maps is desirable, to provide space for annotation. Field boundaries are essential and the more landmarks the map shows, the better.

If fields are small or landmarks closely spaced, morphological detail may be sketched in by eye. More precise location of slope breaks may be provided by pacing and use of a compass and clinometer. If high accuracy is required (for example, when small changes are to be detected), survey techniques involving tape, level, tellurometer or theodolite are required (Bannister & Raymond 1984). These are much more time-consuming, are not worth while if the aim is to produce a map on which the difference in accuracy is within the line thickness plus drafting error.

When field maps are redrawn, even for a thesis or report, some reduction in scale is usually necessary, and this involves generalisation. Figure 2.24, kindly provided by Robert Allison, shows mass movements around Stair Hole, on the coast of Dorset, at three scales: at 1:10 000, a lot of information has been lost; at 1:2500, the main features are clearly portrayed; while at 1:1250 considerable detail is feasible (Allison 1986). Pitts (1979) made photogrammetric base maps of coastal landslides in east Devon at 1:2500, and added detail in the field.

Geotechnical plans are normally mapped at scales of 1:100 to 1:5000 (Dearman & Fookes 1974). Hungarian engineering geomorphology maps are mainly plotted at 1:10 000 (Adam *et al.* 1986). For 'engineering geologic' mapping in Missouri, however, Rockaway (1976) suggested 1:62 500 as an optimum compromise between detail and time and funding limits.

At such broader scales, Boyer (1981) recommended the use of 1:100 000 for (comprehensive) geomorphological mapping of Spain. This would require 297 sheets, compared with 1130 sheets for 1:50 000 (too costly) or 100 for 1:200 000 (significantly generalised). Starting from a 1:50 000 map of 12 × 17 km of the central Huerva valley near Zaragoza, initially produced by air photo interpretation followed by field checks, 1:100 000 requires a limited amount of conceptual generalisation, combining several ages and subunits such as types of glacis. The information loss is estimated as less than one-fifth by van Zuidam (1982), compared with about a half for generalisation to 1:200 000. The latter scale requires much more line generalisation, exaggeration, displacement, combination and omission (e.g. of symbols for chronology), and loses much of the geomorphological landscape character. The three maps are reproduced in colour by both authors, and van Zuidam gives some estimates of times and costs.

When the German Research Foundation agreed a programme of (colour) geomorphological mapping, the scales of 1:25 000 and 1:100 000 were chosen (Barsch & Liedtke 1980; see also parts 30, 31 and 35 of *Berliner Geographische Abhandlungen*). Since 1976, 35 sheets have been mapped at 1:25 000, providing samples of 10 of the 14 major geomorphological landscape types in Germany. Ten sheets have been mapped at 1:100 000 but, according to Liedtke (1984), these are so generalised (e.g. for gradients) that their value is mainly for regional geography; a scale of 1:25 000 is required for planning and other practical applications.

Comprehensive maps of geomorphology at 1:25 000 are also produced in Hungary, Belgium and France. Most such maps. however, are produced at 1:50 000; for example, most of the ITC maps and those from Poland (Klimaszewski 1963), Belgium and most from France (Tricart 1965, 1970, 1972) – now 16 maps. A quarter of the Netherlands is covered by such maps (Ten Cate 1983). Maps at both these scales come under the heading 'large-scale' or 'detailed' geomorphological mapping (Demek 1972); they permit reasonable representation of slope processes, slope breaks and minor forms.

At 'medium scales' (Demek & Embleton 1978), considerable generalisation is required. At 1:100 000 or 1:200 000, 'appalachian' or 'jurassic' structural relief is well portrayed, though slope processes are not (Tricart 1970). Perhaps such scales are best suited not to comprehensive maps, but to maps showing distributional patterns of selected groups of forms, e.g. glacial, volcanic or structural, or terraces of major rivers.

Successful national maps have been produced at 1:500 000 for Hungary, Poland (Gilewska *et al.* 1982) and Portugal (de Brum Ferreira 1981). Delannoy *et al.* (1984) have produced a clear map of karst in the western Alps at 1:500 000. This seems to be the best scale for regional geography (in Europe). The British maps reproduced in colour at about 1:800 000 in the *Geographical Magazine* in 1976 were rather overreduced, necessitating the omission of most information on slopeform and process, which leaves regions of subdued relief almost blank. More information is provided for the Lake District, as reproduced here in monochrome (Fig. 2.25). At still smaller scales, such as the 1:2500 000 geomorphological map of Europe, we have moved into the realm of atlas cartography, and the emphasis is on morphostructural regions rather than identifiable landforms (Demek & Embleton 1978).

In producing maps for books and journals, it is necessary to include a scale bar rather than a representative fraction, since maps are commonly photographically reduced. This has the unfortunate effect that various irregular scales are produced; e.g. one to approximately 95, 156, 103, 34, 213, 82, 35 and 133 000 in one chapter by Sala (1984), and comparison between maps is hindered. In fact, it becomes difficult for the reader to gain a sense of spatial dimensions, in contrast to the use of topographic maps at standard scales. If possible, authors should produce maps with round-numbered scale fractions, which fit the page size envisaged so that their size need not be changed and their scale fraction is given in the caption.

2.9.4 Applications

For applications in construction, agriculture and planning, there are three possibilities. Either comprehensive maps are simplified, reinterpreted and perhaps supplemented; or applied maps are produced without a specific application in mind; or maps are produced tailored (in content and in scale) to a particular application. The latter may or may not be useful in more general terms; their input to a GIS requires adherence to standard classifications and tests.

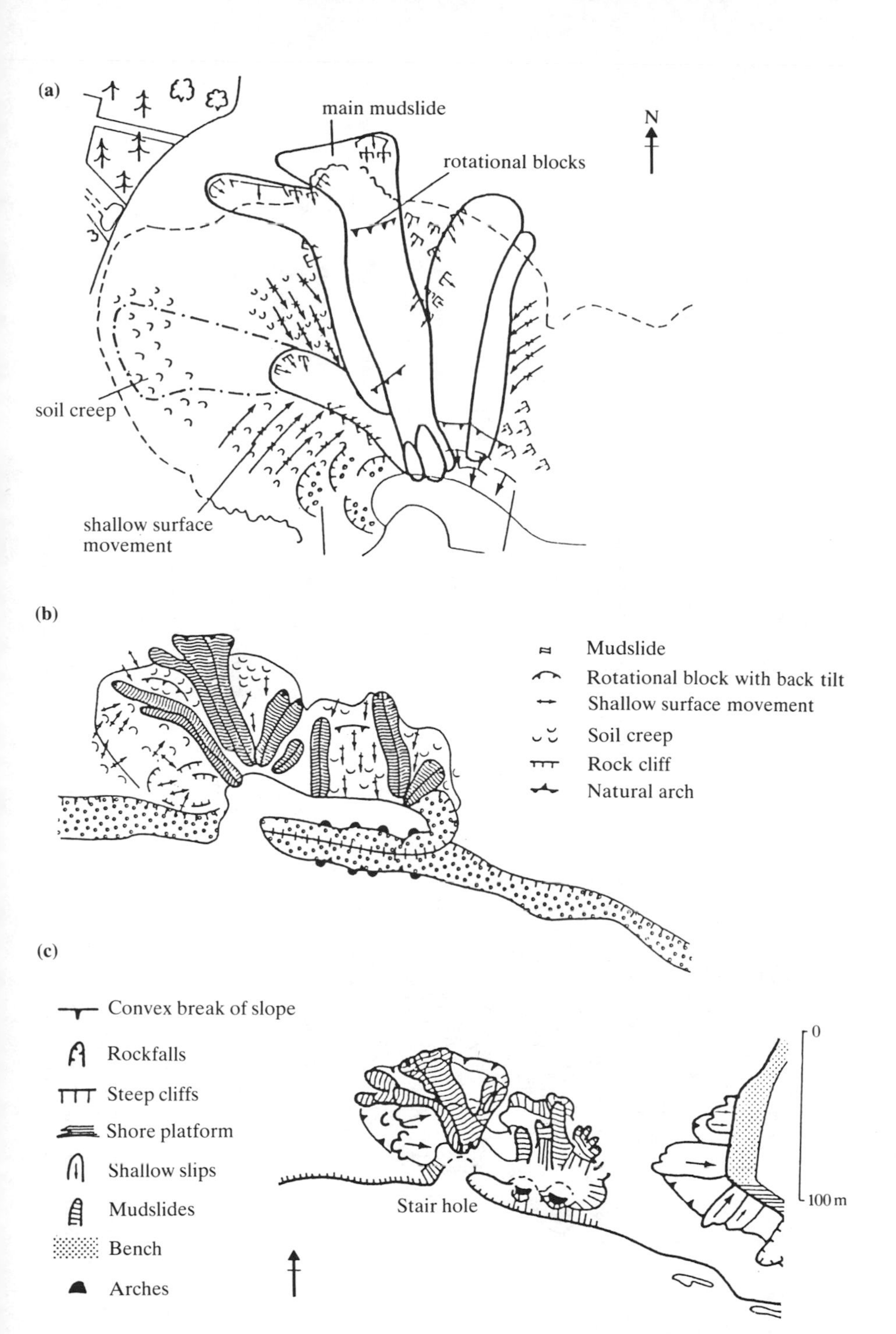

Figure 2.24 Landslides around Stair Hole, Dorset, mapped at three scales: (a) 1:1250, (b) 1:2500 and (c) 1:10 000. (From Allison 1986).

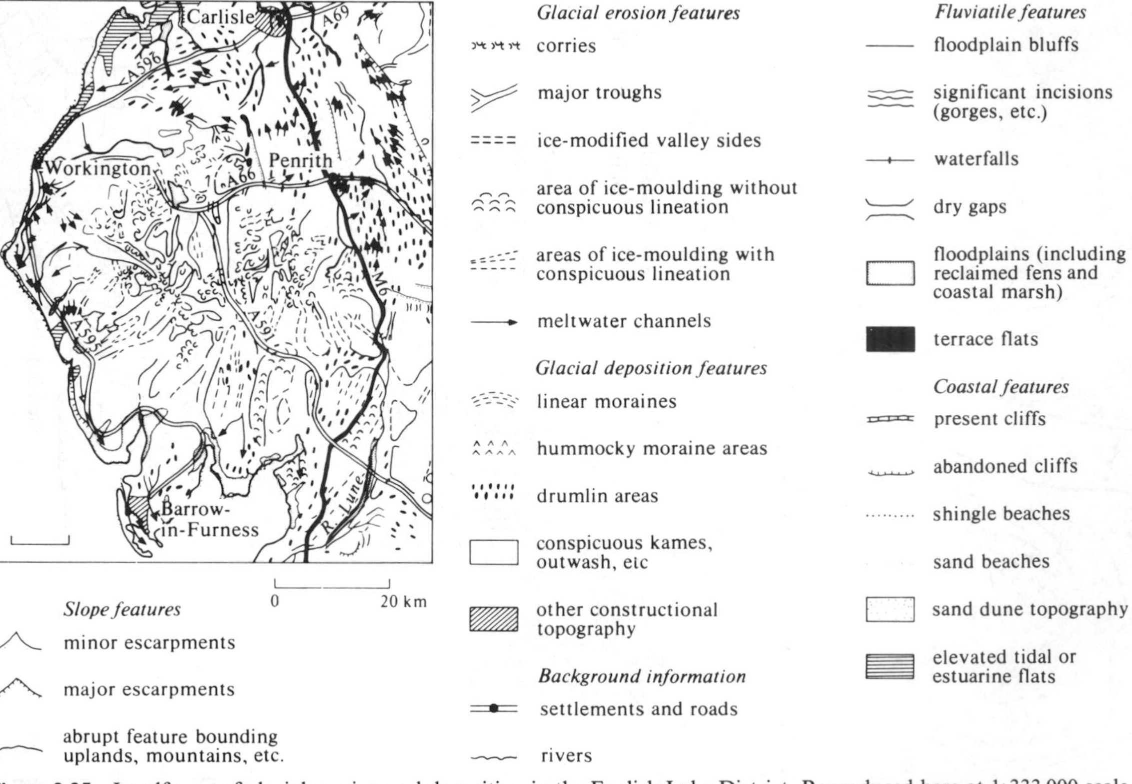

Figure 2.25 Landforms of glacial erosion and deposition in the English Lake District. Reproduced here at 1:333 000 scale after reduction from the coloured map in Waters (1976). Originally compiled by the BGRG at 1:250 000.

In North America, where 'most applied geology is applied geomorphology' (Fleisher 1984), available maps are reinterpreted and simple planning maps are produced. Atlases of single-factor maps are more popular than complex maps. In most applications, materials, slopeform and drainage are of enhanced importance, and morphogenesis and chronology are less important. Cooke and Doornkamp (1974, ch. 14) usefully demonstrated the derivation of geotechnical and hydrogeologic maps (at about 1:60 000) from geomorphological and morphographic/morphometric maps, with additional field investigations. Barsch and Mäusbacher (1979) have discussed the various specialised maps that may be derived from the German comprehensive maps, and listed possible applications.

Brunsden *et al.* (1975) emphasised that geomorphological mapping, at increasing scales, is important at the initial feasibility, the route or site selection, and the site investigation stages. They gave examples of maps at about 1:5000 for three areas of slope instability. Interdisciplinary cooperation is needed for major construction projects, especially with engineering geologists (Dearman & Fookes 1974). Engineering geology maps emphasise material properties such as degree of weathering, jointing, strength and permeability, which are lacking on most geology maps. Increasingly, geotechnical plans (at perhaps 1:1250) are being produced for specific engineering properties. At the reconnaissance stage, however, geomorphological mapping provides a rapid and cheap source of information, which can save enormous resources; for example, by re-routing, by re-siting or by distributing site investigation effort efficiently; see also the examples of Dubai and Suez in Doornkamp *et al.* (1979) and others in Edwards *et al.* (1982).

An important specific application is the production of natural-hazard maps. For avalanches (Mears 1980) these may involve red zones (absolute prohibition), blue zones (uncertain safety: infrequent severe avalanches, or small avalanches every 30 or 40 years; buildings require special design and spacing) and white zones (no hazard). French maps at 1:5000 and 1:2000 on this basis are replacing the 1:20 000 avalanche frequency maps used for broad planning (de Crecy 1980). For landslides, the probability of slope failure might be mapped; most existing maps simply show active and older landslides, though Nilsen *et al.* (1979) map six stability classes based on gradient, geology and landslide deposits. For coastlines, it would be useful to have maps of the expected lifetimes of areas threatened by erosion, and of areas threatened by proposed developments or by likely shifts in offshore banks. Williams (ch. 7 in Townshend 1981) mapped the potential for gullying in relation to future land uses; van Ghelue and van Molle (1984) compared the distribution of soil erosion as mapped with that predicted from slope, erodibility and land use.

On a more positive note, geomorphologic mapping has been applied to the search for mineral resources; for example, placer deposits, which relate to drainage history (Sutherland 1984). Doornkamp *et al.* (1980) predicted sources of aggregates in relation to their survey of Bahrain, while Crofts (1974) discussed the use of large-scale geomorphological mapping in relation to land-use planning. Some applications require three-dimensional mapping of materials, for example to 6 m depth for minor construction projects and domestic septic systems, but to 15 m for sand and gravel resources and for waste disposal. The heterogeneous deposits of glaciated lowlands may require such an approach; letter-coded 'stack maps' are being produced at 1:24 000 in Illinois for such applications, as well as groundwater recharge estimation and general planning (Kempton & Cartwright 1984). Profile-type maps are being produced for alluvia in Belgium (Huybrechts 1985).

A considerable amount of land information can be obtained rapidly by air photo interpretation, preferably before and after fieldwork (Verstappen 1983). At broad scales, the results may be classified into land systems, regions with recurrent patterns of landform and materials (Ollier 1977, Cooke & Doornkamp 1974 ch. 13). All of Papua

New Guinea, much of Australia and many parts of Africa have been mapped in this way. The land systems approach has been developed for enginering applications by Grant and Finlayson (1978), mapping mainly at 1:250 000. For many particular applications it is necessary to map the smaller, more homogeneous components – facets or units. For example, Grant *et al.* (1981) included sample maps at 1:10 000, showing terrain units, while the terrain patterns were mapped at 1:100 000. Alternative air-photo-based schemes for land classification have been used in Canada, leading for example to maps at 1:50 000 letter-coded for material, landform, topography and drainage (Gartner 1984). At broad scales, satellite data are increasingly important (Townshend 1981, ch. 5).

The coming years are unlikely to see the comprehensive geomorphological mapping of whole countries at large scales with standardised legends. Simpler maps, with a certain amount of community in symbol definition, will be produced for both scientific and applied purposes, and the use of GIS will permit flexibility of map content and the production of many more maps, but with shorter lives.

PART THREE
MATERIAL PROPERTIES

3.1 Introduction

Geomorphological explanations are incomplete unless they involve some understanding of the way in which the materials that cover the face of the Earth behave and have behaved. The following discussions are intended as guideposts to those starting out on research paths by presenting some of the main techniques that characterise materials and are useful in geomorphological investigations. Clearly, the authors cannot be all-embracing; full textbooks have been written about individual methods used, so a summary of the method, its principles, instruments and interpretation is given with a selection of references, which lead the way to the more advanced work that the potential user must necessarily move towards.

We have covered the main areas of properties that we believe to be most useful. There are many variations of methods and instruments, which are covered in the referenced work and can be consulted if the intricacies of the problem demand.

Interpretation of results from experiments is where the greatest care is required by the investigator. Rarely will the outcome be unequivocal in some way or another. This is where a full knowledge of the technique used is needed, and this is true whether a 'simple' analysis of particle size is required by sieve or by Coulter counter or of shear strength by field vane or direct shear apparatus. Sophistication is not commensurate with unambiguity, rather the opposite. Detailed analysis frequently gives the precision that is masked by a simpler and more coarse method.

Any investigation, therefore, needs some sort of control and standardisation. This can be achieved in three main ways: external and internal controls and replication. The latter is obvious and is necessary for full statistical treatment, although at times it may not be possible to do replicates. Internal controls are those of the method itself. Care is required in setting up apparatus, at the simplest level of checking connections, etc., but also of making blank runs or trials with materials of known behaviour. Consider a few diverse examples: the output from the two thermocouples of a DTA (differential thermal analysis) apparatus should be identical with a blank run of alumina in each crucible; a standard-concentration solution should give an expected reading within the precision of the instrument. In other words, the calibration of an instrument needs to be checked with the appropriate standard. This should be carried out at intervals according to the manufacturer's instructions, or more often for high-accuracy work.

The second type of control is the procedure or standardisation of pretreatment, preparation and experimental procedure. Some standards, e.g. by the British Standards Institution, have been laid down: others are unofficial but have considerable weight, e.g. the *Soil survey laboratory methods* of the Soil Survey of England and Wales (Avery & Bascomb 1974) and the *Soil Bureau methods* of the New Zealand Soil Bureau (Blakemore *et al.* 1977); and at a third level, authors of papers may recommend a procedure. Which one the investigator takes will depend upon factors that only he or she can decide. It is clearly not possible to discuss all the advantages and problems in the following sections except summarily. Hence, some careful enquiries will be needed before embarking on a method.

3.2 Physical properties

3.2.1 Introduction

A description of the basic properties of geomorphological materials is frequently the most important starting point for an explanation of a geomorphological process. Knowledge of these properties can suggest fruitful lines of experimentation in field or laboratory and may be required in computer simulations or more intensive enquiries, such as in soil mechanical testing.

Most frequently, investigation is of a sedimentary body of some sort and similar problems will be encountered whether this is a hillslope, drumlin or river bank. Figure 3.1 suggests ways in which the properties may be viewed. Geotechnical aspects are covered under Section 3.4 and some aspects of sedimentary structures in Part Four. In this part, a number of the most important basic properties are discussed, some relating to individual clasts and some to bulk character. Following Griffiths (1967), we may write a defining description (D) of properties of a sediment as

$$D = f(C, G, F, O, P)$$

where C is the mineralogical composition of particles $1,\ldots, x$; G is the grading, which consists of sizes $1,\ldots, x$; F is the form, a parameter subsuming shape, angularity, roundness, sphericity and surface texture; O is the orientation of particles $1,\ldots, x$; and P is the complex of factors making up packing (Fig. 3.2). The distinction between bulk and individual properties (Fig. 3.1) is not complete because, for example, packing affects permeability and porosity. Relationships between other parameters are also intricate and some care is needed to work out those necessary for a particular purpose, as well as the best way of obtaining them.

3.2.2 Particle size

Particle size is a parameter having dimension of length and can be defined by:

(a) the width of the smallest square or diameter of the smallest circular opening through which a particle can pass;
(b) the diameter of a circle whose area is equal to the maximum projected area (MPA) of the particle;
(c) the diameter of a sphere whose volume equals that of the particle;
(d) the diameter of a sphere whose settling velocity and density in a given fluid equal those of the particle; as well as
(e) the longest (*a* axis) dimension of the particle.

These definitions are mostly for spherical particles, not usually found in native soils and clays, and therefore methods using these properties will only give approximations of

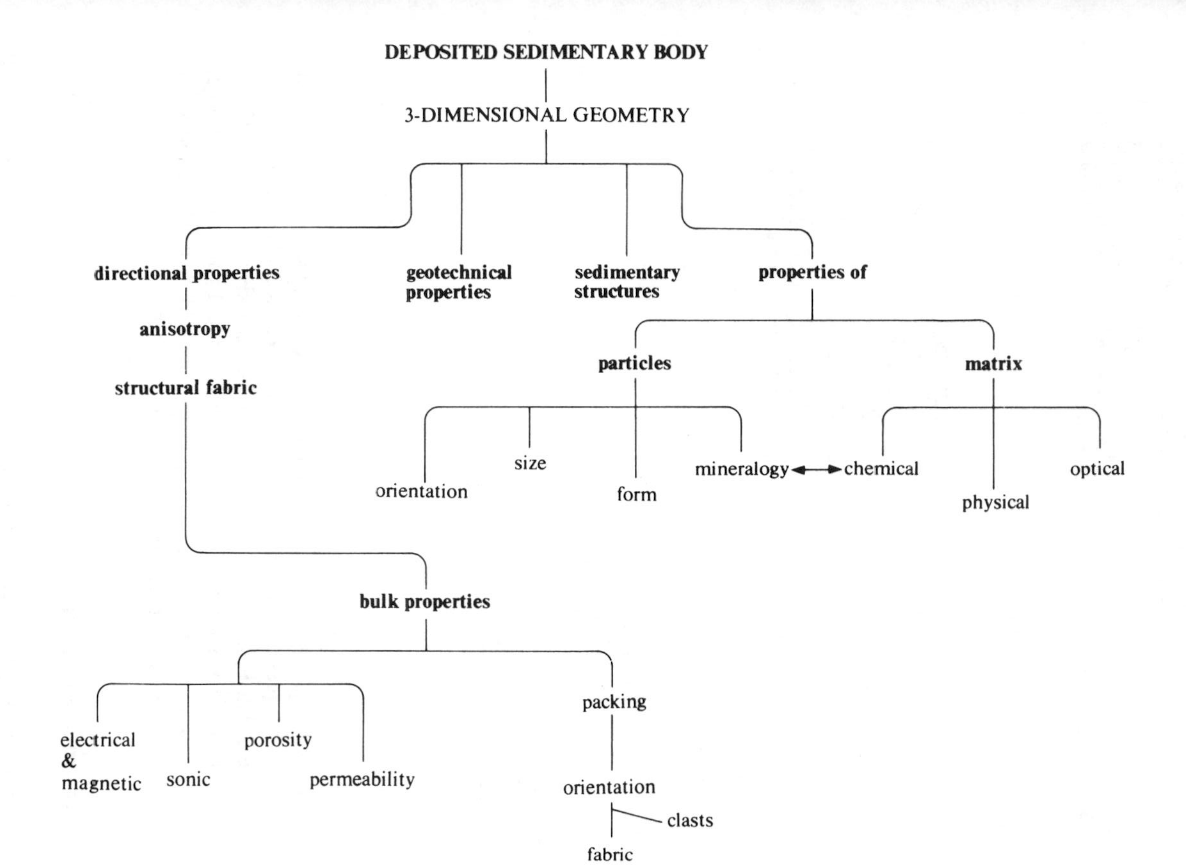

Figure 3.1 Some ways of viewing the properties of a sedimentary body.

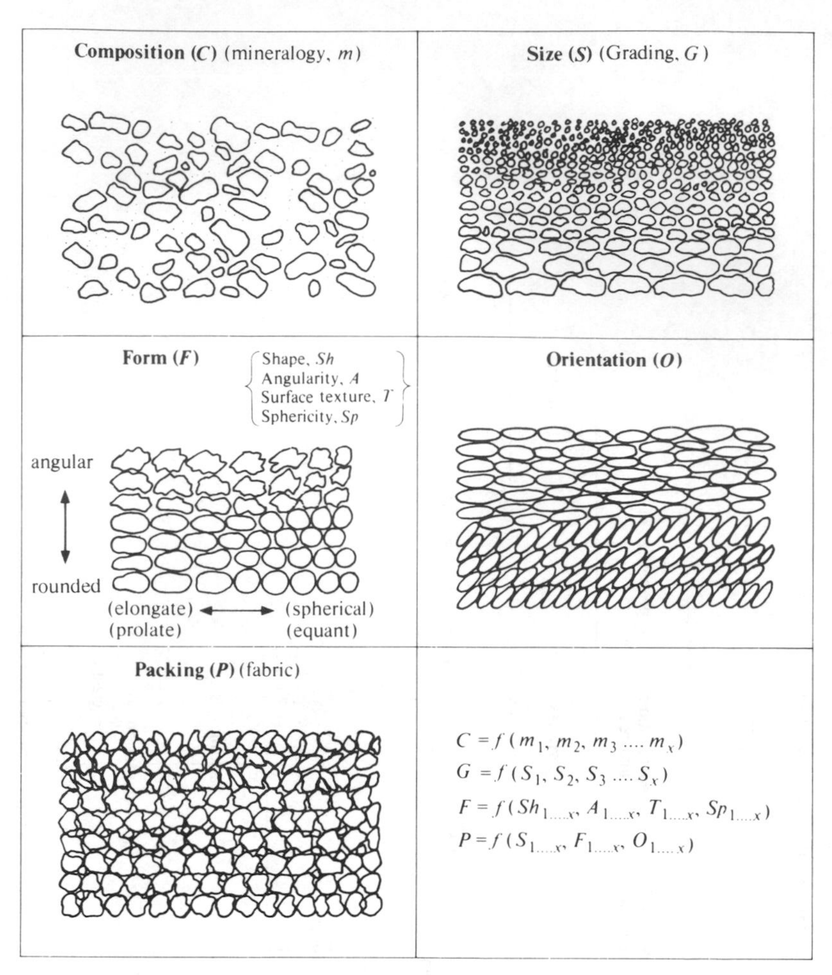

Figure 3.2 Defining equations for a sediment with respect to mineralogy and geometrical properties. (After Griffiths 1967, Pettijohn *et al.* 1972).

particle size. For particles that must be commonly measured ranging from 1 to 10^{-7} mm, a wide variety of techniques must be employed. Only a few of the most common methods can be outlined here. The best general discussion of methods is given by Allen (1981) and the relevant chapters in Carver (1971b) should be consulted for a sedimentological viewpoint. For soil mechanical requirements, books such as Akroyd (1969) and Vickers (1978) provide discussion and procedures for certain methods.

3.2.3 Direct measurement

This can be applied in only a relatively restricted size range (perhaps 1 m to 50 mm) with a ruler, calipers, a frame (Billi 1984) or an electrical device such as that described by Hardcastle (1971). The longest (*a*) axis is easy to determine, but an operational definition needs to be made about the intermediate (*b*) and short (*c*) axes, which contain most of the size information (Griffiths 1967). Such definitions are given and discussed by Griffiths (1967) and Krumbein (1941). For small particles, graduated eyepieces in microscopes or projected images can be employed, although it may not be possible to measure the *c* axis by these means.

3.2.3.1 *Sieve analysis*

This is one of the most widely used methods in size analysis. It is simple but not without limitations – a practicable limit is 0.063 to 16 mm; the nature of the particle and the properties of the sieve also add to these limitations.

Sieve aperture ranges include the German Standard DIN 1171 (1934), the British Standard BSS 410 (1962), the French Series AFNOR, and the ASTM Standard E11–61; only a limited number of sieves are used, so the interval is chosen according to the job in hand.

There are a number of ways in which errors can arise in sieving and these depend on:

(a) the load on the sieve;
(b) the surface properties of the particles;
(c) the particle size range of the sample; and
(d) the method of shaking the sieve.

Investigations on time and load variations have been carried out (see Allen 1975, ch. 5); the recommended sample weight is 0.1–0.15 kg for coarse sand and 0.04–0.06 kg for fine sand and the recommended shaking time is 9 min.

3.2.3.2 *Sample preparation*

The first and most important exercise in sample preparation is the actual sampling itself. The sample must be representative of the batch as a whole. Four stages of sample split-up are given by Allen (1975):

Field sediment	*Gross sample*	*Laboratory sample*	*Measurement sample*
10^nkg	kg	$kg^{1/n}$	mg

This gives an idea of the mass involved in the original sample and that in the final sample used for analysis. Samples should be taken when the particles are in motion, where possible. Handling of samples causes segregation of the fines to the centre of a pile and therefore samples should not be taken from the surface region only.

Splitting down samples to a manageable quantity for analysis must be done with care, as the process may induce substantial errors by giving a non-representative subsample. Quartering the sample is rapid but successive halving by suitably sized riffle boxes is to be preferred (Ingram 1971).

Before dispersal and sieving, the clay cementing agents and insoluble salts must be removed. Carbonates are removed using dilute hydrochloric acid and organic matter with hydrogen peroxide as an oxidising agent (Ingram 1971). The dispersal for subsequent analysis of fines involves replacement of the exchangeable cations present in the clay with sodium ions. Common dispersing agents include trisodium orthophosphate, Calgon and sodium hydroxide. A 5–10% solution of the dispersant is added to the sample suspended in deionised water and shaken overnight. A good comparison of dispersing agents is given by Akroyd (1969). The dispersed sample is then poured onto a 0.063 mm sieve, washed with deionised water until all the fines have passed through and dried in an oven at 110°C. The 'milk shake' mixer type of sample disperser should be avoided with sandy samples as considerable attrition of particles may result.

Dispersal can be carried out rapidly by the use of an ultrasonic probe. The sample and dispersant are placed over the probe for a suitable time (determined by experience), during which the sample is dispersed via high-pressure cavitation bubbles formed in the liquid.

In order to obtain good results from dry sieving, the individual grains must be dry. If surface moisture is as little as 1%, then adhesion forces can overcome the weight of grains below 1 mm (Müller 1967). The analysis is carried out using a nest of sieves of the required subdivisions. The dry weighed sample is poured onto the largest sieve and a lid is placed on top. The nest is shaken for 10 min on a vibration shaker. Full details of this method are given in BS 1377 (British Standards Institution 1975c). If samples are less than 0.063 mm, then an alternative method should be used, since below this level dry sieving becomes inaccurate. If large quantities of fine material are present, then a wet sieving technique should be used. A comparison of sieve methods is given by Wolcott (1978), and Mizutani (1963) looks at theoretical and experimental results. The maximum mass of material to be retained on each test sieve at the completion of sieving is given by BS 1377 (1975) (Table 3.1).

3.2.3.3 Wet sieving

This technique is used for a sample suspended in a liquid. Pretreatment is the same as for dry sieving. The weighed sample is placed in dispersing agent for 1 h with occasional stirring and then poured onto the largest sieve in the subdivided nest. Water is run through the sieves to wash the sample and is continued until that coming from the smallest sieve is virtually clear. BS 1377 (1975) gives full details of this method. Table-top wet sieving units are commercially available and provide a good method of coarse particle sizing.

3.2.4 Sedimentation methods

3.2.4.1 Hydrometer method

The pretreatment of samples is similar to that described above and ultrasonic dispersion is often desirable. This method, which uses about 40 g of sample, records the variation in density of a settling suspension using a specific-gravity hydrometer. It covers quantitative analysis of particle size distribution from the coarse sand size and downwards. If less than 10% of material passes the 63 μm sieve, then the test is not applicable (BS 1377: 1975). Full details again are given of the method in BS 1377 (1975) and in Akroyd (1969) and Allen (1981). The final calculation is derived from Stokes' law

Table 3.1 Table of sieve sizes to British Standards with maximum mass to be retained on each.

	Maximum mass				
BS sieve aperture width (mm)	*450 mm dia. sieves (kg)*	*300 mm dia. sieves (kg)*	*BS test sieve (mm)*	*BS test sieve (μm)*	*Max. mass 200 mm dia. sieves (kg)*
50	10	4.5	3.35	—	0.300
37.5	8	3.5	2.00	—	0.200
28	6	2.5	1.18	—	0.100
20	4	2.0	—	600	0.075
14	3	1.5	—	425	0.075
10	2	1.0	—	300	0.050
6.3	1.5	0.75	—	212	0.050
5	1.0	0.50	—	150	0.040
			—	63	0.025

Source: after BS 1377 (British Standards Institution 1975c).

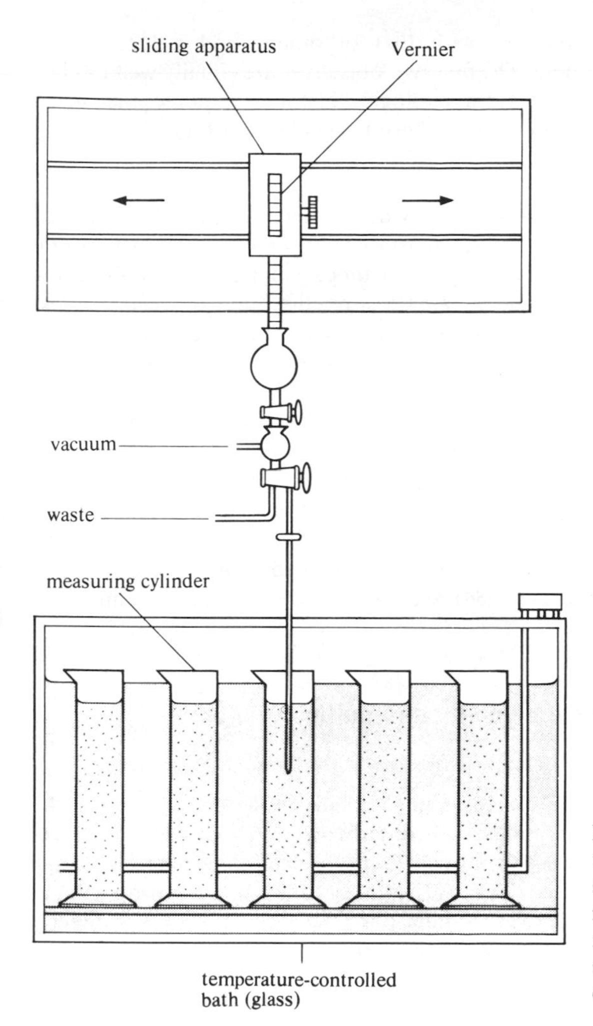

Figure 3.3 Diagram of the apparatus for pipette method of sediment size analysis. The pipette can be moved to each cylinder in turn and racked up and down to allow a sample to be drawn from the desired depth by vacuum.

$$\text{particle size (mm)} = F\left(\frac{\text{effective depth (cm)}}{\text{time (min)}}\right)^{1/2} \tag{3.1}$$

where F is a factor depending on specific gravity and temperature of the sample.

3.2.4.2 The pipette method

This method has the same limitations as the previous one and follows the principle that the changes in concentration in a settling suspension can be followed by removal of definite volumes using a pipette. Details of the method are given in BS 1377 (1975), Akroyd (1969) and Galehouse (1971).

The pipette method is by far the most widely used in sedimentation analysis. The dispersion procedures are the same as those described. To reduce the time for a number of samples to be analysed, this method requires the setting up of an assembly to accommodate 6–10 analyses concurrently (Fig. 3.3). The calculation again is based on Stokes' law.

To compare the methods of size analysis so far, we must look at three things when determining size distribution of samples. These include a physical description, comparison of one sample size distribution with that of others, and a look at the sedimentological history of the sample. The first two objectives are equally well fulfilled by all the methods described, but, when sedimentological history is required, then many workers have found sedimentation techniques superior (Galehouse 1971).

3.2.4.3 Sedimentation columns

In this technique, a cumulative method is involved whereby the rate at which a sample is settling out of suspension is measured. In geomorphological analysis, the incremental methods are mainly used in hydrometer and pipette methods. Cumulative methods show their advantages in that a small amount of sample is required and this leads to less interaction between the particles in the sample. Both methods use inexpensive apparatus but the incremental are more rapid and accurate. Sedimentation methods are likely to become increasingly popular (Galehouse 1971).

The two main types of column are the ICI and BCURA equipment, and a full description is given by Allen (1981). Both types of apparatus are cheap and may be used by semi-skilled workers, allowing a number of analyses to be carried out concurrently. Taira and Scholle (1977) have recently described a modification of the method using a photoextinction tube and Nelsen (1976) details an automated rapid sediment analyser.

Comparisons have been made between various kinds of instrument; e.g. between sieving and settling tube (Komar & Cui 1984), sedimentation and Coulter counter (see below) methods (Stein 1985).

3.2.5 The electrical sensing zone method (Coulter counter)

3.2.5.1 Principle

The Coulter principle depends on particles being suspended in an electrolyte and the passage of these through a small aperture, which contains an electrode at either side. As the particles pass through, they displace their own volume of electrolyte and thus change the resistance in the path of the current. The quantity of resistance change is proportional to the volumetric size of the particle. The number of changes per unit time is directly proportional to the number of particles per unit volume of the sample suspension.

3.2.5.2 Instrumentation

There are a number of instruments in this range, some designed for specific purposes. In the model TA, current is passed through the electrodes and back into the unit preamplifier. A current pulse represents each particle counted and these are detected, amplified and fed to a threshold circuit where the actual sizing takes place according to the previous calibration. The size particle pulses are converted to a voltage representative of the volume of particles in the 16 size ranges. These voltages are then sampled one by one (channel by channel) for differential data and added one by one for cumulative data. The signal from this is then adjusted electronically for an oscilloscope or an x–y recorder.

The diameter sequence for the 16 channels is obtained by starting with 1 μm. All values above 1 μm are obtained by multiplying by $\sqrt[3]{2}$, i.e. 1.26, and all values below by dividing by 1.26, i.e. 1.59, 1.26, 1.0, 0.79, 0.63.

There are several sizes of tubes available from 10 to 2000 μm. They consist of a sapphire orifice let into a glass tube. The particle size range of each tube is between 2 and 40% of the tube's size, e.g. the 100 μm tube has a range between 2 and 40 μm.

3.2.5.3 Operation

The electronic counter in the instrument can be reset either in the count mode or in the manometer mode. In the latter, resetting is achieved when a stopcock is opened to allow a vacuum into the system, and a sample is drawn through the aperture and unbalances the mercury in the manometer. The mercury flows beyond the start contact and the counter is reset. When the stopcock is closed, the mercury travels back to its balanced position and draws a set volume (i.e. 0.05, 0.5 or 2.0 ml) through the orifice. There are start and stop contacts in the manometer, and the distance travelled by the mercury between these contacts determines the volume.

3.2.5.4 Calibration

Before calibration can take place, the aperture resistance must be found, and this can be done by connecting an ohmmeter between the two electrodes, and the matching switch can be set to the appropriate position.

Calibration is carried out by dispersing monosized particles (e.g. latex spheres) in the electrolyte and placing over the tube and stirrer for operation. The size chosen for the tube should lie between 5 and 20% of the tube orifice, e.g. 12.79 μm latex will calibrate a 100 μm tube.

When a larger range of particle size is to be determined and this falls outside the range of one tube, then a two-tube method can be used. This is achieved by overlapping, say, a 100 μm and a 30 μm tube; the overlap must not have a channel shift greater than 5, e.g. channel 11 on the large tube corresponds to channel 16 on the small tube, giving a channel shift of $2^{16-11} = 2^5$. This method is more time-consuming as small-orifice tubes can lead to blockages and also time is spent ‘screening out’ the larger particles before the smaller tube can be used.

The electrolyte itself must be corrected for background and coincidence, i.e. loss in count when two particles go through the orifice together and are counted as one. Coincidence can be kept to a minimum by filtration. Various filters are available and those most commonly used are the membrane type, e.g. 0.2 to 0.8 μm.

3.2.5.5 Presentation of results

The data can be shown in two forms:

(a) a cumulative distribution; or
(b) an incremental distribution.

The data are presented in terms of *d*, i.e. the diameter of a sphere of equivalent volume to the particle

$$d = KV^{1/3} \tag{3.2}$$

where K is the calibration factor determined and is given as $K = (2^w/A)^{1/3}$, w is the peak channel and A is the value for the pulse amplification. The weight per cent frequency distribution may be plotted against particle size on linear axes or on a linear–logarithmic scale.

This technique has added a number of advantages to particle size analysis in geomorphological research. These include speed of analysis and reproducibility. In comparison to other size determination methods, the electronic sensory method is unique in that the size parameter is a sphere having the same volume as the particle being measured. Change in operator and non-representative sampling can cause errors, but this can be reduced through proper training. In comparison to the other methods of particle size analysis, this is among the more expensive. Useful reviews of use of this type of analysis and its comparison with conventional methods are given by McCave and Jarvis (1973), Shideler (1976) and Behrens (1978).

Methods of rapid particle sizing using other than the electrical sensing zone technique have been developed in recent years. These have benefited from developments in electronics and microcomputing power. Thus, the centrifugal methods, used primarily for fine particles ($< 1\ \mu m$), are now much easier to use. In some cases, instruments can combine centrifugal and sedimentation methods in one instrument. Similarly, new optical methods use electronics to count and measure individual particles. In one type of instrument, a collimated light beam detects interruption to the beam with a photodiode as sensor. Alternatively, lasers can provide the light source, with the particles diffracting the light scattered by the particles. Full counting and display and computer facilities are provided by such instruments, and they have the added benefit of being able to analyse wet or dry dispersions of particles. Microcomputers (either integrated with a particle sizing instrument or coupled to it) can analyse the size data and fit them to standard types of distribution. Such software may be included with the instrument package. A recent evaluation of one such laser-diffraction method is given by McCave *et al.* (1986).

3.2.6 Analysis of results

Although it is sometimes possible to use one measurement method for all of a sample, it frequently occurs that more are required. These methods may differ in the way that the size estimate is made. Thus, the finer 'tail' of a sieve analysis is the coarse end of a pipette sedimentation. As the assumptions of the test differ, a smooth transition in the grading curve may not be found. This is discussed by Griffiths (1967), who also suggests solutions to this problem. Replication may be a wise, although tedious, precaution for some methods.

The ways in which results of a particle size distribution may be presented are many and varied. Traditionally, soil mechanics uses a log millimetre abscissa with an arithmetic 'per cent finer by weight' ordinate. This gives a sigmoidal cumulative curve, but is often made to approximate to a straight line by plotting the ordinate on a probability scale. Certain other methods are available to make the visual impression of the line meaningful in some way. A good example is the crushing law distribution of Rossin and Ramler (Kittleman 1964). Sedimentologists most frequently use the ϕ scale of size rather than millimetres. This is given by $\phi = \log_2$ mm; thus 1 mm $= 0\ \phi$ and $2\ \mu m = 9\ \phi$. Other size scales have been used, although not widely accepted.

Other means of presenting data are available, e.g. histograms, triangular diagrams showing three components such as sand–silt–clay or giving percentile values of the distribution. Sedimentological work in quantitative comparisons uses size distribution statistics. These are of two main kinds, which are derived from the approximation of most samples to a gaussian distribution, namely, moment and graphical statistics. Values of mean, standard deviation, skewness and kurtosis are derived from all the data of a grain size analysis according to simple equations. Although the equations are simple, their derivation before electronic calculators and computers was tedious, and more reliance was placed upon graphical methods, which obtained approximations to the moments after the cumulative frequency graph was drawn. There is an enormous literature in this field (McBride 1971, Folk 1966, Moiola & Weiser 1968), especially when concerned with the effectiveness of particular parameters for the interpretation of sedimentary processes (e.g. Friedman 1961, McCammon 1962).

3.2.7 Particle form

3.2.7.1 General form properties (excepting surface texture)

Although particle form is considered as one of the main constituents of sediment description (Griffiths 1967), ranking with particle size as a major contributor to particle resistance in sediment transport, its effective contribution in understanding problems of sediment description has generally been denied relative to the power of particle size discrimination. This stems from both the ill definition of the often illusive nature of particle form, and the concomitant persistent inability of deriving simple measurements by which to characterise form efficiently at all scales of particle size.

The attitude may seem contentious in view of the plethora of methods and indices that purport accurately and precisely to define and measure individual aspects of particle form, but it serves as witness to the debate concerning form definition and measurement, a debate supported by a voluminous literature, which is so interlocked that few papers other than those of the founding fathers of form studies in sedimentology (Wentworth 1919, 1922a, b, Wadell 1932, 1933, 1935, Krumbein 1941) can be cited without injustice to the ranks of form investigators. This complexity underlines the value of the reviews that are available (e.g. Rosfelder 1961, Koster 1964, Flemming 1965, Griffiths 1967, Pryor 1971, Barrett 1980, Beddow 1980).

A major element in form studies is concerned with the resolution of conceptual difficulties presented by the definition of form, and its decomposition into component elements such as shape, sphericity, roundness/angularity and surface textures. A working definition of particle form, based on Griffiths (1967) and Whalley (1972) is given as:

> the resultant of combining those elements which affect the three-dimensional geometrical configuration of the surface envelope that contains the particle mass.

This resultant is affected by the scale at which component elements are represented, such that form is determined principally by shape, then by roundness/angularity, with surface textures adding only marginally to form through micro-scale additions to the particle's surface roughness. None of these elements is independent of each other, or of particle size, despite the continuing debate that attempts to differentiate between them, e.g. shape and roundness (Wadell 1932, Flemming 1965) or shape and sphericity (Whalley 1972). The major difficulty with particle form studies has been the poor operational definition of the form element under examination. Frequently, studies attempt either to gauge one element as the specific characteriser of form without reference to the other elements (Zingg 1935), or, alternatively, to subsume a mixture of elements by one index, e.g. Krumbein's sphericity index (Krumbein 1941).

Citation of sphericity as a fundamental element of form is often considered tautological when it appears alongside shape as a descriptor of macro-form. However, implicit differences between the two concepts make their duality expedient. Expressions of particle shape refer to the closest approximations of the particle's three-dimensional structure by regular solids found on a continuum of oblate–prolate structures, i.e. disc or plate→ blade→roller→sphere. The use of sphericity is a surrogate attempt to monitor form by assessing the differences between the observed particle and a sphere of equivalent dimensions (usually in terms of volume). The use of a sphere as a reference form is based on its limiting position with regard to settling velocity in still water. For any given volume, the sphere has the least surface area of any three-dimensional solid and offers least resistance to flow as it falls, i.e. maximising the settling velocity for that given mass. Shape is a statement of particle similarity, while sphericity is a measure of deviation between the particle and a reference sphere.

Conceptually, sphericity should not be confused with shape except in the limiting case of a true sphere (see Krumbein 1941, figure 5), but there is a tendency for the two items to be considered synonymous. The use of a sphere as a reference index could be superseded by other solids – e.g. Krumbein's (1941) skewed ellipsoid, Aschenbrenner's (1956) tetrakaidekahedron – such that sphericity should not really be considered as a true characteriser of geometrical shape. However, the role of sphericity does highlight an alternative method of describing form through a behavioural or functional approach (Flemming 1965).

This behavioural approach advocates that the full integrated range of form elements offers a collective resistance to fluid activity and that any interest in genetic sedimentation requires behavioural information rather than purely geometrical form. Sneed and Folk's (1958) version of sphericity offers the only standardised index of behavioural form based on the sphericity concept.

This dichotomy of measurement type underlines the need to establish the purpose for which particle form is required. Form should be seen only as a means to an end. Sophisticated form analysis is of little value if the analysis is untenable in the pursuit of sedimentological goals.

By far the largest number of papers are related to the description of form as a pertinent attribute of sediment characterisation. This type of characterisation may be for the purposes of facies analysis and environmental reconstruction (e.g. King & Buckley 1968, Patro & Sahu 1974), or for purposes of environmental gradient analysis, e.g. rates of abrasion in fluvial systems as a function of particle abrasion (Sneed & Folk 1958). Geometrical indices are of use at this stage but, if processes concerning sedimentary genesis are to be elucidated, then behavioural indices should be used.

A second approach is to study the effect of form on particle entrainment. This approach may be based on direct experimentation (Corey 1949) or on an inferential assessment of empirical particle form variation throughout environments and its relation to transport processes (Winkelmolen 1971). In either case, form should always be measured in terms of behavioural indices.

The influence of form on the geotechnical properties of aggregate sediment, e.g. packing and shearing resistance properties, has produced a third, small, strand of particle form research (Lees 1969, Koerner 1970). In this case, standard behavioural properties of transport potential may be inappropriate for static sediment and geometric indices will be of greater value.

3.2.7.2 Methods of form measurement

A universal method of operationalising a standard definition of form is hindered by the range of conditions in which particles are found. The amount of information measured by replicative accuracy of form analysis is minimised in the case of particles observed in thin section under a microscope showing only aggregate, two-dimensional structures, taken in an unknown plane of the particle solid. Information is maximised when disaggregated single particles of a size commensurate with individual axial measurement are available. Between these two extremes lies a range of particles that require an accompanying range of form measurement techniques. The availability of techniques and their efficiency in terms of operator variance is summarised in Table 3.2.

Visual description Methods of this type require a visual estimate of a particle's shape approximation to any regular solid on an oblate–prolate continuum. Allen (1981) gives a British Standards list of shape descriptors viable for very fine sediment. Roundness estimates are not encouraged, as the level of efficiency (replication) is low due to subjective assessments. This type of description should be considered only as a superficial

Table 3.2 Cost–benefit of methods of particle form measurement.

Method	*Technique*	*Type*	*Measuring system complexity*	*Available indices*	*Efficiency*
Visual description	Verbal assessment	Geometric	Low	Shape	Low
Visual comparison	Use of charts	Geometric	Low–medium	Roundness, sphericity	↑ Operator variance
Direct measurement	Axial measures	Geometric/ behavioural	Low–high	Shape, sphericity, roundness	↓
Response measurement	Behavioural measures	Behavioural	Medium	Integrated form	High

field statement when samples are unobtainable for further analysis. Despite the sophistication of required measuring technique, the recognition of surface textures with the aid of a scanning electron microscope (SEM) can be included at this level (see below). Data are recorded only on a nominal scale.

Visual comparison Methods of this type prove to be most popular when dealing with fine sediments (<4 mm; *b* axis) and especially with sand-sized material (0.1–1.0 mm), where direct measurement of particle axes is precluded. Two stages of technique are required: (a) magnification of particles and (b) visual comparison of sample grains' form outline (specifically for sphericity and roundness indices) against charts with known variation of grain form outlines. Several charts are available (Pettijohn 1949, Krumbein & Sloss 1955), but Rittenhouse's (1943) sphericity and Powers' (1953) roundness comparison charts are frequently used. Estimation of sphericity does not prove to be as difficult as roundness estimation, as the interaction of the two variables causes greater confusion with roundness variation than vice versa. Powers gives two sets of roundness outlines at differing sphericity values. Roundness is partitioned into six classes covering a range 0.12 to 1.0. As there appears to be less resolution of roundness differences between very rounded particles, the outlines show a graduated (logarithmic) range. Krumbein (1941) supplied a roundness chart for pebbles, while Powers' chart is supposed to apply to sand grains only. It is unfortunate that a universal size chart has not yet become available.

Principal difficulties with these visual techniques are low efficiency in terms of replication of results due to substantial operator variance (Griffiths 1967), and results application in only two dimensions. Given disaggregated sediment, then it is likely that the section observed is in the maximum projected area (MPA) plane (*a–b* axial plane), but in thin section, unless the sediment has good orientation of grains, the sample plane may not be the *a–b* plane. Data from these visual methods can only be treated on an ordinal scale.

Direct measurement Particle characteristics, e.g. axial values, are measured directly and inserted into formulae that purport to monitor aspects of form. Measurements in three and two dimensions are conventional, the choice of method being based on particle size.

The simplest form methods are based on direct measurement of a particle's three major axes (*a*, *b*, and *c*) taken orthogonally (Krumbein 1941) but not necessarily intersecting at the same point. Humbert (1968) asserts that lack of orthogonality does not affect indices' performance but facilitates speed of measurement.

Given three axial values, then elements of shape and sphericity can be derived.

Roundness indices, depending on the particular index of choice, require additional measurement, usually a radius of curvature for the smallest 'corner' or angle of the MPA outline.

The commonest measures of shape are based on axial ratios. Zingg's (1935) plotting of b/a against c/b proves to be efficient in discriminating between discs, blades, rollers and spheres. A linearised formula for shape on a prolate–oblate scale that achieves success in discriminating between marine and fluvial pebbles is given by Dobkins and Folk (1970).

Sphericity indices for coarse sediment (>4 mm; *b* axis) based on axial ratios have achieved notable popularity since Krumbein's (1941) operationalisation of Wadell's (1932) definition of sphericity (cube root of the ratio of particle volume to circumscribing sphere's volume). Krumbein's index, $(bc/a^2)^{1/3}$ = elliptical volume sphericity, has never achieved the precision of later behavioural indices based on the sphericity concept as proposed by Corey (1949), portance index $c/(ab)^{1/2}$, or its more widely accepted inverse, the sphericity index of Sneed and Folk (1958), $(c^2/ab)^{1/3}$. This latter index as a ratio of the MPA to particle volume contrasts the cross-sectional area of the particle (oriented normal to downward movement in fluids, i.e. resistance) to the particle weight, as a function of volume, which induces the downward component in fluids. Sneed and Folk state that this index correlates highly with the particle's settling velocity ($r = +0.97$), substantiating its behavioural tag and justifying treatment of values as interval-scale measurement, while Krumbein's sphericity can only be considered on an ordinal scale. Cailleux (1947), following Wentworth (1922a, b), suggests a flatness index $(a + b/2c)$, which has a very high inverse correlation with Sneed and Folk's sphericity ($r = -0.981$) (Orford 1975).

Roundness, an elusive quality, is best defined in Wadell's (1933) terms as a planimetric concept, referring to the smooth curvatures of the outline of a projected cross-sectional area, which should represent the MPA. The degree of roundness is a ratio function of the curvature radius of any corner of the MPA to the radius of the maximum inscribed circle in the given area. Wadell's original index is time-consuming to implement and assumes that the MPA is a representative plane of particle abrasion. These problems are not solved by subsequent roundness indices. Both Cailleux's (1945) $(2r/a)$ and Kuenen's (1955) $(2r/b)$ simplifications of roundness indices by restricting curvature measurement to the radius of the sharpest corner (r), and using the *a* and *b* axes respectively as a notional inscribing circle radius, appear to achieve little in environmental discrimination except between distinct differences. Convexity gauges and templates have been devised (Wentworth 1919) to overcome curvature measurement difficulty, but unless a two-dimensional outline of the projected area is available, operator variance will be high, such that roundness indices appear inefficient and poor characterisers. Flemming (1965) asserts that there is little descriptive difference between any of the roundness indices available and that the problem (which relates to convex–concave segment fitting as a function of change in the second derivative of particle outline) is not amenable to these proposed simplistic measures, but requires more complex curve-fitting sequences, such as Fourier analysis.

Alternative methods are required when particles in thin section, or even disaggregated, need magnification for any calibrated direct measurement. At this point, magnification and measuring-equipment cost and complexity may be prohibitive. Despite high cost, efficiency of technique cannot be guaranteed. Such approaches are based on three operations; (a) direct magnification of particle (Burk & Freeth 1967, Pryor 1971), (b) abstraction of particle outline position data, and (c) the use of such raw data to produce a particle form descriptor. A salient problem of all these image-based techniques is the resultant two-dimensionality of particles, thus denying the third dimension that some investigators believe essential in shape studies. However, some shape indices have been adapted for two-dimensional work – for example, Riley's (1941) elongation factor is advocated by Pryor (1971). Most roundness indices previously discussed are feasible at this level.

Over the last decade the greatest change in shape studies technology is associated with the advent of microcomputer-controlled image analysers. The rapid development in microcomputer technology in the 1980s in terms of video-picture capture and digital transformation as well as in data storage and memory have in particular rendered technically obsolete a number of methods that produced form indices from particle sizing methods. This does not deny that the latter methods can still produce valid descriptions of particle form.

Allen (1981) gives an informative review of a number of specific methods of automatic and semi-automatic grain counting, which enable shape indices to be generated from particle size data collected rapidly from large grain samples. Boggs (1957) makes use of a rapid particle size analyser with which direct comparisons between thin-section particle projected areas and areas of adjustable reference circles can be made.

Image analysis techniques decompose a video picture (usually of a magnified grain) into an array of picture elements (pixels), each one of which records a digital value related to the grey scale registered at each pixel. By setting a grey-scale threshold that identifies a two-tone range (black and white), each pixel can then be set to 1 or 0 depending on whether or not the particle's image overlies that particular position. In this way the outline of the particle can be detected at the boundary between 1 and 0 in the pixel matrix.

Most image analysers work on a two-dimensional image – however, grey-level variation could be contoured on a pixel basis and maps of particle surface texture could be prepared. By rotating the grain (e.g. on an SEM scanning stage) a composite view of all dimensions is feasible, though the amount of data required to specify one grain would be very large. Despite the rapid growth in microcomputer memory, it seems likely that microcomputer imaging of three-dimensional information would be concerned only with data capture, while data analysis would be left to mainframe computers. This does not deny the present growing use of image analysers to obtain particle form data in two dimensions (Whalley & Orford 1986).

Once the grain outline is digitally obtained then its transformation into a string of x–y coordinate values, or a string of relative changes in outline edge direction (based on the discrete unit scale of the pixel), follows. This stage used to be done manually by a 'pencil-follow' type of digitiser (Piper 1971) but can now be obtained automatically by applying an edge-finding algorithm to the pixel information (Rosenfeld 1976). Simple form indices of sphericity and roundness can be assessed objectively via iterative algorithms, while more complicated attempts to assess outline curvature variation are now directly feasible with speed and rapidity (Clark 1981). Flemming's suggestion of Fourier analysis of curvature has been taken up by Ehrlich and Weinberg (1970) in particular, likening the grain form outline to a closed loop convenient for Fourier analysis. By decomposing the grain outline into a summed series of differing sine curves, the particle form can likewise be decomposed into the component elements of shape and roundness, which can be related to numerical valuation through the harmonic coefficients associated with decreasing wavelength (Whalley 1978b). The problem of two-dimensional aspect appears less troublesome than first appearances indicate. Whalley (1978b) monitored harmonic changes from a series of digitised particle outlines of the same grain but in different perspective planes to find relative stability of coefficient interpretation especially for the lower-order harmonics controlling particle shape. The use of Fourier coefficients as a means of identifying scale components of form is a technique of considerable importance to form studies (Boon *et al.* 1982). Ehrlich and his coworkers have undertaken considerable research into the analytical basis of Fourier techniques (Ehrlich & Full 1984) in transforming data structures and decomposition of sedimentary samples into shape populations, Ehrlich and his students have pressed claims for Fourier techniques as a key discriminator of depositional environments (e.g. Mazzullo & Withers 1984).

The ability to capture and use grain outline data at all scales introduces the possibility of working with form indices that integrate data at all scales. Form studies are plagued with questions of scale in that indices are often appropriate only to specific scales of size. Fourier analysis is integrative in that all scales from macro-grain form to edge texture can be identified. However, the user is left after analysis with a multivariate data set, which may be or may not be advantageous depending on the user's requirements, e.g. redundant data exist in the data sets and there is a wide choice of the single representative value characterising particle form. An alternative integrated index of form appropriate for any regular or irregular particle is provided by the fractal dimension (Mandelbrot 1982, Orford & Whalley 1983). Although Mandelbrot regards the fractal dimension as being singular for any particle, Kaye (1978a, b), Orford and Whalley (1983) and Whalley and Orford (1986) show that natural and artificial particles may have three fractal dimensions: structural and textural values that reflect the two major scales of edge variation appropriate to macro-form and edge texture respectively, and an overall fractal value that integrates all scale changes of form.

Response measurement These measures are not concerned with form *per se*, but with the integrated effects of form on transport processes; therefore no direct measure of any particle parameter as a function of length ratios is taken. Instead, the particle's reaction in a transport process is obtained. The use of particle settling velocity (w) is the principal response measurement of form that is obtained (Komar & Reimers 1978), although integration of particle mass/size as well means that specific form effects may be difficult to ascertain. The 'Colorado school' of sediment transport investigators has monitored the effects of particle form on settling velocity (Alger & Simons 1968). Their results indicate that w will be an apt behavioural characteriser when sediment genesis is located within the domain $w < U_*$ (U_* = shear velocity) (Middleton 1976). Where $U_* < w$, then bedload transport predominates, in which case w is of limited value in transport characterisation, and methods that monitor the pivotability of particles are of value. Both Kuenen (1964) and Winkelmolen (1971) have made extensive use of the concept of pivotability or 'rollability' in order to explain grain differentiation on a shape basis, in the turbulent swash/backwash zone of a beach face. Winkelmolen has produced a mechanical means of evaluating 'rollability', a property related to the angle of slope in which a grain will roll down. Rollability is indexed by the time it takes for grains to travel down the inside of a slightly inclined revolving cylinder. The time it takes is a measure of the grains' rollability potential. Winkelmolen shows that rollability correlates well with the influence of grain form (given standardised size) on the particles' settling velocity.

There have been few attempts to assess the relationship between geometrical and behavioural indices. Willetts and Rice (1983) is a recent attempt to show that mutually consistent discrimination can be found between the shape characteristics of sand (150–500 μm) in terms of rollability as well as when measured in terms of geometrical sphericity. More research on this comparability question is required.

Choice of method Before undertaking particle form analysis, careful consideration should be made of a number of points that will condition the selection of investigative technique. The choice of form technique is dependent on the answers to the following list of points (modified from Pryor 1971):

(a) What amount of time/expense can be devoted to the analysis? (Form studies are notorious for low returns on expended effort.)
(b) What type of indices are required: geometrical or behavioural? (This point is designed to clarify the purpose of the form analysis.)
(c) What is the mean size of the particle under analysis?
(d) What is the indurate/non-indurate state of the sediment?
(e) What is the availability of sediment samples for replication and statistical design?
(f) What is the availability of magnification and measuring equipment?
(g) What particle parameters can be obtained from sample material?
(h) Are computational facilities for data handling and analysis adequate?

3.2.8 Surface textures of individual particles (scanning electron microscope analysis)

A review by Kuenen (1960a) synthesised the evolution of the general idea that different sedimentary environments gave rise to particular textural features on the surface of quartz sand grains. Such theories have been held, in one form or another, at least since the work of Knight (1924), although the full investigation of the idea was hampered by the poor resolution of light microscopy. However, the development of electron microscopy, first with the transmission electron microscope (TEM) and subsequently, in the late 1960s, with the scanning electron microscope (SEM), enabled these theories to be tested. Despite a series of preparation and examination difficulties, TEM studies (Biederman 1962, Krinsley & Takahashi 1962a, b, c, Porter 1962) were able to differentiate different surface textures that could be considered to have formed in response to a set of energy conditions typical of a particular environment of modification. Thus, 'glacial', 'aeolian' and 'marine' modification processes were identified as resulting in various features on the surface of the quartz sand grains studied. Equally impressive was the discovery that superimposed features upon the sand grains could be identified and, hence, two or more distinct palaeoenvironments of modification could be recognised.

By the late 1960s, the SEM was in general usage by Earth scientists who took advantage of the high resolution and clear, sharp pictures (even stereoscopic ones) that were obtainable. As such, SEM investigations largely superseded TEM analysis in all but specific case studies. This resulted in a proliferation of papers that presented palaeoenvironmental reconstruction of areas derived from quartz grain surface studies. Predominant in this research field was, and is, Krinsley, who together with a number of coworkers (Krinsley & Margolis 1969, Krinsley & Doornkamp 1973, Krinsley & McCoy 1977) has successfully steered the evolution of the research field through facets of environmental reconstruction from known source material to designations of past environments.

The 'look and see' qualitative analysis of this aspect of SEM studies had, of course, a number of built-in problems, not the least of which was the gross generalisation of palaeoenvironment from specific surface textures. A semiqualitative technique devised by Margolis and Kennett (1971), however, enabled researchers to consider environmental reconstruction from the standpoint of surface feature assemblages rather than from individual textures. This satisfied the criticisms often voiced, both generally and in the literature (Brown 1973). Further refinements to this initial catalogue of features followed (Krinsley & Doornkamp 1973, Whalley & Krinsley 1974, Nieter & Krinsley 1976, Krinsley & McCoy 1977, Bull 1978a), so that present studies utilise a whole series of features from which past events can be identified. Summarised in Table 3.3 is a

Table 3.3 Summary table of the main mechanical (M) and chemical (C) textures on quartz sand grains.

1	Conchoidal fractures (large > 15 μm, small < 15 μm)	M
2	Breakage blocks (large > 15 μm, small < 15 μm)	M
3	Microblocks; equivalent to category 2, but probably C	C?
4	Arc-shaped steps	M
5	Scratches and grooves	M
6	Chattermarks; M (Bull 1977), C (Folk 1975)	?
7	Parallel steps; equal to Hogback imbricate structure	M
8	Non-oriented V-shaped pits	M
9	Dish-shaped concavities	M
10	Upturned plates (Margolis & Krinsley 1974)	M
11	Roundness: sub-rounded, rounded, sub-angular, angular; largely M but also C	M/C
12	Relief: (a) low, (b), medium, (c) high	M/C or inherent
13	Facet (Whalley & Krinsley 1974)	C
14	Cleavage flake	Inherent
15	Precipitation platelet	C
16	Precipitation rounding; related to category 11	C
17	Carapace; various forms of adhering particles of comminution debris	C+ inherent
18	Chemically formed V-shaped, oriented pits	C
19	Flat cleavage face; equal to 13?	Inherent
20	Cleavage planes (or microfracture lines, Moss *et al.* 1973)	C+ inherent
21	Silica precipitation: cryptocrystalline/amorphous and euhedral forms	C
22	Silica solution: pits, dulled surface, holes, etc.	C

collection of some of these features compiled largely from Margolis and Kennett (1971), Krinsley and Doornkamp (1973) and Whalley and Krinsley (1974). Features with question marks indicate those textures with controversial origins and, hence, with controversial environmental significance. Table 3.3 is split into mechanically and chemically derived features (M and C, respectively).

The features summarised in the table represent only the major groups of textures that have been identified; within each group there are often a number of specific forms. It can be seen from the complexity of the table, therefore, that great caution must be tempered when deciphering the superimposed features on the surface of a sand grain: combinations of textures are considered, not just individual textures. 'Glacial' features, for example, can be traditionally considered to require the existence of categories 1, 2, 4, 5, 7, 11 and 12c on the surfaces of the sand grains before they can be designated to have derived from glacial conditions. Problems exist, however, in specifying exact environments of modification of glacial action (Whalley & Krinsley 1974) and also of inherent first-cycle quartz features (Schneider 1970). Caution and, perhaps, experience are needed before correct and statistically significant palaeoenvironments can be forecast. An effort is being made in this review to refrain from presenting a list of pertinent surface textures and their related palaeoenvironmental indicators, not because they are unreliable, but rather because the pitfalls of overgeneralisation and a too-lax use of SEM textural analysis has led to unscrupulous and careless application of the technique. Potential SEM analysts are referred to the works of Schneider (1970), Brown (1973), Margolis and Krinsley (1974), Whalley and Krinsley (1974), Whalley (1978a), Tovey and Wong (1978) and Tovey *et al.* (1978) in order that potential pitfalls can be recognised before progression to an application of the technique using the atlas of Krinsley and Doornkamp (1973). More recent reviews of the method can be found in Bull (1981), Whalley (1985), Sieveking and Hart (1986) and Bull *et al.* (1987).

Quartz grains are usually examined because of their ubiquity in the sedimentological record. It is necessary, therefore, to make sure that only such are examined, and so inspection under crossed nicols is advisable before mounting. Alternatively, a rapid method of analysis of the composition of the grain can be achieved using electron probe techniques (energy-dispersive X-ray (EDX) analysis) (see Sec. 3.3.7). Perhaps the most difficult of minerals to distinguish from quartz is feldspar; the similarity of surface texture with weathered quartz is often striking. However, utilising the experience of the trained eye, together with inspection under crossed nicols and EDX analysis eliminates any possible confusion of the minerals.

Although there are methods by which samples can be viewed under the SEM without being covered with a metal conducting film, it is usual to have the samples to be analysed coated with a thin layer of gold or gold–palladium. Vacuum coating is undertaken by a sputter technique on an easily operated machine that is an essential accessory to the SEM. Carbon coatings are required to produce replicas for TEM analysis, although these prove difficult to obtain and are vulnerable to even the most gentle handling. Methods for obtaining such carbon replicas are well documented in the earlier literature and can be found referenced in Krinsley and Takahashi (1962a, b, c) and Biederman (1962).

The selection procedure and the number of grains to be studied from each sample have, in the past, been varied and lacking in any definite instruction. 'Statistically significant' numbers of grains per sample have been designated as low as 10 and as high as 200 (see Bull 1978a for review). Clearly some uniformity must be sought. Perhaps the most significant work relating the number of grains to be investigated is that by Baker (1976), who concluded that 30–50 grains are needed to be sure of obtaining a fair coverage even in the event of bimodality in the sample.

Much of the SEM analysis of quartz grains has been concentrated upon sand-sized particles, i.e. grains with diameters ranging from 63 to 240 μm (with most studies featuring grains greater than 200 μm). However, there has been allied research undertaken using larger grains (Hey *et al.* 1971) and on silt- and clay-sized quartz grains (Smalley 1971, 1978, Moss *et al.* 1973, Bull 1978b, Whalley 1979), the latter studies concentrating largely upon the problems of loess-like sediment development and distribution.

The magnification under which SEM analysis is carried out is as varied as the size ranges of grains investigated. The magnification used generally depends upon the surface textures that are being studied and the size of the grain being examined. As a general but somewhat crude rule, mechanically derived features can be viewed under a lower (30 to 5000) magnification range than chemically derived features (up to 30 000, or indeed to the resolution limits of the particular microscope used). The upper limit of an SEM's magnification is achieved at the cost of picture resolution. Extremely high magnifications (200 000 plus) are, however, possible with the use of the more recently developed STEM (scanning transmission electron microscope), but once again picture resolution of the surface of the quartz grain is often of variable quality. Experience will determine the balance between machine magnification and picture resolution for any particular microscope.

There is much ignorance displayed as to the operating voltage required for SEM mineral grain studies. Many papers advocate machine settings as high as 30 kV although it is quite easy to operate at levels as low as 5–10 kV. The advantages of using the lower setting is that you can be sure of viewing the surface of the grain only; high-voltage (30 kV) operation enables the immediate subsurface phenomena to be seen. If this high kilovoltage is inadvertent, then the picture often contains 'ghosts' of subsurface features superimposed upon the grain surface. This can only serve to confuse the less experienced researcher.

While this review concentrates upon quartz sand grains, some work has been undertaken upon other minerals, although these have been little used for palaeoenvironmental studies (Setlow & Karpovich 1972). Other and more varied applications of SEM analysis include investigations of cathodoluminescence and other secondary electron features (Krinsley & Tovey 1978), stereoscopic picture analysis of sand grains and clay-mineral particle investigations (Tovey 1971).

The great versatility of the SEM and its associated analytical tools has allowed a much fuller understanding of Quaternary environments (and even Ordovician environments; Culver *et al.* 1978). Future developments in SEM analysis should involve quantification procedures and operator variance studies in order that true statistical relevance can be accredited to a previously much maligned technique of environmental reconstruction (Krinsley 1978).

Most recently the development of the scanning optical microscope (Wilson *et al.* 1986, Wilson 1985), which utilises laser technology, provides a mechanism for more specialist research using minerals and, perhaps, biotic associations. The machine has the advantages of requiring no vacuum, no specimen charging, a variety of imaging modes as well as computer storage, processing and even operation! It is most certainly a technique for the future.

3.2.9 Mass

Bulk properties are fundamental physical parameters that help to define the nature of material. Their measurement is often performed quickly, at a low cost and with a high degree of reliability by one operator. Standard methods apply for some of these tests, e.g. British Standards Institution (1975a, c; BS 1924 and BS 1377), Akroyd (1969) and Avery and Bascomb (1974), and good sedimentological overviews are given by Tickell (1965, ch. 3) and Muller (1967).

All materials possess mass, which is defined as the reluctance of a body either to commence or to stop moving. The mass of a body is constant throughout space, in comparison to weight which varies with gravity. Measurement of mass (kg) is by comparison with a standard mass on a beam balance selected in keeping with the size of the sample.

Many types of balance exist and the choice depends upon both magnitude of the mass and the accuracy required, and this will often depend upon the job in hand. Relatively crude beam balances may be sufficient for laboratory or field weighings. Spring balances measure weight rather than mass, although this may not be critical. For moisture content determination, a balance weighing to 10 mg (a ‘two-place’ balance, i.e. 0.01 g accuracy) is required. For analytical chemistry (gravimetry), ‘five-place’ balances may be required. Constructional details vary, but the newest types, the ‘electronic’ balances, have advantages. They use a Moiré fringe-counting technique, which enables a digital value to be displayed directly. Time averaging (to overcome bench movement and air draughts), digital output (which can be interfaced with a microcomputer), robustness as well as a dual range (e.g. 300 g to two places and 30 g to three places on one instrument) and push-button taring are the main reasons why this type of balance is becoming increasingly popular.

3.2.10 Density

The density of a body is the relationship of its mass to its volume (kg m^{-3}). Densities of both rocks and soils vary considerably and are influenced by mineralogy, voids ratio, particle size distribution and particle shape.

Several methods are available for density measurement, and values obtained depend to a certain extent upon the procedure. A standard technique for density measurement is volume displacement using a density can. A sample of material is trimmed to regular shape, weighed, covered in two coats of paraffin wax and reweighed. The volume of water that this sample displaces is collected in a measuring cylinder. A representative sample of material is taken for moisture content determination.

The displacement technique allows the determination of bulk density (γ),

$$\gamma = W_s/V_s \tag{3.3}$$

where

$$V_s = V_w - (W_p/G_p) \quad \text{and} \quad W_p = W_w - W_s \tag{3.4}$$

and W_w is the weight of soil and paraffin wax, W_s is the weight of soil, W_p is the weight of paraffin wax, V_s is the volume of soil, V_w is the volume of soil and wax, G_p is the specific gravity of paraffin wax.

Bulk density may also be determined by weighing a sample in air, in kerosene and finally reweighing the same sample, coated with kerosene, in air. The volume of the sample, and thus density, is determined from the specific gravity of kerosene, i.e.

$$\text{volume of sample} = (W_a - W_k)/G_k \tag{3.5}$$

where W_a is the weight in air of sample coated with kerosene, W_k is the weight of sample in kerosene, G_k is the specific gravity of kerosene. This technique is best employed for gravelly soil.

Density determinations can be made from the specific gravity of soil particles, defined as the weight of a given soil particle to the weight of the same volume of water (at a temperature of 4°C). Specific gravity is measured by either density bottle, for fine-grained soils (BS 1377: 1975), or pycnometer, which is normally used for cohesionless soils. The density bottle method is the most accurate, but in both cases experimental procedures are simple and involve three weighings. Specific gravity of the soil (G_s) is obtained from

$$G_s = \frac{SW_d}{W_d + W_1 - W_2} \tag{3.6}$$

where W_d is the weight of oven-dry soil, W_1 is the weight of container filled with water, W_2 is the weight of container filled with water and solids, and S is the specific gravity of liquid ($S = 1$ if water is used). The relationship between specific gravity and density of soil depends upon void ratio (e), degree of saturation (s) of the soil and unit weight of water (γ_w) as follows:

(a) saturated sample

$$\gamma = \frac{G_s + e}{1 + e}\gamma_w \tag{3.7}$$

(b) unsaturated sample

$$\gamma = \frac{G_s + se}{1 + e}\gamma_w \tag{3.8}$$

(c) dry density, soil heated to 110°C to drive off absorbed moisture

$$\gamma_d = \frac{G_s}{1 + e}\gamma_w \tag{3.9}$$

Density, although simply determined, is an important property in soil and rock mechanics. Knowledge of the density of materials is essential in any analysis of slope form, process and stability. In some soil mechanics calculations, it is sometimes sufficient to use a generalised value for a soil specific gravity, and $G_s = 2.65$ is that most frequently employed.

3.2.11 Porosity and voids ratio

Porosity is the fractional pore space of a rock or soil mass. A dimensionless quantity, porosity (n) is determined from the voids ratio (e) as follows:

$$n = V_v/V \tag{3.10}$$

$$e = V_v/V_s \tag{3.11}$$

$$n = e/(1 + e) \tag{3.12}$$

where V_v is the volume of voids, V_s is the volume of solid, and V is the total volume. As the total volume of a sample equals the volume of solids plus the volume of pores, porosity is evaluated by measuring any two of these properties. Several methods are available for determining porosity; total volume may be measured either by mercury displacement or with a pycnometer.

Similarly, from the determination of the specific gravity (G), dry weight (W_d) and volume of a sample (V), porosity can be calculated as

$$e = (G\gamma_w V/W_d) - 1 \tag{3.13}$$

and

$$n = 1 - (W_d/G\gamma_w V) \tag{3.14}$$

Porosity is of geomorphological significance through its relationship with voids ratio and shear strength of materials, and is one of several factors that influence the permeability of material (see below), although the exact relationship between porosity and permeability is indeterminate. It is also possible, although with more trouble and less accuracy, to determine *in situ* bulk density by replacing a known mass of soil with a measured replacement volume of sand or water (BS 1924: 1975).

The porosity of rocks, determined in bulk or by thin-section analysis, is important in several aspects of sedimentology as well as in the investigation of building stone behaviour. The papers by Haynes (1973) and Fagerlund (1973) (and other sections in this volume) provide a useful introduction to relationships between pore properties, suction and specific surface (see below). Image analysis and quantitative techniques now exist for the characterisation of pore spaces (e.g. Lin & Hamasaki 1983).

3.2.12 Permeability and infiltration capacity

Permeability is the ease with which a fluid can flow through soil or rock. Flow is often complex and is affected by a number of factors: grain shape, water viscosity, density and particle diameter.

Darcy's law relates flow per unit time (q) to the cross-sectional area (A) of sample, coefficient of permeability (k) and hydraulic gradient (i):

$$q = Aki \tag{3.15}$$

The coefficient of permeability is the rate of flow per unit area of sample under unit hydraulic gradient ($m\ s^{-1}$) and thus relates to a particular fluid; for geomorphological purposes this is usually water. The permeability coefficient is in fact a mean velocity for all elements of a sample, which on a smaller scale may have different permeabilities.

Laboratory measurement of permeability is easily performed using a permeameter. The principle of the permeameter is to allow water to flow through a sample, and to measure the hydraulic gradient and rate of discharge. There are two models of permeameter in common use, the constant-head permeameter and falling-head permeameter.

The basis of the constant-head permeameter (Fig. 3.4) is that the permeability coefficient is proportional to the quantity of water per unit time. Distilled water from a balancing, constant-head, tank is allowed to percolate through the previously saturated sample placed in a cylinder of cross-sectional area A. The sample, which rests on a layer of graded gravel and is covered by a gauze, should fill the cylinder from which all air is extracted. Once constant conditions are registered by the manometer tubes, flow is increased until maximum discharge is obtained from the permeameter and conditions are again steady. Discharge Q during time interval t is collected in a graduated vessel and the difference of head H over length of sample l is measured:

$$Q/t = AkH/l \tag{3.16}$$

where H/l is the hydraulic gradient (i).

The accuracy of the constant-head permeameter is dependent upon the measurement of the volume of water. There must be a reasonable flow to be measured. Thus this technique is restricted to fine gravels and sands within the permeability range 10^{-2} to

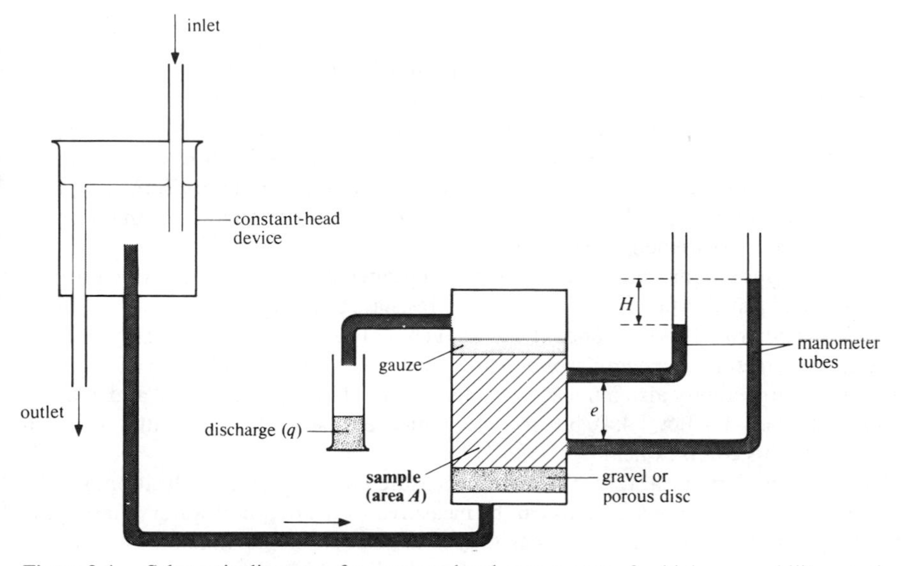

Figure 3.4 Schematic diagram of a constant-head permeameter for high-permeability samples.

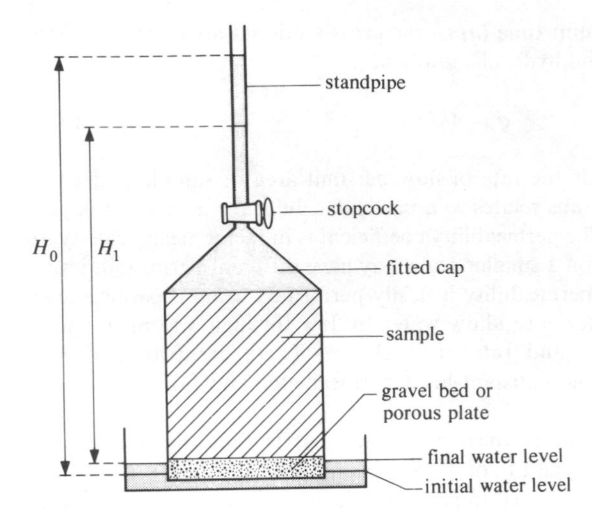

Figure 3.5 Schematic diagram of a falling-head permeameter for low-permeability samples.

10^{-4} m s^{-1}. The falling-head permeability test provides an alternative technique in which the quantity of water flowing into a sample is measured; this test is suitable for low-permeability soils within the range 10^{-4} to 10^{-9} m s^{-1}.

In the falling-head permeameter (Fig. 3.5), the sample is trimmed and placed in a cylinder fitted with a cap. The cylinder is stood in a tank and a vacuum line is connected to the cap. Once saturated, the air-free sample is connected to a standpipe containing distilled water. When the stopcock is opened, water is able to percolate into the container at the base and the time t required for the water to drop to a lower gradation on the standpipe is recorded. The test is repeated several times; a graph of $\log_{10}H$ (H = water head) is plotted against time and the coefficient of permeability is obtained from:

$$k = \frac{2.3al}{A(t_1 - t_0)} \log_{10}(H_0/H_1) \tag{3.17}$$

where a is the cross-sectional area of manometer standpipe, A is the cross-sectional area of sample, l is the length of sample, H_0 is the initial head of water, H_1 is the head of water after particular time period, t_0 is the start time, and t_1 is the time at H_1.

Permeability may also be measured in the laboratory either directly or indirectly during a consolidation (oedometer) test (see below), while a capillarity–permeability test is suitable for partially saturated soils. These tests relate permeability to consolidation, but this is not a precise relationship and is dependent upon other factors. Similarly, field determinations of permeability lack precision as soil and water conditions that exist at the location of measurement are uncertain.

Permeability is a significant geomorphological property, as water percolating through a soil or rock usually takes the line of least resistance or highest permeability. It is possible to construct flowlines and, at right angles, equipotential lines, which yield data on seepage pressures, general groundwater conditions and possible areas of slope instability. Permeability also influences results measured in drained triaxial and consolidation tests (Sec. 3.4.3 & 3.4.4), but soil mechanics texts should be consulted for a full treatment of this complex field.

Associated with permeability, but usually measured in the field, is infiltration, i.e. the flow of water entering the soil. This can be measured as the depth of water entering the soil in a given time, but infiltration capacity is defined as the maximum rate at which rainfall can be absorbed by the soil in a given condition. The last phrase is necessary because antecedent conditions will clearly play an important part in soil-water uptake. Permeability measurements are made when the soil is fully wetted and refer to water flow through it. Infiltration can be associated with very dry soils (i.e. non-Darcy). Its measurement is not especially easy and interpretation of results must be done with care because soil conditions can vary widely. A geomorphological viewpoint is given by Knapp (1978) and other discussions can be found in Gregory and Walling (1973) and Carson and Kirkby (1972). A useful discussion of field determination of permeability of unsaturated soil is given by Weeks (1978).

Two main types of measurement can be made, by flooding infiltrometers and by rainfall simulation infiltrometers. The first type are easier to employ but are more affected by soil variability. They treat the ground essentially as a one-dimensional component (Gill 1976) by allowing water, constrained by a tube or ring, to flood a small area of soil. Water uptake is measured by either lowering of a water head or observing the amount needed to keep a constant head. A plot of water uptake against time shows a levelling off as the infiltration capacity is reached. Slater (1957) gives a critique of the method and Musgrave and Holtan (1964) discuss the relation between this method and sprinkler (rainfall simulation) systems.

Sprinkler techniques are much to be preferred and, although the equipment required is much more elaborate than flooding techniques, some portable devices have been described (Selby 1970). Nozzles or screens provide rainfall whose intensity can be measured, and runoff from the required area is collected and measured. This gives an indirect measure of infiltration. A variety of methods have been used to provide the drops and collect runoff; a useful summary and comparison have been given by Parr and Bertrand (1960).

3.2.13 Specific surface

The specific surface of a particle is the surface area per unit mass. This property is of geomorphological importance, as the magnitude of an electric charge on a soil particle and its influence on the behaviour of that particle are directly related to the surface area per mass of the particle. Surface area may be determined by microscopy for individual particles, but normally surface area is determined for either a known mass or volume of powder. Several techniques of measuring surface area are available and the magnitude of surface area is dependent upon the method of measurement.

In surface area determinations by permeametry, a sample is initially graded by sieving and the surface of each grading is determined independently to obtain the specific surface of the sample. Two types of apparatus are available: the constant-pressure permeameter and constant-volume permeameter. Experimental results are governed by the Carman–Kozeny equation, which holds for low flow rates and grain size greater than 5 μm:

$$S_{w^2} = \frac{1}{K\eta\rho_s^2 u} \frac{e^3}{(1-e)^2} \frac{\Delta\rho}{L} \tag{3.18}$$

where S_w is the weight-specific surface of powder, ρ_s is the powder density, K is an aspect factor, η is the fluid viscosity, u is the average velocity, e is porosity, $\Delta\rho$ is pressure, and L is the thickness of bed.

In the constant-pressure permeameter, the powder is compressed to a known porosity in a permeability cell (cross-sectional area A). Air flows through the powder bed, recording a pressure drop (h_1) on the manometer, and then through a capillary flowmeter, for which another pressure drop (h_2) is recorded. The liquid in both manometers is the same (ρ') and pressure drop small compared to atmospheric pressure. Flow of air through the flowmeter is

$$Q = ch_2\rho'/\eta$$

and pressure drop across the bed is

$$p = h\rho' g$$

Substituting these into the Carman–Kozeny equation with the Carman K value of $K = 5$, then

$$S_w = \frac{14}{\rho_s(1-e)}\left(\frac{e^3 A h_1}{cLh_2}\right)^{1/2} \quad (3.19)$$

where c is a constant for the capillary. Incorporated into some devices is a chart recorder from which specific surface may be read directly.

In the constant-volume permeameter, air is passed through a bed of powder by a pressure difference caused by displacement of oil in two chambers, which are connected in U-tube fashion and to the permeability cell. Air is allowed to permeate through the sample until a predetermined decrease in pressure is attained. This technique is simpler, more robust but less versatile than the constant-pressure permeameter, which can include a provision for varying flowrate.

Specific surface may also be determined by gas diffusional flow. Gas is allowed to flow through a bed of powder, the flowrate being monitored by the changing inlet pressure, which can be recorded as a deflection θ on a chart. A semilogarithmic relationship exists between specific surface and pressure drop:

$$\frac{1}{\theta}\frac{d\theta}{dt} = \frac{8}{3}\frac{A}{lS_u}\left(\frac{e^2}{1-e}\right)\left(\frac{2Rt}{\pi M}\right)^{1/2} \quad (3.20)$$

where A is the cross-sectional area of powder bed, l is the length of powder bed, e is the porosity of powder bed, S_u is the volume-specific surface, R is the molar gas constant, T is absolute temperature, and M is the gas molecular weight.

Liquid permeameters perhaps provide the easiest technique of measuring specific surface but they cannot be used for particles less than 5 μm due to settling and segregation.

Gas adsorption is a widely used technique in the measurement of the specific surface of solids. The principle of this technique is the determination of the amount of physical adsorption of a gas by a solid. Specific surface area (S_w) is given by

$$S_w = N_A \sigma V_m / M_u \quad (3.21)$$

$N_A = 6.0225 \times 10^{23}$ mol^{-1} is Avogadro's number, σ is the area occupied by one molecule of gas, V_m is the monolayer capacity, and M_u is the molar volume. Results are plotted in terms of volume of gas adsorbed (v) as a function of equilibrium pressure; a plot of P/v against V_m yields the monolayer capacity.

Two gas adsorption techniques are available: volumetric methods and gravimetric methods. In volumetric methods, the pressure, volume and temperature of an adsorbate are measured prior to contact with the material and again after contact, once equilibrium has been established. The amount of gas is calculated on each occasion and the difference represents the adsorbate lost from the gas to adsorbed phase.

The basic apparatus consists of a gas (nitrogen) inlet together with a manometer to record pressure. After initial conditions are recorded, a stopcock between the gas inlet and sample is opened and a new pressure is recorded. Pressure difference is proportional to the volume of gas admitted to the sample bulb, and the volume actually adsorbed equals this volume less the dead-space volume, which can be calibrated with helium. The apparatus has an evacuating system to remove gases and vapours picked up on exposure to the atmosphere.

Gravimetric methods are less robust and require calibration and buoyancy corrections. These techniques do, however, have an advantage over volumetric methods in that the amount of gas adsorbed is observed directly by measurement of the increase in weight of the solid sample upon exposure to gas. Details of equipment and other methods are given in Allen (1981). Some methods of specific surface determination can be related to pore properties (e.g. Niesel 1973) and from Coulter counter data; a useful general book is by Lowell and Shields (1984).

3.2.14 Moisture content

Moisture content is the ratio of the weight of water to the weight of solid in a given volume; it is a dimensionless quantity but it is usually expressed in percentage terms.

There are various methods available for moisture content determination but the standard procedure is to determine the loss in weight when a sample is oven-dried to 105–110°C. An appropriate clean container (heat-resistant glass weighing bottles, cadmium-plated metal container or watchglass) is dried and weighed (W_c). A representative subsample of soil (5–10 g) is weighed in the container (W_w) and then with lid/stopper removed the container and contents are oven-dried to a constant weight. The minimum drying period is usually 16 h at 105–10°C. After drying, the container is removed from the oven, cooled in a desiccator, lid replaced and finally container and contents are reweighed (W_d). Moisture content is then determined (as a percentage) from

$$M = \frac{W_w - W_d}{W_d - W_c} \times 100 \qquad (3.22)$$

Rapid methods of moisture content determination utilise the same principle of loss in weight. Results obtained from microwave heating differ little from the conventional oven drying procedure (Ryley 1969) and, moreover, this technique has the advantage that, although water is quickly driven off, soil particles remain cool. Instruments are now available that provide heating and weighing devices in one unit.

Distillation methods are used for moisture content determination if the soil has been stabilised (BS 1924: 1975) but are unlikely to be required for most geomorphological purposes. A variety of methods of moisture content determination are described in Curtis and Trudgill (1974).

Moisture content, although an easy property to determine, is valuable in geomorphological analysis. Soil behaviour and strength vary with moisture content; also moisture content analysis forms the basis of Atterberg limits, which represent changes in soil state (Sec. 3.4.2).

3.2.15 Soil suction (pF)

Soil suction is defined as the negative pressure by which water is retained in the pores of a soil sample that is free from any external stresses. It appears that soil suction represents more than just a capillary phenomenon and osmotic, adsorptive and frictional forces are significant. Soil suction is measured as centimetres of hydraulic head but it is usual to employ $\log_{10}$ of head (or pF) as the scale of measurement. Suction may become very large as a soil dries out.

Two methods are available for measurement of soil suction: the suction plate technique, which is used when suction is less than atmospheric (pF 0–3), and the membrane apparatus, which is suitable for a range of suction pressures from 0.1 to 100 atm (pF 2–5, 0.01 to 10 MPa).

In the suction plate apparatus (Dumbleton & West 1968), the soil sample rests on a sintered glass suction plate, which allows passage of water but not air. The underside of the plate is in contact with a reservoir and water is either drawn to or from the sample depending upon whether the suction of the soil is greater or less than that applied to the plate. This causes movement of the water meniscus in the horizontal tube and suction of the sample is equivalent to that which is applied to the suction plate to prevent movement of the meniscus. Suction may be measured either by manometer or with a gauge. Use of ceramic porous plates has permitted this technique to measure pF up to 3.32. The sample is enclosed in a chamber and air pressure applied; in this form of apparatus, the suction of the soil equals the sum of suction applied to the flow tube together with air pressure in the chamber.

Similar principles apply to the membrane apparatus. The soil sample is enclosed in a cell of robust construction and is separated from a porous disc by a thin cellulose membrane. When the cell is bolted, compressed air is admitted, which causes water to be expelled from the soil through the membrane until equilibrium is reached and applied pressure balances the soil suction (Avery & Bascomb 1974, Curtis & Trudgill 1974).

Suction in many respects is a more fundamental property than moisture content in influencing soil strength. A direct relationship exists between suction and cohesion. The moisture content of a soil in moisture equilibrium with ground water and overburden stresses is determined by suction, and thus has practical significance in terms of stability of cuttings and slopes.

3.2.16 Bulk fabric and structure

The most complex relationships between particles in a sediment are those of packing or bulk fabric (Figs 3.1 & 3.2). Although it is occasionally useful (in paleocurrent analysis) to investigate the orientation of individual particles in space, it is more usual to obtain an estimate of the overall relationships of particles as a statistical value of some kind, and so a bulk method is used. Packing is closely allied to fabric, but it is a general term indicating the way in which particles fit together. As can be seen from Figure 3.2, packing is a function of size and grading, orientation and form. It affects the value of voids ratio, permeability and porosity, but is very difficult to measure in its own right. Most of the following will therefore be concerned with bulk fabric and investigation of internal structure rather than packing *per se*. An excellent review of orientation and isotropy in clays and rocks is given by Attewell and Farmer (1976, ch. 5).

3.2.16.1 Orientation of clasts

The orientation of clasts, sometimes referred to as macrofabric or stone orientation, has been used widely for paleocurrent analysis. The assumption is that a clast will be oriented in a direction that corresponds to an equilibrium condition of an applied force field (gravity, river or glacier flow, etc.) and that analysis of a number of clasts (usually more than 50) will enable the direction of an applied flow regime to be determined. With tills (i.e. till fabrics) an extensive literature has been developed, in both methods used to produce the data, and the means to test the results statistically. A brief résumé is given here but the reader should look at the work of Andrews (1970a) for greater detail. A choice must first be made as to whether three- or two-dimensional measurements need to be made. Clearly, there is more information to be gained from the former, although it presents greater statistical difficulties. In several sedimentological applications where the grains are of sand rather than pebble size, it is more usual to employ two-dimensional methods and this may be the only easy way if thin sections or resin-impregnated sands are to be examined; Bonham and Spotts (1971) give further information and methods. Usually, the long (*a*) axis gives the plunge (below the horizontal), and its direction or azimuth (typically, from true north) on a horizontal plane are determined. A rapid graphical and semiquantitative display can be obtained by plotting both values on an equal-area (Schmidt) projection of a hemisphere with its plane surface as horizontal. Patterns of orientation can easily be visualised and points contoured if necessary. An additional advantage is that structural planes (e.g. joints in till or other significant features) can be plotted on the same diagram (McGown & Derbyshire 1974). It should be noted that the plunge and azimuth of an *a* axis do not fix a particle uniquely in space and that another axis needs to be located. The pole (normal) to the *a–b*, maximum projection, plane is most suitable, as this plane is often diagnostic of processes (Johansson 1965). It is only recently that work has been done on relating processes to fabrics in till (Derbyshire 1978), partly because it is the matrix of clasts that plays an important part. Lilberiussen (1973) has examined till fabrics with X-ray radiography (see also Sec. 3.2.17). In fluvial systems, there are several comprehensive studies of stone fabric and dynamics (e.g. Johansson 1965) and directional properties such as permeability (Mast & Potter 1963), although two-dimensional fabrics were used in the latter investigation.

Fabrics of clasts can be used to help answer three distinct questions, therefore:

(a) identification of a flow direction;
(b) identification or elucidation of a geomorphological process; and
(c) explanation of the mass behaviour of a sedimentary body (e.g. permeability).

Although the measurement of clast orientation may give answers to questions in these categories, it does not necessarily give complete explanations, especially for (*a*) and (*b*). This is because the matrix within which the clasts become preferentially oriented is also affected by the applied shear stress. Indeed, it may be that in matrix-dominated and fine-grain sediments the small particles alone give a true representation of what the shear stresses were. This is akin to the type of analysis used by structural geologists. Thus, we can consider the total fabric (Derbyshire *et al.* 1976) as being made up from the clast fabric and the matrix fabric.

3.2.16.2 Matrix fabric

The definition of matrix will depend upon the size of material but can be conveniently considered as consisting of silts and clays. Indirect methods of investigation could be used, e.g. direction of development of an applied stain. Much more sophisticated methods have been tried to give bulk fabric or a measure of anisotropy, most notably by magnetic susceptibility (Rees 1965, Galehouse 1968). It is common nowadays to use high-magnification photography or the scanning electron miscroscope to obtain a picture, which is analysed subsequently. The difficulty is then to obtain a quantitative estimate of mean particle orientation (which may of course be in one or more directions) or to test against a random model. Two main methods are available, both requiring considerable patience and equipment. Further details can be found in Tovey and Wong (1978).

3.2.16.3 Electro-optical scanning techniques

The 'flying-spot scanner' of Zimmerle and Bonham (1962) was the first semi-automated method in which a line was scanned across a slide and a signal of 'grain'/'no-grain' was compared with other scans at a succession of 10° rotations of the slide; orientation data could be obtained from these signals. A similar, more versatile, but more expensive piece of equipment is the 'Quantimet' image analyser, which has been used for a variety of sizing, form and pattern recognition purposes. Soil structure has been an important area of application (although the limitations are due to the tedious making of sections and photography before analysis can begin). Excellent examples of soil structure analysis can be found in Bullock and Murphy (1976) and Foster and Evans (1974); a general description of the instrument is given in Bradbury (1977). Information can be found about pore structures as well as fabric or orientation of particles, and this depends largely upon the software available for the computing. See also Sections 3.2.7.2 and 3.2.18.

3.2.16.4 Optical methods

Although these methods have been applied to some geological features (McCauley *et al.* 1975) and to rock fabrics (Pincus 1966), much of the work involving these techniques has been in scanning electron microscopy (see below) and a good review of these methods is given by Tovey and Wong (1978). Essentially, optical diffraction patterns are produced from photographs of the soil section to be examined. A coherent beam of parallel light is used with a transparency of the required section as a diffraction grating, the image of which is recorded on the focal plane of a camera. Any preferred orientation of particles, which should be equi-sized (hence clays nearly fulfil this requirement), will show as a particular diffraction pattern, which contains information about orientation and spacing of particles and features in the micrograph. The patterns are complex and their interpretation is somewhat difficult and not easily amenable to mathematical analysis, which will regain most of the information. Fourier analysis can be used for simple patterns and as a means of characterising aspects of form (Kaye & Naylor 1972).

3.2.16.5 Packing

The foregoing discussion has been concerned with fabric or preferred orientation. It is possible to consider fabric as scale-dependent; for example, what shows a strong orientation of large clasts may show a much less defined, apparently random microfabric. Figure 3.2 suggests that packing is a very complex soil parameter. It can be analysed only with difficulty even on a simple level (Smalley 1964) and mathematical or experimental treatment will tend to be concerned with one or other associated or surrogate parameter. Structural studies may be interested in compressibility, permeability, strength, effective porosity, etc., some of which are dealt with elsewhere in this book. In this section, some general methods of structural or packing studies will be discussed.

Soil structure can be examined in two main ways: indirectly by a surrogate such as permeability or directly by microscopy; both may be important in some studies (Hancock 1978). For instance, permeability, porosity and specific surface methods described in this section may be used but, in addition to these standard tests, other useful methods have been tried. A good example is the radial permeameter used in soil (Basak 1972) or rock (Bernaix 1969). These tests provide useful three-dimensional characteristics of a soil mass and can be related to field measurements if required. Although it has not apparently been done with soils, pores in rocks can be examined in three dimensions by impregnating with a resin and then dissolving away the rock. Examination is usually with the scanning electron microscope and has been done successfully with sandstones (Pittman & Duschatko 1970) and chalks (Walker 1978). Such observations could be related to investigation of porosity by the mercury injection method (Diamond 1970), although an investigation of tropical soil structure by mercury injection has been linked to such techniques as X-ray diffraction, thermoanalysis, X-ray fluorescence and Atterberg limits (Tuncer *et al.* 1977), but these authors did not include microscopic examination.

Detailed scanning electron microscopy (SEM) analysis of clay structures has become important in explaining the structural behaviour of clays (and, to a much lesser extent, sands) (Lohnes & Demirel 1978). Much work still needs to be done to relate measured parameters from soil mechanical tests to such detailed observation. One of the major problems is concerned with making sure that what is seen by SEM is not an artefact but is truly representative of the structures developed under given stress conditions. As samples must be examined *in vacuo* in the SEM (although some prototype environmental chambers have been developed recently), drying must be done before examination. Several methods have been tried to minimise specimen shrinkage, and a discussion is given in Tovey and Wong (1978).

3.2.17 X-ray radiography

3.2.17.1 Introduction

X-ray radiography has been applied to the study of structures and inclusions in sediments since the introduction of the technique by Hamblin (1962). This method of testing has found wide use in the field of sedimentology because it provides a rapid, non-destructive and convenient method of recording textural variations, diagenetic products and, to a lesser extent, mineralogical differences in easily destroyed sediment samples. One of the most important aspects of the X-ray radiography method is its capacity to amplify contrasts in sediments that otherwise appear homogeneous, which serves to accentuate subtle sedimentary structures and geochemical inclusions. Studies of Recent sediments (e.g. Calvert & Veevers 1962, Bouma 1964, Coleman 1966, Howard 1969, Roberts *et al.* 1977, and others) have demonstrated the utility of this technique with both terrigenous and carbonate sediments.

The radiography procedure is based on the differential passage of X-ray radiation through a sample onto a film sheet. Variations in textural as well as chemical composition throughout the sediment result in differential attenuation of the incident X-ray radiation before it reaches the underlying film. Therefore, the spatially variable radiation arriving at the film plane creates differences in photographic density.

X-rays can be generated from a number of commercial units. For Recent sediments work at the Coastal Studies Institute (Louisiana), the Norelco (Philips) 150 kV industrial unit with a constant-potential beryllium window X-ray tube has been a durable and reliable system. However, once any adequate X-ray unit is available, this method is relatively inexpensive, non-destructive and time-efficient.

3.2.17.2 Procedure

Many factors affect the quality of radiographs and the exposure times required to produce suitable images. Chemical composition, grain size, packing density, sample thickness, range of densities, water content and other characteristics inherent in a particular sample cause differences in absorption and transmission of X-rays and thus result in differential exposure of the film. Film type, intensity and species of radiation, narrowness of the X-ray beam (focal spot size), source-to-film distance and thickness of the sample are some of the most important variables that control exposure time as well as quality of the resulting X-ray radiograph. Exposure time can be calculated from the following expression:

$$T = FAD^2/S \tag{3.23}$$

where T is the exposure time (min), F is the film factor (from tables), A is the absorber factor (from published curves), D is the source-to-film distance (inches), and S is source activity (mCi).

Although the absorber factor (A) can be found for many industrial products (Richardson 1968), the absorption characteristics of sediments are not so well known. Bouma (1969) provides the most complete source for exposure times versus thickness of various sediments as well as rocks. Roberts (1972) gives the same data for Recent carbonate sediments.

Experience has shown that particular combinations of parameters have proved more effective than others for given sediment types. For clay-rich sediments in the Mississippi River delta, a source-to-film distance of 380 mm, 28 kV, 20 mA, and a small focal spot provide maximum resolution for a minimum exposure time. Most exposures for core slices approximately 10 mm thick are less than 2 min. Quartz-rich sediments can be analysed at these same settings with somewhat longer exposure times. Because of the high absorption characteristics of calcite, aragonite and similar minerals, optimal X-ray radiographs of Recent carbonate sediments demand different power requirements on the X-ray tube. Values of 50 kV and 15 mA for 10 mm slices of carbonate material shot at a source-to-subject distance of 380 mm have proved a successful combination. With these settings, best results are obtained when lead-foil screens 254 μm thick are placed above and below the film sheet to absorb scattered radiation.

In the case of terrigenous sediments, as well as carbonates, it is essential to make radiographs of samples of uniform thickness for the best results. When working with a sediment core, an accepted procedure is to cut the core longitudinally with a wire knife. One core half is then placed in a slabbing frame and sliced to a uniform thickness. Slab thicknesses of approximately 10 mm produce high-resolution radiographs without compounding images.

3.2.17.3 Applications

X-ray radiography is most important in sedimentological studies for accentuating subtle inhomogeneities in otherwise massive sediments. Radiography offers a method of testing that is not only rapid but also non-destructive. Merits of this technique in the study of both primary and secondary sedimentary structures have been extensively discussed by Bouma (1969), Krinitzsky (1970) and others. By varying the intensity of radiation, exposure time and possibly a few other variables, high-quality radiographs can be obtained for most sediment types. In addition to the study of sedimentary structures as shown in Figure 3.6a, cements and many other geochemical products of early diagenesis are made clearly visible. Thus, radiography can be a valuable aid in sampling Recent sediments for geochemical studies.

Recently, this valuable technique has been used in geological hazard studies of Mississippi delta-front sediments. It was found that homogeneous-appearing prodelta clays, remoulded and deformed by mass movement processes, could be distinguished from equally homogeneous-appearing undisturbed shelf sediments (Fig. 3.6b). A recent application of this technique to hillslope hydrology is discussed by Anderson (1979).

3.2.18 Image analysing instruments

A very large range of image analysers are currently available, which enable rapid measurements and counts of many types of data including grains, soil blocks, photographs and other images, ciné film and even materials stored on disc and tape. The type

Figure 3.6 X-radiographs of (a) sedimentary structures revealed in a core cross section (Atchafalayn delta, Louisiana) and (b) an apparently homogeneous clay-rich core from the Mississippi delta front as part of a sub-aqueous landslide. Note the numerous gas expansion and migration structures and the remoulded appearance. The core dimensions are (a) 65×230 mm^2 and (b) 50×120 mm^2.

and cost of the image analyser chosen will determine the various functions and compatibilities of the system. Usually rapid and multiple quantification of parameters such as numerical count, shape, size, length, area, orientation, porosity, perimeter and pattern recognition can be achieved by most microcomputer units. Indeed, microcomputer developments in the 1980s have given rise to powerful video-camera-linked desk-top computers, which are capable of quite sophisticated and relatively inexpensive analyses (e.g. Whalley & Orford 1986). Section 3.2.7.2 gives some further information on these developments.

Perhaps the most sophisticated range of image analysers now available are made by Cambridge Instruments under the name Quantimet. The original Quantimet 720 has been succeeded by the Quantimet 520 and 970 systems. Both are still capable of image analysis from optical and electron microscopes, as well as from photographs, negatives, strip films and from the objects themselves. Images to be analysed are digitised and stored for further reproducibility and indeed for sample comparison as well. All measurement parameters can be undertaken and many statistical tests can be applied from the software incorporated in the machine.

Originally the Quantimet 720 required much operator adaptation for use in a number of different fields of research: remote sensing (Wignall 1977); pedology (Murphy *et al.* 1977); grain-size analysis (Peach & Perrie 1975, Goudie *et al.* 1979); image analysis (Bradbury 1977). The current generation of Quantimet users will need little, if any, operator adaptation to suit their various needs. While the cost of the systems is prohibitively high for individual academics to purchase, many machines will already be functioning in university natural or applied science laboratories. Hourly hire charges, if levied, are extremely moderate, particularly when offset against the speed of analysis and the ability of the system to negotiate most, if not all, measurement and image analysis tasks.

3.3
Chemical properties

Knowledge of the chemical properties of materials is important in several geomorphological contexts, water quality (see Sec. 4.3) being an obvious example, but it is often necessary to know the chemistry of weathering reactions or of sediments and soils. In this section there are two main areas: the analysis of solutions for specific ions and the determination of mineralogy. The latter is particularly important in giving a comprehensive explanation of soil behaviour, whether from a soil mechanical or pedological approach. Both areas of analysis require expensive equipment for detailed examination or where many samples are to be analysed. Although the instruments and methods are outlined here, the procedure for a particular analysis may vary widely and it is vital to be sure of the appropriate method. Suitable reading that covers these points is suggested in each section.

3.3.1 Wet chemical methods

Analysis of water directly or as extracts from soils is not always an easy task, especially when trace amounts of material are to be determined. This field is being extended rapidly in both theory and instrumentation, so that only an overview that discusses the main methods available can be given here. Reference should be made to the basic texts and standards in the appropriate fields such as Rand *et al.* (1980), Golterman and Clymo (1969) and Department of the Environment (1972) for water quality methods, and Black (1965), Hesse (1971) and Allen (1974b) for soils and vegetation. For geochemical analyses of this type, reference should be made to Wainerdi and Uken (1971), Jeffery (1975) and Blakemore *et al.* (1977).

3.3.1.1 Pretreatment of samples

There are certain rules that should be followed when bringing a batch of soils or water samples to the laboratory. The samples should be accompanied by an explanatory note showing relevant data and the analyses required. Transportation time should be a minimum, especially with water samples that undergo rapid changes in chemical composition. Tests like pH and redox potential should preferably be done in the field. Water samples can be carried in vacuum containers at a low temperature (with solid carbon dioxide and acetone). Again, filtration should be carried out as soon as possible using pore sizes between 0.2 and 0.4 μm; pH, alkalinity and dissolved gases should be determined on the unfiltered sample. Further discussion of these topics will be found in Section 4.3.

Soil samples are air-dried in a cabinet below 35°C. They are then removed and divided using a sample divider or riffle box and the homogeneous sample is placed in a grinding machine and passed through a 2 mm sieve (international standard).

3.3.2 Hydrogen ion activity (pH)

This is by far the most commonly made test on soil and water samples and is defined as the negative logarithm of the hydrogen ion concentration. Measurement of pH gives not only the degree of acidity or alkalinity but also element mobility, phosphorus availability and microbial activity. There are two main methods: (a) colorimetric and (b) electrometric. The latter is by far the most commonly used as it is the most accurate and reliable. A glass electrode is usually employed nowadays, replacing an indicator electrode, and this makes the method more flexible for field use.

Measurements are carried out directly on water samples, but for soils a slurry of the sample in water is made by placing 0.01 kg of soil in a suitable beaker and adding 25 ml of distilled water. This is stirred, left overnight and then measured by an electrode and pH meter. The meter must first be calibrated using suitable buffer solution before readings can be taken.

This electrometric method is fast, inexpensive, reliable and the equipment is robust to allow work in the field with only one operator required. More detailed information can be found in Wilson (1970a), Hesse (1971) and Langmuir (1971), as well as manufacturers' literature.

3.3.3 Conductivity

Conductivity is the reciprocal of the resistance of a solution with platinum electrodes, 10 mm square and 10 mm apart, immersed in it. Conductivity is denoted by the symbol K and is given by

$$K = C/R$$

where C is the cell constant and R is the resistance in ohms. It is expressed in microsiemens per millimetre ($\mu S\ mm^{-1}$) in SI units or in reciprocal ohms (ohm^{-1} or mho).

This method is most suited to water samples, as these can be measured directly. At near-neutrality, the effect of different ions at the same equivalent concentration is relatively small, and this gives a useful indication of the ionic strength of the water samples. With soil samples, the test is more laborious, as the sample has to be made into a paste. Soil extracts can also be used, and conductivity can be measured at a soil : solution ratio of 1 : 5.

The principle of all instruments is that of a Wheatstone bridge. The resistance is measured when the cell containing the platinum electrodes is dipped into the sample and current is passed through the circuit. The point of balance is indicated by a zeroed needle or when a 'magic eye' shows a maximum opening. Modern instruments have automatic temperature compensation: this removes the time factor involved in making corrections.

This method, like pH, shows its advantages in geomorphological research in its speed, cheapness, reliability and robustness for fieldwork, with only one operator required. It is possible to record conductivity continuously on a suitable recorder.

3.3.4 Ion-selective electrodes

Analysis by ion-selective electrodes (ISE) is relatively new in analytical chemistry and is probably one of the greatest aids to geomorphological research, especially as it enables work to be done in the field. High-quality instruments and electrodes are vital for the method to work effectively, however.

The principle of this technique is based on the fact that the ionic activity of a solution is related to its concentration. The electrode consists of a membrane that separates two solutions containing the same ion in different concentrations. The membrane is sensitive to this ion and charges are transferred through the membrane with little displacement of the individual ions. The membrane potential and the ionic activity are logarithmically related by

$$E = 2.3\frac{Rt}{nF}\log A \quad (3.24)$$

where E is the net membrane potential, A is the ionic activity, n is the valency and the remainder is the Nernst factor.

3.3.4.1 Instrumentation

Instrumentation is essentially simple and consists of a membrane electrode, reference electrode and a millivoltmeter. Electrodes are relatively expensive and, if given normal laboratory use, will last about two years. There are different types of electrode available, e.g. flowthrough, liquid membrane, gas-sensing, solid-state and combination.

The main difficulty that arises in the ion-selective electrode method is interference effects, e.g. formation of precipitates with ions in the membrane and electrode responses to ions other than those being measured. Such effects and equipment available are discussed further in Edwards *et al.* (1975), which presents geomorphological applications and gives a useful bibliography, and in Bailey (1976), Comer (1979) and Orion Research (1978).

3.3.4.2 Applications

These electrodes may be used to detect the end-points in complexometric and precipitation titrations. Titration methods have led to an increase in the number of measurements and these consist of three basic types:

(a) S-titrations, where the sample (S) species is sensed by the electrode:
(b) R-titrations, where the electrode senses the level of reagent (R) species that has been added to the sample before titration; and
(c) T-titrations, where the titrant (T) species is sensed by the electrode.

More detailed discussion of the use of ISE is given in Midgley and Torrance (1978). Table 3.4 gives a brief idea of the ions that can be measured by this technique which are likely to be of use to geomorphologists. Although ISEs appear to be a cure for many field analyses of ions, they are expensive and may be less reliable in the field than the usual laboratory methods.

3.3.5 Colorimetric and spectrophotometric analysis

The principle of colorimetry and spectrophotometry is based on the absorption of electromagnetic radiation by chemical substances and its measurement. The visible region extends from approximately 380 to 750 nm, and deals with the absorption of light by solutions that are visible to the eye. Ultraviolet (UV) and infrared (IR) spectrophotometry are the same as for the visible, but extended beyond the visible spectrum. Such extended-range instruments are more expensive than for visible light only, but they are sometimes desirable for some elements.

Table 3.4 Summary table of major anions and cations that can be measured with ion-selective electrodes.

Species	*Sample*	*Preparation*	*Method*
Ammonia	Water	Filter, add NaOH	Direct
	Sea water	Spike sample with NH_4Cl, add concentrated NaOH	Known addition
	Sewage	Dilute sample 1 : 10, add concentrated NaOH	Known addition
	Soils	Disperse air-dried < 2 mm soils in 2 M KCl, filter, add NaOH	Direct
Calcium	Water	Add pH 10 buffer to sample	Titration with EDTA*
	Soils	Extract Ca^{2+} with pH 8.2 sodium acetate, centrifuge and dilute supernatant	Direct
Fluoride	Water	Separate system available for fluoride measurements in natural and drinking water	Direct
	Sea water	Dilute with acetic acid/acetate buffer, calibrate electrodes in artificial sea water	Direct
Nitrate	Water (natural)	If low on chloride, no pretreatment	Direct
	Sewage	Dilute sample with silver solution to eliminate chloride interference	Direct
	Sea water	Reduce nitrate to ammonia	Direct as ammonia
	Soils	Disperse air-dried in water, filter through non-acid-washed paper	Direct
Potassium	Soils	Extract with sodium acetate solution	Direct
Sodium	Sea water	Calibrate electrodes with artificial sea water; no preparation necessary for known addition	Direct or known addition
	Soils	Extract with water or $CaCl_2$	Direct
Magnesium	Sea water	Calibrate electrodes with artificial sea water	Direct
	Soils	Add dry ground sample to water	Direct

*Ethylene diamine tetra-acetic acid.

3.3.5.1 Principles

When monochromatic light passes through a solution, it is absorbed, and the degree of absorption depends on the number of molecules or ions in its path. This is expressed as the fraction of light transmitted and is logarithmically related to the thickness and concentration of the solution. This forms the basis of the Beer–Lambert law:

$$I_t/I_o = e^{-kcl} \tag{3.25}$$

where I_0 is the intensity of the incident light I_t is the intensity of the transmitted light, k is a constant, c is the concentration, and l is the solution thickness (path length). This is obeyed for a range of concentrations, and a graph of optical density versus concentration will be linear. The UV region, i.e. below 190 nm, is not used a great deal since atmospheric gases absorb UV light and they must be replaced by a non-absorbing gas.

3.3.5.2 Instrumentation

There are four main parts connected with instrumentation (Fig.3.7):

(a) source of radiation,
(b) monochromator,
(c) sample compartment, and
(d) detection system.

The source is either a tungsten lamp for the visible region or a deuterium, hydrogen or mercury-vapour discharge lamp for the ultraviolet. In modern instruments, these are mounted in the same compartment. The required wavelength can be selected by a prism

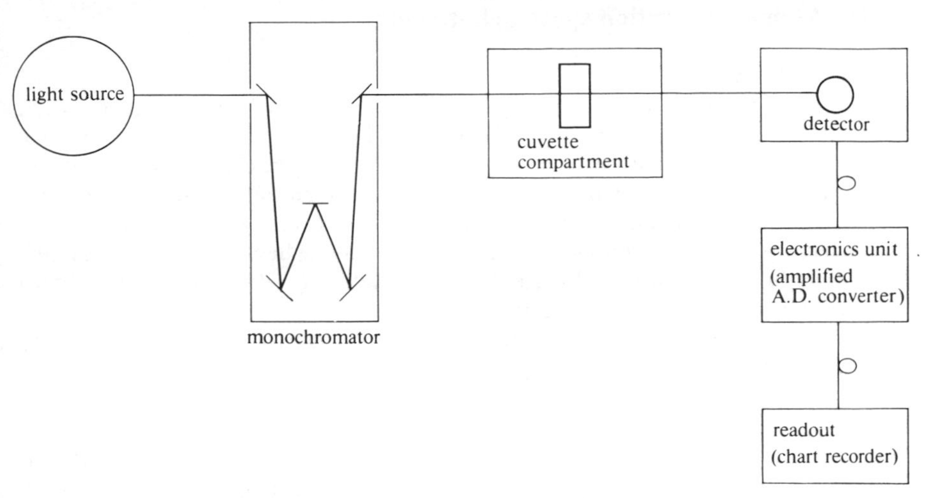

Figure 3.7 Schematic diagram of the main elements of a colorimeter.

or a grating, which when rotated can focus any part of the spectrum on the exit slit of the monochromator. Cells are constructed from glass or quartz and are used according to the wavelength of the transmitted light. The detector may either be a phototube or a photomultiplier, the latter being more sensitive.

Modern instruments have direct concentration readout facilities, which speed up this method considerably, allowing many routine analyses to be carried out each day. They may also have facilities for complete automation. There are a few rules that must be followed when dealing with a reaction suitable for this analysis. The reaction must be simple and free from interferences, have a stable colour, be insensitive to pH and temperature changes, not vary with small changes in wavelength of the incident light, and obey the Beer–Lambert law over a wide concentration range.

Colorimetry is a very suitable method in geomorphological analysis, since it can deal with a large number of samples in a short period of time with the aid of only one operator and at a relatively low cost. The method has become even more widely used because of the progress of automation or 'continuous flow analysis'. Again, there is a minumum of operators used in this method, and junior laboratory staff are capable of setting up and working this type of system, but close supervision by experienced staff should be given in case of malfunction. This shows, therefore, that colorimetric analysis has grown into a fast, economic and reliable method.

When dealing with water samples, this method has the advantage that direct aliquots can be taken for colour development, but, with soils, extracts have to be made. This can be carried out by either extraction or digestion of the soil sample.

There is an extensive literature on colorimetric methods in analytical chemistry, and some texts are available (e.g. Charlot 1964). Parker and Boltz (1971) provide a simple introduction to colorimetric methods in geochemistry and Tuddenham and Stephens (1971) give an introduction to infrared spectroscopy, a related technique.

3.3.6 Atomic absorption spectrophotometry

3.3.6.1 Principles

When a solution containing a metallic species is aspirated into a flame, e.g. air–acetylene system, an atomic vapour of the metal is formed and some of the metal atoms become excited and are raised to a higher energy level where they emit the characteristic radiation of that metal. The greater number of atoms left in the ground state are receptive to light of their own specific resonance wavelength, and if such light is passed through the flame it will be absorbed. The absorption will be proportional to the density of the atoms in the flame. A useful introductory text is Reynolds and Aldous (1970), while most standard analytical books discuss the method in varying degrees of detail, and these should be consulted for specific analyses. Butler and Kokot (1971) discuss its use in geochemistry (see also Sec. 4.3.5).

3.3.6.2 Instrumentation

Atomic absorption instruments are spectrophotometers with a burner compartment rather than a cell (Fig. 3.8). The sources are hollow-cathode lamps that emit radiation corresponding to the emission spectrum of the element to be analysed. The required line is then isolated by the monochromator. The sample enters the flame as a spray, and this is done in the nebuliser system of the burner. The larger droplets are removed by baffles and run to drain. The detector system is similar to those in colorimetry. Today's instruments are usually double-beam, whereby a rotating mirror splits the beam so that the radiation alternately passes through and passes by the flame. This modulates the flame and the detector circuit can adjust for changes in source intensity.

3.3.6.3 Technique

To obtain good results from this method, certain rules should be followed. The lamps must be run at the proper currents, the correct nebuliser–flame system should be used, solutions to be aspirated must be free from solid matter and regular flushing with solvent and checking of the zero standard should be carried out at the end of each run. The flame conditions, gas flowrates and burner alignment are very important factors towards accurate results.

3.3.6.4 Application

When dealing with soil analysis, there are two groups: (a) total elemental analysis and (b) cation-exchange analysis.

In total elemental analysis, the soil sample is digested in a strong acid mixture, e.g. $HClO_4$–HNO_3–HF, to convert the silicon to the tetrafluoride form. The residue is dissolved in hydrochloric acid, and, if the calcium response by aluminium is to be suppressed, a complexing agent (e.g. lanthanum chloride) is added and the solution is made up to a suitable volume. The sample and standard solution are aspirated into the flame for analysis.

In exchangeable cations, the soil is leached by a suitable solution, e.g. 1 M ammonium acetate. This method releases the exchangeable ions. e.g. Ca, Mg, Na, K. After leaching, the solution is made up to a suitable volume with the leachate and aspirated directly into the flame with standards that have been prepared in the leaching solution.

In the case of water samples, much less time is consumed since these can be aspirated directly into the flame after filtration. Lanthanum chloride again can be added to suppress the depressions caused by phosphate and aluminium on the alkaline earths and by iron on the absorption of chromium.

When the element is present in a very low concentration in a sample, then an extraction method is used and this releases the metal into an organic solvent. The most useful method employs chelation with ammonium pyrrolidine dithiocarbonate (APDC) followed by extraction into methyl isobutyl ketone (MIBK).

Over recent years, there have been extensive developments in this field and the most important include the flameless atomisation system and electrodeless discharge lamps. In the flameless system, the entire sample is consumed by ignition to very high temperatures on a carbon furnace and this produces greater detection limits.

The main disadvantage is lack of reproducibility and stability, which is partly due to the background attenuation or absorption caused by certain samples absorbing or scattering radiation from the source. This effect occurs since the volatilised concomitants may exist as gaseous molecular species, salt particles, smoke, etc.

Two methods of correction are available, a continuum source and a Zeeman-effect corrector. The former type uses radiation from a primary source lamp and from a continuum source lamp, passing each alternately through the atomiser. The element of interest absorbs only radiation from the primary source, while background absorption affects both beams equally. When the ratio of both beams is taken electronically, the effect of background is eliminated. The Zeeman-effect corrector uses a magnetic field to split the atomic spectral line into three or more polarised components. Since only atoms and non-solid particles and most molecules that cause non-specific absorption show a line splitting, the Zeeman effect is ideally suited for background correction in atomic absorption spectrometry. Background up to more than 2 Å can be compensated for by this technique.

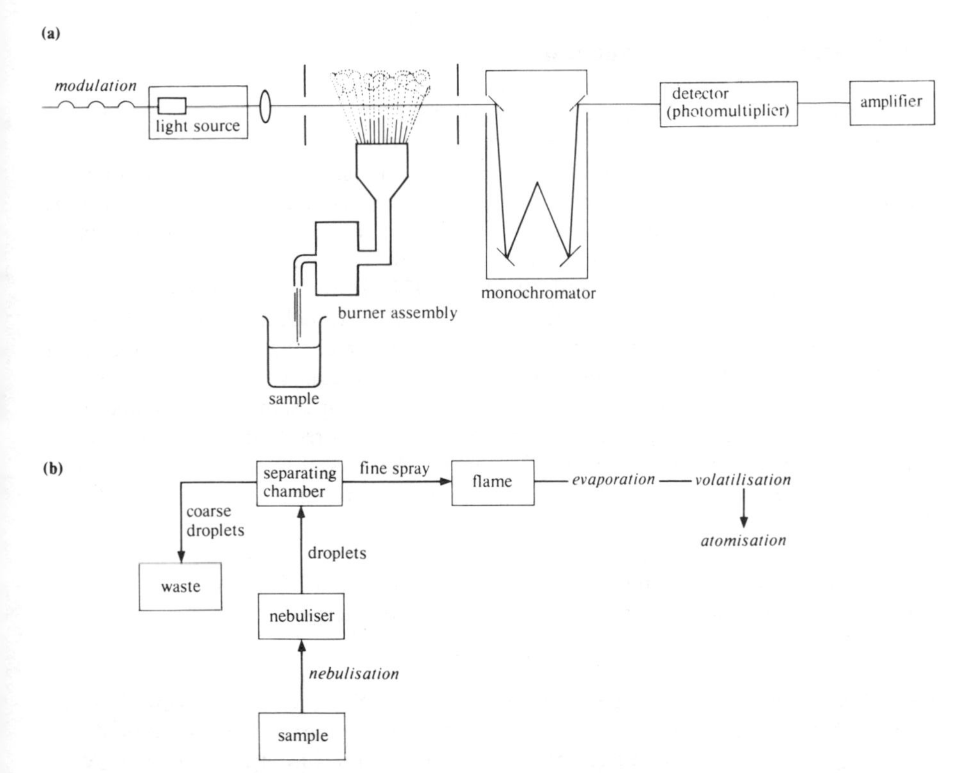

Figure 3.8 Schematic diagram of an atomic absorption spectrophotometer.

Today's atomic absorption spectrophotometers are relatively costly, but this is balanced out by their time saving, reliability and accuracy. Again, one operator is all that is required and this method can be automated to deal with large numbers of samples. By comparison with colorimetric analysis, the main feature, as mentioned above, is its higher degree of accuracy and reproducibility from the flame system. Colorimetric methods are liable to positive error due to colour change and mechanical loss. Atomic absorption preparations are much simpler and less liable to error, and if errors do arise they are easily spotted by the analyst.

3.3.6.5 Inductively coupled plasma spectrometry

A new approach to atomic spectrometry has been brought about by the use of inductively coupled plasma (ICP) emission spectrometry. Today's instruments provide the quality of data required for the vast majority of applications. The advantages of this technique over other emission methods is its low background signals, more precise sample introduction, absence of chemical interferences and better detection limits, allowing for a more comprehensive analysis. All the elements when present in aqueous solution can be measured simultaneously in a matter of minutes, making this the most rapid of all analytical methods.

The ICP is a stream of argon atoms heated inductively by a radiofrequency (RF) coil. Inductive heating of the argon as it flows through the RF field strips electrons from the argon atoms and produces a plasma of ions, which can operate at a temperature of 6000–10 000 K. The cost is below that of X-ray fluoresence systems and the cost per analysis should be very competitive.

3.3.7 Mineralogy: introduction

The determination of mineralogy, especially clay mineralogy, is of increasing importance in geomorphological explanations. Examples range from soil and slope behaviour, to duricrusts, rock weathering and erosion rates. Modern methods used for these analyses are complex, and several of them rely upon electron beams and radiation from various parts of the electromagnetic spectrum, X-ray and infrared wavelengths especially. Excellent texts discussing several of these methods as well as other petrographic techniques are Hutchinson (1974) and Zussman (1977). Infrared spectroscopy is similar in experimental equipment to visible-light spectroscopy, although the graph resulting has peaks of absorbance (or transmittance) that correspond to chemical groups rather than concentrations of elements. Fingerprinting techniques can be applied to identify minerals, and their use in clay mineralogy is described in, for example, Tuddenham and Stephens (1971). The uses of X-radiation are essentially three-fold. The first is X-ray radiography, as described in Section 3.2.7, and in Arthurs (1972) for soil mechanical purposes. The second is X-ray diffraction, to be outlined in Section 3.3.8; and the third is X-ray fluorescence spectroscopy, described in Section 3.3.10. To aid understanding of the last two effects, a brief résumé of the characteristics of X-radiation is necessary. X-rays lie between 0.01 and 10 nm (visible light 400–700 nm) on the wavelength spectrum of electromagnetic radiation. Associated with this is an energy E given by

$$E = h\nu \tag{3.26}$$

where h is Planck's constant and ν is the frequency. Since $\nu = \lambda c$, where λ is the wavelength and c is the velocity of light, the energy equivalent of an X-ray photon of λ = 0.1 nm is 2×10^{-15} J. It is conventional to express energies in electronvolts (1 eV = 1.6×10^{-19} J), because the X-rays are emitted when high-energy electrons (produced by a potential in the region of 20–100 kV commonly) or X-rays strike a metal target. For electrons accelerated by 20–50 kV, the emitted X-rays produce a continuous spectrum of energies, but above about 70 kV, characteristic lines are seen in the spectrum (these are similar to, but far fewer than, the lines seen in visible-light emission spectroscopy). These discrete monochromatic beams have characteristic energies (wavelengths) and varying intensities, and from them the elements can be identified. The groups of lines (denoted by the letters K, L, M and N) are due to rearrangement of electrons in the atomic shells in the excitation process. The characteristic radiation lines are detected in the energy-dispersive X-ray (EDX) analyser, often fitted to scanning electron microscopes (see below), and the electron probe microanalyser (EPMA), which is a common geological tool for quantitative mineralogy.

If a beam of monochromatic X-rays falls onto the regular and periodic array of atoms – a crystal lattice – then diffraction will occur in certain directions. Bragg's law

$$n\lambda = 2d \sin \theta \qquad (3.27)$$

governs the relationship between the wavelength λ used, its incidence angle θ and the interplanar spacing d (repeat distance) of n, which is the (whole-number) order of reflection. This is the basis of both X-ray and electron diffraction and is also used in X-ray fluorescence spectroscopy.

3.3.8 X-ray diffraction of clay minerals

3.3.8.1 Introduction

X-ray diffraction (XRD) of clay minerals is now widely accepted among geomorphologists as an extremely useful method for the study of landform materials, especially in the case of research on weathering processes and mass wasting. There are many published papers using this technique (e.g. Keller 1957, Young & Warkentin 1966, Yatsu 1966, 1967, Gillot 1970, Prior & Ho 1971, Prior 1977, Bentley & Smalley 1978), and these have stressed the importance of the mineralogical and chemical properties of soils to understand their behaviour.

It is inevitable that in this limited space only a guide to these techniques can be given and reference should be made to the existing literature (e.g. Brown 1961, Whittig 1965, Carroll 1970, Dixon & Weed 1977, Jenkins & de Vries 1970a). Those entirely new to X-ray diffraction may profitably consult some textbooks dealing specifically with basic theories in this field (e.g. Cullity 1956, Andrews *et al.* 1971).

Although SI units are generally used in this book, the ångström (1 Å = 10 nm), not an SI unit, is used in this account because of its long-standing use in X-ray methods and in the texts referred to.

3.3.8.2 Clay and clay minerals

The definitions of clay are slightly different in various sciences, e.g. ceramics, agriculture and civil engineering. At present, in the field of clay mineralogy, clay is defined as (a) aggregates of particles finer than 2 μm, (b) adhesive substances and (c) natural inorganic materials mainly composed of silicon, aluminium, iron, magnesium, alkali-metal elements, alkaline earth elements and water.

The constituents of clays may be called clay minerals. Quartz and feldspars usually observed in clay material, however, are not named clay minerals, even though they may be very fine particles. A wide deviation exists among the same species of clay minerals and this is especially the case with phyllosilicates (Fig 3.9). Since the mineralogically identical species of clay minerals are formed by different processes, such as hydrothermal

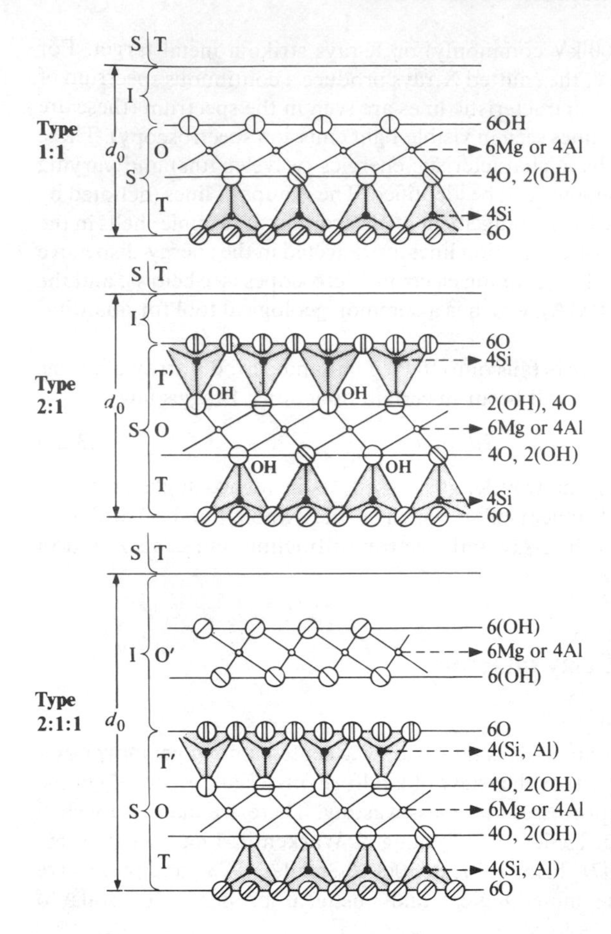

Figure 3.9 Types of phyllosilicates. T, tetrahedral layer; O, octahedral layer; S, unit layer (sheet unit); I, internal layer; O, brucite or gibbsite layer.

reaction, diagenesis and weathering, they usually retain slight differences in their chemical compositions and structures due to these genetic processes. This fact, in turn, carries the significant implication that clay minerals are formed by alterations not completed under low-temperature and low-pressure conditions.

Clay minerals include phyllosilicates composed of sheet structures, fibrous palygorskite and sepiolite, as well as some amorphous substances, e.g. allophane. However, amorphous silica, iron oxide and alumina gels are not called clay minerals. The definition of clay minerals is somewhat ambiguous here.

3.3.8.3 Identification of clay minerals by X-ray diffraction

X-ray diffraction analysis involves the measurement of physical values concerning crystal structures. It is impossible to evaluate exactly the value of interlayer electric charge or to determine whether a clay mineral is dioctahedral or trioctahedral. The intensity of X-ray diffraction and the angle at which diffraction occurs are, however, to some degree related to chemical composition. Therefore, proper use of diffraction methods and appropriate interpretation of the results make it possible to identify the species of clay minerals.

Preparation of specimens For X-ray diffraction analysis, diffractometer or camera methods may be used. At present, the former are common and the specimens are usually powders of clay particle size (e.g. according to the International Soil Science Society method, finer than 2 μm), which may be collected by elutriation (separation by washing). In this procedure, application of dispersing chemicals is not recommended, because clay minerals adsorb the ions derived from these chemicals, and often alter themselves. Clay particles thus obtained will be used for oriented or unoriented specimens.

Oriented specimens Principally, clay minerals are phyllosilicates and have tabular forms. If the clay particles are placed in parallel, i.e. the (001) surfaces of each particle are parallel to each other and to the specimen holder, the specimen is called parallel, 'preferred' oriented, or oriented for short. This is sufficient to measure the (001) spacing, which is the key to the identification of clay minerals.

To prepare oriented specimens, about 30 mg of clay powder and 5–10 cm^3 of water must be mixed in a test tube. Then 1 cm^3 of the dispersion liquid may be taken by a pipette, dropped onto a slide glass and spread over an area of about 1×2 cm^2. It is then allowed to dry in air at room temperature. Another method is as follows. Clay powder and water are mixed in a mortar to form a paste. A quantity of the paste is placed between two coverglasses, which are slid over each other; the film of oriented clay particles adheres on the surfaces of both coverglasses. The coverglasses, with an oriented clay particle film, are dried and used as specimen holders. The measurement of (001) spacing, with some special treatment if necessary, enables classification of the type and group to be made (Table 3.5).

Unoriented specimens In unoriented or randomly oriented specimens, clay particles are not placed in parallel, i.e. the directions of the (001) surfaces are random. As the shape of phyllosilicate minerals is tabular, it is rather difficult to prepare completely unoriented specimens for these minerals. If clay particles mixed with some powdered amorphous material, e.g. silica gel or cork, are packed in a specimen holder of aluminium, then acceptable unoriented specimens may be obtained.

In order to determine subgroups, such as dioctahedral and trioctahedral, or to distinguish the polytype of species, unoriented specimens are indispensable.

Measurement and identification The *d* spacing (molecular plane repeat distance) of each peak on the diffractogram must first be obtained by using 2θ–d tables (which are usually compiled in the manual for the X-ray diffractometer) and it is convenient for identification of clay minerals to insert the *d* spacing value at each diffraction peak on the diffractogram.

The type and group in Table 3.5 are classified by the (001) spacing, i.e. the spacing of the crystal lattice in the direction normal to the surface of the sheet structure. Using an oriented specimen, one must measure the (001) spacing, hereafter expressed as $d(001)$. If the values $d(001)$, $2 \times d(002)$, $3 \times d(003)$, ... are identical, $d(001)$ is the spacing of the sheet lattice. If not, specimens are randomly interstratified minerals or a mixture of two or more kinds of clay minerals. In the case of fibrous palygorskite and sepiolite, the effect of orientation does not appear; this fact is a useful criterion of the detection of these minerals.

In some cases, the identification of groups needs the pretreatment of samples with substances such as ethylene glycol or heating: the changes in the diffraction pattern, i.e. (001) spacing and intensity before and after treatment, provide the diagnostic key to the distinction of groups (see Table 3.5). The recognition of subgroups, i.e. dioctahedral and trioctahedral, is usually based on the diffraction of $d(060)$, using the diffractogram of unoriented specimens. The $d(060)$ of dioctahedral clay minerals is usually 1.49–1.51 Å, while that of trioctahedral subgroups is 1.51–1.53 Å. In the case of vermiculite, the

Table 3.5 Diagnostic characteristics of clay mineral species in their identification by X-ray diffraction.*

Type	d(0 0 1) (Å)	Ethylene glycol treatment	Heat treatment	Group	d(0 6 0) (Å)	Dioctahedral	d(0 6 0) (Å)	Trioctahedral
1 : 1 $Si_2O_3(OH)_4$	7 (*Note 1*)	(*Note 1*)	(*Note 1*)	Kaolinite–serpentine	1.49	Kaolinite†	1.53	Chrysolite, antigorite
						Halloysite	1.60	Greenalite
2 : 1 $Si_4O_{10}(OH)_2$	9.3	No expansion	No change	Pyrophyllite–talc	1.49	Pyrophyllite	1.53	Talc
	10	(*Note 2*)	(*Note 2*)	Mica	1.49–1.50	Muscovite‡	1.53	Biotite‡
						Paragonite (sodium-mica)‡		Phlogopite‡
						Margarite‡		
					1.51	Glauconite (*Note 2*)		Trioctahedral illite
2 : 1 : 1 $Si_4O_{10}(OH)_8$	14.2	(*Note 3*)	(*Note 3*)	Chlorite	1.50	Donbassite	1.53–1.56	Penninite
						Sudoite		Ripidolite
								Chamosite
2 : 1 $Si_4O_{10}(OH)_2$	14.3–14.8	No expansion	(0 0 1) becomes 10 Å	Vermiculite	1.50	Vermiculite	1.53	Vermiculite
	15	(0 0 1) expands to 17 Å	(0 0 1) becomes 9.5 Å	Smectite	1.49–1.50	Montmorillonite	1.52–1.53	Saponite
						Beidelite		Hectorite
					1.52	Nontronite		Stevensite
Interstratified minerals §	24.5–25.5	(0 0 1)– expands to 26–27 Å	(0 0 1) becomes 19.5 Å	Regular interstratified minerals	1.50	Rectorite (Allevardite)	1.54	Biotite–vermiculite regular interstratified minerals § mineral
	29–30	(0 0 1)) expands to 30–32 Å	(0 0 1) becomes 23.5–24 Å	Regular interstratified minerals	1.49–1.50	Alushtite	1.52–1.53	Corrensite
						Tosudite		
	10–15	Changes to be noticed	Changes to be noticed	Irregular interstratified minerals	1.49–1.51	Illite–montmorillonite irregular interstratified minerals	1.50–2.54	Chlorite–montmorillonite irregular interstratified minerals
		(*note 4*)	(*note 4*)			Kaolinite–montmorillonite irregular interstratified mineral		Biotite–vermiculite irregular interstratified mineral

*The table is arranged according to (0 0 1) spacing.
†So-called 'kaolinite-D' is included among the species of kaolinite: it is not certain whether kaolinite-D is acknowledged as one species.
‡The species name of mica mineral is that used in the classification of phyllosilicate minerals, and it is open to question whether or not the name must be retained in the classification of clay minerals, especially because mica clay minerals have less than one cation per two unit cells.

The name of trioctahedral illite is not listed in the classification of phyllosilicates.

Although margarite is dioctahedral brittle mica ($x \simeq 2$) in the classification of phyllosilicates, it is so listed in the table because it has the same nature as other mica clay minerals with respect to X-ray diffraction.

§Interstratified minerals are defined as those which have interstratified structures. There are two different types: regular and irregular. In the former case, the stacking along the *c* axis is a regular repetition of different layers, and the unit cell is equivalent to the sum of the component layers. In the latter, the interstratified structure is a randomly irregular interstratification of different layers. Some explanations may be given as to the species of interstratified minerals:

Rectorite synonymous with allevardite, mica–montmorillonite regular interstratified mineral.
Tosudite di-,trioctahedral chlorite–montmorillonite regular interstratified mineral.
Alushtite dioctahedral chlorite–montmorillonite regular interstratified mineral.
Corrensite trioctahedral chloric–trioctahedral smectite regular interstratified mineral.

Note 1. The *d*(0 0 1) of hydrated halloysite is 10 Å, and it becomes 11 Å on treatment with ethylene glycol and 7 Å on heating at 200°C for 1 h.
Note 2. Some sericites and illites with *d*(0 0 1) of 10–10.2 Å change the (0 0 1) spacing on ethylene glycol and heat treatments. Most glauconites have a similar property.
Note 3. The *d*(0 0 1) of swelling chlorite becomes 15.5–17 Å on treatment with ethylene glycol, and *d*(0 0 1) of unusual chlorites (Kimbara *et al*. 1973) moves, on heating, in a different way from usual chlorite: it beomes 12 Å on heating at 700°C for 1 h.
Note 4. In the case of illite–montmorillonite irregular interstratified minerals, *d*(0 0 1) splits into two peaks on treatment with ethylene glycol, and becomes 10 Å on heating at 300°C for 1 h.

following chemical treatments are applicable. By mixing vermiculite with 1 N solution of NH_4NO_3 in a test tube and heating it in boiling water for 30 min, the diffraction peak of trioctahedral vermiculite at 14.3–14.8 Å is displaced to 10 Å. Dioctahedral vermiculite does not show the displacement by the above treament, but it does by a treatment of boiling with KOH + KCl solution for 5 h (Brown 1961). If all cation positions in the octahedral layer are occupied by divalent ions, e.g. Mg, Fe(II), Mn. etc., these minerals are called trioctahedral, and if two-thirds of them are occupied by trivalent cations, Al and Fe(III), the minerals are named dioctahedral. Therefore, the distinction is essentially chemical.

The determination of species is carried out by means of unoriented specimens. The measurement of $d(hkl)$ diffraction affords the means of distinguishing the species in the case of kaolinite and serpentine minerals where their polytype is the basis of classification. The different ways of stacking the unit cell of crystals are referred to as the polytype. Figure 3.10 presents the polytype of mica. As the classification of species of mica minerals is based on the cations in the tetragonal and octahedral sites, it is usually difficult to identify their species by X-ray diffraction. This is the case with chlorites and smectites, in which the species are distinguished mainly by chemical analyses. The determination of polytype of a species, e.g. 1M, $2M_1$, $2M_2$ ... of muscovite, may be conducted by X-ray diffraction.

In general, the identification of clay minerals is made by comparing the *d* spacings and intensities of diffraction with those of minerals that are already known. These data have been assembled by the American Society for Testing and Materials (ASTM) as a 'powder file'. Identification of minerals is not always straightforward and some experience is required. The books by Andrews *et al.* (1971) Brindley and Brown (1980), Carroll (1970) and Griffin (1971) provide a good grounding.

3.3.8.4 Quantitative estimation of clay minerals in mixtures

To achieve this aim, intensities or peak areas of X-ray diffraction must be measured (see Fig. 3.11). Direct comparison of these values for the individual minerals in mixtures does not give even a rough estimate of their quantities, because of the differences in mass absorption coefficients, crystallinities, particle sizes, and so on, of individual minerals. The precise evolution of quantities of individual minerals in a mixture is really impossible; therefore, their estimation must be at best semiquantitative. Even so, it can still be useful for many purposes.

If the mass absorption coefficient of one kind of clay mineral whose content is to be estimated is different from that of the other portion of the mixture, the content of the clay mineral in question is not in proportion to the intensity of its diffraction. The following procedure, however, can be applied. Adding a precise weight of some standard material (e.g. 2 mg of clay-sized powder of fluorite or calcite) to a definite quantity of the mixture

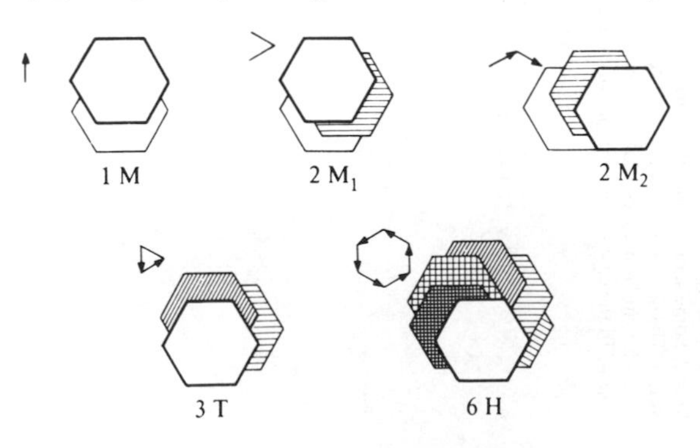

Figure 3.10 The polytype of mica minerals. The arrow shows the displacement of unit sheet stacking.

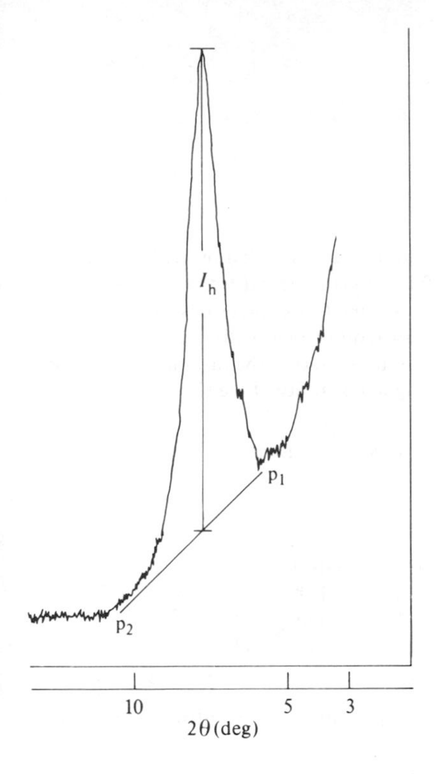

Figure 3.11 Intensity and peak area of X-ray diffraction. I_h is the 'intensity' and the peak area (I_a) is enclosed by the curve and the baseline p_1p_2. The peak area is recommended for quantitative analysis.

(e.g. 50 mg) in which the clay mineral in question is contained, then the ratio of the intensity of X-ray diffraction of the clay mineral to that of the added standard material is measured. Then, one must choose a pure clay mineral of the same kind, with the same particle size, and with a similar degree of crystallinity as the clay mineral in question. Various mixtures of the pure clay mineral mentioned above, other clay minerals and the standard material are X-rayed, and the ratios of intensity of diffractions due to the standard material and the pure clay mineral for each mixture are calculated and plotted against the weight percentage of the pure clay minerals on a graph. Using the rating curve constructed in this way, the percentage content of the clay mineral in question can be estimated. This technique is called the internal standard method (Klug & Alexander 1954).

The small amount of a clay mineral species in a mixture may be estimated by the addition method in which a pure clay mineral of the same species is added successively by a fixed, small quantity, and the increase in intensity or peak area of diffraction is plotted against the weight of pure clay mineral added. If the points lie on a straight line, it must be extended to the abscissa. The distance between its intersection with the abscissa and the point for the original sample without any addition will give the value of clay mineral in question (see Fig. 3.12).

If the (001) diffractions of two or more kinds of clay minerals do not overlap each other, their content ratios can be roughly estimated by the following procedures. Imagine that a clay sample contains A, B and C, and their intensity ratios of (001) diffraction are $I_A:I_B:I_C$. When they are present in equal amounts in the mixture, their intensity ratios are measured to be $a:b:c$. Then the content ratios are estimated to be $I_A/a:I_B/b:I_C/c$). If the diffractions do overlap each other, corrections need to be made. An example is presented here according to Sudo *et al.* (1961). Table 3.6 shows the intensities of

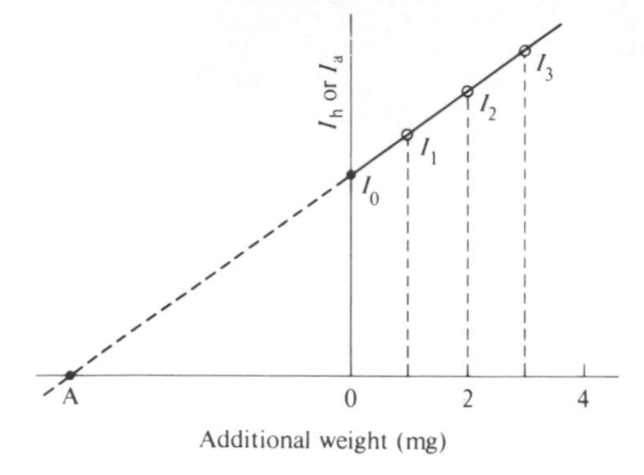

Figure 3.12 The addition method of quantifying a clay mineral from a diffractogram. I_1, I_2 and I_3 are intensities (or peak areas) with addition of an extra 1, 2 and 3 mg of the clay in question. The distance A0 gives the original quantity of the clay.

Table 3.6 Intensities of diffraction peaks in a mixture of montmorillonite, mica and chlorite.

	Intensities		
Treatment	*14–15 Å*	*10 Å*	*7 Å*
Non-treated	168	161	90
300° C, 1 h, heating	25	170	77
Ethylene glycol treatment	16	72	49

diffraction peaks of a sample, containing Al, di-mica mineral, montmorillonite and chlorite.

The peak of montmorillonite at 15.3 Å is displaced to 10 Å by heating at 300°C for 1 h, and, after treatment, rehydration of montmorillonite does not usually take place. By treatment with ethylene glycol, it moves to 17 Å. In the table, therefore, intensity values 25 and 16 at 14–15 Å are due to (001) diffraction of chlorite, and 77 and 49 at 7 Å are due to (002) diffraction of chlorite only. In the case of chlorite, the ratio of (001) diffraction to that of (002) should be $25:77 \simeq 16:49 \simeq 1:3$. Therefore, the intensity value 168 at 14–15 Å is due to montmorillonite as well as chlorite under the non-treatment condition, and one portion of this value of 168, i.e. $90 \times \frac{1}{3}$, should be the (001) diffraction of chlorite in the mixture. Finally, it must be concluded that

$$I(15, \mathrm{Mo}):I(14,\mathrm{Ch}):\mathrm{I}(10, \mathrm{Mi}) = (168 - 90 \times \tfrac{1}{3}):(90 \times \tfrac{1}{3}):161 = 138:30:161$$

where $I(15,\mathrm{Mo})$ is the intesity of diffraction at 15 Å due to montmorillonite, $I(14, \mathrm{Ch})$ is the intensity of diffraction at 14 Å due to chlorite, and $I(10,\mathrm{Mi})$ is the intensity of diffraction at 10 Å due to mica.

On the other hand, their relative intensities are measured as follows, using pure standard species;

$$I(15, \mathrm{Mo}):I(10, \mathrm{Md}):I(14, \mathrm{Ch}):I(10, \mathrm{Mi}):I(7, \mathrm{K}) = 3.6:0.9:0.7:1.0:1.0$$

The notations are the same as above. The symbols Md and K indicate dehydrated montmorillonite (300°C, 1 h, heating) and kaolinite, respectively. By means of the above data, content ratios in this sample can be estimated as

(montmorillonite):(chlorite):(Al, di-mica clay mineral)
= (138/3.6):(30/0.7):(161/1)

Other examples can be found in Griffin (1971) and Jenkins and de Vries (1970b).

3.3.9 Thermal methods

Thermal analysis is a general term for a group of related techniques that, as defined by the International Confederation for Thermal Analysis (ICTA), measure the dependence of the parameters of any physical property of a substance on temperature. Parameters measured can include temperature, weight, energy, gas, radiation, sound, magnetic susceptibility, resistivity and length, but the methods commonly employed in geomorphological analyses involve either temperature (differential thermal analysis, DTA) or weight (thermogravimetry, TG). General discussions of these methods are to be found in Daniels (1973), Hutchinson (1974) and Mackenzie (1970, 1972). Differential thermal analysis records temperature differences between a sample and reference material, against either time or temperature, at a controlled rate. The differences arise from the sample either absorbing or evolving heat as reactions take place during heating.

A DTA assembly (Fig. 3.13a) consists of a light-weight furnace normally wound with nichrome and with a platinum or 13% rhodium–platinum control thermocouple. The vertical positioning of the furnace allows the two specimens (the unknown and an inert reference) to be arranged symmetrically and a temperature control ensures a constant heating rate over a given range of temperature. Sample and reference are contained in separate platinum crucibles placed in a ceramic or metal block, the type of material and design of which affects the sensitivity of the apparatus. A thermocouple in the block assembly measures temperature differences between the two samples, and when one junction is at a higher temperature than the other a small electromotive force (e.m.f.) is set up. This current is amplified and fed into a y–t chart recorder. Temperature is recorded in the x direction, temperature differences (ΔT) in the y direction.

DTA is not necessarily run in an air atmosphere and furnaces are available that allow different atmospheres to be used.

ICTA defines thermogravimetry (TG) as a technique in which the mass of a substance is measured as a function of temperature while the substance is subjected to a controlled temperature programme. Differential thermogravimetry (DTG) is the first derivative of the TG curve with respect to either time or temperature. The basic TG assembly (Fig 3.13b) consists of a furnace, temperature control, electronic or mechanical thermobalance and a recording unit. This system, although simpler, has similarities to that for DTA.

A microfurnace, constructed, for example, from mineral-insulated nichrome wire, sheathed in inconel alloy, is mounted in the body of the unit. The furnace is water-cooled and a platinum resistance sensor acts as the temperature controller. Provision is made for analysis in either air or gas atmospheres. In one type, an electronic balance electromagnetically balances the torque produced by the sample mass. A current flows through the head unit in direct proportion to the applied mass and operates an indicating meter, which provides the electrical output for the chart recorder. The usual convention is to record temperature along the x axis and mass in the y direction (Fig. 3.14). DTG units (Fig. 3.13c) are either available as an accessory or may be incorporated with the TG system.

The validity of thermal methods, particulary DTA, is dependent upon the method of pretreatment. Un-pretreated samples can be analysed by thermal techniques but it is

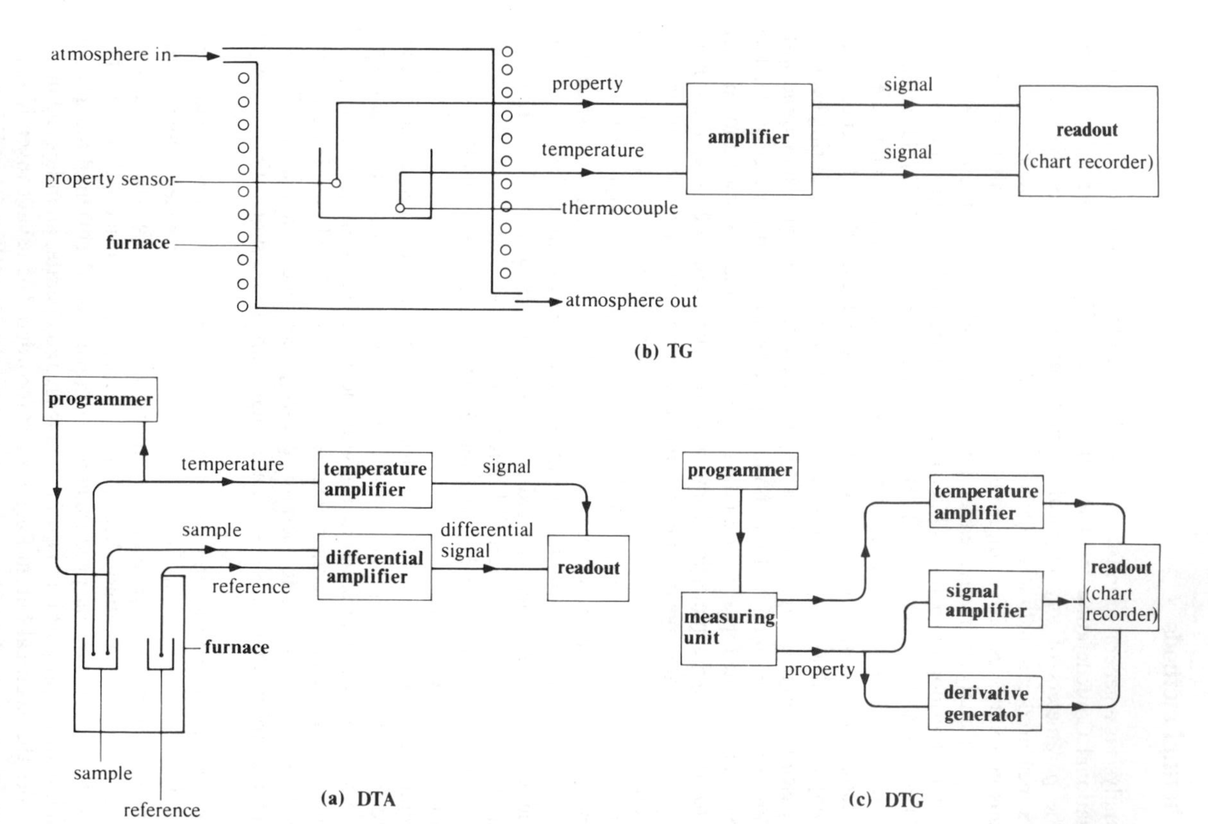

Figure 3.13 Schematic diagrams of (a) differential thermal analysis (DTA), (b) themogravimetry (TG) and (c) differential thermogravimetry (DTG).

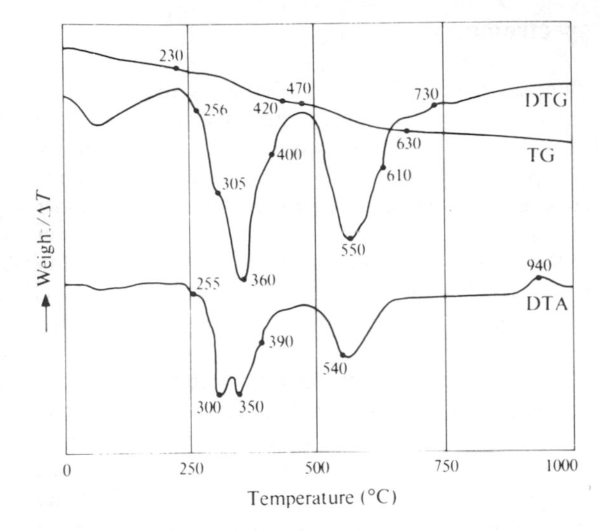

Figure 3.14 Specimen trace of a laterite analysed by thermogravimetry (TG) and its derivative (DTG) and differential thermal analysis (DTA).

desirable to undertake at least some pretreatment, usually removal of organic matter and dispersion of sample. After dispersal, the sample can be divided into different size fractions to analyse for different primary and secondary minerals.

A number of factors influence the shape of thermoanalytical curves, these include method of pretreatment, particle size, heating rates, density of packing, weight of sample and diluent, sample container, atmosphere and sensitivity of instrument used.

DTA curves are characterised by a baseline and positive and negative peaks, which represent exothermic and endothermic reactions, respectively. Both qualitative and quantitative interpretations can be made from DTA curves, but the latter is extremely difficult. Qualitative interpretation is dependent upon reaction temperature and peak characteristics (size, shape and direction), while quantitative assessments depend on the area between the peak and baseline. Examples can be found in Warne (1977).

TG curves show a series of mass losses separated by plateaux of constant mass. Where two reactions overlap, there is no plateau region, and they can only be identified by a change of slope in the curve. Evaluation of the decomposition stages, temperature range and fractional mass loss of each stage can be found from the TG curve, but if interpretation is difficult the corresponding DTG curve may be recorded. The DTG curve consists of a series of peaks corresponding to the various stages of the decomposition. TG provides a useful means of quantification as the mass losses between the plateaux indicate the amount of volatile reaction products and so permit the calculation of the concentration of water, carbonates, etc., within the sample.

The primary use of thermoanalytical methods in geomorphology is one of interpretation and identification of minerals and clay minerals within weathered materials, soils and clay. In many respects, the technique is complementary to X-ray methods, and both sets of analyses shoud be used in conjunction with each other in the identification of mineralogy. However, thermal analysis can be easily performed by a single operator, but speed of testing is dependent upon type of pretreatment, heating rate and instrumentation, the cost of which is relatively expensive. Thermogravimetry is perhaps the most valuable of the thermal methods due to its qualitative and quantitative nature; also it can be carried out more rapidly than DTA since shorter cooling periods and more rapid heating rates can be employed. It has been relatively little used in geomorphology, although some analyses of loess (Seppäla 1972) and quickclay (Smalley *et al.* 1975) have shown its value.

3.3.10 X-ray fluorescence spectrometry

The principle underlying X-ray fluorescence spectrometry (XRFS) is the excitation and emission of fluorescent X-rays, of characteristic energy and wavelength, when an atom of an element is bombarded with a beam of X-rays (see above and Jenkins & de Vries 1970a). This technique is capable of detecting over 80 elements with atomic number greater than 8, but, if electron excitation is used, elements with atomic number 6 or greater can be determined. Elements with concentrations as low as a few parts per million of sample can be detected. Detailed discussion of the method can be found in Allman and Lawrence (1972), Hutchinson (1974) and Zussman (1977). Because elements are determined, it is of most use for elemental ratio determination such as in weathering profiles.

Instrumentation required is an X-ray fluorescence spectrometer, which both disperses X-rays and measures their intensity. The spectrometer (Fig. 3.15) consists of a source, collimators, analysing crystal, detector and readout facility. An X-ray tube powered by a high voltage between cathode and anode is the source and, on bombardment of the anode with fast electrons, the X-ray spectrum is emitted. X-rays are both absorbed and scattered by the sample, and the emitted spectrum is collimated and directed towards a rotating, analysing crystal, which disperses X-rays according to the Bragg equation (Eqn 3.27).

Since wavelength varies, a series of crystals with a wide range of *d* values may be required for analysis of a sample. After dispersion, the beam of selected wavelength is collimated and passed to a rotating detector, normally a scintillation counter for best discrimination. Both analysing crystal and detector are attached to a goniometer and the angle setting by the goniometer is such that the detector moves through an angle twice that moved by the crystal. Pulses from the detector are integrated and for visual display are counted on a scaler, or alternatively a recorder, digital meter or printout is used. The method of counting is either number of pulses in fixed time (counts per second, c.p.s) or the time for a predetermined number of pulses: counting losses due to dead time need to be considered when X-ray intensities are required.

In qualitative analysis, samples may be placed in a X-ray spectrometer in virtually any form and with simple preparation, but, in quantitative analysis, it is necessary to have compacted self-supporting samples. Powders are pressed into holders or pellets, and for light elements the sample is dispersed into a lithium borate glass and cast into discs for direct insertion into the spectrometer. Alternatively, a sample may be digested and introduced in liquid form for which special plastic cells of low atomic number are available.

Care needs to be taken with preparations to eliminate matrix effects. It is also necessary to compare the X-ray response of an unknown sample with a standard of similar but precisely known chemical composition. Where a major difference occurs between the standard and the sample, either dilution of both or addition of a fixed amount of heavy absorbing element is necessary, except for trace-element analysis when synthetic mixtures may be used.

Identification involves measurement of 2θ values for the radiation emitted from the sample; with *d* spacing known, wavelength can be calculated. X-ray spectra are easily recognised, though peaks may not always correspond to those listed in tables due to misaligned equipment, crystal spacing ($2d$) and wavelength variation. The K spectum is obtained from elements of atomic number up to 82 and is commonly used for identification as it is more intense that the L spectral lines, which are excited from all elements. However, the $L\alpha_1$ line is used for heavy elements from which $K\alpha$ spectra cannot be excited. The M spectra are weak and of little interpretative value.

Calculation of the concentration of unknown samples is obtained by comparison with a standard using the following formula:

$$a = \frac{A(p - b)}{P - B} \tag{3.28}$$

where a is concentration of unknown, A is concentration of standard, p is peak intensity of unknown (c.p.s), P is peak intensity of standard (c.p.s), b is background of unknown (c.p.s) and B is background of standard (c.p.s).

X-ray fluorescence spectrometry is a rapid, often non-destructive, method in qualitative analyses, but for precise quantitative measurements it is not non-destructive and considerable effort is required in sample preparation. However, it has a distinct advantage over wet chemical methods in terms of both time and accuracy due to reduction in operational error. In terms of geomorphological research, XRFS is a valuable technique as it can be applied to quantitative, qualitative and trace-element analysis for a wide range of materials including soils, rocks, minerals and water samples (e.g. Watts 1977).

3.3.11 Heavy minerals and insoluble residue

Heavy minerals have been used by geomorphologists for tasks that have included till discrimination, the establishment of sediment source areas, and the determination of the extent of chemical weathering in a profile (see Sec. 4.1), as for example in the case of red-bed formation (Walker 1967).

The heavy-mineral content of a sample is classified as having a specific gravity of 2.85 or greater. This scale is arbitrary and is based on the specific gravity (2.87) of bromoform (tribromomethane, $CHBr_3$), which is used as a separating medium.

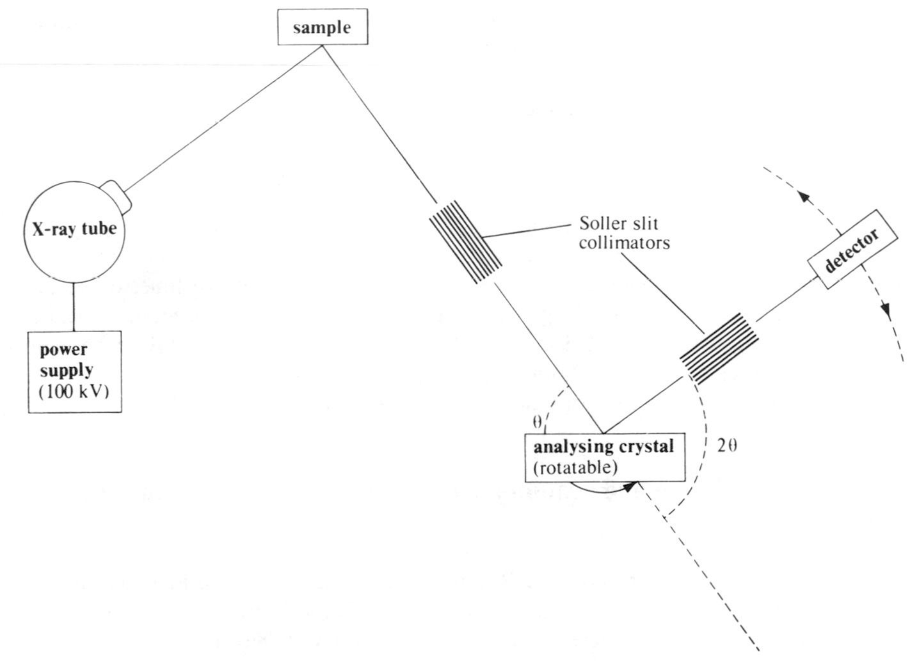

Figure 3.15 Basic components of X-ray fluorescence equipment.

Pretreatment of samples prior to separation is dependent upon whether the sample is rock or unconsolidated material. Rock samples require a two-stage pretreatment of disaggregation and sieving. There is no common disaggregation process, but the chosen technique must be designed to ensure that sample texture is not destroyed. Other procedures, for example, digestion with an acid or alkali, may also be required.

Separation of heavy minerals from rock samples follows a sequence of stages: crushing, washing, sieving and separation. Crushing ensures that specific particle sizes are composed of one mineral type; washing removes particles less than 50 μm, which are not separated efficiently; and sieving divides the sample into groups of uniform grain size.

A number of separation procedures are available but separation with bromoform is a standard method. Approximately 200 ml of bromoform is placed in a separating funnel (a watchglass reduces evaporation), and a representative portion of a sample is weighed and added to the bromoform. By gravity separation, the heavy liquid fraction sinks and is separated from the lighter fraction by a layer of clear liquid. The heavy portion is run off and collected in fluted filter paper; the funnel is removed and the heavy grains washed with alcohol or acetone, dried and mounted. Further separation of the heavy fraction may be undertaken using either a liquid of higher specific gravity or a magnetic separator.

In the case of unconsolidated sediments, the sample is examined under a microscope to determine the approximate percentage of mineral types. An alkali or acid digest may be required, but, if little material adheres to the grains, boiling of the sample in water is sufficient. A number of sieves are selected on the basis of the initial microscope observation and a 20 g sample is removed from the sieve containing the greatest concentration of required minerals. Sample separation then proceeds in a manner similar to that described for rock samples.

Separation of minerals using heavy liquids is a convenient, rapid and efficient technique of recovering minerals of close density. The application of this technique in geomorphology is primarily qualitative as it permits the isolation and subsequent identification of individual minerals from a heterogeneous sample. In quantitative analysis, considerable preliminary work is required to ensure representative samples. Procedures are discussed in detail in Allman and Lawrence (1972), Carver (1971a) and Hutchinson (1974).

In the study of carbonate rocks and carbonate-rich sediments, the determination of the insoluble-residue content has proved to be of some utility; see, for example, Small and Fisher's (1970) study of the 'marl' content of the English Chalk and its bearing on the development of the secondary escarpment. The method is quick and simple, and involves dissolving a known dry weight of rock or sediment in a dilute acid such as 10% hydrochloric acid. Details of the method are given in Carver (1971b). The study of the insoluble content, and many other geomorphologically important properties of limestones, can also be undertaken by means of the petrological microscope using thin sections (see, for example, Sweeting & Sweeting 1969). It has been found that the natures of the carbonate matrix (e.g. whether it is a micrite, bomicrite or a sparite) and of the insoluble residue have an influence on the morphology of limestone pavements and intertidal platforms.

3.3.12 Transmission and scanning electron microscopes and the study of clays

The scanning electron microscope (SEM) has been mentioned several times in this section, mainly as a means of obtaining high-magnification and large-depth-of-field photographs of materials of interest (see papers in Whalley 1978a). It is, however, very much more than a 'supermicroscope'. In the transmission electron microscope (TEM), a beam of electrons is 'defocused' to a small spot by electromagnetic lenses. The energy of the electrons produced may be 50–100 kV, although 1 MV is used in some instruments. The electrons pass through a thin slice of the material to be examined (usually mounted on a specially prepared grid) and an image is produced on a flat viewing plate or photographic emulsion. The types of material examined are restricted to very thin films or slices because the electrons are easily absorbed. The scanning electron microscope is a means of examining solid objects and is therefore most suitable for geological specimens. In this instrument, the defocused electron beam is made to scan a small area of the specimen (a raster). The incident beam gives reflected and secondary electrons, which are detected and modulated with the raster to form an image on a television screen or cathode ray tube. The object image is formed because of surface topography scattering as well as atomic-number differences in the elements in the object. Magnification is changed by reducing or enlarging the raster size on the object. Descriptions of the instrument are given in Thornton (1968) and Swift (1970).

A variety of image processing equipment can be fitted to enhance detail or examine some part of the detected signal. Most importantly in a geological context, the emitted visible light, cathodoluminescence (Grant 1978), back-scattered electrons (e.g. Goldstein *et al.* 1981) and X-rays can be detected. The energy-dispersive X-ray (EDX) analyser is most commonly fitted and this may be used as a semiquantiative tool for elemental analysis by detecting emitted X-rays, the characteristic lines of which can be used for elemental identification. For a detailed discussion, reference should be made to Goldstein *et al.* (1981) and Jones and Smith (1978).

The manual of the electron microscope, supplied by its manufacturer, should be consulted as to the preparation of specimens and operation of the instrument itself. McKee and Brown's (1977) explanations are especially useful for a guide to sample preparation of clay minerals for electron microscopy, both of transmission (TEM) and scanning (SEM) types.

Electron microscopy has permitted the precise determination of the shape and size of various clay minerals. The thickness of an individual particle is also inferred from the images and the angle of shadowing. Numerous authors have published electron micrographs of clay minerals, Beautelspacher and Van der Marel (1968) being the most important. Sudo (1974) has summarised the shapes and the range of particle sizes of various clay minerals, as shown in Table 3.7.

Although SEM has had a great impact upon the morphological studies of fine particles, TEM still has an advantage to permit electron diffraction in the electron microscope. This method is called selected area diffraction, which is very much analogous to X-ray diffraction. Clay minerals usually have the shape of flakes, and thus yield nearly identical hexagonal spot patterns. Thus, it is very difficult to distinguish the various clay minerals. The electron diffraction technique, however, has provided much information regarding the structure of halloysite and chrysotile. Zvyagin (1967) conducted the structure analysis by means of inclined fibre patterns. Nishiyama and Shimoda (1974) applied the electron diffraction to determine the polytype of mica clay minerals. Reference should be made to the discussions of electron diffraction given by Zvyagin (1967), Gard (1971) and Andrews *et al.* (1971). Together with X-ray diffraction, TEM and SEM, is recommended the application of EPMA (electron probe microanalysis), which allows chemical data to be obtained from individual particles as well as from a minute portion, by measuring the secondary X-rays produced by electron beam bombardment. High-resolution TEM (HRTEM) with ultramicrotomy techniques provides valuable images, which can be interpreted as structure images or projections of the atomic arrangments in the sheet silicate structure.

SEM analysis of clays, perhaps aided by simulataneous EDX elemental analysis and

Table 3.7 Electron microscopy of principal clay minerals – shape and size

Shape	*Clay minerals*	*Normal range of size* (μm)
Well or poorly developed hexagonal flakes	Kaolinite	Length 0.07–3.51, width 0.05–2.11, thickness 0.03–0.05
	Dickite	Largest diameter 2.5–8, thickness 0.07–0.25
	Mica clay minerals	
Elongated, lath-shaped unit, bundles of laths, fibrous or needle-shaped	Nontronite	Length 0.0567–0.057, width 0.0108–0.014, thickness 0.0014–0.003
	Saponite	
	Hectorite	Maximum length 0.9, maximum width 0.1, thickness 0.001 (sample, occurred in Hector, California)
	α-Sepiolite, kaolinite, mica clay minerals	
	Attapulgite	Maximum length 4–5, width 0.1 (sample, occurred in Attapulgus, Georgia)
Rolled tube	Halloysite, metahalloysite, chrysotile	Outer diameter 0.07, inner diameter 0.04, thickness of tube 0.02 (on average)
Irregular flake	Montmorillonite, pyrophyllite, kaolinite, mica clay minerals, β-sepiolite	Thickness 0.02–0.002
Fine grain or sphere	Allophane, allophane–halloysite sphere	
Bundles of thin fibres	Imogolite	
Wavy flake	Antigorite	

Source: Sudo (1974).

prior XRD, DTA or DTG results, is mainly by means of their appearance. This is partly a matter of experience, but reference should be made to Mumpton and Ormsby (1978) for zeolites, Waugh (1978) and Whitaker (1978) for a variety of authigenic minerals, Keller (1976, 1978) for kaolins and Tovey (1971) for a number of clay minerals. Examination of clays in their natural state presents difficulties (Tovey 1971), but some success has been achieved with special environmental cells (Kohyama *et al.* 1978).

3.4 Strength of materials

3.4.1 Introduction

In recent years, there has been a trend towards using the behavioural properties of materials, especially soils, to explain geomorphological features. Good examples can be found in Carson and Kirkby (1972), Carson (1971a), Whalley (1976) and Statham (1977). In this section, various ways of investigation of these strength parameters will be examined. The techniques discussed here relate directly to Part Four, especially to slope processes. The methods are primarily for use in the laboratory but some are suitable for field use and require inexpensive equipment. For further reading, the books by Akroyd (1969), Bishop and Henkel (1962), Smith (1978), British Standards Institution (1975c) (BS 1377) and Vickers (1978) should be consulted.

The tests described below are those of the engineer. It should be borne in mind that they do not necessarily provide all the data necessary for a geomorphological interpretation of the phenomenon being studied. The interpretation of laboratory and field estimates of material strength is the second stage of investigation, and perhaps the most difficult. In particular, the time element is especially difficult to study (e.g. Statham 1977).

Although the relationships governing soil strength are essentially simple, the ways in which they are measured may be quite complex, depending on how the stresses in the soil are modelled by the apparatus. The fundamental relationship is the Mohr–Coulomb equation

$$\tau_f = C + \sigma \tan \phi \tag{3.29}$$

where τ_f is the shear stress at failure – the shear strength (measured as a stress in kN m^{-2} or kPa) – C is the cohesion of the soil due to clays (in Pa), ϕ is the angle of internal friction due to granular materials (in degrees), and σ is the normal shear stress on the soil element (in Pa). The problem is to find a value of τ_f that can be used in stability calculations through a knowledge of σ from the geometry of the slope and C and ϕ from tests. The first group of properties is mainly concerned with classification and identification of general behaviour (although it may be possible to approximate a value of the shear strength of the soil, e.g. Worth and Wood (1978)), the complexity and the amount of information about the soil increasing for each test. It is usually necessary to know other basic soil properties such as voids ratio, moisture content and bulk density; these have been discussed in Sections 3.2.9 to 3.2.11 and 3.2.14.

3.4.2 General shear strength measurements and classification tests

There is frequently a need to obtain estimates of soil behaviour away from the laboratory where comprehensive tests can be carried out. In some cases, many samples need to be analysed, and relatively quick and easy methods provide a complement to interspersed, more complete, tests. The methods described in this section provide such results. It should be noted from the outset that, with the exception of the Schmidt hammer, they are primarily for cohesive soils, that is, soils like clays and silts that have little granular material. If the soils do contain clasts larger than sand size, then anomalous readings may result. For granular materials, more elaborate tests are required. Standards are laid down for several of these tests (e.g. BS 1377: 1975). The following classification tests provide simple ways to provide a behavioural description of a soil so that its characteristics can be assessed readily. Particle size range and grading clearly play an important part in the behaviour of (especially granular) soils and reference should be made to Section 3.2.2 for methods of grain size determination.

3.4.2.1 Atterberg limits

These tests provide good indications of three states of many soils: solid, plastic and liquid being delimited by the shrinkage limit (*SL*), plastic limit (*PL*) and liquid limit (*LL*) (Whalley 1976) (Fig. 3.16). The *SL* and *PL* are expressed as moisture content percentages of dry weight. For the British Standard (BS 1377: 1975), all material used in these tests should be finer than BS 425 μm sieve. Reference should be made to this standard for a comprehensive procedure and details of equipment.

Shrinkage limit This is defined as the moisture content of the soil at which the soil stays at constant volume upon drying. A paste of soil (< 425 μm sieve) is made at about the liquid limit (see below) and placed in a semicircular mould 140 mm long by 25 mm diameter and smoothed flush (Fig. 3.17a). The paste is dried, initially at 60°C then at 105°C, the sample length is measured after cooling, and the *SL* is calculated from

$$SL = \left(1 - \frac{\text{length after drying}}{\text{initial length}}\right) \times 100 \qquad (3.30)$$

and expressed as a percentage.

Plastic limit This is the delimitation between a brittle solid and a plastic soil. It is simply (if arbitrarily) found by taking a small piece of soil (10 g) and gently rolling it on a glass plate with the palm of the hand. When the soil thread is 3 mm thick and is just about to break up, then the moisture content is the *PL*. If the sample is too dry then the thread crumbles before the required thickness; it will roll thinner if too wet. The soil is wetted or allowed to dry accordingly. Three determinations are made and the mean value taken.

Liquid limit This is the moisture content at the transition from a plastic to a liquid state. A more complicated test is required for this. The usual, Casagrande, device is as seen in Fig.3.17a. The soil is moulded wet and placed into the bowl and levelled across parallel to the base. A special tool is then used to make a groove and separate two soil masses as shown. The handle is then turned and a cam allows the bowl to fall a predetermined 13 mm twice per second onto a standard hard rubber base. After several turns, the two parts of the soil mass will flow together and join for at least 25 mm. When 25 turns are just required to achieve this, then the moisture content of the soil is the *LL*. It is rare to achieve this exactly and some practice is required to estimate the consistency of the soil near to the moisture content which gives 20 to 30 turns. The test is made easier by taking four moisture contents that correspond to the number of turns in the range 10 to 50 by adding drier soil or water as appropriate. These moisture contents are plotted (as ordinate) against log (number of turns) and the moisture content at 25 is read off from this straight line relationship.

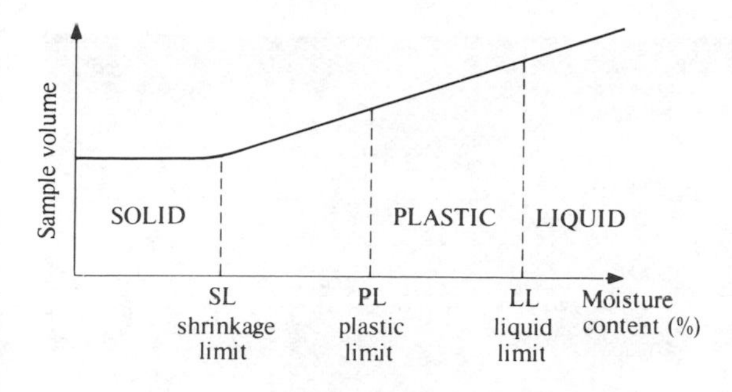

Figure 3.16 Relationship between sample volume and moisture content, showing shrinkage, plastic and liquid limits.

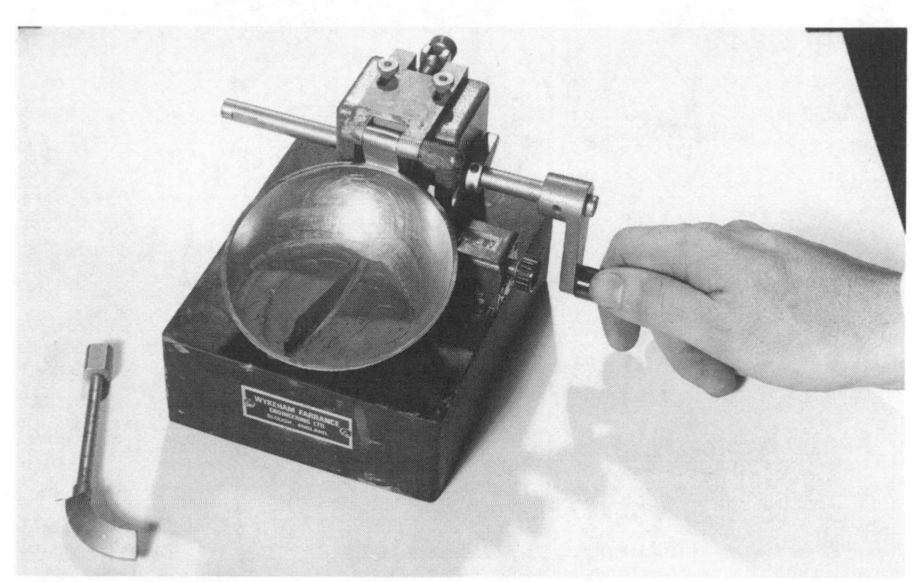

Figure 3.17 (a) Casagrande liquid limit apparatus with groove-making tool. (b) Cone penetrometer in position for start of test. (c) Field shear vane with alternative size vane and dummy probe.

An alternative, and preferred, method of determining the *LL* is by using a cone penetrometer similar to that shown in Fig. 3.17b. The standard cone is placed point down just onto the soil, levelled into a metal container. The cone is released and allowed to penetrate for 5 s and this distance is measured with a dial gauge. Two falls are required with no more than 0.5 mm difference in penetration; if there is, then more trials are required. The moisture content is taken (with 10 g of soil) when this is satisfactory, and the test is repeated at different moisture contents; at least four runs are needed so that a graph of penetration against moisture content gives a linear graph. The *LL* is the moisture content when the cone penetrates 20 mm into the soil.

Generally speaking, this is a quicker and more reproducible way of obtaining the *LL* than the Casagrande apparatus, although the latter is more commonly used. A useful paper discussing the method is by Davidson (1983).

Plastic and liquid limits are useful for many soils, the latter especially in describing phenomena like mudflows where there is no 'shear strength' but a viscous material of complex rheological behaviour.

A useful parameter of liquid behaviour is the plasticity index (*PI*) defined simply as *LL* − *PL*. This is a useful parameter, especially when combined with the liquidity index (*LI*), which relates the field moisture content of the soil (m) thus:

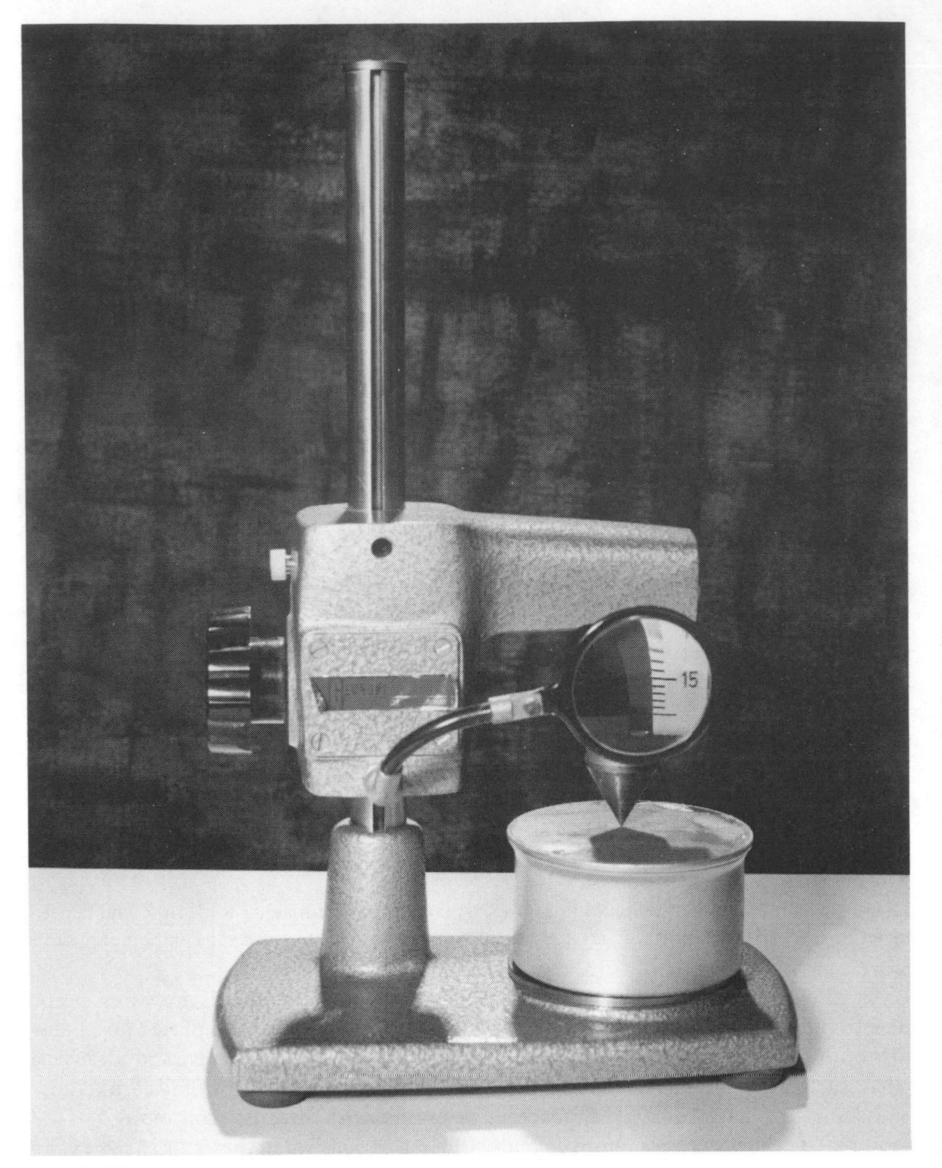

Figure 3.17b

$$LI = \frac{m - PL}{PI} \tag{3.31}$$

Two further relationships are of use in basic soil characterisation. The first is of value when there is a small amount of clay minerals and the value of the PL and LL are difficult to determine, thus: $PI = 2.3 \times SL$. The *activity* of a soil

$$\text{activity} = \frac{PI}{\%\ \text{weight finer than } 2\mu\text{m}} \tag{3.32}$$

is due to Skempton (1953). However, this can be misleading because the cohesion supplied by the clay size fraction depends upon the type of clay.

3.4.3 Laboratory fall cone (cone penetrometer) test for soil strength

This apparatus is similar to that previously described and is shown in Figure 3.17b. The soil to be tested is mixed to a uniform state and placed in a small container directly under the arm, the height of which is adjusted so that the cone point just touches the soil surface. When a trigger is pulled, the cone is released and it falls and penetrates the soil. The distance fallen is measured against a scale on the arm. A value for the shear strength is found directly from tables, which equate it with fall distance for a particular cone. Several cones are required for different types of soils. This method of strength determination is very replicable and compares well with other rapid methods for cohesive soils. The moisture content of the soil needs to be determined at that strength, so a graph of moisture content against fall cone strength can be plotted and the strength at field moisture contents determined. In the field, of course, this strength (and its variation) can be found directly. The use of this method to determine soil strength is described by Hansbo (1957) and has been used to help investigate landslide properties by Caldenius and Lundstrøm (1956) and Prior (1977).

3.4.4 Penetrometer

Large-scale, vehicle-mounted, penetrometers are available but hand-held versions are useful devices to help classification of cohesive soils. A steel rod contained by a sleeve has a line engraved 6 mm from the end, and this rod is pushed into the soil until the line is flush with the soil surface. The rod compresses a calibrated spring in the sleeve and a

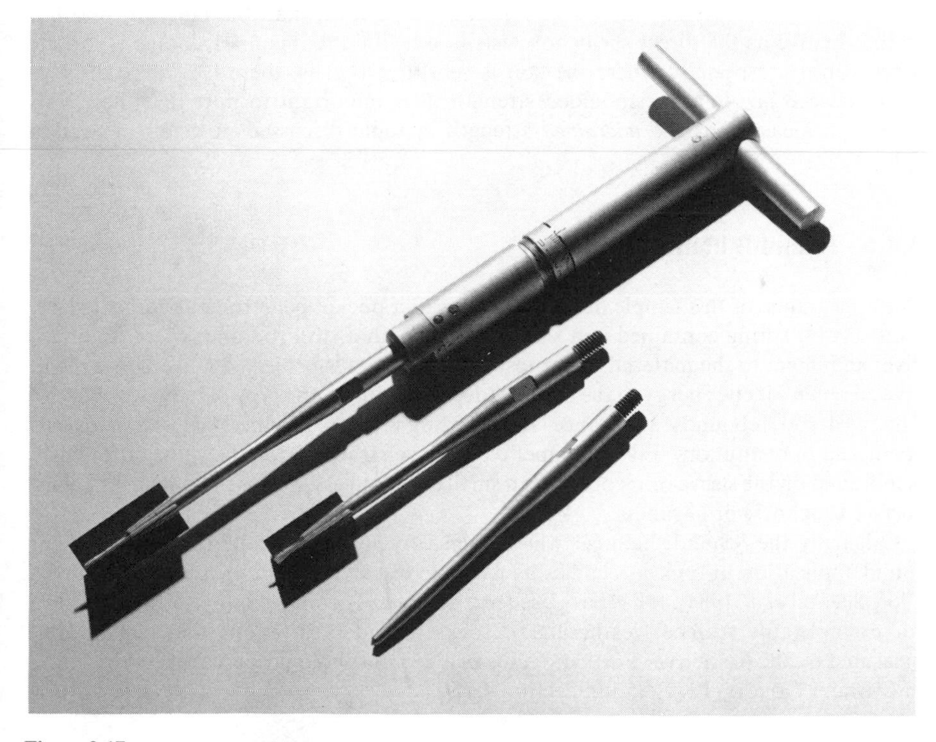

Figure 3.17c

pointer on the rod indicates the 'strength' in kgf cm^{-2} (1 kgf cm^{-2} = 98kPa). A good geomorphological example of a small penetrometer is given in Chorley (1962) for soils and by Krumbein (1959) for beach properties. A full treatment of all types of penetrometers is given in Sanglerat (1972).

3.4.5 Vane test

Versions of this test exist for large-scale field use and allow undrained shear strength determinations up to depths of several metres and within boreholes. Small portable field vanes (Fig. 3.17c) are very useful for preliminary investigations on a site (although, ideally, soil samples need to be collected to determine moisture content). The laboratory version is similar in principle and operation, so a general description will suffice.

Four blades set at right angles are mounted on the end of a central shaft. The vane is pressed into the soil until covered or at the required depth and a torque is applied to the shaft. The torque is increased until, at a maximum value, the soil shears along a cylindrical surface enclosed by the vane. The angular deflection of a calibrated spring is the usual way to determine the torque. A motor is used to apply the torque in the laboratory version and about 9°/min is the rate for the field vane when turned by hand. This torque (T) is converted into undrained shear strength (C_u) when the height (h) and diameter (d) of the vanes (usually a 2:1 ratio) are known from:

$$C_u = T/[\pi d^2 (h/2 + d/6)] \qquad (3.33)$$

The laboratory version of the test is less easy to perform than the fall cone and not as informative as the direct shear box tests (see below). Its main advantage is for site investigations, especially where the soil is sensitive; that is, there is a large discrepancy between *in situ* and remoulded strength. It is important to note that these tests provide a measure of the *undrained* strength, a topic discussed in detail in Section 3.4.9.

3.4.6 Schmidt hammer

The appearance of this simple device is similar to a pocket penetrometer and similarly consists of a spring contained in a sleeve/handle with a steel rod and, when triggered, gives an impact to the material under test. A given mass is projected a fixed distance to give a known kinetic energy to the surface (dependent upon the type of hammer chosen). The steel rod rebounds a distance R according to the 'hardness' of the body (its coefficient of restitution). For a hammer of known characteristics, the value of R (which is indicated on the sleeve or recorded on a small chart) gives a means of comparing rocks' surface toughness or hardness.

Originally the Schmidt hammer was designed for use in concrete testing, but it has found application in rock mechanics as a simple test of strength or durability (Hucka 1965, Sheorei *et al.* 1984). It has been used in a number of geomorphological applications; for example, the surface weathering of rocks provides variations that can be well measured by the technique. Further details in a geomorphological context can be found in Day and Goudie (1977) and in Section 4.1.9.

3.4.7 Other tests of rock strength

While the Schmidt hammer gives an evaluation of surface hardness and has been found to correlate well with compressive strength (Hucka 1965), the testing of rock strength provides many difficulties. Rocks will usually fail under shear or tensile stresses and these are almost always higher than for soils (exceptions are weathered shales and rotted granites, for example): so equivalent strength tests to those mentioned in the next sections need more substantial equipment. Such tests are for small pieces of rock and, although rock masses can be considered as a continuum, as soils usually are, there is a strongly held view that rock masses should be considered as discontinuous bodies (e.g. Trollope 1968). Discussion of such topics, important to geomorphology, although hardly yet impinging upon it (Whalley 1976), must lie outside the scope of this volume.

Some other simple tests relevant to rock behaviour can be outlined, which are of value nevertheless.

The 'Brazilian' test is an indirect measure of uniaxial tensile strength. A cylinder of rock is compression loaded at its centre across the diameter. The load at failure can be related to the geometry of the cylinder or disc to give a measure of the tensile strength. Opinions are varied about the value of this test, although Hawkes and Mellor (1970) suggest that results are valid (at least for comparative purposes) for brittle rocks.

The compressive strength of a single grain of rock has been found by Moss (1972) to be a useful means of characterising the behaviour of sediments undergoing attrition in a stream. Likewise, tumbling experiments have been used by a number of authors to examine the population (Moss *et al.* 1973). A small particle crushing test is used in analysing roadstone in BS812 (British Standards Institution 1975b), although this has not been employed by geomorphologists. A useful summary paper (Anon. 1977) is concerned with evaluation techniques for rock quality from various points of view.

3.4.8 Oedometer (consolidometer) tests

Although this type of test, which examines the effects of loading on the compressibility of soil structure, does not examine soil strength, it is sometimes a valuable indicator of soil behaviour. Primarily, its use is for foundation design and it is described in detail in soil mechanics texts (e.g. Smith 1978). If a soil structure is loaded, then water in the pore spaces increases in pressure (Δu) as a response to the load increment (Δp). Initially, $\Delta p = \Delta u$ but, as drainage takes place. Δu decreases and the increased load is taken by the soil structure. This process is termed 'consolidation' and it depends upon the soil properties, most notably the bulk density and permeability. There may well be a considerable time to regain equilibrium after application of Δp, and for some soils the settlement itself may be substantial. In addition, there will be some change in the soil strength because the pore-water pressure (Δu) changes. It is not possible to discuss the consolidation process in detail, other than to mention factors that are directly related to geomorphological investigations.

Two types of soil are commonly found, which have different consolidation histories: normally consolidated and overconsolidated soils. Normally consolidated soils are those which are at equilibrium under the maximum stress they have experienced (alternatively, the existing effective pressure is the maximum to which they have been subjected). Overconsolidated soils are those which have a present effective pressure less than a previous maximum value. This long-term reduction in stress for overconsolidated soils can be produced by erosion or previous loading (e.g. by ice sheets: Khazi & Knill 1969) or by groundwater variations or a combination of these. The geotechnical properties of

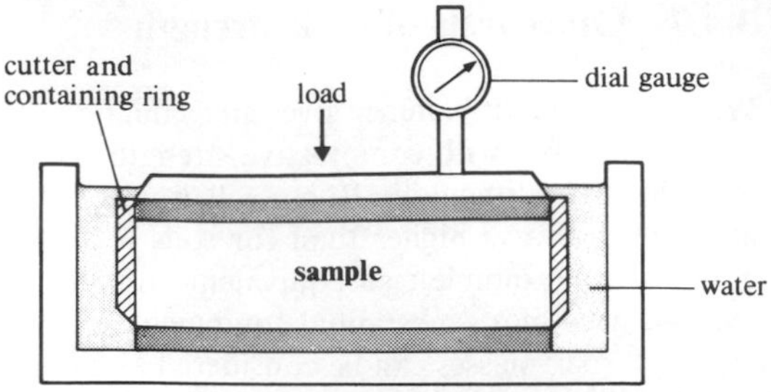

Figure 3.18 Diagram of the main parts of a consolidometer (oedometer).

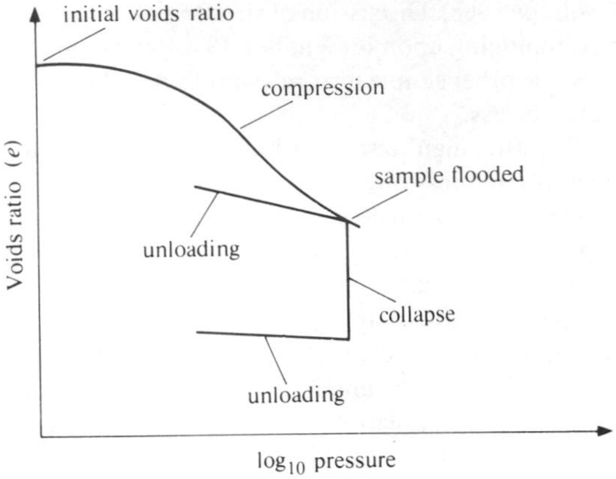

Figure 3.19 Graph of voids ratio against log pressure (applied load) showing the effects of loading and unloading as well as unloading after collapse of a loessic soil.

overconsolidated clays are often important in a geomorphological context. A good example is the mudslide behaviour of many clays (e.g. Hutchinson *et al.* 1974). Of particular value is the way the test can be used to show the collapsing behaviour of loessic soils (Fookes & Best 1969). A sample of the undisturbed soil is cut with the containing collar and carefully trimmed, and placed in a cell on top of a porous disc. Another disc is placed on top of the sample and a load platen on top of that. The apparatus is somewhat more complicated than the diagram shown in Figure 3.18 because of the loading arrangement. An initial load is applied and the compression of the sample is measured at intervals with a dial gauge. This first load is allowed to remain until there is no further settlement or for a fixed time interval of 24 h and then another load increment is applied. For a test on clay, the sample is tested wet with water in the outside container, and drainage is allowed through the porous discs. It is possible to calculate a value for the permeability of the soil with a more sophisticated oedometer, the Rowe cell (Rowe & Barden 1966). To examine the collapse of a soil, however, the test is performed without the water until, at some state (Fig. 3.19), water is added to the cell. Collapse is shown by the trace seen in the e versus log p graph, that is, a plot of voids ratio (e) plotted against $\log_{10}$ (applied load).

Some types of clay soil swell substantially when wetted, and a type of oedometer is available that allows the swelling pressure to be measured as well as a consolidation test.

3.4.9 Direct shear

3.4.9.1 Introduction

Direct shear tests are one of the simplest of all the methods of determining shear strength in the laboratory and consist essentially of tests similar to those used in the determination of the coefficient of friction of two sliding surfaces. In this test, the variation of horizontal strength with variation in the vertical stress is measured. The shear test is similar in that the variation of shear strength, with change in the stress applied normal to the plane of shear, is measured. When the results of the shear tests are plotted, the result is a straight line at an angle ϕ to the horizontal (Fig. 3.20); the intercept of this line with the vertical axis gives the value of cohesion in the Mohr–Coulomb expression (Eqn 3.29).

3.4.9.2 Apparatus

The apparatus used is a standard controlled-displacement, multispeed, reversible shear box, which usually has a 6 cm square sample. The details of construction can be seen in Figure 3.21. The shear box consists of two halves, which can be secured together for consolidation but are free to move during shearing. The sample is placed between the drainage plates. These latter are very important if tests are to be carried out under drained conditions, which permit the free exchange of soil water between test and environment and therefore equalisation of pore-water pressure. The inner box, the shear box itself, is surrounded by a large water-filled box that permits water to enter or leave the test sample at any time. The bottom drainage plate is the bottom pad of the shear box, which is topped by a porous drainage plate and the grid plate. The sequence is reversed at the top of the sample with the pressure pad as the topmost plate. Normal stress is administered to the latter via a ball-bearing contact.

Basically, the box is split in two; the top surface of the soil is loaded with a known weight to give a loading stress usually of 50–350 kPa. Displacement is imposed on the bottom half of the box via a motor-driven ram.

For simple interpretation of the results, the top half of the box should remain fixed but since, in transmitting load onto the proving ring, it compresses the ring, it moves by the amount indicated by the proving ring dial gauge. Therefore, it is the relative displacement that is the lateral displacement.

3.4.9.3 Limitations

It is not possible to produce shear stress if shear displacement is not also applied to the soil, but due to this displacement the area of the plane of shear varies with the amount of displacement and so the amount of shear stress is not uniform. This must be allowed for in any calculations. It is also probable that there is a concentration of shear stress at the edges of the box and that stress progressively increases until failure is reached. Failure of the soil thus starting from the outside gradually moves towards the centre of the sample (Morgenstern & Tchalenko 1967).

The major disadvantage of the direct shear apparatus is the effect of particles of large size upon the resulting value of ϕ. Chandler (1973b) demonstrated the manner in which ϕ is dependent upon particle size until the ratio of sample thickness to maximum particle size (D/d) reaches 10–12. After this ϕ is independent of particle size. Therefore, care is required in designing the experiment to ensure that D/d values are higher than about 10.

3.4.9.4 Drained and undrained tests

Although direct shear tests are simple in their elementary theory and results, it was pointed out by Terzaghi (1936), soon after the introduction of the test, that different results can be obtained by altering the conditions under which the water drains from the sample.

There is no provision for controlling the drainage or for measuring the pore pressure in the sample, and the test cannot be performed with partial pressures. Provided that the test is carried out rapidly, the undrained strength of saturated clays may be determined. However, it is impossible to carry out an undrained test on sand, because of the high permeability, in consequence of which the pore pressures dissipate almost as soon as they are generated by the shear stresses.

Alternatively, drained tests, carried out so slowly that the pore pressure is virtually zero throughout, may be used to determine C' and ϕ', the shear strength parameters with respect to effective stress. The theoretical condition is not possible, but Gibson and Henkel (1954) have shown that at 95% consolidation (i.e. 5% undissipated pore-water pressure) virtually the whole drained shear strength is mobilised. The displacement rate is timed so that 95% dissipation is achieved at failure; therefore, a lower degree of equalisation can be accepted at earlier stages in the test. The time (in minutes) to failure (t_f) that allows 95% dissipation of pore-water pressure in the standard 6 cm shear box can be expressed as

$$t_f = 12.73 t_{95} \tag{3.34}$$

where t_{95} is the time to 95% consolidation derived from Taylor's (time)$^{1/2}$ curve-fitting method (see Scott 1974, p. 113).

3.4.9.5 *Undrained saturated cohesive samples*

In this case, the application of a uniform total stress increment results in an equal rise in pore-water pressure since there is no volume change while the shear stress is being

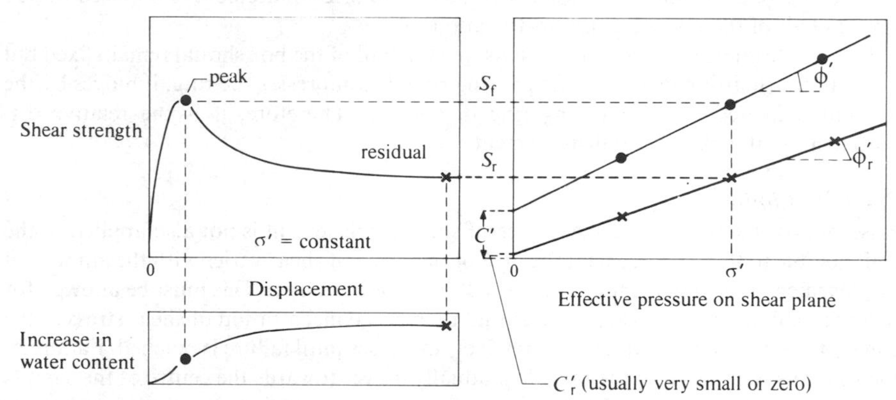

Figure 3.20 Relationship between peak and residual shear (after Skempton 1964) from direct shear tests. The text gives an explanation of the symbols.

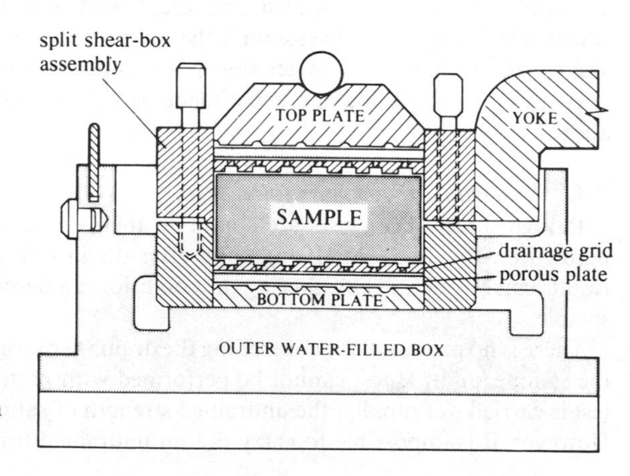

Figure 3.21 Cross section through a direct shear-box apparatus.

applied. The effective stress is therefore unchanged and is independent of the increment in total stress. There is also no way of determining the effective stress since pore-water pressures cannot be measured in the direct shear apparatus. When the soil is loaded to failure in the machine by the application of a principal stress difference (shearing), the shear strength, which is a function of the effective stress, is also constant for all values of total stress. Then, in terms of total stress:

$$t_f = \text{constant} = C_u \qquad \phi_u = 0 \tag{3.35}$$

where C_u and ϕ_u are the shear strength parameters in terms of total stress for undrained conditions. In Figure 3.20 the graph of shear strength against normal stress would be a straight line of zero slope and intercept C_u.

3.4.9.6 Drained saturated samples

When an excavation is made in a saturated soil, the resulting stress relief produces an immediate drop in pore-water pressure in the sides and below the base of the excavation. In this 'short-term' condition, the strength parameters would be the undrained shear strength. With time, the pore-water pressures will reach an equilibrium state, accompanied by a corresponding increase in the water content of the soil around the excavation. The time dependence of the swelling process may be the cause of many delayed or 'long-term' failures and is considered in detail for London Clay by Vaughan and Walbancke (1973). The phrase 'long-term' is used here to (Chandler 1973b):

> indicate that the pore-pressures within the slope have risen from the initial transient low values of the short term condition and have reached or closely approach the constant values that would ideally occur in the long term, given constant boundary conditions.

Experience therefore shows that strength (S) can only be expressed in terms of total stress (σ_n) in a few exceptional cases. Coulomb's expression must generally be restated in the form (Terzaghi 1925):

$$S = C' + \sigma_n' \tan \phi' \tag{3.36}$$

where σ_n' is the effective normal stress and C' and ϕ' are approximately constant parameters (Fig. 3.20) expressing the shear strength in terms of effective stress.

For all practical purposes, it may be assumed that

$$\sigma_n' = \sigma_n - U \tag{3.37}$$

(where U is the pore-water pressure), and if pore-water pressures are kept constant during the direct shear test (i.e. drained conditions), ϕ' and C' may be empirically derived.

This relationship between shear strength and effective stress, first suggested by Terzaghi (1925), is found to be approximately valid for a wide variety of materials and conditions of loading.

3.4.9.7 Residual shear strength

A number of mechanisms have been proposed to explain delayed failure in fissured clays. It has been postulated that drained strength may reduce with increasing loading time (Henkel 1957) as a result of a reduction in the apparent cohesion C'.

Vaughan and Walbancke (1973) have proposed that equilibrium of pore pressure in cutting slopes (i.e. 'first-time' failures) is accompanied by a reduction in the apparent strength, and failure may be delayed by the rate at which swelling can occur. If this is so, it is unnecessary to presume a significant drop in drained strength with time in these 'first-time' failures. However, it is important to differentiate between these 'first-time' failures and the reactivation of old failure surfaces where the apparent strength will be overestimated by laboratory tests unless the residual strength is determined (Skempton 1964).

Skempton (1964) carried out direct shear tests to large displacements on samples of overconsolidated clays (Fig. 3.20). The common method is to use multiple traverses of the direct shear machine, which has the ability to run in reverse as well as forward modes. As the clay was strained, it built up an increasing resistance. However, under a given effective normal pressure there is a limit, and beyond this the sample fails. This limit is known as the peak strength (S_f) and it was normally customary to stop the test after this limit had been reached. Skempton, however, continued the test beyond this peak value and found that a residual shear strength (S_r) existed (Fig. 3.20). The tests were carried out at different normal pressures, and both peak strengths and residual strengths were in accord with the Coulomb–Terzaghi model (Fig. 3.20)

$$S_f = C' + \sigma'_n \tan \phi' \tag{3.38}$$

$$S_r = C'_r + \sigma'_n \tan \phi'_r \tag{3.39}$$

where $C'_r \simeq 0 \neq C'$ and $\phi \geqslant \phi'_r$.

Petley has shown (Skempton 1964, p. 83) that the residual angle of shearing resistance of the Walton Wood Clay, when normally consolidated from a slurry at the liquid limit, is comparable with the value derived from tests on undisturbed samples. Skempton suggests that the residual strength of a clay, under any given effective pressure, is the same whether the clay has been normally consolidated or overconsolidated. The value of ϕ'_r should then depend only on the nature of the particles. When the value of ϕ'_r is plotted against clay content (really the percentage less than 2 μm), a generalised relationship is revealed so that the higher the clay content the lower the value of ϕ'_r (Skempton 1964, Rouse 1967, 1975, Rouse & Farhan 1976). The explanation of the residual shear strength for clay soils involves the orientation of layer-latticed minerals (Skempton 1964 p. 81, Goldstein *et al.* 1961) and the frictional properties of soil minerals (Skempton 1964 p. 84, Horn & Deere 1962). Two very important papers are Morgenstern and Tchalenko's (1967) paper on particle orientation in kaolin samples and Kenney's (1967) paper on the influence of mineral composition on the residual shear strength of natural soils. In this latter paper, Kenny showed that residual shear strength is primarily dependent on mineral composition and that it is not related to plasticity or to grain size characteristics. The massive minerals, such as quartz, feldspar and calcite, exhibit $\phi'_r \geqslant 30°$; the micaceous minerals such as hydrous mica exhibit $\phi'_r \simeq 17°$; the montmorillonite minerals exhibit $\phi'_r \leqslant 11°$. Hence, the natural soils that could be expected to have small residual strengths are those containing montmorillonitic minerals, whereas those soils that contain primarily micaceous minerals and massive minerals could be expected to have larger residual strengths.

Tropical clay soils exhibit unusual geotechnical properties so that they contrast markedly with temperate clay soils (Rouse *et al.* 1986, Rouse & Reading 1985). Tropical clay soils are highly porous with high void ratios and high field moisture contents. Values of ϕ'_r are high, although they vary with clay mineralogy (allophanes 30–38° > smectoids 32–35° > 'dry' kandoids 33–34° > 'wet' kandoids 25–29°). All are above values for temperate clays and are more akin to granular soils. Also none of the tropical clays seem to show any change of τ'/σ'_n as σ'_n increases (Rouse *et al.* 1989).

3.4.10 Triaxial testing of soils

3.4.10.1 Introduction

The triaxial apparatus is the most widely used and most versatile equipment for investigating the shear strength characteristics of soils. The apparatus consists basically of a pressurised chamber in which the sample of soil is subjected to horizontal and vertical compressive stresses. In the most common tests, the horizontal stress is maintained at a constant value and failure is produced by increasing the vertical stress, but it is also possible to perform tests in which the vertical stress on the sample is decreased while maintaining a constant horizontal stress, or to perform tests in which the vertical stress is held constant and failure occurs by increasing (or decreasing) the horizontal stress. The triaxial apparatus includes provisions for varying the conditions of drainage from the sample, and it is this facility which gives it the greatest advantage compared with the direct shear (or shear box) apparatus. Other advantages include the ease with which vertical deformation and volume change can be measured, and the ability to observe pore-water pressures throughout the test. The apparatus is shown diagrammatically in Figure 3.22. The triaxial apparatus consists essentially of:

(a) a triaxial cell, in which the test specimen is placed;
(b) equipment to provide the pressure to the fluid surrounding the test specimen; and
(c) apparatus for applying the increase in vertical load, which produces failure in the test specimen.

3.4.10.2 The triaxial cell

The triaxial cell consists of a baseplate, containing a central pedestal, and a Perspex cylinder, which is attached by tie rods to metal upper and lower plates. It is important that the design of the cell incorporates O-rings in order to avoid leakage when the cell is subjected to hydraulic pressure.

The pedestal contains a central drain, which is used when volume change is allowed during a test or when it is necessary to monitor changes in pore-water pressure. The baseplate also contains connections for the application of the cell pressure and also for a connection to drainage from the top of the sample. The most commonly used sample sizes are 38 mm diameter and 100 mm diameter, but sizes between these two values can be used by changing the pedestal. Some special cells have been manufactured for samples up to 300 mm diameter.

The sample under test is enclosed in a thin rubber membrane, which is sealed onto the pedestal and upper loading cap by rubber O-rings. The sample is therefore totally enclosed, thereby enabling complete control of the drainage conditions to be achieved.

The top plate of the triaxial cell contains a central plunger, which is used to apply the vertical load to the sample.

3.4.10.3 Application of cell pressure

The constant fluid pressure surrounding the sample can be achieved by the use of (a) a constant-pressure mercury pot system, or (b) an air–water cylinder.

The apparatus used must satisfy two requirements – first that the pressure is maintained constant throughout the duration of the test (which may be several weeks), and secondly that the supply system can accommodate changes caused by volume

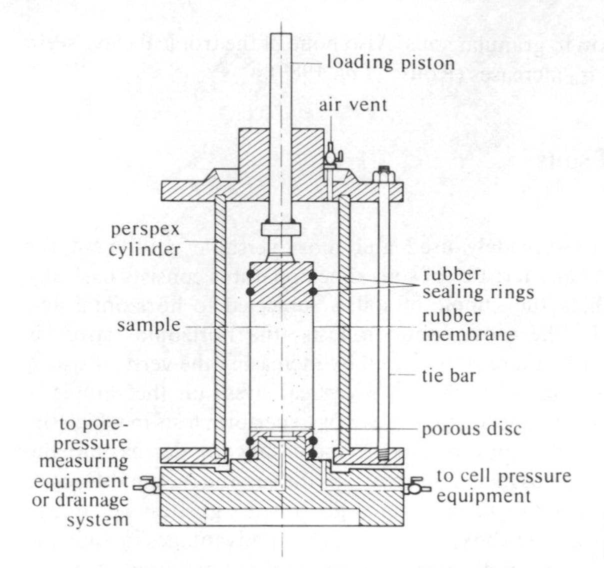

Figure 3.22 Cross-sectional diagram of a triaxial cell.

changes in the sample under test and any leakage. The cell pressure is usually monitored by a pressure gauge connected to the cell pressure lines, although in some circumstances (for example, where low pressures are used) it may be more accurate to use a manometer.

3.4.10.4 Application of vertical load

The most widely used form of triaxial test is one in which the sample is failed by gradually increasing the vertical load on the sample. The load is transmitted to the sample by the plunger fitted into the top plate of the triaxial cell. The increase in vertical load is usually achieved by using a loading frame, which is fitted with an electric motor and gearbox, enabling a wide range of compressive strain rates to be obtained. This type of test is frequently referred to as 'strain-controlled'. As an alternative to the loading frame method, the vertical load on the sample can be increased incrementally, with the size and frequency of the increment being controlled. This testing arrangement produces a 'stress-controlled' test.

3.4.10.5 Preparation of triaxial specimens

Tests may be performed on either undisturbed or remoulded specimens. The undisturbed specimens are normally obtained by extruding the soil from a sampling tube, in which it has been stored, and trimming to the required specimen size. After weighing the specimen and recording its dimensions, the specimen is placed on the pedestal in the cell base. If drainage is to be allowed from the specimen during the test, it is usual to place a porous disc between the pedestal and the test specimen. The top loading platen is then placed in position on the specimen, and the rubber membrane placed around the specimen. A 'membrane stretcher' is used for this operation, the stretcher consisting of a tube of appropriate size with a small pipe connected at its midheight. The rubber membrane is placed inside the stretcher, with its ends folded back down the outside of the stretcher. When suction is applied to the pipe connection, the membrane enclosed in the stretcher can be quickly and easily positioned around the specimen. O-ring seals are then placed over the membrane to seal it to the upper loading platen and base pedestal. In order to ensure that no leakage occurs, it is good practice, particularly when testing large specimens, to use two or more rings at each end of the specimen.

The Perspex cell is then assembled around the specimen, and securely fixed to the base of the cell. The piston is then lowered to rest on the ball seating in the upper platen, taking great care to avoid subjecting the test specimen to any vertical load.

The triaxial cell is then filled with water, from a header bottle or deaerated supply system, with the air bleed valve at the top of the cell open. When the cell is filled, this valve is closed, and the cell is pressurised using either a mercury pot system or an air–water cylinder. After the desired confining pressure has been applied, and assuming that no leaks are apparent, the test can proceed.

3.4.10.6 Types of triaxial test

Failure in the triaxial test takes place at particular combinations of vertical (σ_1) and horizontal (σ_3) stresses. It is important to appreciate that the failure condition is achieved by applying the stresses to the sample in two distinct steps. In the first step, the sample is subjected to the cell pressure (σ_3), which acts equally on all sides of the sample. Secondly, the sample is brought to the failure condition by increasing the vertical stress. Clearly, if failure occurs under a particular combination of stresses, σ_1 and σ_3, the magnitude of the *increase* in vertical stress is $(\sigma_1 - \sigma_3)$, and this is termed the deviatoric stress.

Drainage of water from or into the test sample can be allowed or prevented during either or both of these loading stages, and this means that three distinct types of test can be performed:

(a) Unconsolidated, undrained test, in which no drainage is allowed during either the application of the cell pressure or the application of the deviatoric stress.
(b) Consolidated, undrained test, in which drainage is allowed during the application of the cell pressure but not during the application of the deviatoric stress.
(c) Consolidated, drained test, in which drainage is permitted during both the application of the cell pressure and the application of the deviatoric stress.

The choice of drainage condition is controlled by the practical problem that is being investigated, and it is essential that the correct decision is made at the onset of any testing programme. A detailed description of the application of the different types of test is given by Bishop and Henkel (1962).

Drainage, where it is allowed, is usually monitored by means of a burette, which is connected either to the top or bottom drainage outlet from the triaxial cell. The time required for drainage to be completed following the application of the cell pressure depends on the dimensions of the sample being tested and its coefficient of permeability

3.4.10.7 Application of deviatoric load

Following the application of the cell pressure, and subsequent consolidation, if any, the sample is brought to failure by increasing the vertical load on the sample. The rate at which this is done depends again on the type of test being performed. Thus, an undrained test would be performed at a rate of about 2%/min whereas a drained test, involving the flow of water out of the sample, must be performed at a much slower rate. The appropriate rate for drained tests can be determined using the methods suggested by Bishop and Henkel (1962) and Head (1986).

When the appropriate rate has been chosen, the motor on the loading frame is started and readings of the proving ring or load cell are taken at appropriate intervals together with the reading on the burette (for volume change of the sample) and the dial gauge (for axial strain). The test continues until failure occurs, this being defined usually as a decrease in the deviatoric load; if no decrease occurs, the failure load is taken as that at 20% axial strain.

3.4.10.8 Calculations

The deviatoric stress is defined by

$$\sigma_1 - \sigma_3 = P/A \tag{3.40}$$

where P is the applied vertical load recorded on the proving ring or load cell, and A is the cross-sectional area of the sample. The cross-sectional area A is computed assuming that the test specimen deforms as a right cylinder; thus

$$A = A_o \frac{1 - \Delta v/V_o}{1 - \Delta l/l_o} \tag{3.41}$$

where Δ_v is the change in volume measured by burette (drained tests only), V_o is the initial volume (after consolidation), Δl is the change in axial length of specimen, l_o is the initial length (after consolidation) and A_o is the initial cross-sectional area (after consolidation).

The results are normally presented as a graph of deviatoric stress ($\sigma_1 - \sigma_3$) against axial strain ($\Delta l/l_o$). The value of the deviatoric stress at failure is then used to draw a Mohr's circle of stress (Carson 1971a, Whalley 1976). From a series of tests on identical samples, a failure envelope may be drawn that is tangential to the series of Mohr's circles, hence defining the shear strength parameters.

3.4.10.9 Comments and sources of error

Although the triaxial apparatus is widely used, it is important to be aware of its limitations. Some of these are discussed below:

(a) The aim in all laboratory testing is to reproduce in the laboratory exactly the conditions to which the soil is subjected in the field. This is difficult to achieve in any laboratory apparatus. In the triaxial apparatus, for example, the sample is consolidated under an isotropic stress system, which is unlikely to exist in the field.
(b) The correct choice of the type of test (i.e. drained or undrained) is of fundamental importance.
(c) When undrained tests are performed, it is possible to monitor changes in pore-water pressures using either electronic transducers (see below) or the null-indicator technique devised by Bishop and Henkel (1962).
(d) The rate of drainage from the test specimen can be accelerated by using filter strips between the sides of the specimen and the rubber membrane.
(e) The rubber membrane surrounding the sample has a constraining effect (as do the filter strips, if used) and, in the sophisticated tests, a correction should be made to the results to allow for this effect.
(f) One of the most common sources of error in the triaxial cell occurs due to friction between the loading ram and the top of the triaxial cell. The ram should be carefully polished, highly greased and able to move freely. However, excessive leakage must also be avoided, and this can be achieved by adding a thin layer of oil (e.g. castor oil) to the top of the cell fluid. Clearly, the obvious remedy is to use a load cell mounted inside the triaxial cell.

3.4.11 Residual strength measurements

When a soil is subjected to gradually increasing shear strain, the shear stress on a plane gradually increases until a maximum is reached. If straining continues beyond this maximum value, many soils exhibit strain-softening characteristics, in which the shear stress gradually decreases until ultimately it attains a constant value, known as the residual strength. The process is shown diagramatically in Figure 3.20.

It is generally necessary to apply very large deformations to the sample in order to reach the residual condition. With the great majority of soils, this strain cannot be achieved directly in either the direct shear or triaxial apparatus. In the shear box apparatus, residual strengths have been measured using the 'reversal' technique (Skempton 1964, Skempton & Petley 1967) in which the sample is sheared in opposite directions until the residual condition is reached. An alternative technique involves the use of a preformed failure plane, which then enables the residual condition to be reached with relatively small deformations. Both of these techniques have shortcomings, and possibly the most satisfactory method for determining residual shear strength parameters involves the use of a 'ring shear apparatus', as described by Bishop *et al.* (1971).

The ring shear apparatus contains an annular sample that is subjected to a vertical stress. The shearing deformation is applied by rotating the base of the specimen while maintaining the top in a fixed position. Unlimited deformation can therefore be obtained.

The design of the apparatus is complex in order to eliminate as many potential sources of error as possible, and tests can be lengthy. An additional problem involves the preparation of undisturbed annular specimens.

Published data suggest that the values of the residual shear strength parameters obtained from the ring shear apparatus are generally less than those obtained from tests in the shear box or triaxial apparatus.

3.5 Measuring and recording devices

Many of the methods discussed in this book produce considerable quantities of data, or at least they could do so if a means to take measurements and record them were available. It is particularly important for some tests that a continuous record be obtained (e.g. XRD or DTA). To this end, the manufacturers have provided a recorder of some type, which, when coupled to the instrument, provides a continuous record for subsequent examination. The cost of such instrument–recorder packages may be high, but both are required. Other tests may require only one pertinent value (e.g. a peak stress in a direct shear box test), yet the duration of the experiment could be many hours. It is thus useful to couple a recording device with an apparatus purely for the sake of convenience and labour-saving. Furthermore, it may be possible to obtain more information from a test with many continuous output data rather than from a filtered version based on operator convenience (e.g. for spectral analysis). It is this latter type of data collection that can also be used for field experiments when events are expected but the event time or return period is unknown. A third type of test may produce one answer only but many samples may have to be examined. A good example could be from a range of chemical determinations: the measurement of many cations by atomic absorption spectroscopy or by colorimetry.

Although there may be many differences in the way the tests are made and even the observations recorded, some basic principles can be outlined. Most measurement systems can be considered in three parts, which apply in laboratory or field at high or low cost:

(1) A device that detects the variable to be measured (the *measurand*), which can be a physical (e.g. a position or a pressure) or an electrical quantity (e.g. a voltage). It frequently happens that the measurand can be most easily recorded in some other form to give a *signal* (e.g. conversion of a pressure to an electrical voltage). A device that performs this is called a *transducer*.
(2) A transmission path, which conveys the signal and may incorporate a signal conditioner to amplify or filter it.
(3) A display or recording device.

To these could also be added:

(4) A device for further analysis, statistical or mathematical, which most frequently may be a computer.

To some extent, manual measurement could be identified with these parts, although the examples given previously illustrate the desirability to use some sort of recording system, and it is with this that we will be primarily concerned in this section. From the outset, some thought needs to be given to any instrumentation and recording system; thus:

(a) Are the measurements best recorded automatically?
(b) If the answer is 'yes', is the money available?

Cost determination is governed by instrumental sophistication; thus, another set of questions needs to be asked:

(c) What are the primary variables to be measured? If there is more than one measurand, it may be possible for one system to do all the jobs in (1), (2), (3) and (4) above.
(d) What accuracy and precision are required for each measurand?
(e) What range of the measurand will be required?
(f) How many data points will be required and over what length of time will they have to be made?
(g) What frequency response is required (for dynamic experiments)?
(h) Is the transmission line very long or likely to present mechanical or electrical problems? Various interference or 'noise' problems might appear here and cable shielding or signal filtering may be necessary.
(i) How will the data be processed? (This may determine the type of recorder.)
(j) Are transducers, signal conditioners and recorders available 'off the shelf' or will they have to be constructed?

In any one application or experiment, not all these factors will necessarily come into play, and indeed for some tests (especially the more sophisticated and costly) the manufacturer may have solved these problems – that is why they are expensive. In the long run it *may* be cheaper to buy commercially than to do it oneself. One should not, however, be lured into thinking that even a bought system may do the job perfectly, or as one wants it. One should still weigh up the factors outlined above when reviewing one's requirements. Above all, one needs to be sure that one understands the need for the test in the first place and how it is performed. What works perfectly in an analytical laboratory may not tolerate field operation; chemicals or temperatures may interfere with its operation; geomorphological effects may influence the result it gives. To clarify this last point, consider a sediment whose shear strength is required; if particles larger than 6 mm are a substantial proportion of it, then a 60 mm square direct shear box will probably be inadequate; a larger one is necessary. Again, a colorimeter will be affected by turbidity produced by clays or silts in suspension. Furthermore, one should not expect a definitive answer from any test; full clay mineralogical determinations are often not given by X-ray diffraction alone, and other techniques will be required. We have to consider such factors at many stages, of course, but it is important in the present instance where even a simple chart recorder may cost several hundred pounds.

Before considering recording devices, it is necessary to look at our three other components in the system in some more detail; see also Figure 3.23, which gives a schematic interpretation of possible ways to record data (three examples being given).

3.5.1 Transducers

The type of signal most easily processed in stages (2) and (3) is electrical, although fluid pressure may be used, e.g. as pneumatic pressure in a Bourdon gauge.

Here we will look only at electrical signals. Many devices are now available for converting physical quantities to electrical quantities. Geomorphologically, we may need to sense temperature, position, displacement (or strain), force, pressure or velocity most often. The device selected for any one of these will again have to be done with care. For example, to measure temperature, thermocouples, thermistors, platinum resistances, quartz-crystal thermometers or radiation pyrometers, all of which are temperature–electrical transducers, can be used. Each has advantages and disadvantages regarding accuracy, sensitivity, precision, range, reliability and cost. It is not possible to discuss each of these, still less transducers for other types of variable. In the first place, reference should be made to suitable physics (e.g. Nelkon & Parker 1970), engineering (e.g. Earley 1976) or environmental (e.g. Fritschen & Gay 1980) texts to discover what is available.

Having selected possible types, attention must still be given to the other parts of a system in order to make a final choice.

3.5.2 Signal transmission and signal processing

Having obtained a signal from a transducer, it must be carried to the recording device. In some cases, it may well be some distance away and needs to be carried by a long wire or even sent by telemetry (e.g. from wave buoys or remote river gauging stations). Before we see what the problems are here, there is a need to distinguish between analogue and digital signals. *Analogue* signals are those which are replicas of another signal but in a different form; variations of the measurand are followed by the analogue electrical signals as, for example, a change in voltage. It follows it uniformly (i.e. it is repeatable, subject to hysteresis), although it may not necessarily be linear, as is the case with thermistors for example. Most transducers produce analogue signals. The most convenient form for data analysis (and often to get over transmission and telemetry problems) is a *digital* signal. In digital signals, it is the rate of a series of equal-amplitude pulses that conveys the information.

Some transducers produce digital outputs but, more usually, an analogue-to-digital converter (ADC) is used to produce a digital signal. This conversion is frequently employed in the signal processing part of the sequence (which may be before or after the transmission) and is common nowadays because the recording and processing stage can be carried out most easily by digital techniques (e.g. in a binary coding).

Signal conditioning is frequently employed when there are a number of electrical signals and output, so that a common range of voltages or currents is output. Filtering of the signal may also be done in these units. Another important function is to provide a regulated voltage or current to a transducer which excites it. For instance, a linear variable differential transformer (LVDT), which is often used to provide the signal showing the movement of a shear box or deformation of a proving ring, needs a regulated

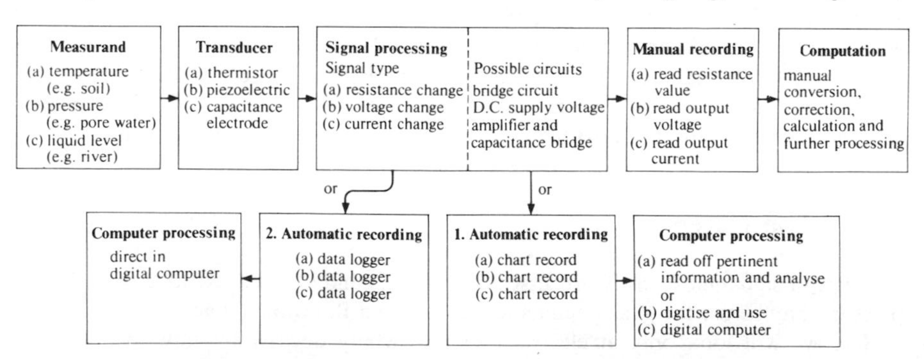

Figure 3.23 The detection, measurement, recording and analysis of a measurand. Three types of measurement are shown as examples.

input voltage of 10 or 12 V. The output of such a device could be in millivolts or volts. Other types of sensor require a different operation, e.g. a thermocouple requires an (electronic) ice reference, a thermistor a bridge circuit.

3.5.3 Display and recording devices

3.5.3.1 Analogue techniques

The signal (analogue or digital) from a transducer usually needs to be monitored visually in some way. It is important that the system be capable of this, even if data processing is carried out directly, because the component parts need to be tested. It is obviously a waste of time if an expensive system fails to record data because the transducer or transmission line does not work. Analogue signals can be used directly to give deflections on a suitable voltmeter or ammeter. For recording such a signal, it must be fed into a suitable device. Two main types of these can be considered: chart recorders and oscilloscopes. Oscilloscopes are devices similar to a television, but where the electron beam is made to traverse the screen in accordance with a selected, horizontal, timebase and its y deflection is a function of the applied signal voltage. Such an instrument is particularly useful for examining dynamic phenomena, but it is of very wide utility, and reference should be made to a suitable text for discussion (e.g. Golding 1971).

Chart recorders come in different types and configurations, of which those most suitable for most 'geomorphological' jobs will be discussed. A chart is moved at a constant speed past the pen, whose deflection is proportional to the analogue signal. Thus the signal trace is given as a function of time (y–t recorder). (In some types, the chart does not move, but one signal fluctuates against another within fixed limits (x–y recorder); such a recorder is used to produce a trace output from a Coulter counter to give a histogram or cumulative distribution curve.) The impression of the pen can be as a normal fluid ink, felt tip, ballpoint or by pressing on sensitive paper. Felt tips are generally to be preferred in that they are easily replaced, trouble-free and different colours can be used. The speed of the chart movement can usually be selected from a range, either electronically or by changing gears manually. The range of deflection (full-scale deflection, f.s.d.) can also be selected on most types; thus output from some transducers may require an f.s.d. of 10 mV, others of 100 mV or larger. Clearly there is a problem if the output is set to a high value, as small variations cannot be resolved, or if a low-voltage f.s.d. is used which misses a value that goes off the scale. To counter this, some recorders have automatic back-offs (chart steppers), which move the pen back to zero if it goes off the scale. Expense increases the greater the range of speeds, voltage scales and other devices that may be fitted (back-offs, integrators, differentiators, etc.), but some recorders can be added to in modular form as and when they are required. This also applies to the number of pens fitted. Two-pen recorders are common, but six is not unusual. To save cost, dot printers are sometimes used. These do not give a continuous trace but dot the chart at selected intervals. This may be sufficiently close to a continuous record for many purposes and coloured inks make reading easy.

There are two main types of chart recorder pen activation, the galvanometric and servo (usually potentiometric). In the former, the pen behaves as if it were the end of a pointer on a voltmeter dial. Although this type is useful (as in an electrocardiogram, for example), it is more usual to use a servo recorder in which the signal is used to drive a motor that positions the pen on the scale according to the signal amplitude.

Chart recorders are relatively cheap and are reliable; some can be used with batteries for field applications. Certain instruments, e.g. X-ray diffractometers (using a single pen), and DTA (using two pens), are always used with such recorders because the experimental

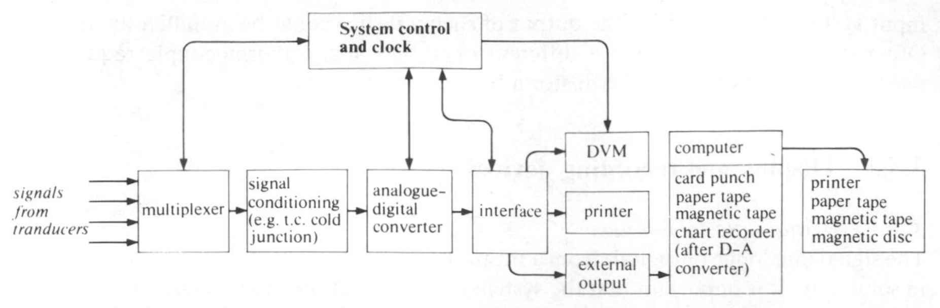

Figure 3.24 Schematic diagram of component parts of an automatic data logger.

time is generally short and it is important not to miss data. Furthermore, the shape of the curve gives a good first impression of data before subsequent detailed interpretation. Other experiments need less frequent sampling but use computer reduction of data for greatest efficiency, so modern techniques are leaning strongly towards digital techniques.

3.5.3.2 Digital techniques

Digital signals can be used to drive a display that shows the value as figures. Thus, a millivolt output can be shown directly on a digital voltmeter (DVM). In its simplest form, it may appear that there are few, if any, advantages over a more normal scale and pointer meter. However, digital techniques allow presentation to be scaled in appropriate units. For example, a digital thermometer can show temperature directly, and the same sort of scaling can also be done for modules that do signal conditioning. The greatest advantages in dealing with digital sources are derived from using a *data logger* (Fig. 3.24) to deal with several inputs almost simultaneously. The capabilities and capacity of the instrument are again a function of cost, but most will do the following: take analogue signals, condition them, produce a visual display for one selected channel on a DVM and output the selected signal to some recording medium. The important item is a crystal-controlled clock, which, with an operating system, selects each input signal at a given time interval and samples it (a multiplexer). This clock also controls the length of the scan period of all inputs as well as the time between successive scans. This can be made almost a continuous process or it might be decided that a sample once every hour or day is sufficient. The output medium may be of several types. For instance, paper tape is often convenient for reading to computer disc file for subsequent analysis. In this case, frequent sampling makes sure that no important points are missed and the automatic analysis reduces manual labour. There are some data loggers suitable for field use; Figure 3.24 shows a schematic layout of a data logging system with the possible output styles.

Some basic considerations of data loggers are given in Armstrong and Whalley (1985).

The advent of cheap, desk- or bench-top computing has changed the logging of data away from rather expensive, purpose-built systems to data acquisition on personal computers. For monitoring a large number of variables there is still likely to be a place for a dedicated laboratory data logger – and the sophistication of such devices has itself increased as the cost of microcomputers has come down. However, as many of the tests described here do not need a large number of channels, it is now cost-effective to buy 'add-ons' for several types of microcomputer to provide data logging facilities, even to the extent of providing one machine per test. When not in use as a logger they can always be used for 'normal' computer use. The main requirement is for an analogue-to-digital converter (ADC) to convert the signal (from appropriate signal conditioning/amplifier) and suitable software. The ADC should preferably be at least 12-bit rather than the 8-bit ADC sometimes supplied in microcomputers; see Armstrong and Whalley (1985) for further discussion. The software should be capable of performing all the necessary sequences required of a test. Similarly, the conversion of logged measurements (usually as voltages) to engineering units is a matter of relative simplicity through software. This may well be a cheaper alternative than providing hardware to convert or to linearise data. Usually, signal conditioning will still be needed to convert the transducer output to a given voltage range.

It is only possible to give a very brief outline of the systems possible and the uses to which they can be put in this section. The many problems that are almost bound to arise (matching inputs, cable capacitance, shielding, etc.) can only be touched on. It is most important that advice be sought from manufacturers and knowledgeable friends. Keeping an eye open for articles is well worth while, e.g. in journals such as *Laboratory Practice* and *Laboratory Equipment Digest*, as they have articles on available equipment and selection of suitable instruments. Manufacturers and distributors often have exhibitions at colleges, etc., so that seeing the capabilities of particular systems and discussing your problems can be a way of both seeing instruments and finding out about the subject. As well as obvious questions of recorder facilities, cost and so on, do not forget to ask if it can be run from batteries and about the current drain if it is likely to be used in the field, and how environmentally protected it is.

PART FOUR
PROCESS

4.1
Denudation and weathering

4.1.1 Introduction

In Part Four of this book we move on to a consideration of the various geomorphological processes that operate on, and sometimes produce, the materials that were the focus of Part Three. In particular we are concerned with the methods that are available for determining rates of geomorphological change.

The study of rates of geomorphological change has a long history, and the Uniformitarian paradigm of the nineteenth century, together with the development of fluvialism, involved basic evaluations of some crucial issues such as the time that was available for a certain degree of landscape change to occur. The first quantitative revolution in geomorphology occurred in the mid-nineteenth century (Chorley *et al.* 1964), at a time when some discharge and sediment load data were becoming available, topographic maps were becoming increasingly accurate, and engineers were requiring precise data on the movement of water and sediment. However, the Davisian geomorphology that followed was less concerned with processes and rates. There were, of course, notable exceptions, and the literature of the US Dust Bowl, for instance, is replete with detailed studies of rates of soil erosion (Bennett 1939). Indeed, erosion plot studies were started in Utah in 1912, and at the Missouri Agricultural Experiment Station in 1917. From 1929 onwards, with L. A. Jones, H. H. Bennett supervised the establishment of 10 further erosion experiment stations in Oklahoma, Texas (two), Kansas, Missouri, North Carolina, Washington, Iowa, Wisconsin and Ohio (Meyer 1982). Another notable exception was in Great Britain, where the Royal Commission on Coastal Erosion (1911) used the evidence of Ordnance Survey maps, together with information provided by local authorities and coastal landowners, to estimate rates of encroachment and accretion.

However, it was not until a quarter of a century ago that there was to be a concerted attack on the question of determining the rate of operation of a wide range of geomorphological processes, and the literature on this subject is now immense, so that 'trying to keep up with current publication is like cleansing the Augean stables' (Saunders & Young 1983, p. 473). A concern with rates of geomorphological processes arose from the mid-1950s in a variety of countries. For example, in Poland, geomorphologists undertook detailed evaluations of processes operation in polar and periglacial environments (e.g. Jahn 1961), and also developed techniques for assessing the intensity of soil erosion processes (e.g. Reniger 1955, Gerlach 1967). Scandinavian geomorphologists were also active at much the same time, and the comprehensive studies at Karkevagge by Rapp (1960a) remain a model for this type of work. Of equal significance was the work by Corbel and coworkers in France (Corbel 1959), and of various teams in the USA (e.g. Leopold *et al.* 1964).

4.1.2 General rates of denudation: problems of assessment

Before moving to a consideration of particular techniques for the measurement of particular processes, it is worth examining some of the general problems that are encountered in estimating rates of geomorphological change.

A first problem of comprehending and comparing different estimates of rates of denudation is that they are often reported in different forms. Thus many of the classic studies of denudation employ data derived from discharge of material by streams and consequently are expressed as the mass or volume of sediment removed from a catchment (e.g. kg, t, m^3, or ppm of material) averaged for a unit area of the catchment in a unit period of time to give a mean rate of land surface lowering. By contrast, data on transport rates of material on hillslopes (Sec. 4.2) are usually defined by the velocity at which the material moves (e.g. m yr^{-1}) or by the discharge of material through a unit contour length (e.g. $m^3\ m^{-1}\ y^{-1}$). One attempt to develop a measure of erosional activity that is applicable to hillslope and river channel processes alike is that of Caine (1976). He proposes that erosion involves sediment movement from higher to lower elevations, and so is a form of physical work. It requires a reduction in the potential energy of the landscape in which there are successive conversions from potential to kinetic energy that is finally dissipated as heat and noise. The change in potential energy through time is defined as

$$\Delta E = mg(h_1 - h_2) \tag{4.1}$$

in which m is the mass of sediment involved in the movement, g is the gravitational acceleration (9.8 m s^{-2}) and h is elevation, with $h_1 - h_2$ being the change in elevation during the interval between time 1 and time 2. The dimensions of work (E) are joules. For most stream channel situations

$$\Delta E = v\rho g(h_1 - h_2) \times 10^3 \tag{4.2}$$

where v is the volume of sediment and ρ is its density (Mg m^{-3}). For slope movements across a unit contour length when the thickness of the moving mantle is known directly

$$\Delta E = v\rho g(d \sin \theta) \tag{4.3}$$

where d is the slope distance (m) and θ is the slope angle.

A second difficulty with evaluation of all rates of geomorphological change is that there is a tendency for workers to undertake investigations in dynamic areas, or to investigate a particular process in a particular location because it is palpably highly effective and important. There is a bias against making measurements where little activity is apparent.

A third general difficulty in comparing data is caused by the storage of material within the system (Meade 1982). In many cases, slope debris may be translocated within a basin but not exported from it, being stored as colluvium in screes and fans or on terraces and flood plains. A scale effect may operate here, for hillslope debris is commonly delivered directly into the channel in a low-order stream but is held in store in higher-order catchments. Thus in a study of river basins in the southeastern USA, Trimble (1977) found that, while upland erosion was progressing at about 9.5 bubnoff units, sediment yields at the mouths of catchments were only 0.53 units. The delivery ratio was thus only 6%, with the difference being stored in valleys and channels. A consequence of this is that different values for rates of erosion per unit area are obtained depending upon whether

Table 4.1 Relations between sediment yield and catchment area (SAY=sediment yield ($m^3 km^{-2} yr^{-1}$), Y=sediment yield (ton $km^{-2} yr^{-1}$), X=area (km^2)).

Source	*Position*	*Basin size (km²)*	*No. of basins*	*Equation*
Scott *et al.* (1968)	Southern California	8–1036	8	$SAY = 1801\ X^{-0.215}$
Fleming (1969)	Mostly American, African and UK basins		235	$Y = 140\ X^{-0.0424}$
Sediment. Eng. (1975)	Upper Mississippi River; *a* varies with 'land resource areas'			$SAY = aX^{-0.125}$
Strand (1975)	Southwestern USA		8	$SAY = 1421\ X^{-0.229}$
	Equations for the upper limits of the silting rate in reservoirs			
Khosla (1953)	Different parts of the world	<2590	89	$SAY = 3225\ X^{-0.28}$
Joglekar (1960)	95 American, 24 Indian, 10 European, five Australian and five African basins		139	$SAY = 5982\ X^{-0.24}$
Varshney (1970)	Northern Indian mountain basins (north of Vindhyas)	<130		$SAY = 3950\ X^{-0.311}$
	Northern Indian plain	<130		$SAY = 3920\ X^{-0.202}$
	Northern India	>130		$SAY = 15340\ X^{-0.264}$
	Southern India	<130		$SAY = 4600\ X^{-0.468}$
	Southern India	>130		$SAY = 2770\ X^{-0.194}$

Source: from various authorities cited in Jansson (1982, table 5).

the study area is large or small. This is, among other things, due to the different transport lengths involved. Therefore, it is hazardous to compare rates of erosion derived from plot studies with those derived from drainage basin studies (Jansson 1982). Table 4.1, which is derived from numerous sources in Jansson (1982, table 5), suggests that, while there is a considerable scatter in the correlations between sediment delivery and basin area, the sediment yield varies with the − 0.04 to − 0.47 power of the drainage area. This inverse relationship may be due to several factors:

(a) Small basins may be totally covered by a high-intensity storm event, giving high maximum erosion rates per unit area.
(b) The relative lack of floodplains in small basins gives a shorter sediment residence time in the basin and a more efficient sediment transport.
(c) Small basins often have steeper valleyside slope angles and stream channel gradients, encouraging more rapid rates of erosion.

A fourth problem is that within a catchment there will be marked local variability in the operation of individual processes, so that detailed sampling may be required if some representative value is to be obtained. As Carson (1967) demonstrated, some soil properties, including infiltration capacity, have a tendency to show a large range over small areas, and this variability will inevitably be reflected in the operation of processes related to it, such as overland flow.

A fifth problem is that as well as being highly variable according to spatial scale and location, the rate of operation of geomorphological processes may be highly variable in time as a response to climatic and other environmental changes and fluctuations. The phenomena responsible for erosion and landscape formation range from those that are normally relatively uniform or continuous in operation (e.g. removal of matter in solution by groundwater flow), through seasonal, periodic and sporadic phenomena to catastrophic phenomena that are abnormal and anomalous (e.g. a major flood caused by an extreme combination of meteorological and hydrological conditions). It is clear that the periodicity and frequency of dominant and formative erosional events in any given area may vary widely, depending on the processes involved and the variables that control thresholds of activity and recovery rates. Thus the reliability and utility of short-term monitoring of rates may be questionable, and there may be advantages to be gained from using historical and archaeological sources (Hooke & Kain 1982). In addition there are the considerations posed by longer-term changes in climate at various scales, such as the Little Ice Age, the Mediaeval Little Optimum, the Holocene neoglaciations, the early Holocene pluvial of low latitudes, and the major glacials, interglacials, stadials and interstadials of the Pleistocene. We cannot necessarily accept present-day measurements, which are taken in an environment that is probably atypical of the Pleistocene and even more atypical of longer timescales, as being representative of geological rates of erosion.

Rates of sediment accumulation are also highly variable through time, and Sadler's (1981) examination of nearly 25 000 published rates of sediment accumulation shows that they span at least 11 orders of magnitude. More importantly, he also found that there is a systematic trend of falling mean rate with increasing timespan. Although measurement error and sediment compaction contribute to such trends, they are primarily the consequence of unsteady, discontinuous sedimentation. Thus, once again, short-term rates of accumulation are a very poor guide to long-term accumulation regimes. The length of the measured time over which processes operate is equally crucial for geomorphological and tectonic processes (Gardner *et al.* 1987).

A major problem with trying to extrapolate present-day estimates of rates of geomorphological change back through time is that 'natural' rates have been modified to an increasing degree over the last two to three million years by the presence of human activities (e.g. fire, deforestation, ploughing, grazing, construction, pollution, etc.) in the landscape.

4.1.3 Gross denudation

In turning to a consideration of specific techniques for calculating rates of geomorphological change, it is logical to start with those methods that have been adopted to calculate overall rates of denudation for relatively large areas over relatively long timespans. The various approaches here are essentially geological in nature.

The basis of Ruxton and McDougall's (1967) approach to estimating rates of denudation in Papua was that, if a dissected volcanic landscape possessed earlier or 'original' forms that could be reconstructed and dated by potassium–argon dating, then, by measuring the volume of material removed, denudation rates could be calculated. Measurement of the volume of material removed was achieved by using generalised contours to reconstruct the original surface of a strato-volcano, and then estimating the amount of ground lowering of concentric sectors distant from the centre by measuring the difference between the present-day and the original cross-sectional areas.

Clark and Jäger (1969) used isotopic dates on biotites in the European Alps to infer long-term rates of denudation. They believed in their model that the apparent ages of biotites from rocks metamorphosed during the Alpine orogeny represent the time at which the biotites had cooled to the point where loss of radiogenic elements no longer occurred. By calculating the required degree of cooling, and estimating therefrom the depths the rocks were at a given date in the past, they were able to calculate rates of denudation. Similarly, in their study of Mesozoic denudation rates in New England, Doherty and Lyons (1980) used the apparent discrepancies between potassium–argon dates and fission-track dates of zircon and apatite for the White Mountain Plutonic–Volcanic Series. The basis of this method is that the K-Ar dates give an initial date for the emplacement of the intrusives and that apatite and zircon have a distinctive temperature range at which tracks will begin to accumulate during cooling. The assumed temperatures are 200°C for zircon and 100°C for apatite. Thus the differences in the three determinations are temperature-related and the temperature is in turn related to depths of emplacement. Assuming a 25–30°C km^{-1} geothermal gradient, it is possible to calculate thereby depths at different times in the past and from that to calculate the thickness of overlying rock that has been removed.

The use of sediment volumes of known age on continental shelves was the technique favoured by Gilluly (1964) to calculate erosion rates over the long term in the eastern USA. By planimetry the area and volume of Triassic and younger sedimentary rocks was calculated. Adjustments were made for the difference in density between source rocks and the deposits derived from them. It was believed, largely because of the absence of calcareous rocks, that the sediments were supplied primarily as suspended and bedload material. Problems were encountered in establishing the source area of the sediments because of tectonic changes, river capture, etc., since the Triassic. Another problem was presented by the reworking of coastal-plain sediments, which meant that some of the measured volume of sediment had been reworked one or more times during its history.

Stratigraphic methods were also used by Bell and Laine (1985) to estimate the degree of erosion achieved by glaciation of the Laurentide region of North America. The technique involved estimating the sediment volumes caused by such erosion by means of studying the large number of seismic studies and Deep Sea Drilling Project boreholes that had become available in the previous 15 years.

Bishop (1985) used a combination of methods in south-east Australia to estimate the late Mesozoic and Cenozoic denudation rates:

(a) Average rates of post-early Miocene incision were determined by dividing the height difference between the present channel of the Lachlan River and the top of the adjacent basalt remnants (age 20–21 Ma), which themselves mark the sites of former channels of the Lachlan and its tributaries.
(b) The post-Oligocene rate of denudation was estimated by calculating volumes of sediment wedges that had been derived from catchments of a known area.
(c) Rates of late Cretaceous and Permian denudation were derived either from sediment volumes or from fission-track ages from granitoid rocks.

In limestone areas it may be possible to obtain measures of channel downcutting into bedrock over long time periods by dating speleothems in relict high-level cave tunnels related to formerly higher base levels. In the Yorkshire Dales of northern England, Gascoyne *et al.* (1983) used the $^{230}Th/^{234}U$ radiometric method (see Sec. 5.2).

Erosion plot studies have been widely employed to determine rates of denudation. Plots are usually of two types, those with constructed boundaries (e.g. of timber, concrete or asbestos sheeting set into the soil) and those without. Difficulties in the interpretation of measured losses from plots can, however, arise from the non-standardisation of plot design and variations in the length of study periods (Roels, 1985).

Finally, increasing use is now being made of isotopic tracers to estimate rates of denudation. On such technique employs beryllium-10 (^{10}Be), which is a cosmogenic isotope produced by cosmic rays interacting with the Earth's atmosphere and surface. Its average deposition rates are known. The affinity of Be for the components of soils and sediments is sufficiently high that it is effectively immobilised on contact, thereby allowing ^{10}Be to function as a tracer of sediment transport. To a good approximation all the ^{10}Be transport out of a drainage system is on the sediment leaving it. The number of ^{10}Be atoms passing a stream gauging station can be determined, if the annual sediment load is known, by measuring the concentration of the isotope in the sediment with a mass spectrometer. The ratio of ^{10}Be carried from the basin to that falling upon it provides an erosion index (e.g. Valette-Silver *et al.* 1986).

The most employed tracer is caesium-137 (^{137}Cs). This is a fission reaction product of atmospheric thermonuclear weapon tests with a half-life of 30.2 yr. It can be used as an indicator of soil loss since 1954 (see, for example, Loughran *et al.* 1987).

4.1.4 Weathering

There are essentially two main categories of weathering processes: those involving decomposition and those involving disintegration. The first category causes chemical alteration of the rock while the second causes rock breakdown without significant alteration of minerals. In most situations, both categories occur together, and the results of weathering normally reflect the combined effects of several different processes. The rest of this section is primarily concerned with the techniques that are available for determining rates of weathering, for simulating physical and chemical weathering, and for describing the nature and degree of weathering. Methods for the sampling and description of soils and weathered material are not included, as these have been reviewed in detail by Hodgson (1978). Three excellent compilations that give an indication of the range of modern studies of chemical weathering and chemical weathering techniques are those of Drever (1985), Colman and Dethier (1986) and Yatsu (1988).

A variety of methods are available for determining rates of weathering. These are classified in Table 4.2. Space precludes that all can be discussed in detail, and so particular attention will be paid to the use of river-water analyses, rock tablets and the micro-erosion meter.

4.1.5 Estimation of rates of chemical denudation by river-water analyses

Of all the methods that can be used to determine rates of chemical denudation or weathering, the most frequently practised has been that of using river quality and flow data.

Various formulae have been developed that can be used to give a figure of chemical denudation, normally expressed either as millimetres per 1000 years (mm/1000 yr) or as tons per square kilometre per year (t km^{-2} yr^{-1}). Williams (1964), for example, developed an earlier formula of Corbel (1959) for use when discharge data are not available. It was developed for use in limestone areas:

$$S = Etn/10d \tag{4.4}$$

where S is the amount of limestone removed in solution (m^3 km^{-2} yr^{-1}; or mm/1000 yr), E is the mean annual water surplus (dm), T is the mean total hardness (ppm, or mg l^{-1}), D is the rock density and $1n$ is the fraction of the basin occupied by limestone.

He also developed a formula for use where discharge has been measured:

$$S = \frac{fQTn}{10^{12}AD} \tag{4.5}$$

Table 4.2 Methods of determining rates of weathering.

(1) *Hydrological observations*
Discharge and water quality analyses

(2) *Instrumental direct observations*
Micro-erosion meters
Plaster casts
Rock tablets

(3) *Archaeological*, through the degree of weathering on:
Tombstones
Buildings
Barrows
Shell middens
Reclaimed land

(4) *Geomorphological*, through the degree of weathering on phenomena of known age:
Drained lake floors
Moraines
Dunes
Volcanic eruptions
Erratics on limestone pavements
Striated limestone pavements or upstanding quartz grains

where S is the amount of limestone removed in solution (mm) for a specified period, Q is the total discharge for the specified period, T is the mean total hardness (ppm, mg l^{-1}) for the period, A is the area of the river basin (km^2), D is the rock density, $1/n$ is the fraction of the basin occupied by limestone and f is a conversion factor: $f = 1$ if Q is in litres, $f = 28.3$ if Q is in cubic feet, or $f = 1000$ if Q is in cubic metres.

Goudie (1970) put forward a formula for situations where both discharge and precipitation data are available:

$$X_a = \frac{QT_1}{A_b} - \frac{PT_2}{A_b} \tag{4.6}$$

where X_a is the net rate of removal (t km^{-2} yr^{-1}), Q is the annual discharge, T_1 is the concentration of dissolved materials in tons per unit area of discharge, A_b is the basin area (km^2), P is the precipitation input in the same unit as Q, and T_2 is the concentration of dissolved salts in tons per unit of P.

Certain general points need to be made, however, in connection with the determination of such rates and in the assessment of previously published rates. First, as has been pointed out by various authors (Janda 1971, Meade 1969, Goudie 1970), allowance must be made for contributions by the atmosphere, which in some cases may be the main source of the dissolved load of streams. This is a difficult procedure for, because of recycling within the basin, it is not easy to differentiate it from material coming from outside the basin. Nonetheless, if data are available on rock chemistry, then some realistic adjustments can be made (Janda 1971). Walling and Webb (1978), for instance, found in their studies of solute loads in Devon, and using their knowledge of rock chemistry, that there were several ionic constituents that were likely to have been derived from non-denudational sources, and these were subtracted from the total to gain a measure of net denudation. They also calculated the theoretical atmospheric contribution to solute loads by making allowance for concentration of the atmospheric input by evapotranspiration:

$$C_s = C_p P/(P - ET) \quad (4.7)$$

where C_s is the theoretical solute concentration in streamflow attributable to atmospheric sources (mg l^{-1}), C_p is the average solute concentration of bulk precipitation (mg l^{-1}), P is the annual precipitation (mm) and ET is the annual evapotranspiration (mm).

The associated non-denudational load component L_n (t km^{-2}) can be calculated as

$$L_n = RC_s/1000 \quad (4.8)$$

where R is the annual runoff, i.e. $P - ET$ (mm).

The results of these two methods were broadly comparable in that both illustrated the substantial non-denudational component of the stream dissolved loads (Table 4.3), though they differ in detail.

Because of the problems created by atmospheric inputs, Moreira-Nordemann (1980) has sought to overcome the situation by employing $^{234}U/^{238}U$ disequilibrium methods. Uranium was employed because it is not to any appreciable extent present in rain, it is soluble and it is a radioactive element with radioactive descendants. The activity ratio of $^{234}U/^{238}U$ measured by alpha-ray spectroscopy in natural waters, rocks and soils can be used to calculate the uranium fraction dissolved during weathering and thereby to calculate weathering rates employing river flow data.

Secondly, the sources of total dissolved solids are complex and human inputs are a particularly serious problem (see Fig. 4.1). These need to be allowed for. Such sources include the addition of nitrate fertilisers, acid rain and municipal and industrial wastes, but there are many problems in assessing their contribution (Kostrzewski & Zwolinski 1985).

Thirdly, much solute material derived by weathering may be taken up by vegetation and then removed from the catchment as litter in streams (Marchand 1971), or, if burning has taken place, as in many savanna catchments, as ash in rivers or by deflation. Likewise, chemical material may be stored within the system as a result of precipitation processes. The deposition of tufa or travertine in limestone environments is an example of this phenomenon.

Table 4.3 Estimates of chemical denudation rates for gauged catchments in Devon

Site	*Gross denudation* (m^3 km^{-2} yr^{-1})	*Chemical denudation** (m^3 km^{-2} yr^{-1})	(%)‡	*Chemical denudation*† (m^3 km^{-2} yr^{-1})	(%)‡
1	35	23	65.7	26	74.3
2	11	0	0	6	54.5
3	24	14	58.3	16	66.7
4	28	17	60.7	20	71.4
5	20	7	35.0	13	65.0
6	35	26	74.3	27	77.1
7	26	15	57.7	18	69.2
8	23	15	65.2	17	73.9
9	39	32	82.1	29	74.4
10	32	22	68.8	22	68.8
11	24	18	75.0	19	79.2
12	23	16	69.6	17	73.9
Mean	–	–	59.4	–	70.7

After Walling and Webb (1978, Table IV).
*Chemical denudation estimated with non-denudational component derived from precipitation chemistry data.
†Chemical denudation estimated with non-denudational component derived from chemical composition of total stream load.
‡Percentage of gross denudation.

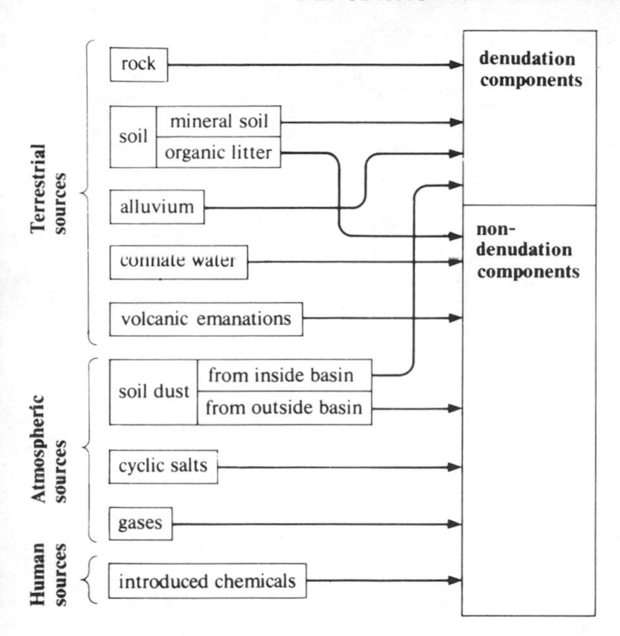

Figure 4.1 Sources of solutes in denudation systems. (Modified after Janda 1971).

Fourthly, the distinction that is normally drawn between the dissolved and suspended loads of rivers can be somewhat arbitrary. For example, colloidal material may be viewed as extremely fine sediment (i.e. less than 45 μm), but in most laboratory procedures will not be retained on the filter medium and so will be included in the solutes represented by the filtrate (Walling 1984).

Fifthly, Beckinsale (1972) has suggested that rates of denudation should perhaps be expressed in terms of a rate per volume of rock (e.g. $m^3\ km^{-3}\ yr^{-1}$), rather than in terms of the surface area of the rock.

4.1.6 Chemical denudation: other methods

The use of direct instrumentation to determine rates of chemical denudation has centred on two techniques: the micro-erosion meter and rock tablets.

The micro-erosion meter (MEM) is an instrument for the direct observation of erosion and weathering on rock surfaces and in its simplest form consists of an engineer's dial gauge that records the extension of its spring-loaded probe on a calibrated dial (Fig. 4.2). This gauge is mounted on a firm metal baseplate, which is supported on three legs (one flat, one a wedge and the other a cone). This enables the instrument to be located on three datum studs, each of which has a stainless-steel ball fixed to it. The three studs are fastened into holes that have been drilled into the rock with a diamond coring bit, and the instrument placed on them. Using arrays of studs it is possible to obtain a large number of measurements over a small near-horizontal area. The micrometer dial on the MEM allows measurements to be obtained to the nearest 0.001 mm or 0.000 01 mm (according to type).

Certain general points can be made about this technique (Spate *et al.* 1985):

(a) Because the measurements are made at a fixed point it is possible to make very selective measurements on a rock surface to determine, for instance, the erosion of different minerals in a coarse-grained rock or to investigate the development of surface micro-relief.

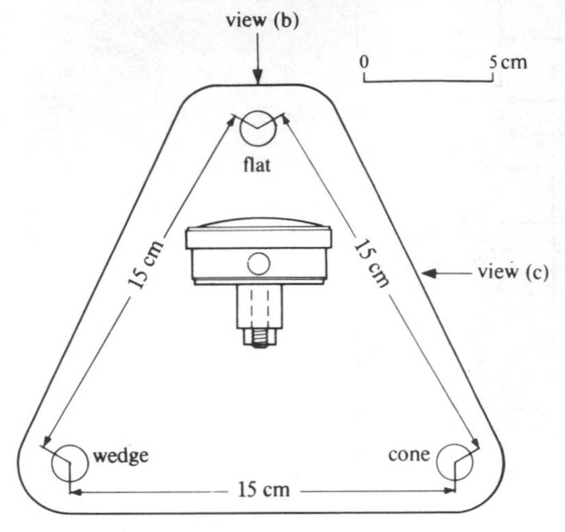

(a) Top view of micro-erosion meter

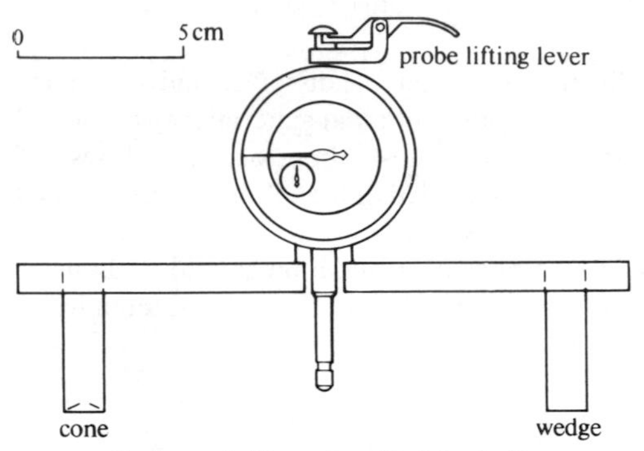

(b) Face view of micro-erosion meter

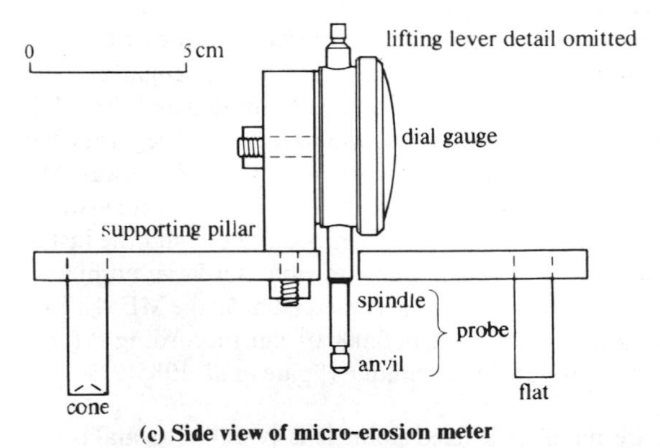

(c) Side view of micro-erosion meter

Figure 4.2 The micro-erosion meter. (After High & Hanna 1970).

(b) As this technique measures the lowering of the rock surface over a period of time, it is a measure of all forms of erosion at the site, and not just of weathering.
(c) On soft or friable rocks it may be difficult to stabilise the datum studs and damage to the rock surface may be caused by the probe, so that the technique is only suitable for firm rock surfaces.
(d) Measurements are made at a point, and therefore to estimate average lowering over an area a considerable number of sites may be necessary. To facilitate this a traversing MEM has been developed from the simple device described above, and with it, by moving the dial gauge within the framework of the three legs, a much larger number of measurements can be made at each site (several hundred readings per 12 cm^2).
(e) The most common long-term problem with the instrumentation is that the steel balls may become detached from their studs.
(f) Problems may be encountered because of the result of the expansion and contraction of the rock surface caused by temperature and moisture changes, and salt and frost heave.
(g) The growth of lichens and other organic processes may interfere with measurement.
(h) The tendency to place sites on flat surfaces, which may be untypical, creates some potential for bias.
(i) The instrument may be affected by temperature changes.

Good descriptions of the construction of the MEM are given by High and Hanna (1970) and Trudgill *et al.* (1981), while a consideration of some long-term results and their implications is provided by Viles and Trudgill (1984). In that paper they express caution about interpretation and extrapolation of short-term data, though their results obtained over the short term on Aldabra (1969–71) gave the same order of magnitude of erosion rates as longer-term observation (1971–82):

1969–71	1971–82	
0.11–0.58 mm yr^{-1}	0.06–0.75 mm yr^{-1}	Sub-aerial sites
0.002–7.5 mm yr^{-1}	0.09–2.7 mm yr^{-1}	Coastal sites

Trudgill (1975) has reviewed the use of rock tablets to detect weathering changes. The basic principle is that a piece of preweighed rock of standard shape is placed in a fine mesh bag (i.e. less than 63 μm)and then suspended in the chosen weathering environment. The fine bag allows no access to abrasive particles, so that when the bag is retrieved from the weathering environment after a period of time, and the rock tablet is reweighed, any weight loss can be ascribed to weathering. The rates of weathering *and* abrasion can be achieved by using both fine and coarse mesh bags, and in this way the tablet method has been used to show the relative importance of limestone abrasion and solution in cave streams and onshore platforms, and to compare rates of solution of different lithologies and under different soil and climate types. The method can also be used for non-calcareous rock types.

Results can be expressed in mg yr^{-1} or as mm yr^{-1} (if the density and surface area of the tablets are known). Trudgill (1977) estimates that if a microbalance is used to weigh the rock tablets a sensitivity of 0.000 001 g can be achieved.

Two main problems have been encountered with this method. The first of these involves standardisation of tablets (Trudgill 1975), though this problem can be minimised by using at least 30 tablets for each combination of site factors and ensuring that as far as possible the drilling of cores minimises variability in rock properties (especially by drilling parallel to the bedding of a sedimentary rock). The second of these problems arises because the emplacement of rock tablets in the soil creates considerable disturbance of pore size and ped arrangement. Disturbance can be reduced by inserting the tablet into the soil directly rather than enclosing it in a nylon mesh bag.

Crowther (1983) has attempted an evaluation of the relative results obtained by the rock tablet and water hardness methods for determining chemical denudation rates on karst surfaces. Field studies in Malaysia showed that the erosion rates determined by the rock tablet technique were between one and two orders of magnitude less than those obtained from runoff data. A possible explanation is that natural rock surfaces come into contact with larger volumes of water than do isolated rock tablets, simply because of their greater lateral flow component. He believes that this casts doubts upon the representativeness, in absolute terms, of chemical erosion rates determined by the weight-loss tablet technique.

Another example of a field monitoring study using preweighed material is Goudie and Watson's (1984) study of salt weathering on sabkha surfaces in Tunisia, using standard blocks of York Stone and concrete, which were emplaced for a period of six years.

Tablets have also been used to monitor accelerated rates of stone weathering in urban atmospheres. Jaynes and Cooke (1987), for example, undertook a monitoring programme in south-east England in which they used stone tablets mounted on freely rotating carousels. The carousel comprised 12 arms extending from a central plastic collar that is free to rotate around a vertical axis. This design solves three problems: it avoids samples standing in water; it prevents possible damage caused if free-standing samples are joggled by the wind; and because the carousel rotates freely, all tablets receive equal exposure.

A variation on this use of tablets is Caine's (1979) technique, which involves crushing rock so that the mean particle size is uniform. For his rhyodacite samples he used an intermediate axis size of 6.2 mm. This crushed material was weighed to four decimal places and then placed in nylon mesh bags (mesh size 1 mm) and left in the weathering environment for five years. Appreciable weathering was detected. One major advantage of this method is that it exposes a larger surface area to weathering attack and so may record detectable changes in a shorter period of time than the tablet method. On the other hand, there will inevitably be some variability in the shapes and surface areas of different samples.

Examination of the degree of weathering on archaeological remains of known age may appear a rather crude way of determining rates of denudation, but it needs to be appreciated that such methods give a representation of the rate of weathering over a much longer timescale than is provided by studies based on drainage-water analyses, the MEM or rock tablets. One of the most widely used artefacts has been the tombstone: a choice occasioned by their widespread distribution and the accurate dates with which they are inscribed. They permit assessment of both rates of weathering and the susceptibility of different rocks to weathering.

Matthias (1967), for example, used the depth of inscriptions in Portland arkose tombstones in Connecticut. Three important decisions were made:

(a) It was assumed that the letters were cut to a standard depth of 10/64 inch (4 mm).
(b) Only measurements for the number '1' of the date were used, in an attempt to eliminate possible differences in depth of cut for different letters.
(c) Only upright stones facing directly north, south, east or west were examined, so that the data could be compared according to geographical orientation.

Depth readings of the inscriptions were made using a standard machinist's depth gauge, calibrated in sixty-fourths of an inch.

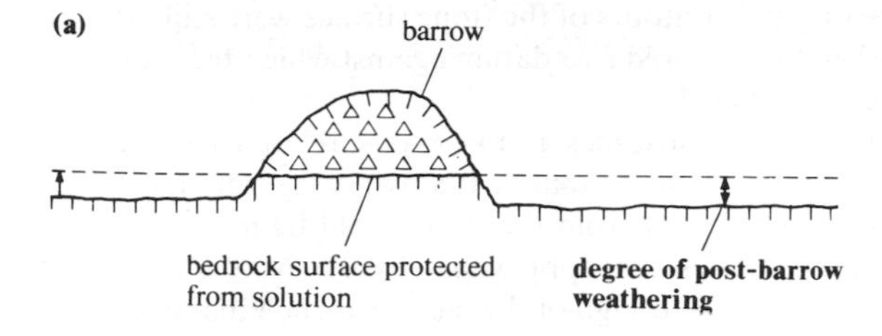

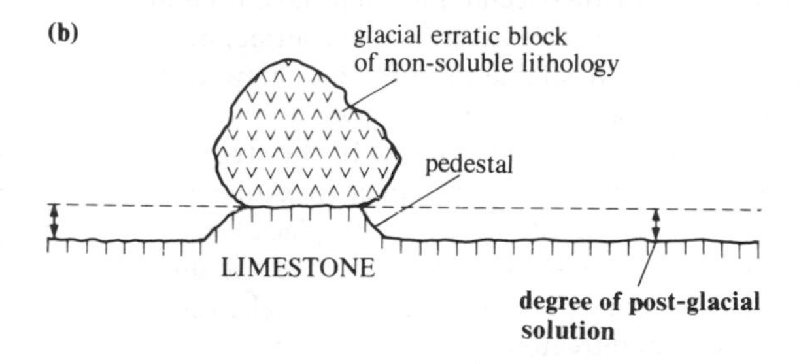

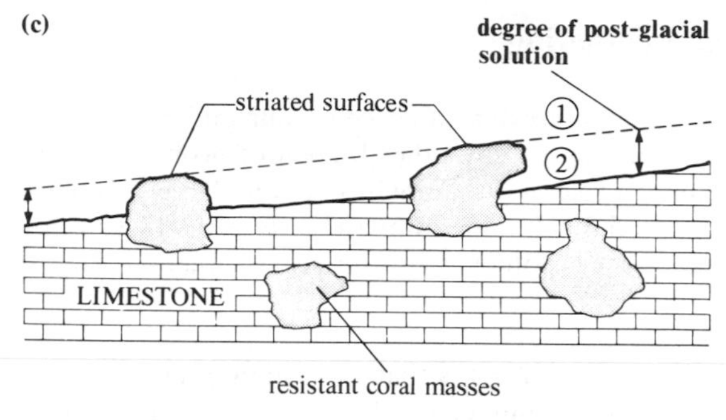

① Initial height of limestone pavement as represented by striations and assuming complete planation by glaciers.

② Height of pavement following post-glacial solution.

Figure 4.3 Methods to determine long-term rates of weathering by means of (a) archaeological remains, (b) perched erratic blocks and (c) preserved glacial striations.

Rahns (1971), on the other hand, used a less ambitious semiquantitative method for determining the degree of weathering. He classified tombstones into six weathering classes:

(1) unweathered;
(2) slightly weathered – faint rounding of corners of letters;
(3) moderately weathered – rough surface, letters legible;
(4) badly weathered – letters difficult to read;
(5) very badly weathered – letters almost indistinguishable;
(6) extremely weathered – no letters left, scaling.

then he collected data on the age of the tombstones and obtained an average degree of weathering corrected to 100 years for different lithologies.

Another approach to weathering rate determination from tombstones comes from Winkler (1975, p. 145). He looked at marble slabs in which there were unreduced quartz-pyrite veins on which the original contours of the stone surface were still evident. The veins stood out and could therefore be used as a datum against which the reduction in the surface of the marble could be judged.

Conceptually related to the use of tombstones is the exposure of rock slabs to weathering processes. An example of this work is that of Reddy (1988), who fabricated stone-exposure racks so that runoff chemistry from the slabs could be monitored and related to precipitation chemistry. The stones were approximately $30 \times 61 \times 5\ cm^3$ in size and were sloped at 30° to the horizontal. The design of the racks was such that one could minimise the influence of processes other than rain interaction with the stone surface.

Finally, mention can be made of three further techniques that have been applied in karstic areas (Fig. 4.3). In the first, a barrow of known age (say Neolithic) has protected the underlying limestone surface from solution, so that the difference between the height of the bedrock surface beneath the barrow and away from it gives an indication of the degree of surface lowering since the barrow was constructed. In the second, the same principle is applied to the effects of an erratic block dumped on a limestone pavement by a glaciation of known age. Thirdly, on some limestone pavements glacial striations may be preserved on relatively insoluble inclusions in the limestone, and the difference in height between the striations and the limestone pavement's general level gives a measure of the degree of lowering since the glaciers retreated.

4.1.7 Experimental physical weathering

The use of experiments has probably had a greater influence on our understanding of physical weathering than it has had on almost any other branch of geomorphology. Appreciation, for example, of the limited role of insolation results very largely from the laboratory experimentation of Griggs and Blackwelder, while as early as the 1820s Brard had shown that salt weathering was at least as potent as any physical weathering process.

4.1.7 Selection of cycles

The advantage of physical weathering experiments is that they allow processes to be speeded up in comparison to nature. This speeding up is achieved either by the removal of what Tricart (1956) calls 'dead time' simply by repeating cycles day after day (whereas, in nature, critical thresholds such as those involved in freeze–thaw might only occur on a certain number of days in the year) or by using shortened and unnatural cycles so that a large number of cycles (possibly artificially severe) can occur in any one day. This latter approach is in general less recommended than the former because of the increased degree of artificiality involved. It is desirable that cycles should approximate those in nature as closely as possible. Different workers have tended to use different cycles even though they may give them the same name. Figure 4.4, for example, shows the 'Icelandic' cycles used by different simulators of frost weathering. When employing the 'Icelandic' cycle, the exact details should be given if truly comparative work is to be possible for subsequent workers.

4.1.7.2 Selection of materials

There has been a tendency in the past for some workers not to pay sufficient attention to the state of their samples. In particular, it is imperative, as shown by Goudie (1974), that size, surface area, etc., be carefully controlled, as they greatly affect the speed at which experimental weathering occurs.

Goudie cut large blocks of quarried Cotswold limestone into rectangles of similar size by weight (45 g) but dissimilar shape (surface area 43.8 to 89.6 cm^3). Thirteen different

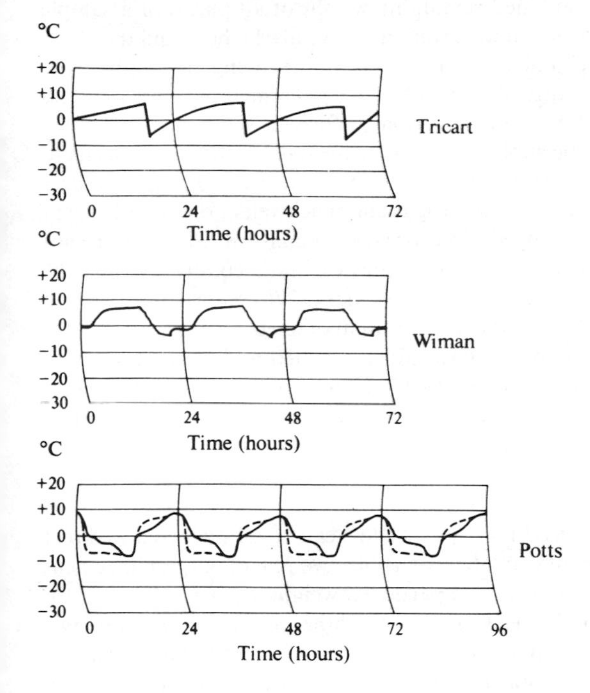

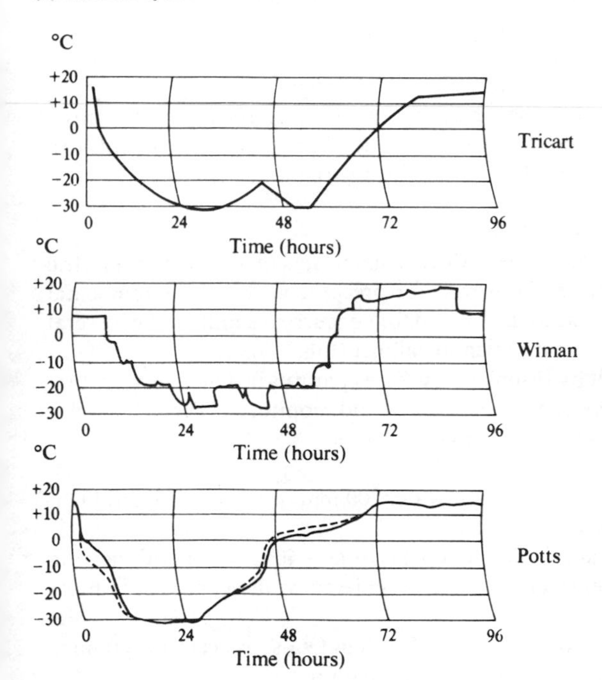

Figure 4.4 Examples of cycles used in frost weathering experiments. (After Tricart 1956, Wiman 1963, Potts 1970).

blocks were then subjected to salt weathering and a significant inverse correlation of $r_s = 0.73$ was found between the combined remaining weight of all pieces of a sample greater than 1 g after 13 cycles and the initial surface area. Similarly, he tested the effect of sample block size by cutting 13 cubes of the same rock ranging in weight from approximately 4 to 400 g. The percentage of initial block weight remaining after 19 cycles correlated against the initial weight gave a correlation coefficient of $r_s = 0.73$.

In selecting the size of sample to be subjected to experimental weathering, it needs to be appreciated that small cubes may respond in a very different way to large boulders encountered in nature. Thus, the conclusion of Griggs and Blackwelder on the efficacy of insolation weathering have been challenged, first, because unconstrained cubes are not subject to the same constraining forces as large rock masses, and secondly, because the larger the specimen, the larger the thermal stress. Thus Rice (1976) reports that a 10 cm diameter rock has 10 times the spalling tendency of a 1 cm diameter rock.

Surface texture may also be an important variable, polished and smooth surfaces disintegrating less rapidly than rough (Kwaad 1970). In the past, also, some workers have used fresh samples and others have used weathered samples. The condition of samples should be clearly stated.

4.1.7.3 Measures of breakdown

Various measures are available to assess breakdown. These can be subdivided into direct and indirect. Direct measures include number of splits, weight of largest fragment remaining, combined weight of all pieces above a particular weight (e.g. 1 g), and so forth. Indirect measures include tests of mechanical strength and other properties (e.g. Young's modulus of elasticity, porosity, etc.) or qualitative comparison of polished surfaces under the microscope. Butterworth (1976) favours the use of dilatometric techniques involving the measurement of the linear coefficient of volume expansion produced by weathering processes. They have the advantage that changes can be detected long before there is any visible damage, which reduces the temptation to increase the severity of the test as a means for obtaining results within a reasonable time. Use of the modulus of elasticity (Battle 1952) suffers from the serious defect that the strength test may destroy the specimens. It may therefore involve the necessity to test one rock sample before freeze–thaw action has taken place and to test another one afterwards. This amounts to a comparison between different samples of what may be a variable material.

4.1.7.4 Hardware for simulation of processes

Some useful simulation work has been done with very simple machinery, such as electric fires, ordinary ovens and refrigerators. However, for precise control of environmental variables, for a reasonable degree of automation and for concurrent simulation of several factors, it is necessary to utilise an environmental cabinet (Fig. 4.5).

A climatic cabinet has been built by Fisons Scientific Apparatus in which temperature and relative humidity can be independently controlled and monitored. The cabinet has the following specifications (Fig. 4.6) (Cooke 1979):

(a) The insulated working chamber is 750 mm wide, 500 mm deep and 750 mm high, and it includes three stainless-steel shelves.
(b) The temperature range obtainable in the chamber is from −40°C to +100°C, heating is provided by electrical elements, and cooling is achieved by direct and/or indirect refrigeration.
(c) Control of temperature is achieved by a West Gardian QG series cam-programme controller. The rotating cam is fashioned to the required shape (e.g. for a 24 h programme) and as it rotates it moves the temperature set point.

Figure 4.5 An environmental cabinet for simulation of weathering.

(d) The relative humidity (RH) range obtainable in the chamber is shown in Figure 4.6 (e.g. at +5°C, a range of 50–90% RH is possible; at 50°C, a range of 18–100% RH can be achieved). The relative humidity is controlled by setting the required 'wet bulb' depression on a control instrument and using this to regulate the amount of water vapour injected into the chamber by an air compressor.

(e) The separate humidity controller operates in the same way as the temperature controller (c).

(f) A Foster Cambridge CP120LD strip-chart temperature and wet/dry bulb temperature differential recorder monitors conditions in the cabinet on a roll of chart paper.

The cam-programmes are capable of accurately reproducing in the cabinet almost all temperature and relative humidity conditions encountered in the ground surface or at the ground/air interface in natural environments, and they can accommodate most naturally occurring rates of change. It is also possible to accelerate most 24 h cycles of temperature and relative humidity and to programme for steady-state testing.

A fundamental weakness of the climatic simulation is that the cabinet can only reproduce given relative humidity and temperature conditions uniformly throughout the chamber. It is therefore not possible to simulate the heat and water vapour fluxes that would normally be found in the 750 mm above a natural ground surface, which affect, for instance, the rate of evaporation. A second weakness is that the chamber is heated indirectly by electrical elements – it is not therefore possible to simulate natural insolation or naturally occurring patterns of light and shadow.

At the CNRS laboratory in Caen, France, a very much larger cabinet has been used for experiments in frost weathering. Indeed, a cold room with a volume of 30 m^3 capable of holding 2000 samples with a weight of 1.2 kg has been utilised (Journaux & Lautridou 1976).

4.1.7.5 Field measurements of weathering cycles

In assessing the frequency of weathering events, and in carrying out experimental weathering studies, accurate data are required on, for example, rock temperatures both at the surface and at depth (see, for example, Roth 1965).

The normal technique used for obtaining such data is some form of electronic temperature sensor such as a thermistor. Smith (1977), for example, used a 'Grant model D' automatic, minature temperature recorder, which measures the temperatures of nine thermistor probes within the space of 1 min, every 15 min. Alternatively, it can be used continuously, so that each thermistor is measured once a minute. The thermistor probes consist of a sphere of 2 mm diameter, housing the thermistor, at the end of a plastic-coated wire 1 mm in diameter. The thermistors have a response time of 1 s and give absolute temperatures to within +1°C over the range +5.0°C to +75.0°C. In some environments, instruments with a different range may be required. Some desert surface temperatures have, for instance, been known to reach 80°C.

The probes can be emplaced in the rock mass under investigation by hand drilling with a tungsten-tipped drill. Smith collected the rock dust and then this was tamped back into place once the thermistor was inserted. Surface thermistors were held in place by small squares of masking tape. Some thermistors were also located in joints between blocks of massively bedded limestone (Fig. 4.7).

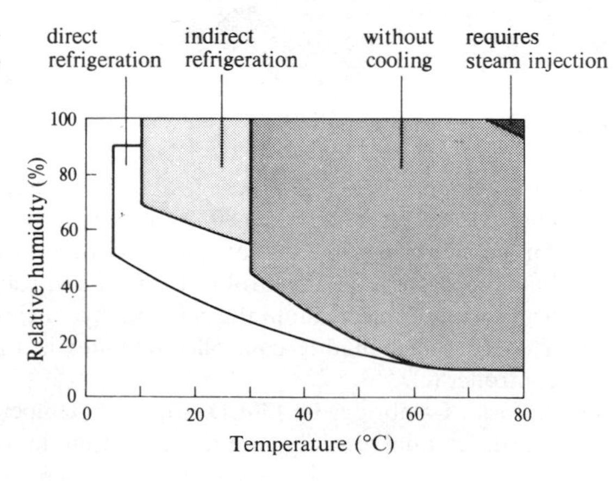

Figure 4.6 Temperature and relative humidity capability of the environmental cabinet.

(a) Set in solid rock

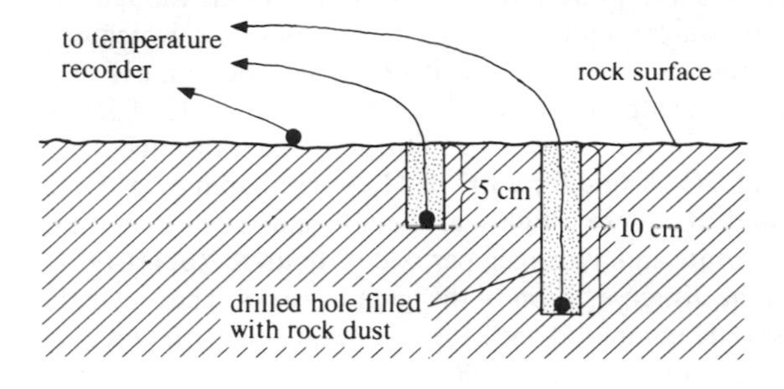

(b) Set in a joint between blocks of massively bedded limestone

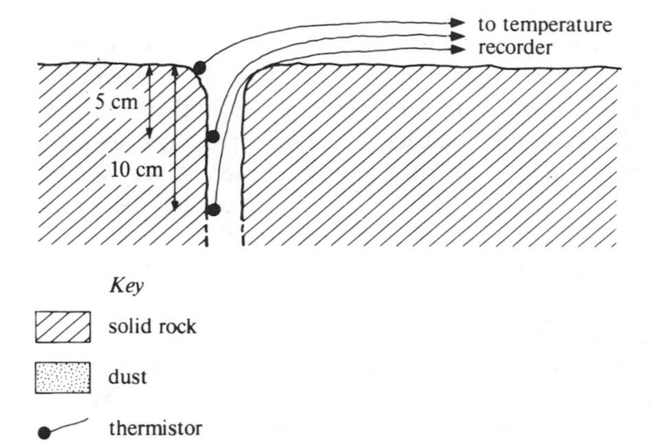

Key

solid rock

dust

thermistor

Figure 4.7 Thermistor arrangements at temperature recording sites. (After Smith 1977).

In the assessment of the frequency and power of frost cycles in the field, it is imperative to measure both temperatures and moisture conditions, as these too are a major control of effectiveness of frost action. Workers who have neglected the moisture factor have thereby tended to overestimate the frequency of events likely to break up rock (Thorn 1979).

A good example of an attempt to monitor environmental conditions and to record the rate of freeze–thaw activity is provided by Fahey and Lefebure (1988). In their field experiment in Ontario, Canada, they monitored temperature conditions in and on rock surfaces with thermistors, and measured available moisture by means of three techniques: hygrothermograph records, collection of seepage water, and sampling of rock samples on a regular basis for pore-water content and degree of pore saturation. The degree of frost-induced rockfall activity was estimated with the aid of a trap installed at the base of the rock slope.

4.1.8 Experimental chemical weathering

In historical terms, experimental chemical weathering has been of value in helping to formulate ideas about the mechanism of decomposition for igneous rocks.

In general, the practice has been to bring a rock sample into contact with a solvent, in either a closed or an open system. Various types of solvent can be used according to what pH, what strength, and what degree of artificiality one wishes to accept. Pedro (1961) queried the value of artificial solvents and preferred to use distilled water, though this itself is never found in nature as rain water contains dissolved gases and salts, which may play an important role in the weathering process (Carroll 1962). To obtain a pH of 3, Correns (961) used a 0.01 N solution of sulphuric acid (H_2SO_4); to obtain a pH of 5.6, he used a twice-distilled solution of water in atmospheric equilibrium; to obtain a pH of 6.6, he used twice-distilled water free of CO_2; while for a pH of 11, he used ammonia. Some French workers, on the other hand, have used highly artificial agents such as sodium cobaltinitrite, sodium tetraphenylborate and hydrogen peroxide (Henin & Robert 1964). Some of these are very efficient in the disaggregation of the igneous rocks. They have the attraction of being particularly speedy compared with some experiments with percolating pure water, which may take several years (Pedro 1961).

The use of water as a solvent can be varied according to whether it is distilled, charged with CO_2, or natural rainfall. Rogers and Rogers (1848), by the comparison of the effects of distilled and carbonated waters, were able to assert the efficacy of the latter when placed in contact with magnesian and calcareomagnesian silicates. They essentially used a closed-system approach where there was no throughput of solvent. This approach was also used by Johnstone (1889), who placed rock chips (orthoclase and muscovite) in various types of water contained in closed containers. He analysed both the altered solvent and the altered rock after the rock chips had been immersed for about a year, and compared them with the initial composition of the rock samples chosen. A very similar closed-system approach was that of Caillère *et al.* (1954), who immersed 100–200 g rock samples in distilled water and wetted and dried them for five months.

The degree of change of the samples and the solvent can be measured using differential thermal analysis and X-ray diffraction techniques. Simpler indications of the degree of change lie in the measurement of pH and the release of sodium and calcium ions. Graham (1949) found that, by experimental weathering in a closed system, he could not only establish the order of weathering of plagioclase as Goldich (1938) had done, but also give an indication of the relative rates of weathering by using those two techniques.

In some ways, the open-system approach is more valuable in that it approximates nature more closely. The apparatus used by Pedro (1961) (Fig. 4.8) enables one to have both drainage and fluctuating vadose and phreatic zones, just as one would expect in the real world. Pedro, concerned not to exaggerate processes, passed distilled water drop by drop through a 300 g column of pieces of rock 10–20 g in weight. After two years had passed, and around 1500 litres had percolated, three horizons had developed in the eluviate. The Soxhlet apparatus pioneered by Pedro has, however, been shown to have certain problems arising from clogging of the siphon tube resulting in inadequate drainage and some waterlogging, and this has prompted Singleton and Lavkulich (1978) to make some modifications to the basic design.

An alternative open-system apparatus is that of Correns (1961). He used a column of very fine mineral powder (less than 1 μm and 3–10 μm sizes), placed some dilute solvent in the cylinder, stirred, and then sucked the solution away through an ultrafilter. More solvent was added till 4 litres had passed through. He found that the changing rate of release of minerals through time, and the total amount of removal of certain ions in solution, gave ideas both about the mechanism of weathering and the relative ease with which different minerals were weathered.

However, such open-system experiments have not been limited to the use of inorganic solvents. Both Schalscha *et al.* (1955) and Huang and Keller (1972) used organic chelating agents as solvents. The results were striking for their speed of operation and did much to bring chelation into serious consideration as a major weathering agent. Likewise, Paton *et al.* (1976) investigated soil profiles that had developed in a matter of

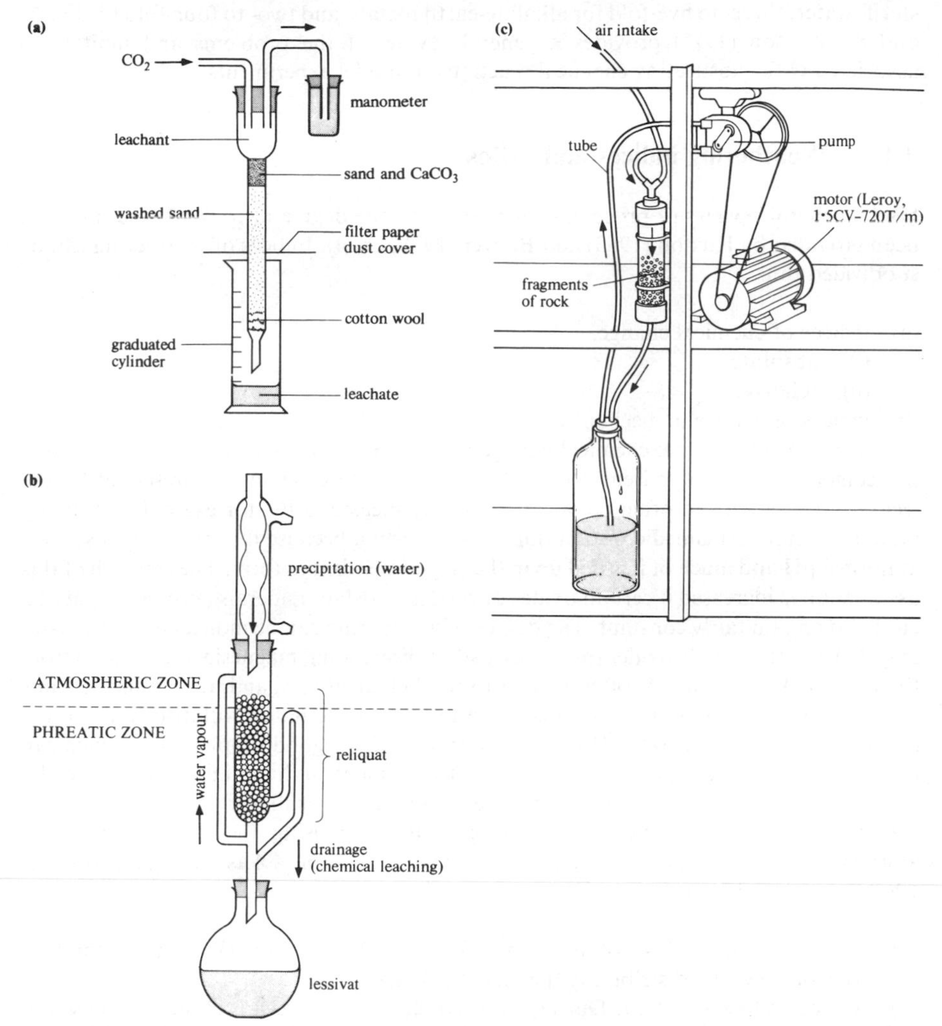

Figure 4.8 Three open-system experimental weathering devices. (a) The column employed by Wells (1965) to investigate the formation of calcium carbonate concretions. (b) Soxhlet extraction apparatus used by Pedro (1961) and Verague (1976). (c) Equipment used by the Centre de Géomorphologie de Caen for experiments involving continuous percolation. (After Birot 1969).

years on dune sand in eastern Australia. They then demonstrated the speed with which the podsol profiles could develop in the laboratory. They prepared aqueous extracts from fresh leaves of plants growing on the dunes and dripped a 5% extract at 1–2 cm^3h^{-1} into the top of a glass tube (0.3 cm internal diameter) packed with homogenised sand. They found that bleaching occurred within hours. Silverman and Munoz (1970) experimented with the role of the fungus *Penicillium simplicissimum*, and also found high rates of weathering, while Lovering and Engel (1967) examined the translocation of silica and other elements from rock into *Equisetum* and various grasses.

Finally, mention must be made of the open-system use of bacteriated water. Percolation tests by, for example, Wagner and Schatz (1967) showed that the dissolution of alkali ions from igneous rocks by bacterial water was six- to eight-fold compared with sterile water, three- to five-fold for alkaline-earth metals, and two- to four-fold for Fe, Al and Si. Watson (1972) provides a general review of the problems and limitations associated with substituting chemical reactions in model experiments.

4.1.9 Weathering indices and ratios

Useful general reviews of procedures for assessing the degree of rock weathering have been provided by Ruxton (1968) and Brewer (1978, ch. 4). Indices of weathering can be subdivided thus:

(a) indices of chemical change,
  (i) absolute,
  (ii) relative;
(b) indices of change in mechanical strength.

The indices of absolute chemical change are based on the assumption that, provided the composition of the parent rock is known, the rate of change represented by the weathered product can often be fairly accurately measured by, for example, assuming alumina to have remained constant during weathering, because it is relatively insoluble at normal pH and much of it is tied up in the clay minerals that form. The strength of this assumption is increased if certain oxides or certain resistant minerals (zircon, magnetite, etc.) also remain fairly constant. The degree of weathering can be indicated by the losses in g/100 g of the mobile oxides (principally silica, lime, soda, magnesia and potash) from the fresh rock. Silica loss is often related to total element loss, and, if alumina remains constant during weathering, the mole ratio of silica to alumina will also indicate the degree of weathering. Ruxton (1968) considers that this a good simple index, which can be used on most free-draining, usually acid environments in humid climates, especially on acid and intermediate rocks that weather to kaolinites or illites.

Calculation of gains and losses during weathering using the assumption about aluminium stability can be carried out using the data in Table 4.4 as follows (Krauskopf 1967, pp. 101 and 103):

(1) Recalculate the analyses of the fresh rock (I) and the saprolite (II) to obtain columns (III) and (IV), by distributing the analytical error.
(2) Assume Al_2O_3 constant. During weathering, 100 g of fresh rock has decreased in weight so that Al_2O_3 has apparently increased from 14.61 to 18.40%. Hence the total weight has decreased in the ratio 14.61/18.40, or from 100 to 79.40 g. The amount of each constituent in the 79.40 g can be found by multiplying each number in (IV) by the same ratio to give the numbers in column (A).
(3) The decrease (or increase) in each constituent is found by subtracting the numbers in column (A) from those in column (III) (treating the latter as g/100 g). This gives the numbers in column (B).
(4) The percentage decrease or increase of each constituent is computed by dividing the numbers in column (B) by those in column (III), giving the numbers in column (C).

The use of stable mineral species entails certain difficulties since no minerals are absolutely immobile and stable and miscellaneous geomorphic processes such as dust input may complicate the situation. Zircon and quartz have been used most frequently because of their relative stability under a wide range of conditions, and because of their widespread occurrence. Brewer (1978) believes that the most dependable stable mineral after zircon is probably tourmaline, while Nesbitt (1979) favoured the use of titanium. In

Table 4.4 Analyses of quartz–feldspar–biotite–gneiss and weathered material derived from it (see text for explanation of columns).

	(I)	(II)	(III)	(IV)	(A)	(B)	(C)
SiO_2	71.54	70.30	71.48	70.51	55.99	−15.49	−22
Al_2O_3	14.62	18.34	14.61	18.40	14.61	0	0
Fe_2O_3	0.69	1.55	0.69	1.55	1.23	+0.54	+78
FeO	1.64	0.22	1.64	0.22	0.17	−1.47	−90
MgO	0.77	0.21	0.77	0.21	0.17	−0.60	−78
CaO	2.08	0.10	2.08	0.10	0.08	−2.00	−96
Na_2O	3.84	0.09	3.84	0.09	0.07	−3.77	−98
K_2O	3.92	2.47	3.92	2.48	1.97	−1.95	−50
H_2O	0.32	5.88	0.32	5.90	4.68	+4.36	+1360
Others	0.65	0.54	0.70	0.54	0.43	−0.27	−39
Total	100.07	99.70	100.00	100.00	79.40	−20.60	

Modified after Krauskopf (1967, Tables 4.1 and 4.2).

all the determinations, numerous samples should be collected at each site to allow for inhomogeneities in the parent rock and weathered materials, and indices need to be determined for a wide range of grain sizes, either by chemical analyses or by grain counts under the microscope.

Ruhe (1975) has proposed two weathering ratios (*WR*), one for heavy and one for light minerals:

$$WR \text{ for heavy materials} = \frac{\text{zircon} + \text{tourmaline}}{\text{amphiboles} + \text{pyroxenes}} \tag{4.9}$$

$$WR \text{ for light minerals} = \frac{\text{quartz}}{\text{feldspars}} \tag{4.10}$$

Eden and Green (1971) used the feldspar : quartz ratio in their consideration of the degree of granite decomposition on Dartmoor. Using coarse sand particles (0.5–2.0 mm), a microscopic examination of prepared slides was utilised. A point-sampled population of more than about 500 feldspar and quartz occurrences was counted on each slide. On the basis of these ratios, and also on the basis of grain size analyses of the growan, they were able to conclude that 'On Dartmoor it is likely that the weathering process has been less effective and less widespread than has previously been implied'.

Two widely used indices of relative chemical changes have been put forward by Reiche (1950). These are based on the assumption that the principal changes during chemical weathering are a decrease in silica and mobile cations and an increase in hydroxyl water (water of hydration above 112°C denoted as H_2O^+). The measures are called the weathering potential index (*WPI*) and the produce index (*PI*). Decreasing *WPI* indicate decreasing mobile cations and increasing hydroxyl water, while decreasing *PI* values indicate a decreasing silica content:

$$WPI = \frac{100(K_2O + Na_2O + CaO + MgO - H_2O^+)}{SiO_2 + Al_2O_3 + Fe_2O_3 + TiO_2 + FeO + CaO + MgO + Na_2O + K_2O} \tag{4.11}$$

$$PI = \frac{100\ SiO_2}{SiO_2 + TiO_2 + Fe_2O_3 + FeO + Al_2O_3} \tag{4.12}$$

Another ratio that is sometimes used is Jenny's (1941) leaching factor (*LF*)

$$LF = \frac{(K_2O + NaO)/SiO_2 \text{ of weathered horizon}}{(K_2O + Na_2O)/SiO_2 \text{ of parent material}} \tag{4.13}$$

A further measure of relative change brought about by weathering is provided by abrasion pH (Grant 1969). Abrasion pH is controlled by the quantity of residual cations released from primary minerals in grinding and by the amount of clay minerals. The procedure is to grind 20 g of crushed rock or saprolite in a 425 ml capacity agate mortar for 2¼ min. Settling is then allowed for 2 min, whereupon a 15 ml sample of supernatant liquid can be extracted for later chemical analysis. The abrasion pH is determined on the remaining slurry.

Indices of change in mechanical strength brought about by weathering have been devised by Melton (1965) and Ollier (1969). The indices are essentially qualitative and rely on very simple equipment (boots, hammers, etc.).

Melton applied his index to volcanic and granitic rocks in Arizona and warned that 'this scale should not be extended to other types of rocks under the assumption that comparable results could be obtained' (Melton 1965, p. 717). His index is determined as follows.

Class 1: Completely fresh fragment showed no oxidation stain on the surface and no visible alteration or weakening of the rock.
Class 2: The surface of the fragment was stained or pitted, but the interior was not visibly altered.
Class 3: The surface of the fragment was deeply pitted or a thick weathering rind was present: the interior showed some staining. The fragment broke after repeated hammering, but was definitely weaker than fresh rock.
Class 4: The fragment was partially decomposed throughout but still cohesive and could be broken by hand or a single light blow of a hammer.
Class 5: The fragment was thoroughly decomposed and not strong enough to resist tough handling or being dropped a foot or two.

Ollier (1969) also developed a five-point scale of friability:

(1) Fresh; a hammer tends to bounce off the rock.
(2) Easily broken with a hammer.
(3) The rock can be broken by a kick (with boots on), but not by hand.
(4) The rock can be broken apart in the hands, but does not disintegrate in water.
(5) Soft rock that disintegrates when immersed in water.

Cooke and Mason (1973) measured the extent to which bedrock was weathered with another rough-and-ready technique. A flat-headed steel punch with a diameter of 0.47 cm was hammered horizontally into bedrock until it was buried by 1.90 cm. The number of hammer blows necessary for this extent of burial was recorded. The punch was hit into the rock in three places at each level.

Changes of mechanical strength brought about by weathering and associated case hardening of limestones and sandstones can be undertaken by the Schmidt hammer (see Sec. 3.4.6, Day & Goudie 1977, Yaalon & Singer 1974, Poole & Farmer 1980). This gives a rapid field guide to surface hardness, which has the advantage over the Cooke and Mason technique that it does not suffer from operator variance. However, extreme care needs to be taken in surface preparation and machine calibration. Results can be presented as shown in Figure 4.9.

Grading of weathering materials for engineering purposes has been discussed by Dearman *et al.* (1978). They have developed a six-fold classification scheme, which is shown in Table 4.5, while Figure 4.10 illustrates Deere and Patton's (1971) terminology for weathering profile classification.

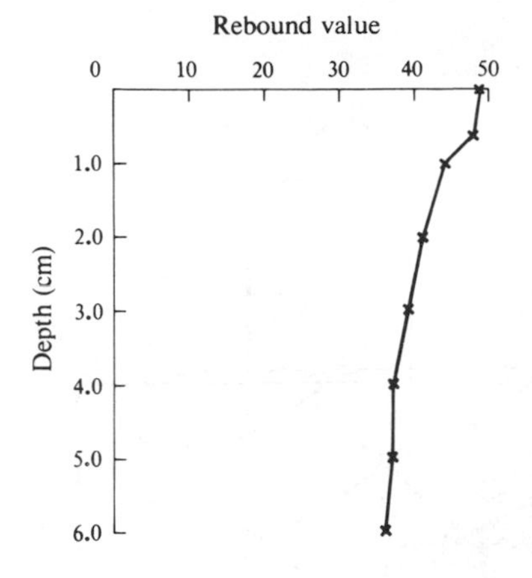

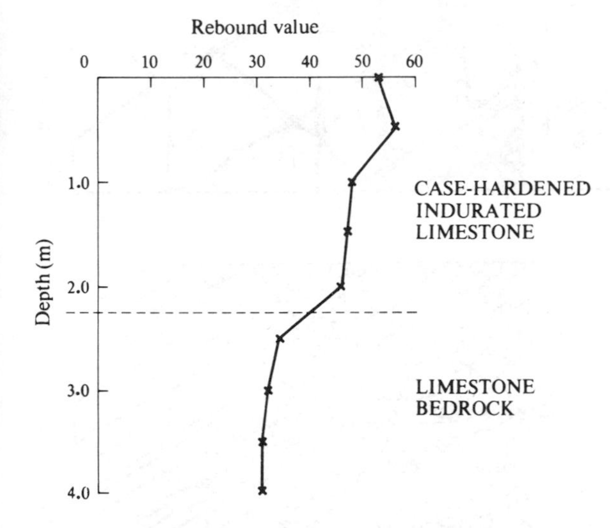

Figure 4.9 Nature of weathering profiles determined by the Schmidt hammer. (From Day & Goudie 1977).

Table 4.5 Scale of weathering grades for engineering description.

Term	*Grade*	*Comment*
Fresh	I	No visible sign of rock material weathering; perhaps slight discolouration of major discontinuity surfaces.
Slightly weathered	II	Discoloration indicates weathering of rock material and discontinuity surfaces. Some of the rock material may be discoloured by weathering; yet it is not noticeably weakened.
Moderately weathered	III	The rock material is discoloured and some of the rock is apparently weakened. Discoloured but unweakened rock is present either as a discontinuous framework or as corestones.
Highly weathered	IV	Some of the rock material is decomposed and/or disintegrated to a soil. Fresh or discoloured or weakened rock is present either as a discontinuous framework or as corestones within the soil.
Completely weathered	V	All rock material is decomposed and/or disintegrated to soil. The original mass structure and material fabric are still largely intact.
Residual soil	VI	All rock material is converted to soil. The mass structure and material fabric are destroyed. There is a large change of volume, but the soil has not been significantly transported. Can be divided into an upper A horizon of eluviated soil and a lower B horizon of illuviated soil.

After Dearman *et al.* (1978).
Note: This scheme has been derived for granite rocks.

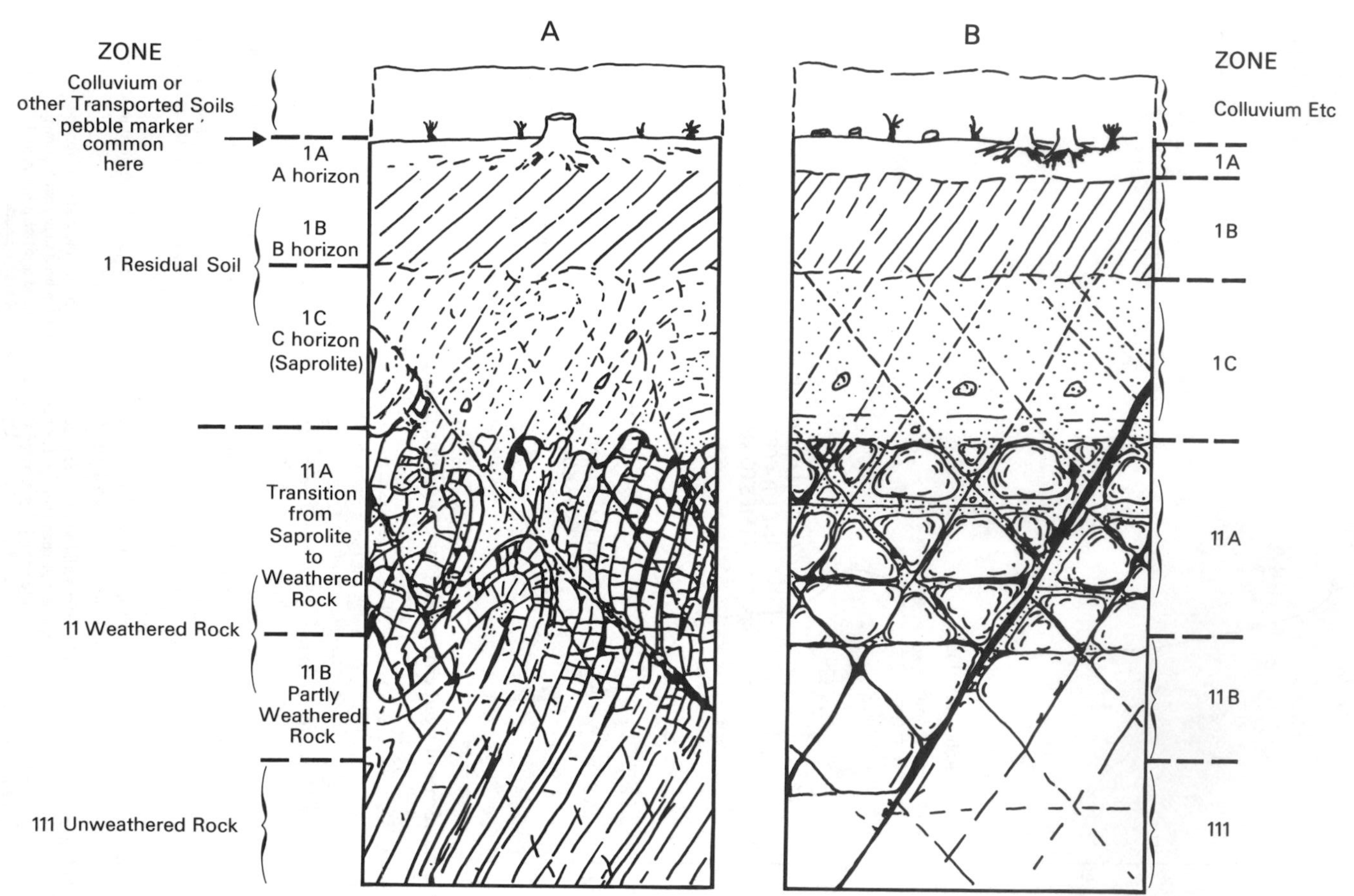

Figure 4.10 Deere and Patton's (1971) categorisation of weathering profiles on (a) metamorphic and (b) intrusive igneous rocks.

Other attempts of a more quantitative type have been used to estimate the extent of physical change brought about by weathering. Ilier (1956), for example, developed a weathering coefficient K, which depends on the ratio between the velocities of ultrasound waves in a fresh and in a weathered rock:

$$K = (V_0 - V_w)V_0 \qquad (4.14)$$

where V_0 is the velocity of ultrasound in a fresh rock and V_w is the velocity of ultrasound in a weathered rock. More recently, Crook and Gillespie (1986) have employed compressional wave speeds.

4.1.10 The measurement of soil carbon dioxide

The production of CO_2 within the soil atmosphere reflects the biological activity of the soil and also, especially in limestone areas, the potential for carbonation. A handy, non-expensive method to assess the soil CO_2 content involves the use of a Draeger device (Miotke 1974). This consists of a small manual pump, which, each time it is pressed, sucks 100 ml of air through a calibrated tube. The tubes come in different ranges and an indicator in them (which turns blue) shows immediately the volume percentage CO_2 of the sampled air.

To extract the air from the soil, Miotke (1974, p. 30) has developed a probe (Fig. 4.11) (C), which has a striking head (A) and a lower tip (D) that can be untwisted. During measurement, the probe is struck into the soil (preferably with a nylon hammer), and the pressure of the soil resistance holds the movable tip tightly against the probe. The closed holes for the air entrance are protected against becoming plugged by soil particles. When the probe has reached the selected depth, the striking head (A) is untwisted and the lower tip is pressed down a few millimetres using a thin metallic stick (H). This opens the air entrances. At this point, the measuring probe (E), which holds a gas test tube (F) in the lower end, is pushed into the drill probe (C). Rubber packings hold the tube tightly within the measuring probe and the measuring probe within the drill probe. The test tube sits directly in front of the air openings. The Draeger pump (G), which is connected by a plastic tube, sucks soil air into the test tube (100 ml each stroke) and the tube changes colour according to the amount of CO_2 present.

Miotke (1974, pp. 31 & 32) provides details of suppliers for both the probes and the Draeger device. An alternative, robust katharometer for field estimation of CO_2 has been described by Poel (1979), and Vincent and Rose (1986) assess the utility of the Jakucs CO_2 microanalyser.

4.1.11 Tracing of karst waters

To comprehend the nature and dynamics of one particular type of landform produced by weathering processes – karst – one further technique needs to be considered: water tracing (Drew & Smith 1969).

Many methods have been developed, but the following properties are generally desirable in the ideal tracer (Sweeting 1972): it should be soluble in both acid and alkaline solutions of water; it should not be adsorbed onto calcium carbonate or onto clay minerals; it should be non-toxic and non-objectionable to both man and beasts; it should be absolutely certain to be determined; it should be detectable in very low concentrations; and it should be cheap and readily available. Few techniques fulfil all these ideals.

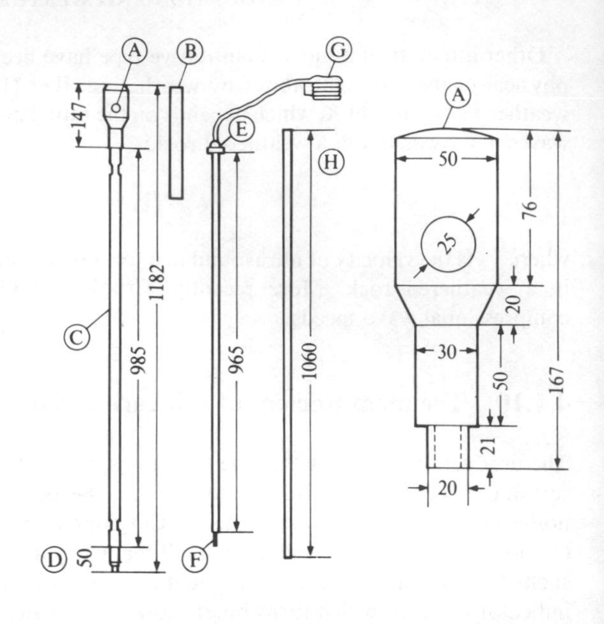

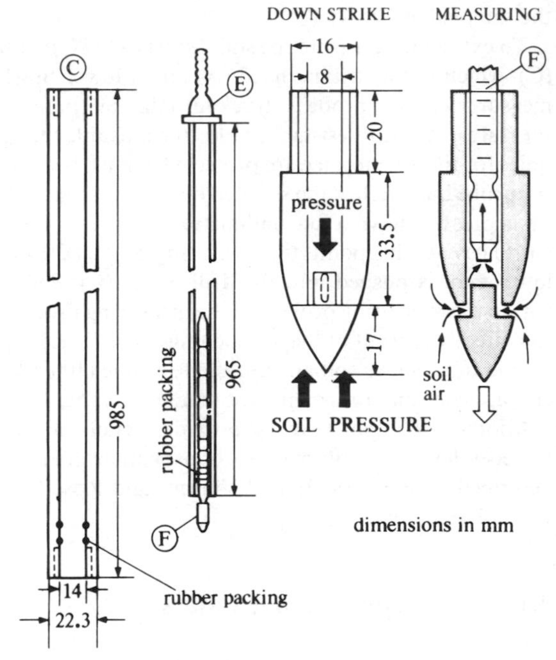

Figure 4.11 CO_2 probe for use in soils. (After Miotke 1974).

4.1.11.1 Fluorescent dyes

The fluorescent dye, fluorescein, has been used since the turn of the century. Initially, its use involved detecting it visually by its coloration of the water being sampled. Plainly, however, the production of bright-green springs by karst geomorphologists is not always welcome to water users. Fortunately, some dyes may be detected below the visible threshold by absorbing them onto granules of activated charcoal held in nylon bags (called 'detectors'), which are suspended in the springs. Alternatively, the dyes may be detected in solution at low concentrations by a fluorometer (Atkinson *et al.* 1973). Fluorescein itself is far from ideal as it may be discoloured by contact with humus, clay minerals and calcite. It also has poor stability under sunlight. Other dyes (e.g. pyranine conc.) have been used with success, though some (such as rhodamine B) have been found to present certain health risks such that their use is not recommended. A full review of the advantages and disadvantages of the available fluorescent dyes is provided by Smart and Laidlaw (1977), while Smart *et al.* (1986) compare this technique with the use of *Lycopodium* spores.

4.1.11.2 Optical brighteners (fluorescent whitening agents)

These are blue fluorescent dyes that absorb light in the ultraviolet spectrum, and are therefore colourless in solution. They have a strong affinity for cellulose fibres, however, which means that they may be used in conjunction with cottonwool or bleached calico detectors to trace water flow in karst areas (Glover 1972). The cellulose material absorbs the optical brightener and this can be detected by using either an ultraviolet light or a fluorometer. Background contamination will be a problem in areas that are in close proximity to high discharges of waste water, but in general is not a major problem.

4.1.11.3 Lycopodium spores

The spores of the club moss, *Lycopodium clavatum*, are small cellulose bodies about 30 μm in diameter. They are commercially available, may be dyed in up to five different colours and, by using uncoloured spores as well, have the advantage over dyes that up to six swallets can be traced simultaneously. These spores also have the virtue that they travel at approximately the same speed as water, but suffer from the fact that they are more expensive than some dyes. The spores are also subject to filtering by sand. Moreover, only relative concentrations can be measured from the recovery of spores as opposed to absolute concentrations using dye. The spores are introduced into the swallet as a slurry in water, and their presence at a spring is detected by filtering the spring water through plankton nets. Samples of sediment are removed periodically and examined for spores under the microscope.

4.1.11.4 Lithium acetate ($LiCOOCH_3{\cdot}2H_2O$)

This may be introduced into a swallet in solution. Water samples taken at springs are analysed for the presence of lithium using a flame photometer. Samples of stream water should be taken before the experiment to determine background concentrations of lithium in the system. Recent evaluations suggest that this is not an ideal method.

4.1.11.5 Polyethylene powder

Polyethylene dip-coated plastic powders have been used as a tracing agent by Atkinson *et al.* (1973), but they report that, as the powder has a specific gravity of 0.96, it is subject to being trapped where cave passages become water-filled.

Some of the methods described so far can be used not only to trace routes from spring to sink but also to calculate the relative importance of each route from sink to spring. If the dye used is conservative (i.e. it is not lost within the system by chemical decay or absorption), and if discharge can be measured at all the springs at which the dye reappears, then fluorometric methods can be used to this end. Smith and Atkinson (1977) propose that rhodamine WT is the best readily available dye for this purpose.

An additional method of water tracing which does not involve the use of tracers in the conventional sense is known as flood pulse analysis. Such a pulse results from a storm or more conveniently may be created artificially by opening sluice gates or breaching a temporary (small!) dam. The pulse is observed at the stream sink and neighbouring springs monitored to establish where the pulse reappears. There is also considerable potential for the use of isotopes as a stream tracing technique. Sklash *et al.* (1976), for example, have used seasonal changes in the oxygen and hydrogen stable isotopic composition of rainfall to this end.

4.2
Slope processes

4.2.1 Introduction

The observation and measurement of processes operating on hillslopes is not new. Many nineteenth-century geologists were well aware of the significance of processes in the development of the landscape, and a casual perusal of much of the early geological literature will reveal many careful observations made on processes. Actual laboratory and field measurements are not quite so common, but some examples do exist; the writings of Davison on soil and rock particle creep readily spring to mind in this context (Davison 1888, 1889). Laboratory and field techniques in slope studies therefore have their roots in the early days of geology. This good start was not maintained through the first half of the twentieth century due to the shift in emphasis towards cyclic models of hillslope geomorphology. This approach, characterised by the work of W. M. Davis, took a rather casual view of processes as the 'agents of erosion' responsible for moulding the landscape. Slope development became the object of intuitive–theoretical discussion, based mainly upon observations of form, at best surveyed but often made by eye. A similar rationale continued into the areal study of landscape development known as denudation chronology, whose aim is to explain present-day landform in terms of the phases of erosion through which it has passed. Cyclic theories of slope evolution and denudation chronology dominated hillslope literature for the majority of the first half of this century, and, with very few exceptions, little progress was made towards an understanding of slope processes.

In the last three decades, there has been a blossoming interest in processes, possibly sparked off by the failure of cyclic models to explain the origin of specific hillslope systems and possibly due to discrepancies between the real world and existing models, which could only be investigated by collecting data. It may also have been related to a realisation that civil engineers and agricultural engineers had been studying slope processes for years. Consequently, there has been a proliferation of process studies and a concomitant growth in the number and sophistication of techniques to make them possible. Process studies set out to explain relationships between slopeform and process, with a view to a better understanding of the origin of specific landscape elements. A long-term view might also be to provide data for a general synthesis of landscape evolution. The main approaches used in hillslope process studies are outlined below.

There are three broad ways in which a slope process may be studied, each requiring a different technical approach. First, the *rate of erosion* on a slope by a process or processes may be measured. It is not absolutely necessary to take much account of the processes involved in order to achieve this. They can be regarded as a 'black box', without any understanding of the way they operate. Rate observations can be made at a point, along a slope profile to show trends in form through time, or they can be bulked for an area of slope. Direct observations require some form of reference, against which repeated observations can be made. Measurements may be quoted as *vertical lowering*, the vertical change of ground elevation, or as *ground retreat*, the retreat of the ground normal to the slope. It is most important to distinguish these two measures of erosion rate. Indirect measurements of erosion rate can also be made, utilising some sort of trap to catch sediment and to gauge overall losses from a definable section of slope. Chemical losses can be estimated from solutes in streams. Indirect measurements require a conversion of collected weight of material to an *in situ* volume in order to estimate ground retreat or vertical lowering. This can be an appreciable source of error in some cases, because it is not always easy to obtain a reliable estimate of *in situ* density.

A second approach to studying processes is to measure their *velocity of operation*. Observations are made of the rate at which particles or masses of soil and rock are displaced downslope. Again, a fixed reference point is necessary to measure surface movements and movement of the entire depth of regolith. Measurement of the velocity of a process and especially the distribution of velocity with depth is often important in the understanding of the process mechanics. It may also lead to estimates of the rate of erosion. One difficulty with measuring process velocity is that some processes operate very slowly and require great precision in measurement, while some take place so quickly that it is extremely dificult to measure them at all. Also, many processes are spatially and temporally distinct events and it is very difficult to locate when and where movement is about to occur. It should be stressed that velocity of operation and erosion rate are not necessarily related at all. For example, the velocity at which a landslide travels down a slope is practically irrelevant in terms of how fast the ground surface is being lowered through time by the landslip process. On the other hand, for slower processes, such as soil creep, there is a relationship between measurement of velocity and direct measurements of ground surface lowering.

A third aspect of processes requiring investigation is the fundamental *mechanics of their operation*. The types of observation and measurement necessary to elucidate process mechanics are not easy to generalise, due to the diversity of modes of operation. However, laboratory and field experiments, designed to simulate the process under controlled conditions, are often used. Some may be very simple, while others require sophisticated experimental design. Knowledge of material properties, for example soil shear strength, is always central to process mechanics. This important aspect of geomorphological investigation is discussed in Part Three.

The three facets of process studies briefly introduced here, rate, velocity and mechanics, can be used to define broad categories of techniques. But it should be stressed that they are not mutually exclusive, and most hillslope process studies would use all three approaches.

In common with other fields of geomorphology, limitations of budget, available time, accessibility and above all the nuances of the natural system demand flexibility in technique and design of equipment. It is rarely possible to specify one technique to measure a particular process which is always superior, since no two projects are identical. It is not even possible to give a definitive design for a particular piece of equipment, for example a sediment trap, because particular circumstances may dictate modifications. The initative of the researcher is therefore always important in meeting the requirements of individual situations, and in the continuous process of technical improvement. This is an important point that should be borne in mind throughout the discussion of techniques. The approach taken throughout this section is to place most emphasis upon simple, low-cost techniques. More sophisticated and expensive devices are discussed briefly and reference is made to more complete descriptions elsewhere.

4.2.2 Process environments on hillslopes

It is simpler to discuss techniques for measuring slope processes with common characteristics and hence similar requirements for study. Accordingly, a very simple classification of slope processes is outlined here to familiarise the reader with basic terms and concepts concerning processes and with their mode of operation. More complete discussion may be found in a number of texts, for example Carson (1971a), Carson and Kirkby (1972), Young (1972) and Statham (1977).

Many classifications of slope processes have been made, but the one used here is essentially embodied in Carson and Kirkby (1972, figure 5.2). Though the figure is entitled a 'classification of mass movement processes', the diagram is more general in its scope and can be adopted as a framework to classify all hillslope processes where material is removed mechanically. It is illustrated in Figure 4.12. The classification is based on three parameters, namely type of movement, rate of movement and water content of the mobile material. These parameters are sufficient to define a slope process classification that has some claim to being genetic, while at the same time generalising processes into a few broad groups. Implicit in the classification, however, is that no clear divisions exist, but that a continuum of processes acts to move material on slopes. Three types of movement are recognised, flow, heave and slide, which correspond to the end-points of rapid mass movements, slow seasonal mass movements and water flow processes, respectively. The main fields of process discussed below are indicated on Figure 4.12 with some examples of specific processes.

Rapid mass movements involve transport of large quantities of rock or soil debris over a fairly well defined rupture surface, or shear plane, within the material. They are widespread but spatially and temporally distinct events, responsible for the observable form of a wide range of slopes in all regions of the world. For example, in Britain, it is true to say that most upland valley sides show evidence of rapid mass movements, and even quite low-angled slopes on clay substrates in lowland Britain are often subject to landsliding. Further, there are even more widespread occurrences of formerly active landslides on slopes down to as low as 1°, which were associated with past periglacial conditions.

Mass movements in hard rocks ('hard' is used here in the broad and not strictly geological sense) are controlled by the intact strength of the rock and by the characteristics of the discontinuities cutting through the rock. Discontinuities include faults, joints, bedding planes and cleavage, and they often form well defined sets with a common orientation and inclination. Small-scale rock failures are almost always entirely controlled by the discontinuities, and so intact rock strength is almost irrelevant. In addition to the geometry of the discontinuity pattern, strength along the discontinuities and the possibility of water pressures within them must be considered in a study of hard-rock failures. Larger-scale failures take place partly through intact rock, but it is probably always true that discontinuties are of major importance in rock-slope stability, irrespective of scale. The smallest-scale process on hard-rock slopes is rockfall, which is the detachment of small fragments from the face along planes of weakness. Rockfall debris may accumulate at the slope base to form a scree slope. Rockfalls often directly follow a thaw on the face or a period of heavy rainfall. But these are only immediate causes of a process that is related to long-term weathering by frost action, loosening by plant roots and other mechanisms. Rockfall is therefore related to weathering back from the face along discontinuities and does not indicate overall slope instability. The style of medium-scale rockfall processes, usually discontinuity-defined, is related to the geometry of the discontinuties with respect to the face. Slab failure takes place along discontinuities dipping towards the face, whose inclination must be less than the face angle but high

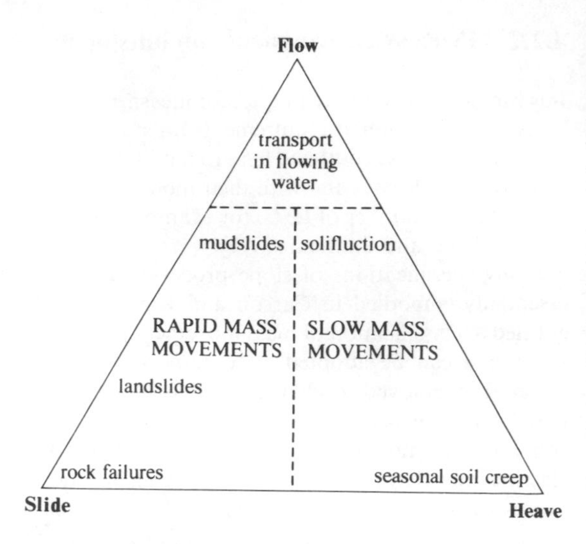

Figure 4.12 A classification of hillslope transport processes. (After Carson & Kirkby 1972).

enough for the strength along the discontinuity to be exceeded. Wedge failure is used to describe a failure on two joint planes whose common intersection daylights in the face. Finally, toppling failure refers to the falling of blocks of rock out of the face along steeply inclined planes, and is often accelerated by basal undercutting. These failures, unlike rockfall, may indicate overall instability of the face if the geometry of the discontinuity net is consistent. Large-scale rockslides do not exclusively follow discontinuities, but cut through some intact rock. Hence, intact rock strength assumes some importance in the stability of large slides, though, as mentioned above, the influence of discontinuities is always appreciable. The mechanics of very-high-magnitude rock failures are poorly understood and opinions differ about the relative importance of structural elements and residual stresses in the rock due to tectonic activity. In most cases, irrespective of scale, rock-slope instabilities operate very quickly indeed and measures of process velocity are difficult to achieve.

Slope failures in soft rocks such as clays, and in regolith materials, may be shallow or deep-seated, depending on whether significant cohesive strength is present in the material. Cohesive soils generally fail in a deep-seated, more or less curved, shear plan and the resulting failure is called a rotational slip. Typical cohesive materials include unweathered London Clay, Lias Clay and most glacial tills. Slope stability on cohesive soils depends not only on material strength and slope geometry, but also on pore-water pressures and sometimes on the presence of fissures and discontinuities. No characteristic slope angle is related to rotational failures since stability is dependent upon overall slope height as well as angle. Rates of sliding on the shear plane are relatively slow, perhaps of the order of a few centimetres to a few metres per year. Zones of repeated rotational failure have a distinctive topography, consisting of rotated slipped blocks, whose slope angle is lower than the original slope and may even be reversed, separated by scars that mark the exposed position of shear planes. Through time, this topography degrades to a hummocky, irregular slope with poorly defined scars. Active rotational slipping is really confined to slopes suffering basal erosion, for only under these conditions can intact, unweathered soil be maintained in an unstable state for a prolonged period. If allowed to weather, cohesive soils and rocks slowly lose cohesion and the slope becomes controlled by the properties of the cohesionless, weathered soil. Cohesionless soils, such as regoliths derived from bedrock weathering, only possess frictional strength. Therefore, only one aspect of slope geometry is significant in stability,

i.e. slope angle, and slides tend to be shallow with the shear plane
to the ground surface over the slide length. This type of failure is k
slide. Pore-water pressures are again important in the stability
cohesionless regoliths. Rates of sliding tend to be faster than coh
relatively slow. Usually, the slide mass breaks up after slid
downslope, leaving a bare scar, unlike a cohesive slide mass, wh
for some time.

Once a landslide has occurred, the slipped debris usually q
position and the process ceases. Sometimes, however, the slipped
into a flowslide, and continue to move for some distance. Th
flowside, differing widely in mechanics of operation. Debris from high-magnitude rockslides can trap compressed air as it falls and move on an air cushion for several kilometres at very high velocity. Flowslides in coarse regolith material can be triggered by very high pore-water pressures and maintained by constant addition of water to the mass. In finer-grained, clay materials, mudflows or mudslides (as they are now known, due to a better understanding of the mechanism) may accumulate under favourable conditions. Mudslides are accumulations of high-water-content clay debris, derived from breakdown on exposed clay cliffs. They move at slow, but sometimes perceptible, rates with most movement taking place after prolonged wet weather. The mechanics of flowslides are poorly understood because, with the exception of mudslides (which are stable forms over long time periods), they tend to take place very rapidly. Consequently, very few observations have been made upon them while in motion.

Slow mass movement processes, often referred to as creep processes in the geomorphic literature, are slow downslope translocations of soil debris, which are imperceptible except to long-term measurement. In contrast to rapid mass movements, creep processes tend to be nearly continuous in operation and to affect very large areas of slope. They also occur on almost any slope angle because generally they do not rely on a critical slope angle for their operation. Strictly speaking, the term ‘creep’ is misleading since the vast majority of slow mass movements are not creep in the rheological sense. Most are due to expansion and contraction cycles in the soil, which are exploited in the downslope direction. Single rock fragments can be displaced over bedrock surfaces by insolation expansion cycles or by freeze–thaw heaving in the underlying rock if it is soft. But the most important slow mass movements are those taking place in soil and regolith, where three main sources of movement can be identified. *Continuous creep* (Terzaghi 1950) is the downslope displacement of a soil layer in response to a gravitational stress and is characteristic of soils whose response to stress incorporates some degree of viscosity. A fairly delicate force balance is necessary for continuous creep to persist because at low stresses its rate decreases progressively towards zero, whereas at high stresses it accelerates towards landsliding. Most slow mass movements can be ascribed to *seasonal creep* mechanisms, which depend on periodic expansions and contractions of the soil due to moisture-content oscillations or to freeze–thaw. Expansion tends to be nearly normal to the soil surface, while contraction is close to vertical, resulting in a net downslope displacement. Seasonal creep is not so restricted to specific types of soil, nor does it require a critical slope angle for continued operation. Accordingly, it is very widespread and can be the most significant transport process on many slopes, especially low-angled slopes in temperate regions where neither landsliding nor slopewash is effective. A third source of creep movements, *random creep*, can be regarded as the net result of all chance, rather than systematic, downslope displacements of soil particles. The collapse of disused animal burrows, worm casting and other similar mechanisms are frequently quoted in this context. Rates of soil creep are usually very low, with ground surface velocities of less than 10 mm yr^{-1} being typical. Faster rates are possible in periglacial regions where the

ed solifluction. However, in that environment it is much more complex and w-angled slide processes such as seasonal creep mechanisms. Only relatively have creep rates been measured with any accuracy, due to the remarkably slow of operation and the necessity of measuring velocity throughout the profile depth. ecause slow mass movements are regarded as important processes, a thriving technology has developed to monitor them, despite the numerous practical difficulties involved.

The last major group of sediment transport processes on slopes is related to the falling and flowing of rain water over the ground surface. The impact of rain water on a slope and the overland flow it generates cause significant transport on bare or poorly vegetated slopes, and is known as slopewash. Slopewash affects large areas of slope but is temporally restricted to periods of fairly intense rain. Raindrop impact dislodges material by undermining, direct impact and as fine-grained sediment in rebounding droplets. It also has an important compacting effect on the soil, reducing its permeability and hence its susceptibility to overland flow. It operates on all slope angles and increases in importance with slope angle. Its efficacy is most pronounced where overland flow erosion is limited. Hence, rainsplash is most obvious close to the divide, where there is insufficient catchment for an erosive surface flow to be generated. As overland flow increases in depth through a storm, it tends to reduce the influence of direct raindrop impact by cushioning the ground. Once overland flow has been generated, it rapidly becomes the dominant transport process. At first, it may be a relatively uniform sheet of water on the slope, but it very soon shows some degree of concentration, however indistinct. Preferred flowlines may be vague, wandering systems related to slope microtopography, or they may develop into well defined rills. Clearly, if rills persist from storm to storm, they may develop into permanent features of the drainage net. However, if they are destroyed by weathering processes and re-form in different places, they can be capable of uniform slope lowering through time. Rates of slopewash are highly variable according to the state of weathering of the surface sediment. During flow, sediment load declines through time due to the rapid removal of the easily transported, weathered material. Weathering may be due to wetting and drying between storms, or to other processes that produce a loose surface layer of soil. Other factors influencing slopewash include soil type, vegetation cover and slope angle.

The brief résumé of slope processes above serves to show their diversity and to summarise the major mechanisms responsible for sediment transport on slopes. One of the major problems with measuring slope processes is of location, in that many are spatially and temporally variable. The extent to which this is a problem depends on the type of process. Slow mass movements are widespread in their operation, and so locating a measuring site to record the process is relatively simple. Rapid mass movements, on the other hand, are often very distinct events in time and space, and it is practically impossible to predict their location prior to failure. Erosion by flowing water on slopes is usually widespread in the areas where it is important, but it operates discontinuously in response to climatic events. Slopewash is also very variable from point to point on a slope, and the extent to which a measurement site is representative of a wider area becomes an issue. Other important general problems concerning process monitoring are disturbance due to the measurement technique and the need for precision when recording very slow processes. Specific problems are discussed under relevant section headings below.

4.2.3 Techniques for the investigation of rapid mass movement in soils and soft rocks

Rapid mass movements in soils and soft rocks include processes such as: rotational slips in overconsolidated clays; translational slides in cohesionless regoliths, weathered and normally consolidated clays; and mudslides in high-water content clay debris. As mentioned in the introductory paragraphs, locating a potential landslide before it takes place is a most difficult task. It is easier if a pre-existing slide zone is under investigation since the spatial location problem is largely solved and time of sliding is usually controlled by high pore-water pressures during prolonged wet weather. But it is seldom that one is fortunate enough to predict a first-time slide and to monitor its behaviour during failure. Unfortunately, to make a detailed stability analysis of a slide, it is necessary to know the conditions within the slide mass *at failure*. Soil strength parameters and slope geometry may be determined with reasonable confidence retrospectively, but the critical factor of pore-water pressure on the shear plane is impossible to measure after failure. Usually, one must be content to make assumptions about pore-water pressures and to perform a retrospective assessment of stability, known as back-analysis. The techniques described below are considered under the headings of rate of erosion and slide movement, and stability analysis of slides. They vary from very simple and inexpensive methods to sophisticated procedures, the latter especially prevalent in stability analysis. Hence, stability analysis receives brief consideration only, with reference made to other works.

4.2.3.1 Rates measured on maps and aerial photographs

One of the simplest and quickest ways of obtaining quite reliable data on rates of slope retreat by landsliding is to examine published maps and aerial photographs. The accuracy of measurements made on maps depends on many factors, such as scale (it should be possible to measure to ±5 m on maps of 1:10 000 scale), date of survey, location, printing technique, and so on. Ordnance Survey maps, published since the mid-nineteenth century, may generally be relied upon, though they may be less accurate away from buildings (Carr 1962). However, older Ordnance Survey maps sometimes suffered stretching in the printing process, and are consequently less accurate. This problem can be overcome by making measurements on the original plates, if these are available. Earlier maps are of variable quality and patchy coverage, but can often be useful if treated with discretion. One may also add purpose-made surveys to published maps for rate measurement. Many studies have made use of map evidence to measure rates of erosion on rapidly retreating slopes and a few examples are quoted here. Practically all are from coastal sites in soft rocks because these locations may show rates of erosion sufficiently rapid to be discernible and irrefutably greater than all possible sources of error. Holmes (1972) examined rates of erosion on maps for a number of sites on the north Kent coast. Warden Point (Isle of Sheppey), for example, has retreated at a rate of about 2.2 m yr^{-1} since 1819 (Fig 4.13). Hutchinson (1970, 1973) has also measured rates of cliff recession on the London Clay of north Kent and has been able to relate slopeform and style of recession to rate of recession. Finally, Brunsden and Jones (1976) have used 1:2500 Ordnance Survey plans to measure average rates of erosion of Stonebarrow Cliff, Dorset. They found the average rate of recession to be up to 0.71 m yr^{-1} between 1887 and 1964, with considerable variation of rate highlighted by intermediate surveys. Where accurate historic maps are unavailable, it may be possible to use datable buildings as reference points to extend the record in time.

Although maps may provide good estimates of overall rates of recession of rapidly eroding slopes, they are of little use in determining the rate at which a specific landslide

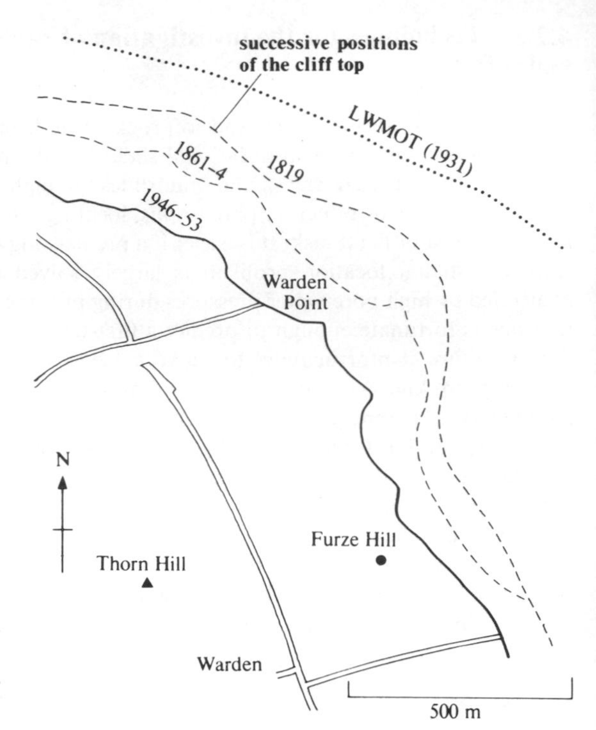

Figure 4.13 The retreat of London Clay cliffs at Sheppey, North Kent, from map evidence. (After Holmes 1972).

mass moves along its shear plane. Even in the case of very large rotational slips, which may take many years to disintegrate, very little information can be gained because the accuracy of surveys in active landslide zones is poor. Brunsden and Jones (1976) have highlighted this lack of accurate detail by comparing contemporary aerial photographs with published maps. Accordingly, to study rates of landslide movement over long time periods one must resort to aerial photographs. Aerial photograph cover varies widely throughout the world, though in Britain cover will usually be available for a number of dates since the Second World War. For example, Brunsden and Jones (1976) had access to photographs flown in 1946, 1948, 1958 and 1969, which allowed very good monitoring of the progress of specific rotational slide blocks. Aerial photographs have a much wider application in geomorphic studies in general, and slope studies in particular, than simply providing information on rate. They are an invaluable source of information for the geomorphological mapping of landslide zones and a very good interpretation of processes is possible from them. They are widely used at the site investigation stage in civil engineering projects as a basic reconnaissance tool, and they should be consulted in a similar manner in any geomorphological project concerned with rapid mass movements. Regarding their use in a reconnaissance context, Norman *et al.* (1975) have discussed some of the factors controlling the success of aerial photographs in detecting old slides. Important in this respect are scale, conditions prevailing at time of photography and, if special surveys can be flown, emulsion type.

4.2.3.2 Colluviation dating

Colluvium is typically fine-grained material that has been brought down a slope to accumulate at a lower position and hence cover up an existing soil layer. This process is known as colluviation. Some authors restrict the term 'colluvium' to material derived only by sheet erosion of the slope and deposited from overland flow, though it has also been applied to landslide debris. Landsliding often takes place in well defined phases, possibly broadly related to climate, during which considerable quantities of debris are moved downslope to cover the formerly exposed lower slopes. The intervening periods of lower activity allow soils and vegetation to develop. Consequently, if the different phases of colluviation can be dated, possibly from materials included in buried soils, and the volume of material involved in each phase calculated, it should be possible to calculate approximate rates of slope recession. This deceptively simple technique is fraught with problems. Suitable material for dating must be found, preferably organic material for ^{14}C isotopic dating, which may be difficult. In the absence of suitable material, it is sometimes possible to estimate age from the relationships between phases of colluviation and datable topographic surfaces, for example river terraces. Clearly, if a colluvial unit overrides a terrace, it is safe to assume a younger age. Unfortunately, such dated surfaces are not always present. It is by no means a simple task to calculate the volume of each colluvium unit. Essentially, this is a resource problem, since the only satisfactory approach is to obtain as much subsurface data from trial pits and boreholes as possible, in order to make a reliable interpretation of the slope sediments. The importance of this point will become apparent from an examination of the published work quoted below. Finally, there is the problem of material lost from the slope by basal erosion. Colluvium on the slope represents only a minimum quantity derived from a higher position, since sediment and solutes may have been subsequently transported away. Despite the pitfalls of the technique, colluvium dating has been used with great success in some studies to infer rates of slope retreat and to reconstruct slopes at specific times in the past. In particular, a recent collection of papers on valley slopes in southern England, published by the Royal Society, shows extensive use of the approach. Two of these papers are mentioned here. Hutchinson and Gostelow's (1976) study of the abandoned London Clay sea cliff at Hadleigh, Essex, revealed seven types of colluvium, and it was shown by dating that four major phases of landslide activity had occurred since 10000 BP (before present). Remarkably good correlation was found between the ages of landsliding and periods known to have had wetter climates. The amount of scarp recession and slope evolution through time are shown in Figure 4.14, which is a simplification of Hutchinson and Gostelow's reconstruction. Chandler (1976) was able to reconstruct a Lias Clay slope in Rutland at various times, on the basis of detailed subsurface investigation of slope sediments and by association with dated morphological surfaces. It is perhaps worth noting that these two studies were able to draw upon better resources than is usual in most geomorphological research. The slope profile investigated at Hadleigh Cliff, for example, was trial pitted for over 110 m (more than 30% of its length) and the pits were supplemented by 22 boreholes. Obviously the quality of the interpretation made using colluvium dating depend upon the availability of sufficient data.

4.2.3.3 Direct measurement of rates of landslide processes

The most reliable short-term technique to measure rate of slope retreat or rate of movement of a sliding mass is to make direct measurements from a fixed reference point. Suitable datum points include stakes hammered into the ground or a prominent, stable natural feature. Very many studies have used direct measurements and a few examples are given here. Lines of pegs across coastal mudslides have been widely used to measure displacement (Hutchinson 1970, Prior *et al.* 1971). Hutchinson (1970) was able to show that movement of a mudslide (previously known as mudflow) was dominantly by translational sliding upon a shear plane, since there was little change in surface velocity from the margins to the centre of the 'flow' (Fig 4.15). This is a clear demonstration of how simple rate observations can yield insight into process mechanics. The use of pegs is

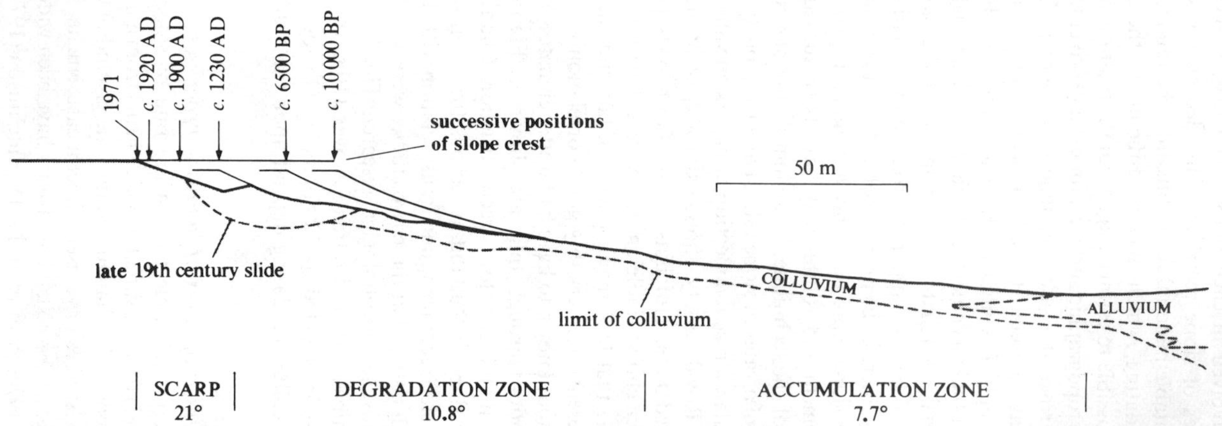

Figure 4.14 Stages in the retreat of an abandoned London Clay cliff at Hadleigh, Essex, based on colluvium dating. (After Hutchinson & Gostelow 1976).

a very simple technique requiring little explanation. However, there are a number of problems regarding their use that require consideration, but discussion of them is deferred until later when the use of erosion pins to measure slopewash is considered.

Although quite reliable estimates of surface movement are possible over a period of months or years from pegs, they give little indication of the precise timing of movement, which is important in showing relationships between the process and external factors, for example rainfall. An ingenious way of overcoming this problem was attempted by Prior and Stephens (1972). They used a modified Monroe water-level recorder to monitor mudslide movements continuously. The recorder was mounted upon stable ground and the steel drive-tape attached to a peg driven into the flow. Despite some problems of underestimating the magnitude of movement over short timespans, some very interesting records were obtained, which demonstrated that the mudslide moved in a series of surges, separated by periods of stability. On an even shorter timescale, Hutchinson (1970) has monitored mudslide movement at the marginal shear plane using time-lapse photography.

So far, only direct rate measurements of surface displacements have been mentioned. Techniques are available for measuring rates of displacement throughout the depth of a landslide, mostly using a flexible insert whose position is subsequently measured by a number of techniques. Once again, Hutchinson (1970) showed that mudslide movement was predominantly by sliding over a basal shear plane. Methods of measuring displacement at depth in a landslide are similar to those for measuring soil creep, and are discussed later.

4.2.3.4 Stability analysis of landslides and landslide-controlled slopes

Stability analysis of landslides forms a very major branch of soil mechanics and it is only possible to give a brief introduction to aspects relevant to geomorphological problems here. Reference is made to more comprehensive works later in the section. A landslide consists of a mass of material sliding over a single plane or a series of planes. Thus, analysing the critical stability conditions for a slide is a problem in calculating the forces tending to cause movement and those tending to resist it, since at failure they are equal. To make a stability analysis, it is therefore necessary to know the geometry of the slide

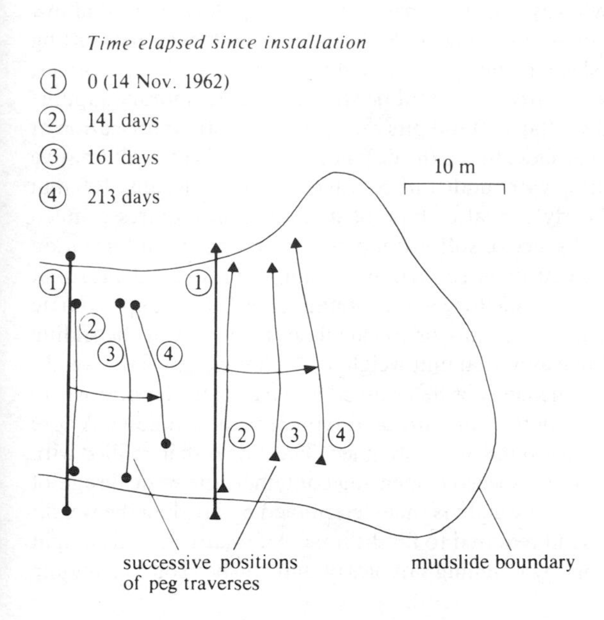

Figure 4.15 Displacement of peg traverses across a coastal mudslide, Beltinge, North Kent. (After Hutchinson 1970).

mass, which is bounded by the shear plane and the ground surface, since many of the forces involved in sliding are components of the slide weight. The parameters necessary for a complete stability analysis can be understood from a basic law of soil mechanics, Coulomb's failure law, which describes the strength of a soil at failure. Coulomb's failure law states:

$$\tau_f = S_f = C' + (\sigma - U)\tan\phi' \tag{4.15}$$

where τ_f is shear stress at failure, S_f is the total strength of the soil at failure, C' is the cohesive strength of the soil, σ is normal stress on the shear plane, U is pore-water pressure, and ϕ' is the angle of internal shearing resistance of the soil. It will be valuable to examine these parameters carefully, in the context of a landslide body. The shear stress (τ_f), tending to cause failure of a landslide along its shear plane, is equal to the downslope component of the slide weight. At failure, it is just equal to the maximum shear strength of the soil material (S_f). S_f is made up of cohesive strength (C'), a soil strength parameter, and frictional strength, which is the product of effective normal stress on the shear plane and $\tan\phi'$ (ϕ' is another soil strength parameter). Effective normal stress (σ') is equal to total normal stress (σ), which is the normal component of the slide weight, minus any pore-water pressure (U) acting on the shear plane at the time of failure. Coulomb's failure law therefore shows us that we need to know the slide-mass weight, the pore-water pressure on the shear plane at failure, and the strength parameters of the soil, in order to calculate all the relevant forces in a stability analysis. Two aspects of stability analysis are discussed here: the precise analysis of a single slide and the less rigorous relating of characteristic hillslope angles to material strength parameters.

Defining the slide mass is the first problem in the stability analysis of a single slide. Surface dimensions can obviously be obtained from ground survey, with very little difficulty. Thus the problem of definition may be resolved into one of locating the shear plane. Success in this matter depends upon soil type and landslide style. Clay- and silt-sized soils take shear planes rather well and they show up as polished, slickensided (striated) surfaces within the material. Coarser soils do not show shear planes nearly so well. They may be discernible as poorly defined zones of disturbed and less dense soil, or they may be zones of percolation and show as wetter horizons. But often they are hard to locate with any certainty. Obviously shear planes are easier to locate in shallow, translational slides than in rotational slides (Fig. 4.16). In shallow landslides, trial pitting may be sufficient to locate the shear plane, and, depending upon available resources, these may be hand or machine excavated. It should be stressed that temporary support should be used to avoid injury by collapse. Trial pits may provide useful information at the toe and backscar of a rotational slide, but in the deeper parts there is little alternative to soft-ground boring techniques, with undisturbed tube samples (usually 100 mm diameter) taken at intervals. Clearly the latter type of investigation requires a much higher input of resources. In the absence of soft-ground boring equipment, information from the trial pits may have to be extrapolated, making assumptions about the form of the shear surface. This is much less satisfactory and is bound to lead to an approximate result. Having defined the shear plane, it remains to calculate the weight of the sliding mass from the product of its volume and total unit weight (γ_t). Unit weight is the weight per unit volume of the *in situ* soil, which may be determined *in situ* or in the laboratory. *In situ* determinations can be made using the standard sand replacement technique. A core of soil is carefully removed from the ground and is weighed. The hole is then refilled with a measured volume of dry sand from a special dispensing container, the basic design of which is illustrated in Figure 4.17. Unit weight is then determined by dividing the weight of the soil core by the volume of sand required to fill the hole. Alternatively, unit weight can be determined in the laboratory by weighing samples of undisturbed soil and finding

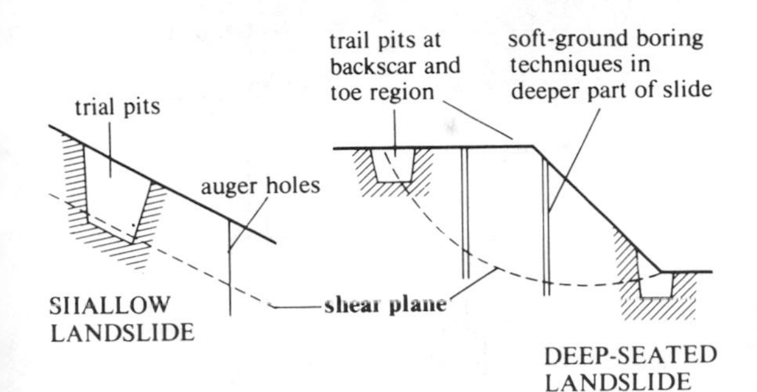

Figure 4.16 Methods for locating the shear plane in shallow and deep-seated failures.

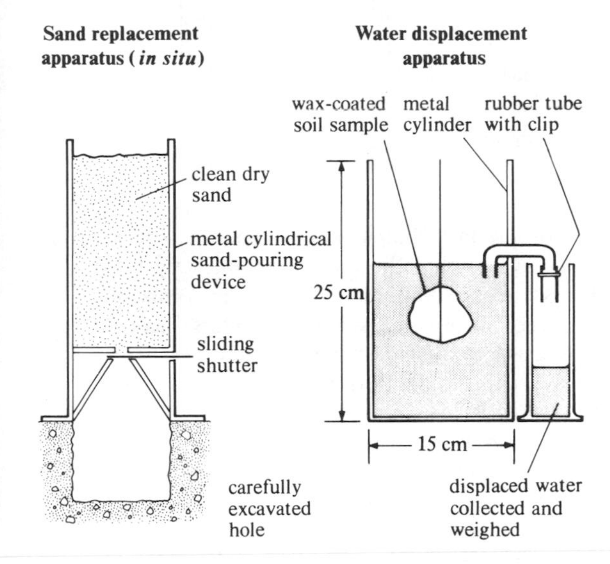

Figure 4.17 Apparatus for the determination of soil unit weight *in situ* and in the laboratory.

their volume by immersion in water. The samples are coated in wax prior to immersion to prevent disintegration, and the volume of wax used must be allowed for. Immersion is best carried out by placing the sample into a special cylinder that allows the displaced water to be discharged into another vessel (Fig 4.17). The displaced water is weighed in order to determine its volume accurately. Full procedural details for both unit weight determination techniques are to be found in BS 1377 (British Standards Institution 1975c).

The second problem in stability analysis is to define the pore-water pressures on the shear plane at the time of failure. This is obviously impossible unless a monitoring system was installed prior to failure, although retrospective observations may be useful in some cases. Pore-water pressures are measured using piezometers. Well-point piezeometers, which are plastic tubes with perforations drilled along their entire length, are used to monitor the position of the water table. Piezometers that are only open at their lower end are used to monitor positive pore-water pressures. Figure 4.18 shows two types. A typical standpipe piezometer consists of an open tube of about 25 mm diameter fitted with a porous ceramic or plastic tip. The tip is surrounded by uniformly graded, clean sand. A cement–bentonite grout seal is placed above the tip in the borehole to ensure that the pore-water pressure measured relates only to tip depth. The borehole is backfilled above the seal and the top set in concrete. A vandal-proof top is usually fitted to the top of the standpipe to prevent deliberate damage and ingress of rain water. Water levels can be

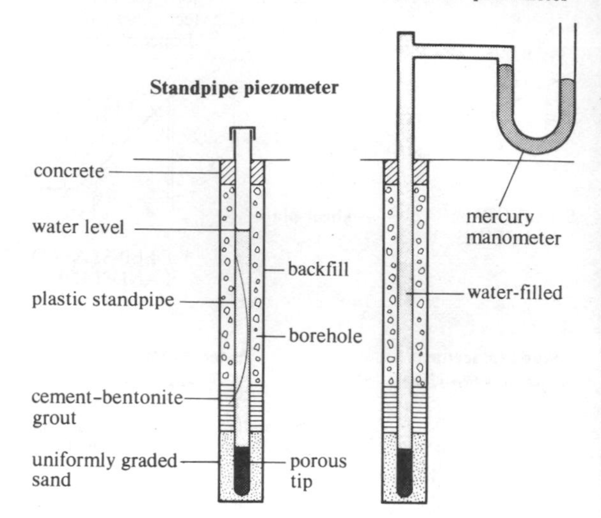

Figure 4.18 Installation of piezometers.

measured in the standpipe using a proprietary dipmeter. Hydraulic piezometers are installed in a smilar way (Fig 4.18), except that the pipe is compeletely filled with deaerated water and is connected to a mercury manometer, or to some other pressure-measuring device such as a transducer. Since some soil tensiometers incorporate a mercury manometer, these may be used as hydraulic piezometers, although there may be some limitation in their use with respect to the depth of installation. Designs for soil tensiometers are given in Webster (1966), Curtis and Trudgill (1974) and Burt and Beasant (1984). Automatic monitoring of pore-water pressures is possible if a pressure transducer is used instead of the mercury manometer. Use of a fluid scanning switch enables a number of piezometers to be logged using just one transducer (Burt 1978). Proprietary systems are also available from suppliers of civil engineering test equipment. Hydraulic piezometers are much more versatile than standpipes because they can measure negative pore-water pressures and are easier to read, especially if automatic monitoring systems are used. Their disadvantages are cost if proprietary equipment is used and problems of leakage and air entry into the system. Whatever technique is used to measure pore-water pressures, it must be stressed that the porous tip of the piezometer should be inserted and grout-sealed to shear-plane depth. It is the pore-water pressure at the shear plane that controls effective normal stress and those recorded at other depths may not give a correct value.

The third aspect of slope stability analysis, that of soil strength determination, is dealt with in the section on soil properties and so it is not covered in any detail here. The basic tests to establish cohesive and frictional strength are the triaxial test and the shear box test, and a wide range of testing procedures can be employed according to the situation under study. Two points of great importance should be noted. First, the distinction between *drained* and *undrained* pore-water pressure conditions in the soil should be made. Drained conditions pertain when the pore-water is moving freely and steadily through the pore space of the soil at a rate proportional to the hydraulic gradient. In natural landslide events this is the more normal situation, since landslides usually take place slowly enough for excess pore-water pressures to drain away. Undrained conditions can occur in places where rapid loading or unloading of a slide mass takes place, leaving a little time for excess pore-water pressures to disperse. For example, Hutchinson and Bhandari (1971) have shown that undrained loading can be important in the study of coastal mudslides, and Skemptom and Weeks (1976) have invoked undrained pore-water pressures subsequent to thawing to explain the instability of periglacial mud-slides on very-low-angled slopes. Strength testing in the laboratory should be carried out to replicate as far as possible the conditions that exist in the field. The second important aspect of strength measurement that has to be considered is what strength parameters are relevant to the field situation. If it is a *first time* slide in intact, unweathered soil, the undisturbed *peak strength* is applicable to a stability analysis. On the other hand, a repeated failure on the shear plane may be related to a lower strength called the *residual strength*. For a fuller discussion of strength parameters and their applicability to different field situations, the reader is referred to Skempton (1964), Skempton and Hutchinson (1969), Chandler (1977) and the section of this book on material properties (Sec. 3.4).

Having collected the relevant data on slide geometry, weight, pore-water pressure and soil strength, one must choose an appropriate model of failure for analysis. This is a very major field in soil mechanics in itself and it is only possible to mention briefly some possible approaches. Rotational slips (Fig. 4.19) are analysed by various circular arc methods, for example the Bishop method of slices (Bishop 1955). The slide is divided into a series of slices, as in Figure 4.19, and the forces acting on each slice are computed and summed for the whole mass. More sophisticated analyses are also in use, not constrained by the assumption of a circular arc shear plane, for example Morgenstern and Price (1965) and Sarma (1973). Shallow landslides may frequently be analysed satisfactorily using the infinite planar slide technique. The slide is approximated as infinitely long, with the shear plane and water table parallel to the ground surface (Fig 4.19). Forces at the edges of the slide mass can often be neglected. It was applied with success to a large translational slide, developed in Coal Measures clays and shales, at Jackfield, Shropshire (Skempton & Delory 1957). The technique is summarised by Chandler (1977). General discussions of slope stability analysis techniques can be found in standard soil mechanics texts, for example Terzaghi and Peck (1967) and Lambe and Whitman (1969). Simpler introductory treatments can be read in Smith (1978) and Craig (1978).

There is no doubt that detailed stability analysis of landlides is an involved task. Considerable data are essential upon slide geometry, pore-water pressure and material strength parameters, and these are not always easy to obtain. However, geomorphologists may be more concerned with answering general questions about the relationships between hillslope angles and landslide processes in an area, possibly requiring less sophisticated means of analysis. Good examples of this type of study are those of Carson and Petley (1970) and Carson (1971b). These studies relate characteristic slope angles on landslide-controlled slopes to material strength parameters, using the infinite planar slide analysis mentioned previously. Many assumptions must be made. It is assumed that cohesion is insignificant as a strength parameter over geomorphological time and that the critical pore-water pressure for failure is related to water flow parallel to the ground surface with the water table at the surface. This general type of analysis seems to explain the limiting stability case of a wide variety of natural hillslopes, and, provided it is used prudently, can yield much valuable information. It is a simple stability framework for relating landslide slope angles to soil parameters, free from the rigours of a complete stability analysis. Furthermore, the technique is directly pertinent to the problems of process–form relationships on slopes and of landscape explanation. The problems and necessary precautions of infinite planar slide analysis are discussed by Chandler (1977).

4.2.4 Techniques for the investigation of failures in hard rocks

In the same way as for soft rocks and unconsolidated materials, it is possible to divide the field problems of studying hard-rock failures into two main issues: namely measuring rates of erosion, and interpreting process mechanics. Once again, the rigorous analysis of process mechanics is a difficult task, which is often not easily tackled on a limited budget. Considerable expenditure on expensive laboratory and field equipment is unavoidable to obtain the necessary data. Practically all important work on rock-slope stability has therefore been done by civil engineers in connection with construction projects. Towards the end of this section, however, some simple techniques to study the rocksliding process are introduced.

4.2.4.1 Rates of movement of rock failures and rate of slope recession on rock slopes

Most rock failures take place very quickly indeed and there is little opportunity to measure rates of movement. Since the rates are so fast they are irrelevant in geomorphic terms anyway, and so the lack of ability to measure rate is of no consequence. Of more importance are the significant, though *relatively* small-magnitude, pre-failure movements that often take place. These movements portend failure and are of interest to the civil engineer, who wishes to monitor slopes for failures, and to the geomorphologist as a means of locating potential slides, as well as giving information about slope recession rate. We have already discussed in the introduction that most rock failures take place along pre-existing joints and other discontinuities in the rock, and a more comprehensive discussion of the interrelationships between failure style and the geometry of the

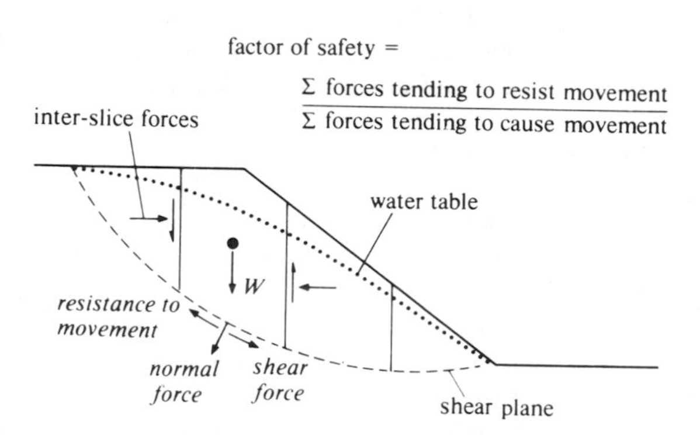

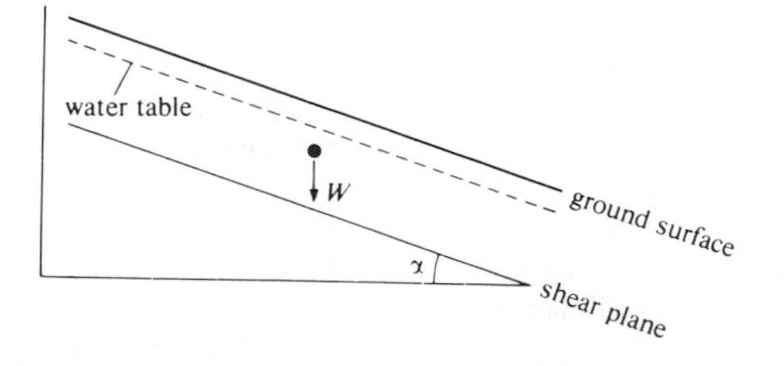

Figure 4.19 The analysis of rotational slip stability.

discontinuity net is to be found in Hoek and Bray (1977). It is hardly surprising that most pre-failure movements in rockslides take place on pre-existing joints and are often to be seen as a progressive widening of joints at the top of the slope through time. The discontinuities at the top of the slope are under tension and as they widen they are usually referred to as tension cracks. Monitoring movement across tension cracks may be carried out using reference bolts inserted into the rock on either side of the crack. A simple calliper device may be used to measure separation of the bolts. If four bolts are installed, two either side of the discontinuity, and changes in separation and inclination between them all are measured, it is possible to record absolute three-dimensional change in position of the moving block. A suitable stud installation for a small tension crack is illustrated in Figure 4.20 and full installation procedure is described by High and Hanna (1970). Many systems have been designed to measure the separation of the bolts, and a suitable instrument can be made very cheaply by modifying standard vernier callipers (Fig. 4.20). It consists of a pair of coned brass cylinders, which can be lock-screwed onto the arms of the callipers. The coned seatings fit precisely over the stud heads. A clinometer or Abney level can be used to measure changes in inclination between the studs. A larger-scale, though essentially similar, instrument to measure greater displacements is described in Boyd *et al.* (1973). A direct-reading system to measure linear displacement only between two points is illustrated in Figure 4.21 (see Hoek & Bray 1977). It consists of an indicator, attached to a wire and running over a pulley to a fixed bolt on the other side of the tension crack. Displacements can be monitored against a scale mounted alongside the indicator.

Other techniques for monitoring rock-slope movements include photogrammetry and electro-optical surveying systems for measuring distance. Both require the establishment of targets on the face as reference points for successive surveys. Photogrammetry has the advantage that a large area of cliff face may be covered by a single observation, while electro-optical distance measurement has the advantage of measuring very accurately the displacement of single points. Both require expensive equipment and are not discussed further here. Brief discussions of these and other techniques, along with a fairly comprehensive list of materials and suppliers, can be found in Franklin and Denton (1973).

Estimates of rates of cliff recession may be made in a number of ways, all requiring estimates of the magnitude and frequency of failures on the slope. This can be achieved by directly observing the processes causing recession or by measuring the amount of debris accumulating at the foot of the slope through time. The most usual direct observation technique has been called the inventory method (Rapp 1960a), in which rockfall events are monitored visually. An area of rockface is watched through the daylight hours and all rockfalls observed are recorded. Where possible, the volume of the fall is estimated, preferably by examination of the fallen debris if this is possible. In order to identify the process controls, precise records of timing, face geometry, meteorological conditions, and so on, should be made. Gardner (1969), working in the Canadian Rockies, has shown that two daily maxima of rockfall activity occur on the 'average' day, one at 0700–0900 hours and the other at 1200–1600 hours. The yearly maximum has often been noted in spring (Luckman 1976, Rapp 1960b, Bjerrum & Jorstad 1968). The last of these studies did not use the inventory technique, but was based on recorded damage to communications in Norway. It is much less easy to relate the rockfalls to specific trigger mechanisms, which are usually assumed to be climatic or tectonic, because of the accessibility problem on the steep rockfaces where the process occurs. Climatic data can be collected, however, and general relationships can be ascertained. Returning to measurement of slope recession, if the volume of rockfalls can be estimated for a section of cliff face, then it is possible to make tentative proposals about the rate of

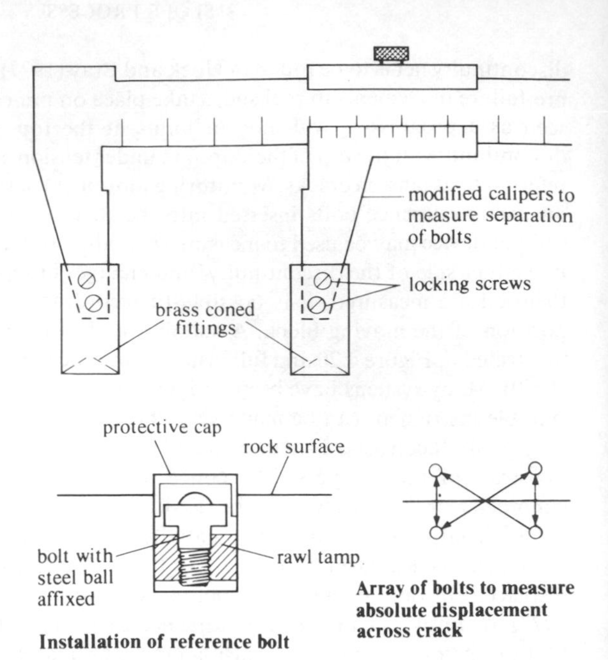

Figure 4.20 An instrument to measure displacement across a tension crack in rock.

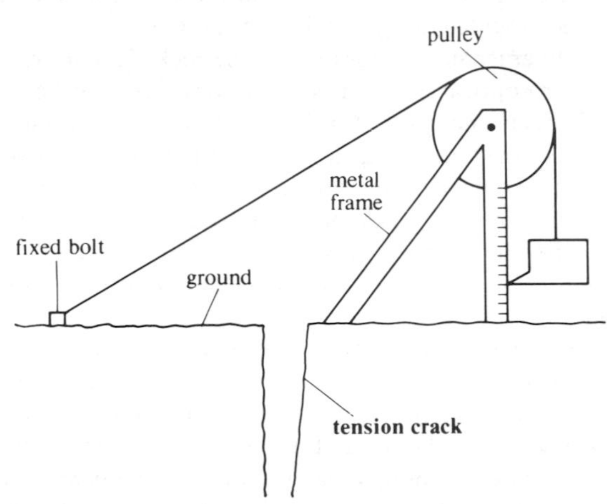

Figure 4.21 A direct-reading instrument to measure displacement across a tension crack in rock. (After Hoek & Bray 1977).

recession. Naturally there are many problems with using the inventory technique. It is fraught with problems of subjectivity, it is easy to miss falls while watching other areas of cliff, estimation of size is difficult, and the collection of the data is very time-consuming. Luckman (1976) presents inventory data for a valley in the Canadian Rockies, makes suggestions about how the data should be analysed and presented, and also discusses the problems inherent in the method.

Various ways of estimating cliff recession over time from debris accumulating at the base of the slope have been used. Debris collection systems are popular; sackcloth squares (Rapp 1960a), wooden collection boxes with wire mesh bases (Prior *et al.* 1971), and polythene sheeting traps (Luckman 1971) are just three methods that have been used, though not necessarily in the pursuit of rate estimates. All small-scale debris traps are problematic. It is difficult to obtain a representative sample from them and they are likely to be completely destroyed or buried under severe conditions. Caine (1969) has estimated the rate of accumulation upon scree slopes from the amount accumulating on and within a winter snow cover. This method could also be used to estimate retreat of the cliff face through the winter season. Caine used 10 m quadrats, laid out at 50 m intervals along the slope profile. Short-term measurements can also be made from stereoscopic photography of a cliff face. Repeated photographs are taken from the same positions, preferably from precisely located tripods, and comparison will show areas of rockfall in the intervening period. Targets of known size should be placed upon the face to give an accurate scale. Longer-term techniques include estimating the volume of scree slopes that can be dated, accumulation upon glacier surfaces (Rapp 1960b), estimating the widening of river canyons (Schumm & Chorley 1966) and estimating the length of time for a major rockfall to move away from a cliff face and to fall (Schumm & Chorley 1964). All of these techniques are subject to serious error, since dating of the initiation of recession is difficult.

4.2.4.2 Process mechanics of rock failures

Analysing the stability of a potential rock failure, or performing a back-analysis upon a failure that has already occurred, is essentially similar to the analysis of a failure in soft rocks or regoliths. The forces tending to cause movement and those tending to resist it must be understood, and hence the geometry of the slide mass, the strength of the material and the pore-water pressures are of importance. Since most small- and medium-scale failures are almost wholly along discontinuities (joints, bedding planes, etc.) within the rock, the geometry of the slide mass is defined by prominent, nearly flat planes. Further, the forces controlling the failure are those acting upon these planes, and so the relevant strength of the material and pore-water pressures are those acting along the discontinuity. Rigorous analysis of rock-slope stability is often difficult, and requires strength testing of the rock discontinuities in a special type of shear box. The excellent text by Hoek and Bray (1977) should be referred to for a comprehensive treatment of this problem.

General assessments of rock-slope stability with respect to specific types of failure present a somewhat easier task. The stability of a rock slope with respect to different styles of failure depends upon the orientation and inclination of rock discontinuities with respect to the cliff face. Thus, if joints or bedding planes dip into the face but at a lower angle, there is the potential for slab failure. Wedge failure is possible when two or more sets of discontinuities are present, whose mutual intersections daylight in the face. Finally, toppling failure may occur if high-angled joints dip away from the face, especially if basal undercutting is present. Thus, assessment of the likelihood of the various failure styles can be made from observations of joint orientation and dip. Such a discontinuity survey can be carried out in a number of ways. First, one may select specific stations on a cliff and record the orientation and dip of every accessible joint within a set distance, using a compass and clinometer. Alternatively, a tape may be laid out along the outcrop or cliff and all joints crossing the tape may be recorded. The joint observations can then be plotted on a stereographic projection as poles to the joint plane. (A pole to a plane is the point at which the surface of the sphere in a spherical stereographic projection is intersected by a line normal to the plane.) Simple graphical techniques, using the plot of poles on a stereographic projection, can be used to assess the risk and style of failure on the rock slope. The general principles and usage of the stereographic projection are described in detail in Phillips (1971) and their specific use in rock-slope stability in Hoek and Bray (1977). Valuable as this general assessment of stability problems on a rockface is, one cannot proceed beyond an indication of *potential* for sliding. To make even a preliminary analysis of whether slides will actually take place, one must measure the strength of the rock discontinuities. This is difficult to do without sophisticated testing equipment, which is discussed in the section of this book on

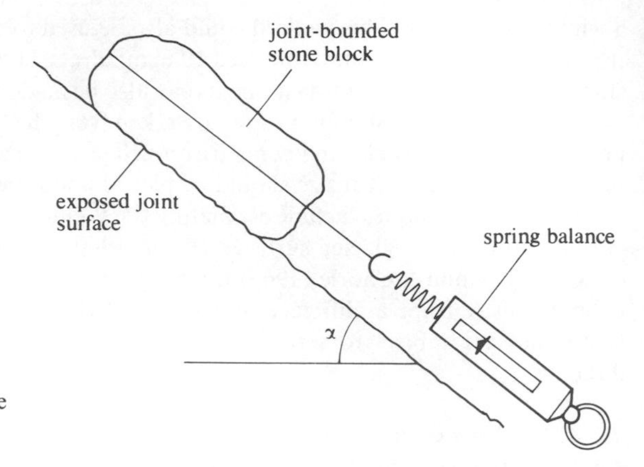

Figure 4.22 A simple technique for measuring frictional strength.

material properties. But a reasonably good idea of the minimum frictional strength may be gained quite easily using a spring balance and rock debris available in the field. The technique, illustrated in Figure 4.22, is to pull a loose block of rock with a joint surface forming its base down an exposed joint plane. The force due to the spring balance should be applied slowly and steadily and the maximum spring balance reading taken when the block just begins to slide. The process is repeated to obtain a reliable mean spring balance reading for the block of rock. The dip of the joint plane (α) and the weight of the block are then measured and the angle of plane sliding friction (ϕ_μ) for the discontinuity is calculated from:

$$W \cos\alpha \tan\phi_\mu = F + W \sin\alpha$$
$$\tan\phi_\mu = \frac{F + W \sin\alpha}{W \cos\alpha} \tag{4.16}$$

where W is the weight off the block and F is the mean spring balance reading. The experiment should be repeated for a number of blocks and joint surfaces to obtain a mean angle of plane sliding friction. If the experiment is carried out in the laboratory using two joint-bounded blocks, the calculation can be simplified. The spring balance is dispensed with and the blocks are placed on a tilting platform, such as that illustrated in Figure 4.23. The platform is slowly tilted until the upper block just begins to slide off the lower one. At sliding, the angle of plane sliding friction (ϕ_μ) is equal to the inclination of the platform (α_{crit}). The technique has been used by Cawsey and Farrar (1976) and Hencher (1976) to measure joint friction.

Thus, simple techniques borrowed from structural geology enable rapid assessments of failure potential to be made, and the frictional strength of joint surfaces can be measured using simple methods. It is therefore possible to make preliminary stability analyses for different styles of rock failure on rock slopes (Hoek & Bray 1977). Difficulties arise when some cohesion exists along the planes and when water pressures in the discontinuities need to be considered. The latter are often important, though difficult to measure.

4.2.5 Techniques for observing and measuring soil creep

All types of soil creep, continuous creep, seasonal creep and random creep, are very slow. Consequently, measurement techniques need to be very sensitive in order to pick up these slow movements accurately. Qualitative indicators of soil creep movement such as curved trees or displaced objects on the soil need to be treated with caution, and the problems have been discussed by Carson and Kirby (1972) and the particular problem of curved trees by Phipps (1974). In practice, although it is possible to define three types of soil creep due to different processes, it is not really possible to measure them separately in the field. All records of slow mass movements are therefore the sum of all processes. Seasonal creep is generally thought to be the most widespread and important process, because it is not restricted to specific soil types or to a narrow range of hillslope angles. Seasonal creep has therefore received theoretical attention (for example, Davison 1889, Kirkby 1967). Lack of agreement between published creep velocity profiles and those predicted from theory suggests that much more work upon the mechanics of seasonal soil creep is necessary. Fleming and Johnson (1975) and Finlayson (1976), for example, have measured quite reliable *upslope* movements over short time periods of a year or so, which are not predicted by any of the theoretical models. The role of lateral expansion and contraction leading to cracking, as opposed to vertical expansions, is obviously important though poorly understood. Understanding the mechanism of soil creep and measuring the rate of downslope soil movement require measurements of the velocity profile of the soil profile, preferably down to the rockhead, if this is possible. A remarkable number of techniques have been designed to measure creep, which is testimony to the importance ascribed to the process, especially on slopes in temperate latitudes not subject to rapid mass movements. It is quite impossible to cover all of the techniques that have been proposed in the limited space available here, and so priority is given to the technically simpler methods for measuring movement of the whole soil profile. Measurements of surface or near-surface movement are briefly discussed, as are more sophisticated methods for whole soil profile measurements.

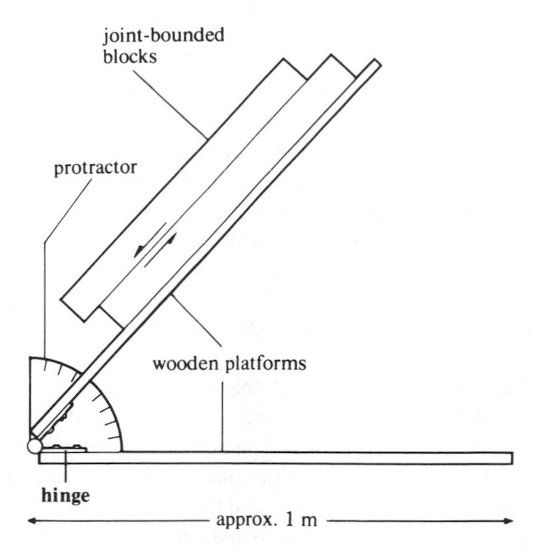

Figure 4.23 A tilting platform device to measure joint friction.

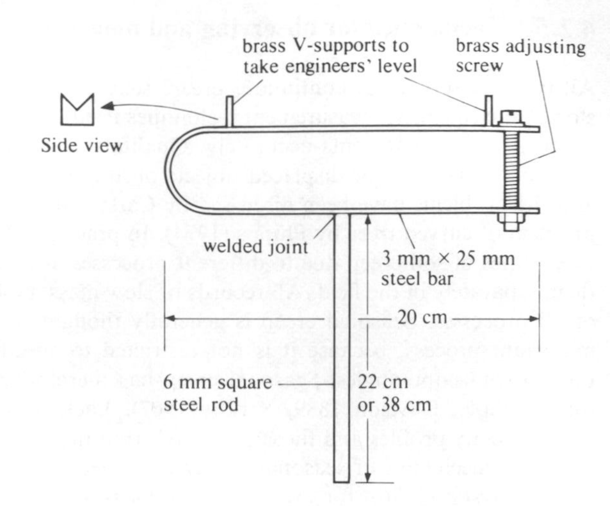

Figure 4.24 Kirkby's T-peg.

4.2.5.1 Surface movements

The simplest method of surface creep measurement is to use pegs inserted a short distance into the soil surface. The pegs may be wooden or metal and should have a reference mark or line engraved upon their tops. Measurement of position from a fixed benchmark, preferably located upon *in situ* bedrock, may be made by theodolite or a steel tape. Painted rocks laid on the ground surface can be used in place of pegs. The main problems with surface markers to measure creep are that they are prone to disturbance and that really precise location is difficult. Hence, they are of most use in remote areas and where rates of creep are relatively rapid, such as in periglacial areas where the process is sometimes known as solifluction.

Surface movements have also been monitored using measurements of angular rotation. A number of systems have been tried but the one described here is basically that of Kirkby (1963, 1967), who used a device capable of measuring very small rotational movements. The apparatus, called a T-peg, consists of a steel rod up to 40 cm in length with a horizontal crosspiece about 28 cm long formed into a U-shape (Fig. 4.24). V-shaped mounting plates are fixed to the crosspiece, which is also equipped with an adjusting screw. The T-peg is inserted into the soil as shown in Figure 4.24 and is levelled with the adjusting screw. An accurate engineer's spirit level, graduated to 10 seconds of arc, is placed onto the V-supports for this purpose. As the soil creeps downslope, it causes the rod to tilt, and the amount of tilt can be measured over quite short time periods with the graduated spirit level. Assuming that angular movement of the bar represents average shear in the soil to the depth inserted, it is possible to calculate relative movement (x) between the depth of insertion (y) and the ground surface from

$$x = y \tan \theta \tag{4.17}$$

where θ is the angle of rotation of the T-peg. In laboratory tests, however, Kirkby (1967) found that the tilting of the T-peg was far too low to justify this assumption, and concluded that there must be some shearing of the soil past the bar. An interesting modification to the approach is Anderson's inclinometer, which allows angular displacements of the inserted peg to be measured in any direction. The apparatus, described fully in Anderson and Finlayson (1975), consists of an adjustable spirit level mounted on a central pivot over the peg. It can therefore be rotated to measure tilt in any direction. The major disadvantage of all techniques to measure angular displacements is that the precise pivot point about which the rotation takes place is not known. Hence it is impossible to convert the observations into reliable estimates of downslope movement. The instruments are also very exposed and are liable to disturbance by animals and humans. These problems are partially solved by the Anderson tube method, discussed under whole-soil-profile movements. Thus, T-pegs and similar systems are best used for comparative measurements between sites over short timespans, where really accurate measurement is necessary.

4.2.5.2 *Movement of the whole soil profile*

Measurement of a creep velocity profile for the whole soil depth requires an insert placed into the ground, whose deformation can be measured after some time has elapsed. Inserts can be divided into those which require re-excavation in order to record their new position and those which can be monitored without being disturbed (e.g. Rashidian 1986). The latter types are much more convenient, in that measurements can be made repeatedly over indefinite timespans. Those requiring re-excavation can often only be measured once, and even if they can be reburied for a further period, they suffer from soil disturbance when they are monitored. Accordingly, if re-excavation is necessary, the insert is best left in the ground for as long as possible before measurements are made. No system to measure soil creep has yet been devised that does not cause some disturbance of the soil when the insert is first placed in the ground. Usually the degree and significance of the disturbance is not known, though inserts that are placed in auger holes are certainly less prone to disturbance problems than those placed in a soil pit. The problem of disturbance is an important one, which is discussed further below. Generally, one should remove as little soil as possible to place an insert, and all holes and pits should be backfilled as near as possible to the original soil unit weight.

A very great number of inserts requiring re-excavation for measurement have been used to monitor creep movements. Original references for those mentioned here are to be found in Anderson and Finlayson (1975). Lengths of 2 mm diameter brass wire or 8 mm lead cable can be placed vertically into an auger hole, which is then carefully backfilled. Wooden or plastic discs can also be inserted into an auger hole. They can be placed inside a steel pipe, which is pushed into the auger hole and then withdrawn, while carefully holding the discs in place with a rod (Fig. 4.25). Two other ways of inserting the discs are illustrated in Figure 4.25. The discs are carefully drilled with a central hole and can then be passed down a vertical steel rod of the same diameter as the central holes, which is then removed. The discs can also be secured together with a cord passing through the central holes, attached tightly to metal pins at each end of the column. The discs are then held rigidly and can be pushed into the soil as a column. After backfilling the hole, the top metal pin is removed, releasing the tension on the cord. A suitable material for the discs is wooden dowelling of about 2 cm diameter, sawn into 1–2 cm lengths. Thin-walled PVC tubing may be inserted vertically into auger holes and their new shape after a period of time can be 'frozen' by filling them with quick-setting cement or plaster ('Cassidy's tubes', Anderson & Finlayson 1975). The tubes are then excavated and the shape recorded. Finally exotic material, such as coloured sand, fine glass beads, or white modelling clay, can be placed into auger holes. Sand or beads can be poured into a carefully excavated auger hole, but the modelling clay must be formed into uniform-diameter cylinders and frozen. They are then transported to the recording site and pushed vertically into the auger hole in the frozen state.

It is most important to place and re-excavate auger hole inserts very carefully in order to minimise disturbance error. The auger hole itself must be vertical and of uniform diameter. A thin-walled tube auger hammered into the ground is the best tool to make the hole, though in stony soils it may be difficult to operate. The insert should also be placed exactly vertically, so that its initial position is exactly known. A plumb-bob is helpful to obtain correct alignment. The auger hole should be backfilled with soil, tamped to original soil density. Alternatively, tamped dry sand can be used, which is easier to place than damp soil. The site should be marked with a steel or wooden peg unless sufficient surface features are present to make relocation possible. To obtain creep measurements, the column is carefully uncovered and its configuration measured from a vertical plumb-line. If the column enters bedrock, it is reasonable to assume that its base has been static. If no fix into stable rock was possible on insertion, one may assume that the base has been static only if no distortion is visible at the bottom of the column (that is, it remains straight and vertical). If distortion is obvious throughout the whole column, one may only report relative displacements of the soil layer throughout the depth monitored.

A widely reported and extensively used technique to measure movement of the whole soil profile is the Young's pit (Young 1960). Short metal pins are inserted into the wall of a soil pit and their new positions recorded after a long time period. A narrow soil pit is excavated in the downslope direction, preferably to bedrock so that stable reference pins can be driven into the rock. One side wall of the pit is carefully trimmed vertical and cleaned, and a plumb-line is suspended next to it. A metal stake is driven into the bedrock at the base of the plumb-line or into the base of the pit if bedrock is not reached. Thin rods of non-corrosive metal or acrylic plastic of approximately 2 mm diameter and 200 mm length are pushed into the pit wall at regular intervals down the line, with their ends flush with the pit wall (Fig. 4.26). Either the pit face is then covered with brown paper to protect the markers and to provide a reference plane when the pit is re-excavated, or small handfulls of sand are placed around the exposed tips of the rods. Finally, the pit is carefully backfilled as close to original soil unit weight as possible, and is marked for future reference. Measurements may be made at any time subsequent to backfilling by re-excavating the pit and suspending a plumb-bob over the bottom stake. Distances from the line to each marker are then measured. In practice, because of the considerable amount of disturbance to the soil profile, the pit should be left as long as possible before re-excavation. While there is no objection to refilling the pit for a further period of monitoring, the possibility of disturbance of the rods by the process of excavation must be borne in mind, and great care is needed. An important asset that the Young's pit possesses over other methods of measuring soil creep is that the vertical component of soil movement may be measured as well as the downslope component

Figure 4.25 Methods of inserting wooden discs to measure soil creep.

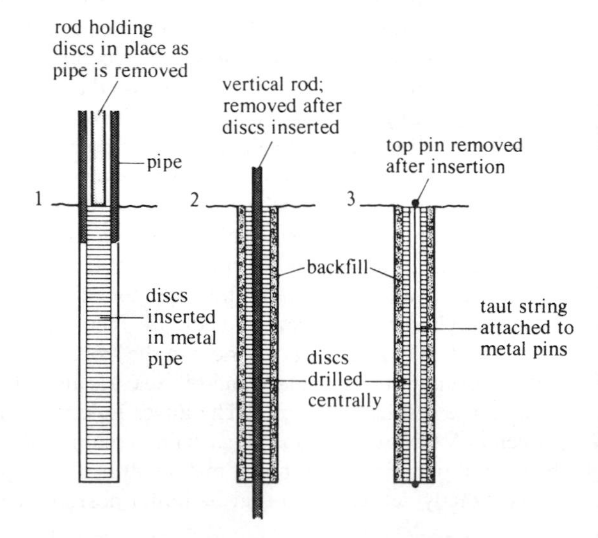

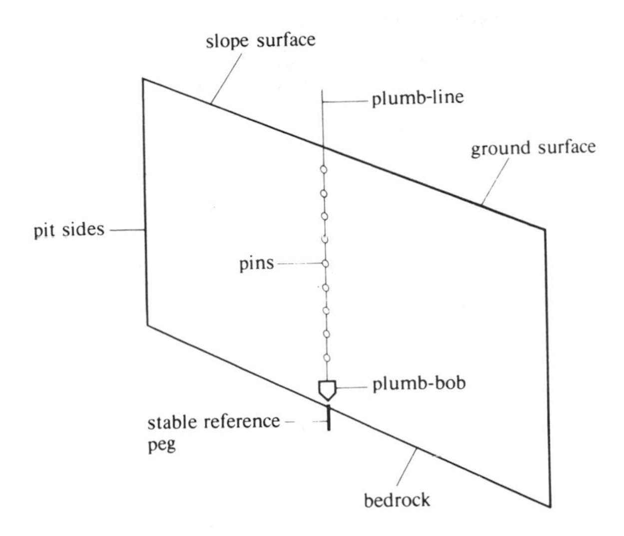

Figure 4.26 The Young's pit.

(Young 1978). This has important implications in estimating the relative dominance of various processes on slopes, and possibly in interpreting whether the soil zone is thickening or thinning through time (Young 1978).

Two other techniques are mentioned here briefly, but are discussed in detail in Anderson and Finlayson (1975). First, small cones made of sheet aluminium are inserted into a soil pit as shown in Figure 4.27. Wires attached to the cones are led to a pipe inserted into the soil. Movement is measured by the change in length of the wires protruding beyond the reference point at the top of the pipe. The wires must be stretched to a constant tension for accurate comparative measurements to be made. The second technique uses a tube inserted into the ground around a central stake driven into bedrock. All the soil is removed from inside the tube and so it is free to move with the soil, with respect to the stationary stake (Fig. 4.28). Measurements are made of the distance between the interior surface of the tube and the stake at four points on the circumference and at two depths, with a pair of inside callipers. This method is a variation upon the angular rotation techniques discussed above, but is more satisfactory in that the pivot point, about which rotation takes place, can be identified.

Recent advances in the measurement of soil creep have centred on the use of flexible tube inserts, which form permanent measuring stations and can be read repeatedly without ground disturbance. Many techniques exist to record the displacement of the tube, but the method of insertion is essentially common to all, and is therefore discussed first. Flexible dust extraction tube, which has a wire coil support, is a suitable material for the tube since it will take considerable deformation without losing its circular cross section. The base of the tube is sealed with epoxy resin before use in the field, and a rubber bung should be pushed into the top to prevent soil falling in during insertion. It is pushed into an auger hole of suitable diameter using a specially constructed insertion tube (Fig. 4.29). This consists of a thin metal tube, slightly greater in diameter than the flexible tube, which has slots cut into it to allow sand to be poured into the auger hole. The sand backfills the gap between the flexible tube and the sides of the auger hole and is tamped into place as the insertion tool is progressively withdrawn. After installation, the tube should be left to settle for some weeks before any mesurements are made.

Many ways of recording the displacement of the tubes have been devised, most of which are rather complex and expensive to build. One simple method, however, is to use

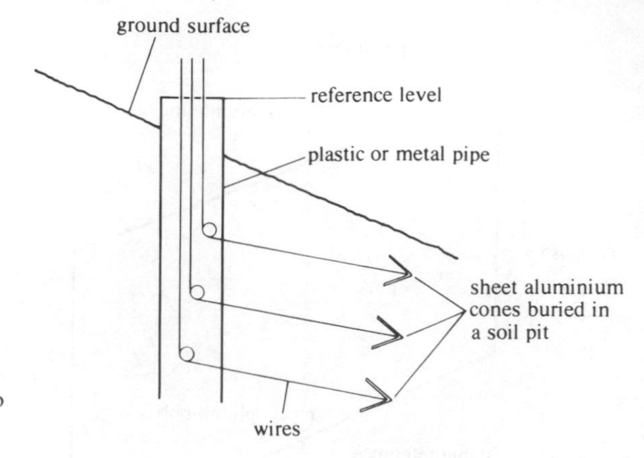

Figure 4.27 Buried cones to measure soil creep.

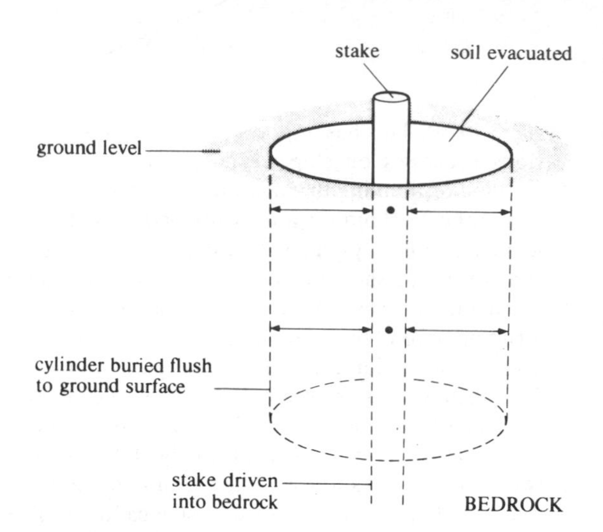

Figure 4.28 Anderson's tube to measure soil displacement.

small glass cylinders, slightly smaller in diameter than the tube, which can be lowered to predetermined depths in the tube. The cylinders are filled with hydrofluoric acid or hot wax and are left in position long enough for an ellipse to be etched on the glass or for the wax to set. The inclination of the tube at that point is then given by the angle between the ellipse and the cylinder wall. One disadvantage of using wax is that the meniscus is often indistinct, though the dangers of using hydrofluoric acid in the field may be great enough to outweigh this problem. A travelling telescope for measuring the displacement of flexible tube inserts has been devised recently (Finlayson & Osmaston 1977). The tube is fitted with crosswires at regular intervals and a 6 V lighting circuit to illuminate them. The tripod-mounted telescope is set up over the tube and the relative displacement of the crosswires with respect to the bottom crosswire is measured. Hutchinson (1970) used a loaded leaf inclinometer to record the inclination of flexible tubes at various depths. The instrument, details of which may be found in Hutchinson (1970), uses standard strain gauges and hence they can be calibrated in terms of inclination. The last two techniques discussed allow quite precise creep measurements to be made, though their expense and degree of technical sophistication are high.

With all systems recording movement of the whole profile down to stable depth, it is possible to estimate abosolute displacement of soil downslope. The area between successive velocity profiles is calculated and can be expressed as a volume of displacement per unit width of slope through time (for example, $cm^3\ cm^{-1}\ yr^{-1}$).

4.2.6 Techniques for the measurement of surface-water erosion processes on slopes

Surface-water erosion on slopes, often called slopewash, results from several mechanisms, already mentioned in the general discussion of hillslope processes. While it is convenient to think of the main mechanisms, rainsplash, unconcentrated flow and rill erosion, as distinct processes for the purpose of explanation, they really operate together and the effects of each are difficult to define. Measurement of slopewash rates is beset with difficulties. Total rates over large areas of slope are difficult to obtain due to the wide variation from point to point. Obviously, erosion along a rill during a storm may be orders of magnitude greater than adjacent unconcentrated areas. Also, many of the techniques have severe operational difficulties, which are discussed below. If rates due to rainsplash are to be separated from those due to surface flow erosion, even greater problems exist. Factors influencing the processes are manifold and include soil, rainfall characteristics, vegetation and slope factors. Often, it is difficult to isolate the factors under natural conditions and more precisely controllable experiments are possible in the laboratory to study process mechanics. The techniques discussed here are those to measure rates of erosion and are considered under the headings of rainsplash, surface flow erosion and total erosion. Brief reference is made to a few simulation studies.

4.2.6.1 Rainsplash

Measurement of rainsplash rates was pioneered by the work of Ellison (1944a, b, 1945). Ellison used a splashboard to measure rainsplash erosion. It consists of a board mounted across the slope, with two narrow troughs sunk flush with the soil surface at its base. The troughs are positioned to catch material splashed onto the upslope and downslope sides of the board respectively in order to estimate net transport (Fig. 4.30). Suitable dimensions for the splashboard are shown in Figure 4.30. After rainfall, material is cleaned from both sides of the boards and the troughs emptied, and the total upslope and downslope catch weighed. Net transport is then expressed in grams per centimetre of slope. Disadvantages of this apparatus include its susceptibility to overland flow entering the troughs and its obvious, though unknown, influence on wind turbulence near to the board.

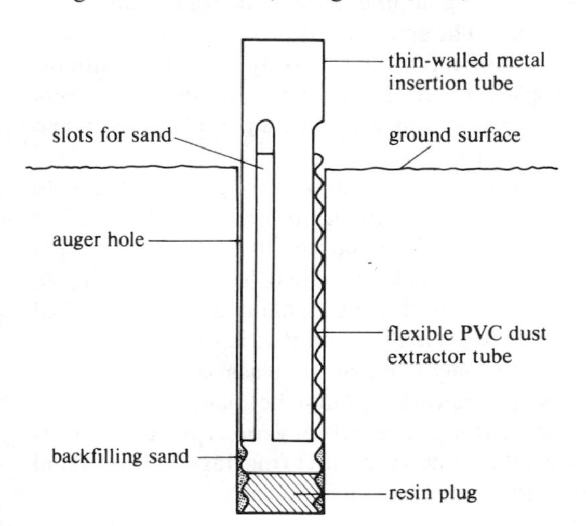

Figure 4.29 Installation of flexible soil creep tubes.

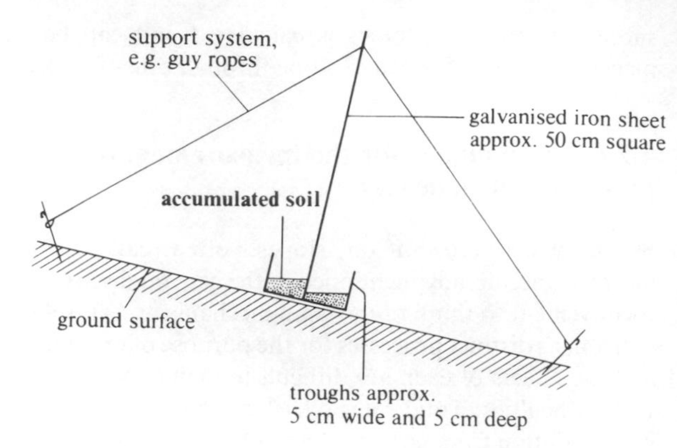

Figure 4.30 The splash trough.

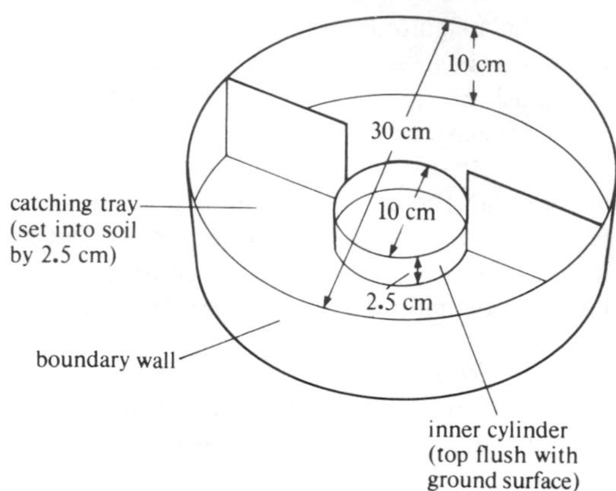

Figure 4.31 Morgan's splash cup.

Morgan (1978) has designed a splash cup to study rainsplash erosion (Fig. 4.31). It consists of an inner tube of 100 mm diameter, which is inserted into the ground until it is flush with the soil surface. An outer circular tray is sunk to a depth 25 mm around the inner tube. This tray catches material splashed from the inner tube and has a central divide so that upslope and downslope transport may be separated. Hence, net transport from the site may be measured and reported in g/m^2 in unit time. Morgan installed two cups at each site and used the average value. The apparatus is protected from overland flow by the outer rim of the collecting tray and it is much less likely to interfere with the wind flow near to the ground than the splashboard. Disadvantages are that only a very small area of soil is sampled by the splash cup in comparison with the splashboard, and that considerable disturbance of the soil occurs on installation.

Bolline (1978) used 52 mm diameter funnels, buried flush with the ground surface, to catch splashed soil. A rot-proof filter is placed into the funnel to catch the sediment. The funnel is weighed before placing in the ground and is removed after each rainfall event for reweighing. The funnel is cleaned on the outside and dried before reweighing to obtain the weight of splashed soil. This apparatus does not separate material splashed downslope from that splashed upslope; nor does it allow soil that has entered the trap from being splashed out again. Therefore, only cumulative splash is measured and estimation of absolute rates of downslope transport cannot be made. Comparative measurements may be made between sites to study the effects of various process controls on rainsplash. Similar to the splashboard, the funnels can suffer from ingress of overland flow, and therefore may not necessarily give a true measure of splash.

A number of other techniques have been tried to investigate rainsplash. Coutts *et al.* (1968) labelled soil particles with a radioactive tracer, ^{59}Fe. The tracer was applied to the soil as a stock solution of $^{59}FeCl_3$. The method was tested in simulated soils and true field conditions, and it was possible to demonstrate the influence of slope and wind direction on soil movement. Unfortunately, the results were not of sufficient quality to allow estimates of rate erosion to be made. Kirkby and Kirkby (1974a) monitored painted stone lines on slopes during rainfall and were able to make interesting observations on the mechanics of the splash process. Generally, marked stone lines can only be used to measure total erosion by slopewash processes, since it is not possible to distinguish splash-displaced particles from overland flow movements without constant observation.

Field studies of controls and mechanics of rainsplash have often been inconclusive, probably due to the great number of factors that exert an influence on the process. Controlled simulations in the laboratory are a way in which this problem can be overcome. Specific controls may be isolated for investigation in the laboratory, which is not possible in the field. An example of this approach is Moseley's (1973) laboratory study of the influence of slope angle on rainsplash. Simulated rainfall is necessary for laboratory studies and it is important that the natural characteristics of rainfall are reproduced as closely as possible. Basically there are two main types of rainfall simulator (Hudson 1972): those that use non-pressurised droppers, and those that use spraying nozzles with pressure. The former suffer from the disadvantage that they need to be very high for the water particles to reach their terminal velocity. The latter suffer from the fact that if the spray is to include drops of the largest size that occur in natural rain, then the nozzle opening has to be large (about 3 mm diameter). Even with low water pressures, the intensity produced from nozzles of this size is much higher than from natural rain, though there are various techniques that have been developed to alleviate this problem. Moreover, as Hall (1970b) has pointed out, if a network of nozzles is used to generate artificial rainfall, an increase in working pressure increases the average intensity but decreases the range of drop sizes within the spray. In natural rainfall, on the other hand, the range of drop sizes increases with increasing rainfall intensity. Mutchler and Hermsmeier (1965), Bryan (1974), Riezebos and Seyhan (1977), Luk Shiu-Hung (1977) and Farres (1980) discuss some of these problems and give details of hardware, while Schumm (1977a) describes the use of a very large rainfall-erosion facility (REF), which can generate whole drainage networks.

4.2.6.2 Surface-water flow erosion

Sediment traps have been extensively used to measure erosion by flowing water on slopes for many years. A box dug into the slope surface with its top flush with the ground surface will catch any overland flow occurring on the slope above it, and most of the sediment being transported will be trapped. The Gerlach trough (Gerlach 1967) has been specially designed for this purpose.

A lid must be incorporated into the design to prevent splashed material from entering if one requires to separate the two processes, and drainage holes should be provided to prevent the trap filling with water and overflowing (Fig. 4.32). If the trap does overtop, a considerable amount of sediment of finer grain sizes would fail to be trapped. Inevitably, some disturbance of the soil must occur on installation of such a trap, which may lead to accelerated erosion near to the upslope lip of the box. However, if care is taken to ensure that the upslope lip is exactly flush with the ground surface, the effect of disturbance will be minimised. It should be mentioned that sediment traps measure net erosion loss for a section of slope above them by catching all sediment in transport at a point. There is often great difficulty in defining the contributing area to the troughs. They do not measure erosion at a point, unlike many of the techniques for total slopewash erosion.

Very few other ways of measuring surface flow erosion *only* have been devised, though many systems exist to measure *total* slopewash erosion, discussed below.

4.2.6.3 Total slopewash erosion

Since the early 1950s, erosion pins, stakes driven into the ground against which erosion can be gauged, have been used to measure slopewash. The technique was pioneered by Schumm (1956) and others, who used wooden stakes. More recently, metal stakes have been used in preference because they can be thinner for the same strength and are generally more resistant to decay. A comprehensive survey of the slopewash literature and summary of the use of erosion pins may be found in Haigh (1977). Many designs of erosion pin have been used since the method was first proposed, but after thorough examination of the literature, Haigh (1977) advocates the use of a design similar to that in Figure 4.33. The pins should be thin (5 mm diameter) metal rods up to 600 mm long, made out of non-corrosive metal if possible. They should be inserted a considerable distance into the ground to minimise risk of disturbance and frost heave, but should not be placed flush with the ground surface. Preferably about 20 mm of peg should protrude to facilitate relocation. Because slopewash rates are so variable between points on a slope, it is advisable to place a number of pegs at each measuring site, arranged in clusters or lines parallel with the contours. Readings are taken of the length of peg exposure using a simple depth gauge (Fig. 4.33) and should be made at least every six months, with the minimum of trampling of the site. If a loose-fitting washer is put round the peg on installation, it should be possible to record maximum and net erosion by the washer position and the depth of sediment accumulated over it respectively. In practice, washers inhibit erosion beneath them and are too smooth to be good sites for sediment deposition. Therefore, pins equipped with washers may give a rather poor indication of erosion and deposition at the site. Some workers have suggested that the washer only be placed over the peg at time of measurement to form a standard reference level on the soil surface, and should be removed after measuring is complete. The difficulties and drawbacks of erosion pins are discussed at length in Haigh (1977). Undoubtedly, there are serious soil disturbance problems on insertion and it is advisable to leave them to settle for some time before commencing observations. The pins also influence flow in their vicinity, though the importance of this problem is not really known.

A variation on the micro-erosion meter (Sec. 4.1.6) devised to measure directly the solution of limestone has been applied to slopewash (Campbell 1970, Lam 1977). Since rates of erosion due to slopewash are much higher than limestone solution, the apparatus need not be so accurate. The so-called 'erosion frame' consists of a rectangular grid of aluminium bars, which can be positioned over four permanently installed steel stakes driven into the ground at the measuring site (Fig 4.34). A depth-gauging rod may be

Figure 4.32 A sediment trap to measure surface-water flow erosion.

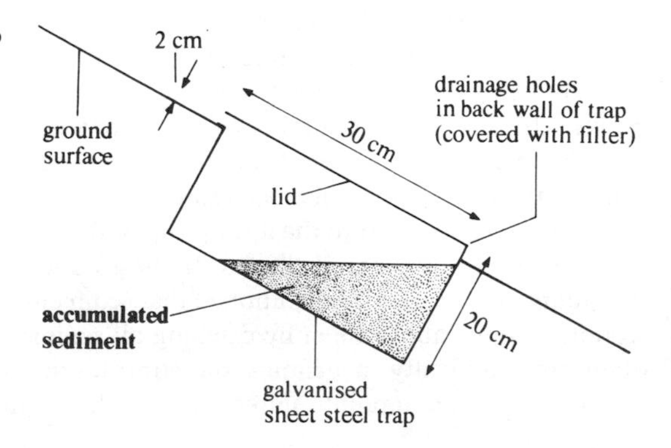

passed through holes drilled into the bars of the frame to measure changes in elevation of the soil surface, with respect to the frame, at a number of points. Several measurement sites may be established along a slope profile to examine changes in erosion downslope. The erosion frame has several advantages over erosion pins. The most significant are that no disturbance of the soil where measurements are made occurs when the site is established, that there is no modification of water flow, and that a number of measurements may be made at the same site.

Painted stones or sediment lines on slopes have already been mentioned in the context of rainsplash observation. When they are not observed continuously during rainfall, they only give information on total slopewash erosion. They have been used by many workers and obviously require little explanation. A paint line is sprayed neatly across the slope surface, or rows of painted stones are placed on the slope. The site is visited after rainfall and displacement of sediment and stones is measured. Particles that have been displaced should also be weighed to make estimates of total transport. Kirkby and Kirkby (1974a) found that the reliability of recovery declined with distance of transport downslope, and that the method was of no practical use at all for parts of the paint line that happened to cross concentrated flowlines, because particles were always lost. This highlights one of the major problem of measuring slopewash; namely rates of erosion fluctuate very widely between areas of concentrated and unconcentrated flow. Kirkby and Kirkby (1974b) have calculated slopewash rates from the degree of degradation of datable archaeological house mounds in Mexico. The calculation was based on the assumption that the transport rate is proportional to slope, on which basis the mound profile approximates to a normal curve whose variance increases linearly with time.

A final point about slopewash that should be made concerns the units in which erosion is measured. Rainsplash instruments and sediment traps catch material, which is weighed, and results are usually quoted in terms of a weight of material displaced from an area of slope, or in terms of a transport rate past a unit width of slope in unit time (for example, $g\ cm^{-1}\ yr^{-1}$). On the other hand, pegs and similar methods of measuring slopewash directly measure ground surface lowering. On this point, it is necessary to be careful to distinguish between vertical ground lowering and ground retreat parallel to the slope. It will be appreciated that these two measures of ground retreat differ by an amount that is a function of slope angle. Comparison of weight measures of transport

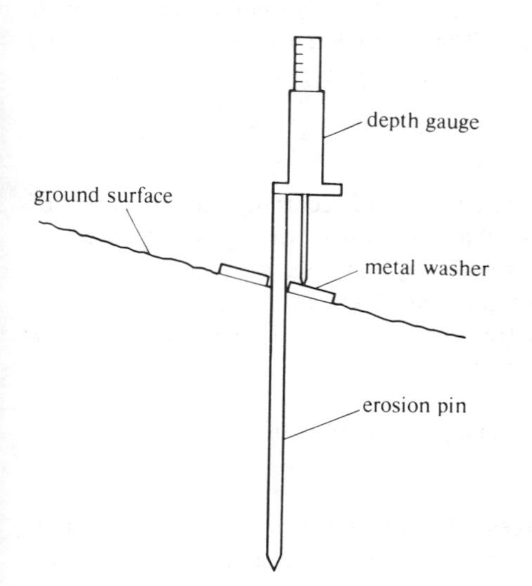

Figure 4.33 A typical erosion pin design.

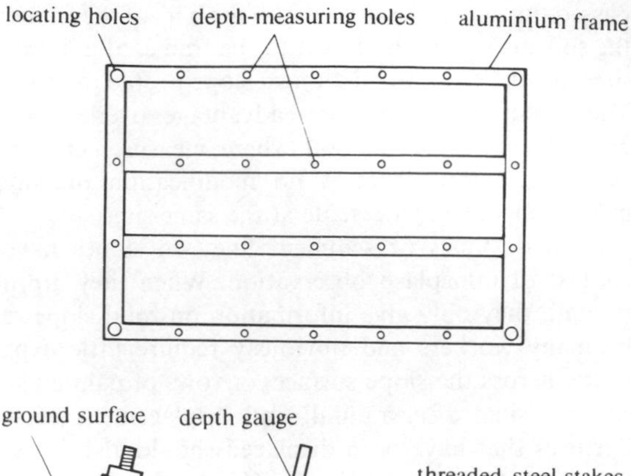

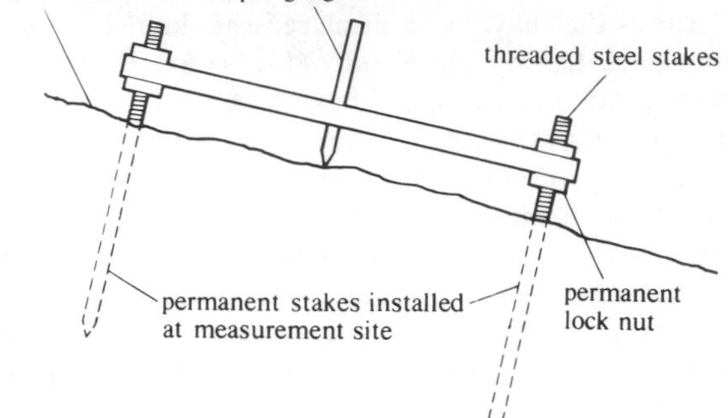

Figure 4.34 An erosion frame to measure slopewash erosion.

with direct surface lowering requires a conversion of the weights of soil into volumes. This in turn requires the unit weight of the undisturbed sediment surface to be assessed.

4.2.7 Frost heave and other ice-related soil movements

The formation of ice within the soil causes a displacement of the soil surface. This process, which is an important cause of sediment movement on slopes, is termed frost heave, and it may be of the order of 30 cm each season in a severe environment, though values of 1–5 cm are probably more typical. The methods available for frost-heave monitoring are summarised in this section. Two main types are recognised: the non-continuous and the automatic.

4.2.7.1 Non-continuous methods

Frame-and-rod instruments These use the principle of recording the movement of rods supported vertically on the soil surface by a metal frame (see Fig. 4.35a). Metal is preferable to wood because of the problems posed by warping. Such instruments can measure heave to an accuracy of 1 mm and are particularly effective where the magnitude of frost heave amounts to several centimetres. Thus, heaving beneath snow can be measured with a cheap, simple and efficient method.

Levelling Levelling can be used to record changes of level with reference to a stable benchmark.

Direct measurement of frost-heave gauges Various types are available (See Fig. 4.35) but James (1971, p. 16) states that buried frost-heave gauges have certain

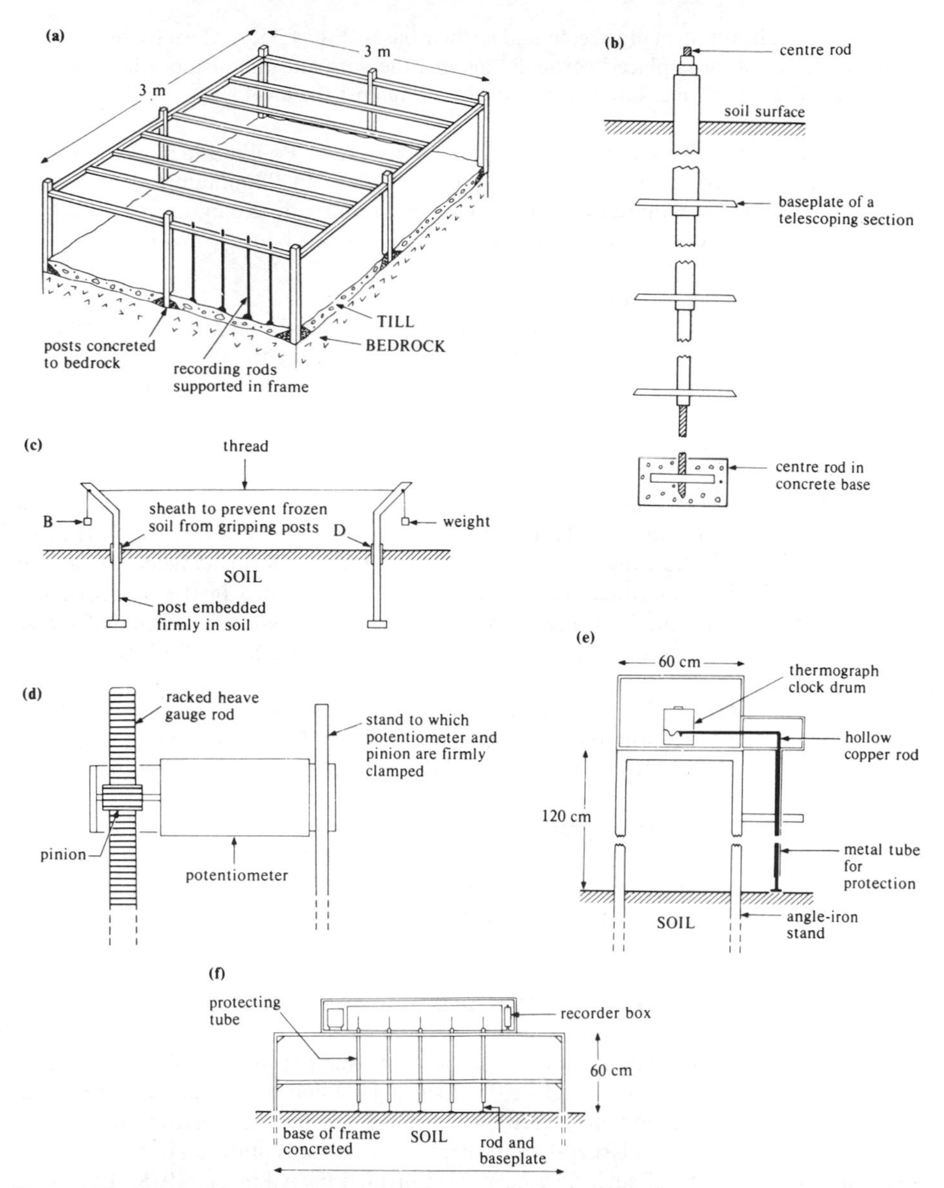

Figure 4.35 Selected instrumentation for the measurement of frost heave. (a) The Schefferville 'bedstead' designed by Haywood. (b) Telescoping ground-movement gauge designed by Baracos and Marantz. (c) The suspended-thread technique employed by Caine. (d) Rack and pinion, and potentiometer designed by Higashi. (e) Recorder designed by Matthews. (f) The Nottingham frost-heave recorder. (After James 1971).

limitations because, first, it seems unlikely that they would be successful in detecting heave of the order of a few millimetres and, secondly, their accuracy is doubtful because of the degree of interference with natural conditions within the soil.

The suspended-thread method Benedict (1970) measured heave with reference to a strand of fine-gauge spring steel wire stretched under a tension of 18 kg between two posts made of heavy steel pipe cemented at their bases (Fig. 4.35c). At each observation, a carpenter's level was placed vertically against the wire and the distance between the ground and wire was measured with a steel tape against the level.

Photogrammetry This can be used (see, for example, Poulin 1962) by placing a specially constructed camera 3–6m above ground level on an aluminium scaffold frame. A contour interval of 6 mm can be determined, but observations can be prevented by a cover of snow. There is the advantage that no ground disturbance is involved.

4.2.7.2 Automatic devices

Ratchet devices Ratchet devices used in association with frame-and-rod instruments of the 'bedstead type' allow upward movement of rods during the frost-heave cycle but prevent their return, thereby allowing the measurement of maximum heave.

Electrical methods The heaving of a soil sample can be recorded by means of such devices as potentiometers (see Fig. 4.35d) and transducers. The following method, developed by Everett (1966), is extremely sensitive, though expensive. The 0.95 cm aluminium rods with aluminium baseplates in contact with the soil were connected to a transducer shaft. The transducers were anchored to a frame of 8 cm diameter, galvanised iron piping, the base of which was cemented into holes drilled in the bedrock. The resistance of the transducers, being proportional to the movement of their shafts, was read on a continuously recording Wheatstone bridge to an accuracy of 0.1 mm.

Clock-driven shafts The bedstead type of frame-and-rod instrument can be modified by the attachment of an arm-and-pen device to the rods. This in turn gives a trace on a clock drum. Details are provided by James (1971, p. 21) and by Matthews (1967).

In periglacial areas, patterned ground phenomena may be associated with frost heave, with the development of ice segregations and with desiccation processes. These can be studied either in the laboratory (e.g. Corte 1971) or in the field by observation of vertical and horizontal particle movements (see, for example, Washburn 1979, p. 80).

4.2.8 The measurement of subsurface water flow on slopes

In recent years the importance of subsurface water movements in the soil zone both as a geomorphological process and as a factor influencing the shape of river hydrographs has become apparent. Such flow may occur either in the soil matrix – matrix flow – or in channels and pipes that are larger than the capillary-size pores of the matrix – pipe flow. Methods for the study of such flow have been divided by Atkinson (1978), on whose work this section is based, into three:

(a) Methods involving interception of the flow, which is then channelled into a measuring device to determine its discharge.
(b) Methods involving the addition of tracers, primarily to measure velocity.
(c) Indirect methods, applicable to matrix flow, in which measurements are made of moisture content and hydraulic potential over the whole of a slope profile or experimental plot.

Interception of flow is achieved by the digging of a pit or trench. It needs to be appreciated that this very act may appreciably alter soil moisture conditions. Normally a series of gutters is installed into the pit or trench face, and to ensure good hydraulic contact between the guttering and the soil the gutters are packed with soil or pea-gravel and protected and sealed at the front by wooden shuttering. Polythene sheeting is dug into the face at an appropriate height for each gutter, to ensure that each draws its flow from only a single layer of the soil (Ternan & Williams 1979). Discharge rates from the gutters can be monitored manually by a stopwatch and measuring cylinder, or automatically using tipping-bucket apparatus. For pipe flow (or tile drainage), flexible hoses can be inserted into the pipe outlet and sealed using bentonite or cement. Discharge can be measured either using small V-notch weirs fitted into cisterns or with tipping buckets. Some suitable designs are given in Jones *et al.* (1984) and in Gilman and Newson (1980, p. 53), who used a combined propeller meter/siphoning tank apparatus. This instrument consists of a propeller meter mounted in a tube, the tube being connected into a tank with a siphon outlet. The outlet is so arranged as to empty the tank when the water in it reaches a given level. A float switch gives two electrical pulses for each filling and emptying cycle. To record flows that exceeded the operating limit of the siphoning tank, the propeller meter was employed. They also used an automatic system consisting of weir tanks with autographic recorders. The weir tank consists of a glassfibre-reinforced plastic cistern with a V-notch plate at one end and a baffled inlet at the other. The stilling well was mounted within the tank, and the level recorder was mounted above its centre. To overcome severe debris blockage problems to both pipe-flow measuring devices, a simple cistern-based sediment trap was placed upflow from both the siphoning tank and the weir tank.

There are two main types of tracer that can be used to measure water flow velocity on hillslopes: radioactive compounds and dyes (see also Sec. 4.1.11). The use of tracers involves injection upslope, or where vertical movement down the soil profile is to be traced, on the soil surface or in the surface horizon. The use of ^{15}N as a tracer is described in Barraclough *et al.* (1984). Smettem and Trudgill (1983) have evaluated some fluorescent and non-fluorescent dyes for the identification of water transmission routes in soils, and describe appropriate microsampling and extraction methods. Examples of field experiments using tracers are given in Trudgill *et al.* (1984) and in Coles and Trudgill (1985). Similar work may be carried out in the laboratory using undisturbed cores cased in resin and glassfibre or in gypsum.

Tensiometers are the prime technique for establishing the direction and rate of moisture movement in the soil (see also Secs 3.2.12, 3.2.15 and 4.2.3).

4.3 Solutes

4.3.1 Solute studies in geomorphology

The study of solutes in natural waters within geomorphology has three major areas of relevance. First, the chemistry of soil water and runoff waters may be used to estimate rates of natural weathering and catchment denudation (see Sec. 4.1.1, Douglas 1973). Likens *et al.* (1970), for example, have noted that:

> if the biotic portion of the ecosystem is at or near a steady state then the net losses of dissolved nutrients (in streamwater) must come from the chemical weathering of primary and secondary minerals within the ecosystem.

Perrin (1965) had earlier used this rationale in proposing the use of chemical analyses of soil drainage waters to assess the rate of soil formation, while Lesack *et al.* (1984) and Sarin and Krishnaswami (1984) have used river solute chemistry to infer continental denudation rates and weathering processes. Cleaves *et al.* (1970, 1974) considered stream-water solute levels to represent the products of certain weathering reactions in a catchment and from this calculated weathering rates, allowing for element incorporation into the catchment's biomass. In earlier studies, Hembree and Rainwater (1961) and Miller (1961) simply used stream-water solute levels and catchment lithology to estimate rates of catchment denudation, whereas Crisp (1966), White *et al.* (1971), Pathak *et al.* (1984) and Peters (1984) incorporated considerations of solute input in precipitation and removal in particulate matter in streamflow and in animals. Estimation of net catchment losses of particular elements, as gauged from solute levels in precipitation and streamflow, has been a major part of the results of the Hubbard Brook Ecosystem Study (Likens *et al.* 1977). Other researchers investigating the actual mechanisms of rock weathering, mineral precipitation and pedogenesis in field and laboratory experiments have used the solute content of soil solutions and mineral leachates to identify the reactions involved (Bricker *et al.* 1968, Deju 1971, Wiklander & Andersson 1972, Waylen 1979, Bryan *et al.* 1984).

The second major use of solute studies has been in assessing the intensity, efficiency and stability of nutrient cycles, often as indicated by the levels and variability of solutes in streamflow (Nelson 1970). Juang and Johnson (1967) and Johnson *et al.* (1969) noted the buffering of stream-water solute concentrations by soil and vegetation cover. Other workers have demonstrated the effects on stream-water quality of disturbance of ecosystems by burning (Johnson & Needham 1966, Watts 1971), deforestation (Bormann *et al.* 1974, Bormann & Likens 1967, Swank & Douglass 1977, Tamm 1974, Davis 1984), agricultural activity (Dillon & Kirchner 1975, Kang & Lal 1981), building construction (Walling & Gregory 1970) and seasonal activation and relaxation of biological activity (Likens *et al.* 1977). Others have simply used precipitation and stream-water solute levels to establish complete nutrient budgets for catchment units (see Likens *et al.* 1977, table 22).

Thirdly, solutes in natural waters have been studied as tracers or labels, both natural and introduced, to indicate pathways for solute transfers and to assess their quantitative significance. Hydrographs have been separated into supposed genetic components on the basis of changing stream-water solute concentrations through time. These changes are attributed to the mixing of runoff waters from different sources that possess different chemical characteristics by virtue of their residence time and/or runoff route (Steele 1968, Pinder & Jones 1969, Johnson *et al.* 1969, Hall 1970a, Walling 1974, Burt 1979). Walling and Webb (1975) studied the spatial variation in stream-water quality in relation to geological outcrops. Others have evaluated human modification to natural background solute levels from urban areas (MacCrimmon & Kelso 1970) and areas of agricultural activity (Minshall *et al.* 1969, Dugan & McGauhey 1974, Kilmer *et al.* 1974). Non-natural solutes have been introduced to a catchment system for the identification of runoff routes (Smart & Wilson 1984) and to assess the timing and magnitude of outflow rates (e.g. Knight *et al.* 1972). Fluorescent dyes and isotopes are most frequently used in tracing experiments (Smart & Laidlaw 1977). Zand *et al.* (1975) and Pilgrim (1976) have used tracer solutions to demonstrate channel mixing and transport processes, and Church (1975) has discussed the use of electrolyte and dye solutions for stream discharge estimation, procedures reviewed by Airey *et al.* (1984) for large rivers in flood. Reference should be made to Trudgill (1986) for a detailed coverage of 'solute studies' in geomorphology.

An obvious requirement in all studies of solutes in natural waters is that these waters must either be sampled for eventual field or laboratory analysis or direct measurements of water quality parameters must be made *in situ*. The following two sections consider the procedures and problems of field sampling of natural waters for laboratory analysis and the field instrumentation necessary.

4.3.2 Field sampling programmes for solute studies

Water sampling is a prerequisite of all solute studies and according to Hem (1970) is probably the major source of error in the whole process of obtaining water quality information. It should therefore be given very careful consideration before embarking upon a solute investigation if worthwhile conclusions are to be drawn from the data collected. The sampling problem is largely one of 'representativeness': whether the samples taken are representative of the spatial variations in the larger body of water from which they were removed and whether these samples represent adequately the temporal variations in that body of water. If the whole volume of water at a site constitutes the 'sample', as may be the case in studies of rain or runoff water quality, then the further problems still remain of field collection and storage of water samples and their transfer to the laboratory in such a way as to minimise the changes in the chemical composition of the water so that the sample in the laboratory may be considered representative of the sample in the field.

The question of the spatial representativeness of water samples depends very much on the particular requirements of the investigation that is being undertaken. In studies of the spatial variation in stream-water solute levels across a catchment, the distribution of sampling points will clearly need to be related to tributary inflows and geological outcrops (Imeson 1973, Walling & Webb 1975, 1980) and other factors that are thought to be significant in affecting this spatial variation such as vegetation patterns (Ternan & Murgatroyd 1984). It is assumed that the selected sampling points will 'represent' certain parts of the catchment with relatively uniform geology, land use, relief, climate, etc. The initial density of sampling points will clearly depend on the individual's preconceptions of the closeness of environmental control and the rapidity of spatial variation in quality, and the number of sampling points necessary to resolve or represent this variation at a level of detail appropriate to the task at hand. A pilot survey is usually a necessary first stage of any such investigation, in which very intensive spatial sampling is made; the redundancies, or lack of them, in the resulting data will suggest the minimum number of sampling points necessary. More than one pilot survey may be deemed necessary to demonstrate any temporal dimension to the 'established' spatial variation, which may confuse later results. The same considerations will also apply to precipitation and runoff sampling; because of the assumed spatial uniformity of precipitation chemistry at the small catchment scale, only one precipitation sampling station is normally thought necessary (Reynolds 1984), though the non-uniformity at a larger scale is well established (Eriksson 1966, Junge & Werby 1958, Granat 1976).

Another aspect of the problem of the spatial representativeness of water samples arises when a single sample at a point is assumed to represent the composition of the total water body, not spatially across a catchment but within a lake or throughout a stream cross section, when these water bodies may not be homogeneous. The problem of the stratification of lake waters, where for example a single surface sample will give a poor indication of mean lake water quality, may well be less important to the geomorphologist than is the variable quality of a stream at a cross section. In small streams with a rocky bed and steep gradient, good lateral and vertical mixing of tributary and groundwater inflows throughout the whole volume of stream water is usually present. In larger and less turbulent streams, however, there may be a definite physical separation of these inflows for distances of several kilometres below a confluence, particularly if the temperatures and dissolved and suspended material concentrations of the tributary and mainstream waters are markedly different. A relatively simple pilot study should establish whether such conditions exist at a selected sampling station, so that its location can be changed or local areas of slack water avoided, or whether a series of samples should be made across the channel so that a discharge-weighted mean solute concentration can be computed. A simple conductivity survey at different stages of flow should be sufficient to establish where, if at all, a single water sample may be taken to represent the quality of the whole stream at that point. It should be noted here that, when continuously monitoring water quality or automatically sampling water, either a sensor probe is fixed in the stream or water is pumped from a fixed intake to an instrument. In both of these cases, a fixed point is being sampled. The 'spatial representativeness' of the samples within the channel should therefore be considered very carefully before installation of such equipment, especially with regard to possible variations in representativeness at different levels of discharge, so that unbiased results or samples may be obtained.

Although continuous monitoring of solute levels largely overcomes problems of deciding *when* to sample, in addition to *where* to sample, the necessary equipment is often unavailable because of its high cost or is inappropriate on grounds of accuracy and reliability. The investigator must then take actual water samples and is faced with the problem of the representativeness of his or her sampling in the temporal dimension in the knowledge that the chemical composition of natural waters at any point is rarely constant but varies in response to seasonal, daily and storm-long fluctuations in both solute and runoff production. Again, it must be stated that the question of temporal frequency of sampling must be answered with reference to the specific requirements and aims of the investigation (whether this is mainly concerned, for instance, with seasonal or with storm-linked quality fluctuations) and also with reference to the nature of the hydrologic system under study (whether rapid fluctuations of solute levels are characteristic, as with precipitation, or whether solute levels are steady for long periods of time, as may be the case for streams supplied mainly from ground water). Claridge (1970), in addition, has noted that:

> sampling according to a standard time series may seriously under- or over-estimate a highly variable quality parameter in streamwater, depending on the nature of the quality fluctuations.

In making general statements concerning sampling frequency, one can say little to fault the comment of Hem (1970) that:

> to determine the water quality regimen of a river at a fixed sampling point, samples should be collected or measurements made at such intervals that no important cycle of change in concentration could pass unnoticed between sampling times.

Exactly what are considered to be 'important' changes will depend on their magnitude, as perhaps established by a pilot survey, and the experimental aims as outlined above, and also more practically on the laboratory facilities and analytical instrumentation available. The Hubbard Brook Ecosystem Study, for example, was mainly concerned with annual and seasonal nutrient budgets and found a weekly stream-water sampling interval to be sufficient for accurate computation of total catchment element output in stream water in conjunction with a continuous record of stream discharge.

Hem (1970) notes that the United States Geological Survey samples rivers on a daily basis, and that this is thought to provide a representative record for most *large* rivers which could be safely used for computation of mean annual chemical composition and annual solute loads. The daily sampling frequency here appears to be largely a convenience, given daily measurement of river discharge, though Rainwater and Thatcher (1960) consider that daily sampling adequately defines water quality for the majority of water uses. The consideration of practical convenience must always form part of the decision to adopt a particular sampling frequency, though obviously it should be secondary to considerations of the necessary frequency to accomplish the particular aims of the project. Walling and Webb (1982, 1985) have clearly demonstrated the profound effect that sampling frequency can have on solute budget calculations.

When the research project is concerned with solute concentration variations during storm events in precipitation, runoff waters or streamflow, then a much more intensive sampling programme is obviously required so that the fluctuations that are of interest are accommodated within the sampling framework so that none 'pass unnoticed between sampling times' (Hem 1970). In such projects, the required frequency of sampling is directly related to the rate of change of solute concentration and, as this rate usually increases during a storm event compared with that between events, distinctly different sampling frequencies are required. Hill (1984) thought it appropriate to sample more frequently in winter than in summer.

Loughran and Malone (1976) sampled every 10 min during the course of a storm hydrograph, and water samplers have been developed that automatically increase the frequency of sampling at a preset water level or discharge rate (e.g. Walling & Teed 1971). By the same token, bulk samples of precipitation or runoff waters that are collected in a single container may conceal significant information regarding temporal quality variations during the storm event. Instrumentation is described below that allows intensive sampling of these waters. Some useful comments on the design of temporal sampling programmes are contained in an article by Walling (1975) in which reference is also made to the consideration of the total duration of sampling necessary in solute studies. He notes that 'several years of record are required if seasonal variations are to be meaningfully distinguished from other more random variations'. Nutrient budget studies will clearly require at least one complete annual cycle to be monitored, though studies of solute response to storm events may be accomplished within one season if enough suitable events occur. Montgomery and Hart (1974) discuss formal statistical procedures for determining the required sampling frequency.

4.3.3 Problems of water sampling for laboratory analysis

Another dimension of the problem of representativeness was noted earlier as being to do with the manner of collection of the sample, its storage and its eventual transfer to the laboratory in such a way that the laboratory sample is representative of the water from which the field sample was extracted. The very removal of a water sample from its field environment will, however, inevitably affect its chemical characteristics (Summers 1972). The degree of sample modification is usually inversely related to the concentration level of its chemical constituents and to the sample size, and is directly related to the time of storage before analysis. Thus, the likelihood of sample modification can be systematically related to the method of sampling, the type of water sampled (i.e. precipitation, soil water or stream water) and the sample treatment prior to analysis.

The type of sample container is most important in influencing the degree of sample modification. If a large sample is taken, then the effects of contamination from the bottle and adsorption of dissolved material by the bottle walls will be reduced compared with those for a smaller sample volume. One-litre bottles are commonly used, partly for ease of transportation and partly because these provide the sample size necessary for eventual analysis by titration and for determination of total dissolved solids concentration. Borosilicate glass (e.g. Pyrex) sample bottles with close-fitting stoppers are often recommended because there is little exchange between bottle and sample by leaching and adsorption, as is expected with other types of glass. But in field use these bottles are heavy and fragile, may not survive transportation and handling in the laboratory and usually do not survive sample freezing, in either the field or the laboratory. While glass bottles allow no gas exchange between the sample and the atmosphere, they are obviously unsuitable for sampling when silica concentrations are of interest. Though Loughran and Malone (1976) used milk bottles and did not find the expected leaching contamination, most popular are screw-top sample containers made from thick, smooth, high-density polyethylene, which are both light and flexible. These may allow some gas exchange and solute adsorption onto the bottle walls, particularly of phosphorus, when samples are stored for longer than a few hours, but these difficulties are usually outweighed by their obvious practical advantages. Polyethylene containers are usually recommended for samples that are to be used in metallic trace-element determinations (Golterman *et al.* 1978).

The need for clean sample containers is fundamental and especially so for waters of low solute content where contamination can completely change the sample's chemical characteristics prior to analysis. New glass containers should be initially cleaned with a hot chromic acid solution to remove contaminants from the manufacturing process, and after later use with a mild detergent, and then rinsed thoroughly, first with tap water and then immediately with distilled or deionised water. Plasticware can be cleaned with concentrated hydrochloric acid or 50% nitric acid or a commercial decontaminant such as Decon. Shaking with a sand and water mixture may help to remove heavy surface contamination and also algae, which tend to grow on some plastic surfaces. Water sampling instruments should also be tested with sample and distilled water 'blank' solutions to check for contamination from this source. Careful attention to this stage of the process of water sampling and analysis is vital if the researcher is to have any confidence in the results of his or her efforts. Lindberg *et al.* (1977) illustrate the scrupulous care that can be taken to avoid, in this instance, contamination of samples by containers. Reynolds (1981) and the series *Methods for the examination of waters and associated materials* (HMSO) are useful guides to good laboratory practice and analytical procedures.

When particular water types are sampled that contain solutes susceptible to immediate chemical precipitation or adsorption effects or to biological uptake, or when samples are to be stored in the field or laboratory for more than a few hours, then some preservative pretreatment is advisable to avoid or reduce changes in the sample. Edwards *et al.* (1975) give an example of the reduction of nitrate concentrations during storage from 5.3 to 3.0 mg l^{-1} in a 72 h period after sampling, probably due to the activity of aquatic organisms. Crisp (1966) stored samples under refrigeration for up to one year, but immediate analysis, if at all possible in the field, is most strongly recommended because the chemical treatments otherwise necessary may well interfere in subsequent analyses or introduce contamination, thus biasing results. If a complete analysis is required to draw up an ion balance and sample treatments are thought necessary, then more than one sample must be taken, to which different preservatives are added. Pretreatments are usually recommended for waters containing solutes in high concentrations that may be precipitated from solution, perhaps under changing temperature conditions, and for waters that must be stored in the field in precipitation collectors or in runoff or streamflow samplers for more than a few hours.

The more common pretreatments are made to reduce bacterial activity and to keep carbonate and iron in solution. The addition of 5 ml of chloroform per litre of sample or acidification to pH 1 with hydrochloric or sulphuric acids will effectively prevent bacterial activity, especially in dark bottles. Minshall *et al.* (1969) added 1 ml l^{-1} of 10^{-5} M 2,4-dinitrophenol to samples and then stored them frozen for more than one year. Iron can be kept in solution by adding 5 ml of 4 N H_2SO_4 to the sample container, though it should be noted that acidification may bring colloids and fine particulate material into solution. If phosphorus determinations are to be made, there are special procedures recommended for the iodine impregnation of polyethylene containers to keep this element in solution (Heron 1962). Gore (1968) found that this practice increased determined phosphorus concentrations by 30% and that the high iodine concentrations present precluded any other determinations.

Whenever possible, the sample bottle should be completely filled to minimise gas exchange with the air, which can particularly affect carbonate equilibria and sample pH and alkalinity. For a similar reason, samples should be shaken as little as possible. Allen (1974b) discusses the usefulness of sample freezing in storage. While storage at between -10 and -15°C will effectively inhibit bacterial modification of solute concentrations, a temperature of just above freezing is easier to maintain and is often recommended to avoid changes in the sample that can be produced by freezing and thawing, this being particularly important at extremely low and high concentrations (Rainwater & Thatcher 1960). Allen (1974b) contains a useful summary of the most popular methods of sample preservation and of types of sample container available. Potential users of sample treatments should consult specific texts on natural-water analysis with regard to interferences with other analyses that a preservative may introduce. Also, chemical preservatives are seldom pure and contamination may be introduced in this way. Above all, researchers should satisfy themselves that the sample treatment does not bias the results of any chemical determinations that are to be carried out by a comparison of combinations of fresh and stored, and treated and untreated samples, together with standards and blank solutions from different container types. This is the sort of pilot study carried out, for example, by Slack and Fisher (1965).

At some stages of flow, stream water may contain higher concentrations of mineral and organic material as fine particles in suspension than are present in solution.

Immediate removal of high concentrations of suspended material is recommended as one of the best and most necessary pre-storage sample treatments. Allen (1974b) suggests that samples should be filtered within 8 h of collection because the field chemical equilibria between solutes and sediment may not be maintained in the laboratory. Others have recommended settling or centrifuging of samples to avoid contamination due to the filtration process (Nelson 1970). The removal of suspended sediment is recommended because it may otherwise interfere with titrimetric and optical techniques of chemical analysis by light scattering and absorption, and also because solution and adsorption may go on in storage between dissolved, colloidal and suspended phases in the sample. There appears to be little quantitative information available concerning this last point, but Hem (1970) emphasises the likely importance of this interaction, particularly in rivers. Sediment–solute interactions are not only important in stored samples but are also relevant to element budget calculations for a catchment. Much 'dissolved' material may in fact be transported adsorbed onto sediment, depending on stream-water acidity and concentrations of adsorbed and dissolved phases (Bradley 1984, Morrison *et al.* 1984). As these conditions vary during storm events, then the sediment–solute interactions are likely to be highly complex (Grissinger & McDowell 1970, Gibbs 1973).

The filter pore size most widely recommended is 0.45 μm, which is about 100 times larger than the smallest colloidal particles, so that some colloidal material will be included in analyses that purport to determine dissolved material concentrations. It is thought that for most purposes this should not trouble the geomorphologist unduly. Golterman *et al.* (1978) have a useful summary of different materials that can be used as filters and a list of manufacturers. They also point out the problems introduced by filtering samples, particularly the long duration of the process without applying a vacuum and the possible contaminations from each filter type. Material may be leached from or adsorbed onto the filter medium, which has been found to increase sample concentrations of N, P and organics, in particular, and cause degassing of samples with high CO_2 concentrations.

The decision of whether or not to filter water samples will depend on the suspended sediment concentration encountered and the particular analyses that are to be carried out. Determinations of pH, alkalinity and dissolved gas concentrations should always take place prior to filtration. If at all possible, filtration is to be avoided as it is time-consuming and a very real additional source of sample modification and contamination. Again, a pilot study is recommended in which both filtered and unfiltered versions of water samples, blank solutions and standard solutions are all analysed for the full range of chemical parameters that are to be routinely determined. Filtration effects should then become apparent, or filtration may be shown to be an unnecessary operation.

4.3.4 Field techniques of natural-water sampling

4.3.4.1 Precipitation

Particularly since the work of Eriksson (1955, 1960) and the publication of Gorham's (1961) classic paper, there has been increasing recognition of the fact that the solute content of surface waters is affected by material derived from the atmosphere, both as dry fallout and as solutes in precipitation. Also, it is now clear that these element inputs from the atmosphere may contribute substantially to the nutrient cycle of ecosystems, particularly on unreactive rocks, in addition to the contributions from catchment denudation by soil and rock weathering (Fisher *et al.* 1968, Likens *et al.* 1977). The estimation of atmospheric element inputs is then very necessary if complete nutrient budgets are to be drawn up so that, for instance, weathering reactions can be evaluated or the effects of land management assessed (White *et al.* 1971). However, the estimation of this input poses considerable problems of sample collection, laboratory analysis and eventual interpretation of results inasmuch as the measured chemical composition of precipitation samples depends very much on the techniques used for their collection. Excellent summaries of procedures for collection, preservation, storage and analysis of precipitation samples have recently been presented by Galloway and Likens (1976, 1978) and reference should be made to these papers. Also Cryer (1986) has summarised recent publications on solutes in precipitation.

A distinction has been drawn between 'wet deposition' of material in precipitation by 'rainout' and 'washout' processes (White & Turner 1970), and 'dry deposition' of particulate material between precipitation events. The catch of a precipitation collector continuously open to the atmosphere, a combination of wet and dry deposition, has been termed 'bulk precipitation' by Whitehead and Feth (1964), who determined that it is 'the most meaningful measure of solute input to a catchment'. The importance of dry deposited material is that it probably represents material of local origin and as such may not be wholly a net gain to the catchment. Also, dry deposition may be inefficiently sampled by precipitation collectors, so that a bulk precipitation collector that collects dry deposition by sedimentation only may give a considerable underestimate of total atmospheric input to a vegetated catchment where impaction is probably a more important depositional process (White & Turner 1970). The contention of Pearson and Fisher (1971) that a continuously open collector receives 'what actually reaches the earth's surface and is incorporated into natural water' may be slightly optimistic. Swank and Henderson (1976) are of the opinion that all estimates of atmospheric element input are likely to be minimum estimates.

Large polyethylene conical funnels have been most frequently used to collect precipitation for chemical analysis. These have usually been 60° funnels between 15 and 30 cm in diameter, though a general rule is probably 'the larger the better' so that any contamination is given maximum dilution in a larger sample; see Stevens (1981a) for collector design details.

Gambell and Fisher (1966) and others have recommended Buchner-type funnels with straight-sided rims, though it is not clear how other funnel types will adversely affect sample representativeness. Fish (1976a, b) used V-shaped plate-glass troughs to collect precipitation. The fact that precipitation samplers collect less than similarly exposed rain gauges of the same orifice size (Crisp 1966) should only prove to be a problem if this difference is due to evaporation loss. A separate standard rain gauge should always accompany a precipitation sampler. Others, like Stevenson (1968), have used glass funnels to collect precipitation, these being manufactured by removing the base from a Winchester bottle. Laney (1965) collected bulk precipitation on a 625 cm square corrugated fibreglass sheet held at 1.5 m above the ground to obtain a representative sample that includes a greater, but still unknown, proportion of dry deposition. The rim height of a precipitation collector is important in that it should be high enough to avoid heavy contamination from splash-in from the ground or from low vegetation. Rim heights of 30 inches (77 cm) (Attiwill 1966) and 34–45 cm (Allen *et al.* 1968) are possibly too low, though the last-named authors considered their collectors to be 'above splash level', The Hubbard Brook project uses collectors positioned at least 2 m above the ground and in a 45° clearing (Likens *et al.* 1977).

The catch of the funnel is led through plastic tubing into a reservoir, which is usually of polyethylene. Likens *et al.* (1967) have recommended the insertion of a vapour barrier, simply a loop in the plastic tubing to prevent evaporation loss, and a bubble trap to allow air to escape as the reservoir fills but to prevent contamination from the atmosphere. The

Figure 4.36 A bulk precipitation collector.

reservoir should be either painted black, buried, or tied inside a dark bag (Anon. 1973) to reduce algal activity during field storage. One such precipitation collector is shown in Figure 4.36.

The low solute concentrations that are usually present in bulk precipitation samples, often less than 1 mg l^{-1}, mean that any contamination is likely to alter profoundly the sample composition, so that extreme care must be taken in precipitation collection. The sampled water is usually 'filtered' at the base of the collecting funnel. This has been effected by a 'replaceable fritted glass disc' (Gambell & Fisher 1966) and also by 0.2 mm (Allen *et al.* 1968) and 0.7 mm (Gore 1968) mesh nylon filters to keep insects, leaves and other debris out of the sample container. Will (1959) used a plug of cottonwool to do this, and Miklas *et al.* (1977) put a fibreglass mesh over the whole collector orifice. Allen *et al.* (1968) found bird droppings to be a source of serious contamination but a row of spikes around the rim proved ineffective in reducing this and samples with a high phosphorus content were rejected. Gore (1968) found that up to 12% of the annual input of some elements was retained in a greasy deposit on the funnel surface and recommended procedures for funnel changing and cleaning and laboratory analysis of the deposit. Swank and Henderson (1976) found that the ratio of dry deposited material leached by acid digestion for later analysis compared with that removed by water varied from 1 for sodium to 2 for calcium, magnesium and potassium, up to 9 for phosphate phosphorus, an indication of the practical complexities that can arise. Carlisle *et al.* (1966) also found it necessary to change reservoir bottles frequently because of the growth of algae that modified sample composition. Because of the necessity for field storage of precipitation samples, the addition of a chemical preservative is usually recommended, though it should be noted that Galloway and Likens (1978) suggest that short storage with no treatment is preferable because every treatment introduces some analytical contamination or interference. Swank and Henderson (1976) added 1 mg of phenylmercuric acetate to the sample collector, while others recommend a few drops of chloroform or mercuric chloride as a preservative. Gambell and Fisher (1966) implemented a sophisticated installation in which samples were stored at a controlled, low temperature before removal to the laboratory. Lindberg *et al.* (1977) have presented a very useful discussion of the careful procedures necessary for precipitation collection and analysis.

In order to estimate separately the size and nature of the dry deposited component of bulk precipitation, more sophisticated sampling devices are necessary unless it is possible to ensure that a wet deposition collector is opened only during rainfall or that a dry deposition collector is covered during that time (Miklas *et al.* 1977, Benham & Mellanby 1978, Schroder *et al.* 1985). The Hubbard Brook ecosystems project developed equipment that accomplished this automatically (Likens *et al.* 1977), and Zeman and Nyborg (1974) include circuit diagrams for their automatic wet fallout collector. Lindberg *et al.* (1977) also have a good description of the electromechanics and operation of a 'dry fallout only' collector. Torrenueva (1975) has described a very simple wet deposition collector consisting of a plastic cover clipped to a return spring and held as a cover over the collector by tissue paper, which breaks when wetted at the start of rainfall, thus releasing the cover. The cover must then be manually reset at the end of the rainfall event. Edwards and Claxton (1964), Rutter and Edwards (1968) and White and Turner (1970) describe dry deposition collectors in which filter paper discs are mounted in tubes that point into the wind on a vane, thus collecting particulate material only. It is perhaps worth mentioning here that many studies have failed to find a significant difference between the catches of bulk precipitation and 'wet deposition only' collectors (e.g. Likens *et al.* 1967). Dust collection is discussed in Section 4.6.2.

The frequency of sampling has been found to affect solute concentrations in precipitation samples, largely because there are concentration variations both within and between precipitation events (Douglas 1972). Galloway and Likens (1978) have, for instance, demonstrated that the accumulated element input from an 'event-sampled' collector is not the same as the input from an identical adjacent 'composite' collector sampled only once over the same period. The sample changes and contaminations during storage are presumably responsible for this effect. The appropriate frequency of sampling will again depend on the objectives of the research project. If this is concerned with catchment element budgets, then weekly (Zeman & Slaymaker 1978) or monthly (Boatman *et al.* 1975) sampling may be appropriate, but if interest is in the particular meteorological factors that affect precipitation quality then event collection will be necessary. In most cases, event sampling cannot be carried out by hand and there are commercial samplers available that can be used to sample during rainfall events. Torrenueva (1975) has described a rotating sampler, powered by the weight of the water collected, which retains separately 5 mm increments of precipitation.

There has been much interest in recent years in the possible widespread increase in acidity of precipitation and the effects of this on the stability of natural ecosystems, and consequently on stream-water chemistry (Anon. 1976, Whittaker *et al.* 1974, Newman *et al.* 1975, Department of the Environment 1976). The published proceedings of the first American symposium on this subject (Dochinger & Seliga 1976) in relation to forest ecosystems contains the most comprehensive and up-to-date collection of papers available on all aspects of field techniques for sampling precipitation and dry fallout. A great number of recent publications concerned with precipitation collection for solute studies has been referred to by Cryer (1986).

4.3.4.2 Throughfall and stemflow

The collection and analysis of throughfall and stemflow beneath vegetation covers has demonstrated the significance of nutrient losses from plants via intercepted precipitation, which then makes a direct addition to the available nutrient pool, no decomposition being necessary (e.g. Reiners 1972). The increase in the solute levels of net precipitation over gross precipitation has been attributed to the leaching of material from leaves and the washing from these leaves of dry deposited material from the atmosphere (Madgwicke & Ovington 1959, Voigt 1960, Attiwill 1966, Carlisle *et al.* 1966, Duvigneaud & Denaeyer de Smet 1970, Eaton *et al.* 1973).

The sampling of throughfall and stemflow presents greater problems than does precipitation because of the considerable spatial variation present in both quantity and quality. Helvey and Patric (1965) and Kimmins (1973) have considered the network design requirements for interception studies and for representative throughfall sampling, respectively. The instrumentation for throughfall collection usually consists of rain gauges or collecting troughs subject to the same operational problems as for precipitation collection except that the possibility of contamination by litter fall is much greater. Stemflow is usually sampled by collars around the base of the tree, as in Figure 4.37, such as those described by Likens and Eaton (1970), Likens *et al.* (1977) and the Institute of Hydrology (1977).

Nicholson *et al.* (1980) contains a useful discussion of the considerable difficulties inherent in estimating the input of solutes from the atmosphere to vegetated surfaces.

4.3.4.3 Soil water

The collection of soil waters for chemical analysis presents special problems of representativeness inasmuch as the very act of sampling matrix water can disturb the existing soil-water equilibrium situation, and the water that is in motion in the soil may not be identifiable with that which is in chemical equilibrium with the mineral and organic phases of the soil. Atkinson (1978) has a critical review of field methods

Figure 4.37 A stemflow collector.

involving interception of downslope and downprofile water flow, which, as well as measuring subsurface flowrates, provide water samples for chemical analysis. Throughflow gutter and trough equipment and tension lysimeters are described by which matrix flow may be sampled, as are flexible hose and siphoning tank arrangements for measuring pipeflow discharge (see Sec. 4.2.8). Trudgill and Briggs (1978) have a recent review of techniques for assessing solute movement in soils.

Atkinson (1978) also reviews methods for tracing flow in soils by the injection of chemical solutions of tracer materials, usually fluorescent dyes or radioactive compounds, though electrolyte solutions can also be used. Sampling at a known outflow point may demonstrate the presence of a labelled inflow and its time of travel, and the tracer concentration of the outflow may give some indication of flow route within the system. Porous ceramic cup samplers (Levin & Jackson 1977, Chow 1977) may be used to recover natural soil water and tracer solutions at points between injection and outflow. These have been critically evaluated in a recent paper by Talsma *et al.* (1980), and Stevens (1981b) has produced a guide for their field use. Tracer solutions have also been used in streams to demonstrate quantitatively the downstream dilution of tracer by subsurface inflows to the channel.

4.3.4.4 Stream water

Because stream water is usually in a continuous flow and is available in large volumes relative to the necessary sample size, the problems of sample collection and storage that are present with precipitation sampling are far less important. Consequently, less attention has been paid in the literature to the details of stream-water sampling, as the procedure is relatively straightforward. In Section 4.3.2 it was suggested that the location and frequency of sampling would be largely determined by the nature and data requirements of the particular research project. The sampling technique that is most appropriate will mainly depend on the size and accessibility of the stream.

When the researcher requires regular but relatively infrequent samples, e.g. at weekly intervals, as may be the case in a catchment element budget study, or when the object of study is the establishment of the spatial variability in stream-water quality variables or background concentrations in relation to geology or land use (Walling & Webb 1975, Webb & Walling 1974), then simple 'grab' sampling may suffice. In streams that can be waded, which are most streams draining catchments of the size most appropriate for the kinds of investigations described in Section 4.3.1, then a sample bottle can be filled by hand. Stilling pools above weir controls should be avoided; Johnson and Swank (1973), for instance, sampled at least 10 m above a control to avoid any effects introduced by temporary storage in the stilling pool and the addition of solutes by the concrete of the structure. The sample bottle and cap should be thoroughly rinsed a few times in the water to be collected and the final sample taken well upstream of any bed disturbance produced by the person sampling. Several authors recommend that the sample be taken from mid-depth, so as to avoid the surface, and Loughran and Malone (1976) have referred to 'depth-integrated hand sampling' in which the sample bottle is moved from the surface to the stream bed during filling. Unambiguous and clear labelling of sample bottles is an obvious though sometimes neglected requirement in sampling, along with full notes of discharge and discharge tendency, water temperature and colour, and time, date and site of sampling.

When the stream is too deep to be waded and when sampling from the bank is likely to be unrepresentative, then the use of a sampling instrument suspended from a bridge or a cableway is necessary. Rainwater and Thatcher (1960) have described depth-integrating samplers (sample bottles in weighted holders, which must be lowered to the stream bed at a uniform rate as they fill) and Golterman and Clymo (1969) describe point samplers for use in deep lakes.

When there are storm-linked fluctuations in both flowrate and solute concentration of stream water, then a sampling frequency appropriate to these fluctuations is necessary to document the details of concentration variation and to calculate accurate solute loads. Because storm events occur randomly (often at night!) it is often not possible to reach and remain at the sampling site for the duration of the event. To overcome this problem, various devices have been developed for continuously sampling stream water or for extracting samples at appropriate intervals of time. The simplest devices continuously divert a small proportion of streamflow into a large container between site visits to provide a composite sample. If the mean concentration of stream water is required, then a time-weighted composite sample is sufficient and can be provided by continuously removing a constant volume from the stream or by making a fixed addition to the composite sample at a constant time interval (White *et al.* 1971). There are usually complications in ensuring a constant rate of sample extraction as water levels vary. If there is a need to know the total mass of a particular solute passing the sampling point, that is the concentration of the total volume of stream water if this were all collected in a large container, then additions to the composite sample should be made at each sampling time in proportion to the stream discharge at that time, i.e. a discharge-weighted composite sample. The main attraction of taking composited samples is that they can be collected with relatively cheap and easily constructed field equipment, which usually allows a good spatial cover at the expense of accuracy. The 'Coshocton wheel' sampler used by Duffy *et al.* (1978) provides a discharge-weighted composite sample by diverting 0.5% of streamflow into a large container. Nabholz *et al.* (1984) provide designs for a flow-proportional sampler system.

Much present-day research in geomorphology focuses on storm-linked variations in stream-water quality, information that is lost when samples are composited. The practice of collecting composite samples has been popular largely as a convenience in the absence of adequate field sampling equipment and of techniques of rapid laboratory analysis. Equipment is, however, now commercially available that allows samples to be taken automatically at preselected time intervals and then stored as separate samples. These automatic samplers may consist of a clock that sequentially trips preset pins, thus releasing a vacuum in a sample bottle, which draws a sample from the stream. Some sampler designs have a pump that fills the sample bottle when activated (Walling & Teed 1971). Two different sampler types are shown in Figure 4.38. An additional feature may be a water-level sensor, which can automatically initiate a sampling sequence when a predetermined rate of rise (O'Loughlin 1981) or a preselected water level is reached during a hydrograph event (e.g. Cryer 1984, Turton & Wigington 1984). Samples are then taken at a fixed interval, usually between 10 and 30 min, the selected interval depending largely on the characteristic variability of solute concentrations and stream discharge previously experienced. Automatic samplers are usually installed at existing gauging points so that a continuous record of streamflow is available for the computation of solute rating relationships and total element outputs. Some samplers have been designed that produce, in effect, a discharge-weighted series of samples by either sampling more frequently at higher discharges (Frederiksen 1969, Johnson & Swank 1973) or by sampling at fixed increments of streamflow volume (Claridge 1970).

Also in frequent use is the rising-stage or single-stage sampler, which consists of a set of sample bottles, each with one long and one short tube fixed with a bung in the neck of the sample bottle. When the water level rises above the preset level of the shorter tube, the bottle fills, the longer tube allowing air to escape. These bottles can be held in a weighted crate within the stream and can sample at any preset stage on a rising limb. Ellis (1975) has a description of a mechanical falling-stage sample collector. Clearly, all the above-mentioned sampling equipment may introduce sample contamination, either

Figure 4.38 Two different types of automatic sequential pumping sampler.

directly or as a storage effect. Therefore, the sampler should be assessed for this effect by taking simultaneous hand and 'device' samples and comparing the quality of each for significant differences.

4.3.4.5 Direct collection of quality data

Such are the problems involved in collection, storage and transport of water samples that the introduction of techniques by which solute concentrations may be measured directly and continuously in the field, without the need for temporal sampling programmes and physical removal of water samples, has been welcomed by those involved in solute investigations. However, the welcome should be guarded because, as Walling (1975) has commented, as yet specific conductance is the only water quality parameter that is generally suited to continuous monitoring in terms of accuracy, reliability and cost. The cost factor is considerable because of the particular instrumentation involved and the need for recording equipment, and because of the usual lack of mains electricity at the sampling station so that battery power must be supplied (Toms *et al.* 1973). Recent developments in micro-electronics have, however, made data logging and data transfer possible using very small dry batteries (e.g. the 'Squirrel' series of data loggers from Grant Instruments).

Specific conductance (see Sec. 3.3.3) was successfully monitored continuously by Visocky (1970), after initial problems of cell fouling by algae and by sediment, such that field values were within 10% of laboratory-determined values. Walling (1975) has used a continuous conductance record to demonstrate the value of this record compared with data from spot measurements at intervals. Ellis (1975) has described an experimental set-up for recording the conductance of urban stream water and has noted the problematically high current demand, which involved the use of heavy-duty batteries.

The concentrations of particular solute species can also be continuously monitored using ion-selective, or specific-ion, electrodes (see Sec. 3.3.4). These are described in most modern analytical chemistry textbooks (e.g. Bauer *et al.* 1978) and also by Edwards *et al.* (1975). These electrodes are initially attractive because they seem to allow the rapid estimation, by means of a dip cell or probe, of solutes that otherwise require lengthy, tedious and usually inaccurate laboratory procedures to produce less useful, non-continuous, data records. Specific-ion electrodes would also obviate the errors and contaminations due to the process of water sampling. Edwards *et al.* (1975) deal specifically with the nitrate electrode and, as well as mentioning significant theoretical and practical problems that seriously reduce the efficiency and accuracy of the probe, even in the laboratory, list several deficiencies of the instrument that make it impossible to use in the field. These deficiencies include the fact that appropriate battery-powered meters were (at that time) unavailable, that frequent standardisation of the instrument is necessary, that chemical interferences are likely to result from common constituents of stream water, and that streaming potentials are produced in moving water which produce aberrant meter readings that cannot be corrected. Bauer *et al.* (1978) have also noted that 'in very dilute solutions slow electrode response may require 15 minutes to an hour for equilibrium', creating obvious difficulties in the continuous monitoring of stream-water quality. The pH electrode is the only specific-ion electrode that has been successfully used for continuous monitoring (Ellis 1975), without very sophisticated pumping systems. In these, water is pumped into an off-channel measuring chamber where automatic calibration of the system and cleaning of the probes are incorporated into the operation (Toms *et al.* 1973).

4.3.5 Laboratory analysis

It is not the intention of this section to provide details of analytical procedures; these can be found in a number of texts, to which reference can be made. The purpose of this section is to provide some comments on the appropriateness, advantages and limitations of alternative methods of analysis. It should be emphasised that the detailed specialised texts should be carefully consulted before a decision is made on analytical procedures. It is often the case, moreover, that each analytical situation presents different combinations of problems, and compromises may have to be reached. This is especially the case if the analyses of several elements are required from a large number of samples. In this case, the operator may have to choose an order of priority of requirements. The detailed texts that should be consulted are Schwarzenbach and Flaschka (1969), Hesse (1971), Allen (1974b), Edwards *et al.* (1975) and Golterman *et al.* (1978), of which the latter is undoubtedly the most useful. Reynolds (1981) has also provided a useful laboratory guide for most of the common procedures.

The solute species discussed below have been chosen because together they comprise almost all the dissolved load of most natural waters. If a complete and careful analysis of a sample is carried out, then the results can be checked for accuracy or completeness by calculating the ionic balance. The sum of the cation equivalent concentrations (expressed in mmol dm^{-3} or milliequivalents per litre (meq l^{-1})) should equal the sum of anion equivalents, where the equivalent concentration is the w/v concentration (mg l^{-1}) divided by the equivalent weight of the particular ionic species. A balance of $\pm 5\%$ is desirable in waters with 25–30 mg l^{-1} of dissolved solids, and a consistently large discrepancy can indicate the nature (positive or negative) and concentration of an undetermined solute species, bearing in mind the likelihood of aggregating or compensating analytical errors. Rainwater and Thatcher (1960) note that balance conclusions are usually negative in natural waters!

Tanji and Biggar (1972) have put forward a model by means of which the conductivity of a solution may be 'generated' from a complete analysis by calculating the contribution of each ionic constituent. This provides the basis of another useful check procedure, though it has been noted that the 'constant' in the equation shows systematic variation with water type. A similar check is the relationship:

$$\text{specific conductivity } (\mu\text{S cm}^{-1}) = \text{sum of cation (or anion) equivalents (mmol l}^{-1}) \times 100 \tag{4.18}$$

(Hembree *et al.* 1964) and again the 'constant' varies around 100 for different water types. Probably only major errors are detectable by checks of this kind.

The presence of dissolved organic material may seriously complicate the checks described above as it *may* contribute to sample conductivity, both as positively and negatively charged species. There are lengthy and often complicated procedures for determining dissolved organic material in natural waters, and Golterman *et al.* (1978) contains an excellent summary of these. Problems of definition and assessment of the dissolved organic component have been discussed by, among others, Thurman (1984), Grieve (1985) and Finlayson (1986).

4.3.5.1 Calcium

Atomic absorption spectrophotometry (AAS) (see Sec. 3.3.6) is probably the most frequently used analytical technique and is usually the most reliable provided that samples are treated beforehand with lanthanum chloride in order to damp interference from other ions. Lanthanum chloride is carcinogenic and should be handled with care. Titration of samples with 0.025 M ethylene diamine tetra-acetic acid (EDTA) using a pH 14 buffer and murexide is an adequate alternative (Douglas 1968, Schwarzenbach & Flaschka 1969) especially when high calcium concentrations are involved (50–100 mg l^{-1} or over); titrations are not as accurate as AAS below 10 mg l^{-1}, even if the EDTA used is diluted. The end-point for calcium is often not very clear and requires skill and practice in determination, especially with high background colour, and, in this case, the use of screened murexide (i.e. with added NaCl) may be preferable to murexide. The use of masks (Bisque 1961) may help with this problem and in any case they should be used when Fe, Al, Zn, Cu and other elements titratable with EDTA are present. The best mask is potassium cyanide, but it is not normally desirable to use it unless the laboratory is equipped to handle poisons, including waste, safely. The use of triethanolamine (masks Fe, Mn and Cu), sodium tartrate (masks Al) and hydroxylammonium hydrochloride (masks Fe, Mn and Cu) in combination provides an adequate alternative, although the latter is poisonous by skin absorption. High concentrations of magnesium can cause clouding as the method relies on the use of a pH 14 buffer to precipitate any magnesium present. Calcium can also be analysed on a flame photometer, but again interference is significant and has to be masked with lanthanum. In addition, frequent calibration checks are desirable. AAS instruments are often calibrated at 10 mg l^{-1} full-scale deflection whereas flame photometers may have a recommended full-scale deflection of 75 or 100 mg l^{-1}. The former are therefore to be recommended for the more dilute solutions. The disadvantage of AAS for high concentrations is the amount of dilution necessary, and unless an automatic diluter is used this can be a source of inaccuracy. AAS is thus recommended for low levels of calcium, and titration or flame photometry for higher concentrations. The advantage of titration is that it can be carried out under primitive conditions, an important consideration if the samples cannot be quickly transported to the laboratory. Degassing and precipitation of calcium can occur in carbonate-rich waters, unless preserved with acid, though this may bring sedimentary particles into solution unless the sample is filtered. Problematically, filtration may also alter carbonate equilibria and calcium exchange can occur with the filter paper. Field titration may be an answer to these problems if rapid AAS analysis is not possible. It has been suggested that AAS determines total calcium, including that adsorbed onto or chelated in organic complexes in solution. Care may therefore be needed in interpretation of results in organic-rich waters.

4.3.5.2 Magnesium

Many of the comments about calcium also apply to magnesium, and again AAS is to be preferred to EDTA titration for low concentrations. This is especially so in the case of magnesium since the titration procedure involves (a) a titration for calcium, (b) a titration for Ca + Mg using a pH 10 buffer and Erio T/Solachrome Black as an indicator, and then the subtraction of (a) from (b) to obtain a value for magnesium. Thus, if an error is made in the Ca titration, this will affect the magnesium result. Sample runs where Ca and Mg show inverse relationships should therefore be scrutinised carefully. The Ca + Mg end-point is much clearer than the Ca one, though sluggish at low temperatures (optimum 40°C). Colorimetric methods for Mg are available though not widely used. The Ca + Mg titration can be usefully used for Ca if it is known that no Mg is present, as the end-point is much clearer.

4.3.5.3 Potassium and sodium

These are most commonly determined by flame photometry or by AAS, and the methods are not subject to any great errors or problems in stream water except that instrument calibration should be checked during sample runs.

4.3.5.4 pH

In general, field pH meters are of insufficient stability to be totally reliable and a laboratory pH meter is to be preferred. However, pH is very liable to alteration in transit and storage, especially if degassing occurs, so that field measurement is preferable. A reliable field pH meter is an important but unfortunately rare commodity; in general, those with large high-voltage batteries are the most stable. Concern with precipitation acidity has prompted a fundamental re-examination of the use of the pH parameter, particularly in dilute natural waters (Marinenko & Koch 1985, Neal & Thomas 1985).

4.3.5.5 Nitrate and ammonia

Autoanalyser methods are probably the most useful and reliable for nitrates or the use of a standard colorimeter or spectrophotometer if automatic facilities are not available. The

Kjeldhal distillation of NO_3-N to NH_4-N is time-consuming, though reliable once running smoothly. Nitrate and ammonium electrodes are available, but vary in their response time and reliability. Frequent calibration is necessary, reproducibility is limited at high concentrations, and electrodes are susceptible to error due to interferences in solute-rich waters. However, reproducible results may be obtained if other background contaminants are present in low and/or constant concentrations. Electrode use has become widespread because of the expense and time-consuming nature of the alternatives rather than because of their more accurate results, and so these results should be checked where possible by other methods. The review by Edwards *et al.* (1975) should usefully be consulted before use of specific-ion electrodes.

4.3.5.6 Chloride

Autoanalysers again provide the best answer where available and running smoothly, though titration is a very adequate alternative. The silver nitrate titration is in practice often difficult to use because of indeterminacy of the end-point; the mercuric nitrate titration (USGS 1970) has a sharp end-point but cannot be used where high concentrations of Fe (10 mg l^{-1}), Cu (50 mg l^{-1}) or sulphate (10 mg l^{-1}) are present. Care should be taken when handling the poisonous mercuric compounds, including the waste.

4.3.5.7 Phosphate and silicon

The ammonium molybdate method has become a standard technique, using a spectrophotometer, colorimeter or autoanalyser. There is a choice of reducing agents and ascorbic acid may be preferred for phosphate analyses where low concentrations occur (Watanabe & Olsen 1965).

4.3.5.8 Sulphate

Sulphate analyses are often effected using a conditioning reagent that converts sulphur compounds to sulphate. The analysis poses several problems in that a turbidimetric method employing barium chloride may be used in which timing of the method is crucial. The transmissibility/absorption varies with time and, unless an arbitrary time period for measurement on a spectrophotometer is chosen, each analysis may take a considerable time in order to measure the peak value of barium sulphate sediment present. If an arbitrary time is chosen, it may well be the case that non-reproducible results will be obtained. The method cannot be used if high organic colour or turbidity is present. The barium chloride is poisonous.

4.3.5.9 Bicarbonate

Analysis should be carried out in the field by means of an alkalinity titration to pH 4.5 with dilute acid. Measurements after storage of more than a few hours are of little use as degassing may occur. Sample preservation with acids is obviously inappropriate, as is the use of chloroform.

4.3.5.10 Specific conductivity

This is the ability of a solution to carry an electric current and is dependent on the concentration and type of charged solutes, and on water temperature. It is usually reported in microsiemens per centimetre (μS cm^{-1}) and at a standard temperature of 25°C. Conductivity is easy to measure in the field, as field meters are usually as reliable as laboratory meters. In carbonate-rich waters, conductivity may be influenced by dissolved carbon dioxide such that decreases in calcium or magnesium levels between samples may be masked by increases in dissolved and dissociated CO_2. Conductivity is a useful parameter to the geomorphologist as it is easy to measure and gives an indication of 'salinity levels' in a sample. Also, if a consistent relationship can be established between conductivity and a particular solute concentration, it may be used as a more easily obtained surrogate for this solute (Steele 1968). Conductivity variations may, however, be less easy to interpret meaningfully because, being the sum of the ionic conductances of the species present in solution, conductivity may not vary in direct proportion to changes in individual ionic concentrations when these concentrations change in proportion relative to each other. Thomas (1985) has referred to this last problem among many others in a review of the use of conductivity in cold, dilute waters.

4.3.5.1 Total dissolved solids (TDS)

This parameter can be determined by evaporating to dryness a known volume of sample on a water bath or in a drying oven at 103°C. The sample size normally used is 500–100 ml. The sample container should be weighed when at room temperature both before evaporation and after cooling in a desiccator, and the weight of dry residue expressed in mg dm^{-3} of sample. There are many attendant errors in this procedure, particularly due to taking the difference between two inaccurate weighings and the loss of dissolved material by volatilisation, mainly bicarbonate, organics and anions in acid waters (Hem 1970). Kemp (1971a,b) states that TDS estimates may easily be up to 25% in error in a routine determination and considers such estimates to be of little value.

If TDS concentration estimates for stream water are to be used in calculating a denudation rate, then a rock density conversion factor (m^{-3}) is necessary to convert a long-term estimate of TDS load to rock equivalents (Walling & Webb 1975, 1978). To avoid a large number of laborious TDS determinations, it is often estimated indirectly from specific conductivity (SC) values by means of a relationship of the form;

$$\text{TDS} = \text{SC} \times \text{constant} \qquad (4.19)$$

based on at least 10 careful TDS determinations, though it should be noted that conductivity is only affected by ionic material in solution. The empirical constant in the above relationship varies from 0.55 to about 0.9, depending on the dominant ions in solution, and must be determined from a plot of TDS concentration against conductivity. Clearly the SC–TDS relationship must be a statistically significant one, as assessed by the correlation coefficient, but in many cases there is a great deal of scatter about the regression line, which is largely due to error in the TDS determination. A highly significant, and therefore usable, relationship is much easier to establish when the range of conductivity values is 500–100μs than when the natural range is less than 50 μs, when non-significant correlations can arise.

4.3.5.12 Calculation of long-term solute loads

It is often necessary to progress from spot measurements of solute concentration to long-term estimates of gross solute loads and catchment denudation rates, and for this both concentration and discharge information are required. As a bare minimum, a single concentration value and a discharge estimate are necessary, with the ideal situation being one with continuous records of both discharge and concentration. Usually, some intermediate level of information is available for load calculation, and some calculation procedures, appropriate to different levels of information, are outlined below (see also Sec. 4.1.3):

(a) mean discharge (ls^{-1}) × mean concentration (mg l^{-1}) × time period (s) = load (mg). (4.20)

This form is necessary when only a few, irregularly spaced samples have been made over a period and when a continuous discharge record is not available.

(b) total discharge ($l\ yr^{-1}$) × mean concentration ($mg\ l^{-1}$) = load ($mg\ yr^{-1}$) (4.21)
This is used when a value for total annual (or monthly, etc.) streamflow is available but only a few concentration values.

(c) $\sum_{i=1}^{52}$ [total discharge for week i ($l\ week^{-1}$) × concentration at end of week i ($mg\ l^{-1}$)] = load ($mg\ week^{-1}$) (4.22)
This is used when there has been regular weekly sampling in combination with a continuous discharge record, such as used at Hubbard Brook (Fisher *et al.* 1968). Given that weekly samples are likely to be in 'time-dominating' low-flow periods and that streamflow is more dilute at high flows, then an overestimate of load will result.

(d) Irregular sampling often reveals a straight-line rating relationship of concentration against discharge (often on logarithmic axes), as assessed by correlation and regression coefficients being significantly different from zero. Such a rating can be used in combination with either regular discharge estimates, a continuous discharge record or a flow duration curve to compute solute yield from a catchment. For example, every hourly discharge value from a continuous flow record generates a corresponding concentration value from the rating curve and the product of these two values, summed for every hour for the period of calculation, gives a precise load estimate. In the biological literature, catchment load is often quoted in units of $kg\ ha^{-1}\ yr^{-1}$. Walling and Webb (1982, 1985) have provided a thorough review of the problems in estimating solute loads and have demonstrated the severe biases that can result from the use of particular procedures for load computation.

All the above procedures give an estimate of gross catchment output of dissolved material in stream water. In order to estimate net output, the atmospheric input of elements must be subtracted from the gross output figure. Using bulk precipitation quality data, an estimate of input can be made thus:

bulk precipitation concentration ($mg\ l^{-1}$) × precipitation depth/100 ($mm\ week^{-1}$) = atmospheric element input ($kg\ ha^{-1}\ week^{-1}$) (4.23)

in this case for weekly sample collection.

4.4
River channels

4.4.1 Introduction

The study of river channel processes both has a long history and is of very broad concern. Since the time of Leonardo da Vinci, and more particularly in the last century, the scientific investigation of river channels has expanded to incorporate such fields as flow hydraulics, bed mechanics, sediment transport and sedimentation. Morphology – and particularly the cross section and plan-form properties of channels – has increasingly been linked to river flow characteristics and related to the properties and quantities of bed materials and transported sediments. A number of recent books can usefully be consulted to show how any geomorphological concern for river channel processes must be set within a broad interdisciplinary context arising through fundamental research nominally undertaken within such subjects as physics, engineering, geology and geography (Graf 1971, Gregory 1977c, Mijall 1978, Richards 1982, Hey *et al.* 1982).

Geomorphologists have begun to explore an increasing number of river-related process study fields in the past few decades. These have usually been *field*-based rather than developed through physical theory or laboratory experiment. For example, geomorphologists' estimates of sediment transport rates, and their 'explanation' of meander properties in terms of gross discharge and dimensional parameters, have not necessarily involved physical or mechanical modelling. In fact, the very term 'process' is endowed with a variety of meanings that derive as much from the historical development of geomorphology as a science as from any deliberate conceptual identification. In the geomorphological literature on rivers, 'process' can refer to the *fundamental mechanics* of grains (sediments) marginal to, moving within and deposited by water flowing in open channels. Such an approach derives especially from theoretical and experimental work by Gilbert, Shields, Bagnold and others, and geomorphologists may be especially concerned with the application and development of their ideas in field situations. This may require, for example, field measurement of water depths and velocities, the calculation of fluid forces, and the measurement of sediment type, shape and size distribution.

Secondly, process studies have involved the empirical establishment of a series of *regime* or *equilibrium relationships* between channel-form characteristics and what are usually hydrological or environmental parameters, or between various measures of form themselves. This has included work on the hydraulic geometry of natural rivers, design criteria for artificial canals and work in laboratory flumes. The technical problem often lies in obtaining a representative sample of field data on forms or their controls.

Thirdly, the term 'process' is used for the *rates of activity*, uneven in both space and time, either with which material is moving or with which forms are changing. The former may involve measurement of sediment concentration in stream waters and the analysis of stream discharges; the latter the repeated field survey of form and the examination of historic maps and photographs. Particular catchments, or parts of catchments, may be identified as possessing high or low activity, while identification of the source or

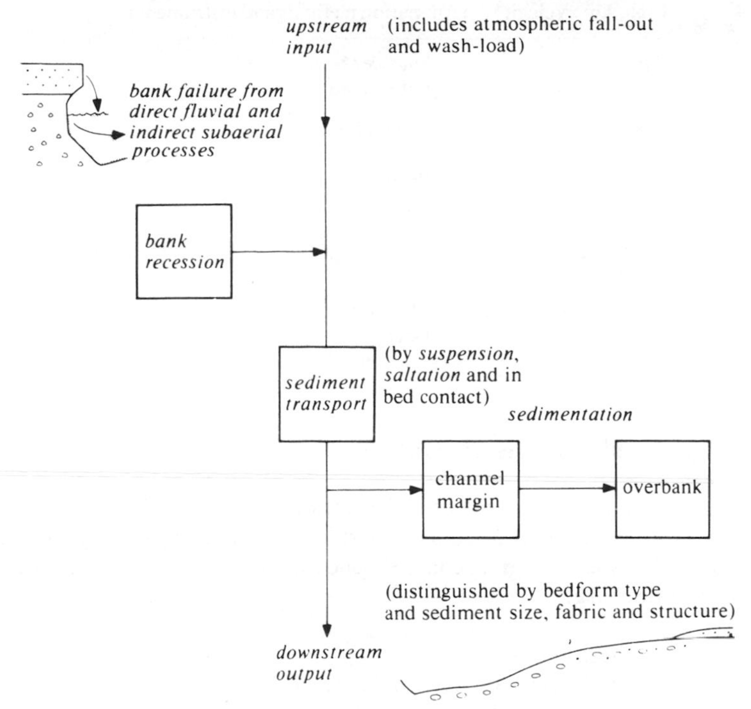

Figure 4.39 The river channel process system.

depositional site for stream-transported sediments may point to the way in which forms are evolving.

In this possibly confusing situation, it is clear enough that a very wide range of techniques may be called upon during an analysis of river processes. Some of them may not be very closely geared to geomorphology in the strictest sense at all, while frequent use is made of climatic and other data that in themselves are not geomorphological. Techniques that the geomorphologist is likely to use directly include flow gauging, sediment tracing and sampling, and the measurement of channel form and its changes by ground survey and from historical sources. Laboratory experimental work is not reviewed here in detail, but at the least the geomorphologist should be aware that many well known physical relationships have first been established under such conditions. Such work often assumes infinite-width channels, and may or may not be relevant to natural-scale and long-timespan developments in rivers, with their characteristically unsteady and non-uniform flows. However, a number of recent experiments should have proved most enlightening to geomorphologists (e.g. Abbott & Francis 1977, Mosley 1976, Mosley & Zimpfer 1978), who are left with the not inconsiderable technical problem of obtaining field data of the same basic kind as those examined in experimental work.

River channel processes are treated here as essentially a matter of sediment transfer (Fig. 4.39). Sediment may be mobilised at channel bed or bank or on land surfaces; this is transported (and modified in transit) by streams, and finally deposited or exported from the stream system altogether. This transfer system may be specified in process terms in the ways described above; in practical terms, this involves the use of techniques for streamflow gauging, sediment movement and channel-form dynamics (see Sec. 2.7 for a discussion of channel form).

Table 4.6 Velocity determination methods and instruments.

Floats	Surface Submerged
Current meters	Pendulum Cup Propeller Electromagnetic
Tracers	
Slope-hydraulic radius relationships	Manning Chézy

4.4.2 Gauging and analysis of streamflow

Measures of water flow may be required for various purposes. These include point measures of velocity in relation to sediment particle entrainment, the establishment of flow structures within a series of river channel bends and the estimation of mean channel velocity and river discharge.

4.4.2.1 Velocity determination

Commonly used methods are summarised in Table 4.6. Surface velocities may be estimated from the travel time of floats over a known distance. This velocity is greater than mean velocity in the vertical, and surface floats may get blown by the wind, but a variety of submerged float designs or ones that have submerged rods or chains have been produced with conversion factors of around 0.80 to 0.95 to estimate mean velocity from observed surface velocity. If an accurate means of calibrating such devices is available, they may be a cheap and useful reconnaissance means of velocity determination, but in general such methods tend to be of limited usefulness.

Velocities at any point within a flowing stream may better be gauged by current meter. Available models vary in design (including pendulum, cup and propeller types), and they may be rod- or cable-mounted; it is important to be aware of the limitations of the particular instrument used:

(*a*) The practicality of using the instrument in the desired place and under the desired flow conditions must be considered. Models may prove insufficiently robust or controllable to be used at times of high flow and high sediment transport rates, or their design may not allow velocities to be gauged sufficiently close to beds or banks, or in shallow enough water. Designs vary in their efficiency in dealing with water movement oblique to the average line of flow.

(*b*) Instruments will require calibration and perhaps (annual) servicing to give accurate results.

(*c*) Designs vary in their sensitivity, and the length of time selected for observations may be important. For example, current meter observations are normally taken for 30 s or more, and mean velocity is calculated as an average. Electromagnetic flow sensors (Kanwisher & Lawson 1975) may be especially useful in that they can give an instantaneous reading of downchannel and cross-channel flow components and they may respond to short-term velocity pulses, and these may be significant for sediment movement.

It is finally important to operate the equipment correctly; this includes standing so as not to interfere with the flow when using a wading rod, and using an appropriate correction factor for true depth with cable-suspended meters. Note that it is extremely hazardous to wade in deep, fast-flowing water on an uneven stream bed. Wading is inadvisable without special precautions where the product of depth (m) and velocity (m s^{-1}) is greater than 1.0.

Average velocity in cross section may be gauged using electromagnetic or ultrasonic devices. These rely on the fact that water, a conductor, flowing through a magnetic field generates an electrical current proportional to its speed of flow; and on the Doppler effect on ultrasonic waves passing through water, which again may be related to water velocity. The practical application of these principles for river velocity gauging is well reviewed by Herschy (1976).

Average velocity for a stream reach may alternatively be estimated using salt tracing techniques (Calkins & Dunne 1970, Church 1975). If a quantity of salt is injected into a stream and allowed sufficient length to mix thoroughly with stream water, its time of travel may be obtained from stopwatch and conductivity readings, because the inserted salt increases the electrical conductance of the stream water.

Finally, mean stream velocity may be calculated from the Manning formula. In SI units this is

$$\bar{U} = \frac{R^{2/3}\beta^{1/2}}{n} \tag{4.24}$$

where $\bar{U}$ is the mean velocity, R is hydraulic radius, β is the sine of the angle of the stream (water surface) slope and n is a lumped roughness coefficient determined empirically, ranging from about 0.02 for straight smooth-walled channels to 0.06 and more for densely vegetated streams containing boulders. Typical n values are given in many texts, including Graf (1971) and Gregory and Walling (1973), and these are illustrated with stream photographs in Barnes (1967). For particular streams, n values are not independent of their discharges, such that shallow water flowing over a coarse bed gives a higher n value than when the bed is deeply submerged.

An alternative, employing the Darcy–Weisbach factor (f) as a flow resistance measure, is given by

$$\bar{U} = \left(\frac{8g\,R\beta}{f}\right)^{1/2} \tag{4.25}$$

Empirical relationships between f and grain size and other channel characteristics have been developed (Hey 1979). It should be appreciated that flow resistance is related to channel plan-form and shape, bedforms and vegetation, the latter varying seasonally.

4.4.2.2 Flow structure

It is known that water movement in rivers involves shearing flows at a variety of scales; these involve small-scale turbulence, jets, meso-scale flow separation (where the major zone of water movement occupies only part of the channel, with low-velocity movement and back-eddies in the rest) and organised secondary motion such as helical flow cells. Field investigation requires instrumentation that sensitively records velocity fluctuations and flow direction, so that field studies have only fairly recently become practicable (see Hickin 1978). It is necessary to record a large number of velocity–orientation readings (Hickin took about 90 for each of 29 cross sections), which may be precisely located in cross section, and from these the directions of both water movement and incident shear stress may be calculated.

4.4.2.3 *Discharge estimation in natural channels*

Methods are listed in Table 4.7. Discharges ($m^3 s^{-1}$) may be estimated by velocity–area methods (in which mean stream velocity, assessed using current meter, electromagnetic, ultrasonic or other methods, is multiplied by the survey cross section of the stream) and by dilution gauging. For very small streams or pipeflows, an easier method may be to catch the entire flow (in a bucket) for a measured length of time; again mean velocity estimated from the equations given above may be used in combination with measurement of channel cross section to give a discharge estimate. This may be helpful for approximate estimation of the magnitude of historical extreme floods whose timing may be obtained in various ways including radiocarbon dating of relict organic material (see Costa 1978).

The velocity–area method using current meter observations is probably the commonest field procedure in the absence of gauging structures. The stream is notionally divided into a series of subsections; the mean velocity of each of these is then estimated, multiplied by the area of the cross section, and the results for each cross section are summed to give total stream discharge. This method requires that three decisions be made: first, the number of subsections or vertical soundings (it is commonly recommended that each should not represent more than 10% of total stream discharge so that perhaps 15 may be needed); secondly, the number of velocity readings in the vertical (one at $0.6 \times$ depth, but more generally two in the vertical at $0.2 \times$ and $0.8 \times$ depth as the minimum number whose average will give mean velocity in the vertical); and, thirdly, the time over which velocity is measured (generally 30–60 s). The accuracy of this method may be low in streams that have irregular beds or high weed growth (see Sec. 4.8.3.1), are very shallow (< 30 cm) or slow-flowing (< 15 cm s^{-1}) or involve strong cross-channel flows.

In these circumstances, and particularly on small streams, dilution gauging may be preferable. Here an aqueous solution of known strength is injected in the stream and its concentration is sampled and measured at a point downstream after thorough mixing with and dilution by natural waters has taken place. The degree of dilution is related to natural stream discharge. This method works well with very turbulent streams in which there are reaches without significant tributary seepage and inflow or water loss. Two techniques can be used: the 'constant-rate' method in which tracer is injected at a uniform rate for a long enough time to achieve a steady concentration in samples taken over time at a downstream point, or the 'gulp' method in which a known volume of tracer is injected and the whole mass of it is accounted for by sampling the pulse of higher tracer

Table 4.7 Methods for discharge estimation.

Volumetric		
Velocity–area	Using methods in Table 4.6 for velocity determinations; also moving boat, electromagnetic and ultrasonic gauging.	
Dilution	Constant-rate	
	Gulp	
Control structures	Weirs	sharp-crested (V-notch, rectangular, compound, etc.)
		broad-crested (flat-V, Crump, compound, etc.)
	Flumes	(H-type, trapezoidal, HRS* steep gradient, etc.)

* Hydraulics Research Station.

concentration as it passes the downstream sampling point. These methods are fully discussed in a Water Research Association (1970) technical paper, complete with 'how to do it' instructions. Special equipment is needed for constant-rate injection (e.g. a Mariotte bottle or André vessel) or for rapid, regular sampling in the gulp method. There may be problems in selecting a suitable reach, and in allowing for background tracer-substance levels and tracer loss (e.g. adsorption). Rather similar problems may occur in using dyes for water tracing; these are thoroughly reviewed by Smart and Laidlaw (1977).

4.4.2.4 Weirs and flumes

There are many designs for quasi-permanent gauging structures for which there is a design relationship between water level (usually gauged upstream of a bed constriction or overflow lip) and stream discharge. These include rectangular or V-notch sharp-crested weirs, broad-crested and compound weirs, and a number of alternative flume designs. The advantage of these structures lies generally in gauging accuracy, particularly with low flows; the major disadvantage is cost. A simple sharp-crested weir can be made relatively cheaply from marine ply and angle iron or from steel decking, and this may provide good discharge data provided that the weir is installed well (without leakage around the plate, at right angles to flow and kept clear of debris and sediment). Installation of a V-notch weir suitable for flows up to about 240 l s^{-1} is shown in Figure 4.40; larger discharges may be gauged using prefabricated flumes in glassfibre or wood (Barsby 1963, Smith & Lavis 1969), which have the advantage that they are at least partly self-cleaning in that sediment may pass through them. Specifications for gauging structures are given in BS 3680 (British Standards Institution 1964), with some useful illustrations in Gregory and Walling (1973).

A particular point to stress is that all structures will be grossly affected by sediment accumulation and debris clogging, and by failure of the structure (either through faulty construction or by distortion during use) to comply with the original design. Normally, the design calibration should be checked by gauging (current metering or preferably dilution techniques). Where possible on larger rivers, it is probably best to profit from the use, by kind permission, of structures already installed at some expense by statutory gauging authorities. These may additionally provide a much longer run of discharge records than could be directly obtained during research. However, again some caution should be exercised over gauge accuracy, particularly during the high flow conditions of especial geomorphological interest. Preliminary estimates of flood magnitude likely to be encountered will help in the selection of structure; this may be obtained by methods in the *Flood studies report* (Natural Environment Research Council 1975).

4.4.2.5 Water-level recording

Fluctuations in river discharge are commonly computed from records of water level (stage) using a rating relationship. Levels may be instantaneously observed from a permanently installed stage board with painted divisions, or using a wire or rod 'dipstick'. Crest stage gauges may record the highest level achieved between observation times, but for most purposes some form of continuous record, or repeated recording at determined time intervals, of stage is needed. This may be achieved via autographic recorders in which a pen produces a trace on a rotating drum. Either drum or pen movement may be activated by water-level changes, usually by a float and counterweight mechanism operating in a stilling well, or alternatively by pressure changes arising from varying water depth using a pressure bulb or the pressure bubbler principle. Time-change recording is achieved through the clockwork movement of pen or chart. Alternatively, digital punched-tape recorders may be used to record levels at set time intervals

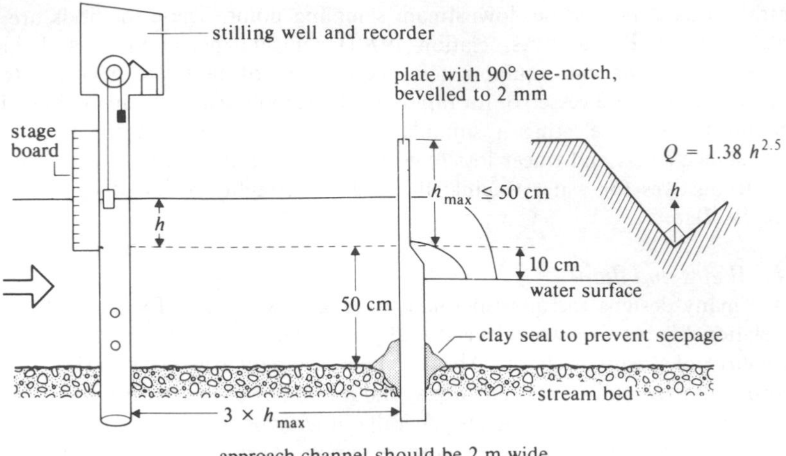

Figure 4.40 Installation of a 90° V-notch weir. Dimensions are given for general guidance.

(commonly 15 min), while interrogable devices are also available that will issue a telephone alarm when set water levels are reached, or will give coded water level when contacted. Three useful memoranda from the Water Data Unit (1967a, b, 1977) discuss commercial instruments available. Multiple stage recording, using a relatively simple home-made arrangement of float with magnet attached operating a vertical array of reed switches, has been described by Isaac (1975).

Major problems in recorder operation include malfunctioning or inaccuracy of the clock mechanism (it is advisable to have a spare ready), poor operation of the pen (which may jam or leave an inadequate trace) and problems with frost or corrosion (affecting especially clocks and the tape connecting float and counterweight, in the author's experience). With tape recorders, clock and punch failures may cause difficulties, while battery and line failures and damage from electrical storms occur with interrogable devices. Above all, many problems arise because of inexperience or casual operation. The installation of instruments in a well designed, wind-secured and weatherproof housing is recommended, so that recorded data can be retrieved looking as unlike used litmus paper as possible.

An alternative to the use of instrumentation solely for level gauging is to use a multichannel data logger, which may be battery-powered to give a data output on magnetic tape. Other channels may record rainfall, temperature, turbidity and real-time data – in fact, any data for which there is an appropriate sensor. The possibilities are illustrated by Butcher and Thornes (1978), who used one channel to record changing water level on a modified 'wave pole' originally developed for use in the coastal environment.

4.4.2.6 River regime and streamflow computations

Water-level recordings may be converted into discharge data via a rating relationship; instantaneous may be averaged to obtain mean daily, monthly or annual values, and flow duration curves may be constructed to give the length of time for which flows of a given magnitude are experienced. Such data may be used to estimate sediment yield (see Sec. 4.4.3), in analysing the frequency of extreme and other events of significance for geomorphological activity (e.g. Gregory & Walling 1973), or in characterising river regime in terms of average, range or variability in discharge (Lewin 1978). Analysis of short-term (storm) hydrographs or of the longer-term (annual) trace of discharge fluctuations may equally be undertaken.

The point for emphasis here is that rather different results may be obtained according to the data used. For example, flow variability will appear considerably less if monthly rather than daily average values are used, and computations of sediment yield using monthly average flow data may be grossly misleading (Walling 1977a, b). In field experiments, it is generally best to use short-time-interval data (15 min) for computational purposes where possible. Computer programs are available for dealing with hydrological data in chart form (Beven & Callen 1978).

4.4.3 Sediment movement

Sediment may move in stream water by turbulent suspension, by saltation, and by rolling and sliding along the bed. Some of this transported sediment may derive from the channel itself (bed material load) and some from sources external to the actual channel, for instance following soil erosion (wash load). Generally, the latter is finer, moving in suspension without being incorporated into the bed and requiring virtually still water to settle in. Bed material may move in a variety of ways depending on its size and the intermittent power of the stream to move it. Such material may therefore move as intermittent suspended load or as bedload (in contact with the bed). It is important to appreciate that the stream sediment load may be distinguished either by source or by transport mechanism, as shown in Table 4.8, and that the processes of sediment transport have a complexity that may be obscured or even misrepresented by simple sampling equipment (see Bagnold 1973). The geomorphologist may be concerned with both source and mechanism, together with the variable rates of sediment movement over time and space. This may involve the gauging or estimation of sediment movement rates over time in conjunction with streamflow observations as outlined in Section 4.4.2, and with the physical conditions for, and location of sediment mobilisation.

4.4.3.1 Mobilisation

Movement of sediment particles as stream competence increases has been detected using acoustic equipment, electromagnetic sensors (Reid *et al.* 1984) and tagged sediment (by fluorescent or waterproof paint, or radioactive treatment) whose movement may be seen and measured after the event. Forces required for initial motion may be considerably greater than those needed to maintain movement once it has started. Such movement has been related to bed velocity, mean velocity, shear stress or shear velocity (see Fig. 4.45), or unit stream power (ω), where

$$\omega = \frac{\gamma g \phi \beta}{w} = \tau \bar{u} \tag{4.26}$$

Where these measures exceed a critical value for a given particle size and density (this value is also affected by water viscosity and bed roughness), then movement occurs. In the classic Shields' diagram, a dimensionless critical shear stress or entrainment function,

$$\frac{\tau_c}{(\gamma_s - \gamma)D} \tag{4.27}$$

where τ_c is the critical boundary shear stress, γ_s is the specific weight of the bed particle and γ is that of the fluid, and D is the particle size (commonly D_{50}), is plotted as a function of shear Reynolds number (Re_*) defined as

$$Re_* = \frac{U_* d}{V} \tag{4.28}$$

where V is viscosity and the other symbols are as given previously. The entrainment function is practically constant at around 0.03 to 0.05 for uniform sediment, but higher for cohesive and variable for mixed sediment.

It should be emphasised that there is a very considerable literature on determining the hydraulic forces (the gravitational, drag and lift forces that must exceed inertial forces) that themselves determine particle mobilisation. Some of the relationships quoted are empirically derived from laboratory work where conditions are not those of streams in the field. For example, many reported experiments have been undertaken with uniform fine sediments, and at steeper slopes, shallower water depths and steadier flows than those obtaining with natural rivers. Particular field problems may be caused by the armouring of the bed (where there is a thin surface layer of coarse particles only, protecting finer materials in a sediment-size mixture beneath) and by the structural packing of the bed materials (see Laronne & Carson 1976). Again, mean stream velocity is often used as an indicator of stream competence, as in the well known curve of Hjulström extended by Sundborg and others. But average velocity may be only poorly related to critical velocity near the bed (e.g. Hickin 1978), while the Hjulström curve itself needs revision for coarser sediments (Novak 1973).

Recently, renewed efforts at the field measuremenht of bed sediment mobilization and transport have been made (e.g. Carling 1983, Ashworth & Ferguson 1986), while there have been several re-examinations of the sets of data available generally (Williams 1984b, Komar 1987). A major factor is the importance of a 'hiding factor' whereby small particles among large ones on a bed are protected from entrainment; by contrast isolated and protruding large ones may be moved more readily than equations developed for uniform-sized bed materials would suggest. All this shows that field observations both of sediment initiation and of the kind of physical data discussed here are needed (see Sec. 4.4.4), but that some caution should be exercised in the prediction of the former from the latter in the absence of direct observations. Nonetheless, data on grain size and channel morphology (including cross section and slope) have been used widely to reconstruct discharges of former rivers (see Gregory 1983).

4.4.3.2 Bedload

Traps of various design may be used to sample the amount of near-bed sediment passing through a section of channel bed in a determined time. Such point samples of transport rate, commonly in the form of weight per unit time per unit stream width, may be related to hydraulic factors or, by continuous use or in conjunction with discharge records, may be used to calculate longer-term bedload sediment yields. This usually amounts only to a small proportion of total sediment yield, even in streams notable for bedload transport, but it is especially important in that it may be related to changes in the form of the channel itself.

Table 4.8 Sediment transport.

According to source	*According to transport mechanism*	
Wash load (from extra-channel sources) →	Suspended load	
Bed material load →	Suspended load	
Bed material load →	Bed load	saltation, sliding, rolling

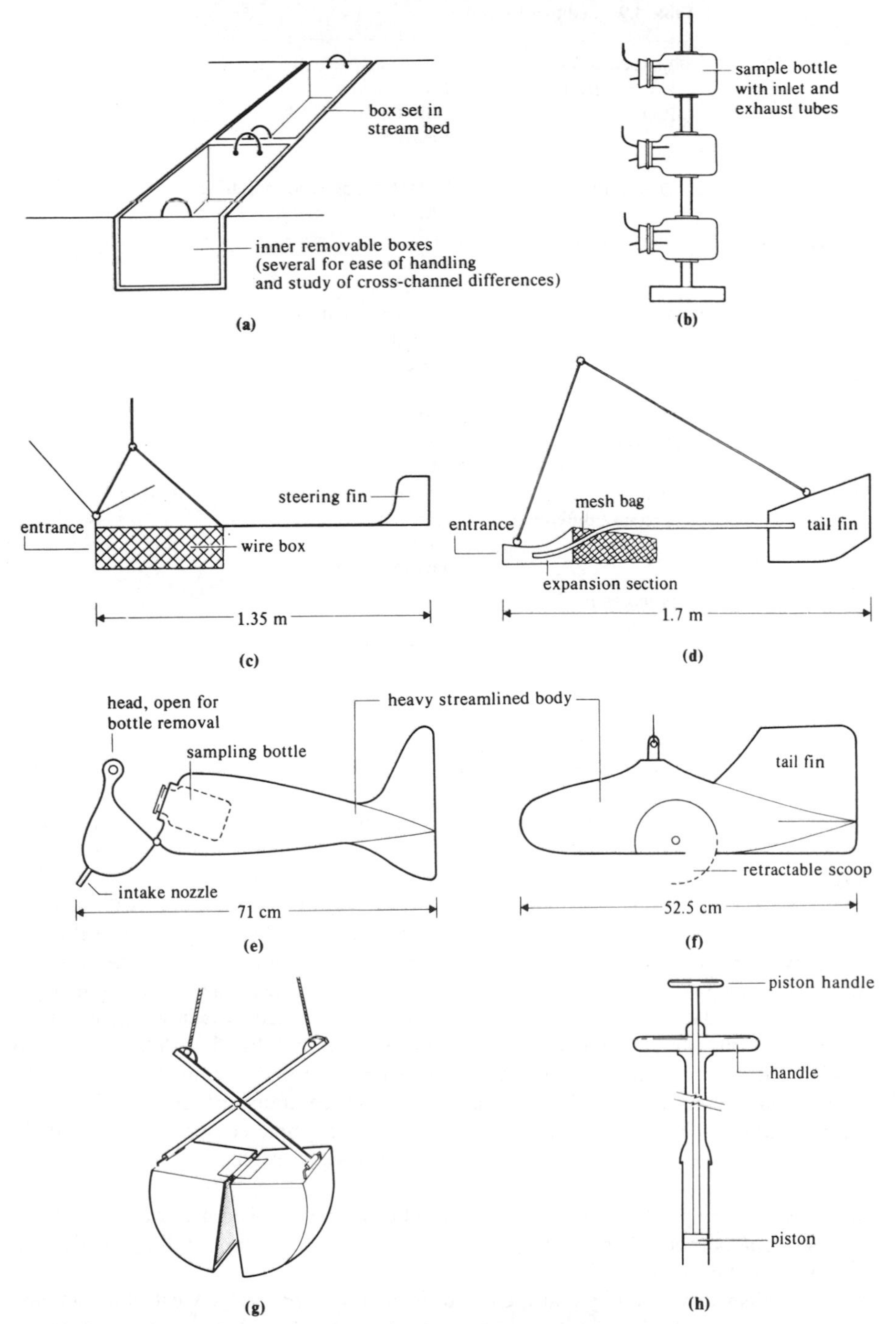

Figure 4.41 Suspended, bed and bed material samplers: (a) design for simple bedload trap; (b) an array of single-stage samplers; (c) Mühlhofer-type basket sampler; (d) Arnhem pressure-difference sampler; (e) USP 61 point-integrating sampler; (f) US BM-60 bed material sampler; (g) bed material grab sampler; (h) US BMH-53 bed material hand sampler.

Table 4.9 Sediment samplers.

Suspended sediments	
Instantaneous	Tait–Binckley
Point-integrating	USP 46, 50, 61, 63
	Neyrpic
	Delft
Depth-integrating	USDH 48, 59 (rod, handline)
	USD 49 (cable)
Single stage	
Pumping	Vacuum
	Peristaltic
Photoelectric	Southern Analytical
	Instanter
	Partech
Bedload	
Pit	USGS East Fork installation
Basket	Swiss Federal Authority
	Muhlhofer
Pan	Poljakov
Pressure difference	Arnhem
	VUV
	Helley-Smith
Bed material	
Scoop	
Drag	US BM 54, 60
Grab	
Corer	US BMH 53, 60

Devices available include pit-type traps permanently embedded across part of the whole stream bed, pervious basket-type or pan-type samplers, which may be lowered by cable (or anchored) to the stream bed, or pressure-difference samplers designed to equate entry velocity into the trap with stream velocity – which other samplers may modify to such an extent as to make unrepresentative the catch of transported sediment that is sampled. Designs are appraised in some detail by Hubbell (1964), and more briefly by Graf (1971) and Bogardi (1974). Several designs are illustrated in Figure 4.41 and listed in Table 4.9.

Many problems arise both in the actual use of samplers and in the interpretation of data they may provide. Without access to expensive equipment, the geomorphologist is likely to use relatively simple basket or pit samplers, which have low but unknown efficiency (ratio of sampled to actual bedload transport rate). But field data are so rarely available that even primitive data may be better than none, and it should be possible, say, to work on seasonal or short-term fluctuations in apparent transport rate (see Fig. 4.42a), or the variations in rate across a cross section using simple equipment provided the limitations of the results are appreciated. Precautions suggested are as follows:

(*a*) Make a comparison between sampled sediment and the bed material present at a site, and between alternative equipment (or bedload formulae). Hollingshead (1971) is a good example.

(*b*) Establish very carefully that the device is operating properly. Most samplers have been designed to suit specific streams and they may not work properly if, say, weight or mesh size is inappropriate elsewhere. Surface-operated baskets can also easily be lowered badly so that they end up at a tilt on the stream bed (allowing scour below the sampler lip, or an unrepresentative sample from a device that lands up oblique to flow), or they may scoop up material during lowering. Neither basket nor pit samplers will continue to operate well when nearly full, and they must be robust or heavy enough to remain correctly positioned at the higher velocities under which bed sediment transport is actually taking place.

(*c*) Note that sampled sediment may only be recoverable between transportation events, and while gross annual sediment yield or relationships with peak event magnitudes may be established, the actual conditions during transport may be unknown. Here the sophisticated apparatus described by Leopold and Emmett (1976), in which sediment passing into a pit-type trap is continuously extracted on a conveyor belt and thence weighed, is to be envied! An alternative is Reid *et al.*'s (1980) permanent sampler where gravel accumulating in a box set in the stream may be continuously weighed using a pressure pillow device. Such technology is particularly valuable in that it allows study of the pulsating nature of sediment transport over time.

(*d*) The samplers described work best where bed sediment is moving as particles or in thin sheets near the bed; in practice, sediment may move in pulses and in the form of discrete dunes or bars, and these may inadvertently be avoided by, or may completely overwhelm, sampling devices. Here, close observation of bedform changes after high-flow events is advisable, together with probing or depth sounding during such events if possible. Volumetric calculation of sediment movement by bedform resurvey may be undertaken.

4.4.3.3 Suspended sediment

Sediment generally supported by fluid turbulence declines in concentration from bed to surface at a rate dependent on grain size. Thus, concentrations are highest at the bed, with coarse sediment found only there, but finer sediment is more evenly distributed in a vertical profile. Samplers may extract mixed water and sediment from streams at determined points within a cross section effectively integrated over a finite time period (time- or point-integrating samplers), or they may be lowered to the bed and raised again

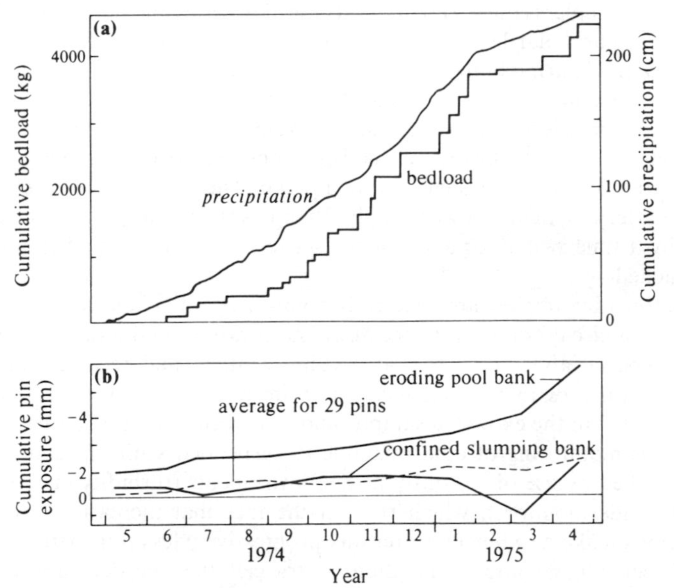

Figure 4.42 (a) Cumulative precipitation and bedload yield, and (b) cumulative bank pin exposure on the River Elan, Wales, in 1974–75. (Based on Lewin & Brindle in Gregory 1977c).

to extract a composite sample representative of concentrations at all depths (depth-integrating samplers). Concentrations vary at a point in time (with turbulence fluctuations), and with depth and position in a stream, so that single instantaneous samples may have dubious meaning. It is also often forgotten that the use of depth-integrating samplers without some accommodation being made for short-term fluctuations (for instance, by bulking several samples if a pilot study shows notable concentration differences between samples taken within minutes of one another), or the use of fixed-point sampling (as in most automatic devices discussed later) without establishing cross-sectional variation in concentrations, will both give misleading pictures of stream sediment content. Most instruments are designed to retain all the water and sediment extracted in a removable chamber, but the Delft bottle is a flowthrough device in which sediment settles out in the instrument. Some sediment is totally lost, but the longer time period for sampling is said to achieve a better estimation of mean conditions.

Samplers may be fixed or movable within the stream, and may be manually or automatically operated. Some rod- and cable-mounted samplers are shown in Figure 4.41. These usually consist of a heavy (cast bronze) streamlined shell within which is a removable sample bottle. The sediment–water mix enters the bottle via an external nozzle, the size of which may be varied, with an air exhaust functioning as the bottle fills. The sampler is kept in alignment with streamflow by a tail fin, and in the case of point samplers some form of opening valve is provided. A series of standard designs is in use (Federal Inter-Agency Sedimentation Project 1963); these are mostly very heavy (the USP 61 illustrated here weighs 48 kg) and need to be operated in conjunction with permanent cableways that are installed at river gauging stations for routine current metering. If workshop facilities are available, a rod-mounted and hand-held sampler can be made of a plastic 500 cm^3 screw-top bottle in which the cap is fixed permanently behind a hemispherical metal head that has been drilled and provided with an intake nozzle, exhaust vent and rod mounting. An even simpler version can be concocted from laboratory stands, clamps and tubing. However, these are very difficult to control in deep or fast-flowing water.

Fixed-point samplers (Fig 4.41) can be arranged in series and activated either as water levels rise to a particular level, or by a time switch. With a suitable system, it may be possible to sample during both rising and falling stages of a flood. Pumping or vacuum samplers can be made (Walling & Teed 1971) or purchased that may be set to sample a fixed amount at preselected time intervals (for example ranging from 0.5 to 2 h). Vacuum samplers may rely on a clockwork mechanism to operate a trip, breaking the vacuum in a series of bottles previously evacuated with a pump. These can thus be usefully operated without a battery or mains power supply. Pump samplers may use a peristaltic pump activated by a time switch, or additionally set to begin sampling when a certain river stage is reached.

These automatic samplers are effectively point-integrating, and for the calculation of loads they must be calibrated, both because concentrations in the stream cross section as a whole may be variably related to that at sampler intake, and because samplers may be inefficient in fully extracting sediment. For instance, larger-size sediment may be underrepresented in the extracted sample, and a correction factor can be applied.

Continuous monitoring of data relatable to suspended sediment concentration has been approached by use of variously named instruments (turbidity meters, silt meters, suspended solids monitors), which rely on the fact that increasing suspended solid concentrations make a sediment–water mix progressively less transparent. If the mix is passed between a light source and a photoelectric cell, the recorded current fluctuations can be calibrated against sediment concentration. Fleming (1969) has reviewed some suitable instruments. Practical problems occur in field situations (for example, due to floating and clogging debris, sediment size variation and variations in daylight), but the results are certainly interesting (Imeson 1977, Walling 1977a). Relationships between X-ray attenuation or ultrasonics and sediment concentration have also been used (Graf 1971).

All the equipment so far described requires some experience in use before good data are obtained. Mechanical or electrical failure, fouling of intake nozzles, failure to maintain the intake orientated properly, and failure to be actually sampling when sediment transport is really happening (or doing the job badly under inclement conditions) can all lead to a lack of useful results or to incomplete data. Damage by vandals and by livestock to equipment left out needs to be minimised by fencing and by locks, though it can be a toss-up as to whether elaborate but obvious security is more effective than unostentatious concealment. Both frost and flood are damaging. Much of the equipment described is heavy to transport and a two-person team is a great help if batteries, crates of sample bottles and depth-integrating samplers have to be carried about. Finally, it is inevitable that a large number of samples will have to be not only collected but also processed, and the data then analysed. It is essential that the *whole* procedure is thought out in advance, or else much physical effort and capital and operational cost will be wasted.

4.4.3.4 Sediment transport formulae

A number of equations for estimating bedload, or total non-wash load, transport rates have been derived through empirical observation and a degree of theoretical reasoning. These may be expressed in terms of tractive force, unit stream power or mean velocity. They usually relate sediment discharge rate per unit width of channel to the difference between actual physical conditions and the minimum critical ones necessary to cause particle movement. Proposed relationships are reviewed by Herbertson (1969), Graf (1971) and Yalin (1977). In view of the difficulties in actually measuring bedload transport in the field that have already been described, the calculation of transporting capacity using data on flow depth, slope, velocity, width, roughness and sediment size would appear to have distinct advantages.

However, as Graf and others have pointed out, the equations developed rely on experimental determination of their coefficients undertaken for the most part in small-scale laboratory flumes. When used outside the range of characteristics in such experiments, it is not surprising that very varied estimates of transporting capacity are arrived at (e.g. Hollingshead 1971). Furthermore, *precise* field determination of such things as flow depth, velocity or energy slope is by no means easy, particularly during floods. The geomorphologist may find use of available formulae inevitable in the absence of any alternative, but more field observations are needed. Worked examples are given by Graf (1971)

4.4.3.5 Tracers

Naturally or artificially 'labelled' sediment may be useful on a broad catchment-wide scale for identifying the source of sediment, or within channel reaches for tracing the paths and timing the rate of sediment movement. Such labelling has to be durable, non-hazardous and inexpensive, and it must allow labelled sediment both to be detected and to operate within the fluvial environment as natural materials would do.

Artificial labelling has been achieved through the use of radioactive tracers, paint and fluorescent coatings and by magnetic tracers; these are briefly reviewed by Bogardi (1974) and in more detail by Crickmore (1967) and Rathbun and Nordin (1971). Another alternative is described by Mosley (1978), who dumped a truck-load of distinctive limestone material into a New Zealand river and monitored its dispersal.

On the grounds of cost, safety and the availability of equipment, the geomorphologist may opt for simple painting of larger sediment particles. Numbering of individual particles and colour coding may allow reasonable tracking especially in channels that are shallow and clear or have flashy discharge regimes allowing for frequent exposure of a high proportion of the channel bed. An experiment along these lines by Thorne is reported in Thorne and Lewin (1979), and the location of 100 numbered pebbles some months after their installation in a line across the stream is shown in Figure 4.43. Recovery rates fall off considerably after a few transporting events. Finer sediments cannot be treated in the same way, and fluorescent coating is preferable.

Natural labelling may be similarly provided by the magnetic properties of transported sediment (Oldfield *et al.* 1979c), by the mineralogy of sediments from distinctive source areas (Klages & Hsieh 1975), by chemical analysis for inadvertently present mining wastes (Lewin & Wolfenden 1978), or indeed any characteristic of transported sediment that in some ways identifies its proximate or ultimate source. For example, suspended sediment derived from surface wash and soil erosion may be different in size and colour from that derived from bank erosion. The colour of suspended sediments (and of the turbid river) may change according to the source dominating at any particular time. Figure 4.44 shows the change in suspended sediment colour in relation to concentration and discharge during a storm in the River Ystwyth, Wales, suggesting an initial dominance of wash (brown) sediment followed by the uptake of channel materials (grey).

4.4.3.6 Sediment analysis

Suspended sediment concentration is usually determined by filtering the sediment–water mix, with sediment concentration expressed as dry weight (mg) per volume of mix (dm^3 or litre); mg dm^{-3} is equivalent to ppm by weight at normal stream sediment concentrations. Vacuum filtering speeds up an otherwise slow process, and evaporation has been used, though the sediment will then include dissolved residue as well. Minimum size of sediment is in effect determined by the filter paper pore size, and may not include colloidal materials, so that filtering materials and techniques should be specified.

Sediment retained on filter papers may usefully be examined for size (especially using instruments like a Coulter counter which require only a small sample), colour, mineralogy, or other attributes. Analysis of the organic fraction is important (see Finlayson 1975) since this may amount to a high proportion of the suspended solids obtained; this may apply to coarser 'sediment' as well.

For bedload traps, the problem may be to cope with the weight and volume of yielded

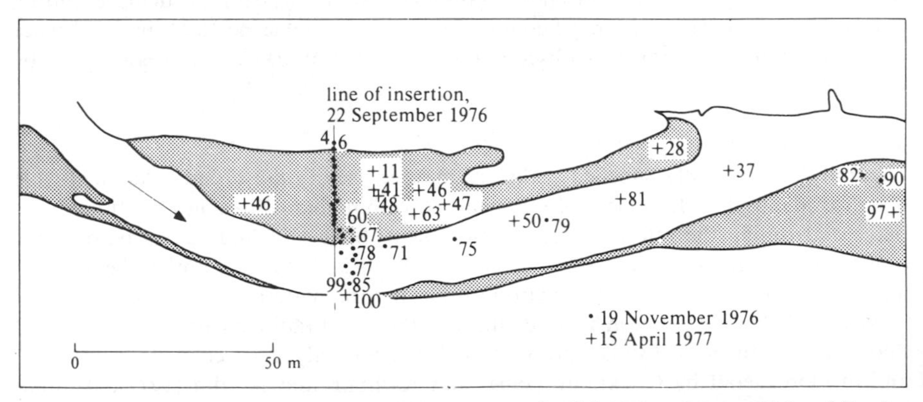

Figure 4.43 Movement of numbered pebbles on a reach of the River Severn, Wales, following insertion on 22 September 1976. (Based on work by Thorne in Thorne & Lewin 1979).

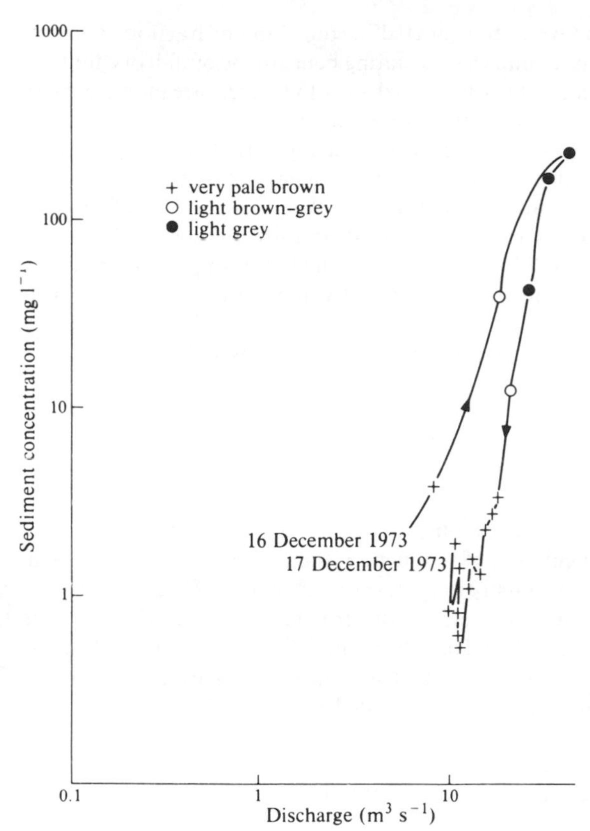

Figure 4.44 Colour of suspended sediment in relation to stream discharge and sediment concentration on the River Ystwyth, Wales. (Analyses by D. L. Grimshaw).

sediment. Some kind of splitter that allows partitioning of the yield without biasing the sample can be helpful, and the results expressed by weight (kg).

4.4.3.7 Sediment yield estimates

Given sample measures of bedload (weight per unit width per unit time) or suspended sediment (concentration per unit volume of stream water), there remains the problem of converting such measures into comparable units, or expressing them in some more desirable form. For estimating sediment yield, this may be resolved into, first, obtaining average sediment transport rate for a stream cross section at a given discharge and, secondly, using a set of such values at different discharges to calculate longer-term sediment yields.

For bedload, sediment discharge in cross section may be obtained by multiplying unit bed width estimates by actual bed width; long-term yields may be obtained cumulatively as long as the sampler is operating. If short-term yields are relatable to discharge magnitude, then longer-term yields can be estimated using magnitude–frequency relationships. In practice, this has rather seldom been attempted.

For suspended sediments, estimates of 'instantaneous' stream yield may be obtained *either* by sampling a number of subsections in the stream that are calculated to be of equal discharge, and summing the concentration × discharge product to give total yield, *or* by moving a sampler at a constant rate through a series of soundings, and determining the concentration of the compound sample to multiply with total stream discharge. Numerous soundings are advisable. An alternative for small streams is to obtain a known fraction of the total sediment and water transported in some kind of fraction collector. Thus, the Coshocton wheel is a horizontal wheel rotating beneath the outfall of a flume; a slot is positioned along the radius, which is carried into the water stream once every revolution, where it samples 1% of total runoff and sediment.

Sediment rating relationships, plots of concentration against discharge, may then be used in conjunction with flow duration curves to estimate yields over time. Such relationships are only approximate, and concentrations relate to a variety of factors as well as discharge (Guy 1964). The computation of yields in this situation has been very thoroughly examined by Walling (1977a, b) and Ferguson (1987). In particular, it may involve the use of separate regressions for summer and winter, rising and falling river stage, and high and low flows. The combined use of rating curves based on a small number of samples and the use of longer-term average flow data (weekly, monthly) is to be avoided.

Alternative methods of sediment yield estimation are provided by surveys of the amount accumulating in control sections, deltas, lakes, reservoirs and estuaries. These may record changes in the nature of the yielded sediment with time, as well as the gross amount if some datable base level can be established. Comparisons between initial surveys of reservoir floors with ones using sophisticated sounding instruments or during reservoir emptying may allow calculation of sediment volume. Here the problems lie in ground survey methods (perhaps of a highly irregular surface that is difficult to cross), in deriving meaningful density values for wet, partly organic sediments that may be identified with fluvially derived material, and of determining the efficiency with which the reservoir has trapped sediment passing into it. This and other methods to estimate sediment yield and rates of general denudation are listed in Table 4.10.

4.4.4. Channel form and channel-form changes

Relationships between river channel dimensions and the discharges and sediments they transmit, or the materials of which the channel is composed, have been looked at in a variety of ways. For example, the desire to provide easily maintained non-scouring and non-sedimenting canals in India led to so-called regime theory as developed by Kennedy, Lacey, Lane, Blench and others. For stability to be achieved, a series of empirical relationships was established, e.g. (in S.I. units).

$$U_c = 0.56\, d^{0.50} \tag{4.29}$$

where U_c is critical velocity and d is canal depth. Critical slopes were also established, while later relations were based on cross-sectional area and hydraulic radius, and include a 'sediment factor' related to particle size. Relationships have constantly been modified as new empirical data have become available.

Relationships have also been established for natural streams between such variables as width, depth, hydraulic radius, wetted perimeter, velocity, discharge and some measure of bed and bank composition. These hydraulic geometry relationships are usually expressed as simple power functions, following the work of Leopold and Maddock and others. This study area has recently been reviewed by Ferguson (1986), who broadly suggests a greater emphasis on hydraulics and bed mechanics in explaining channel dimensions.

Such an approach, based on laboratory experimentation, has been adopted in relating bed geometry (plane, ripple, dune, etc.) to varying flow conditions. Many investigations have also been undertaken concerning the mechanics of channel scour and deposition;

Table 4.10 Selected methods for the estimation of rates of general denudation.

Technique	*Example of method*
River sediment load monitoring	Walling (1978)
Reservoir sedimentation rate	Langbein and Schumm (1958)
Sedimentation behind natural dams of known age	Eardley (1966)
Tree root exposure	Dunne *et al.* (1978) and Section 4.2.2
Erosion plots	Yair *et al.* (1978)
Slope instrumentation (erosion pins, erosion rods, etc.)	Rapp (1960a)
Accumulation on continental shelf	Gilluly (1964)
Isotopic tracing (e.g. Cs-137, Pb-210)	Wise (1977) and Section 5.7
Volume of material removed from land surface of known age	Ruxton and McDougall (1967)
Accumulation on archaeological sites	Judson (1968)

this has involved experimental measurement of shear stress (see Graf 1971), but field geomorphologists have generally preferred to obtain field data on form, flow and materials, and to use these in stability calculations. The technical problem that remains lies in the actual derivation of field data, which may be far from straightforward.

Finally, channel-form process studies have been concerned with establishing the rates and location of form changes. This may involve both erosion and sedimentation, and both short-term direct measurement and the use of historical information. Changes may be linked to broad environmental parameters, to discharge measures, or to mechanical criteria.

4.4.4.1 Channel form

A simple representation of form, flow and material measures commonly used in channel studies is given in Figure 4.45. Only some of the problems concerning the field determination of values of certain measures can be discussed here, and reference to suitable texts (e.g. Graf 1971, Richards 1982) is advised (see also Sec. 2.7).

Channel width can be measured at the water surface, but this is variable along the channel, and it may be desirable to estimate bankfull width, which is often related to bankfull discharge (either measured or calculated as that discharge having a recurrence interval of 1.5 yr) as that responsible for channel formation. In practice, opposite banks are often of different heights: one may be a terrace and the other a surface of sedimentation with an indeterminate boundary. Some of these problems are discussed by Gregory (1976b), who used lichen limits to establish channel margins.

Water surface slope varies with discharge and according to the length over which it is measured. Measurement of high-flow slopes may be difficult at the time, but the waterline may be pegged out for resurvey later. Bankfull slope may be approximated by banktop gradient, but this can give errors where high-flow channels are not uniformly full and banks are uneven in height. Small variations in slope values may be critical in energy or sediment transport calculations and it is important to take measurements carefully.

Mean depth may be obtained from a large number of vertical soundings during current

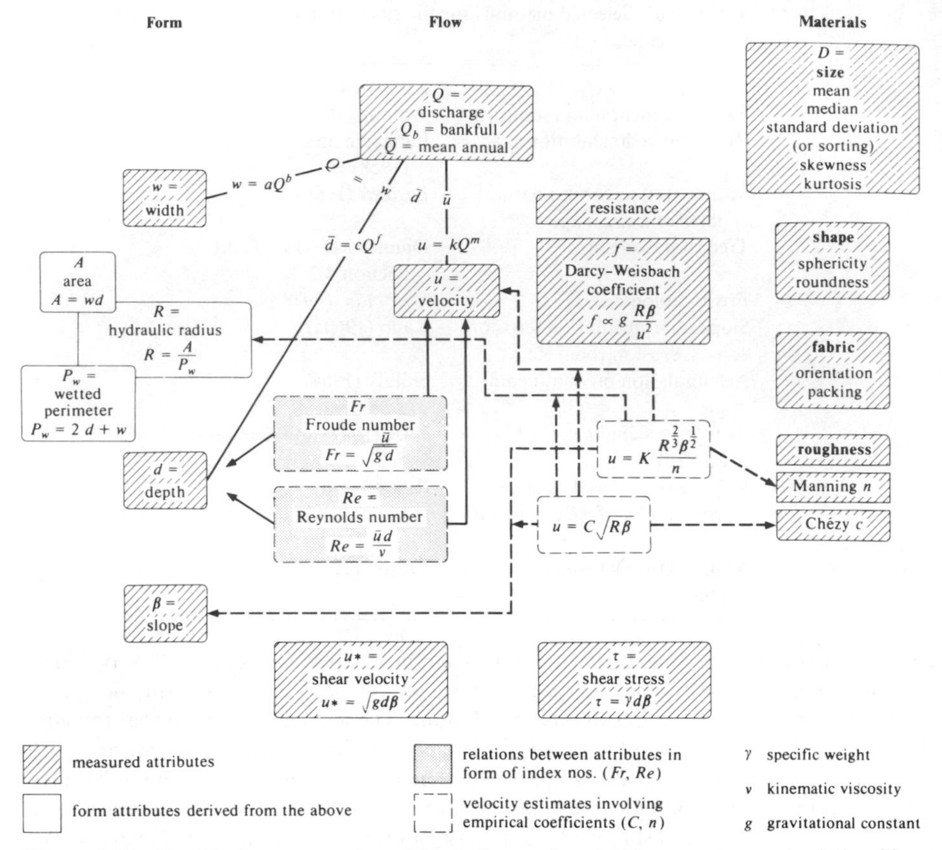

Figure 4.45 Graphical representation of form, flow and material parameters and relationships. (For further discussion see text).

metering, bankfull mean depth from levelling in relation to mean banktop height. However, such measures may misrepresent true scoured depths during floods, and especially in channels with complex bedforms it may not be easy to produce an acceptable value to use in conjunction with relationships established under laboratory flume conditions.

4.4.4.2 Bed material sampling

These materials may be sampled using a scoop, drag or grab bucket, or by using a vertical coring sampler (which may require the freezing of sediment using liquid nitrogen for sediment to be extracted intact) in the case of sampling in standing or flowing water. Surface sampling of exposed river beds may involve the collection of all exposed material within a determined area, and sampling at grid intersections, along a transect, or at random (Kellerhals & Bray 1971). On the whole, grab samplers are useful for finer sediment and for sampling in water, but all methods give inherently different results. Kellerhals and Bray (1971) give weighting factors to apply to the results for different methods. It should be remembered that bed armouring may make the size of sediments found in the surface layer unrepresentative of those beneath, and that sediments may vary systematically with depth so that there is a danger of sampling composite sediment populations if sampling is undertaken indiscriminately.

Properties of materials have been reviewed in Part Three, and it is necessary here only to emphasise that certain bed material characteristics have been used particularly in process work. Thus some measure of size (commonly D_{50} or possibly a value at the higher end of the sediment size range) may be used in sediment transport equations, and the size distribution of bed materials has been used to infer transport modes. On cumulative plots of grain size distribution on log-probability paper, samples may be interpreted as composites of several straight-line segments each representing a specific mode of transport (see discussion by Jackson in Davidson-Arnott & Nickling 1978). Finally, bed particles may be subjected to both local and downstream sorting, and to wearing and fragmentation downstream. These processes may be examined through field sampling, possibly backed up by laboratory experimentation.

4.4.4.3 Bank erosion

This may be measured in a variety of ways depending on the nature of bank failure and the rate and timespan of change. Following Wolman, a number of studies have used bank erosion pins, metal rods upwards of 1mm in diameter and generally ranging from 10 to 30 cm in length, inserted horizontally into the bank. Several pins may be placed in a vertical profile, with a number of profiles positioned along the channel to cover a representative number of sites. Retreat is estimated from progressive pin exposure, generally by ruler, depth gauge or callipers (see Fig. 4.46). In the case of sloping gully walls, pins may be inserted at right angles to the surface, and erosion or accumulation estimated from the amount a washer around the pin has slipped down it or has been buried in between successive observation times. Methods are fully reviewed by Lawler (1978).

This approach relies on the rate of retreat being measurable yet not more than would lead to the loss of the pin during the study period; it also requires that the pin does not interfere with the natural process of bank retreat. In fact, pins may modify bank cohesion locally (creating scour pits around them, facilitating seepage and snagging debris), they may be plucked by scouring iceflows in cold climates, and they may behave in a complex manner where large-scale slips and bank overhangs develop.

It is suggested that pins should always be inserted in large numbers before conclusions are drawn (note the different results in Figure 4.42b). Bent and missing pins may help to locate areas of removal even if precise measurements are impossible; repeat photography and the treatment of bank sections with spray paint may also be helpful in this respect.

Repeat profile surveys on monumented cross sections have additionally or alternatively been used. This can be done using simple levelling equipment, and is useful especially when bank retreat is rapid. An example is shown in Figure 4.46. Monumenting should be undertaken well clear of any likely erosion. Profiling can be undertaken in some detail with respect to the collapse of bank overhangs, though the survey work itself can cause some changes if undertaken clumsily. Long-term data are presented by Leopold (1973). It should be remembered that a permanently positioned profile will become relocated with respect to a shifting channel bend, so that profiles initially positioned orthogonal to the stream at a bend apex may eventually be found oblique to the channel at bend crossovers.

Plan-form changes (including the location of both cut-bank and sedimentation features) can be assessed from repeat plan surveys. Plane table surveys may be compared with already existing maps, plans and vertical air photographs so that patterns of change may be assessed over 100 years and more as well as in the short term (see Lewin in Gregory 1977c). Early plans are commonly generalised and inaccurate, while the accurate comparisons of plots of different scale can be problematical. Here instruments like the Rank–Watta Stereosketch can be useful, while photogrammetric techniques are very useful if equipment is available and photography suitable.

Comparisons of plots can be difficult because river stage is seldom the same for different surveys, so that channel margins are positioned differently even though no morphological change has occurred. Also, insufficient map evidence may be available to document change patterns and rates; an alternative is the dating of floodplain trees on developing sedimentary surface features (Hickin & Nanson 1975).

Data obtained by all these methods may be related to hydrological, meteorological and hydraulic factors. For example, rates of change at a set of individual sites may be related to bank sediment properties, to rainfall or frost incidence, and to parameters of streamflow. A full analysis of hydrological and meteorological variables was undertaken by Hooke (1980). Processes of bank retreat are reviewed in Turnbull *et al.* (1966) and Thorne and Lewin (1979). Somewhat similar techniques may be applied in studies of gully development. For example, Harvey (1974) has combined long-term photography and ground survey with pin-and-washer and sediment trap observations.

Special problems may be encountered in dealing with the particular processes of periglacial environments. Extreme seasonal flood flows of short duration, frost action on exposed river cliffs, the abrasive effects of water-transported ice, and the thermal effects of flowing water on permafrost banks may all produce distinctive bank forms. These may be strongly influenced by ground ice structures; ice melt in banks may produce an extensive thermo-erosional niche back into the bank at water level, and major bank collapse features may be common. These are described by Czudek and Demek (1970). Such processes may be difficult to record as they develop, partly because of the hazards involved, and partly because some of the techniques designed for temperate environments may prove inappropriate. Bank erosion pins, for example, may be unsuitable for permafrost banks and ones liable to rapid scour recession during high-magnitude floods that contain buoyant iceblocks.

Stream gauging during seasonal snowmelt or in proglacial streams may also require unusually robust housing for water-level recorders, the careful reconnaissance of cross sections stable under a wide range of potential flows and strategic planning for times when river crossing may quite suddenly become impossible. The possible range of conditions under which measurements may have to be conducted is graphically shown by McCann *et al.* (1972).

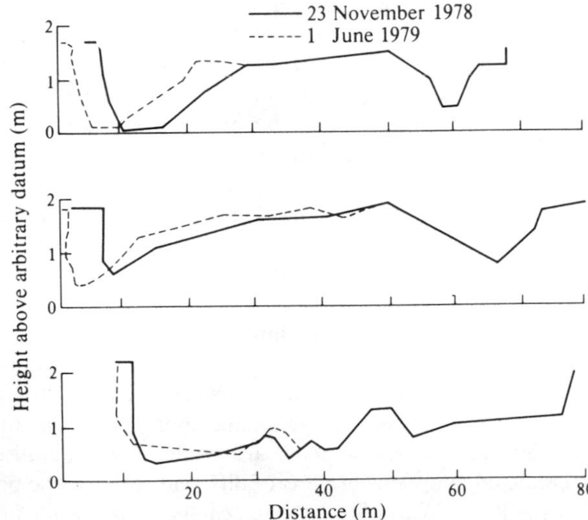

Figure 4.46 Channel profile changes on River Ystwyth, Wales, 1978–79. (Surveys by D. A. Hughes, C. Blacknell and J. Lewin).

4.4.4.4 *Sedimentation*

This occurs where and when stream power is insufficient to allow sediment motion; for the finest washload particles this may take place in virtually still-water pondage in surface lows after floods over a floodplain, whereas the largest sediment may be both mobilised and deposited within a stream during a single flood according to the spatial distribution of velocity. Individual particles may participate in sheet-like movement across the upstream side of major bars, to tumble down an avalanche face, where they are immobilised and buried. Sediment once transported via different modes may become superimposed or mixed on deposition, so that coarse bed sediment may be infiltrated or topped with saltating and intermittent suspension load.

It is rather difficult to observe sedimentation in progress, and studies usually involve the examination of sediments and bedforms in relation to discharges after the event. Experimental work using recirculating flumes (Guy *et al.* 1966) has been extended to field forms (see review by Allen in Gregory 1977c). This may require field measurement of flow characteristics and characterisation of the properties of sediment known to have been laid down under such flows (Jackson 1975). Repeated ground survey of developing bar forms may also be required, and these may be related to flow magnitudes between surveys, or to velocities while sedimentation is occurring (Smith 1971).

4.4.4.5 *Laboratory channels and hardware models*

It is appropriate to conclude a discussion of river channel processes, which has largely concentrated on field methods, with a brief review of laboratory-scale work. This has involved a variety of fixed and mobile bed flumes, particularly for the study of alluvial bedforms and sediment transport, and more extensive channel and network hardware models designed to investigate such phenomena as gullying, the development of channel confluences, or the whole course of 'landscape' development through fluvial activity. Laboratory systems may consist of scaled replicas of particular natural river reaches, or may be taken as universal prototypes from which an understanding of channelled flow and its consequences may be derived. Observable and rapid developments in morphology can stimulate qualitative advances in the general theory of landform development. A valuable recent example is Ashmore's study of braided stream development in a $1.3 \times 39\ m^2$ flume (Ashmore 1982).

Clearly the type of study attempted must be linked to the facilities available. Notable work in the recent past has been carried out on bedforms and sediment transport by the United States Geological Survey (Guy *et al.* 1966, Williams 1970); on channel and network development by Schumm and others (summarised in Mosley & Zimpfer 1978) at Colorado State University, Fort Collins; on sediment transport in curved channels at Uppsala University (Martvall & Nilsson 1972, Hooke 1974); on sediment transport at Imperial College, London (Abbot & Francis 1977); and on channel development at the Hydraulics Research Station, Wallingford (Ackers & Charlton 1970). It may be possible for geomorphologists to pursue similar experimental work if facilities are available, and at the least there is a considerable need for field observations to be undertaken with the results from such experimental work in mind.

4.5 Glacial processes

4.5.1 Introduction

Processes of glacial erosion, debris entrainment and transport, and deposition depend on the behaviour of ice, sediment and rock materials on, within, beneath, around and in front of glaciers. Conditions at the ice/bedrock interface are of critical importance to the operation of glacial processes but, at depth beneath modern glaciers, prove difficult for investigation in comparison with observations that are easily achieved in supraglacial and ice-marginal locations. Because of the problems of gaining access to contemporary glacier beds, glaciologists have relied largely on theoretical considerations, with limited field experiments, in attempts to understand basal processes, whereas geomorphologists have resorted to interpretations of the former beds of ancient glaciers, all too often with little knowledge of contemporary glacial environments. From studies of micro-forms on bedrock and of the texture, lithology, fabric and structure of till, glacial processes have been indirectly inferred from features remaining after the ice has melted or retreated. Even the use of observations around contemporary ice margins necessitates inference, either of present processes or of former conditions responsible for deposits now passively being revealed. Some methods for investigating former fluctuations of glaciers are discussed in Part Five.

Considerable progress has resulted from the application of glaciological theory to basal sedimentation and geomorphological processes and from detailed observations of actual processes in operation under and at the margins of contemporary glaciers and ice sheets. Theories of glacier motion and sliding have been developed extensively (Weertman 1979). Basal sliding where the glacier sole is in intimate contact with a non-deformable bed is well described (Kamb 1970, Nye 1970). Deformation of the bed has also been subjected to theoretical investigation (Boulton 1982). Nye and Martin (1968) extended plastic flow theory to glacial erosion, and a simple sliding theory (Weertman 1964), in which basal drag is due to the resistance of bedrock irregularities, was adopted by Boulton (1975) to demonstrate the relationships between basal ice properties and erosional and depositional processes. The glacial processes controlling the movement of rock inclusions towards the bed of a temperate glacier were treated theoretically by Rothlisberger (1968). Theoretical considerations have also been applied to the processes of abrasion (Boulton 1974, Hallet 1979). The nature of subglacial hydrological systems has also received attention, and thin basal water films (Weertman 1964), conduits in ice (Rothlisberger 1972), channels incised in bedrock (Nye 1973) and water-filled cavities (Lliboutry 1968) have been postulated. Subglacial hydrological conditions and basal thermal regime have been used as theoretical frameworks for processes of erosion, transport and deposition (Boulton 1972). Such theoretical developments have both stimulated field observations and resulted from such investigations. Enhanced opportunities for subglacial exploration have been presented both by tunnels dug for hydroelectric and mining operations (e.g. Vivian 1975, Hooke *et al.* 1985) and as a result of developments in instrument technology for remote monitoring of basal interfacial conditions. Experimental laboratory tests of ice sliding over bedrock have been undertaken, with a view to verification of sliding theory (Budd *et al.* 1979), determination of frictional and adhesive properties of ice (Barnes *et al.* 1971) and simulation of glacial erosion (Lister *et al.* 1968).

A combination of glaciological theory, adapted to include processes of sedimentation, and direct observations of processes in action beneath active glaciers appears to be appropriate to attempts to answer fundamental questions associated with how the physical behaviour of glaciers controls the processes of erosion, transport and deposition. Sediment is supplied to a glacier by rockfall and avalanches from slopes marginal to the ice and becomes buried under the accumulation of snow. Subsequent routing of debris through the glacier depends on surface ablation, internal flow and basal thermal conditions (Fig. 4.47). Further debris is acquired subglacially and all debris subsequently becomes deposited at either the bed or the front of the glacier. During transit along different pathways within the ice, glacial processes modify debris characteristics such as particle size and clast shape and determine the fabric and form of deposits. The routing paths through a glacier are of interest in attempts to determine the relative importance of debris supply sources and of englacial and subglacial transport and deposition processes. It is clear that direct observations of englacial and subglacial processes are essential to the understanding of glacial geomorphology.

Observations of contemporary glacier processes can be made in subglacial sites, on the glacier surface, in englacial tunnels and boreholes and at glacier margins, in crevasses, ice cliffs and recent deposits. While supraglacial observations are useful for studies of depositional processes, other surface measurements, for example of flow velocity, though often the more accurate, are not adequate as surrogates for actual basal determinations. Englacial observations tend to be inconclusive and inaccurate because of the use of inclinometry in boreholes, whereas basal observations are either laborious or limited to certain sites, but are nevertheless most valuable. In subglacial sites, both glaciological and sedimentological characteristics can be measured together simultaneously, and the properties of abraded rock surfaces and tills observed under the exact conditions of their formation. Simple description of sites has proved useful, but several specialised techniques have been developed for quantification. Studies of subglacial processes require a suite of techniques to monitor a moving deforming glacier sole, incorporating rock debris, interacting with deformable basal till, which may contain water, and with irregular bedrock surfaces, in three dimensions. Further measurements of water pressure and of water flowing in channels are needed. The techniques described in this section are generally appropriate for use on cirque and valley glaciers and around the margins of small ice caps.

Most of the techniques used in glacier process investigations are costly, often requiring complex purpose-built apparatus not available commercially, and necessitating considerable logistical field support and teamwork. Theoretical progress in glaciology has advanced ahead of field experimentation, although pertinent observations have been permitted by increasing accessibility to subglacial sites. The techniques described in this section are included on the grounds that they have proved useful in approaching specific questions concerning the actual operation of processes of erosion, transport, deposition and post-depositional change. Details of fabric studies are given in Section 3.2.16.

4.5.2 Access to the ice/bedrock interface

4.5.2.1 Direct access to the bed

Sites permitting direct access to the glacier bed are desirable because detailed observations and measurements of the operation of glacial and geomorphological processes are permitted. Samples of ice and sediment can be collected for subsequent analysis and *in situ* experiments may be conducted. Natural cavities, which exist between bedrock and the glacier sole on the downglacier slopes of transverse rock bars on steep gradients (Vivian & Bocquet 1973) or in the lee of upstanding bedrock protruberances on lesser bedslopes, can be reached from beneath through artificial tunnels in bedrock or from artificial galleries in ice, excavated and accessed from the surface. Conditions in such sites may be atypical and unrepresentative of those existing over most of the basal interface. Cavities at depth may be beneath considerable thicknesses of ice (about 100 m where rock tunnels dug for hydroelectric power purposes penetrated the bed of Glacier d'Argentière, Mont Blanc, France). Although accessible zones of the bed may be of substantial area, about 20 000 m^2 at Argentière, they are restricted to a very small proportion of the total ice-covered area. Accessible sites are dry, either away from the influence of meltstreams or approached during periods of low flow, for example during winter freezing-up of the glacier surface. Where rock tunnels form part of hydroelectric developments, mains power and laboratory facilities are often available outside the cavities. Enlargement of a cavity is possible by melting ice using hot-water jets and sprays on the bottom layers of the glacier. Hot water at a temperature of 40–50°C is pumped through hoses to spray nozzles held in place on the glacier sole by ice screws (Wold & Østrem 1979). Natural marginal tunnels may allow access to limited areas of bed under relatively thin ice cover.

Artificial galleries excavated from the glacier surface also usually expose the bed under only thin ice cover (Haefeli 1951, Kamb & LaChappelle 1964). Tunnels in ice allow englacial and subglacial observations, and can reach bedrock in areas of intimate ice–rock contact or intersect with natural cavities (Boulton 1979). From the glacier surface or margins, manually excavated tunnels are best sited in thin ice and on steep ground, where the approach distance to the bed is minimised (Fisher 1953, McCall 1954). In temperate ice, tunnels should be inclined upwards at an angle of several degrees from the horizontal at the portal to facilitate meltwater drainage. Short-shaft (*c.* 0.7 m) ice picks with heavy heads and sharp points are required for work at the cramped ice face. Ice debris is shovelled using 'avalanche-rescue' aluminium scoop spades into polythene sacks lining a sledge-mounted box for evacuation to the tunnel mouth for disposal. Tunnelling is a slow operation, the rate of advance with two working operators, and reserve manpower, in a tunnel of 2 m^2 cross-sectional area being about 3 m per day (see McCall 1954). Some excavation is required at all subglacial sites in order to gain access to zones of intimate ice–bedrock contact, for the installation of monitoring equipment.

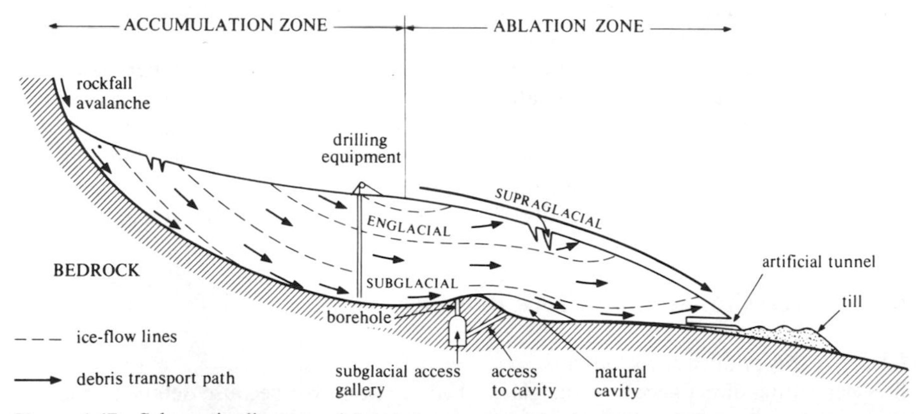

Figure 4.47 Schematic diagram of debris transport routes beneath, within and on a temperate valley glacier. Access to basal sites is possible using boreholes and artificial tunnels in rock and ice.

Apparatus located on the bed in minor access galleries connects with the glacier sole as the ice from immediately upglacier slides forwards.

4.5.2.2 *Remote sensing of the bed*

Remote sensing of conditions at the ice/rock interface can be achieved through boreholes, either in ice from the glacier surface or upwards through drill holes in rock from subglacial tunnels excavated in bedrock. Drill holes in rock have been bored using heavy plant during hydroelectric workings, and have enabled sounding and proving of basal water channels (Wold and Østrem 1979) and measurements of basal water pressures using manometers (Bezinge *et al.* 1973). Such drill holes are usually arranged in lines according to the siting of the tunnels and, while they may emerge at the bed under considerable thickness of ice, only an extremely small area of bed is penetrated.

Advances in ice borehole technology, especially in instrumentation for observations and measurements in small boreholes, have enabled the remote monitoring of englacial and subglacial processes. Boreholes in ice allow the emplacement of sensors that transmit information to the surface. Measurements of glacier sliding have used observations of the tilting of boreholes thought to be drilled to the bed, in combination with determinations of surface velocity (Raymond 1971). Both errors of inclinometry and the uncertainty of the hole actually penetrating the bed limit the utility of this method. An improvement to this system uses borehole photography, and ultimately television, to study the geomorphologically important sliding component of glacier motion (Harrison & Kamb 1973), Engelhardt *et al.* 1978). Remotely sensed, but direct, observations of basal sliding and sedimentation confirm whether a borehole has reached the bed, and allow more detailed measurements of sliding velocities, on a day-to-day basis, in the deeper parts of glaciers. An area of about 60 m^2 of bed only can be observed beneath a borehole, but a net of several holes should provide a reasonable sample of basal conditions. Water levels in boreholes can be measured directly using pressure transducers or membrane gauges, connected to recorders on the glacier surface.

Steam, hot-water and electrical thermal ice drills have been designed in addition to rotary, cable–drill and impacting mechanical devices (Splettstoesser 1976). Borehole diameter, the degree of regularity of hole cross section and circumference required, the verticality of penetration and the depth to be reached are important considerations in the choice of drill type for a particular application. The zone of glacier in which the drilling operations are to take place, the weight of the equipment, likely energy consumption and fuel type are logistical constraints. Drilling equipment is expensive to construct and operate and, with fuel requirements, usually requires helicopter support and two or three operators. Steam-heated ice drills, for which designs have been given by Hodge (1971) and Gillet (1975), are useful only for shallow holes up to 15 m, for the insertion of ablation stakes. Where water is available on the glacier surface, an open-circuit hot-water drill is considered suitable for rapidly and reliably producing vertical wide-diameter, though often irregular, holes to depths of several hundred metres, with reasonable weight and energy consumption. Apparatus designs (Fig. 4.48) are described by Iken *et al.* (1977). Basically, a petrol-engined centrifugal pump forces water through a diesel-oil-burner heating unit into an insulated drilling hose, from the nozzle of which hot water is ejected to melt ice, before flowing back up the drill hole (total weight 150 kg; total fuel consumption 10 $l\ h^{-1}$). The drilling hose consists of flexible pressure-resistant thick-walled tube (ID = 12 mm, OD = 20.5 mm) in a PVC sleeve (total weight 0.45 $kg\ m^{-1}$), ending in a rigid guiding tube 2 m long, at the tip of which is an exchangeable nozzle (ID = 2.5–4 mm). The regular nozzle can be replaced by one at 45° to the drilling axis, which melts an oblique hole into which sand and stones are removed. The hose is carefully controlled during entry to the ice so as to hang freely, preventing

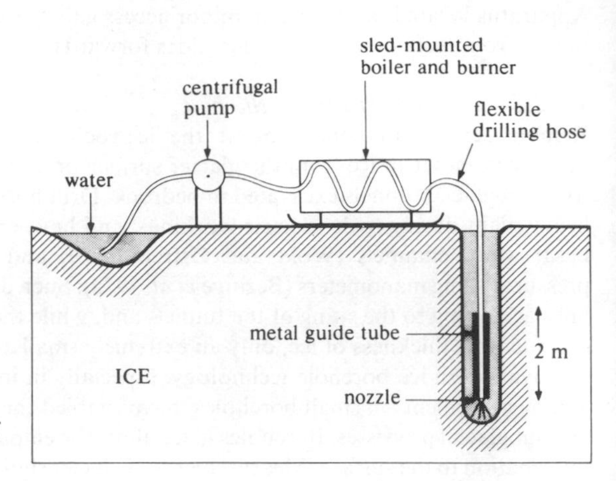

Figure 4.48 Schematic diagram of hot-water ice drill.

bends and spirals. Drilling rate decreases from about 90 m h^{-1} in the first hour to 50 m h^{-1} at depths greater than 200 m. Water flows from the nozzle (ID = 3 mm) at a temperature of 60–80°C with discharge about 0.75 m^3 h^{-1}.

Electrical hotpoint drilling is required for holes in the accumulation zone and for deep holes of regular cross section (for inclinometry). The hotpoint is heated electrically by the Joule effect, with energy supplied along an electric cable from a petrol-engined generator. Small-diameter, *c.* 50 mm, hotpoints with high thermal output descend rapidly and may achieve speeds of 5 m h^{-1}; 50 mm hotpoint drills working from 3 kW generators have proved successful (Hodge 1976). The heating element is cast in the end of a 2 m long drill head. Descent of the hotpoint is controlled by passing the cable over a pulley suspended from a tripod. A potentiometer is attached to the axle of the pulley for the accurate measurement (± 1 cm) of the drill position and rate of advance, so providing a method for recording the internal distribution and thickness of debris bands (reduction in drilling speed) and englacial cavities (sudden abrupt increases). This provides an alternative to the core extraction method for investigation of internal characteristics of glaciers (Splettstoesser 1976). Increased sediment concentration in ice close to the bed may prevent a thermal hotpoint drill from reaching the actual bed. A cable tool, a heavy chisel suspended on a cable, repeatedly raised and dropped from up to a metre above the bottom of the hole will penetrate such dirty ice (Harrison & Kamb 1976). Penetration of the bed may be inferred from the dropping of water levels in boreholes, or from borehole photography.

4.5.3 Techniques for measurement of basal processes

4.5.3.1 Ice motion at the ice/rock interface

Motion of ice at the bed is of fundamental importance geomorphologically, and the nature of the sliding mechanism, involving basal debris, and the rates of movement of ice both relative to velocities of entrained particles and relative to the bed effect the processes of erosion and deposition. Basal sliding velocity can be measured directly in subglacial sites and remotely using borehole photography. Repeated measurements separated by short intervals in time or continuous recording are necessary to show the nature of sliding motion. The simplest technique requires frequent measurements of the displacement of ice screws attached to the sole of the glacier, with respect to fixed points, such as brass screws, cemented into the rock floor of a subglacial cavity. A heavy pointer suspended from an ice screw almost to the bed aids the measurement of the actual distance of movement. An indication of differential motion in the lowest layers of the ice can be obtained by measurements of the displacement of ice screws or pegs inserted about 10 cm apart in a vertical line in a section of basal ice (Kamb & LaChappelle 1964).

A cryokinegraph (Goldthwaite 1973) is used to record variations in the speed of basal sliding (Fig. 4.49). A Stevens, or similar, water-level recorder is anchored firmly on bedrock in a subglacial cavity, with the external pulley aligned in the direction of motion of the sole. The beaded cable, which activates the pulley to move the pen over the chart, is attached to a several-metre length of thin-gauge invar steel wire. The beaded cable is held in its usual position around the recorder pulley with a 0.5 kg weight and the invar wire maintained in a taut catenary curve downglacier in the direction of motion to a 'frictionless' pulley by a counterweight, made up of rock debris ballast to adjust the balance of the system. The pulley is fixed to the glacier sole by a 30 cm long tubular ice screw. Care should be taken to ensure smooth operation of the pulleys. Gearing of the pen movements on the recorder chart to displacement of the wire is adjusted to the observed rate of sliding.

Vivian (1971) has developed the cavitometer as a method of recording basal sliding velocities (Fig. 4.50). The instrument consists of a wheel (0.4 m diameter) held in contact with the glacier sole by a cantilevered arm. Either a tyred bicycle wheel, or a wheel bearing metal studs of small teeth on the rim, ensures that the wheel turns according to sliding of the sole. The rotation of the wheel is continuously recorded using a potentiometer attached to the axle. The arm is 3 m in length, and articulated at a pivot 2.5 m from the business end, counterbalanced by a suitable weight to ensure that contact is always maintained between the wheel and the sole. The cavitometer is fixed on the bedrock floor of a cavity, and the arm directed along the flow axis of the sole in a down-glacier direction. Mains power supply and data recorder are housed separately outside the cavity. Actual velocities of the sole are probably within about 20% of recorded values, with absolute errors of 5–10 mm h^{-1}.

Detailed measurements of short-term variations of sliding and direct observations of processes actually taking place at the bed can be obtained using borehole photography. This method is useful on account of its accuracy and the details of sediment–ice interactions revealed, but is logistically expensive to operate. Borehole cameras, with an outside diameter of 51 mm, have been designed and constructed by Harrison and Kamb (1973) to fit into holes of 60 mm diameter produced by hotpoint drilling (Fig. 4.51). Single-shot cameras use circular discs of film, giving an image 43 mm in diameter. Focal length is fixed (83 or 89 mm), and focusing achieved by adjustment of the film plane.

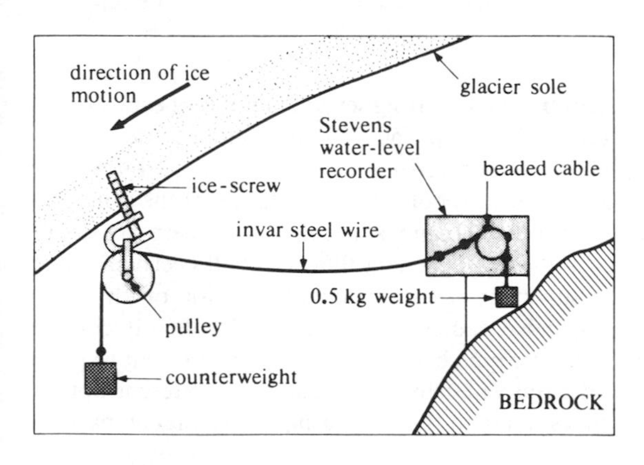

Figure 4.49 A cryokinegraph consisting of a water-level recorder connected to an invar steel wire, attached by an ice screw to the glacier sole.

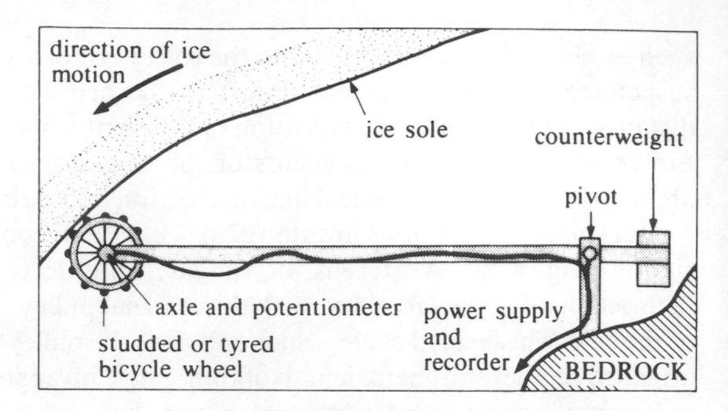

Figure 4.50 The cavitometer for direct measurement of subglacial sliding.

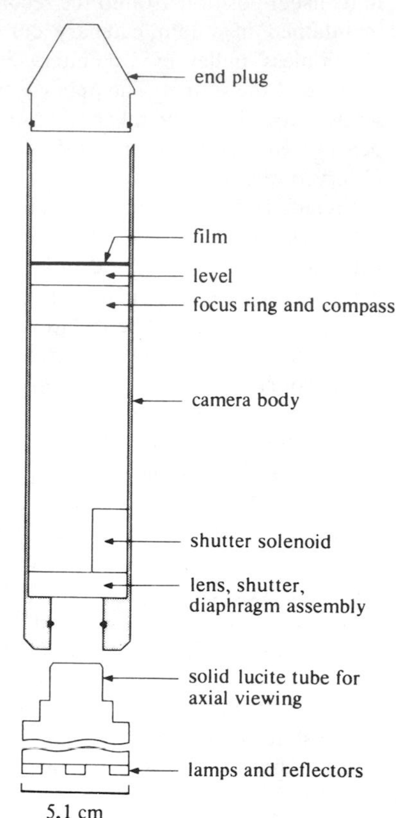

Figure 4.51 Camera for use in glacier boreholes. (After Harrison & Kamb 1973, Fig. 1).

Small apertures (*f*/80) maximise the depth of field, during several minutes of exposure on ASA 400 film. An acrylic viewing tube, of a length (50–250 mm) suited to detailed or wide-angle shots of the bed, bearing lamps and reflectors, is attached in front of the camera, which is suspended on a cable in the borehole. The orientation of the camera is recorded on film from the shadow cast by a small compass needle. The camera can simultaneously function as an inclinometer, as the shadow on the film of a small steel ball rolling on a concave glass plate enables measurement of the angle and direction of tilt of the instrument body. Stability of the suspended camera is maintained by the attachment behind the body of a 40 cm length of plastic pipe brush of a diameter sufficient to press lightly on the borehole walls. Turbid basal water in the borehole may present optical problems. Motion of the basal ice with respect to the bed is measured from the comparison of successive photographs in a temporal sequence. Further, borehole photographs yield information about basal sediments, clasts and rock debris, their relative displacement with respect to basal ice and bedrock, and any ice-bedrock separation or cavitation. Conditions at the actual ice/bedrock interface are revealed, unlike the examination of the sole where it forms the roof of a cavity or tunnel.

4.5.3.2 Debris and debris-laden ice in the basal zone

Observations of properties of basal debris-laden ice may be achieved directly through access to subglacial sites, and indirectly using boreholes. The quantity, concentration, nature and dispersion of particles, their shapes and in-transit modifications, and the direction and amount of their motion with respect to ice are of interest. Subglacial sites enable direct descriptive observations of basal processes, especially of deposition (Boulton *et al.* 1974, Boulton 1974). Some simple measurements, for example of movement of identified clasts, at weekly intervals, with respect to bedrock are useful. Relative clast movement within basal ice can be inferred from observations of small voids preceding clasts in the sole or from tails of fine abraded or crushed debris following clasts that were in traction over the bed immediately upstream of the cavity. Direct observations of clast shapes and sizes and debris fabrics and compaction *in situ* in the environment of their modification or formation are useful for the interpretation of former glacier beds. The vertical and spatial distribution of fine debris inclusions in the lowest ice layers can be investigated by collection of cores obtained using small coring drills such as US Army CRREL ice corers. Cores of about 7–8 cm diameter, up to 50 cm long, are collected from the roof or walls of a subglacial cavity. Alternatively, about 1 kg bulk samples of basal ice can be collected from cavity and tunnel exposures. After division of cores into sections representing layers, using a saw, the ice samples and included debris are melted in graduated cylinders to provide a volumetric measure of sediment content.

Direct and indirect observations of basal ice in cavities have revealed the existence of basal water films and regelation ice. Chemical analyses of ice from basal layers has provided an indication of the nature of basal physical processes (Souchez *et al.* 1973). Crystallographic investigations of basal ice samples, partially melted to take the form of thin sections, provide recognition of regelation phenomena (Kamb & LaChappelle 1964). Samples of ice for chemical analysis can be collected by chipping with a non-contaminating chisel, for example of chromium–vanadium steel previously checked for its lack of alkaline-earth contamination. Samples may also be collected with a PTFE (Teflon)-coated CRREL ice corer. Care should be taken in handling, using clean polyethylene gloves to transfer samples to polyethylene containers previously washed with pure nitric acid and detergent and rinsed in distilled water. A major problem in the treatment of basal ice samples is the chemical interference of particulate material during ice melting, as samples are prepared for instrumental analysis of the resultant water. Souchez *et al.* (1973) warmed ice samples on a water bath, and recovered meltwater immediately (within 30 s) after production using polyethylene syringes. The water was then pressure-filtered through 0.45 μm cellulose acetate filters, acidified to pH 3 and stored in polyethylene bottles before analysis (Sec. 4.5.6).

Temperature is an important variable in basal zones of ice and bedrock, but few measurements have been reported. Accurate thermometers, or copper–constantan thermocouples, inserted into boreholes in ice or bedrock, are allowed to reach thermal equilibrium with their environment. Accuracy of $\pm$ 0.2°C has been obtained in the measurement of ice temperatures using calibrated thermistors with a Wheatstone bridge (Muller 1976).

The forces exerted by the glacier sole on the bed are measured by sensors placed into

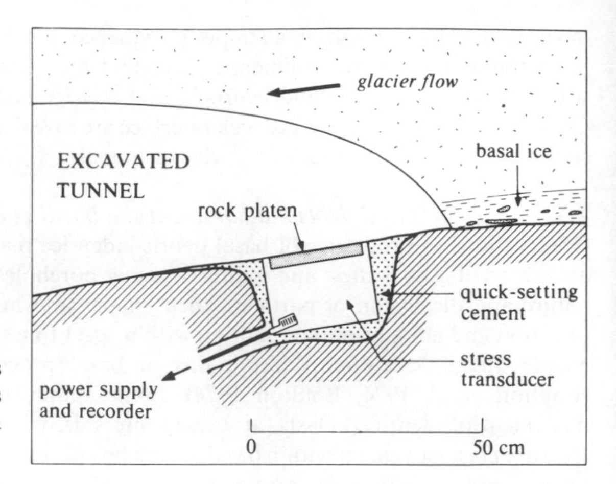

Figure 4.52 Diagrammatic cross section of the location of a stress transducer in bedrock beneath a glacier sole.

holes in bedrock so that the surfaces of the sensors lie flush with the general bedrock surface and act as if they were small areas of the bed (Fig. 4.52). Contact stress transducers, inserted in bedrock, have been used to monitor continuously shear stress and normal pressure changes at the contact with the glacier sole (Boulton *et al.* 1979). Holes are drilled in bedrock in suitable positions, after the overlying ice has been removed. Commercially available stress transducers, using foil-type strain gauges, are located in the holes. A rock platen bonded to the top of the transducer forms the upper surface. Rock platens of known thickness and surface roughness are cut with rock saws and cutters and polished. A force applied to the rock platen deforms internal webs in the transducer, altering the resistance of the strain gauges, to which a direct-current voltage is supplied. Quick-setting cement is used to fill the gaps between the transducer and the rock. Power supply and recorder output cables lead through the rock to respective electronic apparatus in a neighbouring cavity. The rock platen itself becomes subject to glacial wear. After installation in winter, plastic deformation brings ice into contact with sensors, rates of closure of 120–150 mm day^{-1} occurring in tunnels in ice of diameters 1–3 m. Hagen *et al.* (1983) describe an artificial roche moutonnée, made of steel and bolted to the bedrock, filled with reinforced concrete into which pressure sensors were countersunk. A Teflon sheet about 2 mm thick covering the construction was readily removed by abrasion. Three sensors were inserted; on the upglacier slope orientated at 4° dip, on top of the hump arranged parallel to the general rock gradient, and on the steeper downglacier face angled at 65° downslope. The construction was 1.5 m long, 0.4 m wide and had a maximum elevation of 0.3 m at a distance 1 m from the upglacier end. A dome of similar construction with five symmetrically placed pressure gauges was also used. Standard vibrating-wire pressure gauges (0–50 bar range) were calibrated at 0°C to an accuracy of ± 1%. Measurements allow interpretation of individual clast movements, of shear stress–normal pressure relationships, of the role of debris in the frictional drag between basal ice and the glacier bed, and of patterns of stress variations over irregular beds.

Thickness of debris-rich layers in basal ice and beneath glaciers can be determined indirectly from surface boreholes. Performance of the cable-tool drill (Sec. 4.5.2) and recovery of debris by repeated baling using a hand pump (Harrison & Kamb 1976) may be used as evidence, together with penetrometer tests in the subsole drift. A cable penetrometer consists of a sharpened steel rod, of length 30 cm and diameter 1 cm, onto the upper end of which a hollow cylindrical weight of about 20 kg slides on release from above, so driving the rod into the basal sediment. Some measure of differential movement of basal till can be obtained from repeated borehole photography, but the role of deformation of subglacial till, where unlithified material is present between sole and bed, in the total basal movement of a glacier can be determined directly fron subglacial sites. If a section of till is available, welding rods or 1 cm diameter wooden dowels can be inserted horizontally transverse to the direction of movement, in an originally vertical line. Relative movement of marked large clasts in a matrix may also be determined with successive measurements at about weekly intervals in order to ensure sufficient change for accurate measurement. Where subglacial tills are water-saturated, flowage under pressure occurs readily and repeated survey of tunnel floors will indicate the quantity of debris that has flowed. Water pressure can be measured by tensiometer. Boulton (1979) has devised a useful method for use from tunnels in ice 1–2 m above the glacier sole, from which probes are inserted through the underlying ice into the basal till at several locations. The probes consist of a central rod on which 2 cm thick, 1 cm diameter continuous annuli are stacked. After insertion to such depths as permitted by boulders in the till, the central rod is withdrawn by unscrewing it from the tapped bottom annulus. After several days, the site is drained by pumping and then carefully excavated to reveal the final positions of the annuli (Fig. 4.53). Relative vertical motion, net discharge and mean direction of movement of the till can be determined for each site, and a three-dimensional picture of till deformation produced by combination of results from adjacent sites.

4.5.3.3 Measurement of net abrasion rates

Direct experimental investigations are possible given access to subglacial sites. Net abrasion rates may be measured by replacing a small area of natural bed with an artificial plate of metal or rock (Boulton 1974, 1979). Small plates $10 \times 5\ \text{cm}^2$ are smoothly ground, before cementation in holes drilled in bedrock, such that plate surfaces are flush with the bedrock surface (Fig. 4.54). Sites chosen upglacier of cavities provide for examination of products of wear adhering to the sole then subsequently moving over the cavity. Both mean reduction in thickness of plates and hence rates of abrasion with respect to time, and patterns and forms of abrasive wear can be determined from the plates when exhumed. The plates should be left at the bed for at least 12 months to provide sufficient change for reasonably accurate measurement (probably errors of about 10%). The amount of wear can be measured using a precision dial micrometer, traversing and sampling selected points on the plate. The surface roughness can be determined by the use of a profilometer (Rabinowicz 1965). The plates provide a direct measure of the rate of abrasion alone, unlike indirect estimates of glacial erosion by measurements of sediment transport in meltwater streams, which include products from all subglacial processes. Measurements of net abrasion rates are advantageously conducted alongside simultaneous observations of other parameters such as sliding velocity, basal shear stress and sediment concentration in basal ice.

4.5.4 Surface and proglacial process observations

4.5.4.1 Surface observations

Direct observations of the processes of debris incorporation in the accumulation zone, and of melt-out in the ablation area of englacial and subglacial debris to become supraglacial material, are possible on the glacier surface. Supraglacial melt-out and flowage processes can also be investigated directly towards the frontal ice margin (Boulton 1967, 1968). Rates of deposition, rates of movement and their relationship with rates of ablation, ice surface movement and changes in angles of surface slopes are of

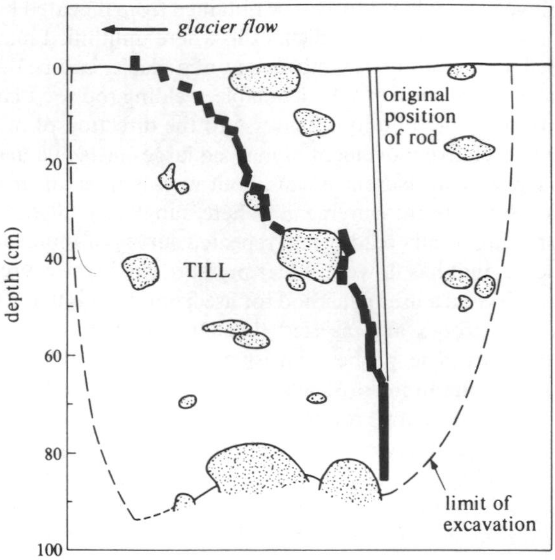

Figure 4.53 Original emplacement and end-of-experiment positions of annuli inserted on a probe from a tunnel within basal ice into deforming subglacial till. (After Boulton 1979, Fig. 7).

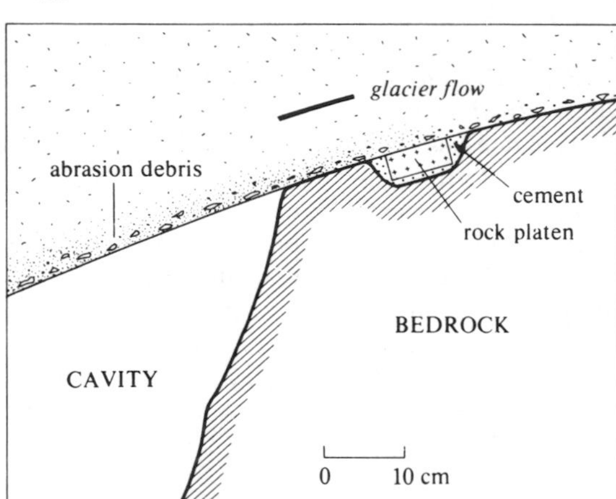

Figure 4.54 Diagram of a rock platen emplacement for *in situ* measurement of net glacial abrasion.

interest in addition to observations of till fabrics developed during deposition by different processes. The character, shape and lithology of surface clasts and moraines can directly provide information about the origins and derivation of detritus and its subsequent routing through the glacier. Quantity, distribution and clast characteristics of annual marginal sediment contributions to the glacier surface are revealed by field survey and sampling of material that lies on and immediately below the surface of snow in the accumulation area at the end of the summer, before burial by following snowfall. Rates of emergence of debris routed englacially can be obtained from debris clearance sites on medial moraines where they appear below the firn line (Rogerson & Eyles 1979). The existing debris revealed after a period of exposure to ablation is collected and weighed. The amount of ice that formerly contained the debris before melting is estimated from ablation stake measurements. Where distinctive rock types occur in a glacierised basin, especially if a lithological unit has no extraglacial outcrop, actual routes of transit and possible subglacial sources of debris can be inferred from surface clasts.

Observations of supraglacial melt-out and flowage processes have usually been descriptive rather than experimental. However, rates of movement of supraglacial tills can be determined from surveys of painted boulders or clasts, or from measurements of thicknesses of till against a network of rods fixed in underlying ice. 'Floating' markers of expanded polystyrene or similar material can be used to monitor flowage. Accumulation of till from the melting of buried ice may be estimated by the insertion of wooden dowels along the ice/till interface from a trench. After emplacement, the trench if refilled, to be re-excavated after the ablation season. The amount of till beneath the dowels is the melt-out contribution (Lawson 1979).

Surface movement and ablation studies use networks of 4 or 5 m long stakes, of aluminium alloy tubing (OD = 32 mm, wall thickness = 2 mm) or ramin wood (25×25 mm^2), firmly located in the holes in the ice. The holes can be bored using a hand-operated ice drill (OD = 35 mm). A robust drill can be constructed from a 1 m length of seamless hardened steel tube, with six sharpened teeth cut into the lower end (Fig. 4.55). Extension rods of aluminium, with brass or bronze couplings, connect the drill with a carpenter's brace (Østrem & Stanley 1969). Holes 4 m deep require 2–3 h drilling time with two operators applying pressure during rotation. Ice chip production and accumulation necessitates frequent removal of the drill from the borehole to maintain efficiency. Ablation lowering of the ice surface is measured relative to the top of the vertical stake, against a graduated pole. Correction should be made for tilting of the stake, and the stake redrilled when ablation has left a length of only 1–2 m buried in the ice.

The same stakes may also act as markers for surveys of glacier surface movement, if small sighting plates are fixed to them. Theodolite stations, cairns built from stones and quick-setting cement, topped with a metal plate, should be located on bedrock as close as possible to the ice margin. The preferred line of sight to the stakes is approximately perpendicular to the direction of movement and two targets fixed to bedrock should lie at the opposite side of the glacier. Velocities are calculated from angular measurements, with three readings of the angle, one with the reversed theodolite position, and the distance between stakes and theodolite, probably 100–250 m. Close to the frontal ice margin, the changing distance from a large lodged boulder or a bedrock marker to a stake in ice can be taped in weekly.

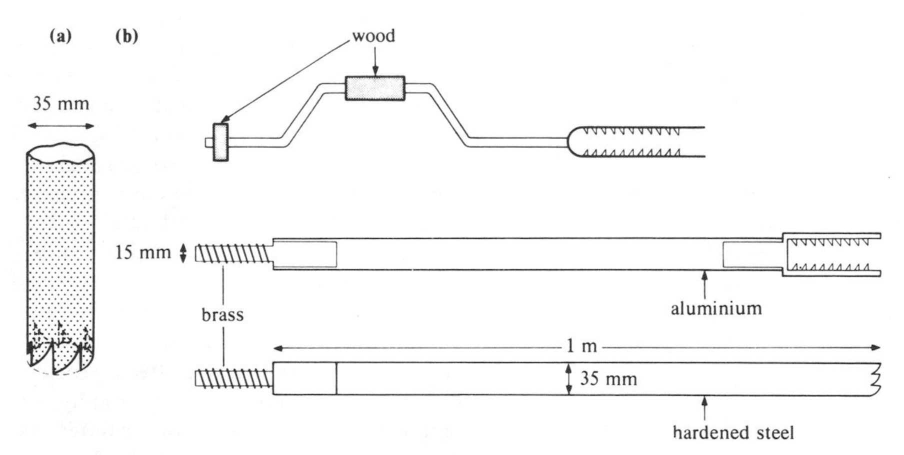

Figure 4.55 (a) Sharpened teeth cut into a seamless hardened steel tube form the cutting edges of a simple hand-operated ice drill. (b) Diagram of drill, extension rod and brace.

4.5.4.2 Ice-marginal observations

Direct observations of active processes such as flow and pushing of debris, and erosion and deposition of sediment, may be made at glacier margins, particularly where glacier recession has ceased to where glaciers are thickening and readvancing. Individual boulders at the ice margin may be identified and painted with (originally) vertical lines. Positions of labelled boulders are subsequently surveyed repeatedly from fixed points on bedrock, and the orientation of paint marks observed in order to ascertain modes of movement and rates of progression in the direction of flow from the glacier (Veyret 1971). Morphological changes, stratigraphic development and ice–sediment dynamics provide information about rates of bulldozing of and sources of sediment composing push moraines (Rogerson & Batterson 1982). Simple measurements of relative motion of ice and sediment might use small ablation stakes, pegs or ice screws. Measurements of debris flows depend on relative displacement of marked blocks and small annuli as described in Section 4.5.3. Sedimentological analyses of structure, fabric and clast dimensions in debris immediately on release from ice margins allow discrimination of which processes are responsible for which sedimentological properties of diamictons in the glacial environment (Lawson 1981). Estimates of quantities of sediment deposited and of rates of operation of ice-marginal and proglacial processes are elusive. Collection of debris sliding from ice surfaces, using polyethylene boxes (dimensions $0.8 \times 0.3 \times 0.3$ m^3) or thicker-gauge polyethylene sheeting spread out at the base of marginal ice slopes, has yet to produce reliable results.

Stratigraphic relationships between debris and ice can be interpreted from descriptions of sections of ice exposed at glacier margins and in crevasses. Englacial and basal debris content of ice can be assessed from samples collected in such sections, although front-marginal sites are probably not representative of conditions throughout the body of the glacier. Transfer pathways and rates of movement of debris through glaciers may also be inferred from observations in exposed sections of ice.

4.5.4.3 Proglacial observations

Subglacial processes affecting deformable sediments rather than rigid beds are not directly observable at depth and have been approached by using theoretical constructs. Deformation of subglacial sediments may be observed in immediate proglacial areas, for example in till 'tails' to roche moutonnées and in fluted moraines that are in the process of emerging from or have just been uncovered by retreating glacier ice (e.g. Boulton 1982). Detailed observation and mapping of the microtopography of recently deglaciated proglacial bedrock areas and of discontinuous deposits of calcite thereon can provide information concerning spatial distribution and geometry of former subglacial cavities and channels. Identification of areal extents of zones subjected to subglacial abrasion is also permitted. Such investigations are of considerable interest in glacier dynamics, hydroglaciology and sedimentation (e.g. Hallet & Anderson 1980).

Processes occurring during post-depositional changes of glacially produced features may be investigated from field descriptive observations (e.g. Price 1969) or from repeated surveys at intervals after glacier recession. A photogrammetric technique for determining rates of ice wastage, meltstream migration and morphological change has been described by Petrie and Price (1966). Special flights across a glacier front provided overlapping aerial photographs of a suitable scale (1:29 000), and photography was available from an earlier date. Ground control is difficult in proglacial terrain, particularly where national grid triangulation stations are absent. Well defined, clearly marked points linked in a terrestrial survey are essential. Petrie and Price (1966) used a Wild B-8 plotter for planimetric triangulation at 1:20 000 scale, plotting all morphological features and contouring at a 15 m interval, although an accuracy for spot heights was as high as 0.02% of the flying height. Retreat of the ice front, melting of buried ice and changes in altitude of esker systems between photography dates were determined for larger areas over a longer time period than would have been possible from field survey. Indirectly, possible processes of esker formation may be inferred from the photogrammetric measurements (Price 1966).

4.5.5 Meltwater characteristics as indicators of subglacial processes

Water has an important role in reworking and removing the products of subglacial erosion. As meltwaters pass through the basal hydrological system of a glacier, their quality characteristics are modified by the acquisition of dissolved and particulate load. Determination of the sedimentological and hydrochemical characteristics of glacial meltwater permits indirect investigation of processes at the ice/rock interface. Meltwaters have sampled basal conditions over those areas of the bed exposed to circulation of water and integrated with channel flow. Measurements of discharge and suspended sediment concentration, bedload transport and solute concentration of evacuation from a glacier can provide estimates of gross rates of denudation by glacial erosion (Østrem 1975) and chemical activity. Some of the water has also passed through the non-glacierised portion of the catchment, and the rate of solute and sediment production in ice-free sites should be evaluated. Studies of rates of process operation assume that the rates of sediment and solute production by glacial erosion and dissolution equal the rate of transport expressed per unit area of bed, that processes operate at a uniform rate over the entire catchment, and that meltwater has access to all zones of the catchment. All are probably unsatisfied.

Shifting beds, standing waves and excessive turbulence prevent reliable calibration of stage–discharge rating relationships at natural cross sections close to glacier portals, where meltwater samples should be collected to minimise extraglacial influences. Accuracy of gauging (see also Sec. 4.4.2) is critical, since errors in discharge measurements are incorporated on multiplication into transport estimates and hence to rates of erosion. Large gauging structures constructed in connection with hydroelectric power schemes are preferred to simple stilling wells on unstabilised channels (e.g. Church & Kellerhals 1970).

Use of rating curves because of incomplete close-interval sampling also increases error in estimates of sediment yield. Records from turbidity sensors produce continuous information, but calibration curves are often imprecise. Variations of sediment concentration again relate to only one point in the cross-sectional area. It is unlikely that suspended sediment yield can be estimated with any certainty from point samples of meltwater at one fixed position in a cross section. Varying water levels and changing bedforms and bank positions influencing turbulence around the sampler nozzle will determine how representative of the cross section is the point of sampling at a particular time.

However, the nature of subglacial sediment-transfer processes can be inferred from temporal variations of suspended sediment concentration in meltwaters (Collins 1979a). In such investigations, indications of relative temporal variations of concentration are more important than representative samples. Either continuous turbidity records (e.g. Humphrey *et al.* 1986) or frequent dip sampling sustained over extended periods can be used to establish patterns of fluctuation in sediment content of rivers. Pumping or automatic vacuum samples may be used to collect 0.7–0.9 l samples from a nozzle at an angle of 45° from the water surface upstream. Effects of the interaction of channels, conduits, basal sediment stores and larger-than-usual flows of meltwater through the

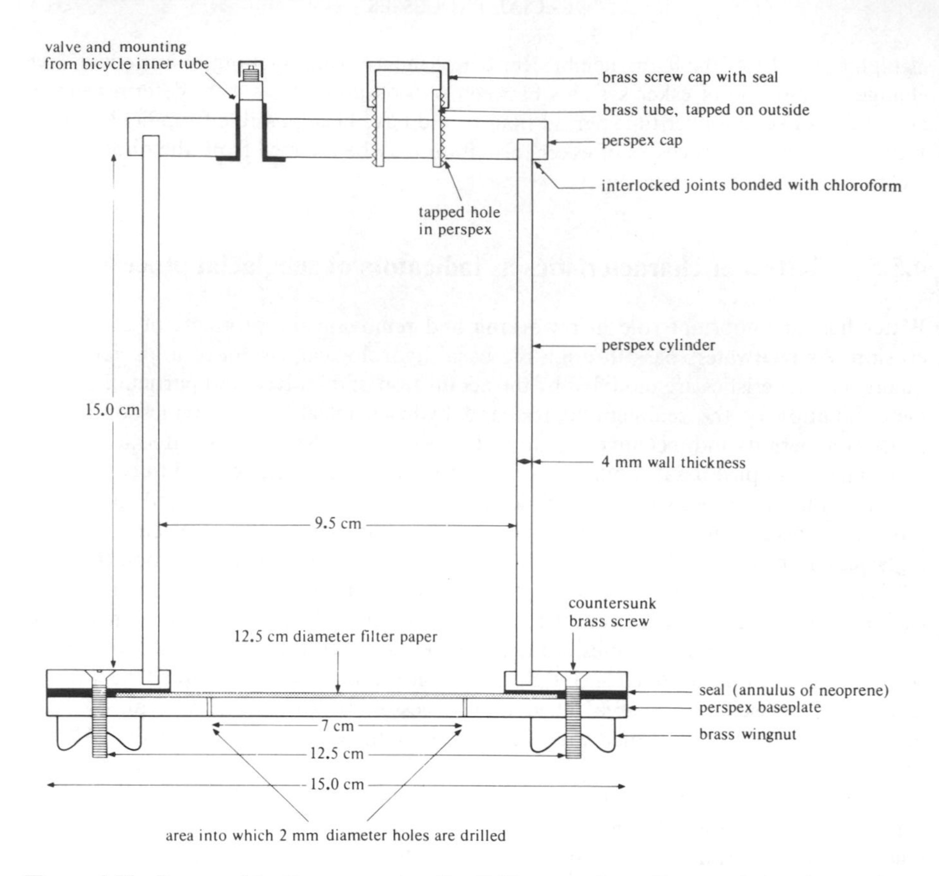

Figure 4.56 Pressure-filtration apparatus for field separation of suspended sediment from meltwater samples.

internal hydrological system can be examined from sediment concentration and discharge records. The volume of a sample of meltwater is determined in a measuring cylinder, and sediments are separated from water in the field by filtration through individually preweighed hardened cellulose filter papers, of initial penetration pore size 2 μm, in Perspex cylinders under pressure from a bicycle pump. The pressure filtration system consists of a basal Perspex plate, in the central area of which 2 mm diameter holes are drilled. The upper is a clear thick-walled Perspex cylinder of internal diameter (9.5 cm) less than that of a filter paper (12.5 cm), together with inlet and valve (Fig. 4.56). Samples on filter papers are returned to the laboratory in sealed polyethylene bags, before oven-drying at 105°C for 8 h, and the quantity of sediment determined gravimetrically; although Østrem and Stanley (1969) preferred to ash filter papers in porcelain crucibles, in a furnace at 600°C for 2 h, before weighing.

Larger debris than that in suspension is evacuated in meltstreams as bedload. Transport of bedload across the width of an irregular boulder-strewn surface may be measured using a 'fence' (Østrem 1975). A net 2.5 m wide with a mesh size of 20 mm woven from 4 mm steel wire is hung vertically across the meltstream and over surrounding banks on vertical sections of steel railway line (3 m long) spaced at 2 m intervals, supported by 25 mm diameter steel cables. The rails are welded to bolts drilled into bedrock and the net supports are similarly fixed. Volume of bedload accumulation is measured at the intersection of a 1 m grid, across the area upstream of the fence, using a dipstick against taut wire levels. The fence design must take into consideration the force, level and volume of discharge and the potential weight of bedload to be supported. The latter can be minimised in a small meltstream by use of a less robust net, though based on a solid supporting structure. Thinner, more flexible, wire is used in the mesh. Hanging from carabiners on the top wire of the fence support, to allow easy detachment, the net is folded at 90° where it meets the bed, and runs upstream for several metres along the bed, to which it is attached flush, using carabiners attached to bolts. At low flow, the net is periodically collected in by rolling from the upstream end, so moving bedload towards the fold, and lifted out of the river for emptying and weighing of the bedload. The net is repositioned for a further collection period, the maximum length of which is determined by rates of bed transport and strength of net. Sediment traps, constructed at intakes to hydroelectric adduction galleries, and periodically flushed out by diverting water and accumulated bedload away from adduction inlets towards the original channel, may provide some information on quantities of bed material moved and relationships between such and discharge (e.g. Beecroft 1983). Purging of bedload from traps may occur automatically or manually, at various levels of refilling of the sedimentation basin with varying packing of sediment. Hence it is unlikely that this type of observation offers an accurate assessment of bedload movement.

Studies of rates of chemical weathering in the glacial environment require sampling strategies that should provide sufficient samples in time to allow accurate identification of seasonal and diurnal variations of solute concentration in melt water. A continuous record of temporal variations of meltwater quality may be provided by monitoring electrical conductivity and collecting samples within the temporal framework so indicated (Collins 1979b). Immediately after collection, *c.* 500 ml aliquots of meltwater should be filtered (in all-polyethylene or plastic vessels to minimise contamination) through nylon membranes (0.45 μm or finer pore penetration size), previously washed in doubly distilled deionised water. Further solution from particles and cation exchange during storage are so prevented. Samples are stored in previously washed polyethylene bottles in dark, refrigerated conditions, or preferably kept deep-frozen (-18°C) before analysis (Sec. 4.5.6).

Relationships between electrical conductivity and cationic concentrations expressed as the sum of cationic equivalents (μEq l^{-1}) are empirically determined (Collins 1983). The sum of products of hourly means of cationic concentration and discharge provides estimates of cationic load and, adjusted for catchment area, of the gross annual cationic denudation rate (in mEq m^{-2} y^{-1}). Subtraction of the atmospheric input component leaves the net annual yield of cations.

An indication of subglacial flucial erosion is provided by direct measurements on trimmed rectangular blocks of rock ($50 \times 50 \times 25$ cm^3) inserted into a proglacial stream bed. Vivian (1975) provided results from blocks forming the artificial bed of a hydroelectric intake after five years of emplacement.

4.5.6 Chemical analysis of ice and meltwater

Collection, preservation, storage and analysis procedures for ice and meltwater attempt to ensure that the final analysis is an accurate determination of the actual field chemical characteristics at the time of sampling, which is made difficult by the presence of substantial quantities of particulate material in very dilute meltwaters. Measurements of pH and electrical conductivity and determinations by titrimetric and ion-selective electrode techniques are possible in the field either before or after filtration (for meltwaters) or after melting (for ice). In the laboratory, ice and meltwater samples, when thoroughly thawed, are analysed by usual instrumental techniques. Reynolds (1971) compared ion-selective electrode determinations of major cations in unfiltered meltwater with those obtained by atomic absorption spectrophotometry, and found significant interference from fine particles that relatively enhanced the results from atomic absorption analysis. However, immediate field filtration at 0.45 μm, following careful melting for ice, overcomes this difficulty and allows high reproducibility of determinations by atomic absorption. Where samples are extremely dilute, major cations may be concentrated above the instrument detection level using a cationic resin column of Dowex 50 or similar ion-exchange resin (Ricq de Bouard 1976). Further details of methods for studies of solutes are described in Sections 3.3 and 4.3.

4.5.7 Laboratory experimental simulation techniques

Several configurations of experimental apparatus have been designed to measure the sliding, friction and abrasional wear of ice on different rocks at various sliding speeds and temperatures, and under different loads (Lister *et al.* 1968, Barnes *et al.* 1971, Hope *et al.* 1972, Budd *et al.* 1979). Experimental simulation should attempt to reproduce ice velocities of $1\text{–}10^3$ m yr^{-1}, normal stresses on the bed of $(5\text{–}50) \times 10^5$ N m^{-2} at temperatures both close to the pressure-melting point and for cold ice. The effects of clean ice, debris-laden ice and rock each sliding over rock, are of interest in the modelling of real ice/rock interfaces.

A simple arrangement consists of a cylindrical block of rock up to 10 cm in diameter, fixed to and rotated on a geared turntable, supporting an immobile ice block loaded with weights (Fig. 4.57), fitting inside a temperature-controlled refrigerator. The ice is prevented from rotation by a spring balance used to measure the torque produced by the sliding face friction (Hope *et al.* 1972). The rock surface is polished before the experiment and wear is measured by a dial micrometer set at different positions on different radii of the rock surface. Debris can be added to the ice during its preparation. Debris may become attached to clean ice during the experiment. Microscopic investigation of the abraded surface and of samples of the products of wear is possible.

Rabinowicz (1965) suggested the use of a localised point of contact for wear experiments so that sliding takes place over a narrow track, at uniform rotational torque with the possibility of a helicoidal path of sliding. A rock pin or button is held, with a strain gauge to permit continuous monitoring of friction, to slide against a rotating ice cylinder or rock cylinder (Riley 1979). Distance of sliding, load velocity and temperature are taken as dependent. Wear is determined by weight loss of the rock stud used, after cleaning off detached material with an organic solvent.

4.5.8 Glacio-lacustrine, glaci-estuarine and glacio-marine sedimentation

Direct observations of processes of sedimentation in ice-dammed and proglacial lakes, in estuaries and fjords and under the sea around tidewater glaciers and ice shelves can be obtained using limnographical and oceanographical techniques. Investigations of such processes also rely on inferences from the interpretation of recently deposited bottom sediments. Frequently repeated measurements of velocities of currents, water density, actual sediment concentration in water or a photo-optical surrogate and of water temperature are necesary in columns of water, which should be spatially well distributed in lakes. From boats, lead-line soundings and acoustic depth-profile techniques are used

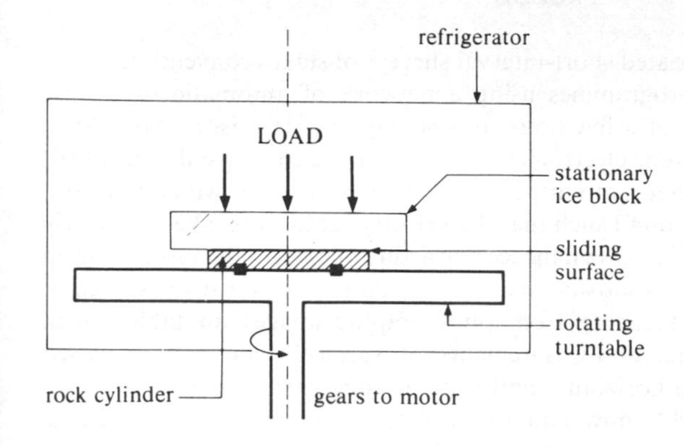

Figure 4.57 Schematic diagram of a simple apparatus for simulation of glacial erosion processes. An ice block rests on a rotating cylinder of rock in a temperature-controlled refrigerator.

for bathymetric surveys. Samples of water from various depths in vertical columns of lake waters can be collected using van Dorn bottles or other types of samplers (e.g. Golterman *et al.* 1978). Thermistor strings suspended beneath floating platforms provide temperature profiles. Photo-optical measurements of turbidity and transmissivity, together with gravimetric determinations on samples, are used to indicate paths of suspended sediment transport (Smith *et al.* 1982); and in lakes, sedimentation rates can be measured directly using traps placed on the bed. Weighted polyethylene containers, with cross-sectional areas of about 0.01 m^2 and locations marked by tethered buoys, are retrieved periodically according to sedimentation rate, providing an indication (in $kg\,m^{-2}\,day^{-1}$) of patterns of sediment deposition. Where velocities are sufficient to initiate rotation of propellers, current movement can be assessed by continuously recording velocity meters, equipped also with direction indicators. Temporal variations in recent sedimentation rates are revealed by analyses of cores up to about 0.5 m in length obtained using gravity corers, or ascertained where beds of ice-dammed lakes subject to periodic draining can occasionally be inspected directly. Similar measurements are possible in fjords, where measurements of salinity and tidal range and current are also necessary for the interpretation of patterns of movement of water and sediment (Elverhoi *et al.* 1983, Gilbert 1982). Shallow-coring, grab-sampling and acoustic characteristics permit inferences concerning facies assocations and stratigraphic units. Measurements of freezing or melting beneath ice shelves are lacking but techniques similar to those described for lakes have had limited use in water columns adjacent to and beneath ice shelves (Carter *et al.* 1981).

4.5.9 Integrated investigations of glacier sliding velocity, subglacial hydrology and meltwater characteristics in streams emerging from glacier portals

Coordinated simultaneous and continuous observations at various transects across glacier surfaces and at portals from which meltwaters emerge are necessary in order to provide information concerning event-based hydrological phenomena and relationships between resulting hydraulic conditions at the glacier bed and glacier motion. Measurements collected during such events are not only of intrinsic interest but also provide data for the testing of hypotheses concerning the seasonal evolution of subglacial drainage systems, the influence of hydrological events on erosion and sediment evacuation, the initiation of surging, and geomorphological effects of surges (e.g. Humphrey *et al.* 1986, Kamb *et al.* 1985, Hooke *et al.* 1985, Iken & Bindschadler 1986).

Surface velocity should be recorded at several transects in order to detect spatial patterns of velocity change. Repeated short-interval surveys of stakes conventionally, or more continuous monitoring programmes using a network of automatic time-lapse cameras, with a time resolution of a few hours to one day (e.g. Harrison *et al.* 1986), provide useful data. Motor-driven, electronically triggered (quartz-crystal-controlled) 35 mm cameras are usually used, mounted in weather-protecting window boxes, mounted about 2 m above the ground such that the velocity vectors of stakes are nearly normal to the direction of view. Targets on the ice consisting of black balls subtending an arc of 35 seconds have been found satisfactory for 300 mm lenses over distances up to 2 km. Data are retrieved by digitising from photographic images on tablets with 0.025 mm resolution. Short-interval measurements of surface velocity have raised questions of surveying errors in horizontal and vertical angles that might result from differential refraction over rock, snow and ice surfaces (Andreason 1985). Seismic activity characterised by high-frequency signals (10^2 Hz) is associated with short-period motion events, and can be monitored using micro-seismometers (Harrison *et al.* 1986). Continuous measurements of discharge and of turbidity or sediment concentration in meltwaters at the portal, and of precipitation over the basin, should accompany the measurements of surface movement. Water pressure is also an important variable, and should be monitored in boreholes.

4.6
Aeolian processes

4.6.1 Airflow characteristics

Recent years have seen a marked revival of interest in aeolian geomorphology, both on Earth (e.g. McKee 1979, Nickling 1986, Brookfield & Ahlbrandt 1983) and on the planets (Greeley & Iversen 1985). To evaluate the role of wind as a geomorphological agent, a knowledge of wind structure on both the synoptic and micro-meteorological scales is necessary

4.6.1.1 Routine meteorological observations

Information on airflow characteristics can come from meteorological records (see Morales 1979). The greater the length of the record, the more likely it is to register rare extreme events, but Striem and Tadmor (1974) suggest that a three-year record reduces the error in estimating the annual frequency distribution to two-thirds and a four-year-record is near the limit of practical improvement

Availability of representative wind data Since windspeed varies markedly with height, it is critical to know the level at which an anemometer has been exposed. Although attempts have been made to standardise anemometer heights at 10 m above ground, there has been a failure in some countries to apply international standards, and this has meant that many observations are made at other heights, so that observations of the wind are not always comparable. The effects of elevation on wind measurement must be corrected to a standard height (Geiger 1965, p. 117).

The timebase for wind data is important to its interpretation, and incorrect conclusions could be drawn in a comparison of stations that used data recorded with different timebase periods. Conclusions based on short records of windspeeds are at best tentative, and can be fallacious. Weighting factors have to be applied to twice-daily observations to make them comparable with five-times-daily observations (Brookfield 1970), while procedural and observational bias has also to be corrected (e.g. Wallington 1968).

Diagrammatic representation of data Two techniques may be found useful: hodograph vector diagrams give a very clear picture of wind regime (Petrov 1976, Warren 1972), while wind-roses may be useful in particular circumstances. Standard climatic tables frequently do not provide the information that is required by the geomorphologist. Where wind directions are variable, frequency studies are the only way to study the regime. The wind-rose provides a graphical indication of the frequency of different ranges of windspeeds with different directions.

4.6.1.2 Non-routine wind observations: introduction

When wind data from meteorological stations are lacking or of poor quality, when interpolations and extrapolations are to be checked, or when wind data are required for heights and locations that are not standard, reliable velocity-measuring instruments are required that are capable of recording local values of mean velocity and the properties of

Table 4.11 Mean velocity anemometers manufactured by Casella & Co. Ltd.

Type of instrument	*Specifications*						*Approximate cost* (£, 1988)
	Principle of operation	*Size* (*cm*)	*Weight* (*kg*)	*Output*	*Range* ($m\ s^{-1}$)	*Accuracy* (%)	
Cup counter anemometer	3 conical cups, 12.7 cm diameter	30.5 high × 23 radius	3.6	Mechanical counter	2–35	±3.1	330
Cup generator anemometer	3 conical cups, 12.7 cm diameter	26.5 high × 22.8 radius	4.5	Dial indicators	2.5–45	±3.5	825
Combined direction and velocity indicator	3 conical cups, 12.7 cm diameter and British Met. Office wind vane	84 high × 51 radius	16.5	Dial indicators	2.5–45	±3.5	2400
Electrical recorders for cup generator anemometer and combined indicator	Transmitter generates AC voltage proportional to windspeed – recorder consists of rectifier and DC voltmeter	*Velocity only*: 30.5 × 21.3 × 24.8	15	Graphical output	—	—	–
		Combined: 35.5 × 33 × 25.4	20	Graphical output	—	—	2500
Portable air meter	Rotation of 8 aluminium vanes about horizontal axis	10 × 10 × 10	0.6	Dial indicators	0.8–16.7	±2.0	–
Hand anemometer	3 metal cups, vertical shaft	15 × 15 × 12	1.15	Dial indicators	0–30	±2.7	–

Table 4.12 Turbulence sensors.

Type of instrument	Manufacturer	Specifications							Approximate cost (£, 1975)
		Principle of operation	Size (cm)	Weight (kg)	Output	Range ($m\ s^{+1}$)	Response ($m\ s^{-1}$)	Accuracy (%)	
Hot-wire anemometer	DISA Elektronik A/S, Denmark	Cooling of heated wire	5 μm diam. × 1 mm long	—	Bridge voltage; voltmeter range 0–10 V; potentiometric recorders	1–150	up to 1000 Hz	—	Probes from £20 up to £100; recording unit > £1000
Three-dimensional anemometer	R. M. Young Co., Michigan, USA	3 propellers measure x, y, z wind components	107 high	3.6	Analogue voltages – 3 bipolar signals from DC tachometer generators; 3-channel recorder or data logger; (−500)–0–500 μA, millivolt or 5 mA galvanometer (3-channel)	—	Distance constant 0.93 m (approx. 2 Hz)	± 1.0; cosine conformity ± 10%	350
Bivane anemometer	R. M. Young Co., Michigan, USA	Rotation of propeller and displacement of vane (Gill)	Height 76; vane length 84; propeller 23	2.7	Analogue voltage proportional to vane angle and proportional to windspeed; 0–1 mA (100 ohm) meter or millivolt or galvanometer type (0–100 mV) or 0–1 mA and 0–5 mA galvanometer	0–32 m s^{-1}; 0–360° azimuth; 50°–0°–50° elevation	0.93 m constant and delay distance (approx. 2 Hz)	± 1.0	350
Hot-wire anemometer with thermometer	Wallace Oy, Finland	Hand-held meter and probe; cooling of heated probe	23 × 10 × 6	1.1	Bridge voltage; hand-held meter or external recorder	2–30	50% response in 2 s	Temperature 0.8°C; Wind 7.0%	150

the turbulent velocity fluctuations. The methods available for recording wind observations may be divided into two groups, anemometry and flow visualisation. The best approach in both field and laboratory is to use both methods to contribute their individual strengths while complementing each other for their inherent weaknesses.

4.6.1.3 Anemometry

A wide variety of instruments is available for measuring windspeed and direction (see Tables 4.11 & 4.12). The choice of a particular instrument depends upon whether it is required to measure mean velocity or turbulent fluctuations about that mean; the dynamic response characteristics of a sensor suitable for registering turbulent fluctuations are quite different from those of sensors for recording mean velocity. Response characteristics should be checked to determine the range of operation of an instrument and any systematic recording errors, so that it is possible to infer the true data from the instrument output. McCready (1965) has outlined a method for examining the response characteristics of anemometers and wind-vanes. It is essential that such an examination be carried out before performing any experiment. The number of parameters needed to give merely a rough description of the atmosphere as a fluid-mechanical system is very large, so that a large number of quantities must be measured simultaneously. Any atmospheric experiment therefore needs many and diverse measuring systems. Since these systems must usually be operated by a limited number of people, operational reliability and simplicity become factors of the utmost importance.

Mean-velocity sensors *Pitot-static tubes* A typical pitot-static tube consists of a 40 mm × 3 mm outside-diameter tube with a 1.55 mm diameter stagnation hole to record the dynamic pressure and five 0.5 mm holes bored 10 mm back from the stagnation hole to record the static pressure; many other instrument designs have been used (e.g. Bagnold 1941, *inter alia*). The instrument is suitable for measuring mean velocities down to 0.15 m s^{-1}. Its main disadvantage is the length of time it needs to stabilise. Where strongly curved streamlines are present or where regions of high turbulence exist, measurements with pitot-static tubes are subject to serious errors. The main disadvantage of these tubes is that they can very easily become blocked with sand in experiments using moving sand. They are also unsuitable in gusty winds. Their main use is in wind-tunnel experiments when no sand is being used.
Vertical-axis rotating-cup anemometers These are more commonly used and have three or four cups mounted on arms free to rotate about a vertical axis. The vertical shaft rotates at a rate proportional to the windspeed, and this rotation is measured in various ways. The convex sides of the cups provide some wind resistance so that the instrument has a poor response to sudden wind changes. The weight of the cup assembly results in high starting speeds and the inertial effects of overshoot. A hysteresis effect exists, the response being quicker to an increase of windspeed than to a decrease (Kaganov & Yaglom 1976). Mean overspeeding of 15% can result from the high inertia in some instruments. Because calibration is usually performed in the steady airflow of a wind tunnel, whereas in the field the wind is turbulent, the mean value of the rotating anemometer reading in a variable wind will be higher than the true mean value of the windspeed.

Very sound reasons can be put forward for using rotating anemometers: they have an essentially linear response between rate of turning and windspeed; their calibration is unaffected by temperature, pressure, density or humidity changes of the atmosphere; and they are reliable for years if reasonably maintained. Cup anemometers exposed with a logarithmic vertical spacing provide the best means of vertical velocity profile measurement. The cheapest and most reliable instrument for this purpose is the Casella pulse anemometer (Sheppard type).

The portable air meter and the hand anemometer are robust, and suitable for checking the exposure and operation of anemometers in fixed locations. The portable air meter can be attached to rods for velocity measurements at various heights above the ground, while the hand anemometer gives direct readings of windspeed when held in the hand at arm's length.

Turbulence sensors The characterisation of turbulence is based on the measurement of the three-dimensional fluctuations or eddying motions of the air and their spectral distribution. Measurements of this type are seldom part of the routine observations made at meteorological stations. There are a number of requirements that must be satisfied by the detecting element if turbulence is to be measured reliably:

(a) The element must be small, so that it does not seriously affect the flow pattern.
(b) The instantaneous velocity distribution must be uniform in the region occupied by the element, so that the detecting element should be smaller than the dimensions of the micro-scale turbulence.
(c) The inertia of the instrument must be low, so that response to even the most rapid fluctuations is practically instantaneous.
(d) The instrument must be sufficiently sensitive to record small differences in fluctuations of only a few per cent of the mean value.
(e) The instrument must be stable, the calibration parameters remaining reasonably constant.
(f) The instrument must be strong and sufficiently rigid to exclude vibrations and motions caused by turbulent flows.

Table 4.12 lists some of the instruments that are used to measure turbulent fluctuations in the atmospheric boundary layer. Hot-film anemometers are not generally recommended for use when sand is in transport. They damage more readily than those shown in Table 4.12.

Precision cup anemometers Only if the rotating anemometer is designed out of lightweight materials with a very low resistance to internal friction will it be responsive to small-scale wind eddies. Starting speed with very light cups can be as low as 0.05 m s^{-1}. In the bridled anemometer, a spring prevents the cups (usually 32) from rotating. Calibration of the instrument allows the force required to keep the anemometer from rotating to be translated into windspeed. The costs of cup anemometers are low, servicing is easy, and the instruments are robust enough to avoid buffeting by the wind, but the response characteristics are usually not good enough to measure anything but the largest eddies of the turbulent air flow.

The Dines pressure-tube anemometer This instrument suffers from several drawbacks. The velocity traces plotted on charts (anemographs) are seldom thin, clear and 'lively', i.e. showing all the details of the minor changes of velocity. The pen requires regular cleaning or renewing to avoid a very thick trace, and excessive pen friction causes a 'stepped' trace, in which the changes of velocity appear to occur in a series of abrupt jumps. Short-period variations of velocity cause excessive broadening of the trace. Regular maintenance and lubrication is necessary. The instrument is difficult to set up, especially when the remote-sensing model is being used. The response of the Dines instrument is about the same as that of the precision cup anemometer: it provides accurate means over periods of 5 s if used with a chart with a sufficiently open timescale.

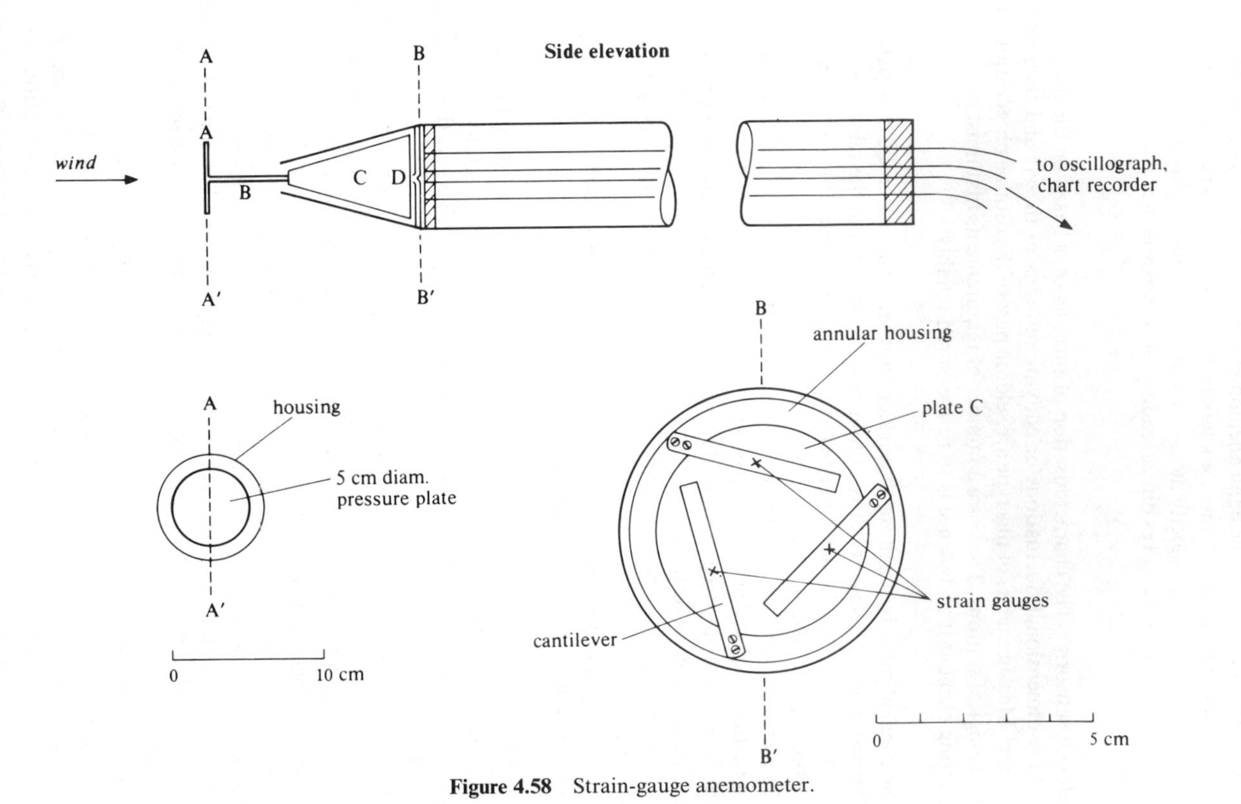

Figure 4.58 Strain-gauge anemometer.

Strong gusts having durations of about 3 s or more are reliably recorded but shorter-period gusts are not fully registered.

Strain-gauge anemometers Chepil and Siddoway (1959) described an apparatus composed of a strain-gauge assembly, electrical amplifier and recording oscillograph for measuring the characteristics of wind turbulence with velocity-fluctuation frequencies up to 120Hz. The anemometer used by the principal author was based on a design developed by Butterfield and his colleagues (personal communication) (Fig. 4.58). It consists of a small circular pressure plate A of tempered aluminium 5 cm in diameter and 0.5 mm thick, mounted on the end of a thin shaft B of tempered steel 1.0 mm in diameter. The other end of the shaft is attached to plate C housed in an annular frame. Plate C is supported within the frame by three steel cantilever strips, which are flexible in a direction normal to the pressure plate but are rigid transversely so as to keep the plate concentric in the frame. The cantilevers are fixed rigidly to the frame of the gauge and have a universal pivot connection to the pressure plate. The load on the pressure plate is measured by recording the strain in each of the cantilever strips by means of two foil-type resistance strain gauges. As changes occur in the resistance value of the strain gauges, they are recorded in direct proportion to the change. The anemometers must, of course, be calibrated before use. Although these instruments do suffer from amplifier drift in harsh environments, they can be reliable over quite extended periods.

Hot-element anemometry The hot-wire anemometer operates on the principles that the electrical resistance of a wire is a function of its temperature, that this temperature depends on the heat transfer to the surroundings, and that the heat transfer is a function of the velocity past the wire. The sensor has the advantage of small physical size and fast response, particularly when operated in the constant-temperature mode, and is sensitive over a wide range of velocities. Because of their small size, hot wires can be used where velocity gradients are large by arranging the probes in an array close to one another. The main disadvantages include the non-linear calibration, and the calibration drift resulting from air temperature changes. A 1°C temperature change in air may produce an error of 2% in measured velocity. These instruments are best used to sense in three mutually perpendicular directions, precision potentiometers being used to read out the real wind (or gust) direction. These instruments need constant checking and recalibration. They are easily broken by saltating sand and costly to repair. Their use in the field is therefore restricted.

Triaxial anemometer This instrument measures the components of turbulent fluctuations in an orthogonal system of coordinates with lightweight helicoidal propellers mounted at the ends of three orthogonal shafts. The polarity of the signal reverses when the wind direction reverses. Their low threshold speed and fast response, and the suitability of the signal for computer data reduction, make these the most practicable instruments for turbulence measurements in the field.

Bivane anemometer This is almost as suitable for measuring turbulence in the field as the triaxial anemometer. It uses the same four-bladed propeller-type plastic windmill (so that the response of the windspeed sensor is the same) (Sethuraman & Brown 1976). The advantage of the bivane is that the propeller is mounted on a Gill vane, which can rotate 360° in the horizontal and ± 50° in the vertical, so that it is always directed into the wind, and both direction and speed can be monitored from a single sensor mounting. McCready (1965) discusses the use and limitations of these instruments. They require a regulated DC power supply or a regulated battery supply for the vane potentiometers; the windspeed signal is self-generating by miniature tachometer generators. One of the few disadvantages of the bivane is that the vane is liable to unbalance during periods of dew formation or precipitation. The bivane is easy to use, rugged, has a low threshold speed (< 0.1 m s^{-1}), a relatively fast response with little overshoot, and the linear calibrations of windspeed and direction do not drift.

Table 4.13 Directional vanes.

Type of instrument	Manufacturer	Specifications							Approximate cost (£, 1988)
		Principle of operation	*Size (cm)*	*Weight (kg)*	*Output*	*Range ($m\ s^{-1}$)*	*Response*	*Accuracy (%)*	
British Meteorological Office wind vane	C. F. Cassella & Co. Ltd.	Horizontal arm with rectangular fin and counterweight	51 × 51	4.5	Direct observation of direction arms or direct-reading dial	–	–	–	55
Bivane (bidirectional wind-vane)	R. M. Young Co., Michigan, USA	Gill vane (as used in bivane anemometer)	76 high; 70 length of vane	2.7	Analogue voltage from linear potentiometers; 0–1 mA (100 ohms) meter or millivolt or galvo (0–100 mV or 0.1 mA) recorder	0°–360° azimuth (5% dead-band) ± 50° elevation	Delay distance 0.93 m (approx 2 Hz)	± 1.0	–

Directional vanes The most successful vane used by the principal author for measuring mean wind direction, where direct visual observations of direction arms was to be made, was made of expanded polystyrene, with dimensions $0.7 \times 10 \times 20\ cm^3$, the response being better than the British Meteorological Office vane and the cost being very low (see Table 4.13). The bivane (Gill vane) is one of the very few vanes suitable for recording turbulent fluctuations of wind direction.

Exposure of instruments The siting of wind-vanes and anemometers is very important if observations made at different stations are to be comparable. If the site is sheltered, the measurements of the windspeed may be unreliable, and eddies created by obstacles may subject the measurements of direction to a large error. The distance from the instrument to any obstruction must be at least ten times the height of the obstruction.

When there are surrounding obstacles disturbing the flow of air, it is necessary to increase the height of exposure above the 10 m standard level in order to obtain an exposure that is virtually clear of the disturbances. Each anemometer is allocated an 'effective height', defined as the height over open, level terrain in the vicinity which, it is estimated, would have the same mean windspeeds as those actually recorded by the anemometer. This effective height should be 10 m above ground level wherever possible if comparisons with routine meteorological data are to be made.

Towers distort the windflow and produce errors in wind velocity measurements (Moses & Daubek 1961, Rider & Armendariz 1966). The errors can be minimised by using two sets of equipment mounted on long horizontal booms on opposite sides of the tower to reduce the effects of leeward wind shadows and acceleration of windflow around the tower. Only the instruments located outside the wake of the tower should be used. Towers are the most reliable way of exposing anemometers for the measurement of velocity profile characteristics close to the surface (logarithmic or power-law profiles, roughness length, shear velocity, etc.) and are usually used for data collection at standard height.

Captive kite balloons can be used when measurements are to be made through greater depths of the atmospheric boundary layer. The principal author used a vane free to rotate about the flying cable of a captive ballon to keep the probes facing into the wind. Altitudes were calculated by multiplying the length of cable paid out by the sine of the angle of elevation to the balloon. Neglecting the bow in the cable introduces less than 2% error under average conditions (Readings & Butler 1972). Alternatively, a double balloon theodolite method can be used. Problems of balloon movements being transmitted to instrumental packages attached to the cable are minimised by flying the probes as far below the balloon as possible, and are controlled for by checking the movements of the probes with theodolite observations.

4.6.1.4 Flow visualisation

Flow visualisation gives qualitative or quantitative Lagrangian information about the wind velocity field. Use is made of a tracer or other indicator introduced into the flow to make the flow pattern visible for photography, or observable by some other suitable detecting apparataus.

The techniques that can be used for visualisation close to the surface include smoke candles or smoke bombs generating a dense, white smoke that shows up particularly well on photographs (Bagnold 1941, Sharp 1966, Tsoar 1978, Livingstone 1986), and neutrally buoyant helium-filled bubbles and balloons. Flow visualisation of upper winds can be achieved with balloons that either rise at a uniform speed (pilot balloons) or float along a surface of constant density (tetroons). The method is fast, reliable and inexpensive when supplies of balloon gas are available. Difficulties of storage of gas and the danger of transporting filled balloons over long distances limits its use to accessible areas (see Knott 1979).

There is evidence that the ordinary pilot balloon does not follow the actual windflow and therefore gives only a semi-Lagrangian measurement (Rider & Armendariz 1966). Aerodynamically induced horizontal motions associated with the balloons may produce a significant error in wind velocities derived from balloon tracking (Scoggins 1964, 1965). The wind data measured by this system are usually so smoothed that it is not possible to measure small-scale variations. The balloon system has a poor response to the rapidly changing wind field, and measures only a small percentage of the total amplitude for wavelengths shorter than 50 cm. A 100% response is only achieved for wavelengths in excess of 2500 m (Scoggins 1965).

The tetroon, a super-pressurised constant-volume balloon (usually inflated with an air–helium mixture), tends to float along a constant-density surface, but is easily displaced from that surface by vertical air motions. The tetroons are towed to flight altitude by a pilot balloon and then tracked by theodolite (see Knott 1979).

If the positions of the tetroon are known with sufficient accuracy, the mean velocity during the period between successive positions can be determined. Similarly, if the velocities are determined with sufficient accuracy, the average acceleration can be evaluated. Hoecker (1975) has developed standardised procedures for this process, and Booker and Cooper (1965) have discussed the theory of tetroon movement.

Measurement of upper winds by balloons depends upon observations of the positions of the balloon at successive intervals of time by photo-triangulation, and tracking with theodolites or radar. The angular position of the balloon is determined by measurement of azimuth and elevation. The photo-triangulation method has many disadvantages.

When tracking radar is used to record the positions of balloons, transponders eliminate the problem of ground clutter when the tetroons are 'skin tracked'. Operation and maintenance costs for radio and radar theodolites are usually too high for general use.

Detailed accounts of manual theodolite methods, including the trigonometrical computation, are summarised in a Meteorological Office handbook (Meteorological Office 1968). They include single- and double-theodolite methods.

4.6.2 Wind erosion

The extent of removal of loose or loosened material by the wind has been empirically expressed in the 'wind erosion equation':

$$E = f(I, K, C, L, V) \tag{4.30}$$

where E is the erosion in tons per acre per annum, I is the soil and knoll erodibility index (a measure of the soil aggregation), K is the soil roughness factor, C is the local wind erosion climatic factor, L is field length (or equivalent) in the prevailing wind direction and V is a measure of vegetative cover (Woodruff & Siddoway 1965).

4.6.2.1 Rates of movement derived from meteorological measurements

The volume of loose material that can be moved by a wind has been calculated in a number of different formulae (see Cooke & Warren 1973, pp. 258–66). It has usually been found that the amount of material Q (usually in tonnes per metre width per year) is proportional to some property of the wind velocity to the third power. However, for low shear (which might move some dust), Gillette (1979) found that the relation was improved if the square of the drag velocity was used.

A useful empirical formula is that of Hsu (1973):

$$Q = (4.97D - 0.47)\left(\frac{0.4(\bar{V}_z - 275)}{\ln[z(gD)^{1/2}]}\right)^3 \tag{4.31}$$

where D is the mean grain diameter (mm), $\bar{V}_z$ is the hourly, averaged, annual velocity from one direction, z is the height of wind measurement (usually 10 m) and g is the acceleration due to gravity.

Fryberger (1978), using WMO 'N-summaries' (held in most major meteorological centres), derived a formula from Lettau's relationship (see Cooke & Warren 1973, p. 265) to compute relative, potential amounts of sand movement in a worldwide study:

$$Q \propto V^2 (V - 12)t \tag{4.32}$$

where V is the wind velocity (knots, at 10 m), t is the percentage of time during which the wind blew from that direction, and 12 is the assumed threshold velocity for sand in the 0.25–0.30 mm diameter range. The use of this formula assumes a relatively flat, dry, sand-covered plain with little or no vegetation. Weighting factors are required (Fryberger 1978).

The values for Q from different compass points can be used to compute and plot 'sand movement roses' whose arms are proportional to the amount of sand moving from the various directions. Annual resultants can be derived from these diagrams either geometrically (Warren 1972) or with Pythagoras's theorem. For special studies, constancy and hourly means can be calculated (Tsoar 1978).

Skidmore (1965) and Skidmore and Woodruff (1968) assessed the direction and relative magnitude of wind erosion on agricultural fields with 'force vectors' r_j for each of 16 compass directions ($j = 0 - 15$):

$$r_j = \sum_{i=1}^{n} \overline{(V - V_t)}^3_{ij} f_{ij} \tag{4.33}$$

where $\overline{(V - V_t)}$ is the mean windspeed above the local threshold for movement (V_t) within the ith speed group (from conventional wind statistics) and f_i is the percentage of total observations that occur in the ith speed group in the direction under consideration: j refers to the direction. The sum of the vector magnitude gives the total magnitude of wind erosion:

$$F_T = \sum_{j=0}^{15} \sum_{i=1}^{n} \overline{(V - V_t)}^3_{ij} f_{ij} \tag{4.34}$$

and the 'relative' erosion vector is given by

$$r'_j = \frac{\sum_{i=1}^{n} \overline{(V - V_t)}^3_{ij} f_{ij}}{\sum_{j=0}^{15} \sum_{i=1}^{n} \overline{(V - V_t)}^3_{ij} f_{ij}} \tag{4.35}$$

such that

$$\sum_{j=0}^{15} r'j = 1$$

This can be used to show seasonal changes as well as annual patterns.

4.6.2.2 *Erodibility of soils*

The erodibility of a surface is a function of a number of measurable factors.

Vegetation cover Plant cover is the most important control on wind erosion. An early study of the role of plant protection was made by Chepil and Woodruff (1963). Bressolier and Thomas (1977) used classificatory and principal-components methods to compare the heights, densities and species composition of different vegetation types with near-ground velocities. They concluded that height is neither the only, nor even the main, factor in controlling the roughness of vegetated surfaces and so near-ground wind velocities and in turn sand movement. Vegetation density was more important than height.

Soil characteristics On bare ground, soil characteristics control the rate of wind erosion. The most important of these are:

(*a*) the size, shape and density of particles, determined by mechanical analysis (Chepil & Woodruff 1963);
(*b*) the mechanical stability of structural aggregates (Chepil 1952); and
(*c*) the soil moisture content (Woodruff & Siddoway 1965).

Chepil and Woodruff (1957) gave a formula for the total soil movement as affected by soil characteristics:

$$X = 400\frac{I}{(RV)^{1.26}} \tag{4.36}$$

where X is the total soil mass moved by a wind tunnel that produced winds of a given U_* (78 cms $^{-1}$), R is ridge roughness or the average depth of furrows (inches), V is vegetative residue (lb acre^{-1}) and I is an erodibility index based on percentage of soil mass in aggregates greater than 0.84 mm. Gillette (1979) modified the formula in the following way:

$$Q = 0.16\frac{LX}{16\,500}\frac{U_*^3}{78} \quad \text{for} \quad \frac{LX}{16\,500} < 1 \tag{4.37}$$

$$Q = 0.16\frac{U_*^3}{78} \quad \text{for} \quad \frac{LX}{16\,500} > 1 \tag{4.38}$$

were Q (g cm^{-1}s^{-1}) is the total soil flux through a surface of unit width that is mutually perpendicular to the wind and to the surface, L (ft) is the length of eroding soil parallel to the wind, and U_* (cm s^{-1}) is the friction velocity corrected for moisture content in the following way:

$$\rho_s U_*^2 = (\rho U_*^2)_{\text{observed}} - 6(w/w')^2 \tag{4.39}$$

where ρ_s is the particle density and w/w' is the 'equivalent moisture'. Here, w is the percentage moisture of the soil in question and w' is the percentage that the same soil holds at 15 atm (see Woodruff & Siddoway 1965). Belly (1962) also examined the effect of moisture on wind erodibility.

The relative erodibility of soil samples can be determined in the wind tunnel by exposing trays of soil (Chepil & Milne 1939). However, the short length of the exposed area of soil trays restricts or prevents the abrasion by the impact of saltating particles that commonly occurs in the field. The quantities of soil erodible in the wind tunnel may therefore be expected to be substantially lower than the quantities in the open field. Condon and Stannard (1957) having classified soils according to their wind erodibility.

The most accurate field method measures the wind and soil from experimental plots. The experimental plot has the advantage that the environmental conditions can be controlled to a large extent, for example, by using a portable wind tunnel (Zingg 1953),

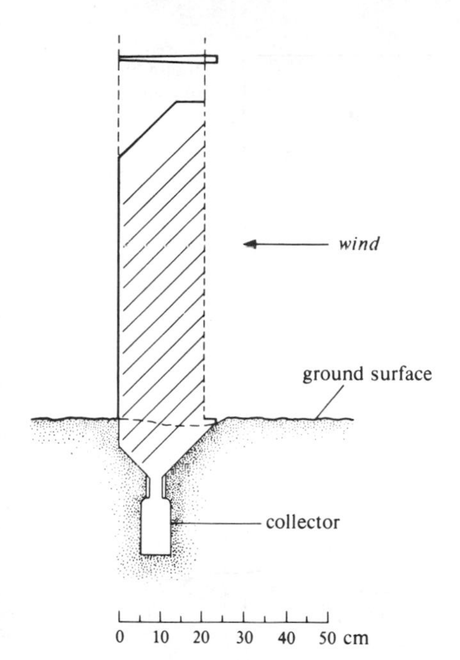

Figure 4.59 Bagnold sand sampler. (After Bagnold 1941).

without disturbing the soil. The main disadvantage is the influence of the plot size on the results, which makes it difficult to extrapolate the results over wider areas: this problem can be overcome by establishing a network of experimental plots rather than basing results on a single plot (Butterfield 1971). Some modern portable wind tunnels are of considerable size and complexity (e.g. Gillette 1978).

4.6.2.3 Monitoring soil loss and soil movement in the field

Sequential surveys Soil loss can be measured against rods (See Howard *et al.* 1977) or pins (Harris 1974). This method is not very accurate because the rods disturb the local velocity field and thus affect erosion rates, and the rods themselves are easily disturbed. Peg displacement, painted stripes or surface lowering fail to differentiate between erosion by runoff, rainsplash or other agencies, and that removed by wind.

Bedload and saltation-load samplers The measurement of the total bedload can only be approximate because the presence of any collecting device interferes with the airstream. The particle collector must minimise airflow disturbance, eddy production and deflection by presenting a thin section to the wind, and must possess a fairly open structure and smooth walls to prevent internal back-pressure.

Bagnold (1941) designed a collector with a narrow vertical slot to catch sand moving by creep and saltation (Fig. 4.59). The design was subsequently modified by Chepil (1945, 1957c), with later versions by Horikawa and Shen (1960) (Fig. 4.60), Crofts (1971) and Kawamura (1951, 1953). The Bagnold–Chepil trap has a low efficiency (only about 20%). Butterfield (1971) and Leatherman (1978) have described more recent and more efficient traps. Butterfield (1971) claims 40% efficiency.

As well as vertical collectors, the rate of sand movement can be determined in horizontal traps, dug into the surface (Bagnold 1941). These are rectangular boxes with small partitions on the bottom to stabilise the trapped sand. The length of the box is such that all saltating sand is trapped when the long side of the box is placed parallel to the wind direction. Figure 4.61 shows the horizontal trap designed by Horikawa and Shen

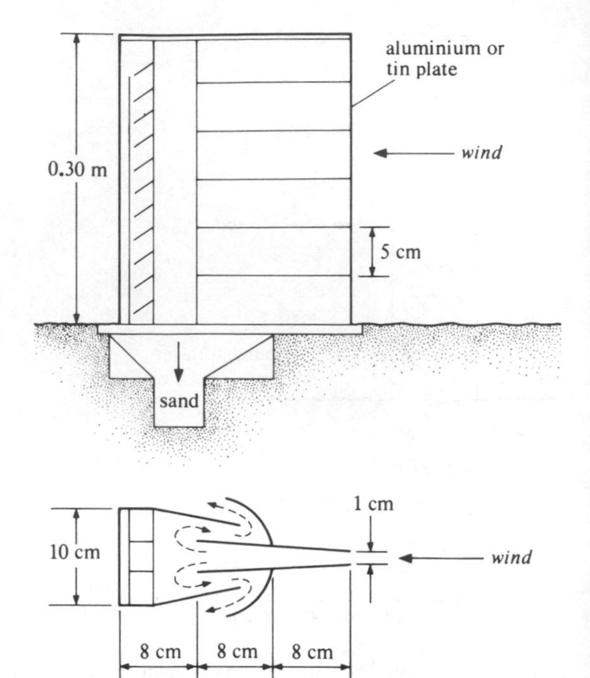

Figure 4.60 Modified Bagnold vertical sand trap. (After Horikawa & Shen 1960).

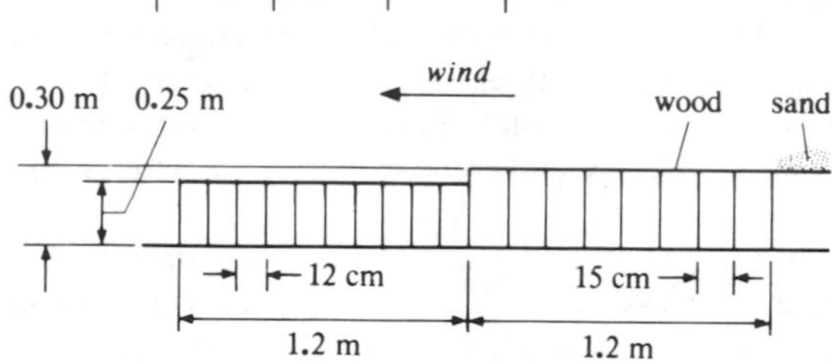

Figure 4.61 Horizontal sand trap. (After Horikawa & Shen 1960).

(1960). The horizontal samplers suffer from the risk of contamination of aeolian samples by sediment derived from other erosion processes, e.g. slopewash material. Sampling in ground-level collections should be restricted to controlled, short-period experiments when other erosion processes do not interfere. The use of a vertical collector fitted with a direction vane eliminates the need for the time-consuming reorientation that is necessary with the horizontal box samplers.

Deposition samplers for dusts The American Society of Civil Engineers (Task Committee on Preparation of Sedimentation Manual 1970) reviewed the measuring sensors and collectors that have been used and the methods of analysis that are applicable to measuring and analysing aerial dusts.

The sedimentation or settling technique is one that collects particles larger than 5μm as they settle out of the air under the action of gravity. Various types of bucket collectors have been designed. Buckets often contain glass or plastic beads, or water, to prevent the re-entrainment of dust. Various deposition samplers are listed in Table 4.14. Engel and Sharp (1958) used a simple 12 inch plastic funnel over a large stoppered bottle to collect dust for trace-element analysis. West (1968a) used a collector surrounded by a box with vents cut into it to minimise the re-entrainment of the dust eddies (Fig. 4.62). Smith and Twiss (1965) used the outer cylinder of the standard rain gauge containing water or antifreeze with 1 mm and 6.35 mm sieves nested and placed in the top to prevent contamination. Chatfield (1967) developed a sampler that can be used for measuring both suspended and settling dust (Fig. 4.63). For deposition sampling, microscope slides on a rotating sampling disc are exposed as they stop beneath the aperture in the lid.

Another type of deposition sampler has been developed by Ganor (personal communication) (Table 4.14). A standard plastic tray is filled with marbles. Dust settles on the marbles and is washed down among them by rain or dew. After exposure, the marbles are washed and the residue allowed to precipitate from the washings. The suspended load can be weighed by evaporating to dryness or by centrifuging.

Dust can also be collected on oiled (e.g. Clements *et al.* 1963) or sticky paper. Vaseline or petroleum jelly can also be used to smear a surface on which the dust collects. Such

Table 4.14 Summary of accumulation techniques for measuring dry deposition.

Type of sampler	*Specifications*	*Principal sketch of sampler*	*Tests of performance*
bucket	HASL: aperture diameter = 300 mm (Steen 1979b)		^{90}Sr against grass
bucket	NILU: aperture diameter = 200 mm		correlations with deposition on moss (Skärby 1977)
bucket	Bergerhoff: aperture diameter = 90 mm turbulence shielded		
bucket	Hibernia: aperture diameter = 250 mm		compared with each other (Köhler & Fleck 1963)
bucket	British Standard deposite gauge: aperture diameter = 300 mm (BSI 1969). Turbulence shielded		BS tested in wind tunnel (Ralph & Barrett 1976)
bucket	Löbner: aperture diameter = 290 mm (Verein Deutscher Ing. 1971)		
Vertical gauge	CERL: tube diameter = 75 mm, opening width = 45 mm	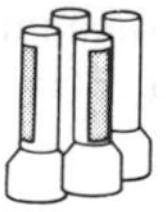	wind tunnel (Ralph & Barrett 1976)
Plane surface	Harwell method: filter under cover (Steen 1979b)		
Natural surfaces	Growing moss, mossbags (Goodman *et al.* 1979; Steen 1979b)		
marbles	Ganor (pers. comm.)		

Modified after Steen (1979b).

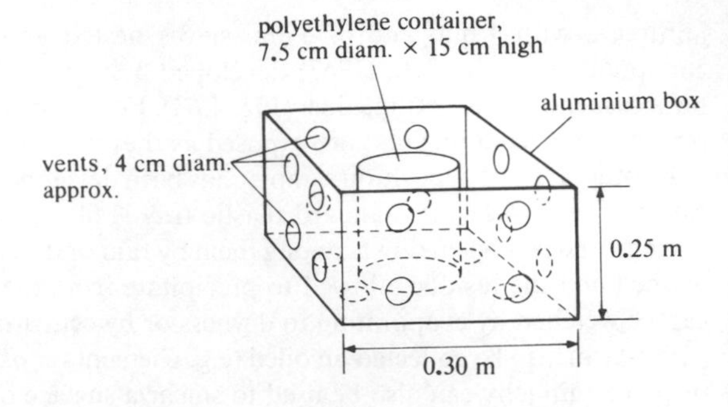

Figure 4.62 Dustfall collector. (After West 1968a).

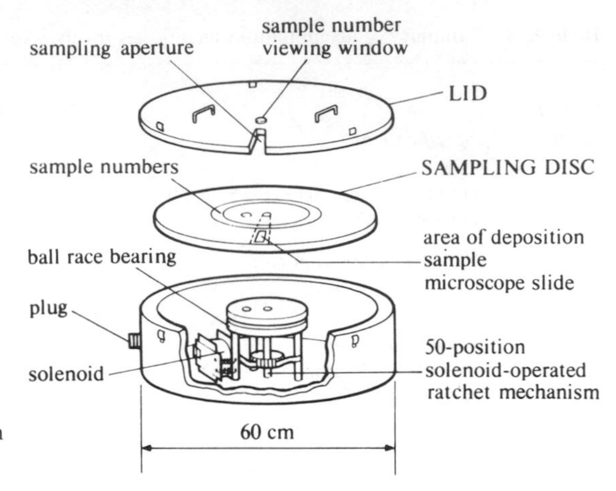

Figure 4.63 Sequential deposition sampler. (After Chatfield 1967).

collectors clearly become less efficient with long exposure, and with Vaseline and petroleum jelly there are problems of recovering the dust.

Mossbags (cleaned moss – usually *Sphagnum acutifolium* aggregate contained in a flat, 2 mm nylon-mesh envelope) can be used to collect aerosols (Goodman *et al.* 1979). They are inexpensive and thus can be used in large numbers to make wide-scale synoptic surveys. Mosses have a high cation-exchange capacity and there is virtually no loss of metallic elements even under periods of intense rainfall.

Suspension samplers for dusts All the devices for sampling airborne particulates (as opposed to settling dust) draw a measured volume of air through the unit usually by means of a motor-driven vacuum pump. Pumping may be for short or long periods. Rahn *et al.* (1979) pumped 2500–7000 m^3 over 1–2 days for their sampling of Saharan dust over the Atlantic.

Filtration Filtration through a suitable material is the most widely used method for collecting aerial dust and has been used for many years (see Chepil 1945). Hendrickson (1968) has evaluated the efficiencies of various filter media, e.g. mineral wool, plastic, glassfibre, asbestos mats, wood-fibre paper, granular media and controlled pore or membrane types. More recently, Sparney *et al.* (1969) reported the use of nucleopore filters, and Gillette and Walker (1977) have used millipore filters.

Impingement Dust impinges on a body when a dust-laden airstream is deflected around it. In measuring devices, dust is made to impinge either on a surface submerged in a liquid (wet impingers) (Fig. 4.64) or on a surface exposed to air (dry impingers)

(Fig. 4.65). Larger particles (> 1 μm diameter) are more efficiently collected than smaller ones.

Thermal precipitation This process separates dust from the airstream on the principle that a thermal force causes suspended particles to migrate from a zone of high temperature to one of low temperature. Electrically heated plates direct the airflow and the suspended particles to water-cooled collection plates. This method has high collection efficiencies for a wide range of particle sizes (0.001–100 μm range) (Task Committee on Preparation of Sedimentation Manual 1970).

Centrifuging This method operates by drawing the air into a tube in which a helical flow generates centrifugal forces that throw suspended particles towards the outside (Task Committee on Preparation of Sedimentation Manual 1970).

Electrostatic precipitation In these collectors, particles passing between two charging electrodes assume a charge, and are transported to a collecting electrode by an electric field. High collection efficiencies are only possible with small particles (0.01–10 μm range) (Task Committee on Preparation of Sedimentation Manual 1970). Steen (1979a) described an electrostatic sampler that might be used for dust sampling over the Sahara (Fig 4.66). It was intended to measure the flux of dust rather than its concentration.

Aeolian deposits The extent of aeolian contamination is best determined in soils and sediments derived from lithologies having distinctive mineralogy (e.g. igneous, volcanic, metamorphic), such that foreign grains are readily recognisable. The percentage of quartz in soils overlying quartz-free mafic (basic) rocks has been used as an index of loess input. Aeolian dust is also an important source of sediment in the oceans (Chester *et al.*

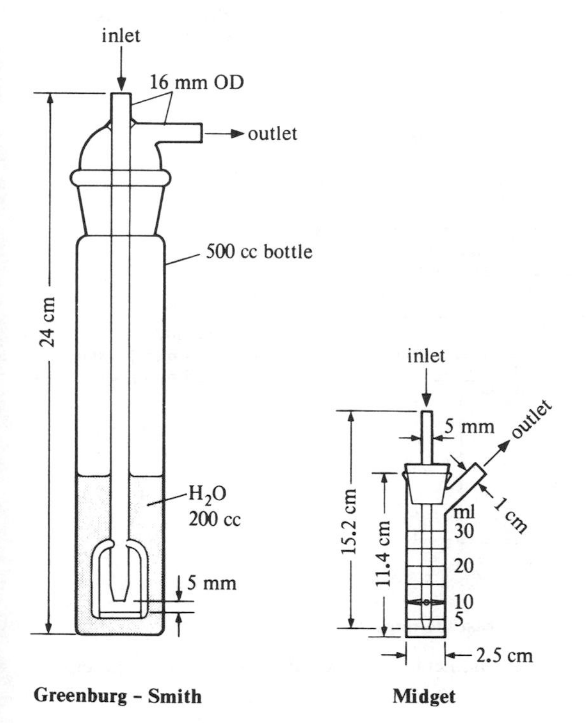

Figure 4.64 Examples of wet impingers. (After Hendrickson 1968, Katz *et al.* 1925).

1972, Eriksson 1979) as, in some cases, is aeolian sand (Sarnthein & Walger 1974). Mineralogical and size analyses of dust accumulated in snowfields, ice sheets and glaciers can be used to distinguish between dust components derived from local sources and dust components derived from distant (global) sources.

The pyroclastic nature of ashfalls and ashfall tuffs is generally discernible in the field, but this depends on the nature of the local lithology and that of the contaminating ash.

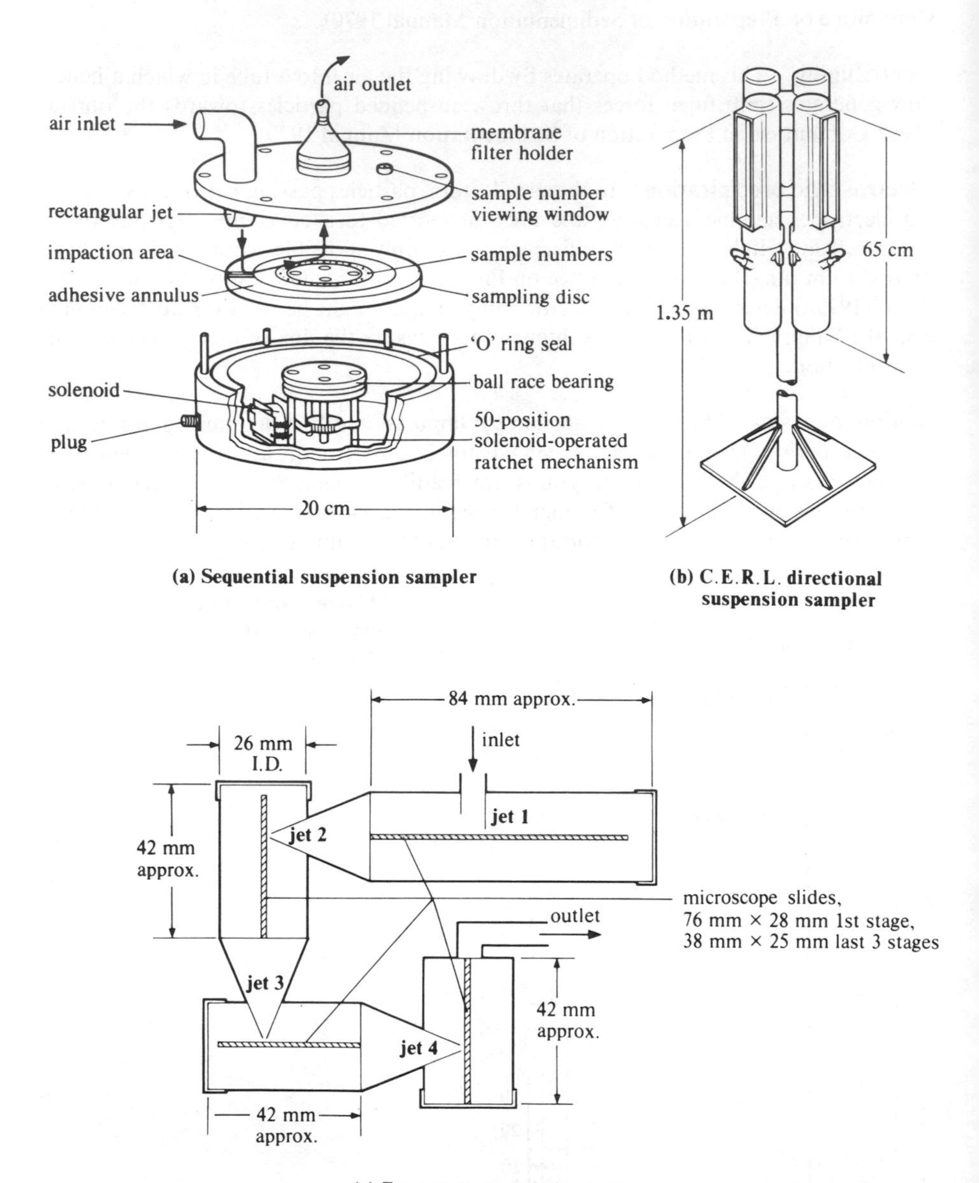

Figure 4.65 Examples of dry impingers. (After Chatfield 1967, Lucas & Moore 1964, Hendrickson 1968).

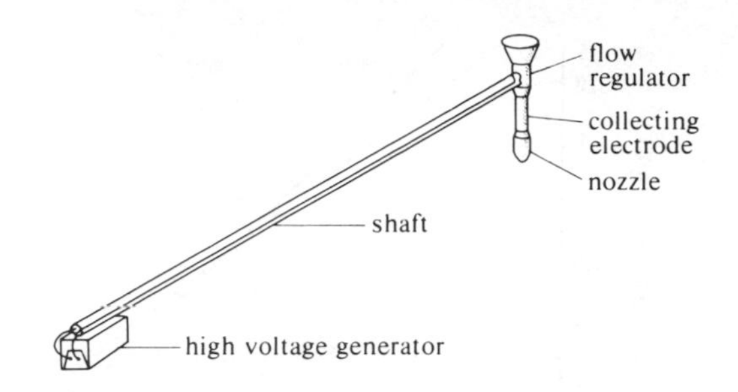

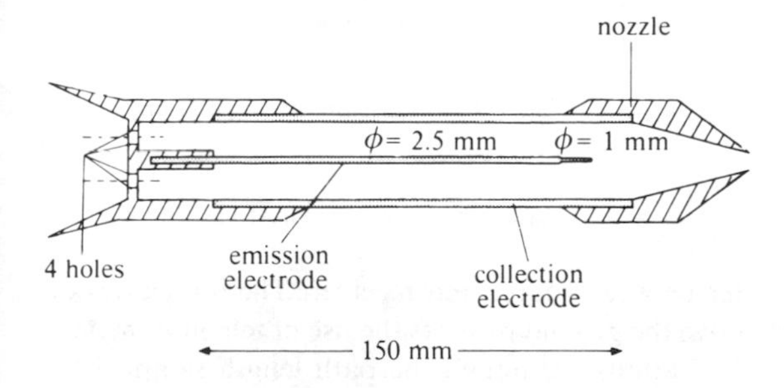

Figure 4.66 Proposed sampler for airborne dust. (From Steen 1979a).

4.6.2.4 Atmospheric dust diffusion

Visibility Particulate concentrations can be estimated by observing visibility (either routine or non-routine meteorological observations). One of the following relationships applies. Chepil and Woodruff (1957) proposed that

$$C_6 = 56.0/V^{1.25} \tag{4.40}$$

where V is horizontal visibility (km) and C_6 is the dust concentration (mg m^{-3}) at 1.8 m (6 ft) above the surface. Robinson (1968) suggested that

$$C = 5.2\rho_s(\tfrac{1}{2}D)/VK \tag{4.41}$$

where ρ_s is the particle density, D is the particle radius and K is the scattering area coefficient. The relationships are plotted in Figure 4.67. Patterson and Gillette (1977) found that visibility V (km) and the mass concentration of dust M (g m^{-3}) are related:

$$MV\gamma = C \tag{4.42}$$

where C is a dimensional constant and γ is a non-dimensional constant.

The relation between mass concentration and visibility must be used with caution because of the variability of C in Equation 4.42, but if fractional size distribution is known, the relation may be calculated with theoretical values of C.

Subjective factors and atmospheric moisture influence the accuracy in recording visibility. Subjectivity is reduced by mechanical instruments such as LIDAR (light detection and ranging) (Patel 1978, Ball & Schwar 1978). Another technique is the telephotometer, which operates on the principle that a perfectly black target viewed from a distance appears grey or white due to light scattering from dust. Yet another technique, the transmissometer, consists of an intense source of light and a photocell receiver on which the transmitted light is focused. The output of the photocell varies with the

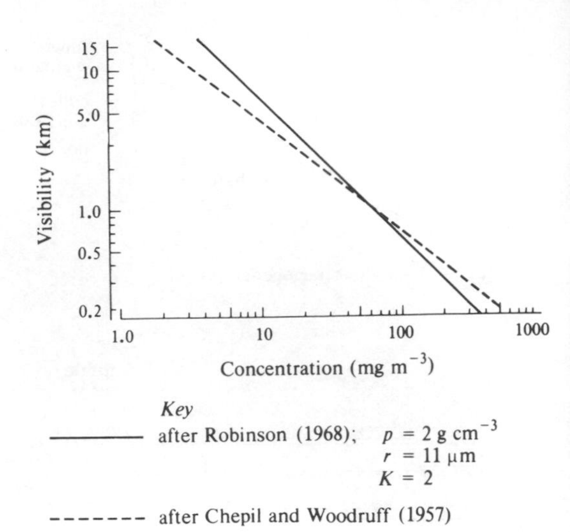

Figure 4.67 Visibility–concentration relationships.

amount and size of particulate matter between source and receiver. The difficulties of introducing targets at some height above the ground restricts the use of telephotometers and transmissometers to surface observations. Although the path length sampled by LIDAR is small, it is more compact than the other systems and the beam can be directed at any angle.

Recent work on the plume of dust emanating from the Sahara over the Atlantic has been based partly on the measurement of Volze turbidity using Sun photometers (Prospero *et al.* 1979). Jaenicke (1979), working on the same problem, used a comparison of astronomically determined sunset with the sunset time recorded on sunshine recorders to study any change in dustiness in the period since recorders were first used in 1880.

Sources and sinks of airborne particles The two most frequent biogenetic components of dust – freshwater diatoms and opal phytoliths – can be used to identify the source area of airborne particles. Because some phytoliths can be assigned to plant species, it is possible to determine whether phytoliths in dust are derived from local or remote sources. A classification of grass phytoliths, based on grass taxonomy, that can be used as an indicator of provenance for dust has been published by Twiss *et al* (1969). Mineralogy can also be used to infer the source of dust (Rahn *et al.* 1979).

There has now been a considerable amount of data acquired on rates of dust accumulation in the oceans by means of analysis of deep-sea core sediments. These cores provide a record for the Pleistocene (e.g. Sarnthein & Koopmann 1980) and, in some cases, as far back as the Cretaceous (Lever & McCave 1983). Records of dust deposition have also been obtained through the analysis of sediments within the major ice caps (Thompson 1977a).

Meteorological observations Meteorological observations provide useful material for investigations of airborne dust, in particular, windspeed and direction, visibility, past and present weather (code 4677 in WMO's *Manual on codes* (WMO 1974)).

Case studies of duststorms can be carried out by means of synoptic weather charts, and the critical windspeed for initiating raising of dust into the air can be determined (Morales 1979, Newell & Kidson 1979). The use of meteorological records to determine duststorm frequencies and distributions has been outlined by Goudie (1983) and analysed in detail by Middleton *et al.* (1986).

Areal extent of dust plumes Parcels of dust-laden air can be delineated on satellite photographs (Carlson & Prospero 1972). Satellite imagery permits a compact view of an entire duststorm at one point in time. Middleton *et al.* (1986) outline the relative merits of various platforms, including geostationary meteorological satellites (e.g. GOES, Meteosat, GMS), polar orbiting satellites (e.g. NOAA-n series, DMSP series) and the Space Shuttle orbiters.

It is possible to visualise duststorms on synoptic weather maps and to follow their migrations and developments by means of isopleths for visibility (Morales 1979). Aircraft soundings incorporating measurements of radon-222 (recorded in units of picoclines per cubic metre of air) provide an effective means of delimiting dust-transporting wind systems (Prospero & Carlson 1970, Carlson & Prospero 1972). The emanation of radon from the ocean is about 100 times less than that over land, so that concentrations within areas of dense dust-produced haze will be markedly higher than in oceanic air parcels. The colour density of the filters used for the radon measurements provides dust concentration values.

Dust-devil dynamics The entrainment mechanisms of particles with a vertical vortex such as in dust devils is quite different to that in uniform boundary layers. Quantitative information on the relation of environmental conditions to the spatial distribution, frequency, duration, diurnal variation, size and direction of rotation of dust devils can be assembled from field observations (Sinclair 1969). Controlled laboratory conditions are necessary to study dust-devil initiation and the complex interactions between relief, the low-level wind field and heat fluxes in the boundary layer. Laboratory simulation with vortex generators presents the most promising approach (Smith & Leslie 1976).

4.6.3 Abrasion

4.6.3.1 Field experiments

Field experiments are usually performed in experimental plots where vegetation cover is sparse and strong winds frequent (Sharp 1964). Large plots (at least 5 m^2) are necessary so that the perimeter fence does not interfere with the wind. Enclosure is necessary to prevent disturbance or vandalism. Natural boulders and stones and artificially shaped blocks (e.g. common bricks, Hydrocal blocks and lucite rods) are exposed in the plot. Targets are exposed at different heights so as to ascertain the height of maximum abrasion (Sharp used vertical 1.2 m lucite rods). Regular observations are necessary to evaluate the shifting, tilting and overturning of the blocks. Rubber strips cemented across the upwind faces of boulders (to protect part of the face from sand blasting) provide datum surfaces from which to measure the cutting rate. Collectors exposed at different heights sample saltating sand and suspended dust so that the distribution of mass within the saltation curtain can be determined for different wind velocity conditions. Wind velocity is recorded at the plot or taken from records at nearby meteorological stations. Wind recording at the plot is preferable, because near the surface the velocity profile is more a function of local roughness than of regional terrain roughness. Field monitoring is difficult because slow, continuous abrasive processes necessitate prolonged exposure to produce measurable results.

The effects of abrasion can be evaluated by studying structures, buildings, ventifacts or yardangs. The nature of markings (pits, flutes and grooves) and their degree of development depend upon lithological characteristics, size, shape and orientation of the face to the wind direction, and projectiles (Sharp 1949). Processes of ventifact formation can be inferred from the study of ventifact morphology (Higgins 1956).

4.6.3.2 Laboratory experiments

Because so many factors enter into abrasion processes, their control in the laboratory is attractive, though care needs to be exercised in the choice of the abrading particles and the samples to be abraded.

Apparatus for studying the abrasion of aggregates, particles or blocks The stability of dry soil aggregates during sieving can be used as an indirect measure of their stability against abrasion (Chepil 1951, 1952, 1962). Revolving cylinder techniques and shaking machines do not simulate the natural aeolian processes and so may give misleading results.

Wind tunnel Abrasion tests can be carried out in a conventional wind tunnel by exposing blocks, clods or trays of soil to various abraders. The abraders are placed on trays upwind of similar trays containing the soil to be abraded. The amount of abrader that passes over the soil is determined either from the difference in weight of the tray containing the abrader before and after exposure or from the weight of abrader collected in traps. Similarly, the amount of soil abraded is determined by the loss in weight of the soil material or the weight of soil (differentiated from the abrader by size characteristics) collected in traps. The relative susceptibility of the soil to abrasion is expressed as the coefficient of abrasion (Chepil 1955).

Circuit tunnel This apparatus, which is used to abrade moving particles, was described by Kuenen (1960b). Figure 4.68 shows its construction. The disadvantages of the circuit tunnel are that (a) the number of jumps per metre of travel is several times greater than in nature because there is less time to gather momentum during each jump; and (c) the concrete floor exaggerates the force of impacts.

Circuit tube This is a vertical loop of hard glass tubing (Fig. 4.69) with a jet at the lower end and an exhaust at the top (Kuenen 1960b). Opposite the jet, the tube is slightly wider so that a 'dune' forms. Sand impacts on the dune and abrasion takes place. The equipment is only useful for low velocities, because if high velocities are used the grains all fly round the tube and no dune forms, so that there is no impacting and no abrasion.

Wind tunnel with belt The apparatus (Kuenen 1960b) that is shown in Figure 4.70 is also used to study the abrasion of moving grains. The main disadvantage of the tunnel is that velocity varies along its length, so that the magnitude of the wind force is difficult to determine. However, the tunnel simulates creep and saltation of particles, which most other pieces of equipment fail to achieve.

Pneumatic devices Pneumatic devices simulate impact forces and not creep and saltation (Fig. 4.71). J. R. Marshall (personal communication) has designed a pneumatic gun apparatus that fires projectiles at quartz plates. The particle velocity, size and impact angle can all be controlled. Crystallographic orientation of the quartz can be varied in the preparation of quartz plates. Bombardment of plates and their examination with an electron microscope have demonstrated that the Hertzian fractures (percussion cones) of Kuenen and Perdok (1962) are the dominant features created by the impact of a spherical body.

Abrasion tests are monitored with electron microscopy of test specimens and projectiles. Examination of electron micrographs of surfaces prior to and following wind blasting permits the identification of characteristics associated with impact (Whitney 1978).

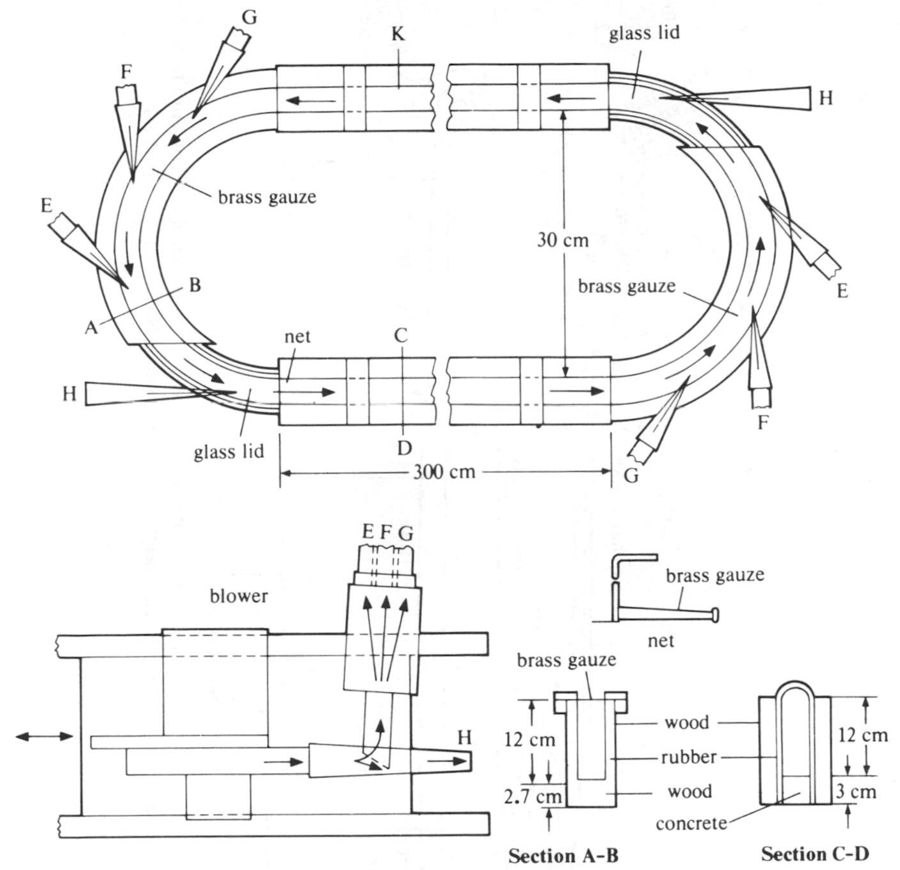

Figure 4.68 Circuit tunnel for studying aeolian abrasion on a concrete floor. The two blowers have a funnel fitting into H. They distribute secondary blasts through tubes connecting E–E, F–F and G–G. (After Kuenen 1960b).

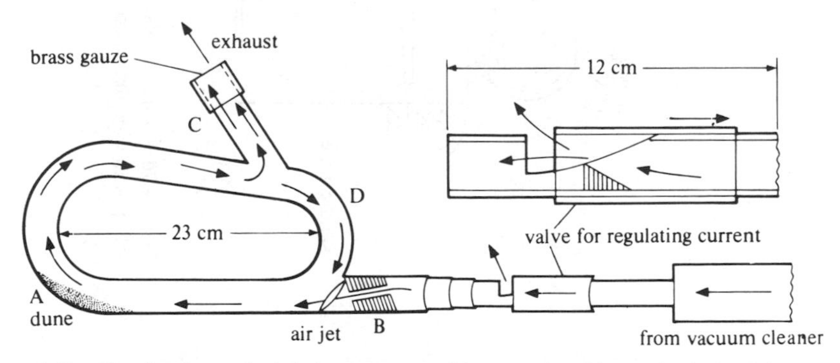

Figure 4.69 Circuit tube used minly for surface markings produced by weak winds. (After Kuenen 1960b).

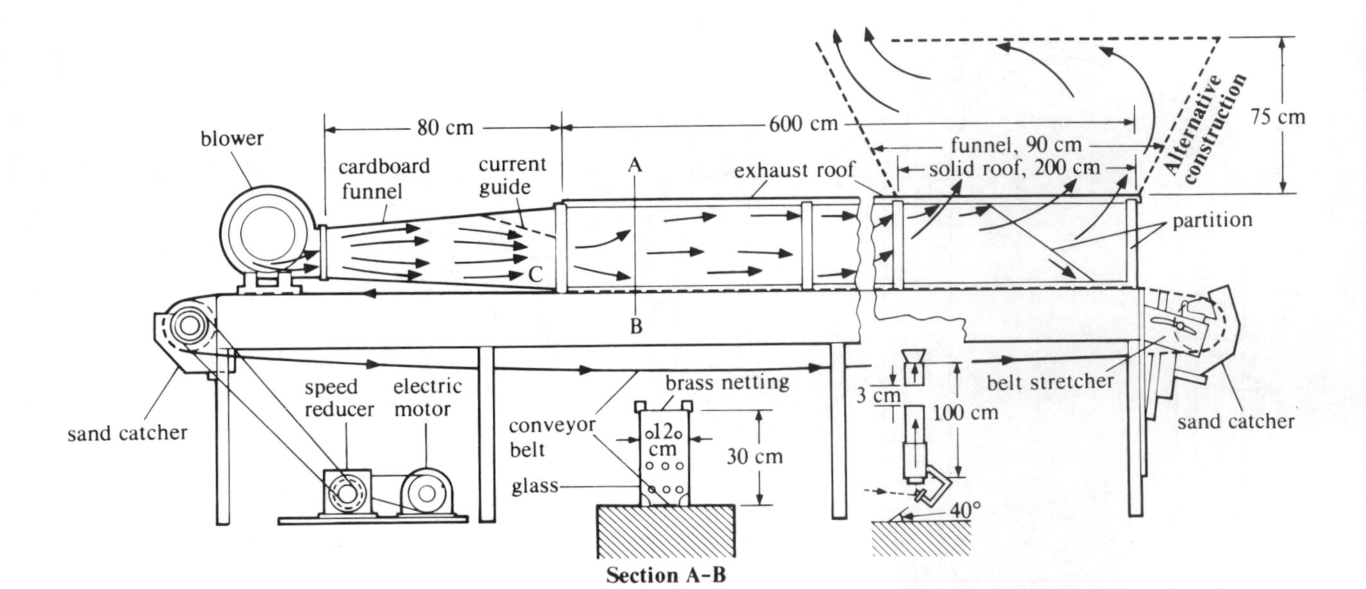

Figure 4.70 Wind tunnel with belt, roofed by wire gauze. The points for measuring wind velocities are indicated in a section. An alternative set-up with wide-funnel exhaust is indicated, used for measuring ricochet heights. (After Kuenen 1960b).

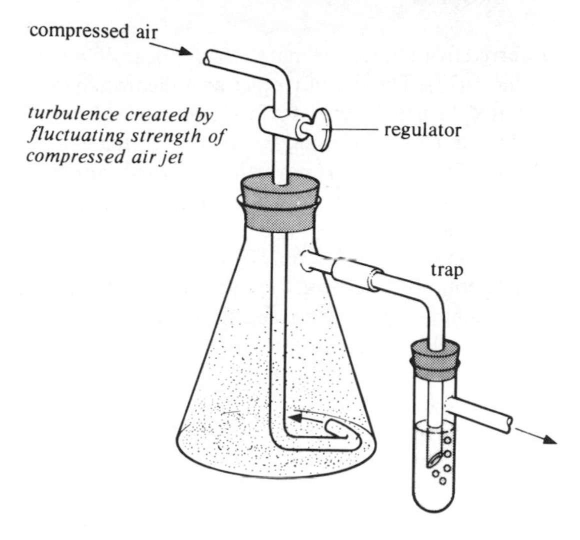

Figure 4.71 Pneumatic abrasion apparatus.

4.6.4 Bedforms

4.6.4.1 Ripples

Ripples are the simplest aeolian bedforms to study both in the field (Sharp 1963, Tsoar 1975) and in the laboratory (Bagnold 1941, de Félice 1955). Field study is hampered by very variable local conditions (on the different parts of a dune) and by gustiness and variability of the wind. Nevertheless, both Sharp (1963) and Tsoar (1975) have developed methods for studying ripples in the field. The $30 \times 1 \times 1$ ft^3 wind tunnels used by Bagnold (1941) allowed much greater control of the ripple environment. A combination of field and laboratory study is ideal.

Ellwood *et al.* (1975) used a computer simulation to make a deductive model of ripple development. Their input data included the impact and rebound velocities of grains of various sizes, calculated by measuring the rebound heights of grains bounding off a bed covered with grains and inclined at 14°. The grains were caught on greased plates.

4.6.4.2 Dunes and megadunes

The explanation of the particular form of a dune (or megadune) entails an understanding of its aeolian and sedimentary environment. This in turn requires study of a number of aspects of the dune: the ambient wind regime (Sec. 4.6.1); the sedimentary parameters of the sand; the flow of air and sand over the dune and between dunes; the rate, direction and mechanism of movement of the dune; its dimensions and place in a dune pattern; and, finally; the internal geometry of its bedding. The results from some or all of these measurements can be fitted into a simulation model to elicit their full significance.

Flow over individual dunes The flow of air over a dune can be simulated in a wind tunnel or in a flume. If simulation is undertaken in a flume there are problems of scaling, but they are not intractable (Rim 1958, Escande 1953, Crausse & Pouzens 1958). If a simple study of very broad similarity is required, one can use simple tanks or even '*trottoirs parisiennes*' (Clos-Arceduc 1966). Rim (1958) used a $10 \times 0.35 \times 0.35$ m^3 flume with a movable 'false' bottom whose tilt and topography could be altered.

Techniques for measuring the velocity and turbulence of the wind in the field are described in Section 4.6.2. Wind direction can be measured with small flags and prismatic compasses (Tsoar 1978) or it can be inferred from the scour marks that appear downwind of stakes sunk in the sand (Howard *et al.* 1977). The use of ripples as indicators of wind direction (Sharp 1966) should be confined to nearly level surfaces; on surfaces with a slope component at right angles to the wind, ripples are said to diverge from a trend at right angles to the flow (Howard 1977). Burrel (1971), nevertheless, used ripples as indicators of sandflow, to map the network of flow in a dune system in Morocco, and analysed it with topological network analysis.

Sandflow can be traced with coloured sand (Tsoar 1978, pp. 80–4). After the coloured sand has been moved by the wind from its point of insertion, the dune surface downwind of the site of release is sampled along straight, regularly spaced transects. The sample instrument is a card covered with Vaseline held in a special holder at the end of a pole (Ingle 1966). In its use, great care is needed to avoid contamination or disturbance of the surface. After sampling and storage, the greased cards are examined in the laboratory under ultraviolet light and all coloured grains are counted. Isolines of particle concentrations can then be plotted on a surveyed plan of the dune to show the pattern of dispersion.

A simpler, but less revealing, technique is to measure the exposure or burial of posts sunk in a regular grid across the dune surface (Howard *et al.* 1977) or the exposure of pins (Harris 1974).

The second author of this section has twice failed to measure sandflow over dunes with painted poles. The first (elaborate) attempt was frustrated by the Libyan revolution of 1969, the second by Bedouin vandals. Several 2 m long round poles were painted repeatedly with layers of different colours. It was hoped that sand blast would remove the layers one by one, so that a series of four-colour photographs (from each cardinal point) could be used to record the height of maximum movement by saltation and its upper limits as well as the principal directions of movement.

Dune shapes, dimensions, patterns and movement The best data about dunes comes from theodolite or plane-table surveys (Verlaque 1958, Sharp 1966, Norris 1966, Hasternrath 1967, Calkin & Rutford 1974, Tsoar 1974, 1978, Howard *et al.* 1977). Tsoar also recorded movements with repeatedly levelled cross sections. Ground survey, though accurate, is very time-consuming, and can seldom be used for a large sample. For dune survey, a scale of 1:2000 is probably sufficient (Verlaque 1958). More replicable but less accurate measurements of dune movement on the ground can come from stakes sunk either in the dune itself, say along the line of the crest of a dune at the start of the experiment (McKee & Douglass 1971), or better, along the easily recognisable base of the slipface (Beadnell 1910, Sharp 1966, Inman *et al.* 1966, McKee 1966, Calkin & Rutford 1974, Tsoar 1974, 1978). Inman was able to cross-check results by referring to small tidelines at the base of the stoss slope of his near-coastal dunes. Kerr and Nigra (1952) also noted similar features on air photographs.

A better sample of dune movements can be taken from maps and air photographs, though duneforms are seldom thought worthy of recording in detail on standard topographic maps (Capot-Rey 1963). Megadunes, however, are beautifully plotted on USGS 1:62 500 maps (Warren 1976) and some data can even be extracted from IGN 1:200 000 maps of the Sahara (Wilson 1970b). Long and Sharp (1964) used sequential topographic surveys but found great difficulty in locating a recognisable point for the comparison of positions; they tried to locate the axial point at the top of the slipface.

Air photographs are undoubtedly the best source of data for measuring dunes and sequences can be used to measure movement, the more sorties the better (McKee 1966). Scales of 1:10 000 to 1:15 000 are adequate to study the plan shape of dunes but not their cross sections (Hunting Technical Services 1977). Scales of 1:30 000 to 1:50 000 are sufficient for some work (Clos-Arceduc 1971).

The main problem with dune measurement is what to measure. In most dune fields, there is an apparent chaos of forms and trends. The problems are least with one of the least common duneforms, namely the crescentic barchan, and most dune morphometric energy has therefore been focused on these. The dune has a clearly defined outline and, being generated by near-unidirectional winds, a clearly defined trend.

Several morphometric studies of barchans have now been made (Capot-Rey 1963, Verlaque 1958, Finkel 1959, Long and Sharp 1964. Norris 1966, Hastenrath 1967. Howard *et al.* 1977, Hunting Technical Services 1977). Figure 4.72 shows a notation for use in such studies, the terms being derived mostly from Finkel and Verlaque.

Where good altimetric data are plotted on maps, as on the USGS 1:62 500 series, dune hypsometry (cf. Strahler 1952) can be computed and used to compare dune types. Warren (1976) found that the hypsometric characteristics of seif-like and barchan-like megadunes were distinctly different. An eyeball classification using air photographs yielded a similar distinction for Tsoar (1974).

Where patterns are regular, dune spacing can be measured on maps or air photographs, but even the most regular patterns provide many ambiguities in measurement. The main difficulty is the choice of exactly equivalent points between which to measure. In many areas, too, the exact trend of the dunes is hard to decide, and may be affected in different parts of the field in response to flow around megadunes. Sampling lines should

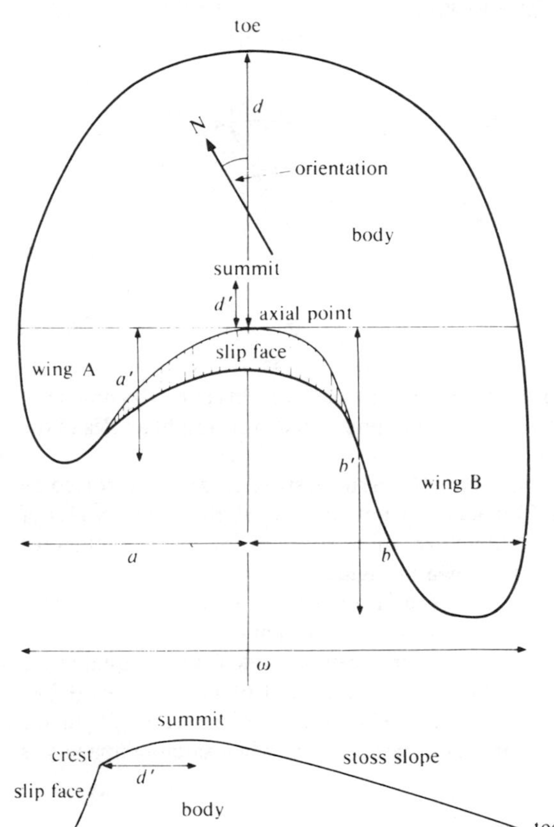

Figure 4.72. A suggested standard nomenclature for measurements of barchans. (Mostly after Verlaque 1958, Finkel 1959).

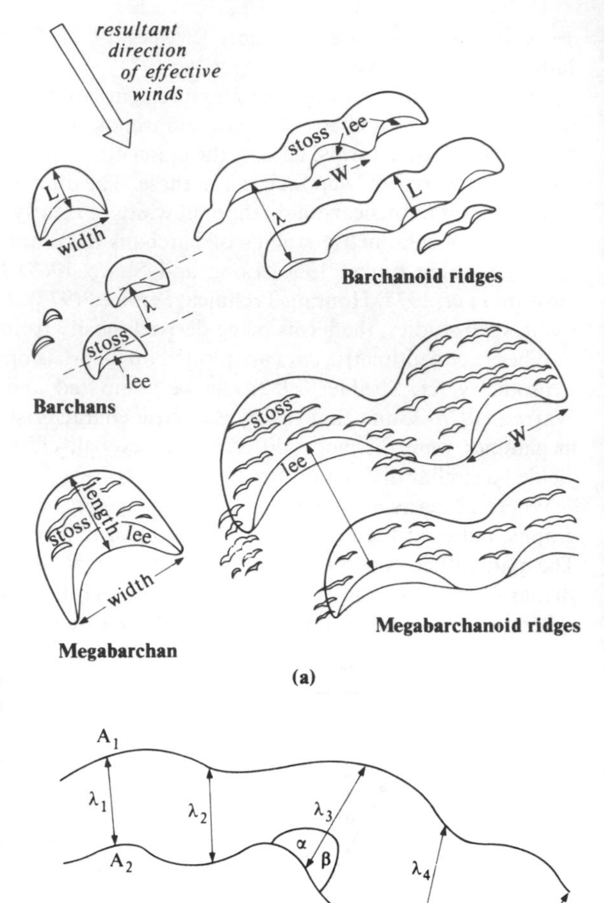

Figure 4.73 (a) Sketches illustrating common forms of crescentic dunes and conventions for measuring width (W), length (L) and wavelength (λ). (After Breed 1977). (b) Measurements of dune wavelengths (λ). The lines (A_1A_2, etc.) should be as near as possible at right angles to the crests, i.e. $(90 - \alpha)^2 + (90 - \beta)^2$ should be a minimum. (After Matschinski 1952).

therefore (a) be short and (b) be replicated in at least three directions narrowly angled to the supposed direction of movement, so that the minimum spacing (at right angles to the true direction) can be estimated.

The main problem of choosing equivalent points on successive dunes is not resolved by choosing only the top of the slipface. Slipfaces may not occur on all the dunes. No set of rules for measuring dune spacing is therefore entirely foolproof. The schemes of Matschinski (1952) and Breed (1977) are shown in Figure 4.73.

In some dune patterns, branching is a conspicuous feature. Goudie (1969) measured such a pattern from the Kalahari. Its interpretation is problematical.

Finally, many significant observations about dunes can be made from the qualitative observation of air photographs and satellite photographs and other imageries (Clos-Arceduc 1971, Mainguet 1972, McKee & Breed 1974, Guy & Mainguet 1975). Image processing can often help to reveal a dune pattern, especially when satellite imagery is being used (Fink & van der Piepen 1974).

Dune volume Several estimates have been made of the volume of sand in dunes. The usual objective is to discover the pattern of volume within a dune field, in order to infer inflow and outflow rates, directions of movement or temporal changes in supply. The most detailed study used a small number of ground measurements and a large number of measurements taken from aerial photographs (Lettau & Lettau 1969). The basal area of individual barchans was multiplied by a simple function relating to cross-sectional shape (both of these were derived from ground survey). Only length and width then needed to be taken from the photographs. Several simple and reasonable assumptions were made about the shape of the dunes. Sand volume was converted to weight by assuming that the bulk density of the dunes was 1.3 g cm^{-3}. Large dunes conform to all these assumptions better than small ones, and because large dunes account for by far the greater proportion of sand, this matters little. Lettau and Lettau (1969) then summed the total volume in dunes in five 600 × 2000 m^2 strips at right angles to the direction of barchan movement (as shown by sand streamers). In each strip, they then calculated the volumetric discharge Q (m^3 yr^{-1}) in dunes thus

$$Q = \frac{\sum cV}{A} \tag{4.43}$$

where c is the mean annual dune displacement (the dune celerity of Bagnold (1941)), V is the total volume of sand in the dunes and A is the area of the sample strip. Then they compared adjacent strips for convergence or divergence. Divergence indicated erosion of the underlying floor.

Sarnthein and Walger (1974) used a modified, but similar, method to study the pattern of sand volume in barchans in the western Sahara, and used their results to predict the future of sand movement and to deduce the history of the dune field.

McCoy *et al.* (1967) used a very generalised technique in a study with similar aims. They wished to discover the age and future of the Algodones dunes in California. Assuming, as do all such studies, that the dunes are composed of aeolian sand throughout, McCoy *et al.* computed the mean height above the nearby dune floor of the four corners of 180 sample squares (each 5000 ft square) taking their measurements from 1:62 500 USGS topographic maps with a contour spacing of 20 ft. A test of this method using a closer grid of points was made by Warren (1976). He first accurately computed dune volume in a stratified random sample of 10 000 square sites using USGS 1:62 500 topographic maps. The area enclosed by each contour above an assumed dune base was multiplied by the contour interval and these figures were summed for the whole dune. Warren's modification of the McCoy method involved the noting of sand depth at 16 grid intersections in the sample squares. The mean sand depth estimated by this second method gave errors of up to 40% when compared with the accurate computation (McCoy *et al.* also checked against the contour method but on a much coarser scale). A better estimate, which can be achieved in a fraction of the time that was needed for the accurate computation, was to average the areas of two cross sections across the sample square and to relate this measurement to the accurate method by a factor relating to dune hypsometric shape (see above).

A fast, but probably much less accurate, method was used by Wilson (1970b) for the Grand Erg Oriental.

Dune bedding The study of dune bedding requires sections of the dune to be cut. Desert dunes are frequently too dry to support cleaned sections so that one can either choose a wet period (Tsoar 1975), artificially wet the dune (McKee & Tibbits 1964), or dig deeply enough to reveal almost permanently moist sand (McKee 1966).

The usual method of recording the bedding is to prepare clean, straight faces. The true dip and strike of beds can be computed from measurements on two orthogonal faces (Warren 1976). The method appears in many standard textbooks of geology. Goldsmith

(1973) inserted an 18 cm disc with a bubble level and then measured azimuth and dip with a Brunton compass. Direct measurement was employed by McKee and Tibbits (1964), Yaalon and Laronne (1971) and Warren (1976). McKee (1966) gridded the sides of his huge faces at White Sands with blue spraypaint, so that various sections could be placed in position. Dune dip is best measured with a Brunton compass.

The study of the distribution of dip angles can use standard linear statistics. Azimuths, however, plot as circular distributions. Goldsmith (1973) suggests the use of the vector summation techniques proposed by Curray (1956, pp. 118–20) and Pincus (1956, pp. 544–6).

If time is limited, or if detailed study of grain orientation is required, a 'peel' may be taken from a cleaned face. Several proprietary sprays (often latex) are available (see McKee 1966, Yasso & Hartmann 1972, Tsoar 1978)). Land (1964) and Sharp (1964) prepared thin sections of dune and ripple bedding after artificial lithification.

Aeolian bedding can also be studied experimentally. In the field, McKee and Douglass (1971) and McKee *et al.* (1971) spread thin marker layers of pulverised magnetite over a slipface at 20–30 min intervals to reveal both the accretion rate and the bedding structures in pits cut subsequently into wetted sections of the face. The rate of deposition was also measured with a rectangular wire mesh screen placed with its long edge parallel to the dip of the slipface. Laboratory studies of dune bedding can be made in a glass-sided tank with a slipface produced by sieving sand over a steep sandy slope (McKee *et al.* 1971). Again, pulverised magnetite can be used for marker layers.

4.7 Coastal processes

4.7.1 Introduction

> I do not know what I may appear to the world; but to myself I seem to have been only a boy playing on the seashore and diverting myself and now and then finding a smoother pebble or a prettier shell than ordinary, while the great oceans of truth lay undiscovered before me.

These sentiments of Newton must find sympathy with many coastal geomorphologists. The coastal geomorphologist often knows exactly what, under ideal circumstances, he or she would like to measure but often lacks the means, due to either technical or financial constraints, to achieve his or her goals. The coastal zone is one of the most rewarding areas of geomorphological research since the relatively high energy input, in terms of waves and tidal currents, promotes responses between process and form over relatively short periods of time. Because of this relatively high energy input, however, the measurement of processes can be a difficult problem, particularly during times of storms and storm surges, which promote the most dramatic form responses. Coastal geomorphology has witnessed over the last decade a rapid development of instrumentation that has few parallels in other branches of the subject. A bonus from this development has been the ability to measure accurately many of the process variables in the coastal zone, although the cost of much of the new instrumentation is beyond many individual researchers. Many of these innovations will be discussed in this section, but simpler and often cheaper methods of measurement will also be described.

Current trends in coastal geomorphology have been reviewed by Viles (1986), Thom (1984), Bird (1985) and Komar and Holman (1986), and particularly useful texts covering wide aspects of this subject are Komar (1976), Pethick (1984) and Trenhaile (1987). This section deals largely with the measurement of processes and the induced movement of sediment. The measurement of morphology is also included since it contains problems unique to the coastal zone. Discussion of process, sediment movement and morphology measurement is confined largely to a consideration of the surf zone. Following Inman and Bagnold (1963), the surf zone comprises that part of the beach over which waves dissipate their energy. Section 4.7.3 deals with the measurement of erosion on coastlines. Emphasis has been placed throughout on the surf zone since the offshore zone enters the realm of the oceanographer and supports a massive literature in its own right.

4.7.2 Surf-zone processes

The most important processes in the surf zone are related to the oscillatory movements and induced currents associated with shoaling and breaking waves. For many years the energy and stresses associated with such waves have been derived from formulae relating parameters of waveform, particularly wave height and wave period, to energy density for either individual waves or wavetrains. Recently, several techniques have been developed to measure the velocities and stresses associated with breaking waves to determine more directly the energy inputs at the water/sediment interface. This section describes techniques available for the measurement of waveform, the analysis of wave data, the measurement of orbital velocities of waves, and the velocities and forces associated with swash and backwash.

4.7.2.1 *Measurement of waves*

Three critical wave parameters are usually measured: wave height, wave period and direction of wave approach. Methods of measuring or estimating wave characteristics fall into two broad categories: direct measurement of waves in the surf zone; and remote measurement or estimation of deep-water waves, which can be used to predict wave spectra incident on adjacent shorelines.

Measurement of waves in the surf zone The measurement of wave height and period will be considered separately from the measurement of wave direction, since different techniques are commonly used in assessing these parameters.

Wave height and period The simplest and cheapest method of measuring wave height and period is by visual observation. Ingle (1966) measured wave height either by comparison with a graduated pole held in the breaker zone or with a graduated pole held at still-water level against which the level of the horizon was measured. Wave period can be determined by noting the number of waves passing a fixed point in a given time interval.

However, to analyse the full wave spectrum incident upon a beach it is desirable to monitor changes in water level at small time intervals over relatively long periods of time using automatic continuously recording instruments. Recently, several electronic wave monitoring systems have been developed, which fall into four categories: wave staffs, wave wires, pressure transducers and ultrasonic devices.

(a) *Wave staffs.* Wave staffs (Koontz & Inman 1967) monitor changes in electrical resistance of a series of bare electrical contacts, which are usually moulded in epoxy resin at equal intervals along the length of the staff. These staffs usually have one of two methods of operation. First, total resistance is monitored as sea water closes the electrical circuit, and resistors in the internal circuit of the staff are selected so that current flow is proportional to the number of submerged contacts or to the submerged length of the staff. Secondly, the number of contacts submerged in sea water are counted during rapid electronic scanning passes, and again an electrical current proportional to the number of submerged contacts is produced. The variable levels of electrical current proportional to seawater level are usually passed along a cable for recording on a suitable device, although some researchers have used telemetry. Williams (1969) gives details of construction, circuit design and calibration of a parallel-type, step-resistance gauge and a relay-type, step-resistance gauge suitable for use in sea water. Greenwood and Sherman (1984) give details of the deployment of an array of seven continuous resistance wave staffs in conjunction with electromagnetic flowmeters to evaluate sediment flux and morphological response in a nearshore system.

(b) *Wave wires.* Wave wires consist of a pair of vertical wires, one bare and the other insulated. The capacitance between the wires varies depending on the length of wire submerged and can be monitored at short intervals to record changes in water level. This technique of wave measurement has proved particularly suitable for recording small waves in laboratory wave tank experiments.

(c) *Pressure transducers and ultrasonic devices*. Pressure transducers are installed on the sea bed and monitor changes in pressure with the passage of waves related to fluctuations in the level of the water column. There are several problems with the use of such gauges, including the fact that the wave crest is not exactly equivalent to the closed water column. Also short-period waves tend to be preferentially filtered out of the record with increasing depth of the water column. The pressure decrease with depth also limits the depth range in which such sensors can be used. A recent review of the applicability of some of these devices, with particular reference to piezoresistive pressure transducers, can be found in Barczewski and Koschitzky (1986).

Analysis of records of any of the above devices allows the determination of the spectrum of wave height and period incident on a shoreline.

Wave direction The simplest method of measuring the direction of an individual wave is to sight along a wave crest using a compass. Measurement of wave direction can also be achieved using an array of three or more of the instruments described above, which should be installed about a wavelength apart. Several instruments have been developed that are capable of automatic measurement of wave direction, including the CERC ultrasonic wave direction recorder and the Delft Hydraulics Laboratory Veldi wave velocity meter, which is suitable for wave periods greater than 10 s. Draper (1967) describes an electromagnetic flow meter that measures the motion of water particles beneath the sea surface to give information regarding the direction of waves. Horizontal velocities in two directions are measured and either a smoothed directional spectrum or the mean directions of waves can be obtained.

Cote (1960) describes the use of photogrammetric techniques for recording the three-dimensional characteristics of a small sea area employing low-flying aircraft covering areas of 813 m × 540 m, from which wave direction data can be derived. Wills and Beaumont (1972) give details of marine surveillance radars that are capable of indicating wave directions over relatively large areas in certain sea conditions. Hails (1974) gives details of a shore-based Decca Type 919 River radar that has been used to indicate wave directions at 3 h intervals.

Deep-water waves Although not as directly applicable to surf-zone research as the instruments described above, several instruments are available for the measurement of deep-water waves in the open ocean that can be used to predict wave spectra incident on shorelines. These instruments include Splashnik, which records wave accelerations, and wave rider buoys, which monitor vertical acceleration and the angles of roll and pitch, and can give useful information including the total spectral energy, mean direction of energy and angular spread of energy.

Several methods are available for estimating wave characteristics on the basis of meteorological data. The methods employed can be divided into those which are semi-empirical and those which are semi-theoretical. An example of the first type is the Sverdrup–Monk–Bretschneider method, which produces estimates of significant wave height and period of predicted waves. An example of the second type is the Pierson–Neumann–James (PNJ) method, in which the emphasis is on the prediction of wave spectrum. For a full account of the various methods, the reader is referred to King (1972), where methods for forecasting waves in shallow water will also be found, and Gardiner and Dackombe (1983). Bertotti and Cavaleri (1986) describe various models that hindcast littoral sediment transport on the basis of atmospheric pressure distribution, which is used to evaluate wind, wave fields, breaking conditions and coastal currents.

4.7.2.2 *Analysis of wave records*

Most wave records generated by continuously recording instruments will be complex and consist of additions of one or more individual wavetrains. Individual wavetrains can be determined using harmonic or spectral analysis (Munk *et al.* 1963). The analysis generates the energy density (the energy per unit frequency interval) for each frequency or period. The total energy in each wavetrain can be obtained by summing the energy densities under each peak generated in the analysis (Komar 1976). Wright *et al.* (1987) discuss the demodulation of wave (and tide) time-series data and the effects of wave groupiness on beach state. There have been many attempts to develop wave climate models based on deep-water wave measurements and offshore bathymetry. A good example is that of Goldsmith *et al.* (1973b) based on a specifically constructed Transverse Mercator projection that is capable of predicting the wave energy distribution of a large area of shelf and coastline in Virginia, which together with wave refraction computations should be able to predict areas of increased erosion and deposition and resulting shoreline changes.

Other simpler parameters are often determined from wave records, the most important of which is significant wave height, which is defined as the average height of the highest one-third of the waves measured over a stated interval of time, usually 20 min. The number of waves to be averaged is obtained by dividing one-third of the time duration of wave observation by the significant wave period, which is defined as the average of the periods of the highest one-third of the waves (Komar 1976, p. 68).

Other parameters that have been used include average wave height, root-mean-square wave height and maximum wave height. Lanfredi and Framinnan (1987) describe programs suitable for a hand-held calculator that relate wave characteristics to coastal morphology. Walton and Dean (1982) describe a computer algorithm to calculate longshore energy flux and wave direction from a two-pressure-sensor array.

4.7.2.3 *Orbital velocities of waves*

Inman and Nasu (1956) found a reasonably close agreement between actual particle velocities in a wave and theoretical velocities predicted by solitary-wave theory, which applies to waves in shallow water. Velocity (C) is determined by

$$C = \sqrt{8}\,(H + d) \tag{4.44}$$

where H is wave height and d is the depth of water below the wave trough. The same authors also produced a graph relating the theoretical maximum bottom velocity with the depth of water for various wave heights.

Any instrument used for measuring the orbital velocity of waves must be capable of both rapid and bidirectional response. The impeller flowmeter generally has a response rate that is far too slow, although several attempts have been made to employ such devices for measurement in the field. Wood (1968) describes a ducted impeller flowmeter that has a velocity range from 0 to 4.757 m s^{-1}, a hyperbolic response in relation to the axial flow and a cosine response to off-axis flow. The instrument was found to have excellent response in relation to the axial calibration, but to assess bidirectional flow two meters must be installed side by side facing in opposite directions.

To overcome the problems of slow response inherent in impeller flowmeters, Summers *et al.* (1971) developed a pendulum-type wave regime indicator that is capable of immediate response to changing velocities and associated stresses. Electromagnetic and small ultrasonic current meters now provide useful devices for measurement of orbital velocities in waves, particularly when installed in orthogonal pairs. Hallermeier (1973) discusses a three-dimensional laser Doppler flowmeter that uses three frequency-modulated light beams and one detector for measurement of the instantaneous velocity vectors in reversing flows.

4.7.2.4 Swash and backwash

Several methods are described in the literature for measuring swash and backwash velocities and energy, ranging from the relatively simple to the sophisticated. Miller and Zeigler (1958) timed floats over measured distances to obtain swash and backwash velocities. Dolan and Ferm (1966) deployed an array of ground control markers in the surf zone and used a chart recorder to assess the velocities of swash fronts. A major disadvantage is that backwash velocities cannot be measured using this technique. Schiffman (1965) and Kirk (1973) developed bidirectional compression spring dynamometers to measure both swash and backwash energies. Both instruments were connected by cable to chart recorders but have the disadvantage that the sensor plates are relatively large compared with the flow to be measured and seriously interfere with this flow. A modified version of the Kirk device is described by Caldwell *et al.* (1982). Various drag meters, usually consisting of strain gauges attached to spheres or cylinders, have been developed that respond to both acceleration-dependent forces and fluid drag forces in the surf zone (Smith & Harrison 1970). Hardisty (1987) describes the use of small ultrasonic current meters to measure the three-dimensional near-bed flow in the swash zone.

Wright (1976) used a photographic technique to measure swash and backwash velocities. Three brass rods, graduated in centimetres, were placed in the beach face a known distance apart and mutually perpendicular to provide control across and perspective into the picture. Film was shot at 16 frames per second, and swash and backwash velocities indicated by surface foam were calculated using a 'Specto' analysing projector. Measurement errors of velocities of $\pm 5\%$ exist due to the effects of wind and changes in the shape of pieces of foam between frames. The velocity direction was measured by tracing on the screen the paths of pieces of foam, corrections being made for perspective distortion. Water depth was measured with an error of ± 0.2 cm by reference to the brass rods.

Horizontal wave wires have been successfully used in studies of swash and backwash velocities, and Waddell (1973) describes a sophisticated system that is both expensive and requires considerable operational support.

4.7.2.5 Swash percolation, groundwater flow and water table

These three aspects of surf-zone processes are extremely difficult to measure and generally involve highly sophisticated and expensive instrumentation. Water table can be assessed by probes or more simply by digging holes or trenches to the depth of saturation. Harrison and Boon (1972) describe a sophisticated system that included 13 wells for monitoring water table and 26 pipe-stations for recording changes in beach elevation. However, this system is both expensive and requires considerable manpower and machinery, particularly for the installation of the wells.

4.7.2.6 Integrated surf-zone process monitoring systems

Several integrated surf-zone monitoring systems have been developed, particularly by research institutions in the USA, which are capable of simultaneous measurement and recording of several process variables. Although most of these systems are beyond the finances of most researchers, they offer considerable promise for the determination of the relationship between form and process in the surf zone. Good examples are to be found in Lowe *et al.* (1972) and Sallenger *et al.* (1983).

4.7.3 Measurement of beach morphology

Beach morphology orthogonal to wave approach is usually defined in terms of a beach profile. The simplest method of measuring beach profiles is the use of tacheometry at low water on beaches that have a large tidal range. Where repeated beach surveys are required, small-circumference wood or aluminium pegs can be installed in the beach face along surveyed profiles. The length of the peg standing above the beach face can then be measured at the required intervals of time. If beach change measurements are required over long periods of time, more substantial markers must be fixed into the beach face, but an allowance must be made for scour around the base of such obstacles. In all cases levels must be related to a datum point that can be related to tide levels. Instruments have been developed to record profiles directly as they are wheeled across the beach but have only been of limited value since they depend on a particularly firm surface.

Beaches with a small tidal range or experiments requiring simultaneous measurement of beach profiles and processes need more sophisticated equipment that is capable of extending a survey into the offshore zone. Inman and Rasnak (1956) describe a method of sounding depth from a boat, which is kept in position by aligning two markers on shore and fixing the distance by horizontal sextant angles to a third fixed point. Using this method the measurements in water up to a depth of 9.15 m fell within 1.52 m radius of the required point for 50% of observations and 80% were located within a 3.05 m radius. Zwamborn *et al.* (1972) describe a surveying method that employs a ski-boat equipped with an echo-sounder, which can travel through the breaker zone. The boat was kept on course by radio communication and two theodolites, one aligned along the proposed line of survey and the other intersecting the position of the boat periodically during the survey. The echo-sounder trace was marked at each intersecting fix to provide reasonable accuracy to within 0.5 m depth after corrections for tidal reductions.

Inman and Rasnak (1956) assessed the relative accuracy of recording profile changes on rods and using echo-sounding equipment. The standard error of rod measurements for four stations in depths varying between 5.5 and 21.3 m was 1.5 cm. The experimental accuracy of echo-sounding was ±15 cm in depths varying between 9 and 21.3 m. Sea sleds of varying designs have also been used to determine beach profiles. Reimnitz and Ross (1971) constructed a device propelled by wave power, which moved parallel to wave orthogonals and automatically recorded profile changes. Other sea sleds are dropped from helicopters and towed by cable to the shoreline through the breaker zone. The advantage of this type of system is that it can be operated under all sea conditions but a major disadvantage is that it requires a relatively smooth sea bed. Fleming and DeWall (1982) describe a package of computer programs for editing, analysing and displaying beach profile survey data.

Both time-lapse and time-interval photographic techniques have been used to assess shoreline changes. Cook and Gorsline (1972) describe time-lapse photography experiments, and Berg and Hawley (1972) describe a time-interval system mounted on top of a 3 m tower to observe both shoreline conditions and wave characteristics. A 16 mm ciné camera, housed in a weather-proof enclosure, was programmed to take film twice each day for a 20 s exposure. Remotely controlled video systems offer a considerable opportunity for the extension of this approach (Hails 1974).

The superposition of a long series of repeated beach profiles allows an assessment of longshore transfer of sediment and of erosion and accretion in the surf zone. The line joining the highest point on all profiles is a measure of the height above which the beach is unlikely to extend. The lowest points are a measure below which material is unlikely to be removed (King 1972). The zone between the upper and lower lines is called the sweep zone by King (1968) and is defined as the vertical envelope within which movement of beach material may take place by wave action.

The concept of the sweep zone is particularly useful since in the short term it indicates the maximum and minimum possible conditions of the beach profile and in the long term it indicates the rates of erosion and accretion, eliminating the effects of temporary variations due to the operation of waves of varying energy levels. The displacement of the whole sweep zone indicates long-term changes usually associated with the longshore movement of material.

The method known as the empirical orthogonal function (EOF) analysis of beach profile data has proved particularly useful for the analysis of both onshore–offshore and longshore patterns of sediment movement in the short term (Clarke & Eliot 1982, 1983). Carr *et al.* (1982) assess the spatial and seasonal aspects of beach stability based on repeated profile surveys, and Caldwell and Williams (1985) discuss the use of profiles to discriminate between different depositional environments. Recent applications of beach profile data to longer-term coastal stability have included Lisle and Dolan (1984) and Fisher *et al.* (1984).

4.7.4 Surf-zone sediment movement

4.7.4.1 Bedload

Although the measurement of bedload transport in the surf zone still depends heavily on the use of natural or artificial tracers, some progress has been made recently on the development of instrumentation to monitor movement of coarser grain sizes. Thorne (1986) describes the development of an instrument that monitors gravel bedload transport by measuring the acoustic self-generated noise produced by interparticle collisions. The instrument has been independently tested by comparison with information collected by underwater TV cameras. Application of this technology in the swash zone is provided by Hardisty (1987) in a study of gravel beaches in southern England.

However, the problem of direct measurement of sand-sized bedload by instruments still remains to be solved, and tracers are widely used in this context. Apart from the fortuitous cases in which naturally occurring tracers can be used, such as the heavy minerals used by Cherry (1966) in his study of longshore movement of sand on part of the Californian coast, tagged natural grains are used as tracers in most studies of beach sediment movement. This method is now well established and employs two main approaches to labelling natural sand: using radioactive isotopes (Caillot 1973) and fluorescent coatings to produce luminophors (de Vries 1973, Ingle 1966, Ingle & Gorsline 1973).

Radioactive tracers Radioactive tracers fall into two categories depending on the method of labelling grains. First, surface labelling involves the placing of a radioactive substance on the surface of natural grains. The most commonly used radio-isotopes are gold, chromium, iridium and scandium, and the choice will depend on the duration of the experiment and the nature, particularly the size characteristics, of the material to be labelled. A major disadvantage with this method of labelling is that particle size distribution must be transformed so that radioactivity is proportional to the mass of the sample (Caillot 1973).

Secondly, mass labelling involves the use of grains of glass, having the same density and hydraulic response as natural grains, which contain a radioactive element incorporated during manufacture. Disadvantages of this method are the cost of manufacture and problems associated with matching artificial grains to the natural population.

Luminophors Teleki (1966) defined luminifors as 'clastic particles coated with selected organic or inorganic substances which, upon excitation of 3560 Å or 2537 Å wavelength ultraviolet light, emit fluorescence of variable wavelengths and intensity in the visible region of the spectrum'.

Coatings are usually applied to natural sediment grains removed from the study site. Techniques for coating grains have not been standardised and a summary of several methods of labelling is given in Ingle (1966). Yasso (1966) also provides details of several coating formulations and the results of laboratory tests concerning thickness, hydraulic implications and durability of coatings. Most coating formulations consist of fluorescent paints in a water base that are not soluble in water after drying and are available in a wide variety of colours. A blue coating should, in general, be avoided since it is difficult to differentiate from skeletal remains in coastal sediments when exposed to ultraviolet light. Some coating formulations employ a resin binding agent to attach the fluorescent agent firmly to the sediment grain and to improve the durability of the coating, but this technique can cause serious aggregation problems during the production of the tracer.

The following technique for labelling grains is a variation of the coating formulations of Yasso (1966) and has the advantage that it is cheap, can be accomplished on site using a minimum of equipment and uses readily available materials. The natural sand to be coated must first be removed from the study site, air dried, and then mixed with three parts Day-glo spray or brush fluorescent paints and two parts xylol solvent. A small concrete mixer is advantageous for the mixing process, particularly when large quantities are involved, and avoids aggregation problems. After thorough mixing the luminophors must be dried before reintroduction into the beach system. Yasso (1966), using a similar coating method, found that the thickness of a coat was 5.94×10^{-4} mm on grains between 0.84 and 0.99 mm in diameter, and that simulated abrasion over a 18 h period produced a percentage weight loss of only 0.15%. In field experiments it has been found that coatings applied by this method can survive a period in excess of two weeks on a low-energy beach.

Experimental design Methods of employing sediment tracers in the coastal environment can be divided into three broad categories.

(a) *The time integration, or Eulerian, method.* This method involves the release of a known quantity of tracer at a point source and a continuous sampling of the moving bedload at a given distance downstream from the source. The mean velocity of tracer grains is determined by noting the time elapsed between the injection and the movement of peak concentration past the sampling point. An example of the use of this method can be found in Brunn (1968).
(b) *The dilution method.* Tracer grains are injected at a constant rate over a given interval of time sufficient to allow the stabilisation of the rate of tracer flux at a point downstream from the source. The velocity of grain movement can be calculated from the distance between the source and the sampling point and the time taken for the tracer concentration to become stabilised. The theoretical aspects of this method of tracer study are outlined in Russell *et al.* (1963).
(c) *The space integration, or Lagrangrian, method.* A known quantity of tracer grains are released at a point source, which is followed by periodical area sampling, usually on a grid, of the surrounding beach to establish spatial variations in tracer concentrations with time. Isolines of tracer concentrations can be constructed over the area to allow the determination of a centroid or centre of gravity of the tracer cloud in a given time period. The mean velocity of grain movement can be determined from the

distance of the centroid from the point source over a given time period. Examples of the use of this method are contained in Komar and Inman (1970)

Although the choice of method will depend on the nature of the problem, a few points of comparison can be made between the three approaches to tracer studies in the beach environment. A disadvantage with both the time integration and the dilution methods is that rates of sand movement can only be determined for a limited spectrum of vectors of potential sand migration. The time integration method allows only the velocity of sand movement, rather than the rate of movement, to be determined because of the lack of information regarding three-dimensional tracer distribution in terms of the mobile layer. There are also serious constraints in both the time integration and dilution methods since they involve assumptions regarding negligible loss in an offshore–onshore direction under oscillatory fluid flows. Both these methods also involve release and sampling in water, although this constraint also applies to the space integration method on beaches with a small tidal range. However, on beaches with a large tidal range the space integration method can be used to assess sediment movement over a tidal cycle when both release and subsequent sampling can be made under 'dry' conditions.

Release of tracers The method of release of tracers will depend in part on the choice of experimental design. In all cases, however, care must be taken to introduce tracers as bedload and not into the water column. A wetting agent, such as domestic liquid detergent, is particularly desirable in this respect. The method of release should keep disturbance to the natural system to a minimum.

Several methods of tracer injection in water satisfy these criteria and include the release from a container or plastic bag, which can be either soluble or cut with a knife. An ingenious method of tracer injection is to freeze the tracer in the form of a tile a few centimetres thick, which is placed on the beach surface or sea bed and rapidly melts to release the tracer. Tracer material has been injected down boat-mounted pipes for use with the dilution method.

The amount of tracer released depends on the objectives of the study, the duration of the study, the dimensions of the area in which the samples are to be collected and the sea state at the time of the experiment. Long-term studies of general patterns of sediment movement have used thousands of kilograms of labelled grains, whereas several tens of kilograms are used in short-term quantitative studies of sand transport.

Sampling tracer material *Radioactive tracers* Radioactive tracer concentration can be measured relatively easily and quickly using either Geiger or scintillation counters, although account must be taken of the fact that activity measured on the sea bed or beach may not be a direct measure of the mass of material present (Pahlke 1973).

Luminophors The method of collection of fluorescent tracers will depend, in part, on the qualitative or quantitative nature of the study. For quantitative assessment of littoral sediment movement, it is necessary to collect three-dimensional bulk samples, which include the mobile layer of the beach, by means of a core sampler. Kraus (1985) used tracers to establish the thickness of the mobile layer and discussed methods for the determination of mixing depth. Other methods of collection involve the sampling of surface grains only and employ equal-area grease-coated cards or plastic strips. Portable 12 V ultraviolet lamps can be used for *in situ* counting of the fluorescent grains. If the space integration method is used, it is advantageous either to sample during the hours of darkness using a portable ultraviolet lamp or to use a shaded light in daylight hours so that the spatial extent of the tracer cloud can be defined prior to retrieval of samples.

Determination of tracer concentrations In the case of radioactive tracers a measure of concentration is gained directly from the Geiger or scintillation counter readings. The most commonly used method of determining fluorescent tracer concentrations is to count the number of tagged grains by eye when the sample is exposed to ultraviolet light against a dark background. Ingle (1966) reports estimates that suggest that fluorescent particles are detectable by eye in concentrations of 1 in 10^6 unmarked grains.

Equal-area samples can be counted directly but bulk core samples require further treatment. To assess variability of concentration with depth, the sample can be impregnated with resin and sliced. Alternatively, after drying, the sample can be passed over a sheet of black paper sprayed with adhesive and excess material removed, leaving a layer of sand approximately one grain thick, which can then be exposed to ultraviolet light and viewed using protective glasses. Several electro-optic counting devices have been developed (de Vries 1973) in which samples are passed across the ultraviolet source and reflections are photomultiplied before being electronically counted.

Interpretation of results The time integration and dilution methods give direct measurements of velocities and rates of transport respectively. As noted previously, however, both these methods generate limited information regarding vectors of sediment transport.

Display of data generated by the space integration method is commonly in the form of a subjective hand-contoured map representing equal concentrations of tracer per unit of area, volume or weight. More objective contour maps can be produced using computer algorithms and tracer patterns can be studied using two-dimensional trend surface analysis. In terms of contour maps, estimations of the direction and rate of sediment transport can be made by calculating the centroid, centre of gravity or weighted mean centre of the tracer cloud.

Several attempts have been made to describe tracer transport in terms of stochastic probability and dispersion models, for example de Vries (1973). Quantitative analysis of rates of sediment transport using tracers has, however, proved a difficult problem in view of the uncertainty regarding three-dimensional attributes of movement associated with the mobile layer. A recent attempt to define the depth of disturbance based on marker stakes and rings is reported in Matthews (1983). The work of Komar and Inman (1970) still remains one of the most successful attempts to date to produce empirical relationships between synoptic measurements of the direction and flux of wave energy and the transport rate of sand in the beach environment using the space integration method. In a review paper dealing with nearshore currents and sand transport on beaches, Komar (1983) points to the scarcity of reliable field estimates of sand transport across the surf and breaker zones and the need for further research using these techniques. Price (1968) has argued that many tracer experiments are misleading, particularly in terms of onshore/offshore movement of material, and has also suggested that quantities of sediment movement should be measured at various levels on the beach rather than *in toto*.

Comparison of radioactive and fluorescence techniques Radioactive tracers have the major advantage that concentration levels can be measured rapidly and remotely using Geiger or scintillation counters, bearing in mind the problems of relating radioactivity to the mass or volume of the material present. Another major advantage of radioactive tracers is that all grain sizes, including silts, can be effectively labelled. Major disadvantages with radioactive labelling are the cost and the inherent health and pollution risks, although Courtois and Hours (1965) have suggested a safe code of practice for radioactive tracers.

There are several advantages in using the fluorescent labelling technique, which include the following:

(a) Naturally occurring clastic grains from a study site can be readily marked and the coating operation, in some cases, need only take a matter of minutes and does not need special laboratory facilities.
(b) The majority of coatings present no legal or health hazard and the cost of coating is relatively low.
(c) The solubility of the binding medium can in some cases be adjusted so that the coatings will adhere for periods of several days to several months, facilitating repeated surveys at one site in short time intervals or long-term experiments.
(d) Different colours can be used simultaneously on different size fractions or to differentiate between successive tests at one locality.

Estimation of initiation of sediment movement and mass bedload transport from wave data *Initiation of sediment movement* There are several empirical relationships for predicting the initiation of sediment movement on beaches under the action of waves, but probably the most widely used is that of Komar and Miller (1975), which relies on the determination of U_t, the near-bottom threshold velocity:

$$U_t = \frac{\pi d_0}{T} = \frac{\pi H}{T \sinh(2\pi h/L)} \tag{4.45}$$

where d_o is the orbital diameter, T is the wave period, H is the wave height, h is the water depth and L is the wavelength. Two equations relate threshold values for different grain diameters:

$$\frac{\rho U_t^2}{(\rho_s - \rho)gD} = 0.46\pi\left(\frac{d_0}{D}\right)^{1/4} \tag{4.46}$$

for grain diameters larger than 0.5 mm and up to 5.0 cm; and

$$\frac{\rho U_t^2}{(\rho_s - \rho)gD} = 0.21\left(\frac{d_0}{D}\right)^{1/2} \tag{4.47}$$

for grain diameters less than 0.5 mm, that is medium sands and finer fractions, where ρ is the density of water, ρ_s is the density of sediment (for quartz, 2.65 g cm^{-3}) and D is the grain diameter. Komar (1976) produces a graph relating wave period and near-bottom orbital velocity to grain diameter for quartz sand.

These equations are limited in applicability, however, in that they assume sinusoidal wave motion and do not take into account asymmetric wave motions in deep-water and shallow-water waves. In any event, results derived from the two equations are likely to be conservative since the interaction of wavetrains of differing periods and roughness elements on the surface of the beach or sea bed (for example, ripples) can generate high instantaneous bottom velocities.

Onshore/offshore sediment transport Inman and Bagnold (1963) produced an indirect model of sediment transport, which is based on the concept of an equilibrium beach slope governed by the asymmetry of energy dissipation in onshore–offshore directions. The orthogonal effect of swash and backwash processes are considered such that the equilibrium slope is a function of the ratio of local onshore and offshore dissipation rates (c). At balance, there must be an equilibrium between the amount of sand carried up and down the beach under wave action. Frictional drag on the swash and water percolation into the beach removing water from the return flow of the backwash produce a net shoreward movement of sand. This is opposed by local beach slope, which through the effects of gravity aids backwash in moving material offshore and produces a balancing offshore tendency. The relationship is:

$$\tan \beta = \tan \phi \left(\frac{1 - c}{1 + c} \right) \tag{4.48}$$

where $\tan \beta$ is the local beach slope and $\tan \phi$ is the coefficient of internal friction in shearing of granular media and is approximately equal to the angle of repose of the sediment. Inman and Frautschy (1966) suggest that:

$$c \simeq \left(\frac{U_m \text{ (offshore)}}{U_m \text{ (onshore)}} \right)^3 \tag{4.49}$$

where U_m is the maximum value of horizontal orbital velocity, evaluated at the breaker zone. This model of equilibrium beach profile appears to agree with the overall character of profiles of sand and pebble beaches, but it remains to be carefully tested (Komar 1976).

Hardisty (1984) developed the beach sediment transport equations of Bagnold (1963), which relate near-bed flows to shoreward and seaward sediment mass transport per wave:

$$J_{in} = (k_1 U_{in}^3 t_{in})/(i + s)$$
$$J_{ex} = (k_2 U_{ex}^3 t_{ex})/(i - s) \tag{4.50}$$

where J_{in} and J_{ex} are the shoreward and seaward mass transports (kg $m^{-1}s^{-1}$), U_{in} and U_{ex} are the maximum horizontal components of shoreward and seaward orbital velocities, t_{in} and t_{ex} are the durations of the shoreward and seaward flows (s) and i is the tangent of internal friction of the sediment. It appears that k_1 and k_2 have similar mean values of 12.78 kg m^{-4} s^{-2}. Jago and Hardisty (1984) combined these equations with the ideas of Inman and Bagnold (1963) discussed above to provide refined predictions of beach gradients. Another example of the consideration of the morphodynamics of a beach is contained in Wright *et al.* (1982), which considers the more complicated situation in a macrotidal environment. A comprehensive review of morphodynamic variability of surf zones is provided by Wright and Short (1984).

Longshore sediment transport The chief causes of longshore sand movement are longshore flow currents generated by oblique wave approach. One of the most frequently used equations to predict longshore sediment transport considers the relationship between longshore wave power and wave energy flux at the breaker zone:

$$Q_s = 6.85(ECn)_b \sin \alpha_b \cos \alpha_b \tag{4.51}$$

where Q_s is the rate of littoral drift (m^3 day^{-1}), $(ECn)_b$ is the wave energy flux at the breaker zone (W m^{-1}) and α_b is the angle of the wave breaking at the shoreline (Komar 1976, 1985). Komar (1985) notes that this model is restricted to conditions where the longshore sand transport is related solely to waves breaking obliquely at the shoreline. If other nearshore currents are involved, then the equations of Inman and Bagnold (1963) and Bagnold (1963) must be employed.

A full discussion of empirical and theoretical formulae for estimating longshore transport rates from wave data can be found in Komar (1976, 1985).

4.7.4.2 Suspended sediment

Measurement of suspended sediment in the surf zone is a particularly difficult problem since concentrations fluctuate rapidly with the oscillatory movement of waves. Several hand-held devices, including siphon bottles, can be used and time integration devices include pumping and modified siphon samplers (Noda 1967). Most of these instruments, however, are unsuitable for long-term or short-interval recording, and recent efforts have concentrated on the development of automatic electro-optical meters. Brenninkmeyer (1971, 1973, 1976) has developed an almometer that is capable of monitoring simultaneously the concentration of material in suspension and saltation throughout the water column and the instantaneous level of the beach surface. The almometer consists of 64 photoelectric cells spaced at 1–2 cm intervals mounted vertically in a watertight acrylic cylinder. At a distance of 30 cm from this tube is another containing a high-intensity fluorescent lamp, which provides a light source. The instrument depends on the interference of light by sediment particles passing between the source and photoelectric cells. Sternberg *et al.* (1986) describe a system of miniature nephelometers and suspended sediment samplers, which is designed to monitor the benthic boundary layer but may find application in the surf zone.

A major drawback with these instruments is the cost, and other cheaper, though in some ways less satisfactory, instruments have been used in the field. For example Cook and Gorsline (1972) developed small sediment traps that consisted of vials arranged in tiers above the beach surface to a maximum height of 30 cm. Such an instrument cannot trap all sediment particles in motion but can collect representative aliquots over a continuous range. Allen (1985) describes the use of a multi-elevational sediment trap of similar application. Hails (1974) reports the development of a sediment flux meter, which consists of suspended solids sampling nozzles that can monitor sand fluxes at given depths and measure changes in concentration of solids in different size fractions. Other examples of the measurement of suspended sediment in the beach environment can be found in Thornton (1972) and Fairchild (1972).

4.7.5 Coastal erosion

4.7.5.1 Erosion of solid rocks – shore platforms and cliffs

Processes Trenhaile (1987) provides a comprehensive review of the geomorphology of rock coasts. There are four main processes by which the sea achieves erosion of solid rock: chemical weathering, corrasion, attrition and hydraulic action. Since erosion usually occurs during high-energy conditions associated with storms, observation is extremely difficult and few techniques have been developed for the measurement of processes. Hydraulic action is probably the most effective agent of erosion, and Mitsuyasu (1967) describes a method using dynamometers to measure shock pressures of breaking waves. Empirically and theoretically derived relationships between wave characteristics and shock pressures are summarised at length in Trenhaile (1987).

Rates of erosion Robinson (1976, 1977) describes a modified version of the micro-erosion meter (MEM) developed by High and Hanna (1970) for use in the littoral environment, which is appropriate for measuring rates of erosion on shore platforms. Using this device on a shore platform in north-east Yorkshire, Robinson (1977) found that corrasion was the main erosive process on the platform ramp, and rates varied between 3.94×10^{-3} and 1.13×10^{-3} cm per tide depending on the wave energy and depth of overlying beach material. Refinements of the MEM and discussion relating to the interpretation of the derived data are contained in Gill and Lang (1983) and Shakesby and Walsh (1986). Donn and Boardman (1986) describe a profiling method for determining erosion and accretion of intertidal rock surfaces.

4.7.5.2 Erosion of unconsolidated cliffs

Wave action plays both a primary and a secondary role in the development of cliffs in unconsolidated materials. Cambers (1976) notes *in situ* attrition of glacial diamictons forming cliffs on the Norfolk coast during storms and storm surges. The retreat of cliffs in unconsolidated material, however, is primarily due to mass movement and other slope processes, which are assisted by waves through the removal of debris at the base and the maintenance of a high, unstable angle due to undercutting.

The measurement of the rates of erosion and retreat of unconsolidated cliffs usually relies on comparison of maps and aerial photographs. Examples of this method include Steers (1951) on the Suffolk coast, Agar (1960) on the Yorkshire coast and Valentin (1954) on the coast of Holderness, who estimates accuracy of map comparison to be ± 5%. Chambers (1976), working on the coast of East Anglia, compared 1:10 560 Ordnance Survey maps to assess cliff retreat over a 70–100 year period and used aerial photographs to update top positions. A general assessment of the use of photogrammetric techniques in the study of coastal changes is contained in Kidson and Manton (1973).

4.7.5.3 Historical data on coastal change

Historical maps, records, manuscripts and archaeology all provide useful evidence for assessing long-term trends in coastal erosion and accretion (Fulton & Olfe 1981). In terms of maps, Robinson (1962) discussed the art and history of the development of marine cartography. One of the major drawbacks is the inaccuracy of many early maps, particularly those produced before the Ordnance Survey began publishing material at the beginning of the nineteenth century. Of the early maps relevant to coastal change, those associated with harbour approaches and entrances are the most reliable and accurate. A good example of the use of historical maps in assessing coastal change is the work by de Boer and Carr (1969) on Orford Ness in Suffolk. The earliest cartographic evidence is a chart dating from the reign of Henry VIII. A long series of maps from the sixteenth century to the present time showed the great variability of the position of the distal end of the spit, which reached its maximum length between 1811 and 1813.

4.8
Biological aspects of process measurements

4.8.1 Introduction

This section describes a number of ways in which biological factors affect measurement of geomorphic processes and suggests ways of quantifying and measuring these effects. Almost all geomorphic processes are influenced by biological factors of some kind, so that the material considered here cuts across the subdivisions of Part Four, implying that similar techniques can be used. On the other hand, biological influences are sometimes extremely complex or concern very particular aspects, and they require specially adapted measurement techniques. Therefore, only if some method is already established or if it is applicable in a wider range of research is it described; if this is not the case, or if methods from preceding sections are suitable with slight modifications, the text is confined to drawing the reader's attention to the effects that the biotic community may have on geomorphic processes.

Also, gross differences in erosion resulting from man-induced accelerated erosion or, for example, from plagues of locusts are not considered here, because they can easily be detected by methods included in preceding sections (e.g. Secs 4.2 & 4.4) and in various soil conservation handbooks. It must be remembered, however, that, although not considered in this section, accelerated erosion induced by man is by far the most important factor controlling erosion rates and sedimentation over most of the world. The main object of this section is to draw attention to a number of biological factors that might influence hillslope and channel transport and erosion rates at the scale of the instrumented hillslope or small catchment, in the hope that by including a consideration of such factors in the experimental design, the interpretation of erosional and transport rates might be assisted. Consequently, the effects of, for example, coral reefs, mangrove forest, peat formation and macroclimatological effects of vegetation are not discussed here.

The whole field of biogeomorphology is discussed by Viles (1988), while Table 4.15 provides a brief guide to the literature on the great diversity of organic influences in geomorphology.

4.8.2 Hillslope processes

4.8.2.1 Vegetational effects on precipitation and soil moisture

Apart from the obvious reduction, through interception, of the amount of rainfall reaching the ground, the drop size distribution, intensity, momentum and areal distribution of rainfall are affected by vegetation. If the vegetation cover is low and complete, the above-mentioned effects lead to a considerable reduction in erosive power. If not, some floral properties need to be taken into consideration.

Table 4.15 Some organic influences in geomorphology: selected reading

Influence	*Authors*
Elephant digging	Laws (1970)
Porcupine and isopod activity	Yair and Rutin (1981)
Penguin striations	Splettstoesser (1985)
Gopher mounds	Hickman and Brown (1973)
Prairie dog burrows	Sheets *et al.* (1971)
Gannet turbation	Blakemore and Gibbs (1968)
Shearwater erosion	Evans (1973)
Ant mounds	Cowan *et al.* (1985)
Worm activity	Lavelle (1983)
Termite constructions	Goudie (1988)
Game wallows and pans	Alison (1899)
Bat erosion of caves	Hooper (1958)
Algal stabilisation of dunes	Van den Ancker *et al.* (1985)
Vegetation and sand mobility	Ash and Wasson (1983)
Vegetation and river channels	Hickin (1984)
Bacteria and travertines	Chafetz and Folk (1984)
Fungi and desert varnish	Taylor-George *et al.* (1983)
Bacterial attack on crystals	Bennett and Tributsch (1978)
Lichen weathering of limestone	Ascaso *et al.* (1982)
Cyanobacterial rock attack	Danin (1983)
Lichens, stromatolites and calcretes	Klappa (1979)
Rainforest nutrient cycling	Jordan (1985)
Tree throw and tree fall	Mills (1984)
Debris dams on rivers (log steps)	Marston (1982)
Rock grazing gastropods	McLean (1967)
Boring molluscs	Craig *et al.* (1969)
Boring sponges	Acker and Risk (1985)
Goose-induced marsh erosion	Dionne (1985)

Source: table compiled by A. S. Goudie and H. A. Viles.

Effects of plant physiology on drop size distribution At low rainfall intensities, water drops collect on leaves to form larger drops, which flow off their surface and attain, if they are not intercepted again, a certain velocity depending on their fall height, or break into smaller droplets if they have a critical size compared to their terminal velocity. Generally, at low rainfall intensities, the number of large drops is increased by the presence of leaves; at high intensities, rain contains a relatively higher proportion of large water drops, which tend to become reduced in size by the presence of vegetation cover. These changes in drop size are reflected by changes in energy or momentum of the drops. Chapman (1948) concluded from a study in a red pine plantation that the kinetic energy of a rainstorm was greater under the canopy than in the open at intensities of less than 50 mm h^{-1}. Of course, the momentum or energy of the drops as they reach the ground is controlled by the distance they can fall from the canopy.

According to Trimble and Weitzmann (1954), a distinct seasonality is present in the protection afforded by vegetation. High rainfall intensities (above approximately 12 mm h^{-1}) are reduced less by a summer canopy due to the 'drooping position' taken up by leaves when they are beaten by rain of high intensity, causing practically all water to run off. On the other hand, high intensities are reduced more by a winter canopy.

Besides this, vegetation type also influences drop size distribution; for example, drops escape more easily from needles than from leaves, resulting in a comparatively smaller drop diameter in a coniferous forest (Geiger 1975), where seasonal differences are also less pronounced.

The dynamic force of falling raindrops can be registered directly with, for example, a (piezoelectric) quartz-crystal sensor, a microphone, a transducer, etc. The measurement technique is still more or less in an experimental phase, but for practical reasons it is advisable to mount the sensor in some kind of rain gauge to ensure simultaneous recording of total precipitation or, if possible, precipitation intensity.

If it is not possible to make direct measurements of drop size distribution and/or kinetic energy, measurements should be made of the crown height of trees (a simple procedure is to use a clinometer) and fall height (the distance from the lowest foliated branches or otherwise dense canopy to the ground can be substituted for this). These two parameters yield the crown depth, which is related to the obliteration of rainfall and thus to interception and drop size distribution. Furthermore, estimates must be acquired of the leaf cover (a very crude estimator is the time of the year; better indicators are vertical photographs taken either from the air or the ground, from which the cover can be evaluated, or estimation of cover with the help of the 'Charts for estimating the percentage of mottles and coarse fragments' from the FAO (1975)).

Effects of vegetation on areal distribution of precipitation Linskens (1951, 1952) made numerous precipitation measurements under isolated trees. It appeared that under foliated conditions precipitation concentrated around the periphery of the tree ('gutter' effect) and along the stem (stemflow). The inclination of tree branches (towards the tree, e.g. beech and oak; away from the tree, e.g. willow; or nearly horizontal, e.g. maple) seems to determine the distinctness of the maxima, which can reach values of more than 150% of the rainfall in the open. Apart from stemflow, rainfall distribution in winter conditions under a single tree is fairly evenly spread. In a closed stand, the spatial distribution of throughfall is irregular, but nevertheless a pattern somewhat like the isolated tree case can be found for every single tree in the stand. So the best way to gauge precipitation, for most research purposes, is (a) to collect stemflow from several trees (e.g. by attaching a circumflexing plastic collar around the tree trunk) and (b) to collect throughfall with the help of several series of troughs (see, e.g. Geiger 1975) of perhaps 100 cm length and 10 cm width, positioned in line from the stem to the rim of the projected crown area, or beyond that if the crown cover is not continuous. A trough of the size specified gives an equivalent of one litre of water to one centimetre of rain.

Only the throughfall part of precipitation can have a direct influence on splash erosion, provided that the drops attain a momentum above the particular threshold value needed to dislodge and transport soil particles or aggregates. On the other hand, the stemflow may very well generate overland flow, even over permeable forest soils, by the concentration of water on a relatively small area near the trunk. To give an example, estimates of Gersper and Holowaychuck (1970) indicated a five-fold amount of water reaching the ground there, compared to the open ground.

Influences of vegetation on water in and on the soil The water reaching the ground usually has to pass through a layer of decaying leaves and other organic residue before it can enter the soil. The presence of this litter has several effects. The water is prevented from reaching the ground directly, so that some of it can easily evaporate. The litter layer absorbs the impact of falling drops and also has an insulating function. Furthermore, it nourishes and protects the soil flora and fauna, which permeate the soil, stimulating infiltration. The water flow through the litter cover will have a component of movement parallel to the soil surface if the land is sloping. This litter flow can occur over a range of distance from a few millimetres up to several metres depending on factors such as the thickness and disturbance of the cover and the slope angle. Occasionally, hardwood leaves may pack tightly if the leaves are lying parallel, and, since dry organic matter is often hydrophobic, water may flow over the leaf cover.

Litter interception may be estimated by collecting litter samples during various stages of drying and wetting, and weighing to determine the moisture content. This method permits natural drainage and is therefore to be preferred to those using leaf samples in containers that drain to storage tanks. Litter flow is monitored with the help of troughs or overland flow traps, installed in such a way that the litter cover is not disturbed.

The water reaching the litter cover from either stemflow or throughfall has usually had a residence time on the leaves or branches of the plant and subsequently contains a much higher content of elements than rainfall in the open (see e.g. Gersper & Holowaychuck 1970). The major source is thought to be aerosols adsorbed on the organic surface. The chemical composition of the water is changed again during passage of the litter cover, since it leaches the organic fragments (see also Sec. 4.3).

The redistribution and spatial variation in amount and composition of the water reaching the soil under a vegetation cover naturally influences soil-forming processes and organic activity. There is not much general knowledge on this subject, but it seems that the pH of a soil is lower close to the tree trunk and that there is a decrease of total nitrogen, organic carbon and exchangeable cations from the tree trunk outwards, reflecting the influence of stemflow, which adds the bulk of elements to the soil. Furthermore, since soil moisture storage and movement are intimately related to the occurrence of mass movements, the withdrawal of soil water for transpiration by the vegetation will have a stabilising effect.

4.8.2.2 Effects of vegetation and litter cover on transport of material

Direct influences There are several events that initiate or cause the transport of material and are directly related to the flora. The most important processes will be considered.

The falling of trees, occurring especially in undisturbed tropical rainforests, but also in humid temperate environments due to gales, may happen through the breaking of the trunk above the ground surface, or by uprooting. In the latter case, the roots are thrown up into the air, usually together with an appreciable amount of soil material. First, the area denuded in this way is exposed to raindrop impact; secondly, since the trees usually fall in a downslope direction, the uplifted material is displaced downslope when it is washed to the ground again. In recent years the importance of this process in terms of slope movement has become more apparent, and there is now a substantial literature dealing with the frequency, magnitude and distribution of its effects (Table 4.16).

When the forest stand is dense enough, one falling tree may detach others. Cailleux (1957) mentioned a frequency of 14 clearly recognisable cases of uprooting per hectare in Guyana. Lutz (1960) reported transport of material due to this effect over distances up to 3.5 m in vertical or horizontal sense. Measurements should involve the number and frequency of occurrence, the volume or weight of the translocated material, and the distance of transport.

Denny and Goodlett (1956) tried to quantify the effect of tree fall on hillslope erosion. They estimated the volumes of the mounds/pits produced by assuming them to be half a prolate spheroid, so that the volume of each is given by

$$V = 0.667\pi ab^2 \tag{4.52}$$

Table 4.16 Some recent studies on the importance of tree throw by wind.

Location	*Authors*
Virginia, USA	Mills (1984)
Southern Appalachians, USA	Lorimer (1980)
Ithaca, NY, USA	Beatty and Stone (1986)
Massachusetts, USA	Veneman *et al.* (1984)
Sri Lanka	Dittus (1985)
South Africa	Versfeld (1980)
Himalayas, India	Pandey *et al.* (1983)
New Zealand	Cremer *et al.* (1982)

where a is half the major axis of the spheroid (i.e. the length of the mound) while b is half the minor axis. Their method has, however, been criticised by Mills (1984).

Another process is mechanical breakdown and transport by the expansion of growing roots. The penetration depth of roots in fractures and fissures can sometimes be considerable, depending on the nature of the rock or soil (e.g. in permafrost or some peaty areas 20–40 cm; in dry sandstones to 15 m). The expanding roots may detach soil or rock material, but on the other hand they may also bind or stabilise (Pitty 1971). There are no generally accepted measurement techniques for these effects. The factors to be measured depend strongly on the aim of the research: for engineering purposes, the pressure exerted by the expanding roots could be measured; to evaluate binding forces a measurement of rooting density and penetration depth might be valuable.

The process of litter transport can be significant in two ways for the transport of material. First, the organic and inorganic material accumulated in the vegetation is returned to the soil when leaves are shed or when the plant dies. This uptake and subsequent deposition may lead to a net downslope transport of material on a sloping surface, like the transport of material induced by frost creep. Quantification of this process should include estimates of the weight or volume of every fallen tree, the direction in which the trees are fallen (upslope or downslope), and the production of litter, using, for example, large collectors. Those collectors should be suspended somewhat above the ground surface, to prevent blowing in of loose surface litter material, and they should be drained since the presence of water promotes decomposition. For whole trees, the weight or volume can also be estimated by finding the correlation coefficient between the diameter of the stem at breast height (*DBH*) and the dry weight (Bunce 1968). The inorganic residue of the materials can be assessed by chemical analysis of samples. It must be mentioned that litter fall and production are very variable spatially as well as temporally. For more elaborate discussions on this subject, reference can be made to Bray and Gorham (1964) and Dickinson and Pugh (1974).

Secondly, the organic material accumulated on a forest floor may be in movement on steep slopes, due, for example, to wind action or animal activity. This transport causes a non-random redistribution of the (in)organic material, but more importantly it also appears that mineral soil material is usually attached to the organic fragments, in particular leaves, because it is splashed onto them. In this way, an appreciable quantity of material is sometimes transported downslope (van Zon 1978). Measurements of this transport can be performed with all kinds of collectors, provided that they do not modify the wind influences too much, and that the attached material cannot be washed away by rain after collection.

The presence and distribution of patches of exposed soil under a forest cover greatly influence geomorphological processes. For instance, splash erosion may be more active there. The causes of bare ground exposure can be manifold: trampling, burrowing, overland flow, litter transport, etc. In general, exposure of bare ground occurs more frequently at the forest boundary, on steep slopes and at places with poor productivity rates (e.g. a young plantation). Gross differences in the amount of bare ground exposed can be attributed to differences in type of vegetation, soil and climate and to the related rates of mineralisation and decomposition. Sometimes even the entire forest floor may be cleared from litter shortly before the annual leaf fall.

Exposure may be estimated from plots by using the 'Charts for estimating the percentage of mottles and coarse fragments' from the FAO (1975), or by measuring the surface actually exposed on many small plots. It is also possible to perform registration at regular intervals along a transect. It must be borne in mind, however, that bare ground exposure is extremely variable within short distances and within short timespans.

Indirect influences Indirectly, the vegetation and litter cover have an influence on sediment transport via the effects that they have on the soil. Among these, the effects on soil structure, infiltration and erodibility are the most important. It has already been mentioned that vegetation controls to some extent the distribution of moisture in a soil, and this has its impact on soil-forming processes. Also, the shading effect, higher humidity and other microclimatological differences, as well as organic-matter production, rooting, soil chemistry, etc., are induced or influenced by the presence of vegetation cover. Vegetation is thus accepted as an important soil-forming factor.

Vegetation can be quantified in many ways. Most methods relevant to the geomorphologist will consist of measurements of vegetation density and species composition. Floristic analysis usually relies on sampling techniques, and the most well known are the transect method (line transects or belt transects) and the 'quadrat' method. In the former method, every specimen that touches or shades a straight line or belt is recorded, together with the relative distances along the line. This method is recommended for rather homogeneous communities. The latter method consists of sampling in a delineated area, which can have any form. Its size is adjusted to the phytocenoses; it depends on the physiognomy and structure of the plant community and also on its floristic homogeneity (e.g. in dense grass 1 m^2 will suffice; in an open forest 100 m^2). Species are recorded by name and coverage, i.e. the area covered by its vertical projection on the ground; see Küchter (1967), Brown (1968) and Mueller-Dombois and Ellenberg (1974) for more detailed discussions concerning vegetation sampling techniques.

In general, it can be stated that under a permanent vegetation cover, and especially under forest that possesses an additional protecting litter cover, infiltration rates are typically higher than on cultivated land on comparable parent materials. These high infiltration rates are the result of the spongy nature of the organic top horizon, the comparatively thick and porous weathering mantle usually present under forest, and the ample presence of organic material and soil fauna and flora, favouring aggregate formation and stabilisation (Baver *et al.* 1972) and permeating the soil. Consequently, Hortonian overland flow hardly ever occurs under forest, and the forest soils are also less erodible compared to their cultivated counterparts due to this better aggregation and permeability.

On the other hand, surface horizons of some forest soils have a definite water-repellent nature, especially in the stage of initial water contact. Plant litter accumulations are clearly correlated with water repellency; in general, repellency is greatest when the soil is dry and it seems to be associated with the presence of fungal mycelium; also, sandy soils are more repellent than soils with a relatively high clay content (Letey *et al.* 1975). Water repellency is increased by high temperatures (e.g. from burning), which affect the surface litter. It has a definite influence on infiltration rates and often interferes with the measurement of this parameter.

Water repellency can be measured in various ways. A simple procedure is the 'water drop penetration time' (*WDPT*) (Letey *et al.* 1975), which consists of placing a water drop on the surface of the soil and measuring the time needed for it to penetrate. If the contact angle between liquid and solid is less than 90°, water is drawn into the soil by capillary forces; if not, negative capillary forces predominate, and in theory the water drop should remain on the surface for an infinite time, though not in reality because reactions would take place between the water and the soil surface, lowering the contact angle value. So the *WDPT* is related to the initial water repellency. More sophisticated methods for measuring water repellency are based upon comparison of capillary rises (Letey *et al.* 1962) or on contact angle–surface tension relationships (Watson & Letey 1970).

4.8.2.3 Effects of animal activity

The burrowing, trampling, tunnelling and other activities of (soil) fauna can have a pronounced influence on geomorphic processes. Darwin (1881) was probably the first to recognise these effects, when he investigated the role of the action of worms in the denudation of the land, and ever since it has been shown that animals influence erosion (see e.g. Pitty 1971). Jońca (1972), studying the erosion induced by mole activity, introduced the term 'zoogenic erosion', defined as the transport over slopes of superficial zoogenic forms. Animal activities are manifold and diverse, so a selection must be made in the discussion of their effects, which might otherwise range from the production of guano to the wallowing of elephants, which are only of local importance.

Worms and other lower animals The effects of these are important enough to justify separate treatment. Worms bring soil material to the surface in the form of casts, which may be subsequently transported downslope by rain, which liquefies the material so that it flows downslope, or by rolling of the (sometimes towering) dry indurated casts down the slope as a whole or in parts (Darwin 1881). The effects may be quantified by counting the number of worm casts per unit area on a representative plot and by measuring and weighing their liquefied remnants after rain, or by registration of the rolling of casts during dry periods on differently inclined slopes. Darwin (1881) estimated the rate of deposition of casts on English grasslands to be between 1800 and 4400 ton $km^{-2}yr^{-1}$. The burrows produced by worms raise the permeability of the soil and induce homogenisation, also because leaves or other organic fragments are often pulled into them by the worms. Especially when burrows are not vertical, subsurface downslope movement of material will take place upon collapsing. Stones lying on the surface may slowly subside due to the burrowing action of worms. Burrows are not easy to trace, however. Pouring in of some liquid that hardens upon drying (like plaster of Paris) may be a promising method (Imeson, pers. comm.). Worms may burrow to a considerable depth (down to 2.5 m). Their burrows do not branch very often. Enormous numbers of worms can be present in a soil, though their activity depends on the time of the day, the season (temperature) and so on. Counting worm casts may give a good general impression of total activity, but not of the total number of worms present.

Apart from erosion and homogenisation, worms seem to promote aggregate formation (Baver *et al.* 1972). In general, every disturbance of the soil, even by microbial action, causes fragmentation of clay skins and the incorporation of the fragments into the interior of aggregates; also any rhizosphere microorganism plays directly or indirectly a role in aggregation through polysaccharide formation.

Other small animals in the soil comprise insects, of which some are notorious mound builders, with the termite as an excellent example. It influences the soil in many ways (Lee & Wood 1971), the mound being their most conspicuous product. Most of the large mounds in tropical areas are built by termites. They usually consist of topsoil material that is devoid of stones (in general of particles larger than 1 mm), may have a textural composition different from that of the original material (Pomeroy 1976, Watson 1976), while possibly the chemical composition is changed (reflected by an increase in exchangeable cations, pH, electrical conductivity and organic carbon). Some opposing views can be found in the literature, probably because of differences in environmental factors and insect type. Hesse (1955) states that chemical changes are unimportant; he maintains, for example, that a better growth of some plants on termite mounds is related mainly to better drainage and improved structure. Omo Malaka (1977a) reports that the crust of some mounds may be poorer in chemical constituents. He also argues (Omo Malaka 1977b) that high bulk density of termite mounds might explain the absence of vegetation on them.

The effects of termite activity include the removal of litter, which is digested thoroughly, leaving few end-products. The selection of fine soil particles for mound construction and the bareness of the mound surface make them susceptible to erosion, as does the comparatively low infiltration capacity of the mount (Omo Malaka 1977a). The eroded products of the mounds may cause a stone-free layer to be deposited on the surface. Williams (1968) mentions deposition rates of 2–3 cm/1000 yr. In some cases, it is suggested that a stone line at some depth in the soil may represent a base of termite activity. Also the termite mound provides a kind of 'umbrella' for the underlying soil, retarding leaching and shedding rainfall (Watson 1976). In humid areas, ant mounds may serve similar functions. Baxter and Hole (1967) reported on ant activity on a Wisconsin prairie, the mounds showing a concentration of organic materials and consisting of 85% of B horizon material. Heights and diameters of mounds were measured with callipers. On average, mounds were 15 cm high and 40 cm diameter and occupied 1.7% of the ground surface.

The activities of other lower animals are not widely accepted as being of appreciable significance for erosional processes. Nevertheless, they may be of importance locally. Examples are the reworking and sometimes sorting of coastal sediments by benthic animals like crabs and aquatic worms, the attack of indurated materials by a variety of mechanical and/or chemical borers (Pitty 1971) and gastropods and chitons that rasp algae from the surface of rocks. Evans (1968) estimated an erosion rate of some 12 mm yr^{-1} mainly due to the action of the rock-boring clam *Penitella penita* in certain localities along the Pacific coast of North America. He used stainless-steel screws, placed flush with the rock surface, for his measurement of normal erosion, and his estimates of zoogenic erosion were derived from regression analysis between the deviation of the dimensions of the burrows from the normal shape due to erosion and the (known) growth rate of the clam, the latter providing a timescale. Van der Pers (1978), investigating the bio-erosion by *Polydora* along coasts, collected rock fragments by scuba-diving and with dragnets. The samples were stored in running sea water, and the number and sizes of animals and their holes were determined. Size of holes is usually recorded as depth and entrance diameter.

Mammals Burrowing animals bring a lot of material to the surface, which is heaped up in mounds near the entrance of the tunnels. The main effect of this is that material may be exposed above a protecting litter layer, subjecting it to rainsplash, while it may also be moved downslope in the course of digging. Besides, the material, particularly at the top of the mound, may come from lower soil horizons, and may be more erodible than other exposed superficial material (Imeson 1976, Imeson & Kwaad 1976). To evaluate these effects, the amount of excavated material and the area affected must be recorded. For this purpose, Imeson (1976) used 22 plots of $5 \times 15\ m^2$ situated along a slope profile. On these plots all newly formed mounds were recorded, measured and numbered each month. Other mounds were measured and weighed, and their composition was determined. Still other mounds were marked with erosion pins, or splashing from their surface was registered (Imeson & Kwaad 1976). The form of fresh mounds and their situation relative to the tunnel gives an indication of the displacement of material due to burrowing. On sloping surfaces, mounds are usually elongated and the largest portion of the material lies downslope of the tunnel entrance. The disturbance and subsequent sediment transport as recorded by Imeson and Kwaad were mainly afforded by moles and voles. They mention a raise of material of $1940\ m^3\ km^{-2}\ yr^{-1}$ in a small catchment in Luxembourg, and this material is for the most part subsequently transported downslope by splash until the mounds are protected by freshly fallen or transported leaves (see also van Zon 1978). The activity of moles in digging new tunnels and the direction of this digging seems to be related to the abundance or lack of food in the soil (see e.g. Godfrey 1955). Thus the extent of the moles' burrowing activities is also controlled partly by land use and/or vegetation type. Furthermore, Abaturov and Karpachevski (1966) also mention the fact that the mounds show lower porosity than the surrounding surface soil, and that mole activity changes the soil moisture regime, the passageways draining the top layers and bringing about a deeper wetting of the soils along the edges of the tunnels and the formation of dry areas under their central part.

Several animals other than moles make large mounds and excavate large tunnels, for example, ground squirrels, prairie dogs, rabbits and other rodents. The activity of rabbits in dune areas can cause the onset of serious wind erosion. The ground squirrel especially is a notable earth handler. From personal observations in Africa, it became clear that this squirrel also sometimes plugs its hole, presumably to protect itself against danger and/or bad weather conditions. These plugs are very hard and compact, and often cannot be broken by hand.

Other animals, such as pigs and deer, may also be relevant to sediment transport, by rooting, wallowing and scraping of the A horizon. They directly displace the soil material, disturb the litter layer, expose and compact the soil, and exert a negative influence on the vegetation, by keeping the understorey open. These effects can be quantified by counting and measuring the affected areas and evaluating the change in soil properties such as hydraulic conductivity. Indirectly, an evaluation can be made by estimating the number and kind of animals present through consideration of grazing or browsing patterns, through determining the density or frequency of their droppings, or through use of aerial survey techniques (see e.g. Norton-Griffiths 1978). Apart from aerial surveys, it must be realised that these methods do not indicate the number of animals directly; to obtain the latter, additional estimates of (re)growth of vegetation and decay/production rates of droppings are required. Otherwise the methods yield relative values.

Overgrazing causes severe erosion, and many methods suitable for investigating accelerated erosion are discussed elsewhere in this manual. The same applies for the accelerated erosion induced by man. However, some effects of human activity are often overlooked, and one of these is the erosion induced by recreational pressures. The impacts of recreational activities are, according to Seibert (1974), mainly: (a) the treading or trampling on sites, causing damage to plants, compaction of the soil, changes in the soil organic material and changes in the microclimate, (b) eutrophication, mainly by pollution, and (c) other damage caused by playing games (e.g. skiing), digging, removal of plants or parts thereof, fire, etc. All these cause an acceleration of erosion, but it is difficult to make quantitative statements on these or to recommend methods of measurement, because the highly diverse and complex nature of the subject does not allow generalisations. A simple, easily applied general condition classification is presented by Frissell (1978). He distinguishes five classes, based on use-related site changes:

Class 1: Ground vegetation flattened but not permanently injured: minimal physical change except for possibly a simple rock fireplace.
Class 2: Ground vegetation worn away around fireplace or centre of activity.
Class 3: Ground vegetation lost on most of the site, but humus and litter still present in all but a few areas.
Class 4: Bare mineral soil widespread, tree roots exposed on the surface.
Class 5: Soil erosion obvious, trees reduced in vigour or dead.

It is clear that, for research purposes, important variables to be considered are the ability of the site to resist pressure, and the duration, type and intensity of the applied pressure. To give an example, Bryan (1977) reports the degradation of land caused by mountain hiking. He found that severity of trail damage is related to soil properties and profile characteristics where use intensity is uniform. Use intensity is difficult to establish, but information of local tourist offices, the presence of trails on tourist maps, their identification by markers and signposts, and the effort needed to follow them may offer support.

Bryan (1977) recognised some stages of footpath erosion, which can be generalised to, first, vegetation disturbance and disappearance of sensitive species, then exposure of soils to erosion, resulting usually in a quick removal of the organic horizon, after which the underlying compacted materials generate increased runoff and erosion, leaving a lag deposit of stones and boulders. Gullies may develop, eroding laterally or vertically depending upon the nature of the underlying materials. The speed with which these stages succeed one another depends on the vulnerability to erosion of the materials, but also on trail direction with regard to the slope of the land, use intensity and practice (e.g. concentrated in wet or dry season), and the sensitivity of plant species to treading. A crucial point is that, after the vegetation has been disturbed, the footpath erosion may proceed even without human intervention. On the other hand, if footpaths are used continuously, walkers avoid the strongly eroded areas, promoting further erosion on nearby spots. Trail conditions can be measured with all kinds of devices, but stereo photography appears particularly suitable for this.

4.8.3 Channel processes

4.8.3.1 Effects on water flow

The influence of vegetation on water supplies is treated extensively in the literature; the determination of water balances, measurements or estimates of evapotranspiration and precipitation, etc., fall well outside the scope of this text, however. Nevertheless, it must be realised also that minor animals present in large numbers may influence water supplies albeit in an indirect way (see Bethlamy (1974), who reports increased streamflow after an insect epidemic), as do small numbers of larger animals (see Sartz and Tolsted (1974), who detected differences in streamflow due to normal grazing practice in only parts of a small catchment).

A direct effect on water flow is induced by the presence of aquatic weeds in a stream channel. This reduces the channel cross section and increases its roughness, thereby changing stage–discharge relationships and thus interfering with measurements, causing errors up to 50% at low discharge in extreme cases. Allowance can be made for this effect by the use of roughness factors in discharge computations (like Manning's *n*; see also Brooker *et al.* 1978) or by the construction of stage–discharge relationships for various growing stages of the weeds. It must be kept in mind that these effects of vegetation may be highly seasonal (Dawson 1978) and also local. The distribution of plant stands can be established from site visits along the river; a method for estimating the biomass might be an optical technique (Owens *et al.* 1967) or sampling.

The supply of organic material to the river from its sides may also interfere with flow, depending on the size of the organic fragments. Leaves and the like will not have much effect, but roots penetrating the channel from the side, or branches hanging in the water, will cause notable reduction of flow. Most conspicuous is the influence of logs and heavy branches that have fallen into the stream and that act as small dams and barriers, often incorporated into the river bed. Heede (1975) reports a rather regular spacing of these obstructions, random mortality of the trees probably being responsible for this, and a considerable reduction of streamflow energy in small mountain streams.

Locally, also, animals may influence streamflow; the dams built by beavers are famous. Also, fish may be of importance, for example, through removing aquatic vegetation (e.g. the grass carp).

4.8.3.2 Effects on sediment transport

One of the most prominent effects of obstruction to flow by vegetation, beaver dams, etc., is the entrapment of sediment. Silt may be deposited between vegetation, and may even remain after decay and subsequent removal of the organic material (Dawson 1978). Especially upstream of coarse organic fragments (logs and heavy branches), notable sedimentation takes place, while further adjustment of the stream is generated through gravel bar formation. A stepped longitudinal profile is then created (Heede 1975) with pronounced dissipation of energy below each step so that scour holes may develop. Clearing the river of these organic obstructions or otherwise interfering with their function (e.g. by cutting and removal of dead and dying trees) will result in a rapid removal of the stored sediment from a catchment. Also the removal of aquatic weeds (by fish, or by man through mechanical clearing or use of pesticides) will increase the sediment yield.

Animal activity may be well related to sediment production; the trampling of mammals on the side of, and in, the river may cause a momentary peak in sediment transport. Besides this, the activity of burrowing animals is sometimes greatest just alongside the river (Imeson & van Zon 1979), giving rise to high sediment production rates.

An often forgotten factor in the measurement of sediment discharge is organic material. Organic fragments floating in the river may contribute to this sediment output. Van Zon (1978) estimated that 2% of the total annual production of leaves in a small catchment in Luxembourg was transported by the river. More important in this context is the fact that sometimes a considerable amount of sediment may be attached to the floating leaves, branches, etc., and so escapes from sediment sampling. Also, the importance of fine organic residue or dissolved organic compounds is often neglected (Arnett 1978). They can play a significant role in forming stable complexes, altering the chemical and physical characteristics of sediment, but organic material may also form a considerable part of the sediment output. According to Arnett (1978) determination of the organic fraction may be done by quantitative wet oxidation using dichromate, and for particulate sediment by filtering through precombusted glassfibre filters and heating to remove oxidisable material. The filtered water samples must be evaporated at 85°C to prevent chemical changes or spontaneous oxidation. From subsequent quantitative oxidation, the consumed oxygen can be determined, from which an equivalent of organic matter can be obtained. Arnett recommends a general figure of 1.43 as the ratio of weight of oxygen required for complete oxidation to the weight of organic matter.

Finally, it must be appreciated that the chemical and physical composition of natural waters is influenced by aquatic life, either by the presence of certain specific species or by prolific growth. Furthermore, recreational use, such as swimming and boating, also modified water quality.

4.8.4 Biological indicators for geomorphological processes

To recognise geomorphological processes in the field, every researcher consciously or unconsciously makes use of indicators, since the real activity is not often encountered. Most of these indicators are manifest and to some extent already mentioned in the foregoing sections, for example presence or absence of vegetation, animal droppings, worm casts, browsing patterns, or height of residual vegetated soil pedestals. Some other less obvious indicators will be discussed here. One of these is the exposure of shallow roots, for example following heavy forest range grazing use. Erosion intensity can be estimated from the age of the tree (or the time elapsed since a major change in land use) and the elevation of the original ground surface, correlating the inflection point between the stem and the first root outgrowths (see Sec. 5.5). The age of a tree may be calculated if the number of growth rings developing each year is known. An interesting fact that may be useful in this analysis is that some tree species (e.g. spruce) have their root system often spread almost parallel to the soil surface. However, it must be kept in mind that using trees as indicators is often hazardous for many reasons: the roots increase in thickness as the tree grows (but remain generally in the same place), the root system may develop a microrelief by itself, and the position of the inflection point or of roots above the surface does not necessarily indicate a lowering of the soil surface. Before measurements are made, the roots of trees from the same species as used for analysis should be examined in neighbouring flat and non-eroded areas; attention should be paid to the average depth of the basal flare, roots growing above the surface, and so on. For a detailed discussion on using tree root exposures as measures for erosion, reference can be made to Dunne *et al.* (1978) for a fine practical example.

Another use of vegetation is described by Olson (1958) in the context of dune development. Excavations of buried stems of marram grass on dunes revealed several former levels at which the base of leaf sheaths were found; these marked the previous positions of the soil surface. In the annual growth cycle, several sheath positions occur, and their relative distance along the stem indicates the speed of deposition in the various periods. Olson mentions also several grasses, shrubs and trees that may be useful as a tool for estimating erosion and/or deposition in dune development.

Another (crude) indicator may be species composition. Some plants or plant communities are characteristic for stable conditions, while others act as pioneers in recent deposits. The development of the plant cover (either single-specimen or the community as a whole) can also be indicative of geomorphic process type and intensity.

A final point worthy of mention is that animals and plants interfere not only with geomorphic processes but also sometimes with the measurement thereof. It is important to realise that, although these interferences are usually experienced as a nuisance, they are actually an indication of the effects of biological activities.

4.9
Neotectonic processes

4.9.1 Introduction

Thus far in Part Four we have been discussing the rates of change that have been wrought by the operation of such sub-aerial processes as weathering, mass movement, fluvial transport, glacial action, wind abrasion and marine attack. However, in addition to such *exogenetic* processes, there is another group of important geomorphological processes associated with crustal movements, to which the term *endogenetic* is applied.

The study of rates of tectonic processes has gained great momentum in the last two decades because of the plate tectonic paradigm, and because of a widespread interest in neotectonics – the study of late Cenozoic deformation (Fairbridge 1981, Vita-Finzi 1986). A large range of techniques has now been developed to permit an assessment of rates of both vertical and horizontal movements (Doornkamp & Han Mukang 1985). Some of these techniques are based on long-term evidence and some on short-term evidence (Table 4.17).

4.9.2 Historical, archaeological and chronometric techniques

One of the simplest methods for inferring the distribution and rates of tectonic change is by present-day or historical accounts of specific seismic events. Although there are inevitable problems with eye-witness reliability after a major catastrophe, and although in some cases earthquake-induced landslide scars may be confused with earthquake-induced fault traces, careful analysis of observations can be a productive source of information. Historical records of earthquake activity have some duration, with the Persian seismic record going back to the Assyrian period in the thirteenth century BC (Ambraseys 1971, Ambraseys & Melville 1982).

Long-term changes in emergence or submergence of land can be conducted using archaeological remains of known age (see, for example, Flemming 1969). The classic example of this approach was provided by the temples at Pozzuoli in Italy, where Babbage (1847) was able to show, from analysis of the local geology and molluscan borings in some of the columns of the Temple of Serapis, that the ruins had subsided into the sea and then been gradually uplifted. Examination of the displacement of archaeological sites can also demonstrate faulting, as in the case of the Great Wall of China (Deng *et al.* 1984) and the Hisham Palace in Jericho (Reches & Hoexter 1981). Evidence of recent anticlinal uplift is demonstrated by the long profiles of Iraqi canals, while arched river terraces may be dated by archaeological means, e.g. at El Asnam in Algeria (Vita-Finzi 1986).

Various other dating techniques have also been employed with some success to date tectonic events and degrees of change. Thus faults, which can be identified on the ground or through remote sensing, can be related to geomorphological features of known age. Relative fault ages in unconsolidated materials may be gauged by looking at their morphology. This may be necessary because mineral or organic material that can be

Table 4.17 Techniques for determining rate of tectonic movement.

	Nature of movement
Long term	
Altitude of dated marine sediments or landforms (e.g. coral reefs, beach rock)	Vertical
Altitude of archaeological remains (e.g. harbour installations)	Vertical
Deformation of sediments of known age (e.g. cave sediments, fans, terraces, raised beaches and platforms)	Horizontal
Palaeomagnetism of submarine basalts	Horizontal
Historical analyses of earthquake records	Vertical
Short-term	
Paired gauges on lakes	Tilt
Tide gauge levels	Vertical
Geodetic levelling	Vertical
Direct observations of earthquake traces	Vertical
Electronic distance measurement and laser ranging	Horizontal
Gravity changes	Vertical
Tiltmeters and creepmeters	Horizontal

radiometrically dated is rarely associated with these fault scarps. Various measures of scarp erosion can be employed: degree of dissection, amount of rounding of the crest of the scarp, and slope of the face. Bucknam and Anderson (1979), for example, found that for young scarps in western Utah there is a clear-cut relationship between scarp height, scarp slope angle and geomorphic age.

It is, however, more satisfactory to use more precise chronometric dating methods when these are available. In California, Sieh and Jahns (1984) used the charcoal present in two ages of alluvial fan and two ages of channel alluvium to estimate the ages of channel offsets of measured length caused by Holocene movement along the San Andreas Fault. Similarly, Clark *et al.* (1972) used the degree of deformation of dated lake deposits that were originally broadly horizontal as the basis for reconstructing an episodic fault displacement chronology for the last 3000 years in the Salton Sea area of California. Likewise tree-ring analysis may permit the identification of trees that have been stressed by fault movement, and may also provide a means whereby such stress can be dated (Sieh 1978). Lichenometric dating may be applied to fault scarps (Nikonov & Shebalina 1979). On a longer timescale, Williams (1982) employed an ingenious geomorphological method for estimating uplift in New Zealand's South Island, where Quaternary marine terraces sometimes cut across limestones that contain caves formed in association with past sea-level-controlled groundwater levels. Uplift, followed by water-table lowering and abandonment of cave passages by active streams, permitted speleothem deposition. Dating of these deposits by $^{230}Th/^{234}U$ and by ^{14}C gives a minimum age for the cave levels and hence also for any terrace to which they may be related, while successive dates at different levels yield an uplift rate for the area. Even longer-term rates of uplift can be obtained by fission-track analysis of minerals like apatite, as has been attempted by Zeitler *et al.* (1982) in the Karakorams of Pakistan, and by Miller and Lakatos (1983) in the Adirondacks of the USA.

Warping of geomorphological features of known age may also be identified. The classic example of this is the suite of warped shorelines caused by hydro-isostatic forces around Pleistocene Lake Bonneville (Crittenden 1967). However, the tilting of marine raised beaches of Pleistocene age may also be inferred from their morphology, because they may be steeper than modern platforms and because the discrepancy becomes greater with age (Bradley & Griggs 1976).

4.9.3 Recent vertical and horizontal movements

Recent vertical crustal movements (RVCM) are those which have been occurring during the present century and have been determined instrumentally (see, for example, Pavoni & Green 1975, Vyskočil *et al.* 1983). The most common techniques for their investigation involve geodetic levelling and relevelling over transects or networks. This can provide data on rates of subsidence or doming over periods of some decades, and can help to explain certain aspects of fluvial morphology (Burnett & Schumm 1983). It is important, however, to recognise some of the limitations and inaccuracies of some past geodetic data (Bilham & Simpson 1984).

Tide gauge records provide a means to determine rates of vertical movement with respect to sea level, though allowance has to be made for such factors as eustatic changes and changes in water level brought about by meteorological and other circumstances. Tidal analyses have been used extensively (see, for example, Lisitzin 1974, Grant 1980) because gauges are in existence in many parts of the world. Possibly the greatest scope for their use at present is found in seas such as the Mediterranean, where the record may be short but the tidal range is small and where there are sheltered bays in which storm effects can be neglected. Similar techniques can be applied to lakes, the extent of tilting over a period of time being obtained by comparing water levels on pairs of gauges at different parts of a lake basin (Wilson & Wood 1980).

A whole range of new survey techniques is permitting the determination of phenomena such as geoidal changes and movements of plates. Satellite altimetry now enables continuous monitoring of the geoid (Horam 1982) while laser ranging is beginning to supply measurements of interplate movement. For example, Bender and Silverberg (1975) discussed the feasibility of determining the motion of the Pacific Plate with respect to North America by using a laser beamed onto the Apollo retroflectors on the Moon. An accuracy of 1 cm yr^{-1} for the change in separation between two stations might, they believe, be achievable. Laser ranging can also be used with respect to satellites, and Smith *et al.* (1979) have derived the distance between two points on opposite sides of the San Andreas Fault by laser tracking of near-Earth satellites as part of an experiment to estimate the motion along the plate boundary. Very-long-baseline interferometry (VLBI) can also be employed (Shapiro 1983, Kroger *et al.* 1986).

Faults and folds of vigorous activity are tempting subjects for direct gauging with such devices as wire-type creepmeters and electronic tiltmeters. Tiltmeters can be installed in boreholes, but may be adversely affected by rainfall events and by groundwater fluctuations (Edge *et al.* 1981). Such techniques, nonetheless, provide a useful means for determining rates of movement along faults (Bolt & Marion 1966, Johnston *et al.* 1976).

A full appreciation of current ground deformation probably requires the use of a variety of techniques in combination. For example, Chadwick *et al.* (1983) used five different methods to monitor ground deformation associated with the eruption of Mount St Helens:

(a) trilateration and triangulation around the volcano to monitor changes in shape;
(b) long-ranging distance measurements to points in the crater made from a point 8.5 km from the lava dome;

(c) measurements of short horizontal and vertical distances across active cracks and thrust faults on the crater floor;
(d) slope-distance and vertical-angle measurements from sites on the crater floor to targets on the floor and dome; and
(e) measurement of ground tilts on the crater floor by means of inexpensive and expendable electronic tiltmeters and precise levelling.

PART FIVE
EVOLUTION

5.1
Radiocarbon dating: principles, application and sample collection

In Part One of this volume reference has been made to models of landscape evolution, and of necessity these require consideration of timescales measured in years. Geomorphology does not have a monopoly of the methods involved in determining ages of past events, and it can be argued that these techniques are primarily the armoury of Quaternary geology. Indeed in their classic work *Fluvial processes in geomorphology* Leopold *et al.* (1964) made the assertion (p. 468) that 'much of geomorphology is stratigraphic geology'. However, in the British Isles, at least, Quaternary geology has been largely ignored by its parent over the last three decades, and research in this field has been dominated by geomorphologists. Naturally the wisdom of this situation is debatable, and Clayton (1985) has suggested that those parts of geomorphology and Quaternary studies that do not have strong linkages with the geographical core should be unloaded onto geology forthwith.

Thus it is not appropriate in the present context to attempt a fully comprehensive overview of geochronology and allied matters. Rather, a selective approach has been adopted and themes of particular relevance to geomorphology have been identified. Since these techniques *per se* are often beyond the capabilities of the average geomorphologist, the actual implementation of the measurements involved is often the province of specialists in other sciences: for instance, the determination of the radiocarbon activity in a sample is essentially the task of a physicist. As a consequence a good deal of the following material is more generalised than that in the preceding sections. A primary objective has been to provide sufficient background in order that a meaningful discourse can be established with specialists in other areas. Clearly the primary role of the geomorphologist is going to be in the field.

5.1.1 Introduction

In this outline of radiocarbon dating we shall concentrate on the field and interpretative aspects, for it is with these that the geomorphologist is able to play a primary role. First, however, it is necessary to outline the physical basis of the method and assay procedures, for invariably liaison between the sample submitter and physicist is desirable and this is helped if the fieldworker is aware of the theory behind the technique.

5.1.1.1 Principles

Carbon-14 (^{14}C, radiocarbon) creation is a continuous process in the upper atmosphere where neutrons (n) produced by cosmic rays react with nitrogen according to the equation:

$$^{14}N + n \rightarrow {}^{14}C + p \quad (5.1)$$

The ^{14}C so formed is radioactive and therefore subject to ultimate decay to a stable form, and when this happens a beta-particle is emitted:

$$^{14}C \rightarrow {}^{14}N^{+} + \beta^{-} \tag{5.2}$$

Meanwhile the newly produced ^{14}C is incorporated within the global carbon cycle as its chemical behaviour, except for isotopic fractionation, is the same as that of stable ^{12}C. It soon combines with oxygen to form CO_2, which is thoroughly mixed with the inactive atmospheric carbon. In turn the CO_2 is either dissolved in water and later incorporated into aquatic life or it becomes absorbed by plants through photosynthesis. In addition, feeding organisms will continually take in food containing ^{14}C. The net result is a global ^{14}C dynamic equilibrium maintained by production and decay. But when an organism that has been in equilibrium with its environment dies, its ^{14}C content is no longer capable of being replenished and there commences a progressive and irreversible process of ^{14}C diminution in accordance with the exponential decay equation:

$$I = I_o e^{-\lambda t} \tag{5.3}$$

where I = activity of sample, I_o = original activity of sample, λ = decay constant, e = base of natural logarithms and t = time elapsed since death.

Decay of radioactive substances eventually reaches a level where only half of the original concentration remains, and the time elapsed is known as the half-life. The disintegration behaviour of ^{14}C achieves its half-life after some 5730 years and, since the decay function is exponential, the concentration is down to a quarter of the original after 11 460 years and so on. Because of the exponential characteristics of the decay rate, the average life of a ^{14}C atom is almost 8300 years, somewhat longer than the half-life.

In his pioneer work in the late 1940s, W.F. Libby realised that this process could be used as a dating technique if the ratio of the activity of an unknown sample could be compared with that of contemporary material in dynamic equilibrium with the environment. But certain assumptions had to be made if the technique was to have a validity. These were:

(a) the biospheric carbon reservoir was in equilibrium,
(b) the cosmic ray influx had been constant, and
(c) no disturbance had occurred to the dynamic carbon reservoir.

Also, the dating potential was dependent upon the technological feasibility of measuring the modern and residual ^{14}C activities. The fundamental problem here is that even the modern level of ^{14}C is exceedingly small, the natural abundance being only one part in 10^{12} parts of atmospheric carbon dioxide. Analytical chemistry was unable to tackle the measurement of such minute amounts and therefore recourse was necessary to an indirect approach that involved the accurate counting of the rate of ^{14}C disintegrations as registered by the beta-particle emissions. This decay, it has to be emphasised, is purely a random process and consequently it is necessary to try to establish what the long-term average rate is in any given case. Barker (1970) has drawn an apt analogy by likening the procedure to an attempt at determining a human population by short-term considerations of the death rate. An inevitable consequence is that age calculations using ^{14}C are dependent upon counting statistics and cannot, therefore, give an unique result but rather the deviation from a mean value.

5.1.1.2 Methods of radiocarbon dating

Currently, two different approaches to the measurement of beta emissions are widely utilised; both initially involve the preparation of CO_2 from the sample whose age is unknown. This is done by combustion if the sample is organic or acidification if inorganic. The long established technique is known as gas proportional counting and 30 years or so ago this method was almost universal. The common gases used are CO_2, CH_4 and C_2H_2, and these are introduced into the counter itself, which consists of a container in the centre of which is a wire carrying a high voltage. This attracts the particle and each neutralisation is registered electronically. However, since the early 1970s most of the newly established ^{14}C laboratories have adopted the alternative method of liquid scintillation counting from foundation, this method being favoured since much of the equipment can be bought 'off the shelf'. Indeed many older laboratories primarily concerned with routine dating work have switched to the latter technique. The key process is the conversion of CO_2 gas to C_6H_6 (benzene) followed by mixing with a non-radioactive scintillator that is activated by beta emissions. The light pulses so produced can be electronically counted. Most laboratories, irrespective of the method they use, have the capability of detecting residual ^{14}C levels that correspond to ages from the present back into the 35–50 ka range.

Predictably there are instances where such an age limitation is not desirable, although samples with true ages in excess of 50 ka are rather rare. In theory, increasingly low levels of residual ^{14}C persist in samples for at least 100 ka but the detection of this presents formidable problems. Nevertheless, instances of field problems that involve ages of this magnitude do occur and some attempts have been made to probe this timespan (e.g. Worsley 1980). The basic approach, pioneered in the Netherlands by H. de Vries, is to enrich isotopically the ^{14}C present in a sample by thermal diffusion so that the ability to discriminate decay rates from background radiation is within the range of gas counters. However, this procedure is both time-consuming and expensive, and as a result it has been restricted to a very few specialised laboratories, e.g. Groningen and Seattle, which seem to be able to attain something like a 75 ka older age limit.

In 1977 the first demonstration was made showing that an entirely different approach to the detection of ^{14}C was a practical possibility. This alternative method involving direct atom counting has come to be generally known as accelerator mass spectrometry (AMS). The original work utilised existing large Van de Graff accelerators in nuclear physics laboratories, which were capable of electrostatically accelerating high-energy beams of carbon ions derived from a sample source in the form of graphite (this, interestingly, is the same form as used by Libby in his pioneering measurements of beta activity). Within electrical and magnetic fields ions of the three carbon isotopes are split into three separate beams according to their atomic mass. The beam of ^{14}C ions is routed into a detector, which is capable of absolute counting. This new approach yields in 30 mins or so an overall efficiency that is equivalent to conventional beta-decay counting of the sample for a period of some 80 years. A significant and valuable consequence of this is the greatly reduced sample mass necessary for dating, amounting to approximately one-thousandth of that required by the conventional method. An additional bonus is that the cosmic-ray background can be ignored since radioactive decay *per se* is not being assayed.

Over the last decade a few specially designed small accelerators have been constructed for dedicated radiocarbon dating. A major constraint has been a very high capital cost, and despite early promise a number of technical problems have dogged the commissioning of AMS-based dating services. One of these concerns the precision accorded to each measurement, for an accuracy of about 1% is necessary to be comparable with conventional dates. However, sufficient progress had been made by 1984 to permit routine age determinations of materials dating from the last 30 ka. In theory AMS ought to be capable of detecting residual ^{14}C back to some 100 ka, but even if the accuracy of AMS can be improved the old bugbear of contamination cannot be eliminated. Hence the immediate prospects for extending the ^{14}C timescale are not encouraging, and the main justification for AMS work is going to lie with the advantages accruing from drastically reduced sample sizes, e.g. it is feasible to date a single cereal grain.

However, the growth of AMS dating techniques must not eclipse equally important technological innovations at some of the 'conventional' laboratories (Otlet *et al.* 1986). This concerns the perfection of miniature gas proportional counters capable of dating, like the AMS method, very small samples, although AMS generally operates with samples an order of magnitude smaller. Essentially, the theoretical approach is identical to that of 'normal' gas counting but design refinements have led to the availability of equipment with a significantly lower intrinsic background, electronic stability and multicounter capability. Predictably this does have a negative corollary, namely a greatly extended count time from a former 1–3 days to the 10–100 day range, the specific time being dependent upon the amount of carbon available. Despite this the dual benefits of miniaturisation and allied microchips has enabled the routine operation, since 1983, of small counter dating services, which are able to handle concurrently 20 plus samples at a financial cost that is minor in comparison to that of its AMS counterpart. Indeed the sample preparation costs are actually less than most demanded by liquid scintilllation counting.

Most laboratories have adopted the practice of routinely assessing the degree of isotopic fractionation that prevails in a ^{14}C sample. This is because different materials incorporate different proportions of carbon isotopes. Most materials have lower $^{14}C:^{12}C$ ratios than that of the atmosphere, and this produces an inherited initial age. The reverse holds for materials from the oceanic environment. The degree of fractionation (expressed as $\delta^{13}C‰$ value) is established by comparing the $^{13}C:^{12}C$ ratio of the sample with a modern standard and by accepting the calculation that the $^{14}C:^{12}C$ ratio is almost twice the $^{13}C:^{12}C$ ratio. Since the $^{13}C:^{12}C$ ratio of abundance is about 1:100 its determination is a straightforward measurement using a mass spectrometer. On average the degree of fractionation will introduce a small error in the establishment of the ^{14}C age of the order of ± 80 years and clearly is important for younger materials. A reported ^{14}C age will acknowledge the fractionation factor either by an adjustment in the age where it is known or by an increase in the uncertainty value.

5.1.1.3 Date interpretation

After the rate of decay of the ^{14}C remaining in a sample has been established, it is then possible to convert this statistically based physical measurement into a radiocarbon date, if the previously noted assumptions are accepted. Invariably the date communicated by the laboratory is in the form of:

> age $\pm$ error in radiocarbon years BP; laboratory identification code and sample number (such a date should be described as a conventional ^{14}C age)

It is *not* a calendar date and the BP relates to years before the present, which is adjusted to a zero at AD 1950. Archaeologists often prefer to use ages in terms of ^{14}C years on an AD–BC scale, but this is not recommended for geomorphological contexts. The error factor largely arises from the random nature of the emissions associated with the disintegration process and the allied experimental uncertainties of establishing their frequency. Unfortunately, there is no universal interlaboratory agreement on precisely how this error is to be determined. Standard practice is to cite each laboratory's doubt as

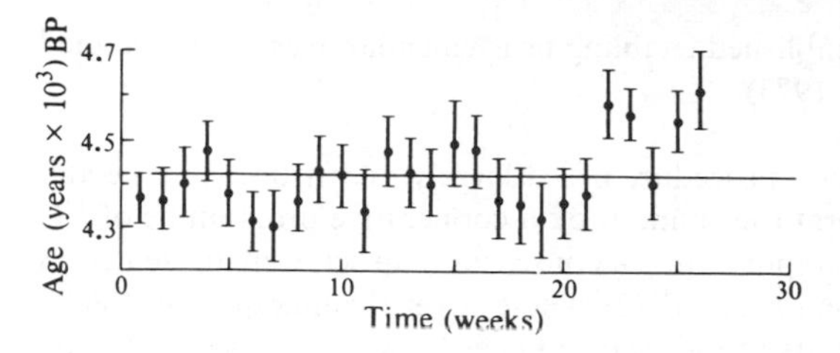

Figure 5.1 Results of 26 replicate measurements, by the liquid scintillation technique, of a single sample over a period of six months. (After Barker 1970). The bars represent one standard deviation about the mean.

one standard deviation (1 SD) on either side of the mean value (about a two-in-three chance of the 'age' lying within the timespan so defined). To increase the likelihood that the true result lies within the error band, the age ought to be quoted in terms of two standard deviations (19 times in 20) and some laboratories do this as routine. Because of the random nature of the decay process and uncertainties in the measurements, replicate counts on the sample preparation are very unlikely to yield precisely identical results, and this is nicely illustrated by the experiment undertaken at the British Museum. In this case 26 standard counts were repeated on the same sample and the variation obtained is shown graphically in Figure 5.1.

In practice, the count times required to achieve the 1 SD level are inversely proportional to the square root of the number of disintegrations counted and hence directly related to the antiquity of the sample. The older the sample material, the longer the count time involved simply because the increasingly low levels of ^{14}C are harder to detect as levels distinct from the background radioactivity.

5.1.1.4 Age calibration

Normally, if the conventional ^{14}C age is to be applied to a field problem, which ideally ought to be expressed in absolute time and not just simply on a relative scale, then the 'raw' age is in need of some adjustment. This arises because of the following factors:

(a) *^{14}C half-life*. At the onset of ^{14}C dating in AD 1950, the half-life was determined as

$$5568\ (5570) \pm 30\ ^{14}\text{C yr BP}$$

and subsequently has come to be known as the Libby half-life. More recent attempts to establish the half-life value have shown the Libby result to be a slight but significant underestimate and the currently accepted best known value is

$$5730 \pm 40\ ^{14}\text{C yr BP}$$

In order to avoid confusion arising from the adoption of differing half-life values in calculating dates, there is an international agreement to continue using the Libby value when reporting conventional ^{14}C ages. Therefore, to convert to the most accurate ^{14}C age such dates need multiplying by 1.03.

(b) *Secular ^{12}C:^{14}C reservoir fluctuations*. Comparison of ^{14}C age results from samples of known ages has revealed that one of the basic tenets of the techniques – a constant level of the dynamic equilibrium – is incorrect. This is known as the de Vries effect, being named after its discoverer. Here the dendrochronological work on the giant sequoias and the bristlecone pine was of paramount importance for it yielded wood samples in 20 year bundles extending over the last 8000 years. This revealed a maximum discrepancy of some 800–900 years in the 6000–7350 'calendar' period. In the 1970s there was no universal agreement on the exact calibration of ^{14}C dates although several tables were published enabling near-calendar ages to be obtained (see, for example, Ralph *et al.* 1973).

In order to rationalise the calibration procedure based largely on bristlecone pine and giant sequoia, a group of five laboratories undertook a cooperative programme of ^{14}C activity measurements. The results, representing a consensus data set, were presented as a series of calibration tables by Klein *et al.* (1982). These cover the timespan of 0–8000 calendar years BP, which corresponds to radiocarbon ages between 10 and 7240 years BP. The tables were easy to use: the standard laboratory ^{14}C date and associated uncertainty value rounded off to the nearest 10 years were found in the tables and adjacent to them was an age range representing a 95% confidence interval containing the 'true' calendar age. Provided that the calendar age was greater than 1000 years BP, each standard date yielded a single age range. However, for ages younger than 1000 years BP, a supplementary calibration table was often necessary since some samples within this period corresponded to multiple calendar age ranges. Although the same effect occasionally occurs in the pre-1000 year period, the uncertainties associated with measurement and calibration lead to an inability to discriminate separate periods.

Calibration procedures attained a landmark with the publication by Stuiver and Pearson (1986) and Pearson and Stuiver (1986) of comprehensive graphs and tables covering the radiocarbon timescale of 0–4000 BP. This period in terms of dendrochronologically determined time corresponds to the last 4500 years. The calibration data result from a collaborative project between the Belfast (liquid scintillation counting) and Seattle (gas proportional counting) laboratories. Both have highly developed systems that yield 'high-precision' output with the uncertainty and errors quoted in terms of ± 20 years rather than the ± 80 years normally associated with standard dating. The wood samples used were western North American (Douglas fir and sequoia) for the first two decades and thereafter western European oak. An absence of bristlecone pine material in this most recent calibration exercise is explained by the need for large decadal or bidecadal samples in high-precision work. Those new calibration data are likely to be the definitive means for deriving near-calendar years from radiocarbon dates for the foreseeable future.

Beyond the current 8000 BP limit of calibration tables the degree of calendar and radiocarbon age divergence is problematic. Earlier estimates placed reliance upon the Swedish varve timescale and indicated that ^{14}C and calendar ages were coincident by about 10 ka. It is now known that there is not a sufficient degree of accuracy in the varve chronology to permit such an assessment. Fortunately, Stuiver (1978) has presented good independent geochronological evidence that suggests that the maximum deviation back to 32 ka is unlikely to exceed 2 ka. On the whole, estimates of earlier biospheric reservoir fluctuations suggest that it has remained within ± 10% over the last 50 ka and necessitates at least an 800 year uncertainty.

5.1.2 Field sampling

Any substance that once formed part of a living organism invariably contains carbon from its contemporary natural environment and, if within the limits of detection, is theoretically suitable for ^{14}C age determination. Naturally enough, any material that the fieldworker considers to have a useful ^{14}C dating potential must have a direct bearing upon a geomorphological problem. Elementary though this statement may seem, it is not unknown for insufficient thought to be given to the relevance of possible sample data to the problem being considered. On occasions the acquisition of a 'date' *per se* has been seen as a justification in itself. In the earlier years of date availability, short papers that often placed total reliance on single results were the vogue and they tended to dismiss other evidence if it did not seem to fit the ^{14}C age. Owing to the problems of derivation and possible contamination, ^{14}C dates should be used as supplemental data rather than in a primary capacity. Further, each date requires the expenditure of manpower and plant (at 1988 prices the average processing cost is some £200 per sample) apart from the resources consumed in collection and transport.

With outcrop sampling the first procedure will be to clean the exposure. This is, of course, often done prior to the discovery of material suitable for dating. Usually it is necessary to cut back the face so as to see clearly the stratigraphical relationships at the site. Often in natural exposures the roots of vegetation masking the slope permeate the organic matter and mass movement processes cause secondary displacements from an original stratigraphic position. As the face is cut back the influence of these two factors often diminishes. Once a clear section has been obtained certain questions ought to be posed, for these can only be satisfactorily answered in the light of field data:

(a) Is the sample in its true stratigraphical context and is there a close relationship between the death of the sample material and the event whose age is sought?
(b) What are the chances of sample derivation?
(c) Did the sample remain uncontaminated after death – what is the incidence of root penetration, recrystallisation and deposition of organic compounds of different origins and activities?
(d) Is it meaningful to collect more than one sample from the same horizon, bearing in mind that single dates are always difficult to interpret?

Following this, the size of sample to be collected needs to be considered. Here the nature of the material is paramount for the carbon content is dependent upon the physical nature of the sample. In Table 5.1 a guide to sample weights is given, and generally the higher the desirable weight the less desirable the material for dating purposes. If any doubt exists with respect to sample size, then it is advisable to err on the excess if plenty is available, for often it is not practicable to return to the sampling site because of costs or, indeed in the case of temporary exposures, survival. It is particularly important to note that the weights given are for dry materials; peat for example can be notoriously misleading because of a water content of some 90% and most of what was obviously a weighty sample at the time of collection may literally evaporate on drying. The method of extraction of the sample from exposure will vary according to the nature of the material to be dated. Concentrated materials, such as *in situ* trees or peat sequences, are easily removed with the aid of a sharp knife or saw. Where the material is dispersed, e.g. charcoal, shells, etc., individual pieces may have to be picked out of the face. However, the general rule is to handle the material as little as possible and it is better if the sample does not come into contact with bare hands. Disposable plastic surgical gloves and metal tweezers are useful in this respect.

Ideally, packaging materials should consist of clean, totally $^{14}C-$ deficient materials. Probably the majority of ^{14}C samples are simply collected in polythene bags and there is currently no reason to suspect their suitability. Obviously, those with thicker gauges are best in that they are more durable and the risk of puncture is less. Where only thin versions are available, they can be used in multiple. If available, aluminium foil is probably the best wrapping material since it is readily adaptable to varying sample size and is effectively self-sealing. Alternatively, aluminium sample tins are very useful especially for fragmentary material. Sometimes it is desirable that comparative reference material is collected, and both this and the sample for dating can be removed in metal or

Table 5.1 Sample size for conventional ^{14}C dating.

Substance	Residual weight after pre-treatment (g)	Weight of raw sample (g)
Wood	16	100–200
Charcoal	10	50–100
Peat	30	200–500
Shells	50	75–100
Bones	collagen extract	300–1000
Soil organic	10	200–1000

plastic monolith boxes that have first been pressed or hammered into the section and then excavated with a knife or trowel. The former type of boxes can be readily manufactured from 18 gauge (about 1.23 mm) sheet and riveted together, while the latter can be made by cutting sections from standard square plastic drainpipe. All sample materials should be oven-dried (85°C) as soon as practicable after collection, for if this is not done there is a risk of mould developing on the materials. If the collecting is undertaken at field sites away from civilisation, then it may be better to attempt drying at the site. It is important to note that the sample must not be treated with any form of preservative agent whatsoever. Care must be taken to ensure sample integrity and identity. All samples should be labelled in such a way that the labels do not decay nor become detached from the sample package. Index cards or indeed luggage-type labels with strong ties are suitable and a supply of adhesive tape is useful. The package, whether it be a polythene bag, aluminium foil, etc., should finally be placed in a clear plastic bag so that the label can be readily seen. Each sample label should have a reference number alongside basic site data. In a field notebook the same information will be recorded along with a much fuller description of the site. In the field it is useful to have a copy of one of the ^{14}C laboratory sample information sheets (each laboratory seems to have its own requirements but all are basically similar). These sheets have to accompany any sample submitted to a dating laboratory and the presence of one in the field ensures that all pertinent factors are recorded in the field notebook on site. An example of one of these is given in Table 5.2. Although the sample data sheet is adequate guidance for the laboratory when processing the sample, many laboratory staff are more than willing to discuss the particular environmental context of a sample, particularly with respect to pre-treatment, and direct personal contact is to be encouraged.

Naturally, to maintain a constant workload, dating laboratories like to have a queue of samples awaiting processing. Even if a submitted sample was passed directly to the production line, the result would not be available for a month or so. Hence the fieldworker should be prepared to be patient and expect a wait of at least several months.

Upon acceptance of a sample for dating, the initial issue concerns the nature of the ^{14}C in it, i.e. does the ^{14}C content solely reflect the appropriate degree of radioactive decay since death. If the dating is to be meaningful, this must be the case when the sample enters the counter. As alluded to previously, samples submitted for ^{14}C dating may be affected by what is termed natural contamination. Such a situation involves ^{14}C that left the exchange reservoir at a time different to that of the specimen, and this may be of two kinds, material older than and younger than that of the sample proper. These additions have opposite effects: the first will increase the resultant apparent age, while the second will shift the date towards the present, thereby giving a minimum possible age (see Fig. 5.2). Provided that the sample submitter has considered the contamination problem and

Table 5.2 A ^{14}C sample description record.

Section 1
(a) Submitter
(b) Address of affiliated organisation
(c) Sample collector and collection date

Section 2
(a) Collection locality – geographical name
(b) Site and/or sample reference numbers, if any
(c) Precise geographical location: latitude, longitude (National Grid reference)

Section 3
(a) Sample composition – popular name and scientific name
(b) *In situ* environment – lithology
(c) Stratigraphical position – depth of burial
(d) Natural contamination – record any possible sources

Section 4
(a) Collection method – e.g. from a core, natural or temporary exposure
(b) Treatment and storage – washed, sieved, etc.; conditions since collection
(c) Possible contamination – during recovery or subsequent storage

Section 5
(a) Estimated age (years BP) and basis of this estimate
(b) Purpose of ^{14}C age determination
(c) Publications and previous ^{14}C measurements on or relating to the sample

communicated the conclusions to the laboratory, then certain pre-treatment operations can be undertaken, which can help to eradicate or identify sources of potential error in the resultant age. These will now be considered in conjunction with notes on the various materials normally dated.

5.1.2.1 Wood (including charcoal)

This is generally considered to be the optimal dating material. Care must be taken to assess the age of the wood at death for this may be significantly later than the initiation of growth. Often small fragments of twigs, etc., are better for dating a geomorphological event (e.g. a flood), as their preservation potential is likely to be less than for logs and hence they are less able to survive at the surface or withstand recycling.

5.1.2.2 Peat

Although not as good as wood, peat is suitable material provided the problem of later root penetration can be recognised and handled accordingly. Care is necessary when sampling since the rate of accumulation can be extremely variable. An extensive horizontal area should be exposed and then only a thin layer per sample collected.

5.1.2.3 Shells

These are notoriously susceptible to environmental contaminants and should only be dated if there is an absence of alternative material. Three environments – marine, freshwater and terrestrial – all yield shell material in geomorphological contexts and each has its own specific problems. As a general rule, monospecific samples are preferable to those which are multispecific and similarly whole shells are better than fragments. Ideally the samples should be submitted to a laboratory with experience in the specific type of shell concerned, especially where they are marine. The key problem is the addition of ^{14}C

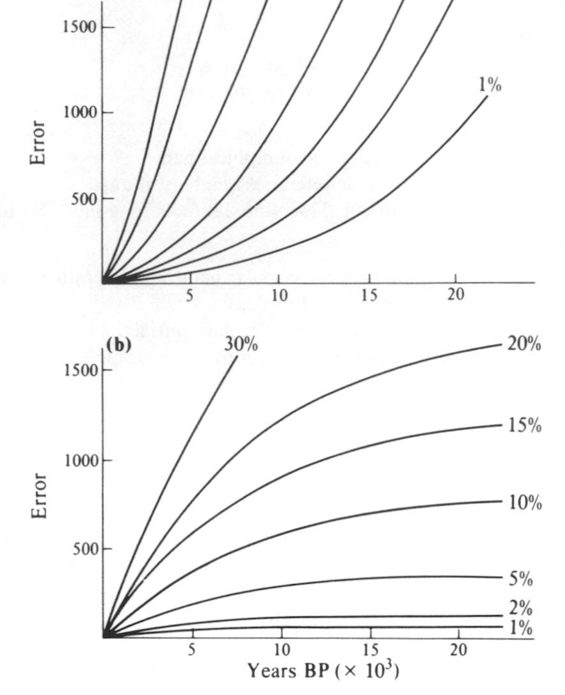

Figure 5.2 Graphs to show the influence of contamination on a sample's apparent ^{14}C age. The upper graph relates to 'infinitely old' carbon and the lower one to contemporary carbon. (After Olsson 1974).

after the death of the shell. Normally, the outer surface (the most susceptible part) of the shell is leached away by acid in the hope that the contamination has not affected the shell core. It has been demonstrated that successively deeper layers of the shell material give increasingly older ages and hence it is best to regard all shell dates as minimum values. With marine material there is also a need to take into account the 'age' of oceanic water. This arises from the long turnover time of the oceanic currents, and because of this modern shells often yield ages of several centuries BP.

5.1.2.4 Bone

Bone material suffers from serious subsequent infiltration by younger ^{14}C as does shell material, but difficulties are increased by a low carbon content. However, provided an ample sample is available it may be possible to separate the collagen (organic protein) content, which is not liable to contamination. This extract can then be satisfactorily dated.

5.1.2.5 Aqueous carbonates

Aqueous organisms (planktonic or plants) assimilate soluble bicarbonate from the water in which they live, and thus carbon is incorporated into their structure. Although the ^{14}C from the water may be in equilibrium with the reservoir, in catchments that include limestone bedrock the latter will invariably contribute some totally non-radioactive carbon as a solute. This has the result that the organisms will be to some degree deficient in ^{14}C with respect to the contemporary biospheric level. This factor is known as the 'hard-water error' and it imposes an older age limitation to dates derived from such materials.

5.1.2.6 Soil organic matter

This probably has the lowest desirability of all ^{14}C-containing materials from the laboratory standpoint, for the organic carbon content is low and consequently a large amount of material has to be processed, a costly and time-consuming procedure. An effective preparation/extraction technique has been developed by Kihl (1975), but before its application by the submitter, it is best to contact the dating laboratory since sometimes the laboratory would choose to 'control' the total pre-treatment operation. The above three materials, namely wood, peat and soil organic matter, may be affected by humic acid and carbonate contamination. Usually these can be circumvented by removing the acids in an alkali wash (e.g. 2% NaOH) and carbonates by treating with acid (e.g. 1 N HCl) followed by repeated washing in distilled water and drying. NaOH treatment of peat requires careful control, especially when the peat lacks a fibrous structure, for it is easy to dissolve the entire sample.

During the initial years of ^{14}C dating the journal *Science* published dating results but this soon became a task beyond the resources of a broad-based publication. With the objective of collating standardised date lists from the various ^{14}C laboratories and giving brief reports on their practices, a new journal, *Radiocarbon*, devoted exclusively to this task was founded in 1959. At first it was a supplement to the *American Journal of Science* and hence it is not uncommon to find it shelved immediately after that journal in libraries. This is an extremely valuable data source but alas it is not all-embracing and some laboratories, especially the better established ones, have tended to neglect the tedious task of submitting regular reports of progress. All the dates reported in *Radiocarbon* are calculated using the Libby half-life value and this permits world-wide immediate comparison. In some archaeological circles such conventional calendar dates have been followed by lower-case abbreviations (e.g. bp) and 'calibrated' dates by upper-case abbreviations (e.g. BP). Since universal agreement and understanding of correction factors has yet to be achieved, this practice is not recommended for adoption by geomorphologists, although it is clearly very important to distinguish between radiocarbon years and near-calendar years. Every few years, there are international meetings of those directly involved in the production of ^{14}C dates and these are invariably followed by proceedings volumes. These give up-to-date reports on the latest developments in the field, but until 1979 the onus for arranging the publication was on the organising institution/country; hence there has been no standard format, publicity, etc., and library availability is often restricted. Following the 1979 Berne meeting the proceedings are published as part of the journal *Radiocarbon*.

It should be noted that the raw sample weights are for ideal situations and laboratories can manage with quantities down to 20% of those given, but in such circumstances additional count times are required, which inevitably increase costs. The advent of AMS and miniature counters has revolutionised the dating potential of situations where only very small quantities of field material are available. For minicounters the sample weights are usually in the 1–5 g range and for AMS 10–200 mg. However, the number of laboratories operating a dating service is small and direct liaison between collector and laboratory is advisable. Also the fieldworker should bear in mind that these specialised facilities are significantly more expensive than 'conventional' dating services. Similarly 'high-precision' conventional dating is some 50% more expensive than the standard dating.

5.2 Uranium-series disequilibrium dating methods

There are two natural parent isotopes of uranium, ^{238}U and ^{235}U. The half-lives are 4.49×10^9 and 7.13×10^8 yr respectively, so that they and their daughters survive as common trace elements in igneous rocks. Upon aqueous weathering, uranium is relatively soluble as carbonate and phosphate complexes of the uranyl (UO_2^{2+}) ion. These are transported in fresh and salt waters and may be precipitated in derived rocks; see Gascoyne (1982) for geochemical details.

Uranium species decay radioactively by emission of α-particles (helium nucleii) and β-particles (electrons) to stable end-products; Ivanovich (1982) outlines physical details. Relevant portions of the decay schemes are shown in Figure 5.3. Uranium-238 (^{238}U) decays through the long-lived daughter isotopes, ^{234}U and ^{230}Th (thorium-230 or ionium), to a series of short-lived isotopes unsuitable for long-range dating. Uranium-235 (^{235}U) similarly decays to an initial long-lived daughter, ^{231}Pa (protactinium-231), and then to a short-lived series.

The decay schemes offer two classes of methods of dating secondary rocks in which U species have been precipitated. In the daughter excess (DE) class, the daughter nuclide is present initially in excess of the amount of its parent either because of preferential leaching of an isotope (e.g. ^{234}U) during weathering of source rocks or because of preferential precipitation of a daughter isotope (e.g. ^{230}Th) from an aqueous medium. In a system that is then closed after initial precipitation, a state of secular radioactive equilibrium will be approached with time. Potentially the most powerful method of this class is the $^{234}U/^{238}U$ method, measurable to at least six half-lives of the daughter (= 1.5 million years). It is described by the equation

$$\frac{^{234}U}{^{238}U} - 1 = \left[\left(\frac{^{234}U}{^{238}U}\right)_0 - 1\right] \exp(-\lambda_{234}t) \tag{5.4}$$

where $^{234}U/^{238}U$) is the present-day activity ratio, $(^{234}U/^{238}U)_0$ is the initial ratio at deposition, λ_{234} is the decay constant or half-life of ^{234}U (2.48×10^5 yr) and t is the time elapsed since deposition.

In sea water, $^{234}U/^{238}U = 1.14 \pm 0.02$. Thurber *et al.* (1965) attempted to use this disequilibrium to date marine corals, and Broecker and Ku (1969) applied ^{230}Th and ^{231}Pa DE methods to deep-sea sediments. Harmon *et al.* (1978b), Osmond and Cowart (1982) and many others have shown that the ratio, $^{234}U/^{238}U$, may vary substantially both in space and time in ground waters and rivers; therefore, the method is not applicable in any simple fashion to freshwater deposits. Gascoyne and Schwarcz (1982) suggest that progressive leaching of carbonate samples in weak acid, with measurement of the relevant isotopes in both leachates and residuum (the 'hot atom' method), may yield estimates of the sample ratios. Ford (in Gascoyne *et al.* (1983) in Yorkshire meteoric water caves, and in work in progress in a thermal water cave in South Dakota) has proposed that where the two standard deviation (2σ) about a mean local or regional initial ratio is determined by daughter deficiency (DD) methods to be low in samples younger

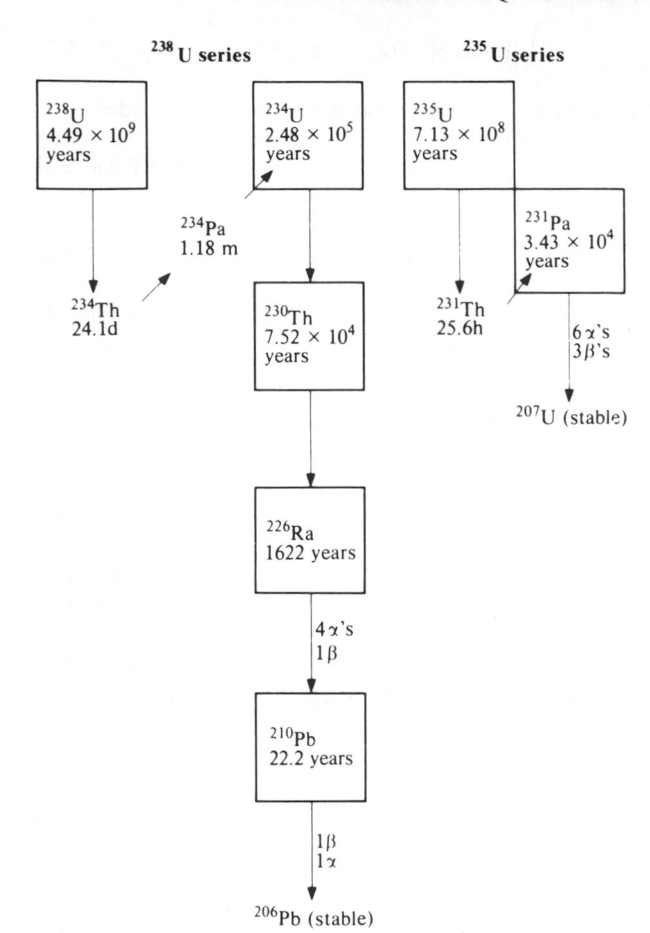

Figure 5.3 The radioactive decay series for ^{238}U and ^{235}U. The longer-lived isotopes of interest are outlined in boxes. Note that the activity of ^{235}U and its daughters is 1/22 that of the ^{238}U-series isotopes.

than *c*.350 000 yr BP, then the mean $(^{234}\mathrm{U}/^{238}\mathrm{U})_o$ value may be taken with 2σ error margins to apply the method to *c*.1000 000 yr BP ('regional uranium best estimate' or RUBE method). Many spring-deposited travertines also exhibit $(^{234}\mathrm{U}/^{238}\mathrm{U})_o$ ratios that are invariate for the range of ^{230}Th/^{234}U dating. Dating such travertines up to a limit of about 1 Ma depends upon the initial ratio; larger ratios allow us to go farther back in time because it takes longer for the ^{234}U/^{238}U ratio to become indistinguishable from unity. Mass spectrometric analysis of ^{234}U/^{238}U will extend the time-range from such deposits even further.

The second class of methods, daughter deficiency (DD) methods, presumes that the parent isotope is deposited without its daughters or with a daughter deficiency of known amount. Age is determined from the extent of growth of the shorter-lived daughter into equilibrium with its parent. As noted, uranium species are readily soluble but the important daughters, ^{230}Th and ^{231}Pa, are effectively insoluble in sea water or in fresh waters of normal pH range. Precipitated rocks therefore may contain abundant uranium but no thorium isotopes at time of deposition. The principal method is based on measurement of the $(^{230}\mathrm{Th}/^{234}\mathrm{U})$ and $(^{234}\mathrm{U}/^{238}\mathrm{U})$ ratios:

$$\left(\frac{^{230}\mathrm{Th}}{^{234}\mathrm{U}}\right)_t = \frac{1-\exp(-\lambda_{230}t)}{(^{234}\mathrm{U}/^{238}\mathrm{U})t} + \left(\frac{\lambda^{230}}{\lambda_{230}-\lambda_{234}}\right)$$

$$\left(1 - \frac{1}{(^{234}\mathrm{U}/^{238}\mathrm{U})_t}\right)\{1 - \exp[-(\lambda_{230} - \lambda_{234})t]\} \tag{5.5}$$

Figure 5.4 is the graph of Equation 5.5. It is seen that dating may be extended to *c.* 600 000 yr BP in favourable circumstances (where there is considerable $^{234}U/^{238}U$ excess at time of deposition) but hitherto the common limit has been between 300 000 and 400 000 yr BP.

$$(^{231}\mathrm{Pa}/^{235}\mathrm{U})_t = 1 - \exp(-\lambda_{231}t) \tag{5.6}$$

Ages up to 300 000 yr BP may be obtained by this method. It is less widely applicable because of the comparatively low natural abundance of the parent isotope, i.e. $^{238}U/^{235}U = 137.9$.

The natural abundance of uranium in carbonates and other common precipitates is found to vary from less than 0.01 to about 100 ppm. The $^{230}Th/^{234}U$ dating method is possible where abundance exceeds 0.01 ppm; the $^{231}Pa/^{235}U$ method requires U concentrations greater than 0.5 ppm.

To obtain an age by these methods, the sample is first dissolved and a spike of artificial U and Th radionuclides (e.g. ^{232}U, ^{228}Th) added to trace the efficiency of succeeding extraction steps. Uranium species are separated and concentrated by ferric hydroxide precipitation and a series of ion-exchange steps; Lally (1982) and Edwards *et al.* (1987) provide full technical details.

There are two means of counting the species and so determining their ratios. The standard method estimates their abundance by counting decay products (α-particles) on silicon surface-barrier detectors. It is practicable to extend measurement to *c.*100 000

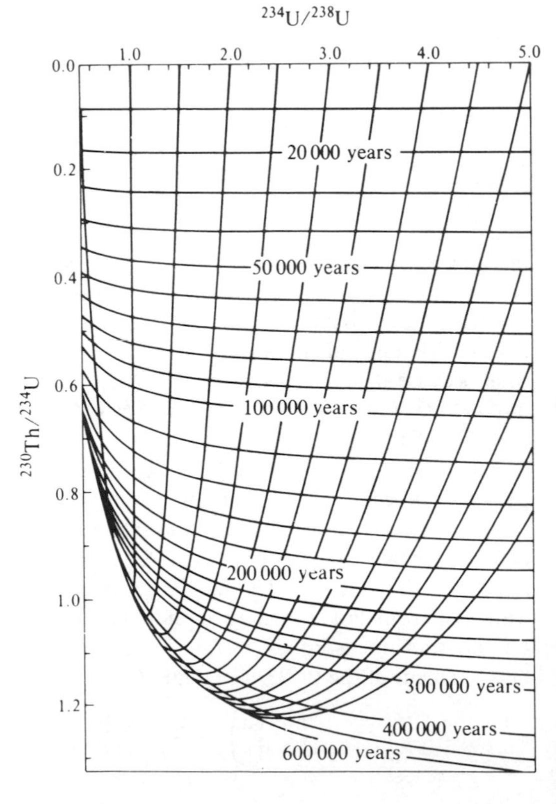

Figure 5.4 Relation between $^{234}U/^{238}U$ and $^{230}Th/^{234}U$ ratios for closed systems initially free of ^{230}Th. The steep curves are decay paths for samples initially deposited with $^{234}U/^{238}U$ ratios equal to left-hand ($^{230}Th/^{234}U=0$) intercepts. The subhorizontal curves are isochrons, in years.

counts. One standard deviation (1σ) error margins in the counting statistics typically are $\pm$ 5–10% in the last 150 000 yr, increasing to 20% or more as the 300 000–400 000 yr limit is approached. The second method counts isotopes directly in a solid-source mass spectrometer. This has long been used to measure U isotopes. Edwards *et al.* (1987) announced the first success with Th species. They worked only with coral. Such isotope dilution mass spectrometry permits 2σ counting error margins as low as 2‰ because of the much higher atom counts obtained (10^6–10^8 compared to 10^4). In principle it extends the dating range by 500 000–600 000 yr BP for the $^{230}Th/^{234}U$ method. Samples of 10–100 g (depending upon U content) are required for the α-particle method; this is reduced to a few grams for young deposits measured by mass spectrometer and to less than a gram for old deposits, a further advantage. Isotope dilution mass spectrometry currently is being evaluated for application to materials other than corals.

Two major problems are frequently encountered in application of the methods. The first occurs where there is significant addition or subtraction of uranium species after initial deposition, i.e. the closed-system assumption does not hold. This is quite common where circulating ground waters are able to penetrate and partly redissolve a precipitate or evaporite. There is a preferential loss of uranium, giving relative thorium enrichment, and the calculation of too great an age. The problem is most notable in deposits of high permeability or where recrystallisation has occurred. It may be recognised by visible evidence of solution damage or failure of the law of superposition in a sequence of dates. A converse difficulty is encountered with teeth and bone, or with calcite or aragonite shell samples. Uranium may be taken up from ground waters into tissue residues in the material *post-mortem*. Concentrations as high as 1000 ppm have been recorded (Schwarcz 1982). Further, the additional U may not be distributed uniformly in a bone, i.e. the age obtained may depend upon the portion that is sampled.

With the DD methods, a second problem is that of ‘detrital thorium’. Although effectively insoluble, thorium species released by weathering of a source rock readily bond to clay colloids, etc., and so may be carried in suspension and deposited as a portion of the insoluble residue present in most chemical precipitates. Perfect separation of such residues is difficult. Also, some of the U and Th present in the detritus is leached by acid during dissolution of the calcite or phosphate being analysed. Presence of detrital thorium is recognised through the observation in the α or mass spectrum of a peak corresponding to ^{232}Th. This is a primordial radioisotope, not derived from U. It is not found in freshly deposited, authigenic carbonates, etc., and therefore it can only derive from detritus. If the modern $^{230}Th/^{232}Th$ ratio is less than 20, it is not possible to infer the age directly from its $^{230}Th/^{234}U$ ratio and a correction must be made. One such ‘correction’ assumes an initial $^{230}Th/^{232}Th$ ratio of 1.25. Most soil and surficial carbonates and other evaporites (spring travertines and tufas, caliche, calcrete, etc.) and speleothems from cave entrances or frequently flooded sites suffer from this problem. Attempts to overcome it are mentioned below.

Reviewing the success of U-series dating, Ku (1982) concludes that the $^{230}Th/^{234}U$ and $^{231}Pa/^{235}U$ methods are the most useful and that coral and cave interior speleothems have proven to be the best materials. With respect to determining or approximating the ages of specific landforms, Bloom *et al.* (1974) and others have dated raised coral strands in New Guinea and elsewhere in the Pacific to estimate ages of high sea stands and rates of coastal uplift. The age of the last interglacial high sea stand in the Caribbean has been determined as 125 000 yr BP with considerable precision. Dating of submerged speleothems is confirming suppositions concerning the last two glacial low sea stands (e.g. Gascoyne *et al.* 1979). Lacustrine marls have also been dated satisfactorily.

On the continents, speleothem ages have been used to determine the mean maximum rates at which vadose channels have been incised in limestone caves (Gascoyne *et al.*

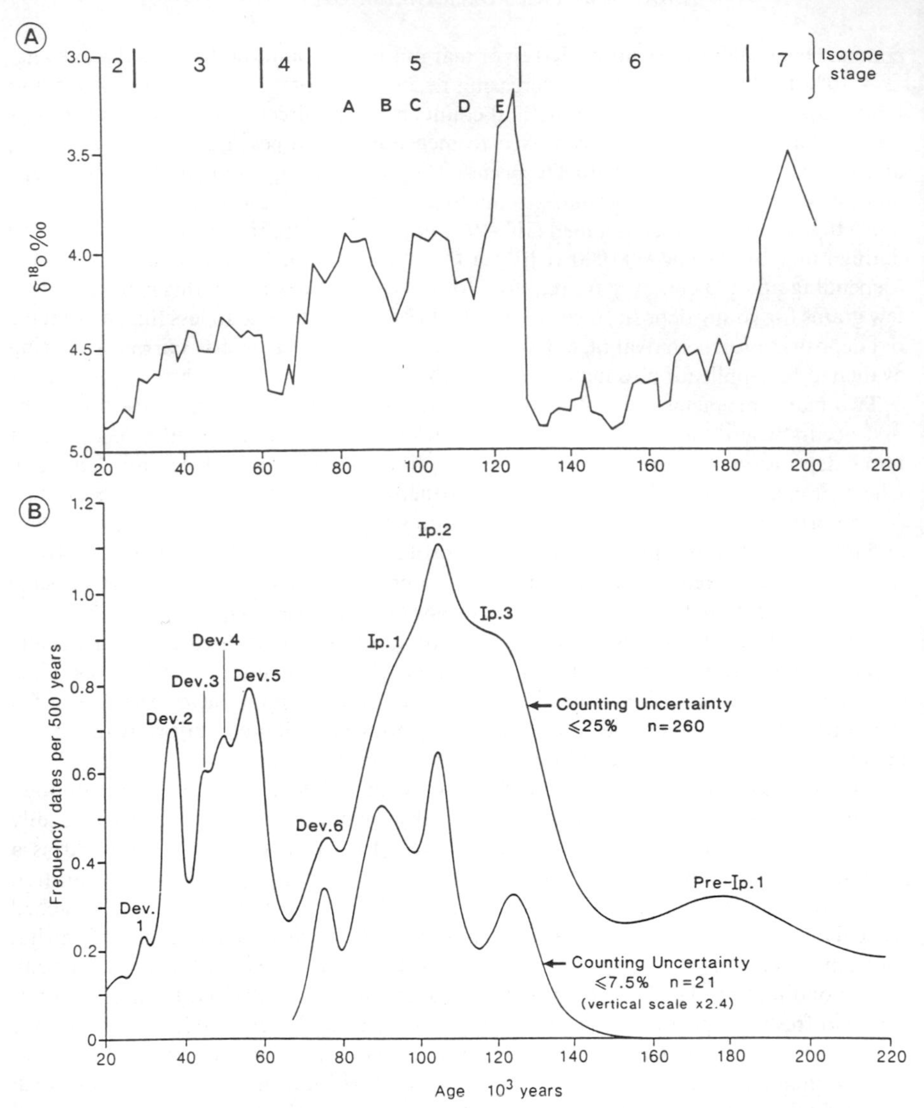

Figure 5.5 (A) A portion of the oxygen isotope record from deep-sea core V19–23 (Ninkovich & Shackleton 1975) plotted using the timescale of Denton and Hughes (1983). (B) For comparison, growth frequency curves (aggregated and smoothed) from UK speleothems dated by the ^{230}Th : ^{234}U method with α counting; see Gordon *et al.* (1988) for details.

1983) or at which phreatic caves have drained (Bakalowicz *et al.* 1987). Extrapolating from the caves to the exterior, they have yielded approximations of the ages of the glacial relief of the Rocky Mountains and the Yorkshire Dales (Ford *et al.* 1981, Gascoyne *et al.* 1983) and of uplift and entrenchment rates in antecedent river canyons in the Mackenzie Mountains (Ford 1973). Many sequences of alternating clastic and calcite deposition in caves have been dated, including important archaeological materials (e.g. Gascoyne *et al.* 1981a, Blackwell *et al.* 1983).

Uranium-series speleothem dates may be used in the aggregate as in Figure 5.5 where they indicate the timing and something of the intensity of interglacial and interstadial episodes in Britain during the past 150 000 years. They are used to measure rates of secular variation of the Earth's magnetic field, as preserved by the inclination and declination of magnetite grains in the calcite (e.g. Latham *et al.* 1986). Uranium-series methods are applied to help to determine ages and sedimentation rates for deep-sea sediment cores from which oxygen isotope ratios of foraminifera establish the fundamental climatic record of the Quaternary (Shackleton & Opdyke 1973). Uranium-dated oxygen isotope palaeotemperature profiles of speleothems have been published from southern France (Duplessy *et al.* 1970) and a wide geographical range of sites in North America (Harmon *et al.* 1978b, Gascoyne *et al.* 1981b).

Systematic investigations of the less tractable deposits such as surficial carbonate, bones and shells continue. Where detrital Th is abundant but closed-system conditions can be assumed, the plots of $^{230}Th/^{232}Th$ and $^{238}U/^{232}Th$ versus $^{234}U/^{232}Th$ yield straight lines whose slopes give the $^{230}Th/^{234}U$ and $^{234}U/^{238}U$ ratios of the chemically precipitated calcite component, respectively. Ku and Liang (1984) show that these slopes can be obtained from analyses of the leachate and the totally dissolved detrital residue (two-point isochron), while Przybylowicz and Schwarcz (1988) show that multiple leachates can yield these slopes and that the residue may not lie on the same line. This technique has proved useful where there is a wide variation of U content in a deposit.

Uranium trend dating has been attempted on a wide range of Quaternary sediments (alluvial, aeolian, glacial, etc.) where it must be assumed that the system remains open to isotopic exchange. It is presumed that U disequilibria will vary with depth due (a) to aging and (b) to U and Th fluxes that may be approximated by an empirical function (see Szabo & Rosholt 1982). Rosholt (1980) has published ages ranging from 4000 to 720 000 yr BP for US loesses, tills, paleosols and tuffs. Causse *et al.* (1987) measured wood from last interglacial sediments preserved beneath clay-rich tills in eastern Canada and obtained $^{230}Th/^{232}Th$ versus $^{234}U/^{232}Th$ isochrons with correlation coefficients higher than 0.97, i.e. the system has behaved in a closed manner here. The prospects for further improvements of techniques and for widening of the scope of U-series dating methods thus appear to be good.

5.3
Application of stable isotopes in waters, sedimentary deposits, and fossil plants and animals

5.3.1 Introduction

Stable isotopes of the light elements are useful tracers of a variety of processes that occur at the surface of the Earth. If two chemical species containing isotopes of the same element (e.g. calcium carbonate and water) have attained isotopic equilibrium with each other, then the difference in their isotopic ratios is a function of the temperature at which they equilibrated (O'Neil 1986). Variation in abundances of stable isotopes can also be used as natural isotope tracers of the sources of the respective elements (oxygen, carbon, etc.) The first practical application by Epstein *et al.* (1953) made possible the determination of the temperatures of growth of some fossil marine organisms (and hence of the sea in which they lived).

In geomorphology, stable isotopes offer a number of fruitful applications, particularly with regard to determining the climate of the recent past. This brief summary touches on only a few of the applications.

5.3.1.1 Notation and technique

Isotopic data are always presented in the delta (δ) notation, e.g. for oxygen, enrichment of ^{18}O with respect to ^{16}O is given by

$$\delta^{18}O_x = \frac{(^{18}0/^{16}0)_x - (^{18}0/^{16}0)_s}{(^{18}0/^{16}0)_s} \times 1000 \qquad (5.7)$$

where x indicates sample and s indicates standard, and the units are per mil (‰). For carbonates the standard is PDB, while for waters Standard Mean Ocean Water (SMOW) is used. Techniques of isotopic analysis are described in Hoefs (1987) and Faure (1986). Precision of analysis is generally ± 0.1‰ except for hydrogen where ± 0.5‰ is commonly obtainable.

5.3.1.2 Deep-sea sediments

Perhaps the most important application of stable isotopes that is relevant to landform studies is the oxygen isotopic analysis of deep-sea sediments, which has yielded the principal record of long-term climate fluctuations in the Pleistocene and late Tertiary. The record has been dated absolutely to yield a chronology for continental glaciations that extends more than 1 Ma. The isotopic compositions of the tests of benthic and planktonic foraminifera are controlled by (a) $\delta^{18}O$ of sea water and (b) temperature of growth, through its influence on the fractionation between calcite and water. Most foraminifera precipitate calcite at or close to equilibrium with sea water. Isotopic analyses of foraminifera from deep-sea sediment cores reveal characteristic, cyclic variation in $\delta^{18}O_{ct}$ with depth in the core (Fig. 5.6) with a cyclicity of the order of 0.1 Ma;

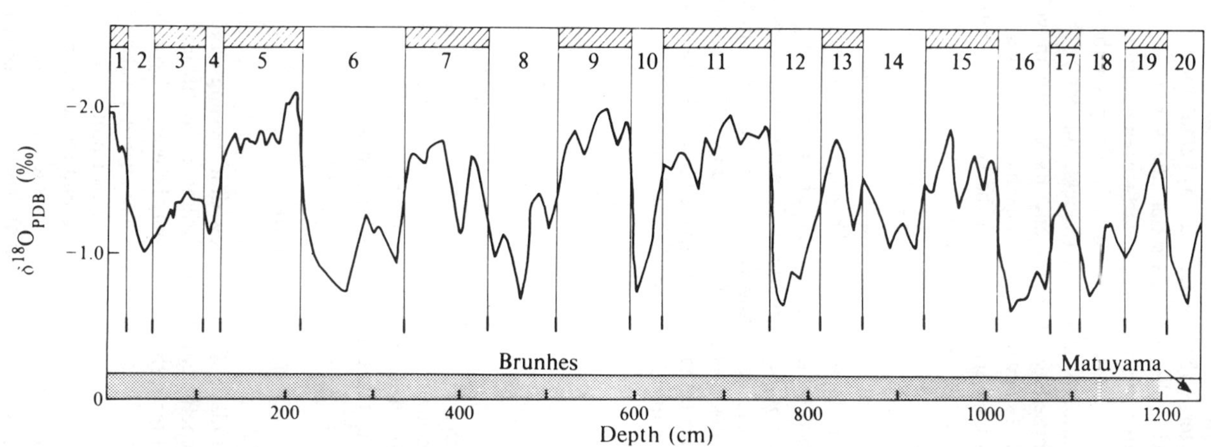

Figure 5.6 Variation in the oxygen isotopic composition of the planktonic foraminifer *Globigerinoides sacculifer* with depth in the deep-sea core V28–238, from the equatorial Pacific, representing the Brunhes normal magnetic period and the beginning of the Matuyama reversed period. These isotopic variations are due to changes in the volume of continental ice. Isotopic stages according to Emiliani (1966) are marked with numbers along the top. (From Shackleton & Opdyke 1973).

the cores are dated either by U-series methods (Ku & Broecker 1967) or by magnetic stratigraphy (Shackleton & Opdyke 1976). The maximum amplitude of the cycles is about 1.8‰ and the minimum (warmest) $\delta^{18}O_{ct}$ values are the same throughout the record and about equal to that of modern foraminifera. The similarity of this variation in foramanifera from bottom waters (where water is always cold) and from surface waters (whose temperature varied with climate) shows that the variation largely represents changes in $\delta^{18}O$ of sea water rather than changes in surface temperature, as first suggested by Emiliani (1966). The $\delta^{18}O$ of sea water reflects, in turn, changes in the volume and, to a lesser extent, isotopic composition of glacial ice (Shackleton 1988). This ice is 20–30‰ 'lighter' than sea water (Johnsen *et al.* 1972) and its growth in volume causes a corresponding increase in $\delta^{18}0$ of sea water. Thus, the curve of $\delta^{18}O$ versus time (Fig. 5.6) has become a master curve for relative changes in the volume of continental ice, and therefore for the sequence of glacial and interglacial stages through the Pleistocene. A similar, but much shorter, record is given by changes in $\delta^{18}O$ (and δD) in ice cores from Greenland and Antarctica (Johnsen *et al* 1972).

The ice-volume record generally suggests slow growth of ice followed by rapid deglaciation (Broecker & van Donk 1970). The cyclicity has been shown to be controlled by variations in the orbital parameters of the Earth (Martinson *et al.* 1987). Following Emiliani's (1966) notation, interglacials are denoted by odd-numbered stages (1,3,5,. . .) and glacials by even numbered stages. Decimal-numbered substages have been suggested by Martinson *et al.* (1987) and other substages are widely used. For example, the generally warm stage 5 is subdivided into substages 5a to 5e; the last interglacial as recorded in continental deposits and maximum high-sea stands is assigned to stage 5e.

5.3.2 Meteoric water

Isotopes of hydrogen and oxygen undergo fractionation during evaporation and condensation. The result is that precipitation becomes progressively depleted in the heavy isotopes ^{18}O and deuterium (D or ^{2}H) as water vapour moves further from its source. This depletion is intensified as temperature falls or as precipitation occurs at progressively higher elevation. There arises a correlated shift in $\delta^{18}O$ and δD in all meteoric waters:

$$\delta D = 8\delta^{18}O + 10$$

In coastal regions and on islands, $\delta^{18}O$ increases with increasing temperature by 0.7‰ $°C^{-1}$ (Dansgaard 1964); this coefficient is more variable in continental regions.

As a consequence of these effects, the isotopes may be used as hydrologic tracers over short timescales such as a single flood event or the passage of a water year. They can be used to discriminate between sources of water such as runoff, snowmelt, glacier-melt, etc., and can be used to determine residence times and storage capacities of aquifers. Evaporation from surface waters causes a characteristic enrichment of stable isotopes with a slope of $\delta D/\delta^{18}O \simeq 5$ so that evaporated (e.g. lake-derived) waters can be recognised. Ancient ground waters, dated by ^{14}C, have shown variations in δD through the Pleistocene (Gat & Isaar 1974) although in some areas surprisingly little variation has been observed (Smith *et al.* 1976). Fluid inclusions in speleothems, which trap samples of recharge water, display decreases in δD of about 12‰ in the northeastern USA during glacial times (Harmon *et al.* 1979b).

5.3.3 Organic matter

When carbon is assimilated from atmospheric CO_2 by photosynthesis, ^{12}C is preferentially taken up (Smith & Epstein 1971). The degree of enrichment depends on the photosynthetic mechanism used by the plant. So-called C4 (or Hatach–Slack cycle) plants, which live largely in warm, semi-arid environments, fix carbon that is 5‰ lighter than atmospheric CO_2, whereas C3 (Calvin cycle) plants of more humid habitats fix carbon that is 19‰ enriched in ^{12}C with respect to CO_2. This isotopic 'label' is transmitted in turn to both fossil organic matter in sediments and to the tissues of animals that eat these plants (De Niro & Epstein 1976). Collagen from fossil bone can be used to infer the proportion of C3 to C4 plants in an environment, from which palaeohumidity can thus be inferred (Chisholm *et al.* 1986).

Variations in δD and $\delta^{18}O$ of cellulose in plants reflect changes in these isotopes in meteoric waters, as well as humidity (which influences the amount of evaporative fractionation by leaves). Isotopic analysis of fossil cellulose in wood and peat can thus be used to reconstruct past variations in temperature (through its effect on $\delta^{18}O$ of rainfall) and humidity (Epstein & Yapp 1976, Edwards & Fritz 1986). Similar climate reconstructions are possible from analysis of $\delta^{18}O$ in phosphate of terrestrial herbivores (Luz *et al.* 1984) and of δD in collagen of bone (Cormie & Schwarcz 1985).

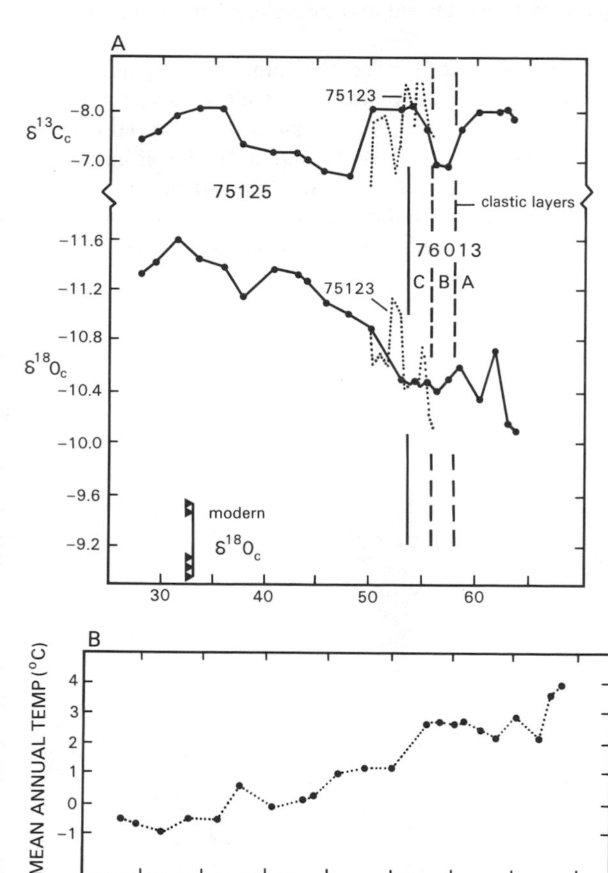

Figure 5.7 (A) Variation in isotopic composition and inferred palaeotemperature of stalagmites from Cascade Cave, Vancouver Island, British Columbia (B) Shifts in temperature, inferred from $\delta^{18}0$ of calcite by assuming that meteoric water varied by 0.7‰ °C^{-1} and that $\delta^{18}O$ of rain water at any stage was offset by an amount equal to the shift in $\delta^{18}O$ of sea water as inferred from Figure 5.6. (From Gascoyne *et al.*, 1981b).

5.3.4 Continental freshwater carbonates

Calcium carbonate is chemically precipitated in various forms: spring-deposited tufa and travertine; lacustrine marls; calcareous zones in soil profiles; and speleothem in caves (stalagmites, stalactites, flowstone, etc.). As well, organically precipitated carbonate occurs as algal marls and mollusc shells. Both types of calcite may be formed in isotopic equilibrium, so that temperatures can be inferred from their $\delta^{18}O$, although $\delta^{18}O$ values of many surficial carbonates largely reflect changes in $\delta^{18}O$ of surface waters or humidity. For example, Cerling (1984) has shown that $\delta^{18}O$ of calcretes and local meteoric precipitation are correlated. Such variations in fossil calcretes, marls, etc., should reflect changes in temperature and humidity (Edwards *et al.* 1986).

Speleothems deposited in the deep interiors of karstic caves, where relative humidity is close to 100%, are formed in oxygen isotopic equilibrium with seepage waters (Hendy & Wilson 1968). The temperature at such sites is constant and close to the average annual surface temperature. The $\delta^{18}O$ of the parent seepage water of ancient speleothem can be obtained by analysis of minute inclusions of water trapped in these speleothems (Schwarcz & Yonge 1983). Fractionation between the calcite and seepage water has been used to infer palaeotemperatures for various cave sites in North America (Thompson *et al.* 1976, Harmon *et al.* 1978b, Gascoyne *et al.* 1981b); see Figure 5.7. These studies have demonstrated changes of up to 8°C between glacial and interglacial stages in the mid-continent.

Carbon isotopes in terrestrial carbonates largely reflect the influence of organic matter on soil CO_2, and are in turn controlled by the relative prevalence of C3 versus C4 plants, and variation in aridity or soil productivity (Cerling 1984, Cerling & Hay 1986, Goodfriend & Magaritz 1988). The latter authors show that $\delta^{13}C$ of soil carbonate decreases with increasing amounts of local rainfall, in Israel. Speleothem precipitated from recharge water that has passed through the soil will also have an isotopic signature acquired from the plants growing in that soil, and that will also change with climate.

5.4 Additional dating methods

5.4.1 Introduction

The importance of reliable independent dating control to place landscapes and events in a secure absolute timescale has become more apparent in recent years. The need for improved dating capabilities has attracted the attention of physicists and chemists who otherwise would not be involved in geomorphological research. Their contributions, coupled with improved instrumental capabilities, have led to significant advances in existing methodologies and to the development of promising new techniques. A number of state-of-the-art summary papers have appeared in the last five years that treat specific Quaternary dating techniques (see, for example, the special issue of Quaternary Science Reviews, **7**, nos. 3 and 4, 1988), and the general topic has been treated in recent books (Mahaney 1986, Easterbrook 1988) as well as substantive chapters in several of the Geological Society of America's DNAG volumes. Problems and potential of dating techniques have also been the topics of symposia at many recent national and international meetings.

The techniques are presented below according to their governing principles:

(a) those methods dependent on the decay of radioactive nuclides that proceed at a fixed rate, for the most part independent of environmental parameters, include radiocarbon and uranium-series dating covered previously, as well as potassium–argon and fission–track dating;
(b) methods such as thermoluminescence and electron spin resonance dating require radioactive decay reactions, but are also dependent on environmental constraints;
(c) chemical reactions, including obsidian hydration and amino-acid racemisation, are controlled by temperature, and therefore are implicitly dependent on regional climatic and microclimatic parameters;
(d) a final category, stable isotopes, are not strictly dating methods but correlation tools; $^{18}O:^{16}O$ (covered elsewhere) and $^{87}Sr:^{86}Sr$ fall in this category.

Radioisotopes are also used as tracers in natural systems. Tritium (^{3}H), ^{137}Cs and ^{14}C were all produced in the upper atmosphere during nuclear weapons testing, mostly in a 10-year interval centred on the middle 1960s. These isotopes were subsequently incorporated into the lower atmosphere, biosphere, hydrosphere and eventually into sediments. They form an important class of compounds that can be used as tracers to determine the rates and pathways of natural processes acting over time periods of a few years to a few decades.

5.4.2 Radiometric methods independent of environmental variables

5.4.2.1 Potassium–argon (K–Ar) and $^{40}Ar:^{39}Ar$ dating

The K-Ar method is applicable to suitable rocks throughout the range of geologic time. Application of the method to rocks of the Quaternary system represents a specialisation of K–Ar dating, and requires specific pre-treatment and detection facilities. Recent advances in grain-discrete laser fusion of individual mineral grains has provided a quantum leap by increasing precision while dramatically reducing sample size requirements.

In naturally occurring potassium-bearing rocks, radioactive potassium (^{40}K) accounts for about 0.01% of the potassium isotopes, whereas the stable isotopes ^{39}K and ^{41}K account for 93% and 7% respectively. ^{40}K decays by electron capture to ^{40}Ar (half life 1.2×10^{10} yr) and, less significantly for the dating of young rocks, by beta decay to ^{40}Ca (half life 1.5×10^9 yr). Argon–40 (^{40}Ar) is the only radioactive isotope of argon, although the half-life is sufficiently long (1.3×10^9 yr) that it can be considered stable in late Cenozoic materials.

In principle any rock or mineral can be dated by K–Ar provided that K and Ar are sufficiently abundant to measure analytically, and that three assumptions are satisfied:

(a) no inherited argon – the material must contain no radiogenic argon at the time of formation;
(b) closed system – the sample has remained a closed system, having neither gained nor lost either argon or potassium, except as a result of the decay of ^{40}K;
(c) at the time of formation the ratio of radiogenic ^{40}Ar to ^{36}Ar was similar to today (*c*.296), and there was no significant fractionation of the potassium isotopes.

These conditions are best met in minerals that have formed by cooling from magma; both intrusive and extrusive igneous rocks are suitable. The occurrence of such rocks in Quaternary deposits is, however, generally restricted to regions adjacent to plate boundaries and areas of active tectonics. Suitable minerals include hornblende, biotite, muscovite, high-temperature plagioclase and potassium feldspar, pyroxene, nephaline and leucite. In the case of volcanic rocks, for which mineral separations are not feasible due to the fine-grained nature of the matrix, whole-rock analyses can yield reliable results (Dalrymple 1964, Everden *et al.* 1964). The most suitable minerals for dating Pleistocene deposits are those with high levels of potassium and low atmospheric argon. Sanadine and potassium-rich whole-rock samples are the most widely used. Some criteria for evaluating suitable samples are given by McDougall (1966), Cox *et al.* (1968) and Naeser *et al.* (1981).

The limitations of the method in young rocks are related to or associated with measurements of very low levels of argon, and the failure of many minerals to satisfy assumption (a); even trace amounts of inherited ^{40}Ar cause major uncertainties when dating young rocks. Argon, being a noble gas, is also more likely to escape from the system than is potassium. Consequently K–Ar dates are more likely to be too young than too old.

Measuring the $^{40}Ar:^{39}Ar$ ratio is an alternative way to estimate the decay of ^{40}K; applications of the method to rocks of Quaternary age are summarised by Hall and York (1984). Prior to isotopic analysis, the sample is placed in a nuclear reactor and exposed to a beam of fast neutrons, which convert some of the ^{39}K to ^{39}Ar. The efficiency of the process is calibrated using standards. A focused laser beam is used to heat discrete mineral grains, releasing trapped argon; the $^{40}Ar:^{39}Ar$ ratio is measured by mass spectrometry. The ^{39}Ar, a short-lived radioisotope that is not present in the atmosphere, can be used as a measure of the amount of ^{40}K in the sample; ^{39}Ar and ^{40}Ar are measured from the same position in the crystal lattice repeatedly as the sample is heated. Multiple analyses allow high-precision dating, even of very young samples, and routine dating of late Pleistocene samples overlapping with radiocarbon is now feasible (e.g. Drake *et al.* 1988). Reliable, precise dating of young samples is enhanced if the sample contains a high potassium content.

Dating bulk samples of volcanic ash by K–Ar or $^{40}Ar:^{39}Ar$ has met with limited success, primarily because the incorporation of detrital grains produces bulk ages that are too old, and weathering of the ash shards promotes the loss of argon. The development of grain-discrete laser fusion techniques allows the problem of detrital grains to be circumvented.

Error estimates typically increase with decreasing sample age; Pliocene and early Pleistocene samples commonly have a precision of better than $\pm 5\%$, middle Pleistocene samples are generally $\pm$ 10–20%, whereas late Pleistocene samples commonly have even greater uncertainty. Replicate analyses of young samples can significantly reduce the standard error of the measurements.

5.4.2.2 Fission-track dating

Fission-track dating was developed as a complementary tool to K–Ar dating, specifically as a means of dating air-fall tephras not suitable for the K–Ar method. Fission tracks are created by the spontaneous fission of ^{238}U nuclei (decay constant $7.03 \times 10^{-17}\ yr^{-1}$). In natural dielectric solids, each fission event is recorded by a zone of intense radiation damage resulting from atomic displacements. Spontaneous fission follows the laws governing radioactive decay, and hence the number of spontaneous tracks (track density) is a function of uranium concentration and time. The uranium concentration is measured by irradiating the sample with slow (thermal) neutrons that induce fission in ^{235}U; tracks that result from the fission events are then counted. The concentration of ^{238}U is calculated from an assumed ratio of ^{235}U to ^{238}U. It is also assumed that:

(a) fission tracks are produced with 100% efficiency (although this can be relaxed if the natural and induced tracks are created under similar efficiencies);
(b) fission tracks are retained with 100% efficiency;
(c) the uranium concentration has remained constant.

Reliable fission-track dates require that all assumptions be met, and that there is a sufficient density of tracks in the mineral being counted. This is particularly important for young deposits. Only high-uranium-bearing minerals contain sufficient uranium to produce a measurable number of fission tracks in less than 10^6 yr. Suitable minerals are zircon and sanadine, as well as apatite and volcanic glass, although the latter are susceptible to annealing.

The application of fission-track dating to young geological events is summarised by Naeser and Naeser (1988). The timeframe is limited primarily by the time required to develop a statistically significant number of tracks; this is generally at least 100 000 yr. Uncertainties are added if the tracks are lost due to subsequent heating or alteration (weathering).

5.4.3 Radiogenic methods with environmental dependences

5.4.3.1 Thermoluminescence dating

Some of the most promising new research in Quaternary geochronology has been in the development of thermoluminescence (TL) dating and the closely related electron spin resonance (ESR) dating (below). Although initial research on the possibility of TL to date geological materials was pioneered in the USA, the recent impetus was initiated in the Eastern European countries in the early 1970s and shortly thereafter picked up by a number of laboratories in Western Europe. The last five years have witnessed a remarkable growth in TL literature and increased understanding of the collecting and analytical strategies that provide the most reliable dates.

TL dating offers the tremendous advantage of having the potential to date sediment directly, rather than being dependent on the fortuitous location of fossil materials. The method is based on the observation that grains of quartz and feldspars act as tiny dosimeters, in which the exposure of the mineral grains to ionising radiation displaces electrons from their normal orbits into lattice-charge disequilibrium traps within the grain. Upon heating, some of the electrons released from their traps move to luminescent centres where they give off light. The amount of radiation received by a mineral grain can be determined by heating the grain in an inert atmosphere in the laboratory and using the intensity of luminescence as a proxy for the number of displaced electrons, which is a function of the accumulated radiation dose; hence the term, 'thermoluminescence'.

The other important observation that allows TL dating of sediment is that electron traps can be cleared not only by heating but also by exposure to light. The exposure of quartz and feldspar to sunlight rapidly reduces (bleaches) the TL level to a small residual value. The efficiency of the bleaching process is a function of exposure time, spectral bandwidth and intensity. Bleaching of quartz is sensitive primarily to the ultraviolet (UV) component, whereas feldspars are bleached by visible light as well. As a general rule, exposure for 5–10 h of even indirect sunlight is adequate to bleach most grains. Aeolian sediments are most likely to be fully bleached prior to deposition, whereas fluvially transported sediments are commonly less fully bleached.

Assuming mineral grains are fully bleached prior to burial, the intensity of the TL signal in a geological sample (generally referred to as the *equivalent dose* (*ED*)) is a function of the age and the total radiation dose. Radiation is derived primarily from the decay of radioactive isotopes of uranium, thorium and potassium, present in trace amounts in nearly all sediment. If the annual *dose rate* (*DR*) can be calculated, the TL signal can be converted to an absolute age.

The optimal methods for calculating the *ED* and *DR*, overviews of the assumptions inherent to the technique and potential applications of the TL method are well covered in several recent publications (e.g. Wintle & Huntley 1982, Aitkin 1985, Berger 1988, Mejdahl 1988, Forman 1989). Although aeolian sediments have been the dominant focus of TL research over the past decade because of their excellent opportunity for thorough bleaching, sediments from other depositional environments also show promise. Buried soils, colluvial sediments and littoral sediments all show extensive bleaching. In certain settings fluvially transported sediments may also be well bleached. Adequately bleached sediments can provide absolute ages by the TL method for the time range of a few hundred to about 100 000 yr. The upper limit of the technique has yet to be well defined, and there are suggestions that deposits as old as 500 000 yr may be datable by TL.

Closely related to thermoluminescence dating (TL) is optically stimulated luminescence dating (OSL), an approach that was first demonstrated by Huntley *et al.* (1985), and which has great potential application to the dating of quartz grains of wind- and water-borne origin. The luminescence emission caused by optical excitation has the advantage that it is only derived from the light-sensitive sources of luminescence from within the grains, i.e. those sources that are most likely to undergo re-setting – 'bleaching' – in the process of being transported and deposited, as compared to the spectrum of both light-sensitive and light-stable traps that are sampled during a TL analysis (Rhodes, 1988).

5.4.3.2 Electron spin resonance dating

First proposed as a potential dating method in the late 1960s (e.g. Zeller *et al.* 1967) electron spin resonance (ESR) dating has only in the last five years emerged from the developmental stage into the applied realm. Hennig and Grün (1983) and Ikeya (1985) provide up-to-date summaries of the underlying principles and applications of the method. Electron spin resonance is a method of measuring the paramagnetic defects in minerals or the skeletal hard parts of organisms; the ESR signal is generally considered to be a close proxy of the total number of defects. These defects are created by penetrating radiation from radioactive elements within, or in the sediment surrounding, the sample.

The method shares many of the fundamental principles with thermoluminescence (TL) dating, being essentially a measure of the accumulated radiation dose acquired by a crystalline solid since its burial. The ESR signal differs from that measured by TL; ESR measures a specific level of trapped electrons rather than all those electrons that emit light in their transition phase, as is the case for TL. An additional advantage over TL is that ESR measurements are non-destructive; the electron distribution remains unchanged.

Determination of the ESR age requires (a) measurement of the ESR signal, and (b) determination of the average dose rate received by the sample. Of these parameters, determining the average dose rate is perhaps most tenuous, at least in part due to the possibility that the modern dose rate may not be representative of the average rate since deposition. Nevertheless, Radtke *et al.* (1988) show a close correspondence between ESR and U-series dates on unrecrystallised corals from Barbados. ESR dating has been successfully applied to corals, molluscan fossil and tooth enamel; ESR dating of bone remains tenuous (Grün & Schwarcz 1987).

The time range of applicability depends on the dose rate and mineralogy; in general a maximum range in carbonate fossils is about 400 000 yr, with an uncertainty of ±10–20%.

5.4.4 Non-radiogenic isotopic methods

5.4.4.1 Strontium dating

The Sr isotopic ratios of sea water are known to have varied over geologic time; several studies have suggested that this ratio can be used to provide absolute dates for Neogene marine carbonates (e.g. Burke *et al.* 1982, DePaolo 1986). The oceans act as a reservoir for strontium, with the isotopic composition dependent on the relative balance of the two primary sources, fluvial transport from the continents and alteration of seafloor basalts. The isotopic composition of strontium derived from these two sources differs; continental runoff carries strontium enriched in ^{87}Sr relative to that derived from oceanic crust; mid-oceanic ridge basalts have a ^{87}Sr:^{86}Sr ratio close to 0.704, whereas rivers are more variable, but generally are above 0.710. The flux of strontium delivered to the oceans is balanced by removal, as strontium is incorporated into the carbonate skeletons of marine organisms, so that the concentration of strontium in sea water remains constant.

The mean residence time of strontium in the oceans is about four million years, much longer than the mixing time, so that the ratio is uniform around the globe. Marine organisms incorporate strontium in their skeletal hard parts without fractionation. The ^{87}Sr:^{86}Sr ratio in carbonate fossils has been used to reconstruct seawater ^{87}Sr:^{86}Sr ratios through the Phanerozoic (e.g. Elderfield 1986); for the Cenozoic, these curves show a nearly monotonic increase in the ^{87}Sr:^{86}Sr ratio, reflecting a higher rate of terrestrial over seafloor sources. Consequently, once this curve is established and independently dated, the ^{87}Sr:^{86}Sr ratio can be used to date marine precipitates. Resolution varies from 10^5 to 3×10^6 yr depending on the rate of change in the global curve. The total change within the Quaternary is small and high-precision ^{87}Sr:^{86}Sr measurements are required to detect differences. Recent advances in instrumentation have provided the precision required to develop a ^{87}Sr:^{86}Sr curve for the Quaternary (e.g. Capo & DePaolo 1988); resolution is *c.* 0.2×10^6 yr.

Although the method must be calibrated by measuring isotopic ratios on samples of known age, and is therefore also susceptible to uncertainties in the ages of the calibration samples, the potential contribution of strontium dating to geomorphic research is large. The method holds promise for providing absolute ages of marine materials of late Pliocene and early Pleistocene age, a time period for which there are few established techniques of wide applicability.

5.4.5 Chemical methods

Chemical methods differ fundamentally from physical methods in that the reaction rate is not a physical constant, but is dependent on one or more environmental parameters, of which temperature is generally the most influential. Chemical reactions respond exponentially to temperature; consequently evaluation of the thermal term is complex. It includes not only the integrated thermal history of the sample since it was deposited, but also a term related to the amplitude of the temperature fluctuations. Samples experiencing high-amplitude temperature fluctuations have an *effective temperature* greater than the arithmetic mean temperature. Only those samples that have experienced similar thermal histories can be directly compared. Cave material, samples buried by at least 2 m of overburden and lake/marine core material have low-amplitude annual temperature oscillations and are preferred sampling sites. Complete evaluation of the thermal history of a sample is usually complicated by uncertainties in glacial–interglacial climate changes.

The equations that define chemical reactions include terms related to both time and temperature. Although chemical methods can be used directly as relative age indices for samples with uniform thermal histories, it is generally not possible to calculate a specific value for either of these terms without independently knowing the other. The most commonly used strategy is to calibrate the thermal term by analysing samples with secure independent dates, and to define a rate constant based on the calibration sample that can be applied to nearby samples from the same thermal regime. In this scheme, it is not necessary to know the precise changes in temperature at each site, only that all samples have experienced a similar thermal history. Alternatively, if the reaction parameters are adequately defined, independently dated samples can be used for palaeothermometry. Because the temperature sensitivity of the rate constant is highly dependent on temperature, palaeotemperature calculations are inherently more precise than are absolute dates; temperature changes of *c*. 1°C can be detected.

5.4.5.1 Amino-acid geochronology

The possibility of using systematic changes in protein preserved in skeletal hard parts for geochronology was first suggested more than 30 years ago by Abelson (1955) and subsequently developed by Hare and colleagues for carbonate systems (e.g. Mitterer & Hare 1967, Wehmiller & Hare 1971) and for bone by Bada (e.g. Bada 1985). Of the various diagenetic reactions affecting proteins, the most widely exploited in geochronology has been the racemisation reaction. Effectively all of the amino acids in structural proteins preserved in geological samples are initially of the L optical configuration. As the protein is degraded, hydrolysing into lower-molecular-weight polypeptides and eventually releasing free amino acids, the L amino acids invert to their D configuration. The extent of the reaction is monitored by measuring the proportion of D and L isomers of a particular amino acid, the most commonly used amino acids being isoleucine (D-alloisoleucine to L-isoleucine, aIle:Ile), leucine (D:L) and aspartic acid (D:L). At death of the organism (t_0) the proportion of D to L amino acids is effectively zero (in practice a small amount of the D amino acid is created during the hydrolysis step required prior to analysis; the D:L ratio in a modern sample is *c*.0.014). The reaction proceeds to racemic equilibrium, at which time the number of D amino acids inverting to their L amino acid configuration is exactly balanced by the reverse reaction (D:L = 1.00, except for isoleucine for which the equilibrium aIle:Ile ratio is 1.30). Unlike radiocarbon, in which the parent material decreases with increasing time, and measurement of the activity of the sample becomes increasingly susceptible to contamination, there is only minor loss of amino acids over time, and the precision of the D:L measurement is the same for modern and very old samples. As in all chemical methods, the reaction rate (i.e. rate of racemisation) is highly dependent on temperature. In high arctic regions (mean annual temperature $MAT = -15°C$) equilibrium requires more than 20 million years, whereas in laboratory experiments at 160°C equilibrium is attained in a few hours. At mid-latitude sites ($MAT = +10°C$), *c*. two million years are required for the reaction to reach equilibrium.

A secondary control on reaction rate is related to the nature of the initial protein, both the actual sequence of amino acids and the cross-bonding between protein chains. In carbonate fossils, these variables are related to taxonomy; to compare directly D:L ratios from molluscan samples, the analyses must be on individuals of at least the same genus, preferably the same species. A similar constraint should be applied to other carbonate fossils.

Although amino-acid geochronology has potential for nearly any system that preserves these organic molecules over geologic time, in practice only bone and carbonate fossils have been routinely utilised. The reliability of D:L ratios for dating bone has been questioned due to the porous nature of the inorganic matrix and the rapid solubilisation of the dominant organic constituent, collagen. Bone preserved in hot climatic zones ($MAT > +15°C$) is frequently devoid of collagen within 5000–10 000 yr, whereas in arctic regions ($MAT < -10°C$) collagen may remain essentially intact for more than 60 000 yr, and well preserved even in samples more than one million years old. Because of the porosity of bone, free amino acids are rapidly leached from the matrix. Consequently, for certain stages of protein diagenesis, the racemisation reaction may follow reversible first-order kinetics. However, the complexity in defining the kinetic pathways has limited the utility of bone in amino-acid geochronology.

In contrast to bone, the integrity and natural buffering effects of the carbonate matrix found in the shells of molluscs and foraminifera provide a more reliable system for geochronological applications. Recent studies have shown that the eggshells of at least the African ostrich (*Struthio camelus*), and possibly other birds, closely approximate a perfect closed system, retaining all amino acids, even in the later stages of diagenesis when the protein has been degraded to low-molecular-weight free amino acids and small peptide chains (e.g. Hare *et al.* 1984, Miller & Wendorf 1986). In these samples, the coefficient of variation in measured D:L ratios is significantly lower than in molluscs or foraminifera. Foraminifera show only minor losses of amino acids over time, whereas molluscan fossils lose a slightly higher proportion of their indigenous amino acids. Even in molluscan samples these losses are relatively small; Cretaceous shells from clayey sediments have been found that contain amino acids at nearly modern concentrations.

Amino-acid D:L ratios can be used directly as correlation indices over regions of uniform thermal history (aminostratigraphy; Miller & Hare 1980), but the conversion of D:L ratios to absolute age is complicated by difficulties in evaluating the integrated thermal history and uncertainties in racemisation kinetics. McCoy (1987) has recently reviewed the precision of the method for applications in geochronology and palaeothermometry. He concludes that, without calibration, racemisation-derived absolute ages have an uncertainty of $\pm 50\%$, but this is reduced to $\pm 15\%$ with a secure local calibration. The effective temperature for an interval bracketed by samples of known age can be calculated to at least ±3°C, whereas the change in temperature between two intervals bracketed by dated samples can be calculated to within 1–2°C. Determination of D:L ratios for correlation purposes is generally reliable within ±5–10%. Although the precision of racemisation-derived absolute ages based on interpolation between calibrated dates is estimated at better than ±20%, due principally to uncertainties in reaction kinetics, extrapolated ages are less precise.

Several recent studies have utilised marine molluscs to develop regional relative chronostratigraphies (aminozones and aminogroups), which, when tied to one or more calibrated sites, have been used to provide estimates of absolute ages for the relative stratigraphies (e.g. Kennedy *et al.* 1982 (USA west coast), Miller & Mangerud 1985 (NW Europe), Bowen *et al.* 1985 (UK), Hearty *et al.*, 1986 (Mediterranean), Wehmiller *et al.*, 1988 (USA Atlantic Coastal Plain)). Terrestrial gastropods have also been utilised, but in general have somewhat less precision than do marine molluscs. In general, the method has shown clear utility in identifying last interglacial deposits in mid-latitude sites; a plot of D:L ratios in moderate-racemisation-rate molluscs from last interglacial sites in NW Europe and the Arctic (Fig. 5.8) demonstrated the temperature sensitivity of the reaction.

Several recent summary articles provide reviews on the use of amino-acid D:L ratios (Miller & Mangerud 1985, Wehmiller 1984), and on the precision of the method and sampling strategies to maximise the potential results (McCoy 1987, Miller & Brigham-Grette 1989).

5.4.5.2 *Obsidian hydration dating*

Fresh rhyolitic volcanic glass (obsidian), when exposed to the atmosphere, will absorb water to form a hydrated rind that increases in thickness with increasing time. The hydration reaction rate is strongly temperature-dependent, increasing exponentially as the temperature rises. The reaction rate is also moderately dependent on the chemical composition and physical properties of the glass (in decreasing order of significance: silica content, refractive index, CaO and MgO content, and interstitial water content). If the exposure of the obsidian surface can be related to a specific event in time (e.g. construction of a projectile point, glacial abrasion, etc.), then the rind thickness provides a measure of time elapsed since that event. Trembour and Friedman (1984) have recently summarised the factors affecting the rate of obsidian hydration.

Rind thicknesses can be used directly to develop relative chronostratigraphies across regions of uniform thermal history. Conversion of rind thickness to absolute age requires quantification of the relationship between hydration rate and temperature for the specific obsidian composition of the sample. This relationship can be derived empirically from ^{14}C-dated sites with obsidian artefacts and pyrolysis experiments at elevated temperatures. Friedman and Long (1976) combined both types of data to show that, at any given temperature, hydration proceeds at a rate proportional to the square root of time, and that the rate is independent of the vapour pressure of water (for all but truly anhydrous settings). The Arrhenius equation derived from their data has an activation energy of *c.*20 kcal mol^{-1}, significantly less than for amino-acid racemisation (*c.* 29 kcal mol^{-1}).

Obsidian hydration dating has been applied most frequently to archaeological samples. The hydration rind thickness on excavated obsidian artefacts may be used to estimate the age of the site, or conversely, if the site is already dated, it is possible to calculate the effective diagenetic temperature of the site since construction of the artefact. It is tacitly assumed that the manufacture of the tool removed any pre-existing rind, although multiple-aged implement surfaces have been recognised.

Geological applications require that the exposure of a fresh obsidian surface can be equated with a specific geological event. One application is the measurement of

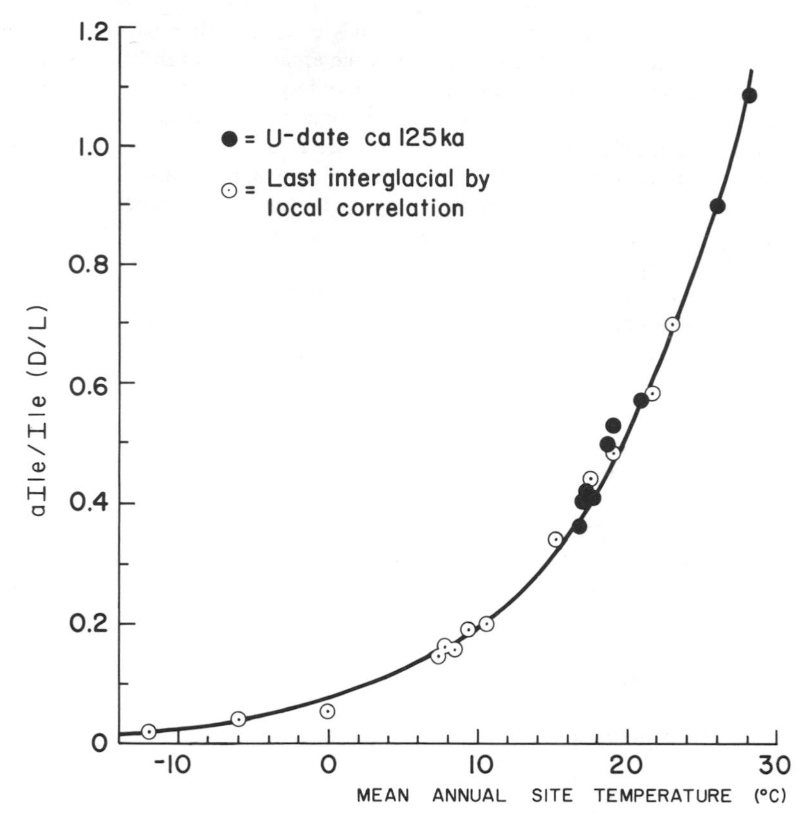

Figure 5.8 The D-alloisoleucine:L-isoleucine ratio from moderate-racemisation-rate molluscan taxa of last interglacial age plotted against current mean annual site temperature. Sites for which a last interglacial age is inferred from correlation or faunal considerations are shown as open circles; whereas those with U-series coral dates are shown with full circles. (Modified from Hearty and Miller 1987).

hydration rinds on obsidian in terminal moraines (Pierce *et al.* 1976). Stress exerted on obsidian pebbles during glacial transport frequently exposes fresh surfaces or fractures. Subsequent hydration on these surfaces or fractures is proportional to the time elapsed since deposition in the moraine. Obsidian hydration has also been used to date volcanic events (e.g. Friedman & Obradovich 1981).

The age range of obsidian hydration dating ultimately depends on the thermal history of the sample. For sites between +5 and +10°C, the useful range of the method is from a few thousand to more than one million years. The precision is generally *c.* ±10%, whereas the accuracy of the dates is somewhat less, *c.* ±30%, due to uncertainties in the thermal term.

5.4.6 Other methods

Brief reference must now be made to certain other techniques that have been employed by geomorphologists and other Earth scientists to obtain dates. One such method is to use the presence of layers of volcanic material of known age in a particular deposit. This is termed tephrochronology and has been well reviewed by Westgate and Gold (1974).

Tephra, which are airborne pyroclastic materials ejected during explosive volcanic activity, may create a blanket of sediment over large areas, which can be considered as more or less instantaneous on a geological timescale. Layers of tephra can be dated by such methods as fission track or K–Ar, and, provided that the dated layer can be uniquely identified in different areas, can then be used as a chronostratigraphic marker horizon to provide limiting dates on the sediments with which it is associated. In some areas, such is the frequency with which tephra ejection events have happened in the past (e.g. in the western USA) that tephrochronology can provide highly detailed dating information (Self & Sparks 1981).

There are also a range of methods that give relative dates, and these include, for example, the use of archaeological materials to give a broad chronology for Pleistocene and Holocene materials (see, for example, Allchin *et al.* 1977). Pottery, coins and other metal objects can provide reasonably precise dates for deposits up to about 2500 yr though care needs to be taken to make sure that they have not been derived from earlier deposits. Stone artefacts tend to be more problematical, and Catt (1988, p. 193) has expressed a wise note of caution:

> At one time it was thought that the typology of stone artifacts could be used to date earlier Quaternary deposits in much the same way as evolutionary changes in fossil groups can help date them. However, it is now known that the methods of tool-making and the styles of different tool types migrated quite slowly from place to place, and some were introduced more than once to certain areas at quite different times. Also some cultures co-existed for long periods in some regions. At present it is therefore more appropriate to date the artifacts of stone age cultures using other dating methods than to expect the artifacts themselves to provide very precise dates.

Likewise, there have been miscellaneous attempts to use the degree of soil formation, mineral alteration and weathering rind development to date such features as neoglacial moraines (see, for example, Porter 1975, Birkeland 1973). This can supplement the information derived from lichenometry, which is outlined in Section 5.6. However, the degree of rock weathering or soil formation that can occur in a given period of time depends on a variety of factors (geomorphic situation, lithology, etc) so that if determination of the time required for a given degree of weathering to occur is the prime object of study then sites need to be chosen that have reasonably similar conditions of parent material and slope. Some of the difficulties posed by this constraint can be overcome by studying the degree of etching of certain heavy minerals, and by selecting sites on the crests of moraines.

5.5 Tree-ring dating (dendrochronology)

5.5.1 Tree growth cycles

In temperate latitudes, tree-ring series and, in some instances, patterns of width variation provide a direct means of establishing timespans of direct relevance to geomorphologists. During growth, trees regularly increase their size by the addition of secondary woody tissue. In a given growth period this material is initially produced as large, thin-walled, cells, but is then followed by a gradual transition to the production of small thick-walled cells. Since the growth pattern is cyclic, with active periods being followed by dormancy, discontinuities, delimited as abrupt changes in cell size, occur between successive cycles, and these constitute the familiar tree-ring series. The growth cycles are normally seasonal in character and are determined by the regular alternation of winter and summer climatic conditions, and hence the rings record annual events.

Although they are often only considered as two-dimensional phenomena, tree rings are in reality long, tapered, cone-shaped bodies, which stack one upon the other (see Fig. 5.9). Once produced, a given growth increment remains constant both in size and in position throughout the existence of the tree. The addition of successive cones of growth in each succeeding year naturally increases the diameter and height of the trees. Tree rings are particularly striking in gymnosperms since they possess a pronounced colour difference between the dark late wood and the following season's light early wood cells on the other side of the discontinuity. In contrast, angiosperm tree rings are not so clearly developed because of a more complex cell structure, but they are nonetheless decipherable.

Two basic kinds of tree-ring series based on gross morphology may be recognised, but it should be remembered that a complete transition occurs between each end-member. At one end of the spectrum, a tree-ring series may occur where, along a given radius, the rings are essentially uniform in thickness. Such a series lacking any marked irregularites in ring width is termed 'complacent'. Conversely, at the other extreme, a tree-ring series may occur where there is a marked variability in the successive ring widths along a given radius, and this type is referred to as 'sensitive'. In all cases, both sensitive and complacent, there is detectable an overall decrease in ring width that is related to increasing tree age. However, the criteria for assigning a tree-ring series to either complacent or sensitive types are independent of this latter factor.

Complacent tree-ring series arise when the environmental conditions are such that there is usually an adequate supply of moisture or a sufficiently high temperature in the growing season. On the other hand, sensitive tree-ring series are often associated with trees that have grown in localities at or close to the tolerance limits for growth. Under these stress conditions, the rings are invariably narrow and possess pronounced year-to-year variations. For example, they may be drought-sensitive and in a year of extreme moisture stress little or no growth will ensue, resulting either in a very narrow tree ring or even, in severe cases, no ring at all. Alternatively, the stress factor may be

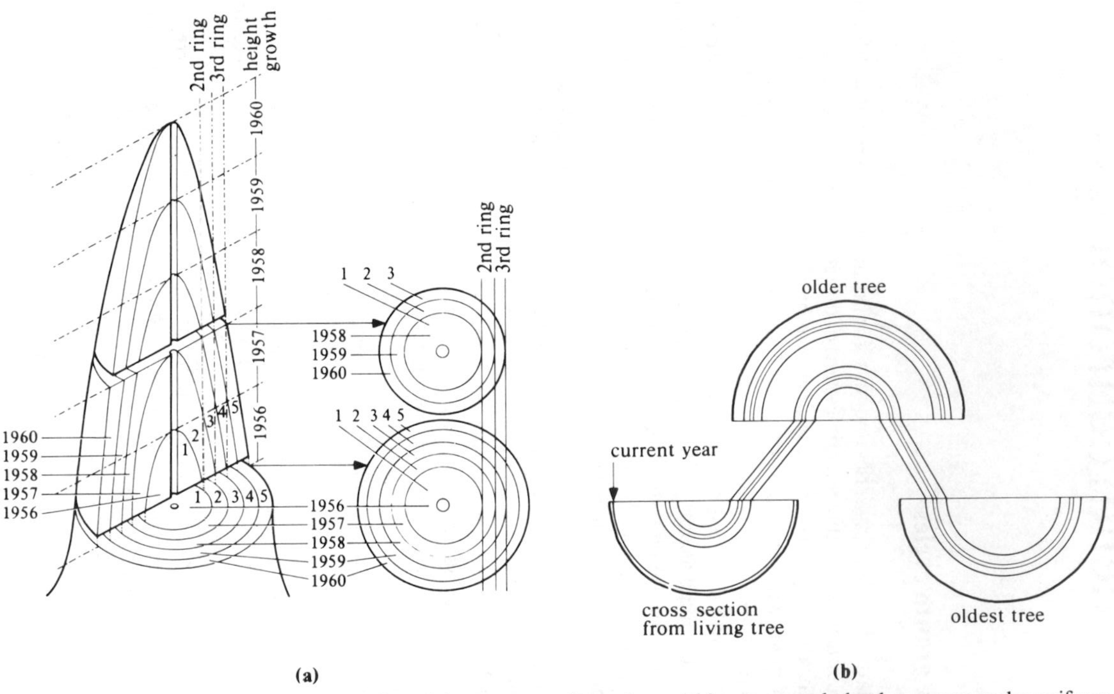

Figure 5.9 (a) Diagrammatic representation of the structure of tree rings within a tree trunk that has grown under uniform environmental conditions. (b) Diagrams to show the principles of cross-dating whereby the ring record in a growing tree can be extended to older and very old wood samples.

related to temperature, with narrow rings reflecting lower-than-normal summer temperatures. Within a given geographical area, contemporaneous rings will display pronounced similarity in relative thickness when compared. It is not surprising, therefore, that the total pattern of ring variability or 'signature' will also be fairly closely replicated in the ring record of adjacent trees. Clearly, in these circumstances, a parameter of environmental character, such as moisture supply or temperature, which fluctuates on an annual basis, is reflected in tree growth response.

5.5.2 Ring counting

Inspection of a horizontal surface cut cleanly across a tree-stump top reveals a concentric series of tree rings, which increase in diameter radially outwards. These are merely a truncated stack of the growth cones mentioned above. If it is assumed that a complete record of the annual growth cycles is present and that the date of tree cutting is known, then the number of years of tree growth after it had reached the height of the stump top can be readily established by simply counting backwards from the outermost (newest) ring to the core. Indeed, the age of any growth ring may be precisely determined. Normally, if none of the rings are very small, then it is unlikely that any rings are missing and a full growth record exists. But, of course, the converse applies. This procedure for estimating the age of a tree has been practised at least since the time of ancient Greece.

Felled trees and subfossil trunks enable the collection of complete disc-shaped cross sections using a hand-operated or powered saw. Where this approach is not practicable, a technique has been developed whereby a tree-ring core is extracted from the trunk without harming the health of the tree. Similarly, with timbers incorporated into buildings, a coring device is often essential. This method utilises an increment borer, consisting of a hollow drill bit that is screwed into the tree trunk and at the same time a pencil-sized, undisturbed core is left in the centre of the borer. The core may easily be extracted by drawing out a semicircular tube from the borer when the centre of the tree is penetrated. The core is then available for ring counting. Although direct ring counting is possible in the field, laboratory examination and measurement are vital. Prior to ring width measurement and counting, the samples, whether discs or cores, are sanded to enhance the ring patterns. Waterlogged material is best deep frozen, planed and then allowed to thaw. Besides the examination of large-sized material, thin slivers of wood removed by a sharp knife are also sufficient. Usually a $\times 10$ binocular microscope is used for ring measurement (to 0.02 mm) and counting. Specially constructed travelling stages facilitate the radial traverses of the sample material. These are linked to a microcomputer, which stores the measurement data. A range of software is available for plotting and cross-dating.

5.5.3 Tree-ring series and cross-dating

The correlation procedure requires painstaking analysis of scores of tree-ring series, since all anomalies in annual ring formation must be identified before a precise chronology is possible. Since sensitive ring patterns are, by their very nature, sometimes associated with absent, partial and double (false) rings, it is essential that such pitfalls are identified, and this is only possible after persistent cross-checking between different tree records, leading ultimately to an internally consistent pattern of growth that constitutes the master chronology for a specific region. Once such a standard is determined, new unknown samples can be compared with the master record and, at least in theory, dated.

The Belfast CROS program (Baillie & Pilcher 1973) is a widely used method for cross-dating and chronology building. Overlaps of 60–80 years between the component series are recommended.

In most environments, trees do not live for more than several centuries, and consequently the direct count method is limited to within such a timescale. By and large, this is the limit to which dendrochronology can have a direct bearing upon contemporary geomorphological problems. Yet there are exceptions, and these are exemplified by the south-west USA. There, two very different tree families occur that have exceptionally long lifespans. The first known was the giant sequoia, *Sequoiadendron giganteum*, a Tertiary relict species that is confined to isolated groves on the western slopes of the southern Sierra Nevada at heights between 1500 and 2100 m. These constitute the largest known trees in the world. Ring counts have demonstrated that this species can live up to 3400 years. In the 1950s it became clear that certain species of *Pinus*, inhabiting high-stress environments at or close to the tree line on the higher summits in the Great Basin area, attained considerable ages despite relatively diminutive sizes. One particular species – the Great Basin bristlecone pine, *Pinus longaeva* – growing at heights between 2700 and 3400 m, attained an age of more than 4000 years and one example proved to be 4500 years old and was still growing.

The facts that sensitive tree-ring series possess a characteristic 'signature', and that within a common geographical region trees growing concurrently will respond to environmental fluctuations by building similar growth rings, have also been noted above. However, it should not be assumed that a cross-dating potential will be universally present, for only certain trees are likely to be occupying habitats where they are under sufficient stress to develop good 'signatures' in their growth records. In the first edition of this book it was stated: 'Similarly, cross-dating is often restricted to specific regions, and interregional correlation is invariably not practicable'. This prognosis has proved to be unduly pessimistic and correlations are now possible between some regions of Western Europe (see later).

5.5.4 Master chronologies

In the pioneer work on chronology construction in the earliest part of this century undertaken by A.E. Douglas in Arizona, cross-dating studies of modern ponderosa pine, wooden beams incorporated within historic buildings and charcoal enabled a precisely dated master chronology to be compiled extending back to the thirteenth century. Concurrently, a floating chronology had been assembled utilising the tree-rings records preserved in the timbers of prehistoric buildings, which had clearly dated from before the Spanish conquest. Ultimately, in 1926, the two series were linked after the excavation of a beam that permitted the recognition of a partial overlap between the floating and absolute sequence.

The Laboratory for Tree Ring Research of the University of Arizona – utilising both subfossil bristlecone pine remains, which, because of their high resin content, are extremely durable after death, and the exceptionally long records from still-living trees – has produced a cross-dated master chronology for *Pinus longaeva*. This extends for more than 8680 years for material from the White Mountains (Inyo Range) of southeastern California (Ferguson & Graybill 1983). Both the giant sequoia and the bristlecone pine were to play a vital role in the initial phase of investigation of the relationship between radiocarbon dates and calendar ages (see Sec. 5.1).

In Western Europe two separate projects, one located in Belfast, Northern Ireland, and the other at Hohenheim-Stuttgart, south Germany, have been engaged in creating a long absolute chronology independent of the south-west USA data. Both projects have utilised subfossil oak material in their chronology building through cross-dating, although in contrast to the long spans of the bristlecone pine, each oak ring series is only 150–250 years long. Thus a large number of individual ring records have to be correlated if long chronologies are to be built from oaks.

Although the two projects both worked on subfossil oaks the material came from quite different stratigraphical contexts. The Irish workers used oaks found in the peat bogs that abound in Ireland supplemented by oak timbers obtained from archaeological and historical sites for the more modern period. In contrast, the German investigators utilised subfossil oaks obtained from gravel pits working the Holocene gravels beneath modern floodplains, especially that of the River Main, although latterly bog oaks from north Germany were also used. Calibration of the radiocarbon timescale was the primary objective of the Belfast project from the onset, whereas the Stuttgart work was initially concerned with alluvial chronology and only later did the calibration issue become important to the work. Happily, the two projects converged and the various floating chronologies could finally be linked (Pilcher *et al.* 1984) to produce a European absolute timescale extending back 7272 years. Correlation over such a long distance was aided by the long records that it had been possible to compile, making the statistics of cross-dating and matching much easier.

5.5.5 Geomorphological applications

Tree-ring studies have been applied to a variety of geomorphological contexts, and an excellent comprehensive review is given by Alestalo (1971). Glacial geomorphology affords examples of both direct and indirect application. First, glacier recession below the tree line can influence the pattern of tree growth directly by determining the age of primary colonisation. Thus the oldest tree on any surface, e.g. a moraine ridge, will give a minimum age for the constructional event. In some instances, moraine ridges are covered by forest in the terminal areas, whereas laterally the same features climb above the tree limit, and patterns of lichen colonisation can be examined in conjunction with tree rings on a synchronous landform (Luckman 1977). A further possibility is that ice-marginal activity may disturb an already established tree by abrading or tilting it, producing an asymmetrical structure to the growth ring pattern, which enables the event to be precisely identified in the growth record (see Fig. 5.10). Secondly, glacier variations may be indirectly related to tree growth via the climatic environment. It has been shown that summer temperature is often critical in controlling tree growth at or near the tree limit, and this same factor can also be vital in the determination of glacier mass balance. Thus, it is frequently possible to demonstrate good correlations between glacier fluctuations and tree-ring widths provided the trees have grown in a stress environment (e.g. LaMarche & Fritts 1971, Matthews 1977).

Although an absolute master chronology covering some 6000 years eventually emerged from the River Main floodplain study, this should not be taken as a guarantee that such a successful outcome will be possible in other areas. By way of example the study of River Trent floodplain gravels at Colwick, Nottingham, England, by Salisbury *et al.* (1984) may be cited. Gravel workings covering about 1 km^2 yielded 124 subfossil oak trunks of which 64 were judged suitable for ring measurement with a view to cross-dating. Of the latter only 38 correlated and these produced two separate floating chronologies, one of 576 years (27 trunks) and the other 280 years (11 trunks). This left 26 that could not be correlated, presumably because their ages related to timespans outside the limits defined by the two floating chronologies. Radiocarbon dating suggests that the

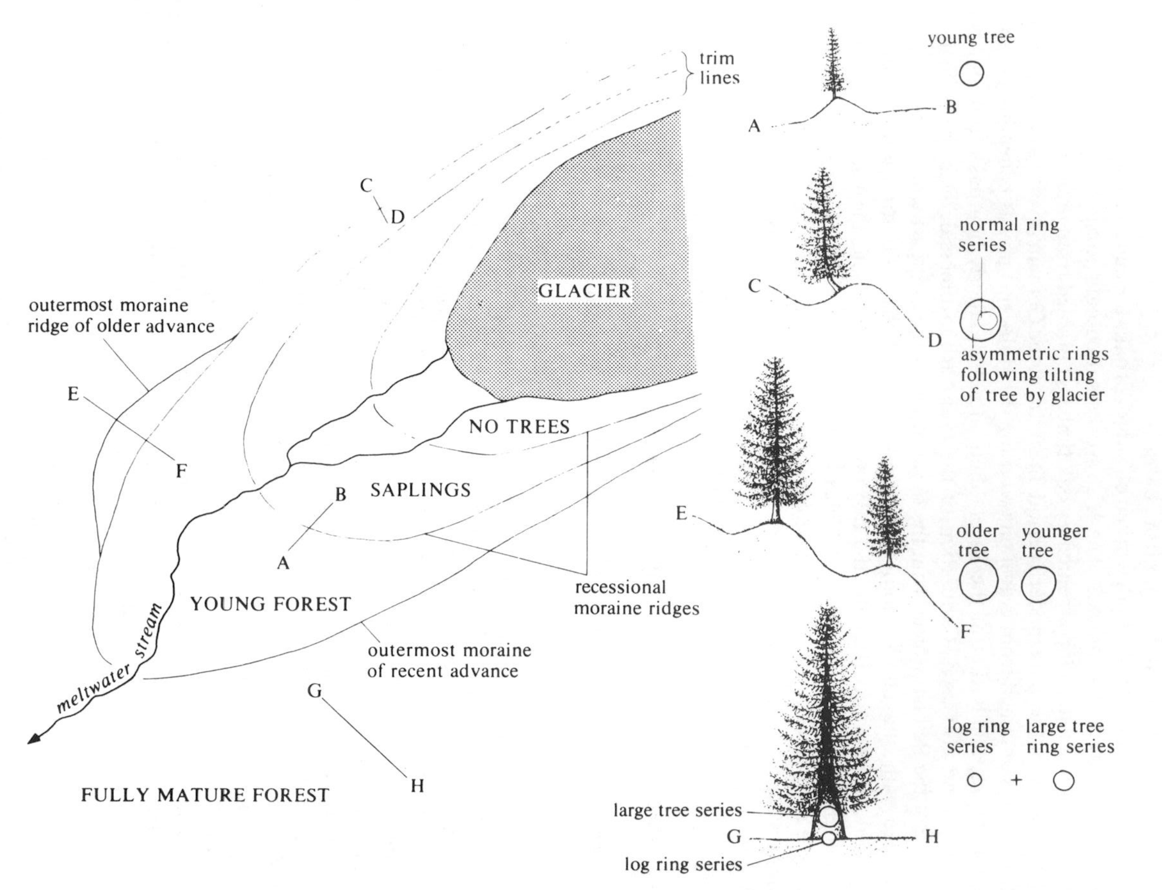

Figure 5.10 Diagram to show how tree-ring studies may be used to reconstruct a chronology of valley glacier variations.

576-year chronology is in the 5000–5500 ^{14}C yr BP range and the 280-year sequence includes 3400 ^{14}C yr BP. A single radiocarbon date on a randomly selected tree from the population of 26 that could not be cross-dated yielded an age just outside the younger floating chronology, suggesting that more material might enable the production of a much longer chronology and possible linking of the two floating sequences. In such circumstances an alternative approach to absolute dating is to seek a correlation of local floating sequences of rings with master chronologies. In the Colwick case attempts to cross-date with the Belfast chronology were unsuccessful despite the fact that some local chronologies from northern England had been satisfactorily linked to the Belfast record. However, correlations with the floating chronology of the Somerset Levels area of south-west England were possible, and the failure of link-up with the Belfast master sequence may be explained by (a) the influence of regional growth differences and (b) the local chronologies not having sufficient length to achieve inter-regional matching.

Some extremely interesting applications of tree-ring studies in recent years have been to determine both the dates and the rates of various types of mass movement. Schroder (1978) has carried out successful investigation of the timing of slope failures, while Carrara and Carroll (1979) have inferred rates of surface lowering from the exposure of tree roots of known age. It is essential that the trees selected are those that have roots that are known to develop in their entirety beneath the surface (LaMarche 1968).

5.5.6 Conclusion

It is recommended that geomorphologists seek the advice of dendrochronological specialists prior to any major use of the technique. Access to a dedicated laboratory would naturally be advantageous. An excellent outline of tree-ring dating may be found in Stokes and Smiley (1968), although this is now dated in some respects. Baillie (1982) gives a good modern introduction with an archaeological slant but from the perspective of European experience. Future development may see much greater usage of X-ray microdensitometry, a technique that exposes wood samples to soft X-rays such that photographic film displays the variability of optical density at high resolution. This can then be related to wood specific gravity and annual growth patterns (Bircher 1986), the minimum density being displayed by the early wood and maximum density by the late wood. Although this approach has been successful with modern trees, doubt is cast on its applicability to subfossil material (Ward *et al.* 1987).

5.6
Lichenometry

5.6.1 Introduction

The technique that has become known as lichenometry has been adopted by geomorphologists, glacial workers in particular, since its introduction by R.E. Beschel in a series of papers dating from 1950. Some workers have claimed that the method has an ability to provide an absolute as well as a relative chronology for substrata exposed over timescales of up to several thousands of years from the present, although the norm for absolute dating is probably restricted to the last 500 years (Innes 1985a). It is especially useful where the utilisation of other techniques such as dendrochronology, radiocarbon dating and palynology are either impossible or difficult due to inclement environmental factors or the lack of suitably preserved material. The most important potential is in those enviroments where glacier fluctuations during the Holocene have long been subject to debate. Lichenometry does, however, have other geomorphological applications such as shoreline dating, river flood frequency studies, mass movement chronology and volcanic activity patterns.

A further attraction of the technique is that its application is apparently straightforward, not requiring the user to undergo any intensive training nor the use of sophisticated instruments save a ruler. On face value, this situation is rather surprising since lichenometry embraces what is probably the least known and most difficult part of botany in terms of taxonomy, ecology and biosystematics. Despite rapid acceptance and a vast proliferation of literature describing its application in many localities, until the 1980s there was a relative shortage of critical assessment of the method itself. Jochimsen (1966) wrote an important thought-provoking review of the technique but the use of German seriously restricted its availability to many English-speaking workers. Fortunately, it is now more widely available in translation (Jochimsen 1973). Yet even so, Webber and Andrews (1973), the editors of a special lichenometry issue of the journal *Arctic and Alpine Research*, which published the translation, had to comment 'she questioned the biological basis of the method and most of her challenges remain unanswered'. Despite this recognition they concluded that the doubts were insufficient to invalidate the use of the technique since it appears to work and give reasonably acceptable results when applied to contrasting environments. This attitude is typical of many of those who have used lichenometry as a dating tool and frequently the caveat 'judicious use' will be appended to comments on the conceptual basis of the method. More recently there has been a growing awareness of the methodological problems posed by the application of lichenometry, and good reviews are to be found in Locke *et al.* (1979) and Innes (1985a).

5.6.2 Conceptual problems

There is little doubt that the most authoritative and comprehensive summary of the fundamentals of the technique are Beschel's papers (Beschel 1957, 1961), and many subsequent workers have simply cited his contribution but not progressed far beyond a restatement of the principles that he established. These may be summarised under four main headings, although it has to be appreciated that a considerable degree of overlap occurs.

(a) Lichen establishment:
  (i) The environment is lichen-free on exposure to sub-aerial conditions.
  (ii) Rates of colonisation and establishment are known.
  (iii) Increasing thallus size, usually represented by a diameter measurement, indicates increasing thallus age and can thus be used to establish a relative chronology for the exposure of the surfaces on which the plants are growing.
(b) Lichen growth and development:
  (i) For any common and competitive species, the size of the largest individual will most closely indicate the absolute period over which its substratum has been exposed to sub-aerial conditions, i.e. for large individuals diameter will be a direct function of age.
  (ii) The rate of growth is known, or can be established, for one or more closely related species.
  (iii) For the timespan under consideration, that growth rate is essentially linear with time.
(c) Environmental factors:
  (i) Environmental differences can be minimised in an area by the selection of the largest thallus, which is likely to be growing under optimal conditions.
(d) Sampling technique and the establishment of lichen growth rates:
  (i) The sampling technique adopted gives significant and reliable results.
  (ii) A reasonably accurate temporal framework can be established for the species involved either by direct measurement of individuals over time, by measurement of thalli on surfaces of known age, or by the calculation of ratios between species with 'known' growth rates and those whose growth rates are 'unknown'.

The following discussion will focus on glacial geomorphological applications, for these are dominant in the applied field.

5.6.3 Lichen establishment

Although occasionally material with pre-existing lichen growths will survive redeposition, e.g. from surficial debris associated with medial moraines (Griffey & Matthews 1978), generally till and bedrock immediately adjacent to a retreating ice margin are found to be lichen-free. This situation certainly applies to areas marginal to temperate glaciers, but as yet there are insufficient data to be entirely confident that this applies to polar glacier margins in all instances. When confronted with a mature lichen community it may not be possible visually to ascertain whether or not any 'recycled' lichens occur. Should they be suspected, then the size structure of the population can be examined for homogeneity. Clearly there is a danger if the investigator simply takes the single largest lichen to represent the age of a surface; rather the upper size groups of any population should be taken and several may then be averaged. The rate of establishment is normally related to the time when thalli first become visible and this is usually in the range of 10–20 years. But greater complexity may exist, for colonisation may be inhibited by some factor, e.g. a period during which perennial snow patches were able to survive, of which subsequently no trace will be left save for an anomalous colonisation pattern. In such an

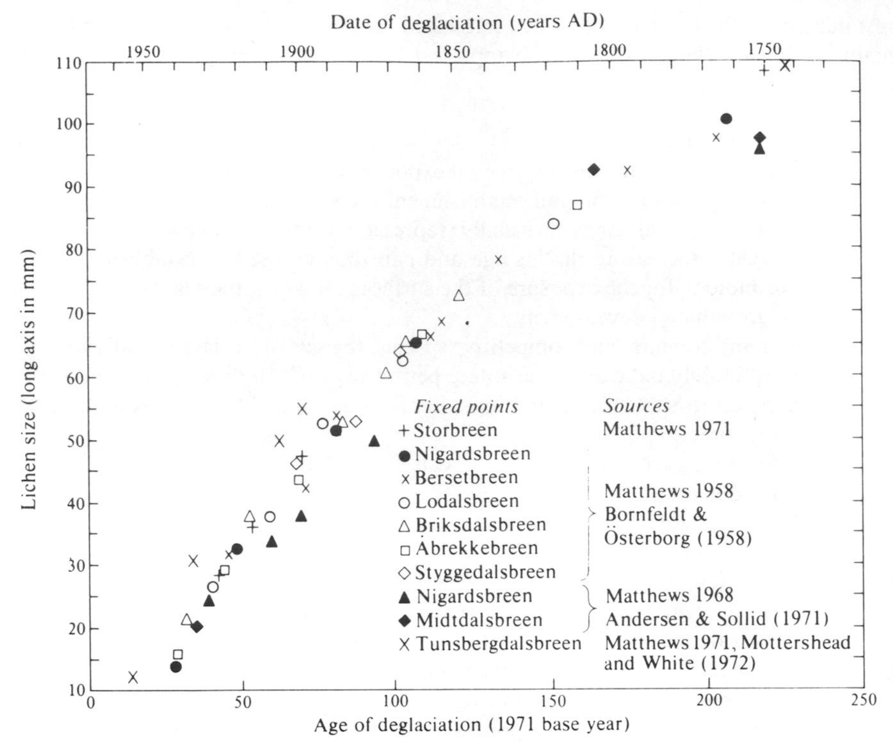

Figure 5.11 A compilation of the results of measuring the longest diameter of the crustose lichen *Rhizocarpon geographicum* on moraine ridges in southern Norway, where the age of ridge construction is known to a high probability (first 100 years) and a reasonable possibility (older dates). Each symbol represents a mean of the five largest lichens, although field techniques do differ to some extent. (After various sources in Matthews 1974).

area the lichen maximum size will underestimate the age of ice recession. This latter situation is especially acute when dealing with older morainic features with ages of many centuries.

Once the lichen is established an initial period of rapid growth ensues (known as the 'great period') but soon this gives way to a regular progressive enlargement of the thallus diameter over time (see Fig. 5.11). The sizes of the largest thalli found on a series of surfaces will hence indicate the order in which these became exposed, and a relative chronology may thus be established. Once growth has progressed to the stage when individual expanding thalli start to make contact with each other, reciprocal effects will be exerted and their subsequent growth may be restrained to a degree, making them unsuitable for age determinations. This factor is frequently overlooked, with the attendant assumption that maturity is never attained. With very old lichens, retrogressive modification becomes a serious problem due to damage by various environmental factors that contribute to the loss of biomass. Finally, the development of a vascular plant cover, especially at or below the tree line, seriously affects the orderly pattern of lichen growth, yet instances occur where the presence of a tree cover is not even mentioned in reports of lichenometric investigations.

5.6.4 Lichen growth and development

Any relative chronology based on maximum sizes on a series of surfaces ideally requires an absolute temporal framework when applying it in geomorphological contexts. Normal practice is to estimage ages for lichen thalli of different sizes and to calculate a so-called average annual growth rate, which enables the ages of lichens growing on surfaces of unknown ages to be estimated. Usually the age–growth relationship is considered to be linear, but control points on calibrated graphs are extremely limited outside the last 100 years or so. Also it is readily assumed that growth as represented by lichen size is a direct function of time, but this assumption is tantamount to claiming that in any generation the tallest or fattest person in a country is the oldest. Clearly size is not just a function of age, and therefore age estimates should theoretically be based upon an assessment of other parameters as well. This is rarely done and great faith is placed on the probability that the largest individual *may* be the oldest and optimally growing individual. Thus acceptance of the regularity of lichen growth is a fundamental but generally untested tenet of the lichenometric creed.

In Europe the most widely used genus is *Rhizocarpon*, which is a distinctive yellow-green-coloured crustose litchen that thrives in environments associated with glaciers. Most workers claim to have identified two common species, *R. alpicola* and *R. geographicum*, although Innes (1985b) has recommended a classification that has an intermediary level between species and genus called section. He states that field identification only to section level is normally possible, and hence the above two should be referred to as *Rhizocarpon* section *Alpicola* and *Rhizocarpon* section *Rhizocarpon*. Conformity to his suggestions will, in the long term, enable more meaningful comparisons between the results of the lichenometric investigations.

5.6.5 Environmental factors

Naturally the environmental influence on lichen growth is a complex of interrelated variables and most studies have contained some statement acknowledging these on the lichens in any given study area. Not infrequently, however, these assessments are derived from Beschel's work, resulting in comments on the avoidance of snow-patch localities, moist hollows, stream courses and the need to ensure lithological substrates. To these may be added possible differences arising from such variables as slope, aspect and exposure. The biological literature includes a wealth of data on lichen metabolism, and moisture appears to be a major controlling influence in this respect. On the other hand, temperature extremes seem to be unimportant, and lichens are able to survive on rock surfaces in exposed locations where an extremely wide range of temperature values are experienced. The stability of the substrate is influential, for lichens can be destroyed or damaged by mass movement of materials upon which they are growing. However, it is often impossible to evaluate the effects of long-term movement where the resulting pattern of colonisation may be composed of several generations of lichens. In summary, it can be stated that lichen growth responds to a complex of microenvironmental conditions that are not reflected in the macroclimatic statistics, and that an average rate of growth is unlikely to be universally applicable within a given macroclimatic region (Haines-Young 1983). Optimum conditions for different genera and species vary considerably, and this must cast doubt on the efficacy of sampling that attempts to eliminate certain habitats in a subjective manner. It is also unwise to claim that lichen growth rates are sufficiently slow to eliminate the effect of temporal variations in environmental factors, as the spatial differences in lichen growth are demonstrably great and result from relatively small-scale, though steep, environmental gradients.

5.6.6 Sampling technique and the establishment of lichen growth rates

Sampling techniques in lichenometry are almost as diverse as the number of workers who have utilised the method. The types of sampling can be classified into two main approaches:

(a) Based on the use of known areas established at 'lichen stations'. Here the main problem is that the selection of typical sites for sampling is not unbiased, since they are influenced by the preconceived ideas of terrain and vegetation.
(b) Based on unrestricted areas, usually varying in size according to the type of feature being studied. It is not unknown in the literature to find that details of the sample design are omitted, whereas in others some have claimed that they have visually inspected every surface within the area of study. The latter case is undoubtedly true in some instances, e.g. Storbreen, south Norway, but this has effectively developed into a lichenometric 'test site'.

A way round the problem that surfaces of interest will vary widely in size and hence influence the chances of finding a lichen of specified size is that the search areas should be proportional to the area of each feature, so that the number of observations from different populations should be the same wherever possible. It is difficult to avoid the conclusion that in many published studies an unwelcome degree of subjectivity has been present in sample data, especially when unskilled assistants have been used for data collection. This makes comparison of values difficult if not wholly impossible, and only a limited amount of comparative work has been undertaken whereby workers have agreed to interchange their results after examining the same ground. The need for standardisation in both technique and definition is paramount. It has to be recognised that nothing can be concluded about the probability of a lichen of a specific size being drawn from a homogeneous population of all lichens on a surface, or being due to chance, unless consideration is given to the representativeness of the sample data, the sampling fraction and the fact that it should represent an unbiased estimate of the population parameters.

Considerable difficulties arise from the basic issue of thallus measurement. In some early studies attention was focused solely on circular crustose lichens, for it was reasoned that these must have increased their size (diameter) in a uniform manner and hence ought to be reliable indices of relative time. However, few crustose individuals maintain a circular state for the whole of their lives and the tendency is for the largest individuals on progressively older surfaces to become increasingly irregular in shape. Even the diameters of apparently circular lichen thalli may differ by up to several centimetres when measurements are made along different axes, and the circular thalli on old surfaces are rarely the largest. If apparently circular specimens are difficult to represent by a single diameter measurement, it can be readily appreciated that this kind of measure is even less situated to elliptical or irregular thalli. Confronted with irregular-shaped lichen thalli the fieldworker has several options and must include:

(a) measuring the diameter of the largest circle that can be accommodated within the lichen borders – effectively this is the shortest diameter;
(b) establishing the largest diameter, taking particular care to avoid inclusion of coalesced thalli;

(c) substituting surface area as a measure of size, but the tedious process of measurement seems to have discouraged its widepread adoption.

The consensus view arising from detailed considerations of sampling strategy is that the mean of the five largest thalli diameters on a given substrate should be used to characterise the minimum age of a surface. The issue of measurement is even more severe with foliose and fruiticose species; the former have uneven surfaces and curled margins, making accurate field measurement impracticable. Dry weight might be a valid alternative but this has the disadvantage of killing the individual. Fruiticose species characterised by either fibrous or lobe-like forms present similar difficulties. Finally, the accuracy of specimen identification needs to be considered, for even trained botanists emphasise the inherent problems attached to specific determinations of lichens.

When considering timescales measured in centuries it is clear that the direct method of ascertaining lichen growth is largely irrelevant since values thereby obtained can only refer to the period of measurement. Recourse is therefore inevitable to indirect methods for the establishment of a temporal framework. The most common method is to examine surfaces of 'known age' within or near to the study area (e.g. bridges, grave headstones, mine waste, railway cuttings, walls), identify the largest lichen specimen on each and then plot the results to produce a relationship between size and age, which is often mistakenly called a 'growth-rate curve'. As noted already this approach is restricted to the last few centuries at the most; beyond this, control is either absent or other stratigraphic methods are brought into play, e.g. dendrochronology or radiocarbon dating (Reger & Péwé 1969, Griffey & Worsley 1978). In those situations where stratigraphic control is lacking, some have attempted more circular routes to derive lichen growth rates. A good example of this is Andrews and Webber's (1964) paper, which, although subsequently revised in part, was a frequent subject of reference. To derive an average annual growth rate they assume parallel and linear ice-marginal retreat to its present position from the location of the first lichen species. The known age in this case was derived from air photographs (1948–63). By simple division of diameter length by time they derived an 'annual growth rate' and used this to extrapolate back to 1887 and to infer ice-marginal retreat from 1680.

At a global level, southern Norway is probably the most intensively studied area as far as glacially related lichenometric work is concerned. Consequently an outsider might reasonably expect that after two decades a consistent interpretation of glacial chronology based on lichen measurements would be forthcoming, yet this is not the case. This dichotomy well illustrates the continuing problems of lichenometric application to geomorphology. An attempt was made by Erikstad and Sollid (1986) to establish a neoglacial end-moraine chronology at 14 glaciers in three subregions. Of note was their conclusion that in the Jotunheimen the outer end-moraines dated from 1780–1820 AD rather than 1750 AD, a date accepted by Matthews (1977) as applicable to Storbreen. The age difference of some 50 years arose from Erikstad and Sollid's construction of an 'adjusted mountain curve' relating lichen size to age. In the absence of intra-Jotunheimen direct evidence for ice-marginal positions prior to about 1890 AD, they were obliged to extrapolate the slope of a growth curve from a valley in the Jostedalsbreen subregion and use lichen size from mining debris at two sites *outside* the Jotunheimen. Not surprisingly this invoked a critical retort by Matthews (1987) since subjectivity inevitably arises if 'calibration' indices are used when they clearly relate to differing environmental conditions. Disputed chronological interpretations are not restricted to the southern part of the country. In north Norway also there is no agreement concerning lichen size–age relationships. Innes (1984) challenged the conclusions of Karlén (1979) and in one example, Blokkfiellbreen, a predicted age of 1780–1830 AD by Karlén's proposed growth curve corresponded to a 1910 AD marginal position on the basis of photographic evidence. With respect to northern Scandinavia Innes (1984, p. 350) concluded: 'it appears that a careful re-evaluation of these curves is required before either the lichenometric evidence for glacial fluctuation dates is accepted with any degree of certainty or the lichenometric dating of other phenomena in the area is attempted'.

In conclusion it is difficult to avoid the suggestion that the lichenometric dating method in its present form is conceptually unsatisfactory with respect both to some of its basic assumptions and to its method of field application. The principal uncertainties are:

(a) That lichen colonisation does not always proceed in a regular and predictable manner.
(b) That thallus size is dependent upon equally dependent significant variables other than time alone. Included here are the environmental and physiological factors, and the effects of competition and succession.
(c) That the sampling methods are often arbitrary and appear not to produce statistically reliable results.
(d) That the use of a single parameter of the lichen population, a diameter measurement, is an inadequate and often unrepresentative index of the character, or age, of that population.
(e) That uncertainties are implicit in the methods used to establish lichen chronologies and growth rates.

It is urged that future workers in the field bear in mind Stork's (1963) statement: 'not by one isolated method can we get reliable information about the age of the moraine surface. Different methods must be combined to elucidate the problem and to permit any conclusions concerning the age of the deposit to be drawn'.

5.7 Peats and lake sediments: formation, stratigraphy, description and nomenclature

5.7.1 Introduction

Just as the most comprehensive records of long-term geological history are to be found in depositional environments, so, in the study of recently operating geomorphic processes, the fullest record of their operation and of the changes in material flux that they have generated will often be found in the stratigraphy of lake sediments and peats. Peat and sediment studies have at least three major additional advantages. The record of material flux that they contain can often be directly related to evidence for changing vegetation and climate, preserved, for example, in the fossil remains of organisms indicative of past conditions. Thus, a more holistic picture of past environment may emerge, and the insights of the geomorphologist become both more valuable and more interpretable by virtue of the broader scientific context within which they can be set. Peats and sediments are also suitable for a wide range of dating techniques, some of which, especially those dependent on annual laminations, on measurements of the presence and decay of short-lived radioisotopes (^{137}Cs, ^{210}Pb, ^{241}Am, ^{239}Pu), or on the detection and calibration of a palaeomagnetic record, are especially suited to sediments. Finally, both the context of present-day sedimentary environments and growing peat bogs, and the development of new techniques for dating very recent changes, mean that the principle of uniformitarianism, critically employed, can be used to provide not merely a sophisticated interpretational key to the reconstruction of the past but also a uniquely valuable opportunity to unite palaeoenvironmental reconstruction and present-day process studies to give a continuum of insight into the changing nature and rate of material flux through time.

5.7.2 Lake sediments and peat: origin and development

The primary distinction between lake sediments and peats arises from their mode of deposition and their relation to the ground water table. All lake sediments represent sub-aqueous accumulation and most include a significant proportion of derived allochthonous material, which is terrestrial in origin, in the form of detrital inorganic and organic particles. Less often, lake sediments may be largely of lacustrine origin in the form of precipitates or of partially decomposed aquatic organisms, which have settled out from the water column. These latter sediments are usually the result of high aquatic primary productivity, low allochthonous detrital input and retarded degradation of organic material at the mud/water interface, often as a consequence of anaerobic conditions on the lake bed. Peats are formed of predominently autochthonous organic matter, the result of the *in situ* accumulation of partially decomposed remains derived

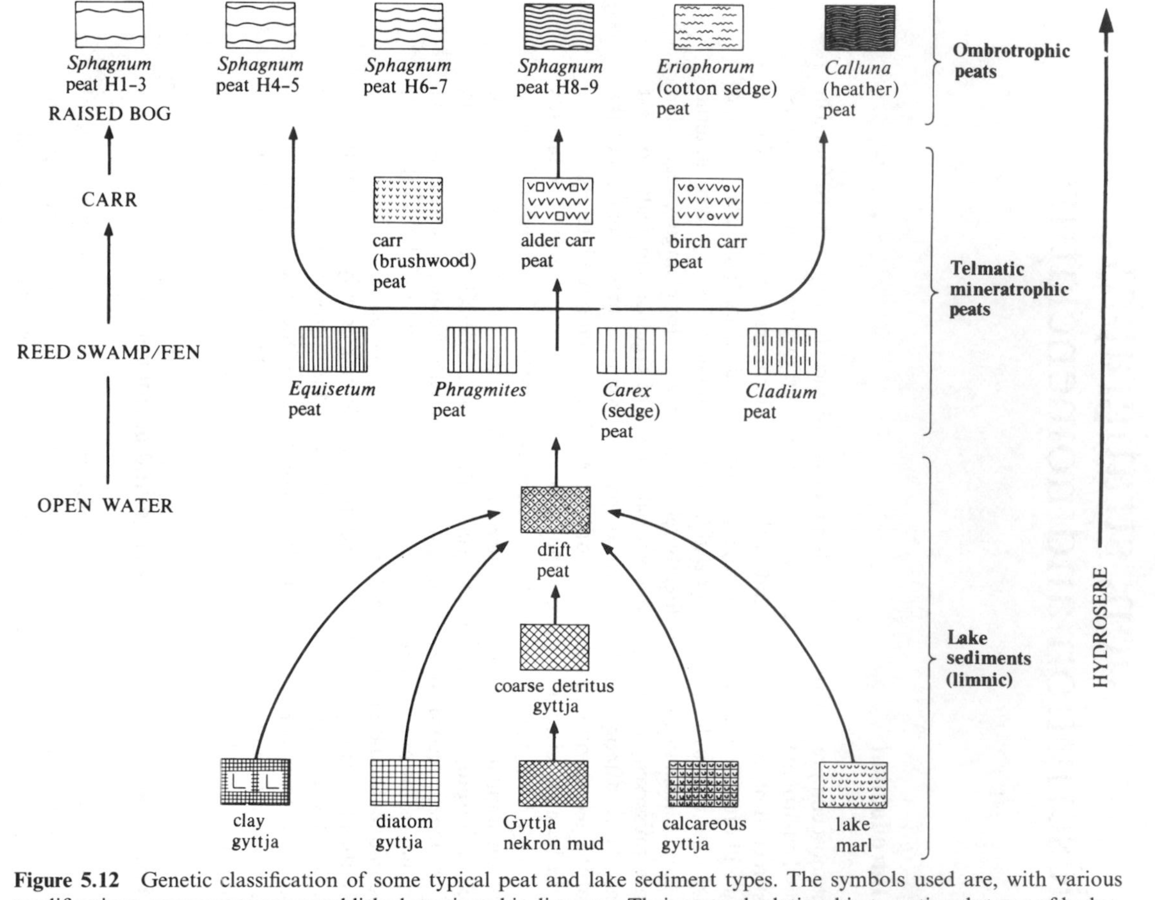

Figure 5.12 Genetic classification of some typical peat and lake sediment types. The symbols used are, with various modifications, common to many published stratigraphic diagrams. Their general relationship to notional stages of hydroseral development (cf. Walker 1970) and related nomenclature (cf. Moore & Bellamy 1974) is shown. The scheme is simplified and illustrative rather than comprehensive. The 'H' scale refers to peat humification: H1–3 is little humified, with fresh recognisable plant remains; H8–9 is almost amorphous dark-brown/black peat. Barber (1976, Fig. 2.1, p. 13) outlines a further version of this genetic classification and documents its origin.

from mire plants growing at or near the site of accumulation. In terms of the classic hydrosere, there is a gradual transition from lacustrine to this form of accumulation in shallow water as reedswamp develops over open-water muds. The nature of peat varies with and reflects the trophic and topographic conditions within which the peat-forming plant species grow. It is often convenient to use a genetic classification of both lake sediments and peats, an example of which is outlined in Figure 5.12.

The genetic approach to peat and sediment classification has the disadvantage of interposing crucial interpretational decisions between the observation of the material and the adoption of a nomenclature for descriptive purposes. Figure 5.13, based on the comprehensive and widely accepted scheme of Troels-Smith (1955), gives an example of a more objective and purely descriptive nomenclature stressing the appearance and content of the sediment rather than its origin or relation to a notional hydrosere. Fuller discussion of seral/genetic aspects of lake sediments and peats can be found in Moore and Bellamy (1974), Walker (1970) and Birks and Birks (1980). Birks and Birks (1980) also provide a good introduction to the Troels-Smith system, and a comprehensive illustration of the effective use of this in a modified form is given in Birks (1973).

5.7.3 Field sampling, cores and coring techniques, conventional land- and ice-based sampling

Where peats and lake sediments have been exposed in section by lowered water tables, erosion or other factors, stratigraphic description and sampling are best carried out from cleaned faces where specialised equipment is seldom required. The specific problems posed by sampling peats in this way are discussed by Barber (1976), and Aaby and Digerfeldt (1986).

Where sampling from an exposure or pit is not possible, a wide range of coring equipment and techniques are available. These have been developed to meet very varied requirements in terms of penetration, sample quality and coring conditions. The design and operation, and the advantages and disadvantages of several types of land-based corer are illustrated in West (1977) and in Barber (1976). Most of these corers can also be operated in reasonably shallow water from a stable boat or from ice where winter climate permits this. Digerfeldt (1978) provides designs for alternative cheap, effective and portable sub-aqueous corers.

For most purposes, a useful combination is provided by a Hiller-type corer capable of penetrating at least narrow bands of relatively consolidated minerogenic sediment (Fries & Hafsten 1965) and a D-section 'Russian' or 'Jowsey' corer (Jowsey 1966), lacking this degree of penetration but capable of retrieving extremely well preserved cores from all but the wettest or most fibrous of soft sediments (Fig. 5.14). The Hiller corer uses a sample chamber opened and closed by the pressure of the sediment on a flange. Penetration is achieved by combining clockwise rotation of the corer with a downward thrust at each turn. Once the depth to be sampled is reached, anticlockwise rotation both opens and fills the core chamber and subsequent clockwise rotation closes it. The corer is then pulled up to the surface, cleaned carefully on the outside and opened. Some degree of contamination is inevitable but this can be minimised by observing the following guidelines:

(a) Use alternate holes for each 50 cm sample (or 30 cm in the case of the small, 30 cm chambered version). This avoids having to sample that part of the sediment distorted by the screw end during the rotation and counter-rotation involved in sampling.

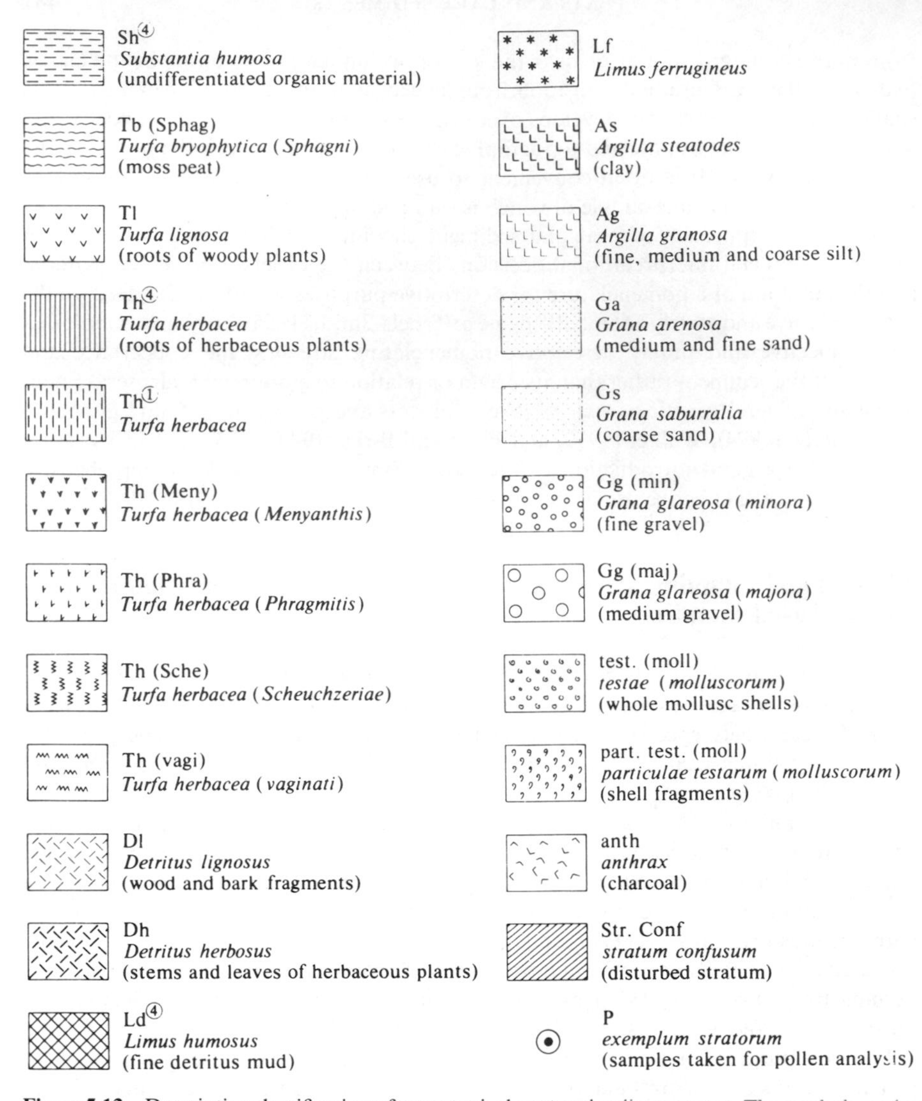

Figure 5.13 Descriptive classification of some typical peat and sediment types. The symbols, codes and descriptions are taken from Troels-Smith (1955), to which reference should be made before the scheme is employed. Graduations of content and colour can be shown by means of symbols shown. Modifications of the Troels-Smith system are given by Aaby and Digerfeldt (1986). (Present illustration taken from Tooley (1978a)).

(b) Keep the corer scrupulously clean and the chamber well sealed before and after each opening of the core chamber for sampling.
(c) Always clean the exposed face of the sediment lying in the sample chamber by gently breaking away surface material with a clean knife; failing this, slice the surface off, but always by cutting across the face and not vertically.
(d) In critical subsampling, avoid sediment in contact with the chamber wall or flange face, look for visual evidence of accidental downcarriage of overlying sediments and avoid material from the very base of the sample chamber.
(e) Reject visually contaminated material and samples from only partially filled sample chambers.

Modifications of the original design of the Hiller corer by Thomas (1964) enable whole cores of sediment to be retained, thus avoiding subsampling in the field.

The Jowsey corer requires only a straight thrust and half-rotation, though it is once more strongly advisable to use alternate holes. The sample retrieved often requires minimal cleaning and can be returned to the laboratory for subsampling by placing each whole 50 cm sample in suitable precut U-section half-drainpipe lengths. A lining of aluminium foil or polythene film allows the sample to be sealed and kept moist and fresh.

Provided care is taken to ensure that the corer is inserted at each successive depth with the flange on the same compass bearing, the flat limb of the D-section sample will have constant azimuth. This allows subsampling for palaeomagnetic measurements (see Thompson & Berglund 1976, Sec. 5.7.6). Extended evaluations of the relative merits and disadvantages of the Hiller and Jowsey corers are given in Moore and Webb (1978, ch. 2) and by Aaby and Digerfeldt (1986).

Another widely used type of corer is the end-filling piston sampler. The Livingstone sampler is the commonest type, together with its various modifications and improvements (Livingstone 1967, Cushing & Wright 1965, Wright 1967, Wright *et al.* 1984). The principles of sediment coring using piston samplers are given by Aaby and Digerfeldt (1986), with more detailed explanations in Wright *et al.* (1965) and Wright (1980).

Corers using heavier apparatus (e.g. Smith *et al.* 1968, Walter & Lowe 1976) are capable of taking samples of greater diameter and have the advantage of providing larger slices of sediment for ^{14}C dating where the material studied is relatively poor in carbon. This reduces significantly those sources of error and uncertainty arising either from

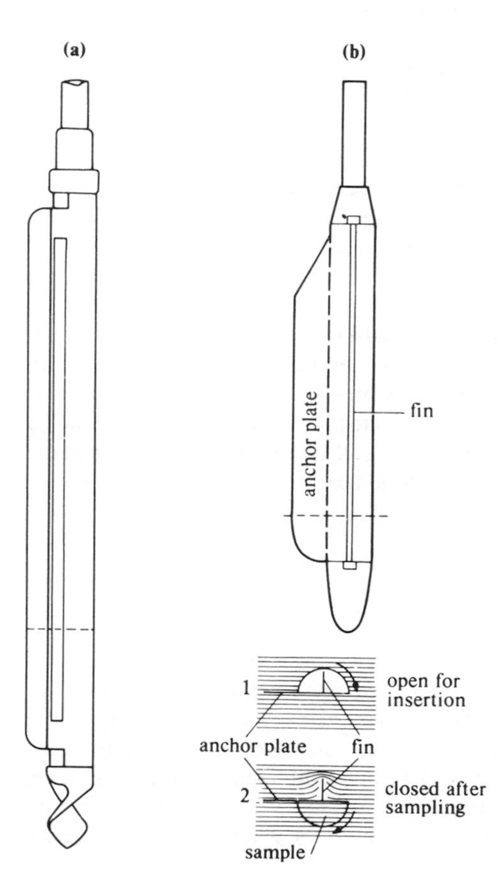

Figure 5.14 The (a) Hiller and (b) Russian or Jowsey corers. An extended discussion of their use is given in the text and also in Moore and Webb (1978, ch. 3). (Diagrams modified from West (1977)).

'multiple-shot' sampling for ^{14}C datable material or from using a thickness of sediment too great to resolve detailed variations in sedimentation rates.

5.7.4 Pneumatic corers

Coring limnic sediments in deep water where raft construction is not feasible and thick ice never forms has until recently proved extremely challenging except where a very high level of capitalisation was possible or short cores obtainable by gravity samplers (e.g. Hakala 1971, Digerfeldt 1978) were adequate. The development of pneumatic coring systems all deriving directly from the designs of Mackereth (1958, 1969) has revolutionised this aspect of sediment study by (a) making feasible the retrieval of continuous cores from 1 to 12 m in length with relatively modest and portable equipment, (b) providing a rapid method of taking undisturbed samples of the mud/water interface with negligible contamination and (c) furnishing cores ideally suited for the range of magnetic measurements outlined in later sections of the text and of increasing importance both for chronology and for palaeoenvironmental reconstruction.

Two types of Mackereth corer have been used as prototypes for complementary apparatus. The early design (Mackereth 1958) uses compressed air not only to drive a core tube up to 6 m in length into the sediment, but also to anchor the corer in an upright position and to raise the corer and full core tube from the bed of the lake once sampling has been effected. Modifications and additions to the original design have led to the construction of 3 m corers using the same principle and to more ambitious designs used successfully to retrieve cores as much as 12 m in length as well as to establish the azimuth of the core during penetration of the sediment, by means of a camera, compass and scribing system. These latter additional features make it possible to give true orientation to measurements of palaeomagnetic secular variation (see Sec. 5.7.6).

The second type of design (Mackereth 1969) uses compressed air for driving the core tube, and all other aspects of operation are carried out mechanically. Consistent recovery of an undisturbed mud/water interface can be achieved under most circumstances. An enlarged modification of the original 'minicorer' taking cores of internal diameter *c.* 10 cm and 1.5 m long has been used successfully to provide a larger bulk of sediment for palaeoecological and radiometric study.

All the above pneumatic corers up to 6 m length can be used safely and effectively from stable inflatable dinghies. Shallowness of water limits operation of the first type to depth no less than the full extent of the core to be taken. Cores have been recovered in water depths exceeding 100 m using both systems. The references quoted include detailed drawings and operating instructions. A film illustrating all aspects of the operation of the Mackereth 6 m corer with orienting and scribing modifications has also been produced (Lucey 1978).

5.7.5 'Freezer corers'

Interest in annually laminated sediments has grown from the demonstration by several workers (see Sec. 5.7.6) that such sediments are not unique to ice-marginal situations but may arise from a range of climatic conditions wherever these impose sufficiently strong and consistent seasonality on depositional processes and where sediment mixing by physical or biological mechanisms is poorly developed. In many cases, deep lakes provide the best laminated sequences (Craig 1972, Swain 1973) and often the high organic content of the sediment has evolved methane bubbles, which expand as the core is raised to the water surface and hydrostatic pressure released. This can destroy the sequence of laminae immediately after recovery of the core. In response to this problem, a variety of 'freezer' corers have been developed (Swain 1973, Shapiro 1958, Huttunen & Merilainen 1978, Saarnisto 1979, Renberg 1981a, Wright 1980). Most of these involve the use of dry ice both to freeze the sediment to the outside of either a sample 'pole' ('icy-finger' corer) or 'plate' ('freezer-box' sampler), and to allow it to be brought back to the laboratory for detailed study. The enhanced chronological control that this provides is becoming increasingly attractive as the limitations and sources of error in the radiometric dating of lake sediments become more apparent (see Sec. 5.7.6). A good summary of freezer coring techniques can be found in O'Sullivan (1983) and Saarnisto (1986).

Projects requiring the recovery of limnic sediments beyond the range of the pneumatic, 'freezer', or piston corers described normally require a drilling platform and much more expensive apparatus. Examples of studies based on cores from sub-aqueous sediments obtained in this way are given in Horie (1972, 1974, 1975).

5.7.6 Peat and lake sediment chronology

5.7.6.1 *Introduction*

In the context of sediment studies, chronologies are vital not merely for dating events but also for calculating rates of material flux (cf. Oldfield 1977).

The dating techniques applicable to peats and sediments provide chronologies on timescales ranging from annual to millennial. The dating principles, the precision that can be achieved and the accuracy and reliability of the techniques vary immensely. Three main types of chronology are outlined in this section, based respectively on annual rhythms, on radiometric decay and on palaeomagnetic variations.

5.7.6.2 *Laminae*

Varved sediments arising from the seasonal rhythm of glacier melt have been used for dating ever since the pioneer work of de Geer (1912). Annually laminated sediments are formed in many other environmental situations and the development of the freezer coring techniques described in the previous section has encouraged studies of laminated sediments in areas of cool temperate boreal and continental climates. In most cases, the laminations arise from the dominance of patterns of sedimentation by the annual rhythm of the seasons. In some cases, this is expressed by an alternation between layers of minerogenic allochthonous sediment representing inwash resulting from snowmelt for example, and more organic autochthonous muds reflecting the summer productivity and subsequent sedimentation of aquatic organisms and their degradation products. In other cases, laminations may arise from seasonal changes in biochemical conditions within the water column of the lake and at the mud/water interface. These lead to annual cycles of, for example, FeS precipitation or biogenic accumulation identifiable in the colour, texture and composition of the sediment. Several recent references outline the origin of some of the different types of laminae and illustrate the value of studies based on this type of sediment (e.g. Swain 1973, Ludlam 1976, Saarnisto *et al.* 1977, Simola 1977, 1979, Cwynar 1978, Renberg 1978, 1981b, Sturm 1979, Dickman 1979, O'Sullivan 1983, Peglar *et al.* 1984). Ideally, for well developed laminations to be preserved, the lake should be both flat-bottomed, to prevent slumping and lateral sediment movement, and deep, so that circulation of the water column due to wind stress and seasonal mixing is at a minimum. The absence of significant bioturbation at the mud/water interface is also critical. In most lakes in maritime climates, conditions are unfavourable for the development of laminae and, as a result of either mixing, exposure or lack of strong seasonality, sediments are less easily datable by purely stratigraphic means. Interesting exceptions are, however, coming to light and it seems probable that in future, in view of the enormous value of annual laminations and the continuing doubts surrounding virtually all other dating techniques, studies based on annually laminated sediments in a wide range of environments will become increasingly prevalent. Recent work in subtropical areas has revealed annual laminations and, as elsewhere, these are likely to be of crucial significance in evaluating both ecological and geomorphic change in the drainage basins from which they derive. In some instances, laminae can now be shown to reflect single major erosional events in the catchment rather than seasonal cycles and, in such cases, the direct links with morphogenesis may often be even more direct (e.g. Digerfeldt *et al.* 1975, Simola 1977, Simola *et al.* 1981, O'Sullivan *et al.* 1982).

In the case of ombrotrophic peats (i.e. those formed above the ground water table), Pakarinen and Tolonen (1977) have recently shown that, in the cool continental climatic regimes to which peat-forming mosses are subject in east Scandinavia, annual growth increments in both *Sphagnum* and *Polytrichum* species allow precise age calculation for layers of peat up to 600 years old.

5.7.6.3 Radiocarbon dating (^{14}C)

Of the radiometric dating techniques available for lake sediments and peats, by far the most commonly used and best known is the radiocarbon method. The principles of this dating method and most aspects of its application are fully discussed in a previous section. There are, however, two particular problems associated with lake sediments that merit special attention here.

The error introduced by the photosynthesis of aquatic organisms in hard-water lakes has been documented since Deevey and his colleagues studied Linsley Pond, Connecticut (Deevey *et al.* 1954). They were able to show that the incorporation of old carbon into the living tissue of the organisms from solution in the lake water produces a level of specific activity lower than that for contemporary terrestrial material. Since these aquatic organisms contribute carbon to the sediment, this gives rise to a dilution of the radioactivity arising from ^{14}C and produces radiocarbon age determinations older than they should be. This problem may not be restricted to areas of carbonate rocks or substrates. In recently deglaciated areas, silt and clay fractions of rock flour may provide a supply of inert carbon that is transported via ground water to lakes, becoming incorporated into the sediments and organisms (Sutherland 1980). Although some authors have attempted to correct for this hard-water error by assuming constant divergence between real and measured radioactivity, using surface sediments as a reference point, this has had to rely on many untestable assumptions about the proportion of organic carbon in the sediment derived from different potential sources within the lake and its catchment. Work in Finland by Tolonen (1978) and by Brugam (1978) in Linsley Pond demonstrates that the divergence between apparent age and true age in a lake subject to hard-water error is not constant. This limitation is therefore of major and, perhaps in view of Olsson's (1974) work, almost general significance in aquatic and semi-aquatic environments.

The second problem with radiocarbon dating specific to lake sediments is that arising from the input to the lake of old carbon derived from stable organic residues in the land surfaces draining to the lake. Many authors have documented this old-carbon error and a review of its impact is given in Oldfield (1977). Most authors have responded to this problem by assuming that, in sections of a core for which radiocarbon dates continue to get older with depth, the dates can be accepted. Only when the relationship becomes reversed have they been rejected. This practice is obviously dubious since the degree of

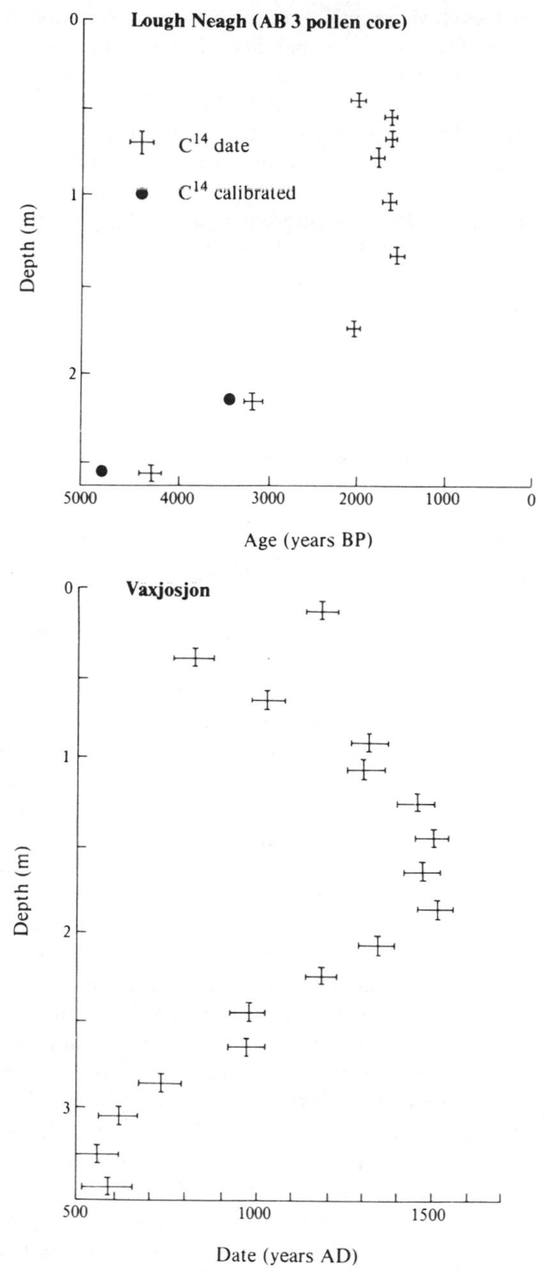

Figure 5.15 Erroneous 'inverted' ^{14}C-dated lake sediment sequences. The Lough Neagh profile is based on O'Sullivan *et al.* (1973, p. 272, Fig. 7) and the inverted age–depth relationship arises from the inwash of old, stable terrestrial organic material from soils and blanket peat in the catchment. The Växjösjön profile is from Battarbee and Digerfeldt (1976, p. 339, Fig. 6). In this instance, recent oil spillage is a possible additional contributory factor.

dilution required to swing the ages older as one moves up a profile must be very excessive and, even below the level where inflection takes place, more modest dilution may have occurred (Fig. 5.15). The strong probability of this is brought out by the work of Thompson and his colleagues, who have been calibrating palaeomagnetic secular variation with radiocarbon dates in a variety of lakes in north-west Europe (Thompson 1975, 1983a, Turner & Thompson 1979). Their demonstration that the radiocarbon age of synchronous swings in secular variations can vary significantly from lake to lake strongly suggests that, even in those sections of radiocarbon-dated lake sediment cores where there appears to be a normal monotonic relationship between radiometric age and depth, there is often significant dilution from old carbon washed in from the drainage basin (Dickson *et al.* 1978). In most cases, this dilution occurs from a point in time when human activity in the lands around the lake has begun to provide surface material available for inwash to the lake through overland flow. During the preceding period of dominance by mature vegetation and soils in early to mid-Flandrian times, such problems are of much less significance. From the point in time when ploughing has become prevalent in a catchment, the dilution effect seems to become rapidly more serious, and it is often at this stage that the inflection in radiometric age versus depth makes dilution obvious.

5.7.6.4 Lead-210 (^{210}Pb)

More recently, alternative radiometric techniques have been developed for looking at changes in sedimentation on a much shorter timescale. Radiocarbon dating is unsuitable for this, in view of the high standard errors associated with even recent dates and the problems arising from atomic testing and the preceding Suess effect (Pennington *et al.* 1976). The principle of ^{210}Pb dating is comparable to that applied in the case of radiocarbon. Lead-210 is a natural radioactive decay product, part of the uranium-238 (^{238}U) series. It has a half-life of 22.26 yr. Lead-210 reaches lake sediments in two forms, as a 'supported' component arising from the presence in the sediment of its parent isotope, radium-226 (^{226}Ra), which precedes ^{210}Pb in the decay series, and as an 'unsupported' rainout component derived from atmospheric ^{210}Pb (Fig. 5.16). The 'supported' component is obviously derived directly from the substrates of the drainage basin. The 'unsupported' component appears to be derived predominantly from the atmosphere, though there is the possibility that soil enriched in 'unsupported' ^{210}Pb has reached the lake as a result of surface erosion. In order to derive dates from ^{210}Pb analysis, it is necessary to determine both the total ^{210}Pb content, and also the supported activity. Measuring total ^{210}Pb can be carried out directly by gamma-ray assay (Gäggeler *et al.* 1976, Durham & Oliver 1983, Appleby *et al.* 1988) or indirectly by measuring the activity either of its daughter isotope bismuth-210 (^{210}Bi) or more often its grand-daughter isotope polonium-210 (^{210}Po) (Eakins & Morrison 1977). Subtraction of the radium-supported activity from the total ^{210}Pb activity allows calculation of the 'unsupported' ^{210}Pb content in the sediment. This will tend to decline with depth in accordance with a negative exponential function derived from the short half-life of the radioisotope. The precise way in which dates are calculated using the ^{210}Pb assays varies with the assumptions made about the way in which ^{210}Pb is incorporated into the sediment. All authors are obliged to assume that ^{210}Pb does not migrate within the sediment column. Beyond this, the assumptions used in the dating model (whether explicit or implicit) are bewildering in their variety (Robbins 1978). Recent evidence (Appleby & Oldfield 1983, Oldfield & Appleby 1984a, b) from a wide variety of lakes suggests that, in very diverse circumstances, the assumption of a constant net flux of unsupported ^{210}Pb to the sediment surface despite any changes in dry mass sedimentation rates that may have occurred as a result of lake or catchment disturbance – the

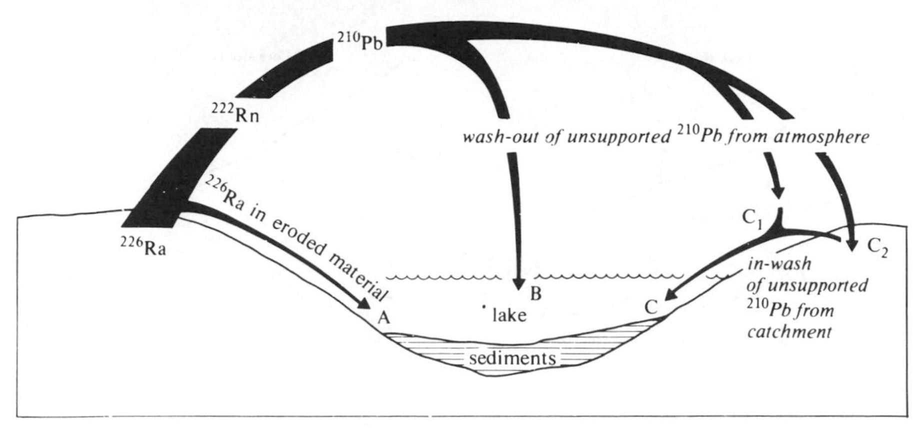

Figure 5.16 Pathways of ^{210}Pb input to lake sediments. The diagram is a modified version of that presented by Wise (1980). A comprehensive analysis of ^{210}Pb pathways is given by Krishnaswamy and Lal in Lerman (1978).

constant rate of sedimentation (CRS) model *sensu* Appleby and Oldfield (1978) – will often provide the best guide to age–depth and sedimentation rate calculations. The assumptions and methodology involved in applying this approach to ^{210}Pb dating are outlined in Appleby and Oldfield (1978). Figure 5.17 shows age–depth curves and dry sedimentation rate profiles from parts of Lake Michigan where the CRS model can be confirmed by testing against external evidence (cf. Appleby & Oldfield 1979, Appleby *et al.* 1979, Oldfield & Appleby 1984a). In summary, ^{210}Pb dating is clearly an exceptionally valuable technique of particular interest to the geomorphologist by virtue of the promise it holds of assisting quantification of material flux during the period of maximum human impact on environmental processes right through to the present day. At the same time, critical evaluation of the technique has lagged behind its application, and its full value will only emerge once it has been tested against independent and unambiguous dating methods in a wide variety of contexts. This applies not only to its use in dating lake sediments but also its application to soil erosion studies, to ice-core chronology and to estimating the rates of saltmarsh accretion (McCaffrey & Thompson 1974, Clark & Patterson 1984), nearshore marine sedimentation (Koide *et al.* 1972, Benninger *et al.* 1976), estuarine sediment accumulation (Brush *et al.* 1982) and ombrotrophic peat accumulation (Oldfield *et al.* 1979a, Aaby *et al.* 1978, El-Daoushy *et al.* 1982).

5.7.6.5 Caesium-137 (^{137}Cs)

Caesium-137, unlike ^{210}Pb, is an artificial radioisotope, the presence of which in the environment is the result of fallout from atomic-bomb testing. Figure 5.18 shows the atmospheric content of ^{137}Cs in the atmosphere of the Northern Hemisphere from the early 1950s onwards. The beginning of detectable concentrations can be dated to 1954 and the peak activity to 1963. The reference curves provide calibration for ^{137}Cs concentrations in sediments once correction has been made for variations in rainfall, delays in hemispheric exchange and the decay constant of ^{137}Cs (half-life 30 yr). As a result, several authors have been able to date recent sediment accumulation and estimate influx variations using the ^{137}Cs dates provided by the 1954 onset and more especially the 1963 peak. The technique is attractive partly because the normal method of assay (Pennington *et al.* 1973) is non-destructive. As with ^{210}Pb, bioturbation and physical mixing of sediments will inevitably introduce error into ^{137}Cs-based dating. Moreover, the evidence for ^{137}Cs migration in the sediment column is much stronger than for ^{210}Pb,

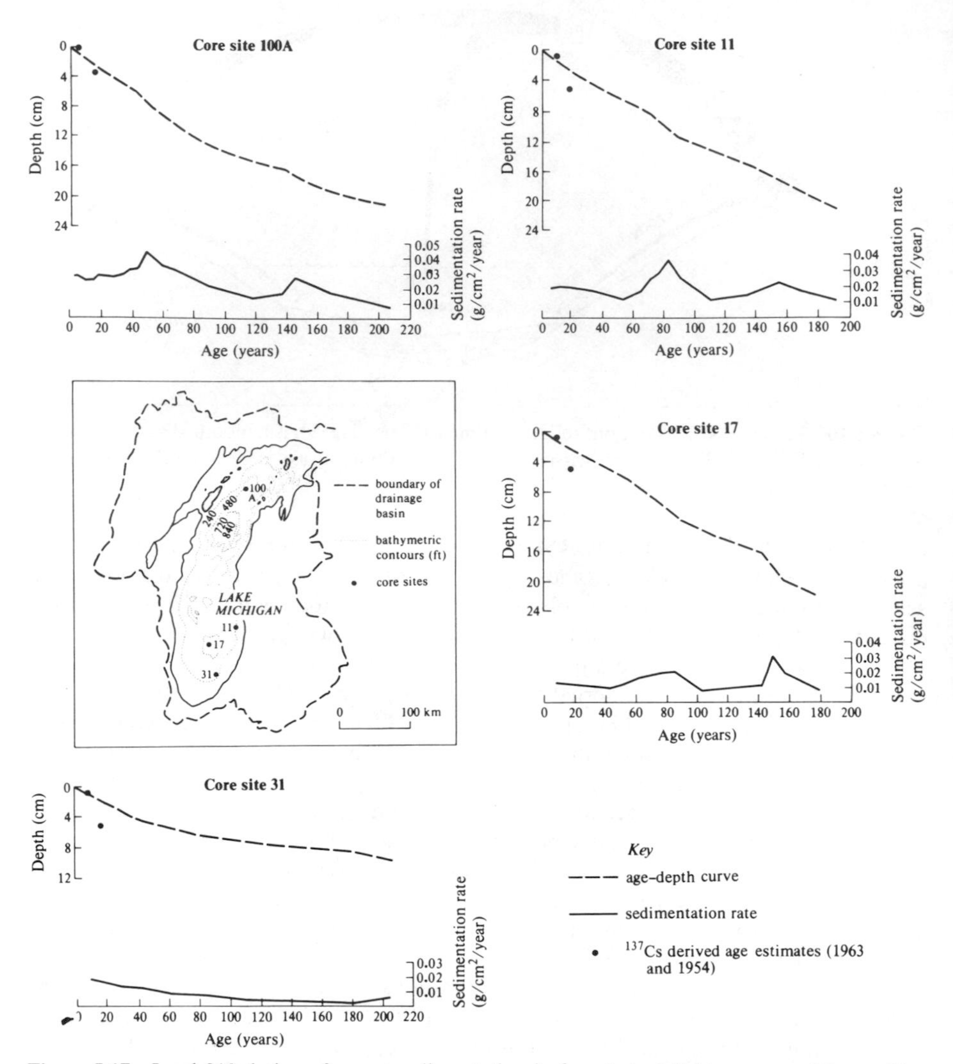

Figure 5.17 Lead-210 dating of recent sedimentation in four Lake Michigan cores. The profiles present recalculations by P. G. Appleby of the original data set published in Robbins and Edgington (1975). Cores 11 and 17 appear to receive relatively high and variable inputs of allochthonous sediment from the large and mainly agricultural south-east corner of the catchment. The profiles of dry mass sedimentation for cores 11 and 17 identify an early nineteenth-century peak associated directly with the *Ambrosia* pollen increase and they parallel the independently derived curve of sedimentation at Frains Lake on the edge of the catchment (Davis 1976). Core 3 lies outside and to the south-west of the zone affected by allochthonous input. Core 100A shows a similar depositional history to 11 and 17 though sediment focusing at the site greatly reduces the reliability of the calculations. A validation of the constant rate of sedimentation (CRS) approach to the Lake Michigan cores is set out in Appleby and Oldfield (1979) and Oldfield and Appleby (1984a).

and it has also been suggested that ^{137}Cs is released from the sediment into the water and taken up by zooplankton (Longmore *et al.* 1983a). Furthermore, whereas inwash of ^{210}Pb from soils in the catchment remains an unevaluated though possibly significant source of uncertainty in ^{210}Pb dating, in the case of ^{137}Cs its importance can be readily confirmed (Fig. 5.18). The combined effect of downward diffusion and delayed input from ^{137}Cs-enriched catchment surfaces provides, for ^{137}Cs dating, conflicting sources of potential error, which require careful site-by-site evaluation. The documented tendency for some of the ^{137}Cs in lake sediments to be derived from the surface soils of the catchment implies that, even where, as a result of this, ^{137}Cs is of only limited value as a dating technique, it can be valuable as an indicator of rates of surface soil erosion and of areas from which there has been significant net soil loss over the last one or two decades (Wise 1980, McCallan *et al.* 1980, Longmore *et al.* 1983b). Caesium-137 dating is not often applicable to peats. Oldfield *et al.* (1979a) have shown that in ombrotrophic mires it is subject both to downward diffusion and to active uptake by living plants. Other nuclear fission products, e.g. plutonium-239 (^{239}Pu) and americium-241 (^{241}Am), have been used as dating tools with varying degrees of success. Generally, measuring their residual activity is more costly than ^{137}Cs assay alone, and only in the case of peats do they seem to hold out the possibility of providing a better indication of age. Locally within north-west Europe, deposition of ^{134}Cs as well as ^{137}Cs resulting from the Chernobyl accident in April 1986 has provided an additional opportunity for dating and tracing.

5.7.6.6 *Palaeomagnetic dating*

During the last few years, geophysicists have provided evidence of chronological value by documenting the way in which the Earth's magnetic field has changed through time and calibrating these changes with the help of appropriate radiometric dating techniques. Reversals of geomagnetic polarity during the geological past are summarised

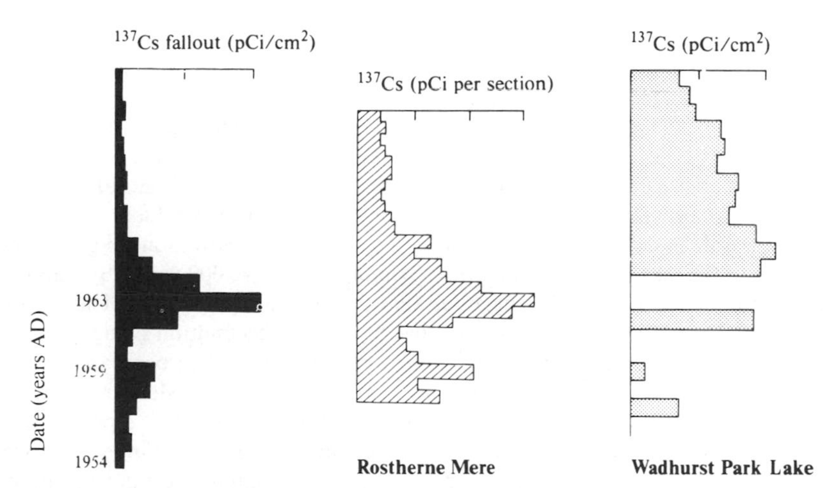

Figure 5.18 Caesium-137 in the atmosphere and its use as a dating parameter and erosion indicator in lake sediments. The black reference histogram shows measured annual atmospheric fallout at Milford Haven, UK, since 1954, and is taken from Cambray *et al.* (1972). The profile from Rostherne Mere is taken from Livingstone and Cambray (1978). This lake is highly productive and has a small drainage basin. The resulting combination of rapid and predominantly autochthonous sedimentation provides an ideal combination for dating by ^{137}Cs, and the ^{137}Cs dates are independently verified by algal stratigraphy. The Wadhurst example (Wise 1980) illustrates delayed input of allochthonous soil-derived ^{137}Cs from eroding catchment soils (cf. McHenry *et al.* 1973).

in, for example, McElhinney (1973). These subdivide the magnetostratigraphic record into epochs lasting between 10^5 and 10^6 yr. Within these, polarity events (shorter-term flips to opposite polarity) have been detected lasting between 10^4 and 10^5 yr. The European Pleistocene spans part of three polarity epochs, the Gauss (normal), Matuyama (reversed) and Brunhes (normal) (Fig. 5.19). Within each of these, polarity events have been detected with varying degrees of confidence. In particular, there are well documented polarity events within both the Matuyama and Gauss epochs. Within the Brunhes epoch, which spans the last 0.7 Ma, various authors have claimed to identify temporary reversals as well as major fluctuations in geomagnetic polarity termed 'excursions'. These latter involve changes in the position of the virtual geomagnetic pole of at least 45° of latitude followed by a return to original polarity within a short (e.g. 10^2–10^4 yr) but often poorly defined period of time. Denham (1976), Smith and Foster (1969) and Barbetti and McElhinney (1972) report well authenticated excursions. More controversial are the claims of Mörner & Lanser (1974) and Nöel (1975) for a recent 'Gothenburg' excursion coincident with some part of the late Weichselian period. These and other authors claim to have identified events and excursions within the recent Brunhes polarity epoch, at a wide range of sites (e.g. Nakajima *et al.* 1973, Kopper & Creer 1976, Vilks *et al.* 1977). Also, there has been a tendency to correlate these on the basis of very poor chronological control. In many cases, the proposed events and excursions coincide with lithological changes that help to explain the geomagnetic features used to indicate the magnetic field variations inferred (Thompson & Berglund 1976). Under these circumstances, it is very important to bear in mind the minimum criteria advocated by such authorities as Von Montfrans (1971) and Thompson (1977c). The whole field of late Pleistocene magnetostratigraphy based on events and excursions remains somewhat open to doubt, and many of the problems are unresolved. There is a tendency, as in many other aspects of Quaternary study, to resort to what Watkins (1971) has termed the 'reinforcement syndrome', whereby coherent but untested reconstructions tend for a while to be self-perpetuating.

Within the context of lake sediment study, for most practical purposes, especially in latitudes where sedimentation processes were interrupted by major glacial advances during the late Pleistocene, greater chronological significance attaches to the ability to detect a record of secular variation within the major epochs. This record is independent of the more extreme events and excursions discussed in the previous paragraph. The first author to decipher a continuous and well dated record of secular variation in magnetic declination was Mackereth (1971), who analysed a 6 m core from Lake Windermere, UK. He was able to demonstrate a sequence of east-to-west fluctuations in geomagnetic pole positions (i.e. changes in declination) spanning the last 10 000 years. By calibrating these with radiocarbon dates, he suggested that they had a regular periodicity approximating to 2700 yr for a full cycle. The most recent westerly declination swing recorded in Lake Windermere and in many other sedimentary profiles can be compared with the instrumentally recorded extreme westerly declination of AD 1815. Mackereth's pioneer work in this field has stimulated a large number of studies from lakes, especially in the Northern Hemisphere, designed to complement and extend his Windermere work. Palaeomagnetic studies of this kind can provide no more than indirect dating but, once calibrated either by annual laminations, radiometric dates or other types of chronology, palaeomagnetic measurements provide an exceptionally attractive dating technique.

Calibration from the sixteenth century onwards can be achieved by direct comparision with observatory records or with the global models of the geomagnetic field reconstructed for 50 yr intervals since AD 1600 (Thompson & Barraclough 1982). For the period between AD 1000 and the late sixteenth century, archaeomagnetic records of declination and inclination provide a reference curve that is well dated, and the

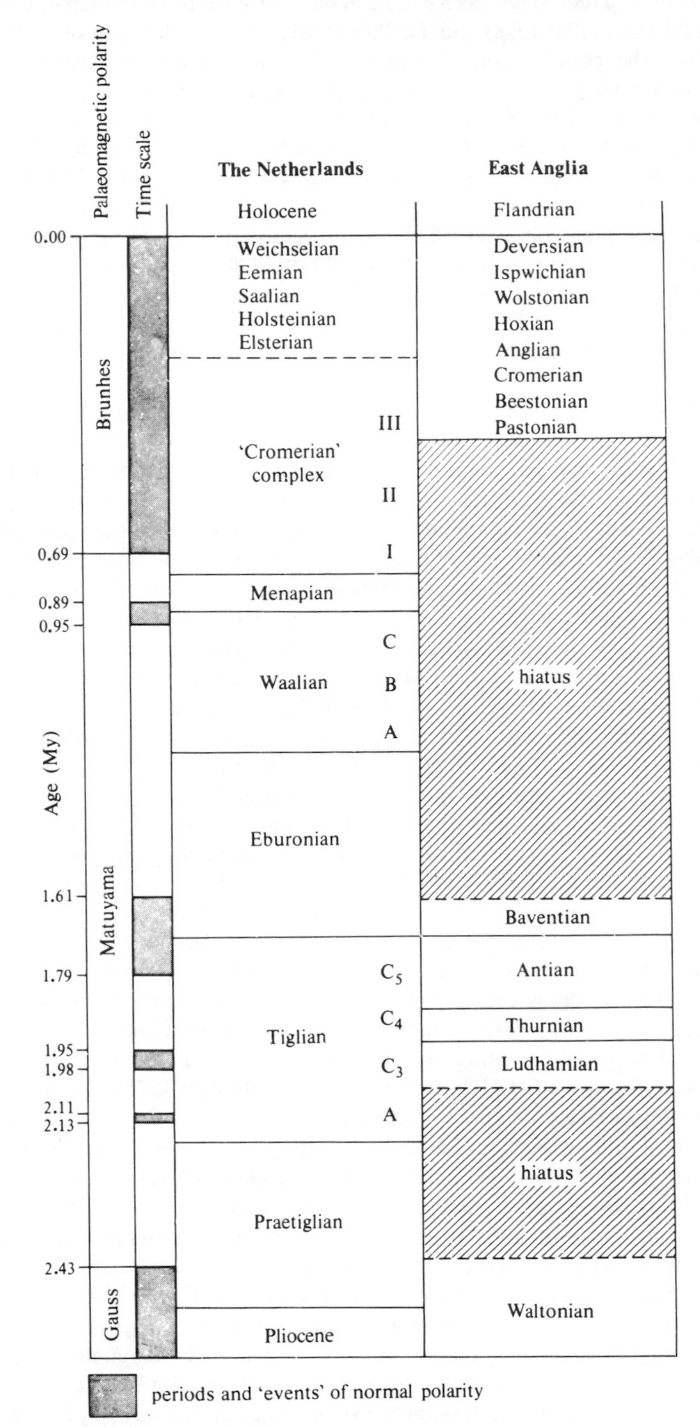

Figure 5.19 Geomagnetic polarity timescale for Pleistocene of north-west Europe and its relation to the major regional stratigraphic units. (Based on Thompson 1975).

archaeomagnetic record has been extended backwards in time, though not continuously, by both Aitken (1974, 1978) and Thellier (1981). Beyond the span of archaeomagnetic calibration, the combination of pollen analytical horizons, tephrochronologies and radiocarbon dates provides approximate calibration, once these latter are corrected for the error established by comparison with dendrochronological timescales. In the light of the most recent studies, the initial hypothesis of Mackereth that declination swings were harmonic seems now to be unjustified (Turner & Thompson 1981, Thompson 1984).

Work carried out at Loch Lomond, Loch Shiel (UK) and several other sites, as well as in the Jura region of France, and Switzerland, shows that the declination and inclination records preserved in lake sediments, once calibrated, provide an exceptionally consistent and detailed clue to the chronology of sedimentation. Figure 5.20 summarises the published evidence from Loch Lomond (Turner & Thompson 1979) amplified by remeasurement of Flandrian sediments from Windermere (Thompson & Turner 1979). It shows, however, that the variations are not harmonic. The task of calibration by means of radiocarbon dates and other methods continued (Thompson & Edwards 1982) and it is apparent from these studies and from others carried out in low latitudes (e.g. Thompson & Oldfield 1978) that palaeomagnetic dating adds a significant new dimension to all studies based on lake sediments. One of the major attractions of the technique lies in the speed with which it can be applied to unopened, continuous, pneumatically obtained cores using the Mackereth systems described in the previous section. Measurements of declination on whole cores can be obtained within a matter of minutes or at most hours using entirely non-destructive techniques and leaving the core tube unopened and ready for all further types of analysis. Inclination measurements require the core to be opened and individual subsamples taken. Nevertheless, the rate at which this procedure can be accomplished compares very favourably with the rate at which radiocarbon dates can be obtained, and the palaeomagnetic trace, when it is stable and smooth, is not subject to serious sources of error such as those arising from the inwash or the photosynthesis of old carbon described in the previous section. The major limitations of the technique arise from the fact that not all lake sediment sequences are suitable for palaeomagnetic dating. At present, it is difficult to predict which lakes will have a good palaeomagnetic trace and which will not, though some controlling factors have begun to emerge. Lake sediments that have predominantly *coarse* minerogenic influx appear to be unsuitable since the palaeomagnetic remanence preserved is not sufficiently well locked into the sedimentary structure to retain a signal from the Earth's magnetic field at the time of deposition. Equally, lake sediments that are predominantly organic and poor in allochthonous minerogenic influx have weak unstable natural remanence and have so far proved unsuitable for palaeomagnetic dating, though this limitation may be overcome by means of exceptionally sensitive SQUID magnetometers (Collinson 1975). Ideally a lake should have a drainage basin yielding relatively large amounts of fine-grained allochthonous minerogenic input at least moderately rich in magnetic minerals and also should be free from either bioturbation or physical mixing by means of currents or slumping at the site of deposition. Under these relatively favourable circumstances, there is an increased probability that the directional properties exhibited by the palaeomagnetic trace from the core will reflect secular variation in the Earth's magnetic field during the course of sediment accumulation. The signal from the Earth's magnetic field in most cases appears to be the result of what is known as post-depositional remanent magnetisation (PDRM). A signal of the past geomagnetic field can also be retained by magnetic minerals forming authigenically near the sediment surface (chemical remanent magnetisation or CRM), though the importance of this has not been demonstrated conclusively for lake sediments, since the early indications of this, suggested by Mackereth (1971), Creer *et al.* (1972) and Thompson (1975), have been shown to be misleading (Stober & Thompson 1977).

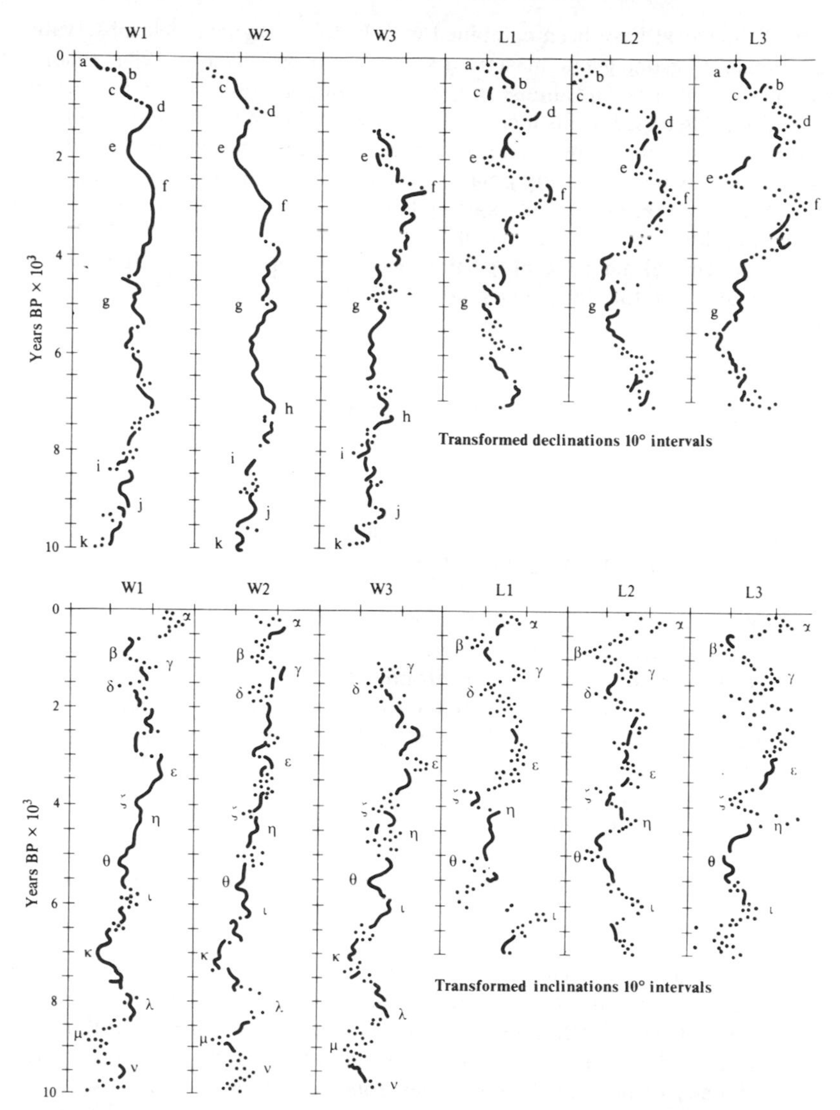

Figure 5.20 Master curves of British geomagnetic secular variation 10 000-0 yr BP. (From Thompson & Turner 1979). W1–3 are cores from Windermere, and L1–3 are cores from Loch Lomond. A total of 23 features (10 declination and 13 inclination) are used. The preferred timescale for the features is given in the accompanying table. Turner and Thompson discuss the calibration of the master curve.

Once a master curve of secular variation in declination and inclination has been established (preferably for several cores from a range of lakes), and then calibrated chronologically by radiometric or other methods, the rate at which whole cores can be scanned and dated is impressive (Turner & Thompson 1979, 1981, Thompson & Edwards 1982). However, since the Earth's geomagnetic field at any place comprises both dipole and non-dipole components, each changing through time in ways not yet fully understood theoretically or documented empirically, the master curve will be applicable to a limited area only. Therefore palaeomagnetic records of lake sediments from seven regions of the world have been combined with historic magnetic field observations to produce regional geomagnetic master curves (Thompson 1983a, b, 1984, 1986). These master curves, which depend almost entirely on the radiocarbon method for their dating, may be applicable to sediments located up to 1000–2000 km from the type sites. In this context it is important to note that geomagnetic master curves relying largely both on timescales identifying and dating polarity epochs, events and excursions as well as on shorter-term, more recent timescales using the sedimentary record of secular variation are being developed from the study of marine sediments (Opdyke 1972). For all the studies of directional magnetic properties, one of several types of magnetometer and ancillary equipment for demagnetisation tests (Collinson 1975, Banerjee 1981) are necessary and widely available.

5.7.7 Palaeoenvironmental reconstruction from peats and lake sediments

5.7.7.1 Pollen analysis

Palaeoecological tools such as pollen analysis have usually been used in geomorphology for one or more of the following purposes:

(a) the establishment of relative chronologies and indirect dating by means of correlation with dated sequences elsewhere;
(b) characterisation of depositional environments at or near the sample site;
(c) reconstruction of palaeoenvironmental (often palaeoclimatic) conditions pertinent to the explanation of geomorphological features or processes; and
(d) establishment of the vegetation and land-use regimes within which features and processes have developed during the recent period when man's impact on all aspects of landscape evolution has been most important.

A simple introduction to the techniques of pollen analysis for dating and environmental reconstruction is provided by Jones and Cundill (1978).

Relative chronology Relative dating and correlation by means of fossil biotic assemblages, and especially pollen zones, is still of major significance in reconstructing stratigraphic sequences for the Pleistocene as a whole (e.g. Turner & West 1968, Zagwijn 1975). For more recent times, and especially most of the Flandrian period, the regional non-synchroneity of most of the old pollen zone boundaries has by now led to little reliance being placed on pollen either as a means of inter-regional correlation or as a dating tool for the last 10 000 years (Smith & Pilcher 1973, Hibbert & Switzur 1976). At present the best available methods are either to obtain approximate chronologies through the biostratigraphic correlation of sequences with well dated, standard regional pollen diagrams from the same area or, alternatively, to estimate dates from published isochron or isopoll maps (Huntley & Birks 1983).

Depositional environments The use of pollen diagrams together with ancillary evidence from macrofossils, in the provision of information about local environments, continues to be of major value in Pleistocene geomorphology. For example, in studies of land and sea-level change, palaeoecological information from peats and associated saltmarsh deposits is often crucial in establishing precise evidence of former sea level (Fig. 5.21) at stages that can also be dated by ^{14}C (e.g. Tooley 1978a, 1985a, b, Devoy 1977, 1979). Similarly the characterisation of local environments during the various

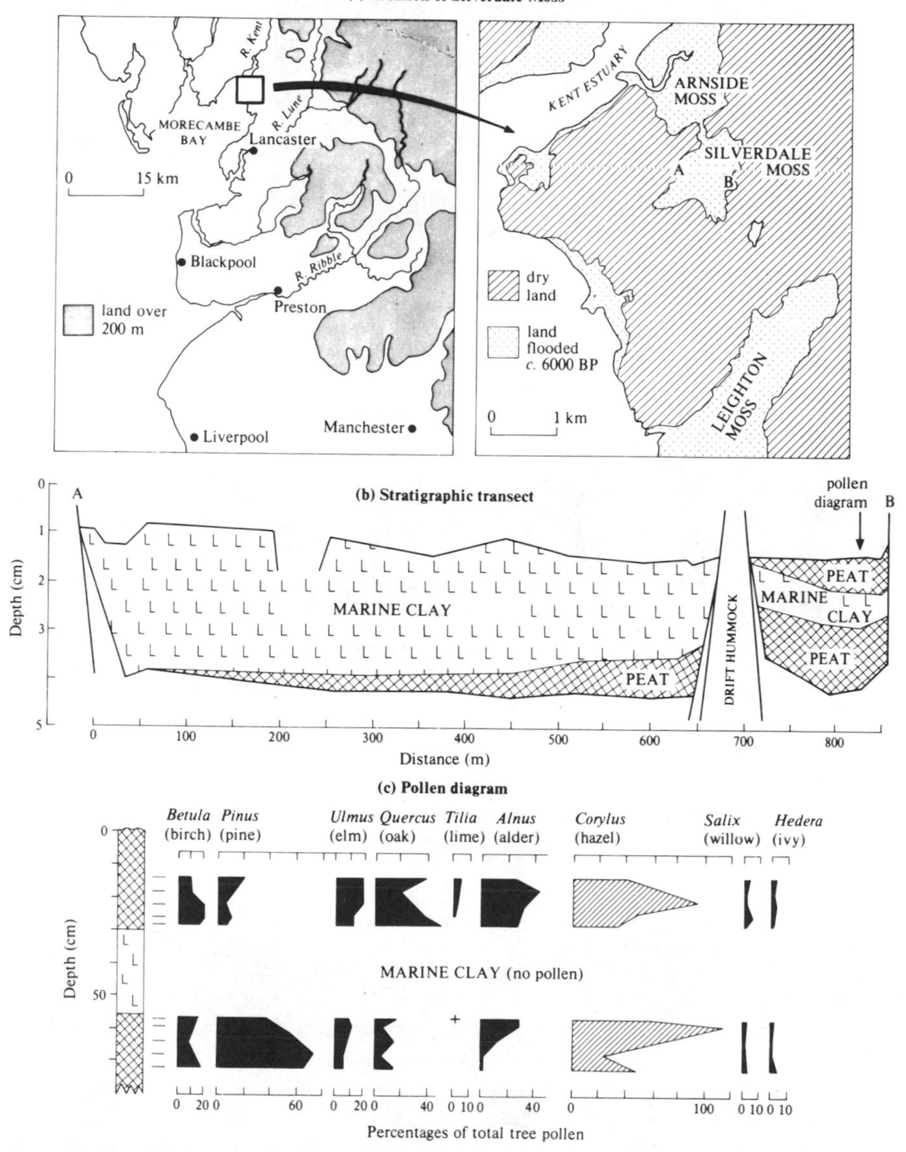

Figure 5.21 Evidence for land and sea-level change in stratigraphic and pollen analytical study, Silverdale Moss. The stratigraphic transect is located on the inset map and the pollen diagram comes from the eastern end of the transect where the marine alluvium resulting from tidal incursion of the basin is sandwiched between freshwater muds and peats. Radiocarbon dates of *c.* 6000 BP associated with the peat and the pollen analytical changes recorded provide an estimate of the age of the marine transgression at the site. (From Oldfield, 1960a, b).

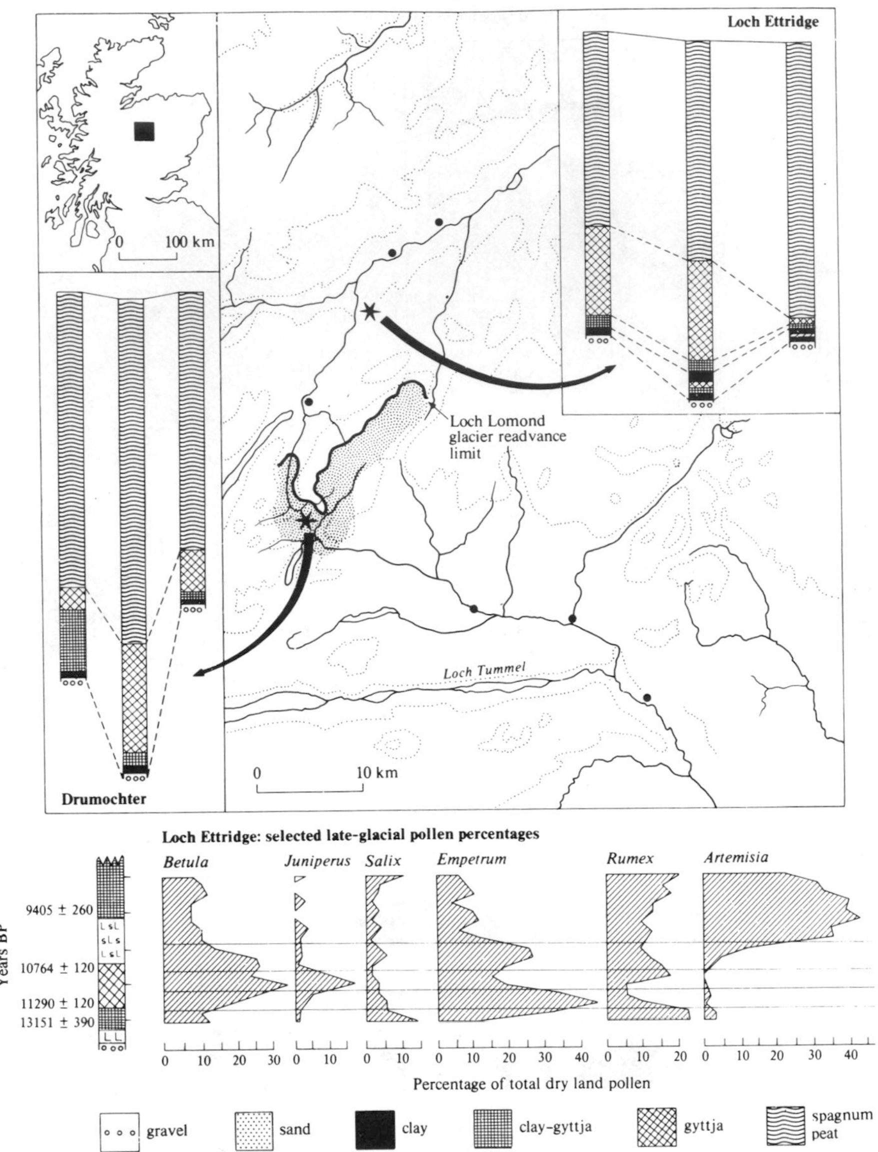

Figure 5.22 Evidence for late-Glacial (late-Devensian) deglaciation sequences in stratigraphic and pollen analytical study, Loch Ettridge and Drumochter in the central Grampian Highlands, Scotland. (From Walker 1975). The sites lie respectively beyond and inside the maximal extent of the Loch Lomond readvance ice. The sites are located, stratigraphic profiles from the centre of each basin are shown and the changing relative frequency of selected pollen types at the Loch Ettridge site are plotted alongside four radiocarbon dates. As indicated by the diagrams, this site, lying beyond the Loch Lomond readvance limit, shows the lithostratigraphic and biostratigraphic changes resulting from late-Glacial climatic changes at the site before and during the readvance. The Drumochter site contains only post-glacial (Flandrian) sediments directly overlying the drift of the Loch Lomond readvance.

substages of the late Devensian (Fig. 5.22) depends on the association of palaeoecological evidence from pollen, macrofossils and coleoptera with stratigraphic and morphological data.

Palaeoclimatic reconstruction Palaeoclimatic reconstruction from fossil biotic assemblages in lake sediments and peats usually implies some reliance on one or more of the following:

(a) identification of so-called climatic indicator species (e.g. Iversen 1944);
(b) reconstruction of ecological communities, the present distribution of which is closely linked to climatic parameters (e.g. Fig. 5.23); and
(c) establishment of numerical transfer functions linking directly the fossil assemblage to the climatic parameters circumscribing its present-day distribution.

Biological evidence for palaeoclimatic reconstruction has tended to focus strongly not only on pollen but on coleoptera within the first two deductive frameworks (e.g. Coope 1975, 1977, Atkinson *et al.* 1986). Currently, increasing emphasis is being placed on the development and application of broad-scale transfer functions for pollen analytical data, originally devised by Webb and Bryson (1972). This latter approach, which basically involves taking statistically derived calibration functions relating modern pollen spectra and climatic variables, and applying them to fossil pollen data in order to derive palaeoclimatic estimates, is reviewed by Webb and Clark (1977), Howe and Webb (1983) and Webb (1985). It has been used to provide climatic calibration of pollen data and palaeoclimatic reconstruction for the Flandrian in central North America (Bartlein *et al.* 1984) though there is still a general need for the approach to be subjected to vigorous tests of compatibility with evidence independently derived from other palaeoclimatological techniques.

Human impact The fourth contribution of pollen analytical and related studies to geomorphology has been somewhat neglected given the consistency with which lake sediments preserve both a history of vegetation and land-use change and a record of variations in material flux in the catchment associated with different patterns and intensities of human usage (cf. Oldfield *et al.* 1980). This theme will be more fully developed below, and it is also a major element in Oldfield (1977).

Faegri and Iversen (1975), Moore and Webb (1978) and Barber (1976) all set out the main steps in the chemical preparation of samples for pollen analysis. Identification requires use of a comprehensive reference collection of contemporary pollen types, though keys such as those published by Faegri and Iversen (1975) and Moore and Webb (1978) and illustrated in Figure 5.24 are of great value. High-quality optical equipment in the form of a compound microscope with mechanical stage and magnification range of $\times 250$ to $\times 1000$ is also essential.

Irrespective of the main purpose underlying the adoption of pollen analytical or similar palaeobiological methods by the geomorphologist, the techniques involved remain demanding and time-consuming. Outlining the many alternative approaches to methodology, listing the numerous limitations of the techniques available and sorting out the actual details of procedure at each stage lie beyond the scope of the present section, and may be gleaned from the references already quoted and the introductory texts of Birks and Birks (1980) and Mannion (1980) and the references contained within.

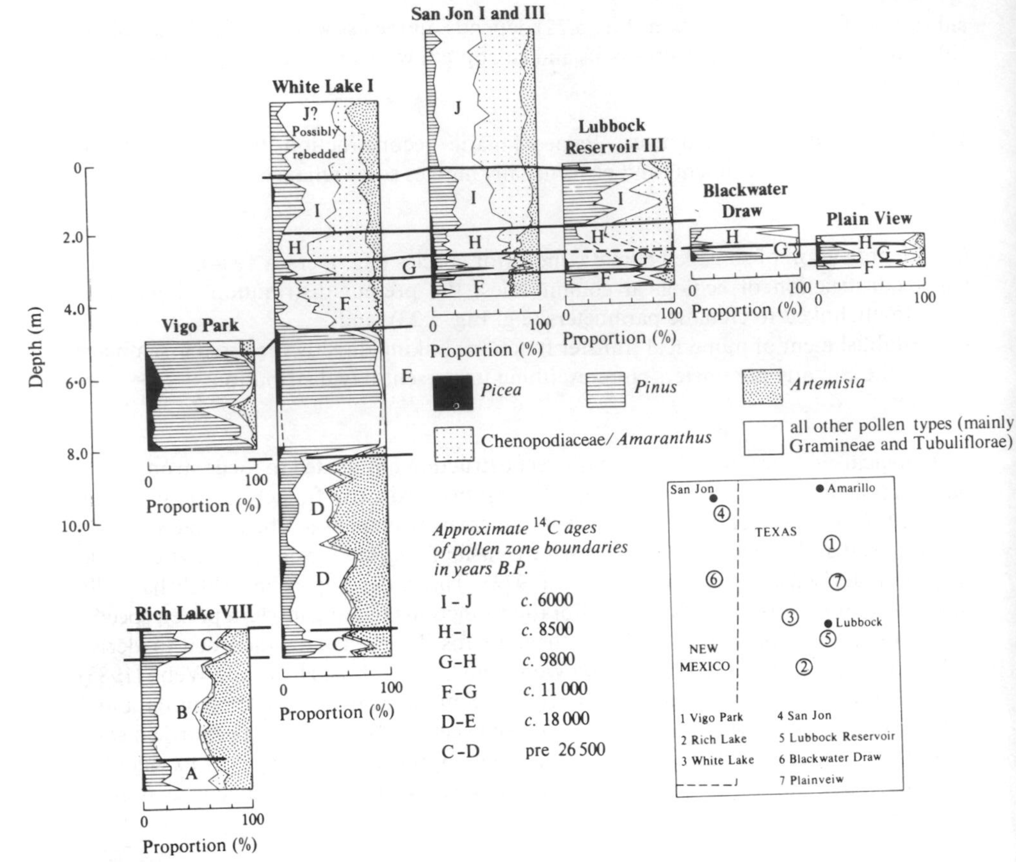

Figure 5.23 Pollen analysis and regional environmental change. The summary diagrams have been compiled from cores and sections taken from major playa lakes (1–4) and from infilled ponds in the former through drainage lines or 'draws' (5–7). The playa sequences portray the major vegetational and climatic shifts preceding, during and after the Wisconsin glacial maximum. In areas of short-grass prairie at the present day, cool sagebrush prairie and savanna gave way to pine or pine–spruce forest, which, during the late Wisconsin, alternated with prairie communities that finally became generally established at the opening of the Holocene. The Vigo Park site comes from a section at the maximum areal extent of the former playa and confirms the coincidence between highest lake level and the Wisconsin maximum. The 'draw' sites record lacustrine deposition only during the relatively brief period between their existence as continuously flowing river valleys during and after the Wisconsin maximum and their desiccation during the Holocene. For detailed discussion of the results, see Oldfield (1975) and Oldfield and Schoenwetter (1975).

5.7.7.2 *Material flux*

Not only do lake sediments contain a palaeoecological record in the fossil remains of organisms preserved within them, they also provide a detailed and continuous record of the changing material flux within the catchment of the lake during the whole period of infilling. The processes giving rise to variations in the material flux are inevitably those which are the direct concern of geomorphologists. Thus by studying the record of material flux contained in lake sediment sequences, we may hope to gain some time perspective on the processes operating on the present-day landscape.

Lake sediment stratigraphy, even studied in the crudest sense, clearly demonstrates many direct links with climatic change and morphogenesis. This is best illustrated by the

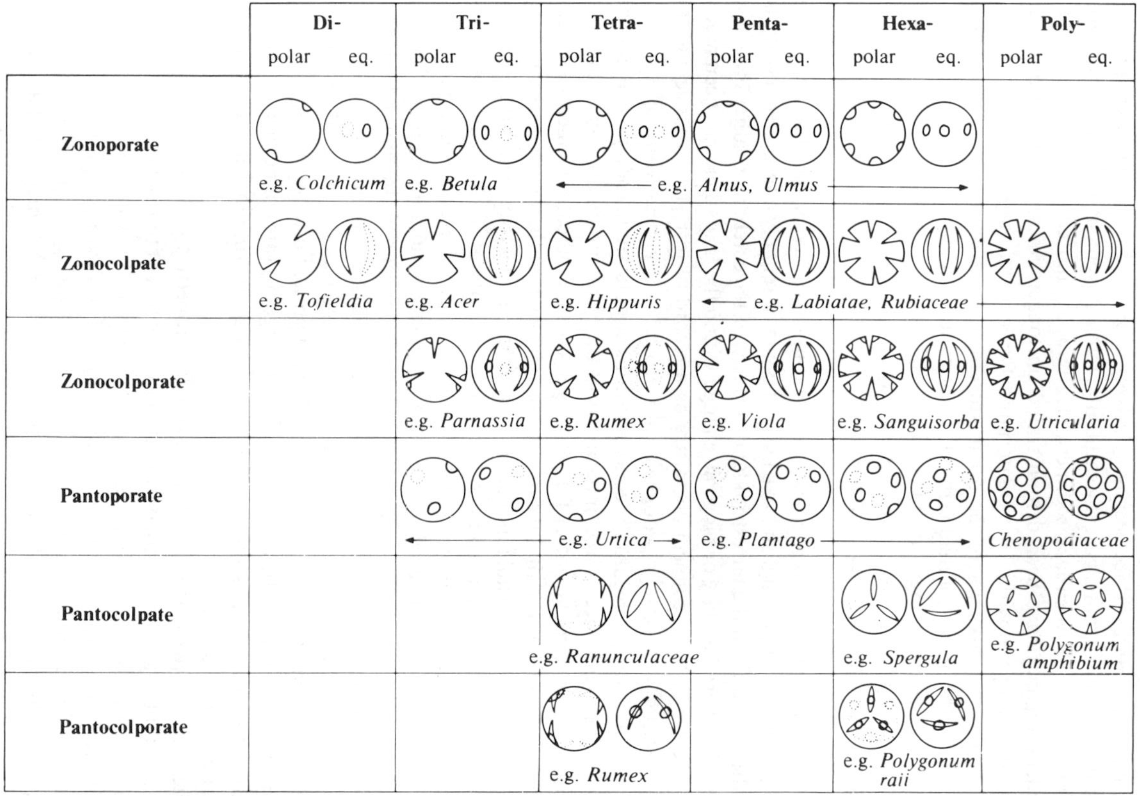

	Di-		Tri-		Tetra-		Penta-		Hexa-		Poly-	
	polar	eq.	polar	eq.	polar	eq.	polar	eq.	polar	eq.	polar	eq.
Zonoporate	e.g. *Colchicum*		e.g. *Betula*		← e.g. *Alnus, Ulmus* →							
Zonocolpate	e.g. *Tofieldia*		e.g. *Acer*		e.g. *Hippuris*		← e.g. *Labiatae, Rubiaceae* →					
Zonocolporate			e.g. *Parnassia*		e.g. *Rumex*		e.g. *Viola*		e.g. *Sanguisorba*		e.g. *Utricularia*	
Pantoporate			← e.g. *Urtica* →				e.g. *Plantago* →				*Chenopodiaceae*	
Pantocolpate					e.g. *Ranunculaceae*				e.g. *Spergula*		e.g. *Polygonum amphibium*	
Pantocolporate					e.g. *Rumex*				e.g. *Polygonum raii*			

Figure 5.24 Morphological features of some of the main pollen types. The arrangement of apertures illustrated is used as a basis for pollen identification and the diagram relates directly to the scheme and key set out fully in Moore and Webb (1978).

many studies of late-glacial sequences where the direct relationship between stratigraphy and changing climate emerges (see, e.g. Fig. 5.22).

Minerogenic sediments often associated with either direct glacial outwash or periglacial conditions and solifluction within the lake drainage basin alternate with more organic sediments representing periods of relative warmth during which soil development took place, the land surface became relatively stable and the accelerated input of mineral matter resulting from physical weathering, mass movement and glacial outwash ceased. Similar links between lake sediment stratigraphy and climatic change (hence morphogenetic regime) have been demonstrated on longer timescales of the order of glacial–interglacial cycles within the Quaternary. On shorter, more recent timescales, there are comparable links where the dominant effect on erosional and depositional regimes has been the impact of man rather than of climatic change.

Studies of the sedimentology and mineralogy of lake sediments are relatively rare, though the papers that have been published suggest that much of interest both to limnologists and geomorphologists may eventually emerge in this field (e.g. Lerman 1978).

Since Mackereth (1965, 1966) demonstrated that much of the chemical record in the sediments of the lakes of the English Lake District could be reliably interpreted mostly as a trace of the output from the terrestrial land surfaces of the catchment rather than as a record of primary productivity or biogenic change within the lake, there has been much interest in the use of lake sediment chemistry as a record of terrestrial change and, in particular, varying weathering, pedogenic and erosional regimes. More specifically, many authors have come to regard the concentrations of potassium, sodium, magnesium and, in some lakes, calcium as direct indications of the changing relative importance of erosion within the lake catchment. All these alleged erosion indicator elements are very easily leached from substrates; thus it can often be inferred that their presence in the lake sediments is a function of detrital input of unweathered material from which the elements have not been previously removed in solution. Interpretations of lake sediment chemistry, with important implications for the reconstruction of past geomorphological processes, have usually been based on the main interpretational guidelines set down in Mackereth's original paper with some modifications, as for example in Pennington *et al.* (1972), Engstrom and Wright (1984) and Oldfield (1977). As indicated in the latter references, it is important to attempt to join chemical and sedimentological analysis to derive timescales accurately so that diagrams, which at present usually represent relative changes in *concentration*, can be transformed into changes in rates of net accumulation. Only by doing this will it be possible to make the fullest geomorphological use of the lake sediment record (Fig. 5.25).

Over the last few years, studies of the magnetic properties of lake sediments have begun to contribute significantly not only to the development of timescales, as outlined in Section 5.3.5, but also to tracing the source of sediments (whether, for example, predominantly from substrate or topsoil material) and to developing firmer empirical bases for estimating the total quantity of sediment input during any given time interval. Several recent publications illustrate this approach, define the magnetic parameters, measuring techniques and units used, and provide illustrations of both aspects of the potential application of magnetic measurements (Thompson *et al.* 1980, Oldfield & Maher 1984, Oldfield *et al.* 1985a, Thompson & Oldfield 1986).

Not only do magnetic measurements provide an extremely rapid method of whole-core logging, which can form the basis of detailed correlation schemes, they also provide a basis for characterising sediment type, since many of the transformations of iron compounds within the environment have as their end-product forms of iron oxide that possess magnetic properties diagnostic of the processes themselves.

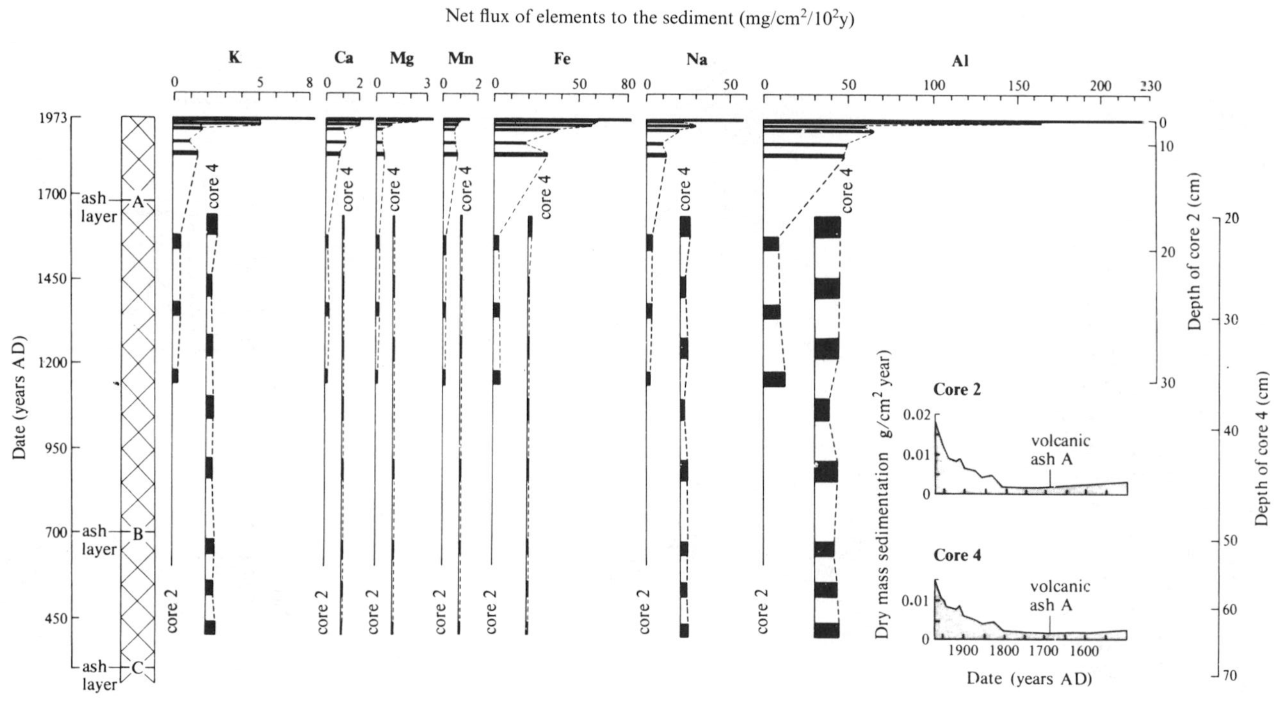

Figure 5.25 Chemical stratigraphy of lake sediments expressed as changes in the net flux of each element to the sediment. The timescales and dry mass sedimentation rates used in calculations (see inset diagrams) are set out more fully in Oldfield *et al.* (1980). The increase in the rate of influx of each element in core 2 since *c*. AD 1700 is seen to be largely a function of accelerated sediment composition.

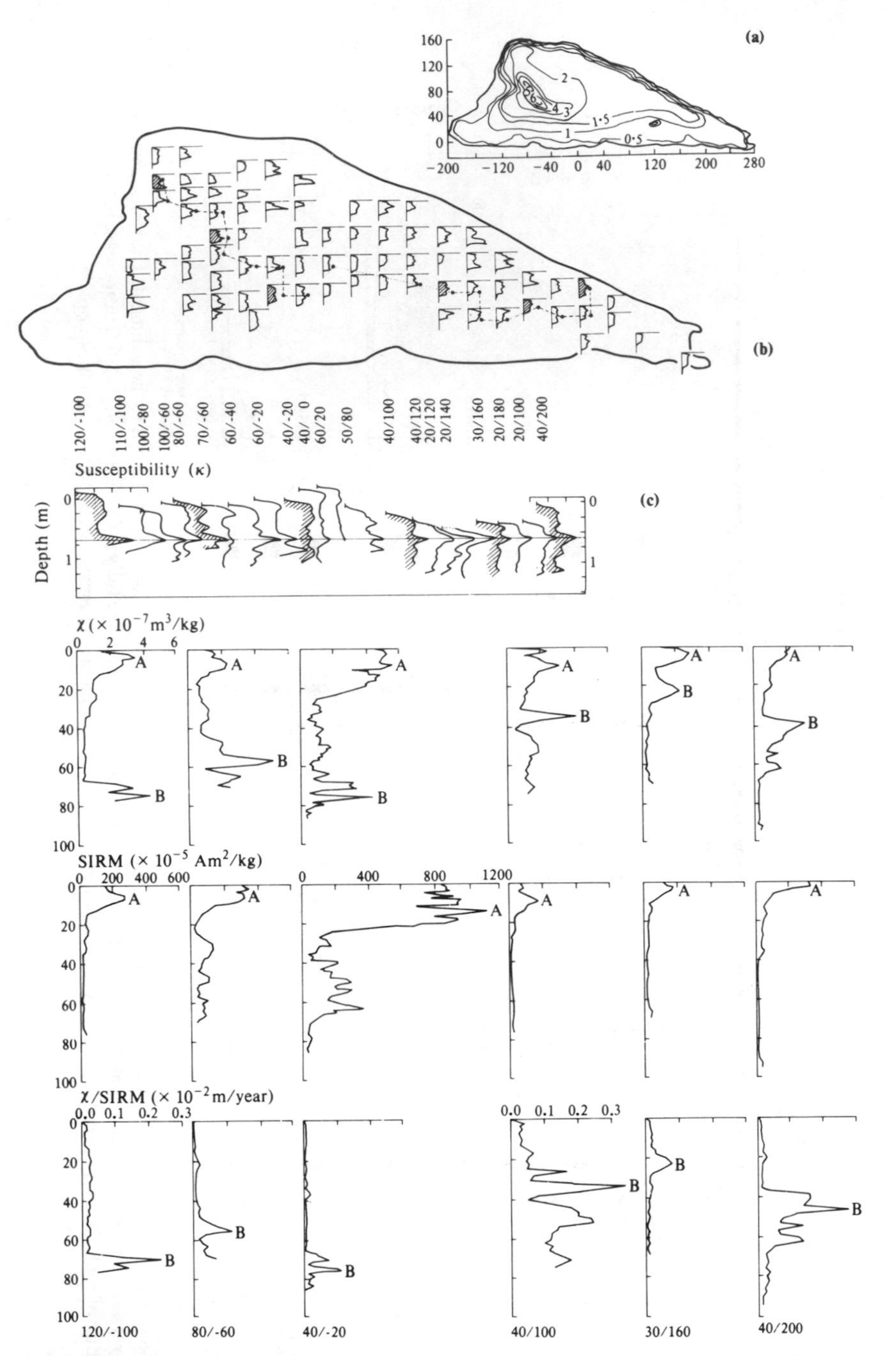

Figure 5.26 Core correlation by magnetic measurements, Llyn Goddionduon, north Wales. The upper part of the figure shows bathymetry and core coordinates (a), plots whole-core susceptibility traces for minicores from the main sampling grid (b), and links the correlated cores along a 'tie-line' for which expanded traces are shown in (c). The cores with shaded traces are those for which single-sample susceptibility (χ), saturation isothermal remanent magnetisation (*SIRM*) and $\chi/SIRM$ ratios are plotted below. In all cases the upper peak (A) in χ and *SIRM* gives low $\chi/SIRM$

Use of magnetic correlations to provide a further basis for quantifying changing rates and forms of sediment influx over time is well illustrated in studies at a number of European sites (Bloemendal *et al.* 1979 (Fig. 5.26), Dearing *et al.* 1981, 1982, Dearing 1983, 1986, Foster *et al.* 1985), as well as from a variety of small farmed catchments in the New Guinea Highlands (Oldfield *et al.* 1980, 1985b). In all cases, the rapid core correlation made possible by magnetic measurements speeds work by many orders of magnitude, and assists quantification of material flux on a whole-catchment/whole-lake basis.

Using magnetic measurements to characterise sediment source depends very strongly on the type of lithology on which the lake catchment has been formed. In lithologies as contrasted magnetically as strongly ferrimagnetic basalt, weakly magnetic or diamagnetic limestone and strongly antiferromagnetic Permo-Triassic substrates, the magnetic transformations that take place during weathering and pedogenesis provide a clear key for identifying the source of the lake sediments and for relating changes in sediment source to processes within the lake drainage basins. In each case studied so far, the dominant control of the sediment source during the Flandrian period has been the changing land-use regimes produced as a result of human activity (Thompson *et al.* 1975, Dearing 1979, Higgitt 1985). Insofar as one can generalise at present from these diverse case studies, there is, during the Flandrian, a shift from predominantly channel sources during early forested periods to topsoil sources as overland flow in a deforested landscape, and especially a cultivated landscape, yields an increasingly high proportion of the sediment output from the catchment.

5.7.8 Integration

There is now immense scope for integrated multidisciplinary studies of lake watershed ecosystems using magnetic properties as a basis for rapid core logging and correlation as well as for sediment source identification, using sediment chemistry as a key to the type and nature of the material being derived from the catchment at different times, using both long- and short-term radiometric dating techniques alongside palaeomagnetic dating for chronological control, and using pollen analytical and related biostratigraphic studies for reconstructing the major palaeoecological variations associated with changes in material flux at the time. Such studies, coming right up to the present day, should help to provide the continuum of insight into geomorphological processes that only depositional environments have recorded.

---

values. The low peak (B) used to link together the whole correlation scheme is characterised by exceptionally high $\chi/SIRM$ ratios. Thus, qualitative magnetic parameters help to confirm the correlation scheme initially derived from rapid whole-core scanning. Peak A results from a forest fire in 1951. Peak B is not yet dated precisely but is probably developed in sediment some 1800–1300 years old. (Taken from Bloemendal *et al.* 1979).

5.8
Raised shorelines

5.8.1 Introduction

Shorelines are isochronous surfaces formed at the interface between land, atmosphere and ocean or lake. They represent morphostratigraphic units that can, in a displaced form, be used to determine mass transfer of water (eustasy) or crust (isostasy), with a resolution determined by the rate of displacement. Because such mass transfers are environmentally controlled, a relationship can exist between specific shoreline types and climatic events. For instance, eustatic raised shorelines are related to interglacial climatic optima, and isostatic marine limits are most commonly related to maximum submergence immediately following glacier wastage. As morphostratigraphic units, displaced shorelines can be related to environmentally sensitive phenomena such as palaeosols or glacial, fluvial, aeolian and periglacial landforms and sediments, so that complex stratigraphic sequences can be constructed, and displaced shorelines can provide a basis for regional correlation. Correlation has been aided by graphical and numerical methods, and shoreline ages can be inferred from stratigraphic position or obtained directly from radiometric dates.

5.8.2 Methods of measuring displaced shorelines

5.8.2.1 Geomorphological mapping

Although regionally extensive surveys, supported by radiometric dates, can provide a general interpretation of shoreline history (Andrews 1985), comprehensive geomorphological mapping should be the first step in any study of raised shorelines. Without a reliable database, subsequent measurements and analyses can produce meaningless results, as emphasised by the early studies of Scottish raised beaches, which were based on single measurements on conspicuous landforms, and produced the well known, but totally spurious, 100 ft, 50 ft and 25 ft raised beach sequence (Sissons 1962). Geomorphological mapping of raised shorelines and associated features is most efficiently carried out by stereoscopic examination of air photographs at scales between 1:10 000 and 1:25 000 followed by field mapping on at least 1:10 000 scale base maps to check and complete the details shown by air photograph interpretation (Gray 1975, 1983, Rose & Synge 1979). In the process of geomorphological mapping, particular attention should be given to form and size and to the relationships between different shorelines and between shorelines and other landforms such as deltas. The material into which erosional features are cut and of which depositional features are composed should be noted (Fig. 5.27). This process is time-consuming, but essential.

In glaciated areas, particular attention should be given to the identification of the marine limit, which is the highest level reached by the sea after deglaciation. In addition to forming readily recognisable landforms, it can also be identified by the upper limit of washed rock, the upper limit of marine-worked superficial deposits, or the lowest level of unstable sediments such as perched blocks (Synge 1969).

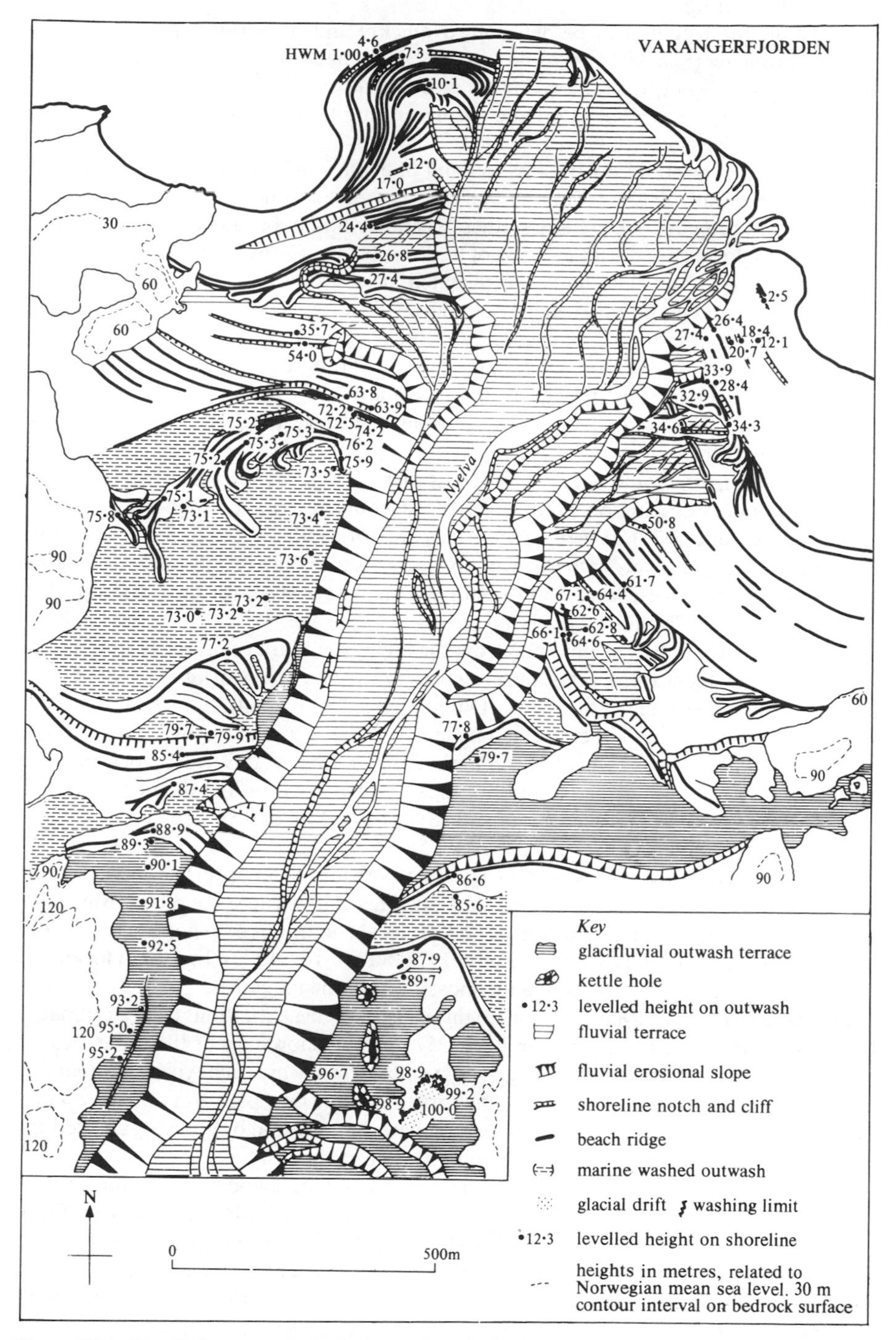

Figure 5.27 Detailed geomorphological mapping of raised shorelines and associated river terraces, Nyelv, south Varangerfjord, north Norway. (After Rose & Synge 1979).

5.8.2.2 Heighting

Raised shoreline heights can be determined quickly and by a single person with an aneroid barometer (Sparks 1949, Sollid *et al.* 1973) or a hand-held level such as an Abney (Sollid *et al.* 1973), but, with the exception of use in reconnaisance or general studies in areas with a wide range of shoreline displacement, these implements create more problems than they solve (Gray 1975, 1983).

The surveyor's level, used with a tripod and Sopwith staff, provides heights that are reliable and accurate (Sissons 1967), and results determined by this method have been the basis of recent progress in shoreline studies. Readings should be taken to 0.01 m and all traverses should be closed with an error of less than 0.15 m (Gray 1975, 1983). Traverses with errors larger than this should be repeated, a task that can be minimised by the creation of frequent temporary benchmarks during the original survey, so that in the case of a resurvey only these need be checked and the section of the traverse with the error located. Although heights can also be obtained by photogrammetry and tacheometry, particularly in remote areas with a wide height range, the predominantly linear distribution of raised shorelines makes levelling the most suitable of all techniques, despite the need for two operators.

Ideally, all heights should be related to benchmarks and so be expressed in terms of the national datum. Comparison between different countries can be made by a simple correction (Jardine 1976). Where benchmarks are not available, mean sea level can be used as a datum by measuring the level of high water under calm conditions on a number of days and relating these levels to those for the same days given for the nearest port in the Admiralty Tide Tables (Synge 1969).

Temporary datums can be obtained from the highwater mark of ordinary spring tides (McDann 1965), the upper limit of the growth of seaweed, *Fusus vesiculosus* (Sollid *et al.* 1973) and the upper limit of barnacles (Donner 1963, Lewis 1964), but ultimately these should be converted to either mean sea level or the national datum (Gray 1983). Connection with a region where benchmarks exist can be established by marking the level of high water on a calm day in both areas, either by recording the upper limit of fresh, drifted sea salad (*Enteromorpha* sp.) (Synge 1969) or by two people marking its position on some convenient rocks (Gray 1975).

Problems exist in the selection of a levelling position on appropriate landforms in order to represent mean sea level at the time of formation, because the relationship with mean sea level varies with landform type and the local conditions of exposure, fetch, tidal range and parent material (Møller & Sollid 1972, Sollid *et al.* 1973). In certain localities, such as the firths of eastern Scotland, most measurements can be made on a single landform (Sissons *et al.* 1966), but, where this is not possible, variations can be estimated by making similar measurements on currently forming shorelines at the same locality (Rose 1978). In practical terms, it is most convenient to measure each type of shoreline at an easily located position, such as the inner break of slope on a platform or the crest of a beach ridge (Table 5.3), and to attempt to eliminate the variation by applying corrections derived from the measurements on currently forming features.

The effects of post-formational modification should be avoided, as in the case of talus from cliffs (Gray 1975), or measured, as in the case of burial beneath peat (Gray 1974) or younger marine sediments (Sissons 1966, 1969). The thickness of peat can be determined by metal rods, and fine-grained marine sediments can be penetrated with hand-operated augers such as the Hiller type. In both cases the form of the buried shoreline must be determined by a series of probes before the appropriate position can be located.

Along the shorelines, heights should be taken at 50–100 m intervals according to local conditions.

Table 5.3 Summary of levelling positions and some indication of measured deviations from mean sea level.

Landform	*Levelling position*	*Deviation from mean sea level** (m)			
		Lofoten, etc‡	*Finnmark* §	*Varangerfjord* **	*Bugöyfjord* ‖
gravel beach	inner break of slope	0.0 ± 0.5			
sand beach	inner break of slope	+ 1.8 ± 0.4	+2.0 ± 0.4	+ 1.5 + 0.5	+ 1.8 ± 0.2
rock platform	inner break of slope	+ 1.5 ± 0.5	+ 1.4 ± 0.6		+ 1.3 ± 0.1
clay platform	highest elevation				+0.7 ± 0.6
beach ridge	crest	+2.5 ± 1.3	+3.9 ± 1.4	+ 1.8 ± 0.8	+2.2 ± 0.1
delta	lowest position on subaerial surface	− 1.5 †		+0.6 ± 0.6	−0.7 ± 0.3
beach ridge around delta	crest				+0.7 ± 0.1
ice-pushed boulders	base of boulders				0.0 ± 0.3

* Mean ± one standard deviation.
† One height only.
‡ Møller and Sollid (1972).
§ Sollid *et al.* (1973).
** (Rose & Synge 1979)
‖ Rose (1978).

5.8.3 Methods of dating and correlating displaced shorelines

5.8.3.1 Morphostratigraphic methods

Elevation is the main morphological criterion for separating different shorelines according to age, and because a relative rise of sea level leads to burial of existing shorelines, this relationship applies primarily to regressive sequences and is the basis of the staircase constraint (Cullingford 1972). The only typical case where a relative rise of sea level can be determined geomorphologically is where a delta terrace is submerged and only partially buried by beach ridges (Rose & Synge 1979).

Displaced shoreline landforms are most commonly correlated on the basis of size and continuity. It is these morphological properties that form the basis for identifying the Main Lateglacial and Perth Main shorelines in Scotland (Sissons 1974, Sissons & Smith 1965) and the 'Main Line' and Main Post-Glacial (Tapes) shorelines in Norway (Marthinussen 1962). Because some morphological properties such as ice-pushed boulder rims in Norway (Synge 1969) or frost-fractured pebbles in Mallorca (Butzer & Cuerda 1962) have a climatic significance, these criteria have also been used as a basis of correlation between sites and to give a general estimate of shoreline age.

Where displaced shorelines are developed in association with deglaciation, their distribution is restricted to the contemporaneous ice-free areas. The result is that, as ice cover diminishes, progressively younger shorelines may cover more extensive areas (Fig. 5.28). This ice-margin constraint (Cullingford 1972) therefore provides a useful morphostratigraphic control. Because certain shorelines may be associated with glacier margins, either by their distribution or by their relationship with contemporaneous outwash, this property has frequently formed a basis for correlation (Donner 1969, Sissons *et al.* 1966, Andrews 1970b). However, this method is subject to misinterpretation if shorelines are destroyed by glacial readvance such as applied to some of the Late-glacial shorelines during the Loch Lomond readvance in Scotland.

5.8.3.2 Lithostratigraphic methods

Stratigraphic sequences, consisting of alternating terrestrial and shoreline sediments, or littoral and neritic sediments, can provide evidence for water-level changes, and in particular for transgressions. In a continuous sedimentary sequence, the change from marine to terrestrial or vice versa is known as the index point (Tooley 1978a) (Fig. 5.29 &

5.33). However, attention has been drawn to the problems of associating changing sediment sequences (for instance, terrestrial to littoral to inner, middle and outer neritic facies, or vice versa) with changes of sea level, rather than changes of water depth brought about by erosion or sedimentation (Tooley 1982). For this reason the terms 'transgressive' and 'regressive' overlap have been introduced to describe the sedimentary sequence, and the term 'tendencies of sea-level movement' is used to refer to the changes of sea level when these can be identified from additional independent evidence (Tooley 1982, Shennan 1983, Tooley & Shennan 1987).

Most typically, and well illustrated by the succession of beach sand, terrestrial peat and estuarine clay in the Forth lowlands of Scotland (Jamieson 1865, Sissons *et al.* 1966), shoreline deposits overlain by peat indicate a regressive tendency, and peat overlain by marine clay represents a transgressive tendency. In other environments, such as the Mediterranean region, the alternation of beach and aeolian sands provides evidence of similar changes of sea level. Inevitably, the occurrence of unconformities complicates the interpretations, and the identification of a continuous succession is of prime importance. For this reason, the evidence is best obtained from low-energy environments such as tidal flats and lagoons, and successions from headland regions or tidal mouths should, at best, be considered suspect (Tooley 1978a, 1982).

Individual lithological properties, such as mineralogically distinctive beach sand or lithologically distinctive pumice (Boulton & Rhodes 1974), can be used to identify particular shoreline sediments and hence provide a criterion for correlation between sites. Pumice is particularly useful in this context because it can be related on lithological properties to dated volcanic events and, because of its capacity to float, it can be spread rapidly over very wide areas such as the North Atlantic and Arctic Oceans (Binns 1967).

When lithological properties can be related to some climatically controlled process, age can be inferred by relation to an established, climatically based stratigraphy. In high

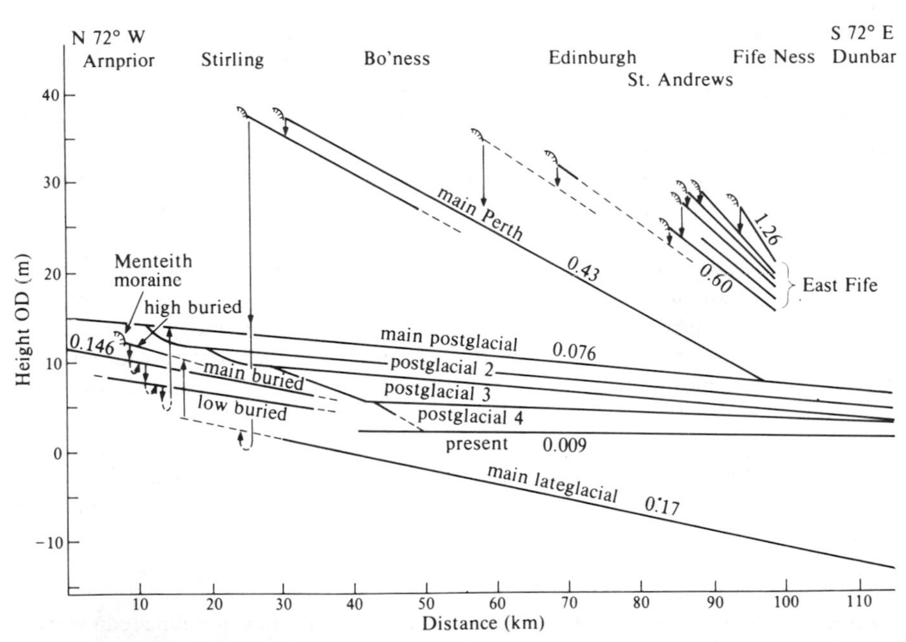

Figure 5.28 Height–distance diagram showing the shoreline sequence in south-east Scotland and the relationship between late-glacial shorelines and contemporaneous ice margins. Figures on the shorelines are the gradients (m km^{-1}). (After Sissons 1974).

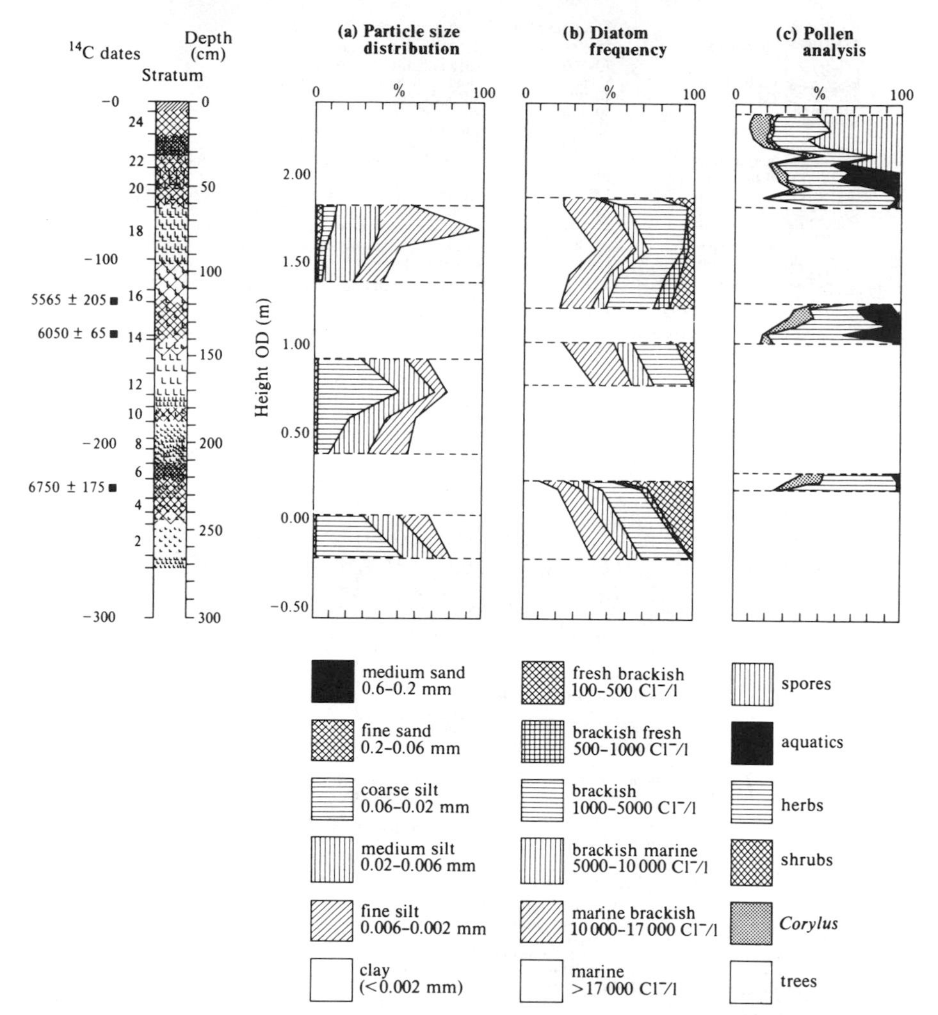

Figure 5.29 Lithological, diatom and pollen assemblages associated with sea-level change in a tidal flat/lagoonal sequence from Downholland Moss, west Lancashire. (After Tooley 1978b).

latitudes, the occurrence of ice-rafted debris has been taken as evidence for formation in a period of relative climatic severity (Synge 1969); and in the Mediterranean region, the occurrence of various types of palaeosol, developed on wind-blown sands, has been used to infer eustatic sea-level changes (Butzer 1975). Thus a *terra rossa* is associated with a full interglacial climate, a *terra fusca* is associated with an interstadial climate and unweathered aeolianite indicates maximum climatic deterioration, so that, in a 'littoral sedimentary cycle' consisting of aeolianite–*terra fusca*–aeolianite–*terra rossa*, beach sand would therefore be interpreted as indicating an interstadial transgression followed by a stadial regression and terminated by an interglacial transgression (Fig. 5.30).

5.8.3.3 Biostratigraphic methods

Biological remains can be used in shoreline studies to provide evidence of contemporaneous water depth, palaeoenvironments with particular reference to change from terrestrial to marine or lacustrine conditions, and a stratigraphic framework for

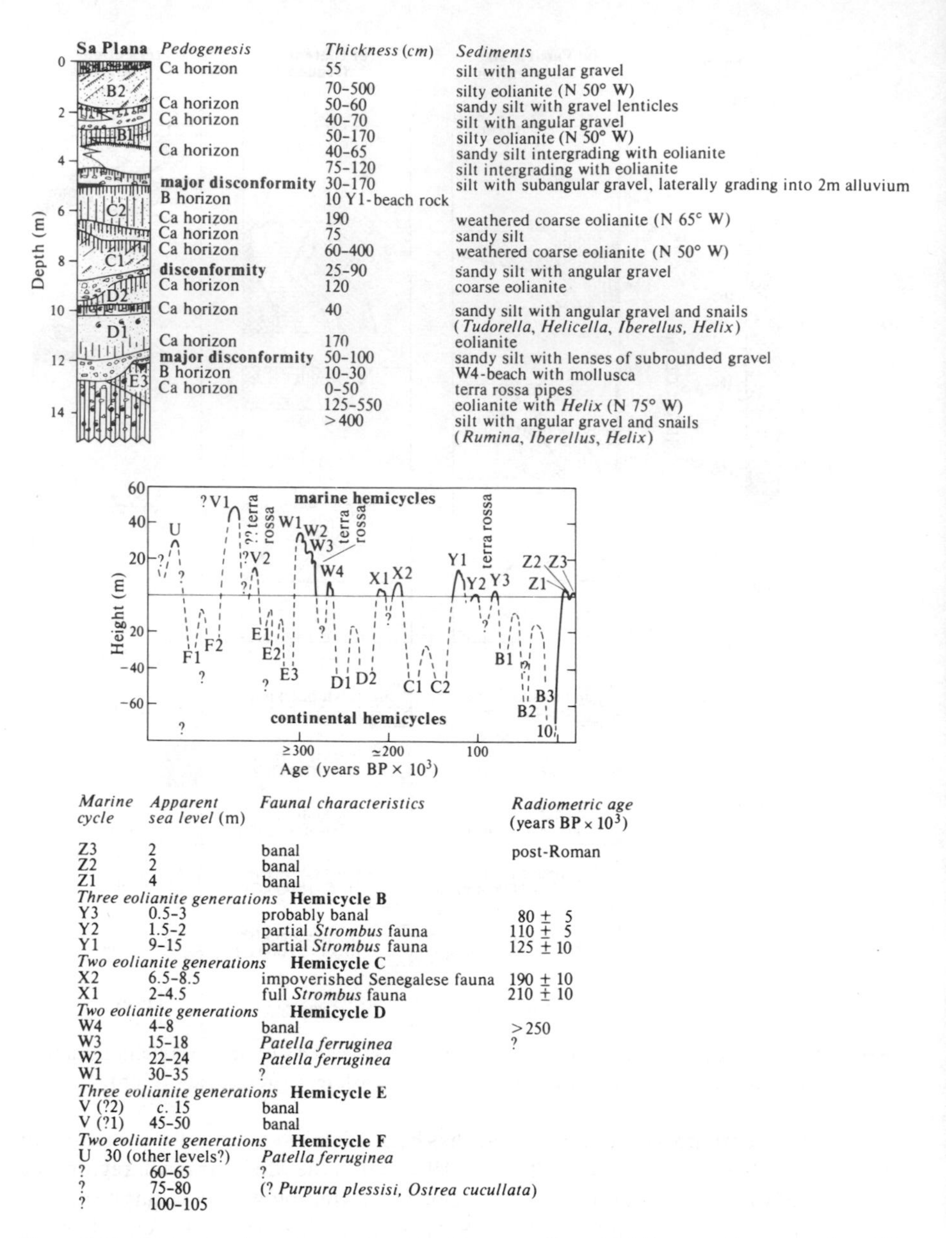

Pedogenesis	Thickness (cm)	Sediments
Ca horizon	55	silt with angular gravel
	70–500	silty eolianite (N 50° W)
Ca horizon	50–60	sandy silt with gravel lenticles
Ca horizon	40–70	silt with angular gravel
	50–170	silty eolianite (N 50° W)
Ca horizon	40–65	sandy silt intergrading with eolianite
	75–120	silt intergrading with eolianite
major disconformity	30–170	silt with subangular gravel, laterally grading into 2m alluvium
B horizon	10 Y1-beach rock	
Ca horizon	190	weathered coarse eolianite (N 65° W)
Ca horizon	75	sandy silt
Ca horizon	60–400	weathered coarse eolianite (N 50° W)
disconformity	25–90	sandy silt with angular gravel
Ca horizon	120	coarse eolianite
Ca horizon	40	sandy silt with angular gravel and snails (*Tudorella, Helicella, Iberellus, Helix*)
Ca horizon	170	eolianite
major disconformity	50–100	sandy silt with lenses of subrounded gravel
B horizon	10–30	W4-beach with mollusca
Ca horizon	0–50	terra rossa pipes
	125–550	eolianite with *Helix* (N 75° W)
	>400	silt with angular gravel and snails (*Rumina, Iberellus, Helix*)

Marine cycle	*Apparent sea level* (m)	*Faunal characteristics*	*Radiometric age* (years BP × 10³)
Z3	2	banal	post-Roman
Z2	2	banal	
Z1	4	banal	
Three eolianite generations		**Hemicycle B**	
Y3	0.5–3	probably banal	80 ± 5
Y2	1.5–2	partial *Strombus* fauna	110 ± 5
Y1	9–15	partial *Strombus* fauna	125 ± 10
Two eolianite generations		**Hemicycle C**	
X2	6.5–8.5	impoverished Senegalese fauna	190 ± 10
X1	2–4.5	full *Strombus* fauna	210 ± 10
Two eolianite generations		**Hemicycle D**	
W4	4–8	banal	>250
W3	15–18	*Patella ferruginea*	?
W2	22–24	*Patella ferruginea*	
W1	30–35	?	
Three eolianite generations		**Hemicycle E**	
V (?2)	c. 15	banal	
V (?1)	45–50	banal	
Two eolianite generations		**Hemicycle F**	
U	30 (other levels?)	*Patella ferruginea*	
?	60–65	?	
?	75–80	(? *Purpura plessisi, Ostrea cucullata*)	
?	100–105		

Figure 5.30 Littoral sedimentary cycles at Sa Plana, south-east Mallorca, with summaries of the sequences of relative sea-level change indicating diagnostic faunal characteristics and $^{230}Th/^{234}U$ dates derived from molluscs. (After Butzer 1975).

correlation either between sites or with formal, established sequences. With all biostratigraphical methods, standard analytical procedures have been established and should, wherever possible, be adhered to when these methods are applied to shoreline studies.

The classical studies of Mollusca (Brogger 1900) and Foraminifera (Feyling-Hanssen 1964) from marine clays in Oslofjord provide evidence of habitat in addition to climatic affinity and therefore give a general estimate of water depth when sediments accumulated on the sea bottom. This type of information is especially useful when shoreline landforms and sediments have been eroded and the only evidence available survives in the form of bottom sediments deposited some distance from the contemporaneous coastline. This is well illustrated by the work of Peacock *et al.* (1977), who, in an integrated study of Mollusca, Foraminifera and Ostracoda, have reconstructed a detailed pattern of sea-level change from a truncated late-glacial succession in western Scotland.

This approach has been applied extensively to the identification of sea-level changes within the early and middle Pleistocene marine 'crags' of East Anglia. Using death assemblages of Mollusca collected from these deposits, Norton (1967, 1970) has drawn analogies with present-day ecotones and made inferences about the life environment and deposition environment. Thus, from a given stratigraphic sequence, sea-level changes can be determined and transgressions have been associated with the Antian and Pastonian temperate stages (West & Norton 1974, West *et al.* 1980) and with the upper part of the Beestonian cold stage (West 1980).

Biological environmental indicators provide the most critical evidence for changes from terrestrial to marine conditions and vice versa. In north-west England, diatom assemblages from tidal flat and lagoonal deposits have been used to identify marine, brackish-water and freshwater environments (Tooley 1978a), and in Scandinavia the same technique has been used to determine the point in a sedimentary sequence from a lake basin at which sea level fell below the lowest col of that basin (Fig. 5.31) (Hafsten 1983). Dinoflagellate cysts (previously called 'hysterix' in Scandinavia and incorrectly used as marine indicators (Anundsen 1978)) can also be used to indicate marine, brackish-water and freshwater environments (Hafsten 1983, Anundsen & Fjeldskaar 1983). Likewise, changes in the frequency of pollen of open-habitat taxa such as Chenopodiaceae, *Plantago maritima, Armeria maritima* and *Ruppia* have been used to show the relative proximity of marine conditions (Tooley 1978a). Pollen spectra have been used for this purpose extensively throughout the Pleistocene, and the technique is clearly demonstrated in the middle Pleistocene of north Norfolk (West 1980), the middle and upper Pleistocene of south-east England (West 1972) and the Holocene of the Netherlands (Jelgersma 1961).

By far the commonest use of biostratigraphy is for correlation, either locally or regionally, using informal or formally established biozones (Hedberg 1976). Mollusca, Foraminifera, Ostracoda and pollen are all used informally, but pollen is used most widely and is the basis of the formal terrestrial biozones into which the Pleistocene has been traditionally subdivided (Turner & West 1968). In Norway, Mollusca from marine clays from Oslofjord (Brogger 1900) to Nordland (Andersen 1975) form the basis of assemblage biozones characterised by such species as *Portlandia arctica* and *Tapes decussatus*. In Mallorca, mollusc assemblage biozones appear to form the only sound basis for differentiating a most complex association of raised beaches (Butzer 1977), with *Patella ferruginea* and *Strombus bubonius* being the diagnostic species.

Both the Norwegian and Mallorcan mollusc biozones can be related to established, climatically defined stratigraphies. In Norway, the *Portlandia* fauna is related to the late-glacial climatic deterioration of Younger Dryas age and the *Tapes* fauna is associated with the Holocene thermal optimum. In Mallorca, the *Strombus* fauna is associated with climatic conditions similar to those off Senegal at the present time and hence considered to be formed during the last two interglacials (Butzer 1975, Hearty *et al.* 1986) (Fig. 5.30). However, pollen analysis provides the most effective biostratigraphic method for linking shoreline events with an established stratigraphy. Although the interpretation of pollen spectra from marine sediments is made difficult by the possibility of reworking, overrepresentation by differential transport rates and the separation of the waterborne and airborne components of the pollen influx, results show a general

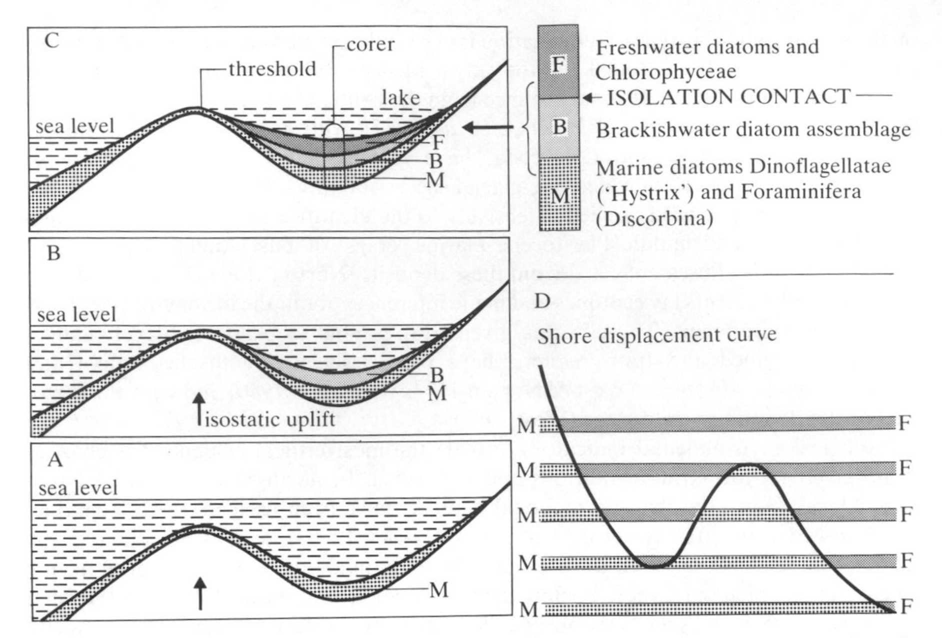

Figure 5.31 Sedimentary sequence formed during a progressive fall of sea level and isolation of a basin as a freshwater lake. (After Hafsten 1983).

consistency (West & Norton 1974, Turon 1984), and most pollen spectra are obtained from associated terrestrial deposits (West 1972, Tooley 1978a). Pollen analysis has been used successfully for past interglacials, where it is often the only method of deriving an age (West 1972, Hollin 1977), and most effectively with sea-level changes following the last glaciation where the chronological resolution is increased by the relatively short duration of the pollen assemblage biozones and the likelihood of analysed material lying close to a known zonal boundary. This has been demonstrated most effectively in Scandinavia (Hafsten 1983), central Scotland (Brooks 1972, Sissons 1972) and north-west England (Tooley 1974, 1978a).

5.8.3.4 *Methods of direct dating*

Shorelines can be dated directly from the equilibration of $^{230}Th/^{234}U$ and the decay of ^{14}C or by floating timescales such as amino-acid diagenesis, glacial varves and palaeomagnetism, which must subsequently be linked to radiometric scales.

Ratios of ^{230}Th:^{234}U have been used most effectively for dating corals (Thurber *et al.* 1965) in areas such as Barbados (Mesolella *et al.* 1969) and the Huon Peninsula, New Guinea (Bloom *et al.* 1974) (Fig. 5.32), and a general review of $^{230}Th/^{234}U$ ages for the last interglacial is provided by Kaufman (1986). The accuracy of the results is dependent upon a correct estimation of initial ^{230}Th:^{238}U ratios in the sample and the assumption that the coral–ocean system remains closed after death. Application of this method to calcitic and aragonitic molluscs has proved less successful, largely due to diagenesis and the relative increases of ^{234}U and ^{238}U. However, a series of dates obtained from molluscs from Mallorcan raised beaches appear to be reliable (Fig. 5.30), apparently due to cementation of the specimens in beach rock and the consequent reduction in diagenetic alteration (Butzer 1975).

Radiocarbon dates can be obtained either from marine shells or from terrestrial biogenic materials associated with shoreline stratigraphy. Errors are introduced into ^{14}C

determinations on marine shells due to an enrichment of ^{14}C caused by isotopic fractionation between carbon dioxide and ocean carbonate, and a dilution caused by variable mixing of the oceans. If isotopic fractionation is corrected to $^{13}C = -0‰$, 410 yr should be subtracted from the derived age (Mangerud & Gulliksen 1975, Sutherland 1983, 1986). Problems also arise due to the comparison of dates from different fauna, as differences of up to 2000 yr have been recorded from individual species of a mixed population derived from different shoreline habitats (Donner *et al.* 1977).

Radiocarbon dates derived from terrestrial material are particularly vulnerable to hard-water error caused by the differential isotopic fractionation of 'old' carbon derived from calcium carbonate dissolved in rivers or lakes (Shotton 1972, Lowe *et al.* 1988). Dates have been shown to be up to 1700 yr too old, and for this reason terrestrial plants provide the most suitable material.

Diagenesis of protein has been used to obtain relative ages of marine Mollusca (Wehmiller 1982). In particular, the extent of isoleucine epimerisation (aIle: Ile ratio) has been used to identify groups of fossils that lived around the same time. The rate of diagenesis is dependent on species, temperature and age (Hare 1974), and dating must be obtained by calibration with other dating methods such as ^{14}C or U-series. Thus, investigations using this method must begin by analysing selected species from a restricted geographical province, although, with careful stratigraphic control, correlation may be extended across a wider area (Hearty *et al.* 1986, Miller & Mangerud 1985). Partly because epimerisation rates are slower in cold regions, early applications of the technique were concerned with shoreline problems in high latitudes such as Arctic Canada (Andrews & Miller 1976). More recent developments have enabled the differentiation and correlation of shorelines of different ages in south-west Britain (Bowen *et al.*

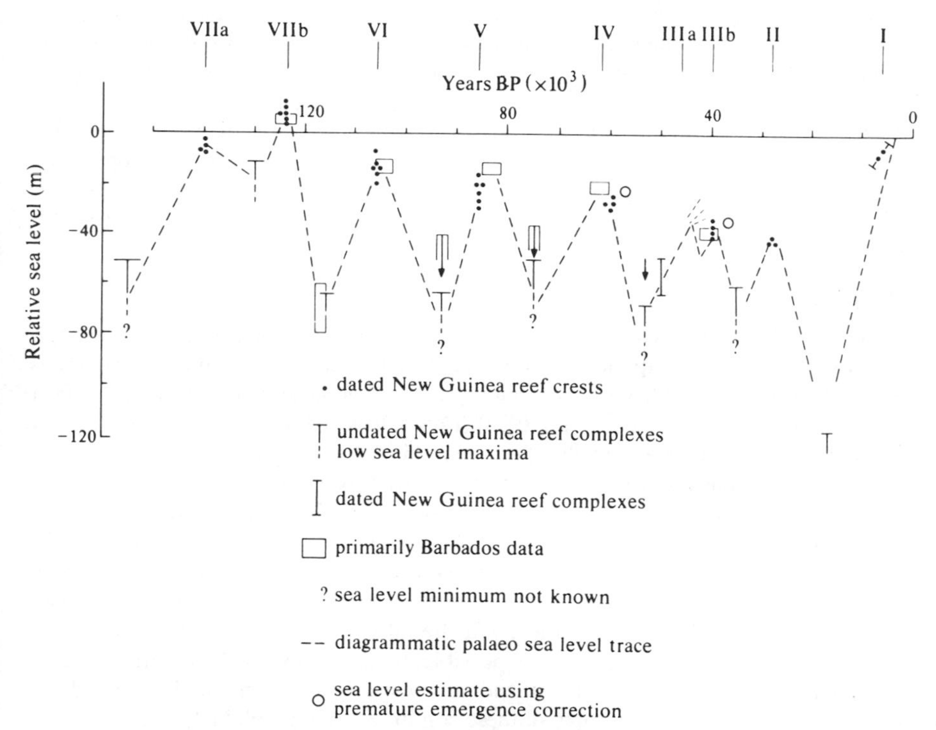

Figure 5.32 Pattern of sea-level change based on $^{230}Th/^{234}U$ dates of coral reefs from New Guinea and Barbados. (After Bloom *et al.* 1974).

1985) and across northern Europe (Miller & Mangerud 1986). Although this technique is still in the early stages of development, its application to relatively abundant fossil material and its wide age range makes it especially useful for shoreline studies.

Palaeomagnetic relative declination, inclination, intensity and susceptibility show synchronous changes through time that can be linked to radiometric timescales (Thompson 1973b). As the technique is best applied to relatively rapidly deposited fine-grained sediments, it can be a most effective method of dating changes in water level if these changes are indicated by other types of evidence. Additionally, palaeomagnetism is capable of indicating changes from marine to lacustrine conditions, as intensity and susceptibility have very low values in marine sediments, which contrast with the normal values for the lacustrine deposits. This has been demonstrated for the Holocene transgression in Loch Lomond (Dickson *et al.* 1978).

Glacial varves are distinctive annual increments of glacio-lacustrine sediments, and as such provide the longest-known method of directly dating lake shorelines (de Geer 1940). Linkage is established between the bottom sediment and the contemporaneous shoreline on the basis of ice-marginal positions, and the technique has proved very effective in deriving a floating late-glacial chronology in the Baltic region (Sauramo 1923). However, the method suffers from the problem of relating the floating timescale to radiometric chronologies (Fromm 1970, Stromberg 1983).

5.8.4 Methods of representing results

Many results of shoreline studies can be expressed numerically and graphically with the aim of representing (a) changes of level over an interval of time, (b) changes of level in different places at the same point in time and (c) rates of change, with the emphasis on graphical and numerical prediction of relationships between shorelines. Without exception, the quality of any representation is dependent upon the quality of the data used, and it is essential to appreciate that results illustrated below are the product of considerable time and effort accurately collecting the data by the methods described in the preceding sections. Each representation can be an end in itself, but can equally contribute to further analysis and a greater understanding of the primary data and their underlying assumptions (Andrews 1974).

5.8.4.1 Shoreline displacement curve

This method shows changes of shoreline level plotted against time. In its simplest and most successful form, it is applied to data from a small area, as the shape of the curve is dependent upon the stability of the region. A predominantly stable area will reflect eustatic changes (Butzer 1975, Tooley 1974) (Figs 5.30 & 5.33), whereas an isostatically or tectonically deformed region will reflect a combination of land and sea-level changes (Hafsten 1983, Sissons 1962) (Fig. 5.34). In regions such as Arctic Canada where isostatic rebound is dominant, the shoreline level can be plotted against age since deglaciation to give a postglacial uplift curve (Andrews 1970b) (Fig. 5.35). Ideally, the data for the shoreline displacement curve should be collected from a small, homogeneous area, which is devoid of major tidal irregularities or tectonic movements and parallel to any isobases of isostatic or geoidic deformation. Ideally, it should also be collected from a depositional environment characterised, as far as possible, by continuous sedimentation so that index points can be identified and linked to biostratigraphy and radiometric dates (Tooley 1978a). It is only rarely that all these conditions can be fulfilled, and often such properties as isobase direction are only discovered as a result of the work. In the construction of the curves, the tidal range can be represented by lines for both mean high

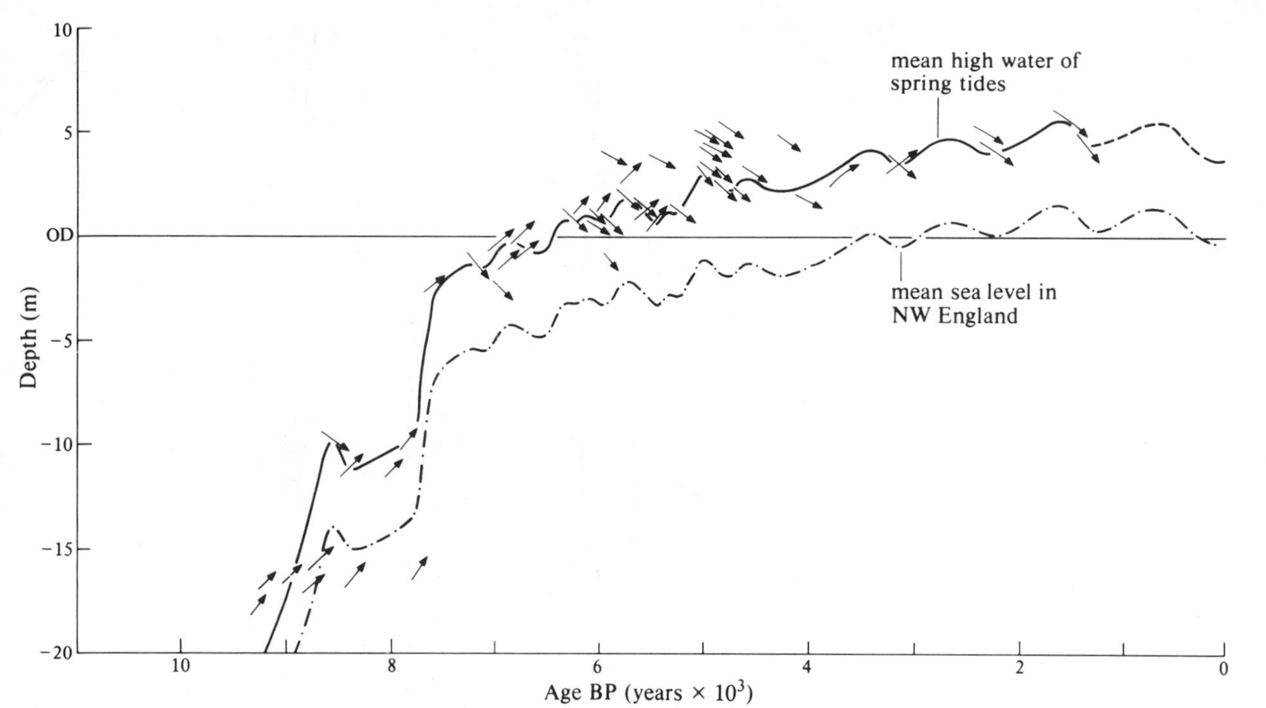

Figure 5.33 Relative sea-level curve for north-west England from 9200 to 0 yr BP. Each point on the graph is a radiocarbon-dated level, the relationship of which to its contemporary sea level has been established. (After Tooley 1974).

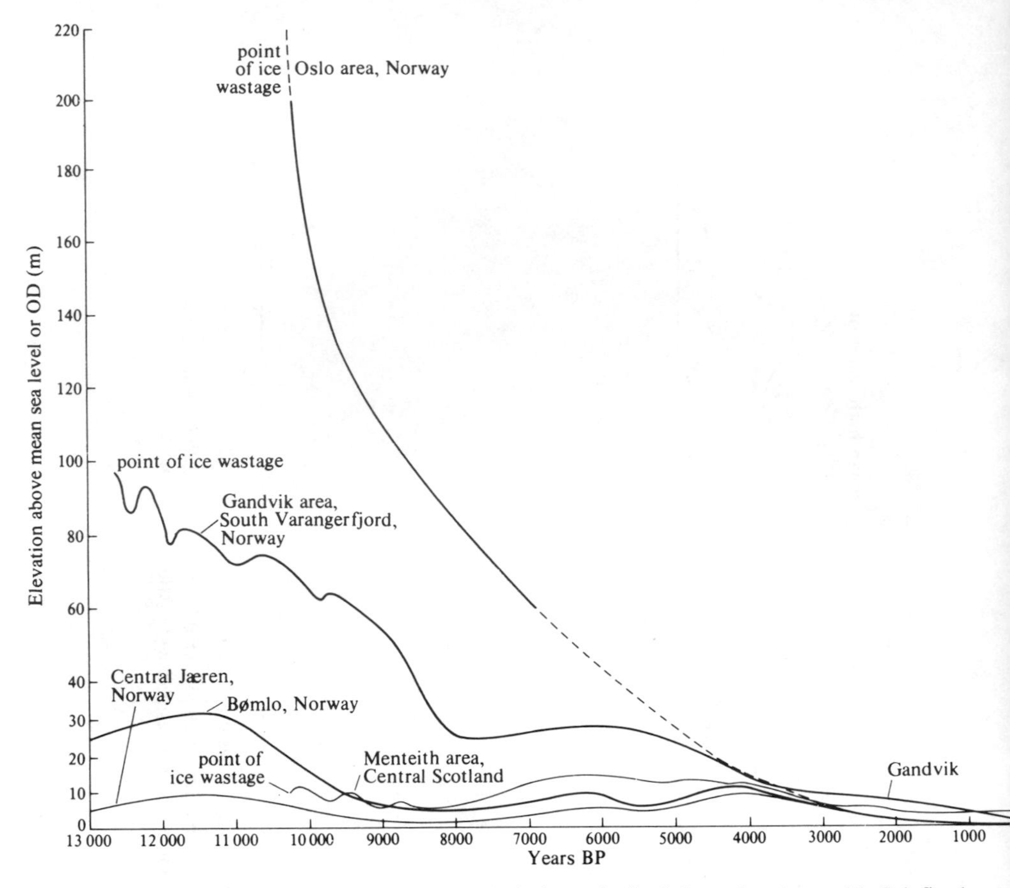

Figure 5.34 Relative sea-level curves from glacio-isostatically deformed regions, with Oslofjord showing the greatest rebound. The relatively high beaches between 10 300 and 8700 yr BP in central Scotland are due to renewed loading caused by the Loch Lomond readvance in the region in Younger Dryas time. (After Sissons 1967, Hafsten 1960, Rose 1978).

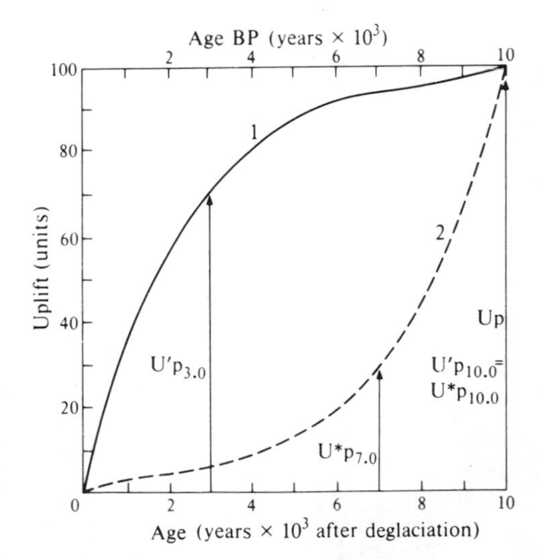

Figure 5.35 Postglacial uplift curves, with $U'p$ referring to the amount of uplift since deglaciation and $U^{*}p$ referring to uplift plotted in relation to the present day. (After Andrews 1969).

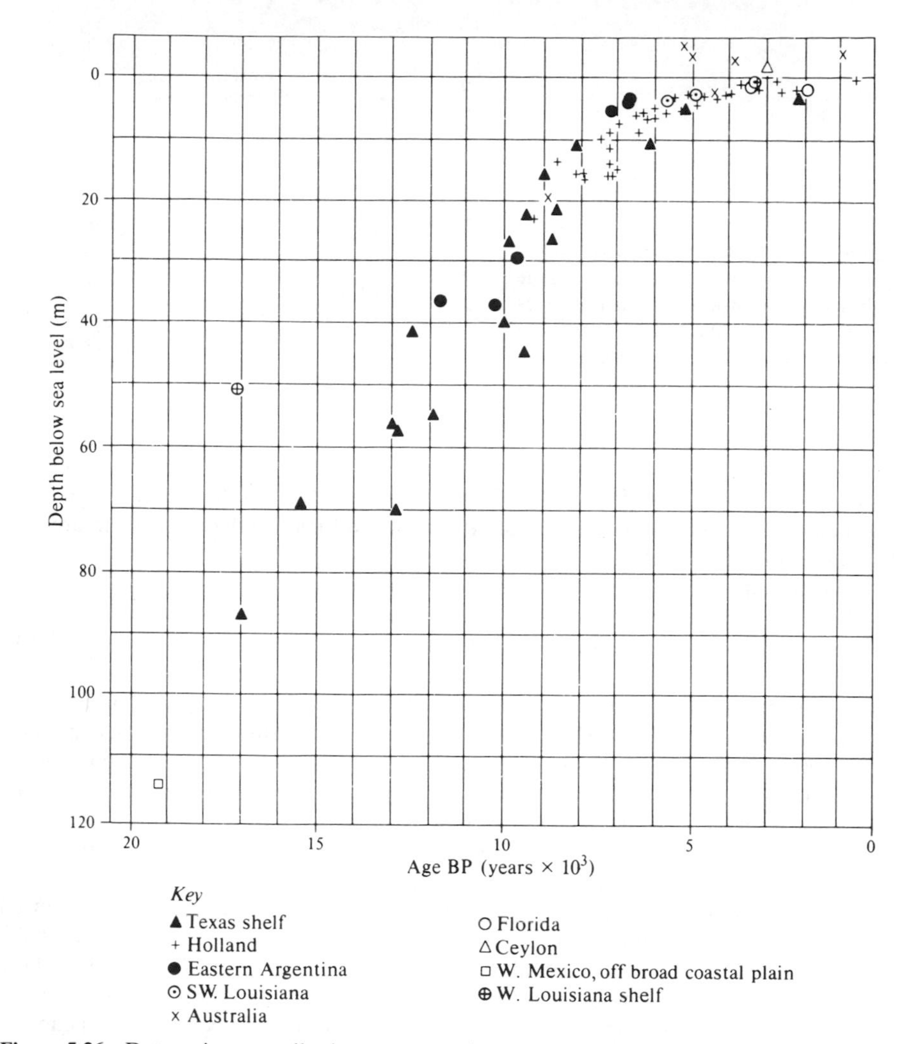

Figure 5.36 Data points contributing to a eustatic sea-level curve taken from a wide range of supposedly stable areas. (After Jelgersma 1966).

water of spring tides and mean sea level for the region (Fig. 5.33).

Shoreline displacement curves have also been constructed with data from a wide range of supposedly stable areas, on the assumption that the result gives a fair estimate of eustatic change (Fairbridge 1961, Shepard 1963, Milliman & Emery 1968) (Fig. 5.36). The accuracy of these results is restricted by the variety of palaeoenvironments used as index points.

In order to overcome the sampling restrictions imposed by the submerged or buried position of much data used in eustatic curve studies, Mörner (1971) constructed a eustatic curve with evidence collected from the glacially isostatically uplifted part of south-west Sweden. This region is normal to the isobases so that, unfortunately, eustatic level cannot be obtained without the prior determination of the isostatic component at a site. The result is that, until the isostatic factor can be satisfactorily predicted, results from such a region will remain suspect.

Uplifted data have also been used from the Huon Peninsula, New Guinea (Bloom *et al.*

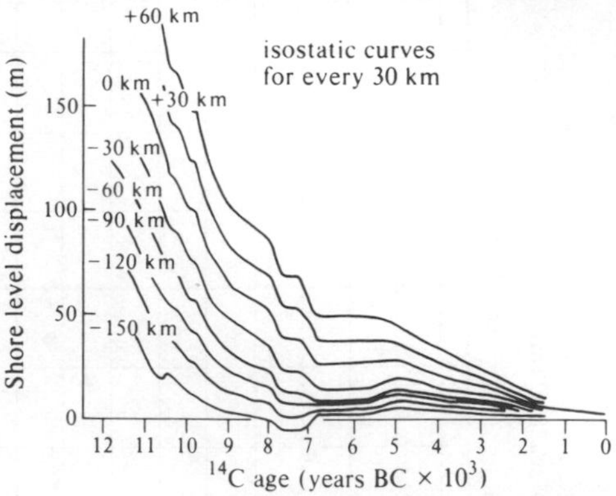

Figure 5.37 Isostatic rebound curves constructed for a transect drawn across the isobases in south-west Sweden. Curves relate to locations 30 km apart, with those closest to the centre of glacier loading showing the greatest rebound. (After Mörner 1971).

1974, Chappel & Vech 1978) where tectonically uplifted coral reef terraces have been dated using $^{230}Th/^{234}U$ dates (Fig. 5.32). However, the results depend upon the assumption that tectonic uplift is constant, which is difficult to demonstrate in areas of known tectonic instability.

Using the equation

$$I = S + E \tag{5.8}$$

where I is the isostatic rebound in relation to present sea level, S is the shoreline level (with + for above and − for below present sea level) and E is the eustatic level in relation to present sea level (with + for below and − for above sea level), it is possible to determine isostatic displacement curves for a locality through a given period of time. Typically, a series of displacement curves are produced for different localities located normal to the isobases, so that the different rates and patterns of isostatic deformation can be demonstrated (Mörner 1971) (Fig. 5.37).

Anundsen and Fjeldskaar (1983) have drawn attention to the importance of renewed crustal depression brought about by glacial readvance. This has been demonstrated for Weichselian late-glacial readvances in Scandinavia (Anundsen 1978, Roise 1978) and the late-Devensian Loch Lomond readvance in Scotland (Firth 1986). In this case the shoreline displacement curve takes on a sinuous form with periods of maximum submergence correlating with glacial readvance (Fig. 5.34, Gandvik area).

5.8.4.2 Height–distance diagrams and isobase maps

These methods are used to describe changes of level of given shorelines across a given area. Height–distance diagrams represent a projection plane, drawn normal to the isobase in the area concerned, on which the elevations of shoreline fragments are plotted. Distance is plotted on the x axis from an arbitrary point of origin and height is plotted on the y axis in terms of the appropriate datum (Fig. 5.38). This method is best used when shoreline fragments have a sound stratigraphic control, but it is usually used as a device for correlating shoreline fragments when stratigraphical control is poor and, at best, morphological (Sissons & Smith 1965, Synge 1969). Regression analysis has been applied to determine the best-fit line for a given set of data points (Cullingford & Smith 1966, Gray 1975), and goodness of fit, which can be expressed by a correlation coefficient, has been used as an objective procedure for correlating shoreline fragments where there is no stratigraphical control or the projection plane is not known (Andrews 1967). In regression analysis, the use of several levelled heights from a single shoreline fragment introduces problems of autocorrelation, which can give spuriously high levels

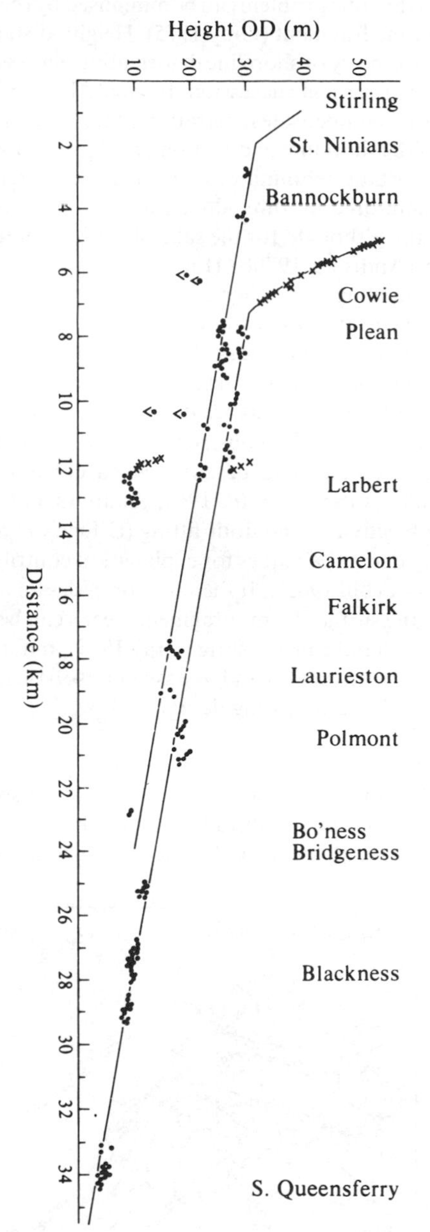

Figure 5.38 Height–distance diagram on the Perth beaches in central Scotland. Raised shoreline heights are projected onto a vertical plane aligned northwest–southeast, and the lines through the shoreline heights are calculated by regression analysis. (After Sissons & Smith 1965).

of significance, and it has been suggested that this problem can be minimised by the use of a single, mean height value for each shoreline fragment (Gray 1975). Height–distance diagrams are most effective in showing the geometry of shoreline deformation across a region and the distribution of shoreline levels in a vertical succession. It should be emphasised, however, that the statistical methods do not compensate for stratigraphical weaknesses, or any false assumptions about the slope of the shoreline or direction of the projection plane.

Isobase maps contour the plane of isostatic rebound of a given shoreline relative to present sea level. They represent, therefore, in three-dimensional form, the same properties as the height–distance diagram, although, for the sake of clarity, contours are usually drawn on only one shoreline (Andrews 1970b) (Fig. 5.39). Like the height–distance diagrams, their reliability depends upon stratigraphical control, but they are usually based on relatively weak morphostratigraphically defined data (Sissons 1967, Marthinussen 1960). Ideally, in order that the direction of projection planes can be determined, isobase maps should be constructed before height–distance diagrams, but most usually the reverse is the case, and the height–distance diagram provides the basis for correlating the shoreline fragments used in the construction of the isobase map (Sissons 1967, Smith *et al.* 1969). Isobase maps can be constructed either manually (Sissons 1967) or by trend surface analysis (Smith *et al.* 1969, Andrews 1970b). The latter technique has the advantage of objectivity in contour fitting (O'Leary *et al.* 1966) and evaluating the goodness of fit (using derived *F* values for explanation contributed by each surface component (Krumbein & Greybill 1965)). It therefore provides an objective method of describing the complexity of the surface form. Fragment means can be used to reduce the problem of autocorrelation in significance testing (Gray 1975), but, irrespective of the statistical devices used in their construction and assessment, isobase maps are only as good as the stratigraphical control and heighting data.

5.8.4.3 *Shoreline relation diagrams*

These are based on the assumption that the rate at which shorelines are displaced is predictable and that there is a direct ratio between the altitude of a shoreline, the speed of uplift and its age. The method is considered most applicable to glacio-isostatically

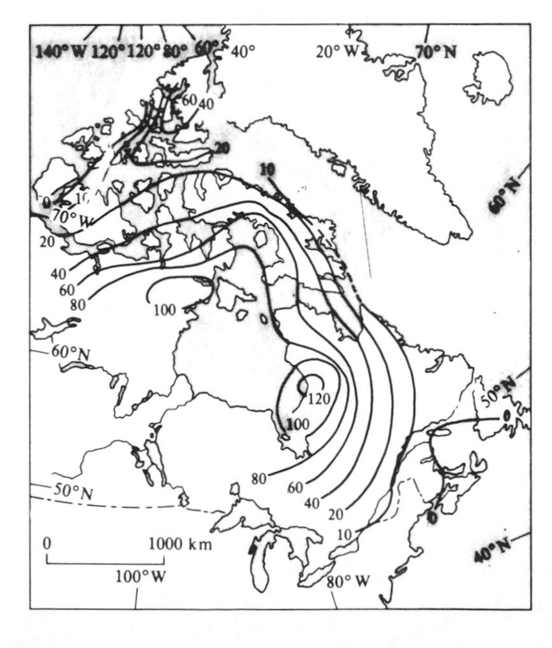

Figure 5.39 Isobase map for northern and eastern North America showing emergence since 6000 BP. (After Andrews 1970b).

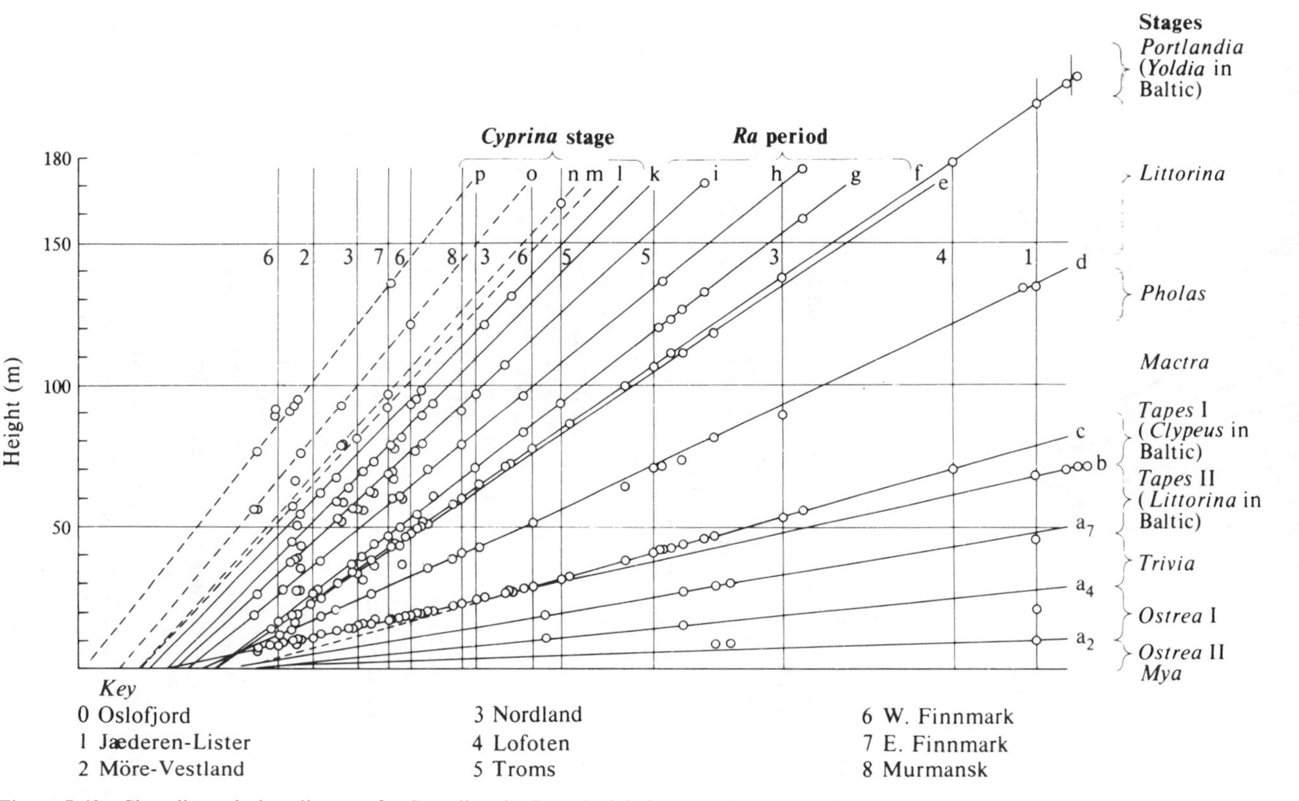

Figure 5.40 Shoreline relation diagram for Scandinavia. Postglacial shoreline b (Tapes II) and late-glacial shore zone e and f are used as reference levels. (After Tanner 1933).

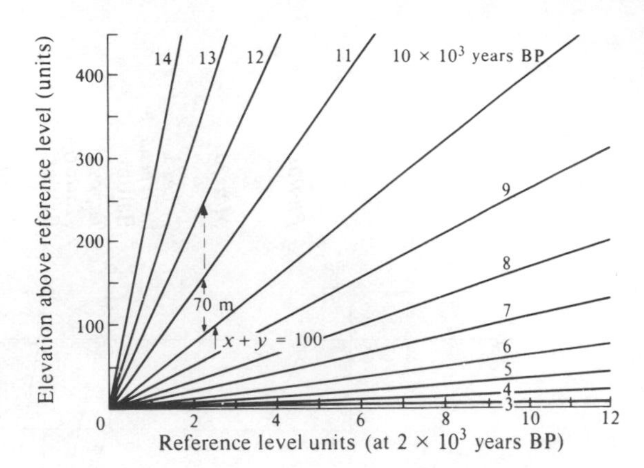

Figure 5.41 Shoreline relation diagram for Arctic Canada based on postglacial uplift. (After Andrews 1969).

uplifted areas where the amount of uplift is related to the thickness of ice and the time since deglaciation. Thus, given the rate of change at any point within an uplifted area, shoreline age can be determined for a measured elevation or shoreline elevation can be predicted for a required age.

As rate of change is proportional to the difference in elevation between two dated shorelines, relation diagrams consist of at least two reference levels plotted with age along the x axis and elevation above or below the reference level along the y axis (Gray 1983). The method was first developed in Scandinavia (Tanner 1933) where prominent shorelines were used as upper and lower reference levels. At this time, geochronometric dating was not available and estimations of age were therefore relative, and the method was used essentially for correlation between different shoreline fragments and with existing mollusc stratigraphies (Fig. 5.40). Using this technique, Tanner contended that for any given locality it would be possible to predict the relative age of a shoreline from its elevation relative to that of the reference level.

The technique has been developed further in Arctic Canada (Andrews 1969), with the lowest reference level plotted, in time units, as the x axis and the elevation above the reference level plotted on the y axis. In this case, the rates of change are derived from a large number of radiocarbon-dated, postglacial uplift curves (Fig. 5.34) explained by the equation

$$U^*p_t = A \sum_{t=0}^{t=n-1} i^t \tag{5.9}$$

where U^*p_t is uplift in relation to the present day, A is the amount of rebound between $t = 0$ and $t = 1 \times 10^3$ yr, t is time in increments of 10^3 yr, i is a constant, equal to 1.521 for Arctic Canada, and n is the length of postglacial recovery in 10^3 yr.

Thus a succession of reference levels can be constructed for any age required, and Andrews (1969) has drawn appropriate lines for each 1000 yr (Fig. 5.41). Thus, given the elevation of a reference level, it is possible to predict from the nomograph the age of the unknown shorelines, and most particularly the age of the marine limit, which, in much of Arctic Canada, coincides with deglaciation.

There are several problems associated with the use of relation diagrams (see Gray 1983). First, reference levels should be corrected for eustatic sea-level movement by subtraction of the appropriate eustatic value (Andrews 1970b). Secondly, and most fundamentally, the rate of change must be a smoothly decelerating function, and, if this is interrupted by renewed loading in a readvance, such as occurred in north-west Europe, the method cannot be applied to shorelines that pre-date the interruption; see discussion regarding renewed ice loading (Sec. 5.8.4.1). Similar interruptions can be caused by shifts in the patterns of isobases during ice wastage. Thirdly, the method is dependent upon accurate determinations of the age and elevation of the displaced shorelines, and it was on this basis that the technique was criticised in Scandinavia, where the reference levels used by Tanner (1933) were of different ages in different locations in the case of the Tapes level and had a wide altitudinal spread in the case of the late-glacial references zone e and f (Marthinussen 1960).

5.8.5 Conclusions

Certain characteristic patterns of shoreline displacement can be recognised. Where eustasy is the dominant control, the pattern of sea-level change reflects global climatic changes (Figs 5.30 & 5.32), and, because the highest sea level in the middle and late Pleistocene appears to have been during the last interglacial, raised shoreline sequences, including those in Mallorca, must have been uplifted by tectonic activity (Shackleton 1975). In areas where isostasy contributes to the sea-level curve, the timescale concerned is restricted to the period since deglaciation, which, with rare exceptions such as the Hoxnian of eastern England, consists of late last glaciation and Holocene time. In some areas where isostasy has been dominant, the effects of eustatic changes are largely obscured and sea-level curves are a simple, smoothly decelerating function (Figs 5.34, Oslo area, & 5.35). However, where isostatic recovery is less pronounced, the interaction between rebound and eustasy produces a series of complex curves characterised by relatively rapid regression immediately following deglaciation, and a transgression in Holocene times (Fig. 5.34, all curves except Oslo area). Often, in this intermediate case, local circumstances can produce a unique pattern of land, lake and sea-level changes such as is associated with glacier thickening in central Scotland in late-glacial time (Figs 5.28 & 5.34, Menteith area) or with the opening and closing of the Baltic during late-glacial and Holocene time (Eronen 1983).

5.9
Palaeosols

5.9.1 Introduction

For several decades, palaeosols have figured strongly in stratigraphical studies of continental European Quaternary sediments, but only recently have they been encountered in published work on such deposits in the British Isles. In many ways, this is perhaps as well, since many problems must be resolved before we can be satisfied that such soils provide acceptable stratigraphic markers and, still more, before they can be used for dating purposes.

In the first instance, we need to define the term 'palaeosol'. If we accept the opinion of the ISSS/INQUA Working Group (1971), then it is a soil formed on a land surface of the past or in environmental conditions different from those of the present day. However, since virtually all soils have been subject to the vicissitudes of the Flandrian climate, it is apparent that such a definition is somewhat meaningless. In the absence of a better alternative, we shall take as a working definition 'soils that have either been isolated from present-day pedogenesis or show features that are unlikely to have been produced in environments similar to those of the Flandrian'. Even this is unsatisfactory, since many soils display such characteristics but the profiles as a whole are dominantly the result of late-Flandrian soil formation. It is appropriate, then, to refer to many soils as possessing 'relict features' (Catt 1979).

5.9.2 Buried soils

Again, such definitions suppose that our perception of what is soil is clear and unambiguous. However, Valentine and Dalrymple (1976) have made it clear that, at least in the case of buried soils, there are few, if any, single characteristics that allow such a positive identification. There are also many features, such as fragipans and void coatings (cutans), which are widely supposed to result from 'soil formation' but whose origins are still open to a variety of interpretations. Cases also arise in the literature where buried horizons that show strong cryoturbative disturbance have been referred to as 'arctic structure soils' (e.g. Rose & Allen 1977). Although such features clearly relate to a period during which non-sedimentary sub-aerial processes operated, unanimous agreement on describing them as pedogenetic would, even now, not be forthcoming. However, the proposition advanced by Valentine and Dalrymple (1975) would gain some support:

> the crucial test of a buried soil landscape must be the existence of a palaeocatena . . . whose original soil properties differ owing to their original landscape position and soil-water regimes.

One has only to look at soil maps of present landscapes to appreciate the marked lateral variations in soil characteristics in relation to these controls. Conditions for the preservation of such extensive remnants are not that frequent. Nevertheless, palaeocatenas have been identified by Valentine and Dalrymple (1975) beneath marine clays and fen peat, and by Paepe (1971) largely within the loess/coversand sequence of Belgium, while extensive lateral sections can be seen in the Dinkel Valley of the eastern Netherlands, revealing buried soils of Allerød and middle Holocene ages (Vink & Sevink 1971).

Many would not regard the existence of a palaeocatena as the *crucial* test and have argued that the possession of an assemblage of characteristics should enable identification as a soil (Fenwick 1985).

It would seem that loess and coversand deposits provide ideal materials for the burial of pre-existing soils. However, this supposes that burial by aeolian material is rapid and leads to an immediate inhibition of soil-forming processes. However, Ruhe *et al.* (1971) showed that aeolian sedimentation during the Wisconsin cold period in southwestern Iowa varied between 1 ft/121 yr and 1 ft/1400 yr. It is doubtful whether such rates could truly isolate the soil from the influence of present pedogenesis even though they could result in considerable depths of accumulated material. Indeed, it is debatable whether any effect would be noticeable, other than the possible 'contamination' of the soil by exotic mineral species.

Despite the difficulties in establishing the existence of buried soils or of relict features, we need criteria with which to make at least *preliminary* recognition possible. For field purposes, these must of necessity take the form of colour and textural characteristics.

As with modern soils, it is horizon colour and horizon assemblages based on colour that permit an initial identification. However, it is rare to find fossil profiles intact and many relict features may be confined to but a part of the present profile. For the most part, we must, therefore, be content with recognising individual horizons that are either buried or anomalous in the context of the modern soil.

In much of western Europe, currently dominated by brown soils, it is the rich 'orange-red' horizons, often at a depth of several metres, that immediately catch the attention. Indeed, they have even taken on an important role in soil classification – clay-rich horizons with hues of 7.5 YR or redder and with chromas greater than 4 being referred to as palaeoargillic horizons by Avery (1980). Concentrations of manganese, often brought about by gleying processes long since ceased, may be suspected by purplish-black dendrites on stones and aggregates.

Many important palaeosols may reflect only limited pedogenesis and are represented only by thin, dark organic horizons. Such are the Allerød soils buried by solifluction and slopewash materials at the foot of the chalk scarps in southern England (Evans 1966) or by aeolian coversands in the low countries (Vink & Sevink 1971, Paepe & van Hoorne 1973). Again, organic staining is the principal primary clue to the recognition of many of the interglacial palaeosols of the American plains (Thorp 1965).

Where fossil soils have been preserved intact, horizon colour assemblages should prompt initial suspicions of a pedogenic origin. For details of modern analogues, the reader should refer to a text on pedology such as Fenwick and Knapp (1981).

In certain instances, colour differentiation may be absent but evidence of relict features may still be provided by textural breaks in the profile. Clearly this could be due to original sedimentary differences, but it could also be brought about by the translocation of, for instance, clays.

Buried palaeosols, if they can be unmistakably identified, are immediately attractive to the Quaternary geologist and geomorphologist since they form recognisable marker horizons. To be used in this way demands that buried soils retain a certain identity of properties over wide areas, and this is not always the case. Indeed, the concept of the palaeocatena outlined by Valentine and Dalrymple (1975) assumes considerable lateral variability. However, occasionally it has proved possible to trace such buried soils, as for instance with the Sangamon interglacial soil in North America. Morrison (1967) has referred to such soils, which can be consistently recognised and mapped, as geosols. But, even where the soil consistently occurs, it is not always possible to identify it solely on its own internal properties and assign it to a particular stratigraphic position. Indeed, Catt (1979) has gone so far as to say that 'the only way of dating soil development in periods too early for radiocarbon assay is by careful stratigraphic interpretation', i.e. the relative dating of buried soils is as dependent on the adjacent deposits as the converse.

Nevertheless, certain geosols do possess very distinctive characteristics that facilitate their immediate recognition. Thus, the Beuningen soil of upper pleniglacial (Weichselian) age in the Dinkel Valley of the Netherlands is notable for the strong development of a fragipan, up to 40 cm in thickness, in what are otherwise loose, structureless sands. Even here, identification is facilitated by the presence of a bed of fine gravel (the Beuningen Gravel Bed) immediately above it (Vink & Sevink 1971). In Britain, the only soils that are readily identifiable and have been recognised over wide areas, thus qualifying as geosols, occur in southern England. They are the Valley Farm rubified *sol lessivé*, described from southern East Anglia, originally by Rose *et al.* (1976) and more recently re-examined by Kemp (1985), and the associated cold-climate soil, the Barham arctic structure soil (Rose & Allen 1977, Rose *et al.* 1985a). In addition, the Pitstone soil (Rose *et al.* 1985b) of Windermere Interstadial age has been extensively recognised, especially on the chalklands of southern England.

5.9.3 Radiocarbon dating

Buried soils do in certain circumstances contain material suitable for radiocarbon assay (see also Sec. 5.1). However, dating of soil organic material poses problems that are unique to soil material. Thus, where O horizons have been buried, even where this may have been sudden, the organic matter at the time of burial would have contained a fraction that had been within the soil for many decades. On the other hand, there would also have been material from very recent incorporation. Accordingly, if a bulk sample is taken, the 'date' yielded will reflect the *apparent mean residence time* (*AMRT*) (Campbell *et al.* 1967) of this material together with the elapsed time since burial.

The interpretation of the mean residence time is complicated by the fact that certain fractions of the soil organic matter turn over much more rapidly than others. In particular, Campbell *et al.* (1967) showed that humin, or fine residual organic carbon, was much more stable than mobile humic acids and fulvic acids. In a chernozemic soil, the comparative *AMRT* values were *c.*1135, *c.*785 and *c.*555 ^{14}C yr respectively. These compared with an *AMRT* for the unfractioned soil of *c.*870 ^{14}C years. Differences between the 'ages' of the three fractions have been attributed by Jenkinson and Rayner (1977) and Stout *et al.* (1981) to normal decomposition, humification and translocatory processes prior to burial. However, in the view of Cook (1969), they could also result from continued microbial activity post-burial.

It would appear, therefore, that in bulk analyses the fulvic acid fraction is likely to give the nearest approximation to the date of burial. Even so, contamination by continuing microbiological activity remains an unknown.

Interestingly, Griffey and Matthews (1978) examined the dating potential of the included rootlets in a soil buried under an advancing glacier. Normally, such material is discarded as likely to be a modern contaminant. In this instance, where the present root mat and the buried horizon were some 1 m apart, they argued that the root material had been alive close to the time of burial and, therefore, could provide a particularly precise indication for the date of glacial advance.

Research by Matthews and Dresser (1983) has, however, opened up the possibility of

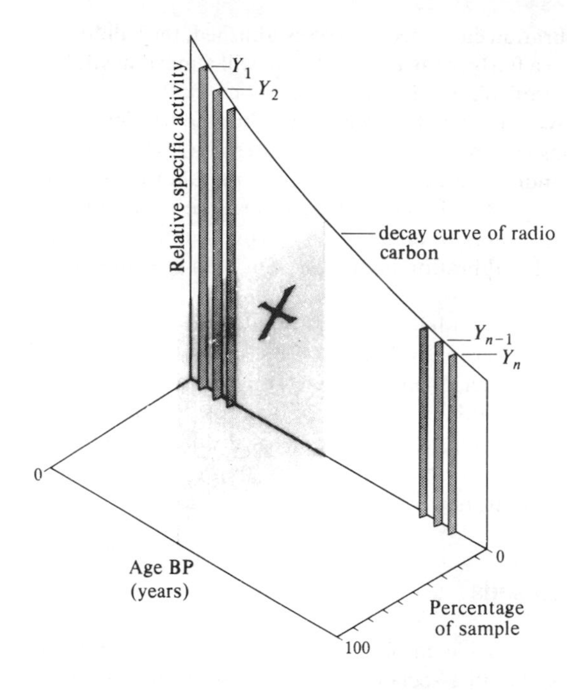

Figure 5.42 Diagrammatic representation of the components of the total ^{14}C activity in a sample from a Bh horizon. The total activity is indicated by the area *X*. This represents the integration of small additions of humus through time represented by $Y_1, Y_2, \ldots, Y_n$, where Y_n is the oldest and, therefore, the least active element. (From Perrin *et al.* 1963).

much more precise dating of buried soils. By intensive sampling of a 15 cm thick Ob horizon from the Jotunheimen, southern Norway, they were able to show that there was a uniform and consistent decrease in age towards the top of the horizon. In this case, they concentrated on the humic acid fraction, although a similar trend appeared to be evident with the other fractions. It becomes clear that little if any correction for *AMRT* is necessary if the date of burial is based on a depth–age relationship.

The quantification by Matthews and Dresser (1983) of the age of different layers within an O horizon also highlights the difficulty of establishing the date of initiation of podsolisation by bulk sampling and dating of Bh horizons. As long age as the early 1960s, Perrin *et al.* (1963) hypothesised that the activity level is in fact a composite of successive increments throughout the period of Bh formation. While the older material will, if rates of addition of organic material to the Bh have remained uniform, inevitably contribute a small element to the total activity, recent additions will account for a much higher proportion (Fig. 5.42).

It is clearly only possible, therefore, to state a latest date for the initiation of Bh formation, and this would also imply that the process ceased immediately thereafter. If it is assumed or can be proved that organic matter translocation has been taking place at a constant rate and is still continuing, then it is possible to establish the date for the initiation of Bh formation based on the *AMRT* provided by the assay.

Attempts to employ radiocarbon dating techniques for pedological purposes hinge very much on the development of satisfactory models of soil carbon dynamics. For a review of these, the reader is referred to Scharpenseel and Schiffmann (1977).

5.9.4 Dating using chronosequences

In recent years there has been a great deal of progress in establishing indices of the rate and magnitude of soil formation. Such indices offer possibilities of inferring the age of soils from their morphological characteristics. In many respects, the technique displays marked similarities with lichenometry. A calibration curve needs to be established; the validity of the calibration is likely to be limited to a fairly local area and there will almost inevitably be considerable variation in the extent of soil development in any one locality.

Some indices (e.g. Harden 1982) have been based on field data. Thus the Harden index compares horizon properties with those of an assumed parent material. Others, such as Rockwell *et al.* (1985), have devised indices based, at least in part, on laboratory data. Using this technique, Rockwell *et al.* suggest, although without offering tested examples, that it should be possible to assign age ranges to undated deposits with intact soils. By various means they were able to provide calibration points for their data over the range from 0 to 200 000 yr BP.

Working with much more recent materials, Mellor (1985) has adopted an alternative approach by obtaining carbon, pH and pyrophosphate-extractable Fe and Al. In an examination of soil development on neoglacial (post–1750 AD) moraine ridges in the Jotunheimen and Jostedalsbreen areas of southern Norway, he was able to demonstrate some significant statistical relationships. In areas where radiometrically dated material is sparse or imprecise, these functions could be of great value in providing a guide to the age of the soils and, conceivably, of the depositional form itself.

5.9.5 Relict features in modern soils

Although buried palaeosols, therefore, provide an obvious and, in some cases, datable horizon in stratigraphic work, many soils still associated with the present landscape also display relict features. If it is possible to establish when such features were initiated, then we may have some control on the dating of the surface and of the sediments in which the soil has developed. Some of these relict features are evident even to a casual observer, although their interpretation is less straightforward.

Thus, in southern England, strongly reddened horizons, where the colouring can be shown not to have been derived from source materials, occur extensively on all but the most recent landscape facets. Usually, such horizons underlie E and B horizons that contain more recent material such as Devensian loess. Furthermore, in Britain, these reddened horizons are often strongly disturbed by involutions and other periglacial activity. There is thus evidence that much of this reddening, or rubification, must pre-date the main late-Devenisan cold phase.

Avery has long recognised the existence of what he has described (e.g. Avery 1985) as palaeoargillic B horizons. Generally these are found beyond the last cold stage glacial limit. With matrix chromas more than 4 and hues redder than 7.5 YR in fine-textured soils, these horizons are often described as reddened. They display not only strong colours but also complex sepic fabrics resulting from a pronounced reorganisation of clay. Even so, Avery (1985) has been reluctant to be more precise than to ascribe such features to 'pre-Devensian pedogenesis'.

Catt (1979) summarised the argument thus:

> as the characteristic colours of palaeoargillic horizons probably result from iron oxides and hydrated oxides derived by weathering of iron-containing minerals, and the extensive reorganisation of their clay component probably results from long-continued clay illuvation and repeated shrink-swell cycles in a seasonally humid climate, it is reasonable to attribute the pedogenesis to interglacial stages of the Quaternary rather than colder periods.

Federoff (1966) argued that in the Paris Basin such red soils did not even generally develop during the Eemian (Ipswichian). Boulaine (in Federoff 1966) went further and suggested that, for red *lessivé* soils to develop, a mean annual temperature of 16°C and an annual rainfall of 800–900 mm would be required – or a climate similar to the western Mediterranean at the present time. More recently, Federoff and Goldberg (1982) have played down the correlation between soil colour and climate. Despite a soil hue of 5 YR, they inferred from micromorphological characteristics that Eemian soils in the Paris Basin formed in 'a temperate climate similar to today's'.

Evidence produced by Schwertmann *et al.* (1982) demonstrated conclusively that reddening has occurred during the Flandrian, with mean annual temperatures as low as 7.5°C. The Bavarian soils on which they reported were notable for their coarse texture. It is possible, therefore, that provided there is a warm pedoclimate, as in such coarse materials, rubification can occur even under cool atmospheric conditions. In general, however, many would probably agree with Folk (1976) that relatively high temperatures and distinct dry seasons occurring over a long soil-forming period provide the optimal conditions for rubification.

Even though the relationship between soil colour and climate may be far from clear, there seems to be a consistent pattern of rubification increasing with age. This was amply demonstrated by Birkeland (1984) in a compilation of data from the southwestern USA. Similarly, Federoff (1966) found that soils developed in sediments older than the Riss (Wolstonian) were always rubified.

Whether the formation of these colours is a result of warmer climates or simply a response to the passing of time, it is not yet possible to say.

Similarly, the occurrence of pseudostagnogleying in *lessivé*-type soils has been seen by several French workers (e.g. Federoff 1973) as indicating formation over a very long period. Federoff (1973) has observed that in Atlantic France soils originating from before the Riss cold period generally show these features. Catt (1979) has placed much emphasis on the colour of these red mottles, in that 10 R and 2.5 YR mottles are found in Cromerian soils of a variety of textures and even in some Hoxnian and Ipswichian soils. However, he is of the opinion that Hoxnian and Ipswichian profiles more frequently contain 5 YR mottles. To be more precise than this in dating either rubified or gleyed palaeoargillic horizons is impossible, since we have little idea at present as to the influence of factors other than climate and time on such soil properties.

While it is evident that it is not possible to attribute isolated occurrences of relict features to any particular climatic episode, it is possible to utilise soil catenas, which may provide stratigraphic markers in this way. Furthermore, micromorphological analyses can, in certain circumstances, provide a record of environmental changes and hence some uncertain indication of the period of initiation of soil formation. The collection and preparation of thin sections of soils is well documented in Osmond and Bullock (1970), as also is the interpretation of several micromorphological features. For a fuller aid to the recognition of such features, the reader is referred to Bullock *et al.* (1985).

In one of the most convincing applications of soil micromorphology to stratigraphic investigations, Chartres (1980) demonstrated evidence of successive phases of soil development. In a study of soil formation on terrace deposits of the River Kennet (S. England) he argued that different phases of clay translocation can be identified by the contrasting colour of the clays – deep red, egg-yellow and orange (when viewed under crossed polarised light). Red clays are confined to the higher terraces (especially the 47 m terrace), while yellow matrix clays can be seen on both this higher terrace and also at the 10 m level. Orange clays, mainly as void cutans, are ubiquitous, even on the lowest 2 m terrace. Whereas the orange clays are *in situ* and undisturbed, the yellow and red clays show severe disruption, although it is not uncommon for them to be present as recognisable elements of strongly oriented clay (papules) suggestive of former void coatings. On the highest terraces, the disrupted red clays are clearly embedded within, and therefore are presumably older than, the yellow clays, which suggests that the former are the younger features. On the basis of this pedochronology he tentatively suggested that the red clays were formed during the Hoxnian, translocation of the egg-yellow clays followed in the Ipswichian with the orange argillans being a product of Flandrian pedogenesis.

For such relationships to be interpreted in terms of environmental changes, and hence to be used as a basis for relative chronology, demands a considerable element of speculation not only on the reasons for variations in the colour of the clay but also for the disruption of clay coatings. Although it is popular to ascribe such disruption to cryogenic processes, it could be that similar forms, for instance, result from stress due to moisture changes or from the effects of tree fall.

For a site in central Oxfordshire close to those of Chartres, Bullock and Murphy (1979) have deduced that a soil developed in the Plateau Drift has experienced clay translocation, gleying and disturbance, as well as erosion. In this case, they felt that the disruption of cutans was probably due to cryoturbation. Often such an interpretation can be supported by the existence of involutions or ice-wedge casts. Valuable though this is, it will be apparent that we are, as yet, not sufficiently secure in our interpretation of relict features to utilise the evidence to the full. Once this stage has been reached, then soil micromorphology offers the prospect of the establishment of a real pedochronology. This, then, is the prospect, but for the present we must beware of rash interpretations.

5.9.6 Conclusions

Perhaps the real value of palaeosols lies in the fact that 'forming remnants of the older buried surfaces, they [fossil soils] are one of the most important unconformities in the geological context' (Paepe 1971). Furthermore, it is well to bear in mind that, for example, in southern England the majority of the Quaternary record is represented by such depositional hiatuses. J. Rose (pers. comm.) has estimated that the depositional record only accounts for some 800 ka out of a total span of 2300 ka. For the rest, we are largely dependent upon the record bequeathed by the relict features within our soils.

References and bibliography

Aaby, B. and G. Digerfeldt 1986. Sampling techniques for lakes and bogs. In *Handbook of Holocene palaeoecology and palaeohydrology*, B. E. Berglund (ed.), 181–94. Chichester: Wiley.

Aaby, B., J. Jacobsen and O. S. Jacobsen 1978. Pb-210 dating and lead deposition in the ombrotrophic peat bog Draved Mose, Denmark. *Dan. Geol.Unders.Arbog*. 45–68.

Abaturov, B. D. and L. O. Karpachevski 1966. Effects of moles on the water physical properties of sod-podzolic soils. *Sov. Soil Sci.* 667–74.

Abbott, J. E. and J. R. D. Francis 1977. Saltation and suspension trajectories of solid grains in a water stream. *Phil.Trans.R.Soc.Lond.A* **284**, 255–54.

Abelson, P. H. 1955. Organic constituents of fossils. *Carnegie Inst. Washington Year Book* **54**, 107–9.

Abrahams, A. D. 1980. Divide angles and their relation to interior links lengths in natural channel networks. *Geog. Anal.* **12**, 151–71.

Abrahams, A. D. 1984. Tributary development along winding streams and valleys. *Am.J.Sci.* **284**, 863–92.

Abrahams, A. D. and A. J. Parsons 1977. Long segments on low-angled hillslopes in New South Wales. *Area* **9**, 124–7.

Acker, K. J. and M. J. Risk 1985. Substrate destruction and sediment production by the boring sponge *Cliona carribbaea* on Grand Cayman Island. *J. Sedim.Petrol.* **55**, 705–11.

Ackers, P. and F. G. Charlton 1970. *The geometry of small meandering streams*. Proc.Inst.Civ.Eng. Paper no. 73285.

Adam, L., M. Pecsi and G. Mezosi 1986. Geomorphological mapping. In *Physical geography and geomorphology in Hungary*, M. Pecsi and D. Loczy (eds), 51–9. Budapest: Hungarian Academy of Sciences.

Agar, R. 1960. The post-glacial erosion of the north Yorkshire coast from the Tees Estuary to Ravenscar. *Proc.Yorks.Geol.Soc.* **32**, 409–27.

Ahnert, F. 1970. An approach towards a descriptive classification of slopes. *Z. Geomorphol. Supp. NF* **9**, 70–84.

Airey, P. L., G. E. Calf, A. Davison, J. F. Easey and A. W. Morley 1984. An evaluation of tracer dilution techniques for gauging rivers in flood. *J. Hydrol.* **74**, 105–18.

Aitken, M. J. 1970. Dating by archaeomagnetic and thermoluminescent methods. *Phil.Trans R.Soc.Lond. A* **269**, 77–88.

Aitken, M. J. 1974. *Physics and archaeology*. Oxford: Clarendon.

Aitken, M. J. 1978. Archaeological involvements of physics. *Phys. Rep.* (Sec. C of *Phys. Lett.*) **40C** (5), 277–351.

Aitken, M. J. 1979. *Physics and archaeology*, 2nd edn. Oxford: Clarendon.

Aitkin, M. J. 1985. *Thermoluminescence dating*. London: Academic Press.

Akroyd, T. N. W. 1969. *Laboratory testing in soil engineering*. London: Soil Mechanics Ltd.

Alestalo, J. 1971. Dendrochronological interpretation of geomorphic processes. *Fennia* **105**, 1–140.

Alger, G. R. and D. B. Simons 1968. Fall velocity of irregular shaped particles. *Proc.Am. Soc.Civ.Eng.* **HY3**, 721–37.

Alison, M. S. 1899. On the origin and formation of pans. *Trans. Geol.Soc.S. Afr.***4**, 159–61.

Allan, J. A. 1978. Remote sensing in physical geography. *Prog.Phys.Geog.* **2**, 36–54.

Allchin, B., A. S. Goudie and K. T. M. Hegde 1977. *The prehistory and palaeogeography of the Great Indian Desert*. London: Academic Press.

Allen, J. R. 1985. Field measurement of longshore sediment transport: Sandy Hook, New Jersey, USA. *J. Coastal Res.* **1** (3), 231–40.

Allen, J R. L. 1968. *Current ripples: their relation to patterns of water and sediment movement*. Amsterdam: North-Holland.

Allen, J. R. L. 1974a. Reaction, relaxation and lag in natural sediment systems: general principles, examples and lessons. *Earth Sci.Rev.* **10**, 263–342.

Allen, S. E. (ed.) 1974b. *Chemical analysis of ecological materials*. Oxford: Blackwell Scientific.

Allen, S. E., A. Carlisle, E. J. White and C. C. Evans 1968. The plant nutrient content of rainwater. *J. Ecol.* **56**, 457–504.

Allen, T. 1975. *Particle size measurement*, 2nd edn. London: Chapman & Hall.

Allen, T. 1981. *Particle size measurement*, 3rd edn. London: Chapman & Hall.

Allison, J. R. 1986. *Mass movement and coastal cliff development of the Isle of Purbeck, Dorset.* Unpubl. Ph.D. Thesis, Kings College, University of London.

Allman, M. and D. F. Lawrence 1972. *Geological laboratory techniques.* London: Blandford.

Ambraseys, N. N. 1971. Value of historical records of earthquakes. *Nature* **232**, 375–9.

Ambraseys, N. N. and C. C. Melville 1982. *A history of Persian earthquakes.* Cambridge; Cambridge University Press.

Amorocho, J. and W. E. Hart 1964. A critique of current methods in hydrologic systems investigations. *Trans Am.Geophys. Union* **45**, 307–21.

Amos, C. L. and T. T. Alfoldi 1979. The determination of suspended sediment concentration in a macrotidal system using Landsat data. *J. Sedim.Petrol.* **49**, 159–74.

Anderson, B. G. 1975. Glacial geology of northern Nordland, north Norway. *Nor. Geol. Unders.* **320**, 1–74.

Anderson, E. W. and B. L. Finlayson 1975. Instruments for measuring soil creep. *BGRG Tech.Bull.* **16**.

Anderson, M. G. 1973. Measure of three-dimensional drainage basin form. *Water Resour. Res.* **9**, 378–83.

Anderson, M. G. 1979. On the potential of radiography to aid studies of hillslope hydrology. *Earth Surf. Processes* **4**, 77–83.

Anderson, M. G. and T. P. Burt 1977. Automatic monitoring of soil moisture conditions in a hillslope spur and hollow. *J.Hydrol.* **33**, 27–36.

Anderson, M. G. and T. P. Burt 1978. The role of topography in controlling throughflow generation. *Earth Surf. Processes* **3**, 331–44.

Anderson, M. G. and A. Calver 1980. Channel plan changes following large floods. In *Timescales in geomorphology*, R. Cullingford, D. Davison and J. Lewin (eds). Chichester: Wiley.

Anderson, M. G. and S. Howes 1985. Development and application of a combined soil water–slope stability model. *Q. J. Eng. Geol.* **18**, 225–36.

Anderson, M. G., K. S. Richards and P. E. Kneale 1980. The role of stability analysis in the interpretation of the evolution of threshold slopes. *Trans. Inst. Br. Geog.* **5**, 100–12.

Andreason, J. O. 1985. Apparent short-term glacier velocity variations. *J. Glaciol.* **31** (107), 49–53.

Andrews, J. T. 1967. Problems in the analysis of the vertical displacement of shorelines. *Geog. Bull.* **8**, 71–4.

Andrews, J. T. 1968. Postglacial rebound in Arctic Canada: similarity and prediction of uplift curves. *Can.J.Earth Sci.* **5**, 39–47.

Andrews, J. T. 1969. The shoreline relation diagram: physical basis and use in predicting age of relative sea-levels: evidence from Arctic Canada. *Arctic Alpine Res.* **1**, 67–78.

Andrews, J. T. 1970a. Techniques of till fabric analysis. *B.G.R.G.Tech.Bull.* **6**.

Andrews, J. T. 1970b. A geomorphological study of post-glacial uplift with particular reference to Arctic Canada. *Inst.Brit.Geog.Spec.Publ.* **2**, 1–156.

Andrews, J. T. 1974. Introduction. In *Benchmark papers in geology. 10: Glacial isostasy*, J. T. Andrews (ed.), 1–17. Stroudsburg, Pa.: Dowden, Hutchinson & Ross.

Andrews, J. T. (ed.) 1985. *Quaternary environments: eastern Canadian Arctic, Baffin Bay and West Greenland.* London: Allen & Unwin.

Andrews, J. T. and R. E. Dugdale 1971. Quaternary history of northern Cumberland Peninsula, Baffin Island, NWT. Part V: Factors affecting corrie glaciation in Oka Bay. *Quat. Res.* **1** 523–51.

Andrews, J. T. and G. H. Miller 1976. Quaternary glacial chronology of the eastern Canadian Arctic: a review and a contribution on amino acid dating of Quaternary molluscs from the Clyde Cliffs. In *Quaternary of North America*, W. Mahanney (ed.), 1–32. Stroudsburg, Pa.: Dowden, Hutchinson & Ross.

Andrews, J. T. and P. J. Webber 1964. A lichenometrical study of the northwestern margin of the Barnes Ice Cap: a geomorphological technique. *Geog.Bull.* **22**, 80–104.

Andrews, K. W., D. J. Dyson and S. R. Keown 1971. *Interpretation of electron diffraction patterns*, 2nd edn. London: Hilger & Watts.

Aniya, M. and R. Welch 1981. Morphometric analysis of Antarctic cirques from photogrammetric measurements. *Geog. Annaler* **63A**, 41–53.

Anon., 1973. Analysis of the atmospheric environment. *NERC Newslett.* **7**, 10.

Anon., 1976. Report from the International Conference on the Effects of Acid Precipitation, Telemark, Norway. *Ambio* **5**, 2.

Anon. (Geol.Soc.Eng.Gp) 1977. The description of rock masses for engineering purposes. *Q.J.Eng.Geol.* **10**, 355–88.

Anundsen, K. 1978. Marine transgression in Younger Dryas in Norway. *Boreas* **7**, 49–60.

Anundsen, K. and W. Fjeldskaar 1983. Observed and theoretical Late Weichselian shore-level changes related to glacier oscillations at Yrkje, south-west Norway. In *Late- and postglacial oscillations of glaciers: glacial and periglacial forms*, H. Schroeder-Lanz (ed.), 133–70. Rotterdam: Balkema.

Appleby, P. G. and F. Oldfield 1978. The calculation of lead-210 dates assuming a constant rate of supply of unsupported ^{210}Pb to the sediment. *Catena* **5**, 1–8.

Appleby, P. G. and F. Oldfield 1979. Letter on the history of lead pollution in Lake Michigan. *Environ.Sci. Technol.* **13**, 478–80.

Appleby, P. G. and F. Oldfield 1983. The assessment of ^{210}Pb data from sites with varying sediment accumulation rates. *Hydrobiologia* **103**, 29–35.

Appleby, P. G., P. J. Nolan, D. W. Gifford, M. J. Godfrey, F. Oldfield, N. J. Anderson and R. W. Battarbee 1986. ^{210}Pb dating by low background gamma counting. *Hydrobiologia* **143**, 21–7.

Appleby, P. G., F. Oldfield, R. Thompson, P. Huttenen and K. Tolonen 1979. ^{210}Pb dating of annually laminated lake sediments from Finaland. *Nature* **280**, 53–5.

Armstrong, A. C. and W. B. Whalley 1985. An introduction to data logging. *B.G.R.G.Tech.Bull.* **34**.

Arnett, R. R. 1971. Slope form and geomorphological process: an Australian example. *Inst.Br. Geog.Spec.Publ.* **3**, 81–92.

Arnett, R. R. 1978. Regional disparities in the denudation rate of organic sediments. *Z. Geomorphol. Supp. NF* **29**, 169–79.

Aronovici, V. S. 1966. The area-elevation ratio curve as a parameter in watershed analysis. *J. Soil Water Conserv.* **21**, 226–8.

Arthurs, J. R. F. 1972. New techniques to measure new parameters. In *Stress–strain behaviour of soils*, R. H. G. Parry (ed.), 340–7. Proc.Roscoe Memorial Symp., Cambridge University. Henley-upon-Thames: Foulis.

Ascaso, C., J. Galvan and C. Rodriguez-Pascual 1982. The weathering of calcareous rocks by lichens. *Pedobiologia* **24**, 219–29.

Aschenbrenner, B. C. 1956. A new method of expressing particle sphericity. *J.Sedim.Petrol.* **26**, 15–31.

Ash, J. E. and R. J. Wasson 1983. Vegetation and sand mobility in the Australian desert dunefield. *Z. Geomorphol. NF* **45**, 7–25.

Ashmore, P. E. 1982. Laboratory modelling of gravel braided stream morphology. *Earth Surf. Processes Landforms* **7**, 201–25.

Ashworth, P. J. and R. I. Ferguson 1986. Interrelationships of channel processes, changes and sediments in a proglacial braided river. *Geog. Annaler* **68A**, 361–71.

Atkinson, T. C. 1978. Techniques for measuring subsurface flow on hillslopes. In *Hillslope hydrology*, M. J. Kirkby (ed.), 73–120. Chichester: Wiley.

Atkinson, T. C., K. R. Briffa, G. R. Coope, M. J. Joachim and D. W. Perzy 1986. Climatic calibration of coleopteran data. In *Handbook of Holocene palaeoecology and palaeohydrology*, B. E. Berglund (ed.), 851–8. Chichester: Wiley.

Atkinson, T. C., R. S. Harmon, P. L. Smart and A. C. Waltham 1978. Palaeoclimatic and geomorphic implications of ^{230}Th/^{234}U dates on speleothems from Britain. *Nature* **272**, 24–8.

Atkinson, T. C., D. I. Smith, J. J. Lavis and R. J. Whitaker 1973. Experiments in tracing underground waters in limestones. *J. Hydrol.* **19**, 323–49.

Attewell, P. B. and I. W. Farmer 1976. *Principles of engineering geology*. London: Chapman & Hall.

Attiwill, P. M. 1966. The chemical composition of rainwater in relation to cycling of nutrients in mature eucalyptus forest. *Plant Soil* **24**, 390–406.

Avery, B. W. 1973. Soil classification in the Soil Survey of England and Wales. *J.Soil Sci.* **24**, 324–38.

Avery, B. W. 1980. *Soil classification for England and Wales (higher categories)*. Soil Surv. Tech. Monogr. no.14. Harpenden: Soil Survey of England and Wales.

Avery, B. W. 1985. Argillic horizons and their significance in England and Wales. In *Soils and Quaternary landscape evolution*, J. Boardman (ed.), 69–86. Chichester: Wiley.

Avery, B. W. and E. L. Bascomb (eds) 1974. *Soil survey laboratory methods*. Soil Surv. Tech. Monogr. no. 6. Harpenden: Soil Survey of England and Wales.

Babbage, C. 1847. Observations on the temple of Serapis, at Pozzuoli, near Naples. *Q.J.Geol. Soc.Lond.* **3**, 186–217.

Bada, J. L. 1985. Amino acid racemization dating of fossil bones. *Annu. Rev. Earth Planet. Sci.* **13**, 241–68.

Bada, J. L. and R. Protsch 1973. Racemization reaction of aspartic acid and its use in dating fossil bones. *Proc.Nat.Acad.Sci.USA* **70**, 1331–4.

Bagnold, R. A. 1941. *The physics of blown sand and desert dunes*. London: Methuen.
Bagnold, R. A. 1963. Mechanics of marine sedimentation. In *The sea*, M. N. Hill (ed.), vol. III, 507–28. New York: Interscience.
Bagnold, R. A. 1973. The nature of saltation and 'bed-load' transport in water. *Proc.R.Soc.Lond.A* **332**, 473–504.
Bailey, P. L. 1976. *Analysis with ion selective electrodes*. London: Heyden.
Baillie, M. G. L. 1982. *Tree-ring dating and archaeology*. London: Croom Helm.
Baillie, M. G. L. and J. R. Pilcher 1973. A simple cross-dating program for tree-ring research. *Tree-Ring Bull.* **33**, 7–14.
Bakalowicz, M., D. C. Ford, T. E. Miller, A. N. Palmer and M. V. Palmer 1987. Thermal genesis of dissolution caves in the Black Hills, South Dakota. *Geol. Soc. Am. Bull.* **99**, 729–38.
Baker, H. W. 1976. Environment sensitivity of submicroscopic surface textures on quartz grains: a statistical evaluation. *J. Sedim.Petrol.* **46**, 871–80.
Baker, V. R. 1979. Erosional processes in channelized water flows on Mars. *J. Geophys.Res.* **84**, 7985–93.
Ball, D. J. and M. J. R. Schwar 1978. New developments in air pollution monitoring techniques. *Clean Air* **8**, 25–7.
Band, L. E. 1986. Topographic partition of watersheds with digital elevation models. *Water Resour. Res.* **22**, 15–24.
Banerjee, S. K. 1981. Experimental methods of rock magnetism and palaeomagnetism. *Adv.Geophys.* **23**, 25–99.
Bannister, A. and S. Raymond 1984. *Surveying*. London: Pitman.
Barashnikov, N. B. 1967. Sediment transportation in river channels with flood plains. *Int.Assoc. Sci.Hydrol.Publ.* **75**, 404–13.
Barber, K. E. 1976. History of vegetation. In *Methods in plant ecology*, S. B. Chapman (ed.), 5–84. Oxford: Blackwell.
Barbetti, M. and M. W. McElhinney 1972. Evidence of a geomagnetic excursion 30 000 yr BP. *Nature, Phys. Sci.* **239**, 327–30.
Barczewski, B. and H. P. Koschitsky 1986. Investigations on the applicability of inexpensive piezoresistive pressure transducers for measurements in water. In *Measuring techniques in hydraulic research*, A. C. E. Wessels (ed.), 97–112. Proc. IAHR Symp., Delft, 1985. Rotterdam: Balkema.
Barker, H. 1970. Critical assessment of radiocarbon dating. *Phil.Trans. R.Soc.Lond.A* **269**, 37–45.
Barnes, H. H. 1967. *Roughness characteristics of natural channels*. US Geol.Surv.Water Supply Paper no. 1849.
Barnes, P., D. Tabor and J. C. F. Walker 1971. The friction and creep of polycrystalline ice. *Proc.R.Soc.Lond.A* **324**, 127–55.
Barraclough, D., E. L. Geens and J. M. Maggs 1984. Fate of fertilizer nitrogen. II. Nitrogen-15 leaching results. *J.Soil Sci.* **35**, 191–9.
Barraclough, D. R. 1974. Spherical harmonic analysis of the geomagnetic field for eight epochs between 1600 and 1910. *Geophys.J.R.Soc.* **36**, 497–513.
Barrett, P. J. 1980. The shape of rock particles, a critical review. *Sedimentology* **27**, 291–303.
Barsby, A. 1963. *A wooden flume*. Water Res. Assoc. Tech. Paper no. 28.
Barsch, D. and H. Liedtke 1980. Principles, scientific value and practical applicability of the geomorphological map of the Federal Republic of Germany at the scale of 1:50 000 (GMK 25) and 1:100 000 (GMK 100). *Z.Geomorphol. Supp. NF* **36**, 296–313.
Barsch, D. and R. Mäusbacher 1979. Geomorphological and ecological mapping. *GeoJournal* **3**, 361–70.
Bartlein, P. J., T. Webb III and E. Fleri 1984. Holocene climatic change in the northern Midwest: pollen-derived estimates. *Quat.Res.* **22**, 361–74.
Barton, C. 1979. Modifications to the Mackereth corer for palaeomagnetic measurements. *Limnol. Oceanog* **24**, 977–83.
Basak, P. 1972. Soil structure and its effects on hydraulic conductivity. *Soil Sci* **114**, 417–22.
Battarbee, R. W. and G. Digerfeldt 1976. Palaeoecological studies of the recent development of Lake Vaxjosjon. I. Introduction and chronology. *Arch. Hydrobiol.* **77**, 330–46.
Battiau-Queney, Y. 1984. The pre-glacial evolution of Wales. *Earth Surf. Processes Landforms* **9**, 229–52.
Battle, W. B. R. 1952. Unpublished manuscript on freeze–thaw processes. Department of Geography, University of Cambridge.
Bauer, H. H., G. D. Christian and J. E. O'Reilly 1978. *Instrumental analysis*. Boston: Allyn & Bacon.
Baulig, H. 1935. *The changing sea level*. Inst.Br.Geog. Publ. no. 3.

Baver, L. D., W. H. Gardner and W. R. Gardner 1972. *Soil physics*. New York: Wiley.

Baxter, F. P. and F. D. Hole 1967. Ant (*Formica cineres*) pedoturbation in prairie soil. *Proc.Soil Sci.Am.* **31**, 425–8.

Beadnell, H. J. L. 1910. The sand dunes of the Libyan Desert. *Geog.J.* **35**, 379–95.

Beardsley, G. F., G. C. Knollman, A. S. Hamamoto and J. D. Eisler 1963. A device for measuring orbital velocities in an underwater wave field. *New Sci.Instrum.* **34**, 516–19.

Beatty, S. W. and E. L. Stone 1986. The variety of soil microsites created by tree falls. *Can.J.For. Res.* **16**, 539–48.

Beautelspacher, H. and H. W. Van der Marel 1968. *Atlas of electron microscopy of clay minerals and their admixtures*. Amsterdam: Elsevier.

Beck, J. R. 1971. *Use of time-lapse photography equipment for hydrologic studies*. USGS Open File Rep.

Beckinsale, R. P. 1972. The limestone bugaboo: surface lowering or denudation or amount of solution. *Trans. Cave Res. Gp GB* **14**, 55–8.

Beddow, J. K. 1980. Particle morphological analysis. In *Advanced particulate morphology*, J. K. Beddow and T. P. Meloy (eds), 1–84. Boca Raton, Fl. CRC Press.

Beecroft, I. 1983. Sediment transport during an outburst from Glacier de Tsidjiore Nouve, Switzerland, 16–19 June 1981. *J.Glaciol.* **29**, 185–90.

Behrens, E. W. 1978. Further comparisons of grain-size distributions determined by electronic particle counting and pipette techniques. *J.Sedim.Petrol.* **48**, 1213–18.

Bell, M. and E. P. Laine 1985. Erosion of the Laurentide region of North America by glacial and glaciofluvial processes. *Quat.Res.* **23**, 154–74.

Bell, S. A. 1983. *Attributes of drainage basin topography: an evaluation of profile and altitude matrix approaches and their hydrological relevance*. Ph.D. Thesis, University of Durham.

Belly, P. Y. 1962. *Sand movements by wind*. Monogr. Univ. Coll. Berkeley, Hydraul. Eng. Labs, Wave Res. Projects.

Bender, P. L. and E. C. Silverberg 1975. Present plate-tectonic motions from lunar ranging. *Tectonophysics* **29**, 1–7.

Benedict, J. B. 1970. Downslope soil movement in a Colorado alpine region: rates, processes and climatic significance. *Arctic Alpine Res.* **2**, 165–226.

Benham, D. G. and K. Mellanby 1978. A device to exlude dust from rainwater samples. *Weather* **33**, 151–4.

Bennett, H. H. 1939. *Soil conservation*. New York: McGraw-Hill.

Bennett, J. C. and H. Tributsch 1978. Bacterial leaching patterns on pyrite crystal surfaces *J. Bacteriol.* **134**, 310–17.

Bennett, R. J. and R. J. Chorley 1978. *Environmental systems*. London: Methuen.

Bennett, R. J., W. J. Campbell and R. A. Maughan 1976. Changes in atmospheric pollution concentrations. In *Mathematical models for environmental problems*, C. A. Brebbia (ed.), 221–35. London: Pentech.

Benninger, L. K., R. C. Aller, J. K. Cochran and K. K. Turekian 1976. Lead-210 geochronology of contemporary near shore sediments: status and problems. *Trans.Am.Geophys.Union* **57**, 931–2.

Benson, M. A. 1960. Areal flood frequency analysis in a humid region. *Bull.Int.Assoc.Sci.Hydrol.* **19**, 5–15.

Benson, M. A. 1965. Spurious correlations in hydraulics and hydrology. *Proc.Am.Soc.Civ.Eng., J.Hydraul. Div.* **91**, 35–42.

Benson, R. H. 1967. Muscle-scar patterns of Pleistocene (Kansan) ostracodes. In *Essays in palaeontology and stratigraphy*, C. Teichert and E. L. Yochelson (eds), 211–42. Lawrence, Kans.: Kansas University Press.

Bentley, S. P. and I. J. Smalley 1978. Mineralogy of sensitive clays from Quebec. *Can.Mineral.* **16**, 103–12.

Berg. D. W. and E. F. Hawley 1972. Time-interval photography of littoral phenomena. In *Proc.13th conf.on coastal engineering* vol. 2, 725–45. Am. Soc. Civ. Eng.

Berger, G. W. 1988. Dating Quaternary deposits by luminescence – recent advances. *Geosciences Canada* **13**, 15–21.

Berglund, B. E. 1973. Pollen dispersal and deposition in an area of southeastern Sweden: some preliminary results. In *Quaternary plant ecology*, H. J. B. Birks and R. G. West (eds). Oxford: Blackwell.

Bergqvist, E. 1971. Runoff and discharge of suspended matter and dissolved solids at Stabby and Sundbro. *Drainage basins in Central Sweden*, Part 4. Uppsala Universitet Natugeographiska Institutionen.

Bernaix, J. 1969. New laboratory methods of studying the mechanical properties of rocks. *Int.J.Rock Mech.Min.Sci.* **6**, 43–90.

Berry, B. J. L. 1964. *Sampling, coding and storing flood plain data.* Agric. Handb., US Dept Agric., no.237.

Bertotti, L. and L. Cavaleri 1986. Hindcast of short-term shoreline evolution. *Estuar., Coastal Shelf Sci.* **23**(4), 499–512.

Beschel, R. E. 1957. Lichenometrie im Gletschervorfeld. *Sonderdr. Jahrb. 1957 Ver. Schutze Aplenflanzen Tiers* **22**, 164–85.

Beschel, R. E. 1961. Dating rock surfaces by lichen growth and its application to glaciology and physiography (lichenometry). In *Geology of the Arctic: proc.1st int.symp.on Arctic geology*, G.O. Raasch (ed.), 1044–62. Toronto: University of Toronto Press.

Best, J. L. 1986. The morphology of river channel confluences. *Prog.Phys.Geog.* **10**, 157–74.

Bethlamy, N. 1974. More streamflow after a bark beetle epidemic. *J.Hydrol.* **23**, 185–9.

Beven, K. J. and J. L. Callen 1978. HYDRODAT – a system of FORTRAN computer programs for the preparation and analysis of hydrological data from charts. *BGRG Tech.Bull.* **23.**

Bezinge, A., J. P. Perreten and F. Schafer 1973. Phénomènes du lac glaciaire du Garner. *Int.Assoc. Sci.Hydrol.Publ.* **95**, 65–78. Symp. on hydrology of glaciers, Cambridge, 7–13 Sept. 1969.

Biederman, E. W. 1962. Distinction of shoreline environments in New Jersey. *J.Sedim.Petrol.* **32**, 181–200.

Bigelow, G. E. 1984. Simulation of pebble abrasion on coastal benches by transgressive waves. *Earth Surf. Processes Landforms* **9**, 383–90.

Bilham, R. and D. Simpson 1984. Indo-Asian convergence, and the 1913 survey line connecting the Indian and Russian triangulation surveys. In *The international Karakoram project*, K. J. Miller (ed.), vol.I, 160–70. Cambridge: Cambridge University Press.

Billi, P. 1984. Quick field measurement of gravel particle size. *J.Sedim.Petrol.* **54**, 658–60.

Binns, R. E. 1967. Drift pumice on postglacial raised shorelines of northern Europe. *Acta Borealia* **24**, 1–64.

Bircher, W. 1986. Dendrochronology applied in mountain regions. In *Handbook of Holocene palaeoecology and palaeohydrology*, B. E. Berglund (ed.), 387–403. Chichester: Wiley.

Bird, E. C. F. 1985. *Coastal change: a global review.* Chichester: Wiley.

Birkeland, P. W. 1973. Use of relative age-dating methods in a stratigraphic study of rock glacier deposits, Mt Sopris, Colorado. *Arctic Alpine Res.* **5**, 401–16.

Birkeland, P. W. 1974. *Pedology, weathering and geomorphological research.* Oxford: Oxford University Press.

Birkeland, P. W. 1984. *Soils and geomorphology*, 2nd edn. New York: Oxford University Press.

Birks, H. J. B. 1973. *Past and present vegetation of the Isle of Skye.* Cambridge: Cambridge University Press.

Birks, H. J. B. and H. H. Birks 1980. *Quaternary palaeoecology.* London: Edward Arnold.

Birot, P. 1969. Etude experimentale de la décomposition des roches cristallines. *Bull. Centre Géomorphol. Caen* **3**.

Bishop, A. W. 1955. The use of the slip circle in the stability analysis of slopes. *Geotechnique* **5**, 7–17.

Bishop, A. W. and D. J. Henkel 1962. *The measurement of soil properties in the triaxial test.* London: Edward Arnold.

Bishop, A. W., C. E. Garga, A. Anderson and J. D. Brown 1971. A new ring shear apparatus and its application to the measurement of residual strength. *Geotechnique* **21**, 273–328.

Bishop, P. 1985. Southeast Australian late Mesozoic and Cenozoic denudation rates: a test for late Tertiary increases in continental denudation. *Geology* **13**, 479–82.

Bisque, R. E. 1961. Analysis of carbonate rocks for calcium, magnesium, iron and alunimium with EDTA. *J.Sedim.Petrol.* **31**, 113–32.

Bjerrum, L. and F. Jorstad 1968. Stability of rock slopes in Norway. *Nor. Geotech.Inst.Publ.* **79**, 1–11.

Black, C. A. (ed.) 1965. *Methods of soil analysis.* Madison, Wisc.: American Society of Agronomy.

Blackwelder, E. 1933. The insolation hypothesis of rock weathering. *Am.J.Sci.* **26**, 97–113.

Blackwell, B., H. P. Schwarcz and A. Debenath 1983. Absolute dating of hominids and paleolithic artefacts of the cave of La Chaise-de-Vouthon (Charente), France. *J.Archaeol.Sci.* **10**, 493–513.

Blair, D. J. and T. H. Biss 1967. *The measurement of shape in geography.* Quant. Bull., Dept Geog., Univ. Nottingham, no. 11.

Blakemore, L. C. and H. S. Gibbs 1968. Effects of gannets on soil at Cape Kidnappers, Hawke's Bay. *NZ J. Sci.* **11**, 54–62.

Blakemore, L. C., P. L. Searle and B. K. Dale 1977. *Soil Bureau Methods. A: Methods of chemical analysis of rocks.* NZ Soil Bureau Sci. Rep. no. 10A.

Bloemendal, J., F. Oldfield and R. Thompson 1979. Magnetic measurements used to assess sediment influx at Llyn Goddionduon. *Nature* **280**, 50–3.

Blong, R. J. 1972. Methods of slope profile measurement in the field. *Aust. Geog. Stud.* **10**, 182–92.

Bloom, A. L., W. S. Broecker, J. M. A. Chappell, R. K. Matthews and K. J. Mesolella 1974. Quaternary sea-level fluctuations on a tectonic coast: new $^{230}Th/^{234}U$ dates from the Huon Peninsula, New Guinea. *Quat. Res.* **4**, 185–205.
Board, C. 1967. Maps as models. In *Models in geography*, R. J. Chorley and P. Haggett (eds), 671–725. London: Methuen.
Boatman, D. J., P. D. Hulme and R. W. Tomlinson 1975. Monthly determinations of the concentrations of sodium, potassium, magnesium and calcium in the rain and in pools on the Silver Flowe National Nature Reserve. *J.Ecol.* **63**, 903–12.
Bogardi, J. 1974. *Sediment transport in alluvial streams*. Budapest: Akademiai Kiado.
Boggs, S. 1957. Measurement of roundness and sphericity parameters using an electronic particle size analyzer. *J.Sedim.Petrol.* **37**, 908–13.
Bolline, A. 1978. Study of the importance of splash and wash on cultivated loamy soils of Hesbaye (Belgium). *Earth Surf. Processes* **3**, 71–84.
Bolster, S. J. S. 1985. The nature of origin of earth hummocks, Breisterdalen, Jotunheimen, Norway. *Swansea Geog.* **22**, 94–115.
Bolt, B. A. and W. C. Marion 1966. Instrumental measurement of slippage on the Hayward Fault. *Seismol. Soc.Am.Bull.* **56**, 305–16.
Bonham, L. C. and J. H. Spotts 1971. Measurement of grain orientation. In *Procedures in sedimentary petrology*, R. E. Carver (ed.), 285–312. New York: Wiley.
Bonnefille, R., P. Cormault and J. Vlembois 1967. Progrès des méthodes de mésure de la Hade naturelle au laboratoire natural d'hydraulique. In *Proc.9th conf.on coastal engineering*, 115–26. Tokyo.
Bonny, A. P. 1976. Recruitment of pollen to the seston and sediments of some Lake District lakes. *J.Ecol.* **64**, 859–87.
Booker, D. R. and L. W. Cooper 1965. Superpressure balloons for weather research. *J.Appl.Meteorol.* **4**, 122–9.
Boon, J. D. 1969. Quantitative analysis of beach sand movement, Virginia Beach, Virginia. *Sedimentology* **13**, 85–103.
Boon, J. D., III, D. A. Evans and H. F. Hennigar 1982. Interpretation of grain shape information from Fourier analysis of two-dimensional images. *Math.Geol.* **14**, 589–605.
Boothroyd, J. C. and D. K. Hubbard 1974. *Bedform development and distribution pattern, Parker and Essex estuaries, Massachusetts*. Coastal Res. Center, Univ.Massachusetts, Amherst, Mass., Misc. Paper no. 1–74.
Boots, B. N. and R. K. Burns 1984. Analysing the spatial distribution of drumlins: a two-phase mosaic approach. *J.Glaciol.* **30**, 302–7.
Bormann, F. H. and G. E. Likens 1967. Nutrient cycling. *Science* **155**, 424–9.
Bormann, F. H., G. E. Likens, T. G. Siccama, R. S. Pierce and J. S. Eaton 1974. The export of nutrients and recovery of stable conditions following deforestation at Hubbard Brook. *Ecol. Monogr.* **44**, 155–77.
Boulton, G. S. 1967. The development of a complex supraglacial moraine at the margin of Sorbreen, Ny Friesland, Vestspitzbergen. *J.Glaciol.* **6**, 717–36.
Boulton, G. S. 1968. Flow tills and related deposits on some Vestspitzbergen glaciers. *J.Glaciol.* **7**, 391–412.
Boulton, G. S. 1972. The role of thermal regime in glacial sedimentation. In *Polar geomorphology*, R. J. Price and D. E. Sugden (eds). *Inst.Br.Geog.Spec.Publ.* **4**, 1–19.
Boulton, G. S. 1974. Processes and patterns of glacial erosion. In *Glacial geology*, D. R. Coates (ed.), 41–87. Binghampton: State University of New York.
Boulton, G. S. 1975. Processes and patterns of sub-glacial sedimentation: a theoretical approach. In *Ice ages: ancient and modern*, A. E. Wright and F. Moseley (eds), 7–42. Liverpool: Seel House.
Boulton, G. S. 1979. Processes of glacier erosion of different substrata. *J.Glaciol.* **23**, 15–38.
Boulton, G. S. 1982. Subglacial processes and the development of glacial bedforms. In *Research in glacial, glaciofluvial and glaciolacustrine systems*, R. Davidson-Arnott, W. Nickling and B. D. Fahey (eds), 1–31. Norwich: Geo Books.
Boulton, G. S. and M. Rhodes 1974. Isostatic uplift and glacial history in northern Spitzbergen. *Geol.Mag.* **111**, 481–576.
Boulton, G. S., D. L. Dent and E. M. Morris 1974. Subglacial shearing and crushing, and the role of water pressures in tills from south-east Iceland. *Geog. Annaler* **56A**, 135–45.
Boulton, G. S., E. M. Morris, A. A. Armstrong and A. Thomas 1979. Direct measurement of stress at the base of a glacier. *J.Glaciol.* **22**, 3–24.
Bouma, A. H. 1964. Notes on X-ray interpretation of marine sediments. *Mar. Geol.* **2**, 278–309.
Bouma, A. H. 1969. *Methods for the study of sedimentary structures*. New York: Wiley.

Bowden, K. F. 1962. Turbulence. In *The sea: ideas and observations*, M. N. Hill (ed.), 3. New York: Interscience.

Bowen, D. Q., G. A. Sykes, M. A. Reeves, G. H. Miller, J. T. Andrews, J. W. Brew and P. E. Hare 1985. Amino acid geochronology of raised beaches in southwest Britain. *Quat. Sci. Rev.* **4**, 279–318.

Box, G. E. P. and G. M. Jenkins 1970. *Times series analysis – forecasting and control*. San Francisco: Holden-Day.

Boyd, J. M., D. V. Hinds, D. Moy and C. Rogers 1973. Two simple devices for monitoring movements in rock slopes. *Q. J. Eng.Geol.* **6**, 295–302.

Boyer, L. 1981. Generalisation in semi-detailed geomorphological mapping. *ITC Journal* **1981** (1), 98–123.

Bradbury, S. 1977. Quantitative image analysis. In *Analytical and quantitative methods in microscopy*, G. A. Meek and H. Y. Elder (eds). Cambridge: Cambridge University Press.

Bradley, S. B. 1984. Flood effects on the transport of heavy metals. *Int. J. Environ. Stud.* **22**, 225–30.

Bradley, W. C. and G. B. Griggs 1976. Form, genesis and deformation of Central California wave-cut platforms. *Geol.Soc.Am.Bull.* **87**, 433–49.

Bragg, G. M. 1974. *Principles of experimentation and measurement*. Englewood Cliffs, NJ: Prentice-Hall.

Brassel, K. 1975. A model for automatic hill-shading. *Am. Cartogr.* **1**, 15–27.

Bray, J. R. and E. Gorham 1964. Litter production in forests of the world. *Adv.Ecol.Res.* **2**, 101–57.

Breed, C. S. 1977. Terrestrial analogs of the Hellespontus dunes, Mars. *Icarus* **30**, 326–40.

Brenninkmeyer, B. M. S. J. 1971. New instrumentation to determine topographic changes, bed load movement, and suspended sediment concentration. In *Abstract volume, second national shallow water research conf.*, 29. Los Angeles: University of Southern California Press.

Brenninkmeyer, B. M. S. J. 1973. Topography changes and sediment movement in surf zone. In *Int.symp.relations sedimentaires entre estuaries et plateaux continentaux*. Inst.Géol.Basin d'Aquitaine, Bordeaux, France.

Brenninkmeyer, B. M. S. J. 1976. Sand fountains in the surf zone. In *Beach and nearshore sedimentation*, R. A. Davis and R. L. Ethington (eds), 69–81. Tulsa: Society of Economic Palaeontologists and Mineralogists.

Bressolier, C. and Y.-F. Thomas 1977. Studies on wind and plant interactions on French Atlantic coastal dunes. *J.Sedim.Petrol.* **47**, 331–8.

Brewer, R. 1978. *Soil fabric and mineral analysis*, 2nd edn. New York: Wiley.

Brice, J. C. 1960. Index for descriptions of channel braiding (abstract). *Geol.Soc.Am.Bull.* **71**, 1833.

Bricker, O. P., A. E. Godfrey and E. T. Cleaves 1968. Mineral–water interaction during the chemical weathering of silicates. *ACS Adv. Chem. Ser.* **73**, 128–42.

Brindley, G. W. and G. Brown 1980. *Crystal structures of clay minerals and their X-ray identification*. London: Mineralogical Society.

British Standards Institution 1964. *Methods of measurement of liquid flow in open channels*. Part 3: *Velocity area methods*. BS 3680. London: British Standards Institution.

British Standards Institution 1975a. *Methods of test for stabilized soils*. BS 1924. London: British Standards Institution.

British Standards Institution 1975b. *Methods for sampling and testing of mineral aggregates, sands and filters*. BS 812. London: British Standards Institution.

British Standards Institution 1975c. *Methods of testing soils for civil engineering purposes*. BS 1377. London: British Standards Institution.

Broecker, W. and T. L. Ku 1969. Caribbean cores P6304–8 and P6404–9: new analysis of absolute chronology. *Science* **166**, 404–6.

Broecher, W. and J. van Donk 1970. Insolation changes, ice volumes and the ^{18}O record in deep-sea cores. *Rev.Geophys.Space Phys.* **8**, 169–98.

Brogger, W. C. 1900. Om de seniglacial og postglacial nivaforandringer i Kristianiafelter. *Nor. Geol. Unders.* **31**, 1–731.

Brooker, M. P., D. L. Morris and C. J. Wilson 1978. Plant-flow relationships in the R. Wye catchment. In *Proc.EWRS 5th symp.: aquatic weeds*, 63–70.

Brookfield, M. 1970. Dune trends and wind regime in Central Australia. *Z. Geomorphol. Supp. NF* **10**, 121–53.

Brookfield, M. E. and T. S. Ahlbrandt (eds) 1983. *Eolian sediments and processes*. Amsterdam: Elsevier.

Brooks, C. L. 1972. Pollen analysis and the Main Buried Beach in the western part of the Forth valley. *Trans.Inst.Br.Geog.* **55**, 161–70.

Brown, D. 1968. *Methods of surveying and measuring vegetation*. Commonw. Bur. Pastures Field Crops Bull. no.42.

Brown, D. A. 1971. Stream channels and flow relations. *Water Resour. Res.* **7**, 304–10.

Brown, E. H. 1950. Erosion surfaces in North Cardiganshire. *Trans.Inst.Br.Geog.* **16**, 52–66.

Brown, G. (ed.) 1961. *The X-ray identification and crystal structures of clay minerals.* London: Mineralogical Society.

Brown, J. A. H. 1972. Hydrologic effects of a bush fire in a catchment in south-eastern New South Wales. *J.Hydrol.* **15**, 77–96.

Brown, J. E. 1973. Depositional histories of sand grains from surface textures. *Nature* **242**, 396–8.

Brown, L., M. J. Pavich, R. E. Hickman, J. Klein and R. Middleton 1988. Erosion of the eastern United States observed with ^{10}Be. *Earth Surf. Processes Landforms* **13** (5), 441–57.

Brugam, R. B. 1978. Pollen indicators of land use change in southern Connecticut. *Quat. Res.* **9**, 349–62.

Brunn, P. 1968. Quantitative tracing of littoral drift. In *Proc. 11th conf. on coastal engineering* vol. 1, 322–8. Am.Soc.Civ.Eng.

Brunsden, D. and D. K. C. Jones 1972. The morphology of degraded landslide sloped in south west Dorset. *Q. J. Eng. Geol.* **5**, 205–22.

Brunsden, D. and D. K. C. Jones 1976. The evolution of landslide slopes in Dorset. *Phil.Trans. R.Soc.A* **283**, 605–31.

Brunsden, D., J. C. Doornkamp, P. G. Fookes, D. K. C. Jones and J. M. H. Kelly 1975. Large scale geomorphological mapping and highway engineering design. *Q.J.Eng.Geol.* **8**, 227–53.

Brush, G. A., E. A. Martin, R. S. Defries and C. A. Rice 1982. Comparisons of Pb-210 and pollen methods for determining rates of estuarine sediment accumulation. *Quat.Res.* **18**, 196–217.

Brush, L. M. 1961. *Drainage basins, channels and flow characteristics of selected streams in central Pennsylvania.* US Geol.Surv.Prof. Paper no. 282F.

Bryan, R. B. 1974. Water erosion by splash and wash and the erodibility of Albertan soil. *Geog. Annaler* **56A**, 159–75.

Bryan, R. B. 1977. The influence of soil properties on degradation of mountain hiking trails at Grovelsjon. *Geog. Annaler* **59A**, 49–65.

Bryan, R. B., A. C. Imeson and I. A. Campbell 1984. Solute release and sediment entrainment on microcatchments in the Dinosaur Park Badlands, Alberta, Canada. *J.Hydrol.* **71**, 79–106.

Bucknam, R. C. and R. G. Anderson 1979. Estimation of fault-scarp ages from a scarp-height–slope-angle relationship. *Geology* **7**, 11–14.

Budd, W. F., P. L. Kleage and N. A. Blundy 1979. Empirical studies of ice sliding. *J.Glaciol.* **23**, 157–70.

Bull, P. A. 1977. Glacial deposits identified by chattermark trails in detrital garnets: a comment. *Geology* **5**, 248–9.

Bull, P. A. 1978a. A quantitative approach to SEM analysis of cave sediments. In *Scanning electron microscope studies of sediments*, W. B. Whalley (ed.). Norwich: Geo Abstracts.

Bull, P. A. 1978b. Observations on small sedimentary quartz particles analysed by scanning electron microscopy. In *Scanning electron microscopy 1978*, O. Johari (ed.), vol.1, 821–8. Chicago: SEM.

Bull, P. A. 1978c. Surge mark formation and morphology. *Sedimentology* **25**, 877–86.

Bull, P. A. 1981. Environmental reconstruction by electron microscopy. *Prog.Phys.Geog.* **5**, 368–97.

Bull, P. A., W. B. Whalley and A. W. Magee 1987. An annotated bibliography of environmental reconstruction by SEM (1962–1985). *BGRG Tech.Bull.* **35**. Norwich: Geo Books.

Bullock, P. and C. P. Murphy 1976. The microscopic examination of the structure of the sub-surface horizons of soils. *Outlook Agric.* **8**, 348–54.

Bullock, P. and C. P. Murphy 1979. Evolution of a palaeoargillic brown earth (paleudalf) from Oxfordshire, England. *Geoderma* **22**, 225–52.

Bullock, P., N. Federoff, A. Jongerius, G. Stoops and T. Tursina 1985. *Handbook for soil thin section description.* Wolverhampton: Waine Research Publications.

Bunce, R. G. H. 1968. Biomass and production of trees in a mixed deciduous woodland. *J.Ecol.* **56**, 759–75.

Bunge, W. 1966. *Theoretical geography.* Lund Ser. Geog. Ser. C no. 1.

Burke, K. and S. J. Freeth 1967. A rapid method for the determination of shape, sphericity and size of gravel fragments. *J.Sedim.Petrol.* **39**, 797–8.

Burke, W. H., R. E. Denison, E. A. Hetherington, R. E. Koepnick, H. F. Nelson and J. B. Otto 1982. Variation of seawater ^{87}Sr/^{86}Sr throughout Phanerozoic time. *Geology* **10**, 516–19.

Burnett, A. W. and S. A. Schumm 1983. Alluvial-river response to neotectonic deformation in Louisiana and Mississippi. *Science* **222**, 49–50.

Burrell, G. J. 1971. The locational and spatial analysis of a dune system in the Ziz valley, Southern Morocco. *Reading Geog.* **2**, 24–41.

Burrough, P. A. 1985. Fakes, facsimiles and facts: fractal models of geophysical phenomena. In *Science and uncertainty*, S.Nasa (ed.), 151–68. London: IBM/Science Reviews.

Burt, T. P. 1974. *A study of hydraulic conductivity in frozen soil*. Unpub. M.A. Thesis, Carleton University, Ottawa.

Burt, T. P. 1978. An automatic fluid scanning switch tensiometer system. *BGRG Tech.Bull.* **21**.

Burt, T. P. 1979. The relationship between throughflow generation and the solute concentration of soil and stream water. *Earth Surf. Processes* **4**, 257–66.

Burt, T. P. and M. Beasant 1984. An improved design of a mercury manometer tensiometer. *BGRG Tech.Bull.* **33**, 43–8.

Burt, T. P. and D. P. Butcher 1985, On the generation of delayed peaks in stream discharge. *J.Hydrol.* **78**, 361–78.

Burt, T. P. and D. P. Butcher 1986. Development of topographic indices for use in semi-distributed hillslope runoff models. *Z. Geomorphol. Supp. NF* **58**, 1–19.

Burt, T. P. and D. E. Walling 1984. Catchment experiments in fluvial geomorphology: a review of objectives and methodology. In *Catchment experiments in fluvial geomorphology*, T. P. Burt and D. E. Walling (eds), 3–18. Norwich: Geo Books.

Burt, T. P., M. A. Donohoe and A. R. Vann 1983. The effect of forestry drainage operations on upland sediment yields: the results of a storm-based study. *Earth Surf. Processes Landforms* **8**, 339–46.

Butcher, G. C. and J. B. Thornes 1978. Spatial variability in run-off processes in an ephemeral channel. *Z. Geomorphol. NF* **29**, 83–92.

Butler, L. R. P. and M. L. Kokot 1971. Atomic absorption. In *Modern methods of geochemical analysis*, R. E. Wainerdi and E. A. Uken (eds), 205–43. New York: Monographs in Geoscience.

Butterfield, G. R. 1971. The instrumentation and measurement of wind erosion. In *Proc.6th geographical conf.*, 125–30. New Zealand Geographical Society.

Butterworth, B. 1964. The frost resistance of bricks and tiles. *J.Br.Ceram.Soc.* **1** 203–36.

Butzer, K. W. 1975. Pleistocene littoral–sedimentary cycles of the Mediterranean Basin: a Mallorquin view. In *After the Australopithecines*, K. W. Butzer and G. L. Isaac (eds), 25–71. The Hague: Aldine.

Butzer, K. W. 1977. Pleistocene sea-level stratigraphy in Mallorca. In *Géologie Méditerranéene*, H. DeLumley (ed.).

Butzer, K. W. and J. B. Cuerda 1962. Coastal stratigraphy of southern Mallorca and its implications for the Pleistocene chronology of the Mediterranean Sea. *J.Geol.* **70**, 398–416.

Caillère, S., S. Henin and P. Birot 1954. Essai d'alteration artificielle de quelques roches. *C.R. Acad.Sci. Paris* **240**, 1441–2.

Cailleux, A. 1945. Distinction des galets marines et fluviatile. *Bull.Soc.Géol.Fr* **5**, 375–404.

Cailleux, A. 1947. L'indice d'émousse des grains de sable et grès. *Rev. Géomorphol. Dyn.* **3**, 78–87.

Cailleux, A. 1957. Etudes sur l'érosion et la sedimentation en Guyane. In *Comm.Quatr.Conf. Géol.des Guyanes*, 50–71.

Caillot, A. 1973. Sediment labelling with radio-active isotopes. In *Tracer techniques in sediment transport*, 169–78. Vienna: International Atomic Energy Agency.

Caine, N. 1976. A uniform measure of subaerial erosion. *Bull.Geol.Soc.Am.* **87**, 137–40.

Caine, T. N. 1969. A model for alpine talus development by slush avalanching. *J.Geol.* **77**, 92–100.

Caine, T. N. 1979. Rock weathering rates at the soil surface in an Alpine environment. *Catena* **6**, 131–44.

Caldenius, C. and R. Lundstrøm 1956. The landslide at Surte on the River Gota Alu. *Sver. Geol. Unders. C* 27.

Caldwell, N. E. and A. T. Williams 1985. The role of beach profile configuration in the discrimination between differing depositional environments affecting coarse clastic beaches. *J. Coastal Res.* **1** (2), 129–39.

Caldwell, N. E., A. T. Williams and G. T. Roberts 1982. The swash force transducer. *J.Sedim.Petrol.* **52** (2), 669–71.

Calkin, P. E. and R. H. Rutford 1974. The sand dunes of Victoria Valley, Antarctica. *Geog.Rev.* **64**, 189–216.

Calkins, D. and T. Dunne 1970. A salt tracing method for measuring channel velocities in small mountain streams. *J. Hydrol.* **11**, 379–92.

Calvert, S. E. and J. J. Veevers 1962. Minor structures of unconsolidated marine sediments revealed by X-radiography. *Sedimentology* **1**, 296–301.

Cambers, G. 1976. Temporal scales in coastal erosion systems. *Trans Inst.Br.Geog.* **1**, 246–56.

Cambray, R. S., E. M. R. Fisher, D. H. Pierson and A. Parker 1972. *Radioactive fallout in air and rain. Results to the middle of 1972*. AERE Report no 7245. London: HMSO.

Campbell, C. A., E. A. Paul, D. A. Rennie and K. J. McCallum 1967. Applicability of the carbon-dating method of analysis to soil humus studies. *Soil Sci.* **104**, 217–24.

Campbell, I. A. 1970. Micro-relief measurements on unvegetated shale slope. *Prof. Geog.* **22**, 215–20.

Cannon, P. J. 1976. Generation of explicit parameters for a quantitative geomorphic study of the Mill Creek drainage basin. *Oklahoma Geol. Notes* **36**, 3–19.

Capo, R. C. and D. J. DePaolo 1988. Plio-Pleistocene strontium isotope record from southwest Pacific Ocean DSDP site 590B. *Eos* **69**, 1253.

Capot-Rey, R. 1963. Contribution à l'étude et à la répresentation des barkanes. *Trav. Inst.Rech.Sahar.* **22**, 37–60.

Carling, P. A. 1983. Thresholds of coarse sediment transport in broad and narrow natural streams. *Earth Surf. Processes Landforms* **8**, 1–18.

Carlisle, A., A. H. F. Brown and E. J. While 1966. The organic matter and nutrient elements in the precipitation beneath a sessile oak (*Quercus petraea*) canopy. *J.Ecol.* **54**, 87–98.

Carlson, T. N. and J. M. Prospero 1972. The large-scale movement of Saharan air outbreaks over the Northern Equatorial Atlantic. *J.Appl.Meteorol.* **11**, 283–97.

Carlston, C. W. and W. B. Langbein 1960. Rapid approximation of drainage density: line intersection method. *US Geol. Surv.Water Resour. Div., Bull.* **11**.

Carr, A. P. 1962. Cartographical error and historical accuracy *Geography* **47**, 135–44.

Carr, A. P. 1983. A new technique for determining the surface level of salt marshes and other uncompacted surfaces. *Earth Surf. Processes Landforms* **8**, 293–6.

Carr, A. P., M. W. L. Blackley and H. L. King 1982. Spatial and seasonal aspects of beach stability. *Earth Surf. Processes Landforms* **7**, 267–82.

Carrara, A., E. Catalano, M. Sorriso Valvo, C. Reali and I. Osso 1978. Digital terrain analysis for land evaluation. *Geol. Appl. Idrogeol.* **13**, 67–127 and maps.

Carrara, P. E. and T. R. Carroll 1979. The determination of erosion rates from exposed tree roots in the Piceance Basin, Colorado. *Earth Surf. Processes* **4**, 307–17.

Carroll, D. 1962. *Rainwater as a chemical agent of denudation.* US Geol. Surv. Water Supply Paper no 1535–G.

Carroll, D. 1970. *Clay minerals: a guide to their X-ray identification.* Boulder, Colo.: Geological Society of America.

Carruthers, J. N. 1962. The easy measurement of the bottom currents at modest depths. *Civ.Eng.* **57**, 484.

Carson, M. 1966. Some problems with the use of correlation techniques in morphometric studies. In *Morphometric analysis of maps*, H. O. Slaymaker (ed.), 149–66. BGRG Occas.Paper no. 4.

Carson, M. A. 1967. The magnitude of variability in samples of certain geomorphic characteristics drawn from valleyside slopes. *J.Geol.* **75**, 93–100.

Carson, M. A. 1971a. *The mechanics of erosion.* London: Pion.

Carson, M. A. 1971b. Application of the concept of threshold slopes to the Laramie Mountains, Wyoming. *Inst.Br.Geog.Spec.Publ.* **3**, 31–49.

Carson, M. A. and M. J. Kirkby 1972. *Hillslope form and process.* Cambridge: Cambridge University Press.

Carson, M. A. and D. J. Petley 1970. The existence of threshold slopes in the denudation of the landscape. *Trans.Inst.Br.Geog.* **49**, 71–95.

Carter, D. J. 1986. *The remote sensing sourcebook: a guide to remote sensing products, services, facilities, publications and other material.* London: Kogan Page/McCarta.

Carter, L., L. S. Mitchell and N. J. Day 1981. Suspended sediment beneath permanent seasonal ice, Ross Ice Shelf, Antarctica. *NZ J.Geol.Geophys.* **25**, 249–62.

Carver, R. E. 1971a. Heavy mineral separation. In *Procedures in sedimentary petrology*, R. E. Carver (ed.), 427–52. New York: Wiley.

Carver, R. E. (ed.) 1971b. *Procedures in sedimentary petrology.* New York: Wiley.

Caston, V. N. D. 1970. Tidal sand movement between some linear sand banks in the North Sea off north-east Norfolk. *Mar.Geol.* **9**, M38–42.

Caston, V. N. D. 1972. Linear sandbanks in the southern North Sea. *Sedimentology* **18**, 63–78.

Catt, J. A. 1979. Soils and Quaternary geology in Britain. *J.Soil Sci.* **30**, 607–42.

Catt, J. A. 1988. *Quaternary geology for scientists and engineers.* Chichester: Ellis Horwood.

Causse, C., A. de Vernal and C. Hillaire-Marcel 1987. Elements of a pre-Late Wisconsinan climatostratigraphy in eastern Canada from Th/U dating and palynological studies of organic-rich deposits. In *Programme with abstracts, INQUA '87*, 142.

Cawsey, D. C. and N. S. Farrar 1976. A simple sliding apparatus for the measurement of rock joint friction. *Geotechnique* **26**, 382–6.

Cerling, T. E. 1984. The stable isotopic composition of soil carbonate and its relationship to climate. *Earth Planet. Sci. Lett.* **71**, 229–40.

Cerling, T. E. and R. L. Hay 1986. An isotopic study of paleosol carbonates from Olduvai Gorge. *Quat.Res.* **25**, 63–78.

Chadwick, W. W., D. A. Swanson, E. Y. Iwatsubo, C. C. Helicker and T. A. Leighley 1983. Deformation monitoring at Mount St Helens in 1981 and 1982. *Science* **221**, 1378–80.
Chafetz, H. S. and R. L. Folk 1984. Travertines: depositional morphology and the bacterially controlled constituents. *J.Sedim.Petrol.* **54**, 289–316.
Chandler, R. J. 1973a. The inclination of talus, Arctic talus terraces and other slopes composed of granular materials. *J.Geol.* **81**, 1–14.
Chandler, R. J. 1973b. Lias clay: the long term stability of cutting slopes. *Geotechnique* **24**, 21–38.
Chandler, R. J. 1976. The history and stability of two Lias clay slopes in the upper Gwash valley, Rutland. *Phil.Trans.R.Soc.A* **283**, 463–92.
Chandler, R. J. 1977. The application of soil mechanics methods to the study of slopes. In *Applied geomorphology*, J. R. Hails (ed.), 157–82. Oxford: Elsevier.
Chandler, R. J. and N. Pook 1971. Creep movements in low gradient clay slopes since the Late Glacial. *Nature* **229**, 399–400.
Chang, M. and D. G. Boyer 1977. Estimates of low flows using watershed and climatic parameters. *Water Resour.Res.* **13**, 997–1001.
Chapman, C. A. 1952. A new quantitative method of topographic analysis. *Am.J.Sci.* **250**, 428–52.
Chapman, G. 1948. Size of raindrops and their striking force at the soil surface in a red pine plantation. *Trans Am.Geophys.Union.* **29**, 664–70.
Chappell, J. and H. H. Veeh 1978. $^{230}Th/^{234}U$ age support of an interstadial sea-level of −40 m at 30 000 yr BP. *Nature* **276**, 602–3.
Charlot, G. 1964. *Colorimetric determination of elements: principles and methods.* New York: Elsevier.
Chartres, C. J. 1980. A Quaternary soil sequence in the Kennet Valley, central southern England. *Geoderma* **23**, 125–46.
Chatfield, E. J. 1967. A battery operated sequential air concentration and deposition sampler. *J.Atmos.Environ.* **1**, 211–20.
Chayes, F. 1971. *Ratio correlation.* Chicago: University of Chicago Press.
Chepil, W. S. 1945. Dynamics of wind erosion. 1: Nature of movement of soil by wind. *Soil Sci.* **60**, 305–20.
Chepil, W.S. 1951. Properties of soil which influence wind erosion. V: Mechanical stability of structure. *Soil Sci.* **72**, 465–78.
Chepil, W. S. 1952. Improved rotary sieve for measuring state and stability of dry soil structure. *Proc.Soil Sci.Soc.Am.* **16**, 113–17.
Chepil, W. S. 1955. Factors that influence clod structure and erodibility of soil by wind. IV: Sand, silt and clay. *Soil Sci.* **80**, 155–62.
Chepil, W.S. 1957a. Sedimentary characteristics of dust storms. I: Sorting of wind-eroded soil material. *Am.J.Sci.* **255**, 12–22.
Chepil, W. S. 1957b. *Width of field strips to control wind erosion.* Kansas State Coll.Agric.Appl.Sci. Tech.Bull. no. 92.
Chepil, W. S. 1957c. Influence of moisture on erodibility of soil by wind. *Proc.Soil Sci.Soc.Am.* **20**, 288–92.
Chepil, W. S. 1962. A compact rotary sieve and the importance of dry sieving in physical soil analysis. *Proc.Soil Sci.Soc.Am.* **26**, 4–6.
Chepil, W. S. and R. A. Milne 1939. Comparative study of soil drifting in the field and in a wind tunnel. *Sci.Agric.* **19**, 249.
Chepil, W. S. and F. H. Siddoway 1959. Strain-gauge anemometer for analysing various characteristics of wind turbulence. *J.Meteorol.* **16**, 411–18.
Chepil, W. S. and N. P. Woodruff 1957. Sedimentary characteristics of dust storms. II: Visibility and dust concentration. *Am.J.Sci.* **255**, 104–14.
Chepil, W. S. and N. P. Woodruff 1963. The physics of wind erosion and its control. *Adv.Agron.* **15**, 211–302.
Cherry, J. A. M. 1966. Sand movement along equilibrium beaches north of San Francisco. *J.Sedim.Petrol.* **36**, 341–57.
Chester, R., H. Elderfield, J. J. Griffin, L. R. Johnson and R. C. Padgham 1972. Eolian dust along the eastern margins of the Atlantic Ocean. *Mar. Geol.* **13**, 91–105.
Ching, R. K. H., D. J. Sweeney and D. G. Fredlund 1984. Increase in factor of safety due to soil suction for two Hong Kong slopes. In *Proc. 4th int. symp. on landslides*, vol. 1. Toronto: International Society for Soil Mechanics and Foundation Engineering.
Chisholm, B., J. Driver, S. Dube and H. P. Schwarcz 1986. Assessment of prehistoric bison foraging and movement patterns via stable carbon isotopic analysis. *Plains Anthropol.* **31**, 193–205.
Chorley, R. J. 1958. Group operator variance in morphometric work with maps. *Am.J.Sci.* **256**, 208–18.

Chorley, R. J. 1959. The shape of drumlins. *J.Glaciol.* **3**, 339–44.
Chorley, R. J. 1962. The geomorphic significance of some Oxford soils. *Am.J.Sci.* **257**, 503–15.
Chorley, R. J. 1966. The application of statistical methods to geomorphology. In *Essays in geomorphology*, G. H. Dury (ed.), 275–388. London: Heinemann.
Chorley, R.J. 1969. The drainage basin as the fundamental geomorphic unit. In *Water, earth and man*, R. J. Chorley (ed.), 77–100. London: Methuen.
Chorley, R. J. 1978. Bases for theory in geomorphology. In *Geomorphology: present problems and future prospects*, C. Embleton, D. Brunsden and D. K. C. Jones (eds). Oxford: Oxford University Press.
Chorley, R. J. and P. F. Dale 1972. Cartographic problems in stream channel delineation. *Cartography* **7**, 150–62.
Chorley, R. J. and B. A. Kennedy 1971. *Physical geography, a systems approach*. London: Prentice-Hall.
Chorley, R. J., A. J. Dunn and R. P. Beckinsale 1964. *The history of the study of landforms or the development of goemorphology*. Vol. I: *Geomorphology before Davis*. London: Methuen.
Chorley, R. J., D. E. G. Malm and H. A. Pogorzelski 1957. A new standard for estimating basin shape. *Am.J.Sci.* **255**, 138–41.
Chow, T. L. 1977. A porous cup soil-water sampler with volume control. *Soil Sci.* **124**, 173–6.
Chow, V. T. 1959. *Open channel hydraulics*. New York: McGraw-Hill.
Church, M. 1975. Electrochemical and fluorometric tracer techniques for streamflow measurements. *BGRG Tech.Bull.* **12**.
Church, M. and R. Kellerhals 1970. *Stream gauging techniques for remote areas using portable equipment*. Dept. Energy, Mines Resour., Ottawa, Tech.Bull. no. 25
Church, M. A. 1984. On experimental method in geomorphology. In *Catchment experiments in fluvial geomorphology*, T. P. Burt and D. E. Walling (eds), 563–80. Norwich: Geo Books.
Churchill, R. R. 1979. A field technique for profiling precipitous slopes. *BGRG Tech. Bull.* **24**, 29–34.
Claridge, G. G. C. 1970. Studies in element balances in a small catchment at Taita, NZ. *Int.Assoc. Sci.Hydrol.Pub.* **96**, 523–40.
Clark, J. S. and W. A. Patterson, III 1984. Pollen, Pb-210 and opaque spherules: an integrated approach to dating and sedimentation in the intertidal environment. *J.Sedim.Petrol.* **54**, 1251–65.
Clark, M. M., A. Grantz and M. Rubin 1972. Holocene activity of the Coyote Creek Fault as recorded in sediments of Lake Cahuilla. *US Geol.Surv.Prof. Paper* **787**, 112–30.
Clark, M. W. 1981. Quantitative shape analysis: a review. *Math.Geol.* **13**, 303–20.
Clark, R. M. 1975, A calibration curve for radiocarbon dates. *Antiquity* **49**, 251–66.
Clark, S. P. and E. Jäger 1969. Denudation rate in the Alps from geochronologic and heat flow data. *Am.J.Sci.* **267**, 1143–60.
Clarke, D. J. and I. G. Eliot 1982. Description of littoral, alongshore sediment movement from empirical eigen-function analysis. *J. Geol.Soc.Aust.* **29** (3–4), 327–41.
Clarke, D. J. and I. G. Eliot 1983. Onshore–offshore patterns of sediment exchange in the littoral zone of a sandy beach. *J. Geol.Soc.Aust.* **30** (3–4), 341–51.
Clarke, J. I. 1966. Morphometry from maps. In *Essays in geomorphology*, G. H. Dury (ed.), 235–74. London; Heinemann.
Clayton, K. M. 1985. The state of geography. *Trans.Inst.Br.Geog. NS* **10**, 5–16.
Cleaves, E. T., D. W. Fisher and O. P. Bricker 1974. Chemical weathering of serpentinite on the Eastern Piedmont of Maryland. *Bull.Geol.Soc.Am.* **85**, 437–44.
Cleaves, E. T., A. E. Godfrey and O. P. Bricker 1970. Geochemical balance of a small watershed and its geomorphic implications. *Bull.Geol.Soc.Am.* **81**, 3015–22.
Clements, T., R. O. Stone, J. F. Mann and J. L. Eyman 1963. *A study of windborne sand and dust in desert areas*. US Army, Natick Labs, Mass., Earth Sci.Div.Tech Rep. no. ES-8.
Cliff, A. D. 1973. A note on statistical hypothesis testing. *Area* **5**, 240.
Cloet, R. L. 1954. Hydrographic analysis of the Goodwin Sands and the Brake Bank. *Geog.J.* **120**, 203–15.
Clos-Arceduc, M. A. 1966. Le rôle determinant des ondes aériennes stationnaires dans la structure des ergs sahariennes et les forces d'érosion évoisinantes. *C.R.Acad.Sci. Paris D* **262**, 2673–6.
Clos-Arceduc, A. 1971. 1. Disposition des structures d'origine éolienne au voisinage d'un groupe de barkhanes a parcours limité. 2. Etude d'un groupe isolé de barkhanes au sud du Tibesti. 3. Evolution des barkhanes sur un parcours limité par deux bandes où la déflation éolienne interdit la présence de dunes. *Photo-interprétation* 71–2. Paris: Technil.
Coates, D. R. 1958. *Quantitative geomorphology of small drainage basins of southern Indiana*. Office Naval Res. Geog. Branch, Tech.Rep.no. 10.
Cody, R. D. 1976. Growth and early diagenetic changes in artificial gypsum crystals grown within bentonite muds and gels. *Bull.Geol.Soc.Am.* **87**, 1163–8.

Coffman, D. M. 1969. Parameter measurement in fluvial morphology. *Indiana Acad.Sci.Proc.* **79**, 333–44.
Coffman, D. M., A. K. Turner and W. N. Melhorn 1971. *The WATER system: computer programs for stream network analysis.* Purdue Univ.Water Resour. Res. Center, Lafayette, Indiana, Tech.Rep.no. 16.
Cogley, J. G. 1986. Hypsometry of the continents. *Z. Geomorphol. Supp. NF* **53**, 1–48.
Cohen, I. B. 1985. *Revolution in science.* Cambridge, Mass.: Harvard University Press.
Coleman, A. M. and W. G. V. Balchin 1960. The origin and development of surface depressions in the Mendip Hills. *Proc.Geol.Assoc.* **70**, 291–309.
Coleman, J. M. 1966. Ecological changes in a massive freshwater clay sequence. *Gulf Coast Assoc.Geol.Soc. Trans.* **16**, 159–74.
Coles, N. and S. T. Trudgill 1985. The movement of nitrate fertilizer from the soil surface to drainage waters by preferential flow in weakly-structured soils, Slapton, S. Devon. *Agric., Ecosyst. Environ.* **13**, 241–59.
Collins, D. N. 1979a. Sediment concentration in melt waters as an indicator of erosion processes beneath an alpine glacier. *J.Glaciol.* **23**, 247–57.
Collins, D. N. 1979b. Hydrochemistry of meltwaters draining from an alpine glacier. *Arctic Alpine Res.* **11**, 307–24.
Collins, D. N. 1983. Solute yield from a glacierised high mountain basin. *Int. Assoc. Hydrol.Sci. Publ.* **141**, 41–9.
Collins, M. B. 1976. Suspended-sediment sampling towers as used on the intertidal flats of the Wash, eastern England. *Estuar.Coastal Mar. Sci.* **4**, 45–7.
Collinson, D. W. 1975. Instruments and techniques in palaeomagnetism and rock magnetism. *Rev.Geophys.Space Phys.* **13**, 5.
Colman, S. M. and D. P. Dethier 1986. *Rates of chemical weathering of rocks and minerals.* Orlando, Fl.: Academic Press.
Colwell, R. N. (ed.) 1983. *Manual of remote sensing*, 2nd edn. Falls church, Va.: American Society of Photogrammetry.
Comer, J. 1979. Modern applications of ion selective electrodes. *Lab. Equip. Digest* **17**, 73–9.
Condon, R. W. and M. E. Stannard 1957. Erosion in western New South Wales. *J.Soil Conserv.Serv. NSW* **10**, 17–26.
Cook, D. O. and D. S. Gorsline 1972. Field observations of sand transport by shoaling waves. *Mar. Geol.* **13**, 31–55.
Cook, F. D. 1969. Significance of micro-organisms in palaeosols, parent materials and ground water. In *Pedology and Quaternary research*, S. Pawluk (ed.), 53–61. Edmonton: University of Alberta Press.
Cooke, R. U. 1970. Stone pavements in deserts. *Ann.Assoc. Am.Geog.* **60**, 560–77.
Cooke, R. U. 1979. Laboratory simulation of salt weathering processes in arid environments. *Earth Surf. Processes* **4**, 347–59.
Cooke, R. U. and J. C. Doornkamp 1974. *Geomorphology in environmental management.* Oxford: Clarendon.
Cooke, R. U. and D. R. Harris 1970. Remote sensing of the terrestrial environment: principles and progress. *Trans.Inst.Br.Geog.* **50**, 1–23.
Cooke, R. U. and P. F. Mason 1973. Desert Knolls pediment and associated landforms in the Mojave Desert, California. *Rev. Géomorphol. Dyn.* **22**, 49–60.
Cooke, R.U. and A. Warren 1973. *Geomorphology in deserts.* London: Batsford.
Coope, G. R. 1975. Climatic fluctuations in northwest Europe since the Last Interglacial, indicated by fossil assemblages of coleoptera. *Geol.J* **6**, Spec.Issue 153–68.
Coope, G. R. 1977. Fossil coleopteran assemblages as sensitive indicators of climatic changes during the Devensian (Last) cold stage. *Phil.Trans.R.Soc. B* **280**, 313–48.
Coope, G. R., A. Morgan and P. J. Osborne 1971. Fossil Coleoptera as indicators of climatic fluctuations during the last glaciation in Britain,. *Palaeogeog.,Palaeoclim.Palaeoecol.* **10**, 87–101.
Corbel, J. 1959. Erosion en terrain calcaire. *Ann. Géog.* **68**, 97–120.
Corey, A. T. 1949. *Influence of shape on the fall velocity of sand grains.* Unpubl. M.Sc. Thesis, Colorado Agricultural and Mechanical College.
Cormie, A. and H. P. Schwarcz 1985. Stable hydrogen isotope analysis of bone collagen has potential for paleoclimatic research. In *Geological Society of America, annual meeting, abstracts*, 553.
Cornish, R. 1981. Glaciers of the L.Lomond Stadial in the Western Southern Uplands of Scotland. *Proc.Geol.Assoc.* **92**, 105–14.
Correns, C. W. 1961. The experimental chemical weathering of silicates. *Clay Min. Bull.* **26**, 249–65.
Corte, A. E. 1963. Particle sorting by repeated freezing and thawing. *Science* **142**, 499–501.

Corte, A. E. 1971. Laboratory formation of extrusion features by multicyclic freeze–thaw in soils. *Bull.Centre Géomorphol. Caen* **13–15**, 157–82.

Costa, J. E. 1978. Holocene stratigraphy in flood frequency analysis. *Water Resour. Res* **14**, 626–32.

Cote, L. J. 1960. *The directional spectrum of wind generated sea as determined from data obtained by S.W.O.P.* Meteor Pap.no.2. New York.

Courtois, G. and R. Hours 1965. *Propositions concernant les conditions particulières d'emploi des radioéléments artificiels pour étudier les movements de sédiments.* Earsotop, Cah. Inf. Commun. 29 Rapport Intern SARS/64–13.

Coutts, J. R. H., M. F. Kandil, J. L. Norland and J. Tinsley 1968. Use of radioactive ^{59}Fe for tracing soil particle movement. Part 1: Field studies of splash erosion. *J.Soil Sci.* **19**, 310–24.

Cowan, J. A., G. S. Humphreys, P. B. Mitchell and C. L. Murphy 1985. An assessment of pedoturbation by two species of mound-building ants, *Camponotus interpidus* (Kirby) and *Iridomyrmex purpureus* (Smith). *Aust.J. Soil Res.* **22**, 95–107.

Cox, A., R. R. Doell and G. B. Dalrymple 1968. Radiometric time-scale for geomagnetic reversals. *Q.J.Geol.Soc.Lond.* **125**, 53–66.

Cox, N. J. 1977. Allometric change of land forms: discussion. *Geol.Soc.Am.Bull.* **88**, 1199–200.

Cox, N. J. 1978. Hillslope profile analysis. *Area* **10**, 131–3.

Cox, N. J. 1983. On the estimation of spatial autocorrelation in geomorphology. *Earth Surf. Processes Landforms* **8**, 89–93.

Craig, A. 1972. Pollen influx to laminated sediments: a pollen diagram from northeastern Minnesota. *Ecology* **53**, 46–57.

Craig, A. K., S. Dobkin, R. B. Grimm and J. B. Davison 1969. The gastropod *Siphonaria pectinata*: a factor in destruction of beach rock. *Am. Zool.* **9**, 895–901.

Craig, R. F. 1978. *Soil mechanics*, 2nd edn. London: Van Nostrand Reinhold.

Crausse, E. and G. Pouzens 1958. Sur la possibilité d'étude hydrodynamique en modélé réduit des dunes désertiques isolées du type de barkhane. *C. R. Acad.Sci. Paris* **246**, 554–6.

Creer, K. M., R. Thompson, L. Molyneux and F. J. H. Mackereth 1972. Geomagnatic secular variation recorded in the stable magnetic remanence of recent sediments. *Earth Planet. Sci. Lett.* **14**, 115–27.

Cremer, K. W., C. J. Borough, F. H. McKinnell and P. R. Corter 1982. Effects of stocking and thinning on wind damage in plantations. *NZJ. For. Sci.* **12**, 244–68.

Crickmore, M. J. 1967. Measurement of sand transport in rivers with special reference to tracer methods. *Sedimentology* **8**, 175–228.

Crisp, D. T. 1966. Input and output of minerals for an area of Pennine moorland; the importance of precipitation, drainage, peat erosion and animals. *J.Appl.Ecol.***3**, 327–48.

Crittenden, M. D. 1967. New data on the isostatic deformation of Lake Bonneville. *US Geol. Surv.Prof. Paper* **454E**, 1–31.

Crofts, R. 1971. Sand movement in the Emlybeg dunes, Co.Mayo. *Irish Nat. J.* **17**, 132–6.

Crofts, R. S. 1974. Detailed geomorphological mapping and land evaluation in Highland Scotland. In *Progress in geomorphology*, E. H. Brown and R. S. Waters (eds), 231–5. Inst.Br.Geog.Spec. Publ. no.7.

Crofts, R.S. 1981. Mapping techniques in geomorphology. In *Geomorphological techniques*, A.S. Goudie, *et al.* (eds), 66–75, London: Allen & Unwin.

Crook, R. and A. R. Gillespie 1986. Weathering rates in granitic boulders measured by P-wave speeds. In *Rates of chemical weathering of rocks and minerals*, S. M. Coleman and D. P. Dethier (ed), 395–417. Orlando, Fl: Academic Press.

Crowther, J. 1979. Limestone solution on exposed rock outcrops in west Malaysia. In *Geographical approaches to fluvial processes*, A. F. Pitty (ed.), 31–50. Norwich: Geo Abstracts.

Crowther, J. 1983. A comparison of the rock tablet and water hardness methods for determining chemical erosion rates on karst surfaces. *Z.Geomorphol. NF* **27**, 55–64.

Crozier, M. J. 1973. Techniques for the morphometric analysis of landslips. *Z. Geomorphol. NF* **17**, 78–101.

Cryer, R. 1984. Two simple devices for stage-activation of pumping water samplers. *BGRG Tech. Bull.* **33**, 33–41. Norwich: Geo Abstracts.

Cryer, R. 1986. Atmospheric solute inputs. In *Solute processes*, S. T. Trudgill (ed.), ch. 2. Chichester: Wiley.

Culling, W. E. H. and M. Datko 1987. The fractal geometry of the soil-covered landscape. *Earth Surf. Processes Landforms* **12**, 369–85.

Cullingford, R. A. 1972. *Lateglacial and postglacial shoreline displacement in Earn–Tay area and eastern Fife.* Unpubl.Ph.D. Thesis, University of Edinburgh.

Cullingford, R. A. and D. E. Smith 1966. Late-glacial shorelines in eastern Fife. *Trans. Inst. Br.Geog.* **39**, 31–51.

Cullity, B. D. 1956. *Elements of X-ray diffraction*. London: Addison-Wesley.

Culver, S. J., H. R. Williams and P. A. Bull 1978. Infra-Cambrian glaciogenic sediments from Sierra Leone. *Nature* **274**, 49–51.

Curran, P. J. 1985. *Principles of remote sensing*. London: Longman.

Curray, J. 1956. The analysis of two-dimensional orientation data. *J.Geol.* **64**, 117–31.

Curtis, L. F. 1974. Remote sensing for environmental planning surveys. In *Environmental remote sensing: applications and achievements*, E. C. Barratt and L. F. Curtis (eds), 87–109. London: Edward Arnold.

Curtis, L. F. and S. T. Trudgill 1974. The measurement of soil moisture. *BGRG Tech.Bull*, **13**.

Curtis, L. F., J. C. Doornkamp and K. J. Gregory 1965. The description of relief in field studies of soils. *J.Soil Sci.* **16**, 16–30.

Cushing, E. J. and H. E. Wright 1965. Hand-operated piston corers for lake sediments. *Ecology* **46**, 380–4.

Cwynar, L. C. 1978. Recent history of fire and vegetation from laminated sediment of Greenleaf Lake, Algonquin Park, Ontario. *Can.J.Bot.* **56**, 10–21.

Czudek, T. and J. Demek 1970. Thermokarst in Siberia and its influence on the development of lowland relief. *Quat. Res.* **1**, 103–20.

Dacey, M. F. and W. C. Krumbein 1971. *Comments on spatial randomness in dendritic stream channel networks*. Tech. Rep.17, Dept Geol. Sci., Northwestern University, Evanston, Ill.

Dacey, M. F. and W. C. Krumbein 1976. Topological properties of disjoint channel networks within enclosed regions. *J.Int.Assoc.Math.Geol.* **8**, 429–61.

Dahl, R. 1967. Post-glacial micro-weathering of bedrock surfaces in the Narvik District of Norway. *Geog. Annaler* **49A**, 155–66.

Dalrymple, G. B. 1964. Potassium–argon dates of three Pleistocene inter-glacial basalt flows from the Sierra Nevada, Calif. *Bull.Geol.Soc.Am.* **75**, 753–7.

Dalrymple, J. B., R. J. Blong and A. J. Conacher 1968. A hypothetical nine unit landsurface model. *Z. Geomorphol. NF* **12**, 60–76.

Damon, P. E., C. W. Ferguson, A. Long and E. I. Wallick 1974. Dendrochronologic calibration of the radiocarbon time scale. *Am. Antiquity* **39**, 350–65.

Daniels, T. 1973. *Thermal analysis*. London: Kogan Page.

Danin, A. 1983. Weathering of limestone in Jerusalem by cyanobacteria. *Z. Geomorphol. NF* **27**, 413–21.

Dansgaard, W. 1964. Stable isotopes in precipitation. *Tellus* **16**, 436–68.

Darwin, C. 1881. The formation of vegetable mould through the action of worms with observations on their habits. Published in 1945 as *Darwin on humus and the earthworms*. London: Faber.

Daubrée, A. 1879. *Etudes synthétiques de géologie expérimentale*. Paris: Dunod.

Davidson, D. A. 1978. *Science for physical geographers*. London: Edward Arnold.

Davidson, D. A. 1983. Problems in the determination of plastic and liquid limits of remoulded soils using a drop-cone penetrometer. *Earth Surf. Processes Landforms* **8**, 171–5.

Davidson, R. N. and P. N. Jones 1985. Initiations in computer cartography: using a Versawriter tablet. *Comput. Geosci.* **11**, 335–6.

Davidson-Arnott, R. and W. Nickling (eds) 1978. *Research in fluvial geomorphology*. Norwich: Geo Books.

Davis, E. A. 1984. Conversion of Arizona chaparral to grass increases water yield and nitrate loss. *Water Resour. Res.* **20** (11), 1643–9.

Davis, M. B. 1976. Erosion rates and land use history in southern Michigan. *Environ.Conserv.* **3** (2), 139–48.

Davison, C. 1888. Note on the movement of scree material. *Q.J.Geol.Soc.* **4**, 232–8.

Davison, C. 1889. On the creeping of a soil cap through the action of frost. *Geol.Mag.* **6**, 255.

Dawson, F. H. 1978. The seasonal effects of aquatic plant growth on the flow of water in a stream. *Proc.EWRS 5th symp.: aquatic weeds*, 71–8.

Dawson, M. 1985. Environmental reconstructions of a Late Devensian terrace sequence. Some preliminary findings. *Earth Surf. Processes Landforms* **10**, 237–46.

Day, M. J. 1976. The morphology and hydrology of some Jamaican karst depressions. *Earth Surf. Processes* **1**, 111–29.

Day, M. J. and A. S. Goudie 1977. Field assessment of rock hardness using the Schmidt Test Hammer. *BGRG Tech.Bull.* **18**, 19–29.

Dearing, J. A. 1979. *The application of magnetic measurements to studies of particulate flux in lake-watershed ecosystems*. Unpubl.Ph.D.Thesis, University of Liverpool.

Dearing, J. A. 1983. Changing patterns of sediment accumulation in a small lake in Scania, southern Sweden. *Hydrobiologia* **103**, 59–64.

Dearing, J. A. 1986. Core correlation and total sediment influx. In *Handbook of Holocene palaeoecology and palaeohydrology*, B. E. Berglund (ed.), 247–70. Chichester: Wiley.

Dearing, J. A., J. K. Elner and G. M. Happey-Wood 1981. Recent sediment influx and erosional processes in a Welsh upland lake-catchment based on magnetic susceptibility measurements. *Quat.Res.* **16**, 356–72.

Dearing, J. A., I. D. L. Foster and A. D. Simpson 1982. Timescales of denudation: the lake-drainage basin approach. In *Recent developments in the exploration and prediction of erosion and sediment yield*, 351–61. IAHS Publ.no.137.

Dearman, W. R., F. J. Baynes and T. Y. Irfan 1978. Engineering grading of weathered granite. *Eng. Geol.* **12**, 345–74.

Dearman, W. R. and P. G. Fookes 1974. Engineering geological mapping for the civil engineering practice in the United Kingdom. *Quarterly Journal of Engineering Geology* **7**, 223–56.

de Boer, G. and A. P. Carr 1969. Early maps as historical evidence of coastal change. *Geog.J.* **135**, 17–39.

de Boodt, M. and D. Gabriels (eds) 1980. *Assessment of erosion*. Chichester: Wiley.

de Brum Ferreira, D. 1981. *Carte géomorphologique du Portugal*. Memorias Centro Estudos Geograficos, Lisboa no. 6.

de Crecy, L. 1980. Avalanche zoning in France – regulation and technical bases. *J.Glaciol.* **26**, 325–30.

Deere, D. U. and E. D. Patton 1971. Slope stability in residual soils. In *Proc.4th Panamerican conf.soil mechanics and foundation engineering*, vol. **1**, 87–170. San Juan, Puerto Rico.

Deevey, E. S., M. S. Gross, G. E. Hutchinson and H. l. Kraybill 1954. The nature of ^{14}C contents of materials from hard-water lakes. *Proc.Nat.Acad.Sci.* **40**, 285–8.

de Félice, P. 1955. Etude de la formation des rides de sable. *C. R. Acad.Sci.Paris* **240**, 1253–5.

de Geer, G. 1912. A geochronology of the last 12 000 years. XIth Int. Geol. Cong., Stockholm, **1**, 241–53.

de Geer, G. 1940. Geochronologia suecica principles. *K. Sven. Vetenskapsakad. Handl., 3rd Ser.* **18**, 1–367.

Deju, R. A. 1971. A model of chemical weathering of silicate minerals. *Bull.Geol.Soc.Am.* **82**, 1055–62.

Delannoy, J.-J., R. Maire and J. Nicod 1984. Karsts des Alpes Occ., Jura mer.et de Provence. *Karstologia* **3** (map): see also pp. 2–11.

Demek, J. (ed.) 1972. *Manual of detailed geomorphological mapping*. IGU Subcommission on Geomorphological Survey and Mapping. Prague: Academia.

Demek, J. and C. Embleton (eds) 1978. *Guide to medium-scale geomorphological mapping*. Stuttgart: Schweizerbart'sche, for IGU.

Demirmen, F. 1975. Profile analysis by analytical techniques: a new approach. *Geog.Anal.* **7**, 245–66.

Deng, O., S. Fengnin, Z. Shilong, L. Mengluan, W. Tichin, Z. Weiqi, B. S. Burchtal, P. Molnar and Z. Peizhen 1984. Active faulting and tectonics in the Ningzia-Hui Autonomous Region, China. *J. Geophys.Res.* **89**, 4427–45.

Denham, C. R. 1976. Blake polarity episode in two cores from the Greater Antilles Outer Ridge. *Earth Planet. Sci. Lett.* **29**, 422–34.

DeNiro, M. and S. Epstein 1978. Influence of diet on the distribution of carbon isotopes in animals. *Geochim. Cosmochim. Acta* **42**, 495–506.

Denness, B. and P. Grainger 1976. The preparation of slope maps by the moving interval method. *Area* **8**, 213–18.

Denny, C. S. and J. C. Goodlett 1956. Microrelief resulting from fallen trees. *US Geol. Surv. Prof. Paper* **288**, 59–66.

Denton, G. H. and T. J. Hughes 1983. Milankovich theory of the Ice Ages: hypothesis of ice-sheet linkage between regional insolation and global climate. *Quat. Res.* **20**, 125–44.

DePaolo, D. J. 1986. Detailed record of the Neogene Sr isotopic evolution of seawater from DSDP Site 590B. *Geology* **14**, 103–6.

Department of Energy, Mines and Resources 1972. *List of map sources*. Ottawa.

Department of the Environment 1972. *Analysis of raw, potable and waste waters*. London: HMSO.

Department of the Environment 1976. *Precipitation acidity*. Central Unit Environ. Pollut., Pollut. Paper no. 7. London: HMSO.

de Ploey, J. 1971. Liquefaction and rainwash erosion. *Z. Geomorphol. NF* **15**, 491–6.

de Ploey, J. and D. Gabriels 1980. Measuring soil loss and experimental studies. In *Soil erosion*, M. J. Kirkby, and R. P. C. Morgan (eds) 63–108. Chichester: Wiley.

Depraetere, C. 1987. *Classification automatique interrégionale à partir de MNT issus de la BDZ.* Paris, Inst. Géog. National, ANGEO/DELI/SCME Rapport 4.

Derbyshire, E. 1978. A pilot study of till microfabrics using the scanning electron microscope. In *Scanning electron microscopy in the study of sediments*, W. B. Whalley (ed.), 41–59. Norwich: Geo Abstracts.

Derbyshire, E. and I. S. Evans 1976. The climatic factor in cirque variation. In *Geomorphology and climate*, E. Derbyshire (ed.), 447–94. London: Wiley.

Derbyshire, E., A. McGowan and A. Radwan 1976. 'Total' fabric of some till landforms. *Earth Surf. Processes* **1**, 17–26.

Devoy, R. J. N. 1977. Flandrian sea-level changes in the Thames Estuary and the implications for land subsidence in England–Wales. *Nature* **220**, 712–15.

Devoy, R. J. N. 1979. Flandrian sea level. Changes and vegetational history of the lower Thames Estuary. *Phil.Trans.R. Soc.B* **285**, 355–407.

de Vries, M. 1973. Applicability of fluorescent tracers. In *Tracer techniques in sediment transport*, 105–23. Vienna: International Atomic Energy Agency.

Diamond, S. 1970. Pore size distribution in clays. *Clays Clay Min.* **18**, 7–23.

Dickinson, C. H. and G. J. F. Pugh (eds) 1974. *Biology of plant litter decomposition.* London: Academic Press.

Dickman, M. D. 1979. A possible varving mechanism for metomictic lakes. *Quat.Res.* **11**, 113–24.

Dickson, J. H., D. A. Stewart, R. Thompson, G. Turner, M. S. Baxter, N. D. Drndarsky and J. Rose 1978. Palynology, palaeomagnetism and radiometric dating of Flandrian marine and freshwater sediments of Loch Lomond. *Nature* **274**, 548–53.

Dietrich, R. V. 1977. Impact abrasion of harder by softer materials. *J.Geol.* **85**, 242–6.

Digerfeldt, G. 1978. *A simple corer for sediment sampling in deep water.* Univ. Lund, Dept Quat. Geol., Occas. Paper no. 14.

Digerfeldt, G., R. W. Battarbee and L. Bengtsson 1975. Report on annually laminated sediment in Lake Jarlasjon, Nacka, Stockholm, *Geol.Foren. Stockholm Forhandl.* **97**, 29–40.

Dillon, P. J. and W. B. Kirchner 1975. The effects of geology and land use on the export of phosphorus from watersheds. *Water Res.* **9**, 135–48.

Dionne, J-C. 1985. Tidal marsh erosion by geese, St Lawrence Estuary, Quebec. *Geogr. Phys. Quat.* **39**, 99–105.

Dittus, W. P. J. 1985. The influence of cyclones on the dry evergreen forest of Sri Lanka. *Biotropica* **17**, 1–14.

Dixon, J. B. and S. B. Weed (eds) 1977. *Minerals in soil environments.* Madison, Wisc: Soil Science Society of America.

Dobkins, J. B. and R. L. Folk 1970. Slope development in Tahiti-Nui. *J.Sedim.Petrol.* **40**, 1167–203.

Dochinger, L. S. and T. A. Seliga (eds) 1976. *Proc.1st int.symp. on acid precipitation and the forest ecosystem.* US Geol Surv.Gen.Tech.Rep. NE-23.

Doherty, J. T. and J. B. Lyons 1980. Mesozoic erosion rates in northern New England. *Bull.Geol.Soc.Am.* **91**, 16–20.

Dolan, R. and J. C. Ferm 1966. Swash processes and beach characteristics. *Prof. Geog.* **18**, 210–13.

Donahue, J. J. 1972. Drainage intensity in western New York. *Ann.Assoc. Am.Geog.* **62**, 23–36.

Donahue, J. J. 1974. Measuring drainage density with a dot planimeter. *Prof. Geog.* **26**, 317–19.

Donn, T. F. and M. R. Boardman 1986. A profiling method for measuring erosion and accretion of intertidal rock surfaces. *J. Coastal Res.* **2** (1), 69–73.

Donner, J. J. 1963. The late- and post-glacial raised beaches in Scotland. *Ann.Acad.Sci.Fenn. A III Geol.-Geog.* **68**, 1–13.

Donner, J. J. 1969. Land/sea level changes in southern Finland during the formation of the Salpausselka end-moraines. *Bull.Geol.Soc. Finl.* **41**, 135–50.

Donner, J. J., M. Eronen and H. Jungner 1977. The dating of the Holocene relative sea-level changes in Finnmark in northern Norway. *Nor. Geog. Tidsskr.* **31**, 103–28.

Doornkamp, J. C. and C. A. M. King 1971. *Numerical analysis in geomorphology.* London: Edward Arnold.

Doornkamp, J. C. and Han Mukang 1985. Morphotectonic research in China and its application to earthquake prediction. *Prog.Phys.Geog.* **9**, 353–81.

Doornkamp, J. C., D. Brunsden and D. K. C. Jones 1980. *Geology, geomorphology and pedology of Bahrain.* Norwich: Geo Abstracts.

Doornkamp, J. C., D. Brunsden, D. K. C. Jones, R. U. Cooke and P. R. Bush 1979. Rapid geomorphological assessments for engineering. *Q.J.Eng.Geol.* **12**, 189–204.

Douglas, D. H. 1986. Experiments to locate ridges and channels to create a new type of Digital Elevation Model. *Cartographica* **23**, 29–61.
Douglas, I. 1968. Field methods of water hardness determination. *BGRG Tech.Bull.* **1**.
Douglas, I. 1972. The geographical interpretation of river water quality data. *Prog. Geog.* **4**, 1–82.
Douglas, I. 1973. *Rates of denudation in selected small catchments in Eastern Queensland.* Univ. Hull, Dept Geog., Occas.Paper no. 21.
Dowling, J. M. 1974. A note on the use of spectral analysis to detect leads and lags in arrival cycles of water quality. *Water Resour. Res.* **10**, 343–4.
Drake, R. E., A. L. Deino, G. H. Curtis, C. S. Swisher, B. Turrin, M. McCrory and T. Becker 1988. Applications of $^{40}Ar/^{39}Ar$ dating of single-crystals by laser fusion. *Eos* **69**, 1502.
Draper, L. 1967. Instruments for measurement of wave height and direction in and around harbours. *Proc.Inst.Civ.Eng.* **37**, 213–19.
Dreimanis, A., G. Hutt, A. Raukas and P. W. Whippey 1978. Dating methods of Pleistocene deposits and their problems: 1. Thermoluminescence dating. *Geosci. Can.* **5**, 55–60.
Drever, J. I. (ed.) 1985. *The chemistry of weathering.* Dordrecht: Reidel.
Drew, D. P. and D. I. Smith 1969. Techniques for the tracing of subterranean drainage. *BGRG Tech. Bull.* **2**.
D'Souza, V. P. C. and R. P. C. Morgan 1976. A laboratory study of the effect of slope steepness and curvation on soil erosion. *J.Agric.Eng.Res.* **21**, 21–31.
Duffy, P. D., J. D. Schreiber, D. C. McClurkin and L. L. McDowell 1978. Aqueous and sediment-phase phosphorus yields from five Southern pine watersheds. *J.Environ.Qual.* **7**, 45–50.
Dugan, G. L. and P. H. McGauhey 1974. Enrichment of surface waters. *J.Water Pollut. Control* **46**, 2261–80.
Dumbleton, M. J. and G. West 1968. Soil suction by the rapid method: an apparatus with extended range. *J.Soil Sci.* **19**, 40–6.
Dunkerley, D. L. 1977. Some comments on stream ordering schemes. *Geog.Anal.* **9**, 429–31.
Dunne, T. 1981. Concluding comments to the Christchurch Symposium on erosion and sediment transport in Pacific rim 'steeplands'. *J.Hydrol. (N.Z)* **20**, 111–14.
Dunne, T. and R. D. Black 1970. An experimental investigation of runoff production in permeable soils. *Water Resour. Res.* **6**, 478–90.
Dunne, T., W. E. Dietrich and M. J. Bruneggo 1978. Recent and past erosion rates in semi-arid Kenya. *Z. Geomorphol. Supp. NF* **29**, 130–40.
Duplessy, J. C., J. Labeyrie, C. Lalou and H. V. Nyuyen 1970. Continental climatic variations between 130 000 and 90 000 years BP. *Nature* **226**, 631–2.
Durham, R. and B. G. Oliver 1983. History of Lake Ontario contamination from the Niagara River by sediment radiodating and chlorinated hydrocarbon analysis I. *Great Lakes Res.* **9**, 160–8.
Dury, G. H. 1961. Bankfull discharge: an example of its statistical relationships. *Int.Assoc. Sci.Hydrol.Bull.* **65**, 48–55.
Dury, G. H. 1969. Relation of morphometry to runoff frequency. In *Water, earth and man*, R. J. Chorley (ed.), 419–30. London: Methuen.
Dury, G. H. 1972. A partial definition of the term 'pediment' with field tests in humid-climate areas of southern England. *Trans.Inst.Br.Geog.* **57**, 139–52.
Duvigneaud, P. and S. Denaeyer de Smet 1970. Biological cycling of minerals in temperate deciduous forests. In *Analysis of temperate forest ecosystems*, D. E. Reichle (ed.), 199–225. London: Chapman & Hall.

Eakins, J. D. and R. T. Morrison 1977. *A new procedure for the determination of lead-210 in lake and marine sediments.* UK AERE Rep. no. 8475. London: HMSO.
Eardley, A. J. 1966. Rates of denudation in the High Plateaus of Southwestern Utah. *Bull.Geol. Soc.Am.* **77**, 777–80.
Earl, M. D. and G. F. Beardsley 1969. A three component drag probe for use in wave fields and tidal channels. *Abstracts volume, American Geophysical Union meeting*, 631. Washington DC.
Earley, B. 1976. *Practical instrumentation handbook.* Scientific Era Publications.
Easterbrook, D. J. (ed.) 1988. *Dating Quaternary sediments.* Geol.Soc.Am.Spec. Paper no. 227.
Eaton, J. S., G. E. Likens and F. H. Bormann 1973. Throughfall and stemflow chemistry in a northern hardwood forest. *J. Ecol.* **61**, 495–508.
Edelman, T. 1967. Systematic measurements along the Dutch coast. *Proc. 10th conf. on coastal engineering*, ch. 30, 489–501. Am.Soc.Civ.Eng.
Eden, M. J. and C. P. Green 1971. Some aspects of granite weathering and tor formation on Dartmoor, England. *Geog. Annaler* **53A**, 92–9.
Eden, R. A., A. V. F. Carter and M. C. McKeown 1969. Submarine examination of Lower

Carboniferous strata on inshore regions of the continental shelf south-east Scotland. *Mar. Geol.* **7**, 235–51.
Edge, R. J., T. F. Baker and G. Jeffries 1981. Borehole tilt measurements: a periodic crustal tilt in an aseismic area. *Tectonophysics* **71**, 97–109.
Edwards, A. C., J. Creasey and M. S. Cresser 1985. Factors influencing nitrogen inputs and outputs in two Scottish upland catchments. *Soil Use Managem.* **1** (3), 83–7.
Edwards, A. M. C., A. T. McDonald and J. R. Petch 1975. The use of electrode instrumentation for water analysis. *BGRG Tech.Bull.* **15**.
Edwards, R. J. G., *et al.* 1982. Land surface evaluation for engineering practice: Report by a working party . . . . *Q.J.Eng.Geol.* **15**, 265–316.
Edwards, R. L., J. H. Chen and G. J. Wasserburg 1987. ^{238}U–^{234}U–^{230}Th–^{232}Th systematics and the precise measurement of time over the past 500 000 years. *Earth Planet. Sci. Lett.* **81**, 175–92.
Edwards, R. S. and S. M. Claxton 1964. The distribution of airborne salts of marine origin in the Aberystwyth area. *J.Appl.Ecol.* **1**, 253–63.
Edwards, T. W. D. and P. Fritz 1986. Assessing meteoric water composition and relative humidity from $^{18}0$ and ^{2}H in wood cellulose: palaeoclimatic implications for southern Ontario, Canada. *Appl.Geochem.* **1**, 715–23.
Edwards, T. W. D., R. O. Aravena and P. Fritz 1986. Comparison of postglacial palaeoclimatic records in southern Ontario from stable isotope studies of terrestrial plant cellulose and lacustrine carbonates (abstract) *American Quaternary Association program with abstracts*, 78.
Ehrlich, R. and W. E. Full 1984. Fourier grain shape analysis – a multi-variate pattern recognition approach. In *Particle characterization in technology*. Vol. II: *Morphometric analysis*, 89–101. Boca Raton, Fl.: CRC Press.
Ehrlich, R. and B. Weinberg 1970. An exact method for the characterization of grain shape. *J.Sedim.Petrol.* **40**, 205–12.
Ehrlich, R., P. J. Brown, J. M. Varus and D. T. Eppler 1980. Analysis of particle morphology data. In *Advanced particulate morphology*, J. K. Beddow and T. P. Meloy (eds), 101–19. Boca Raton, Fl.: CRC Press.
El-Ashry, M. T. and H. R. Wanless 1968. Photo interpretation of shoreline changes between Cape Hatteras and Fear (North Carolina). *Mar. Geol.* **6**, 347–79.
El-Daoushy, F., K. Tolonen and R. Rosenberg 1982. Lead-210 and moss-increment dating of two Finnish *Sphagnum* hummocks. *Nature* **296**, 429–31.
Elderfield, H. 1986. Strontium isotope stratigraphy. *Palaeogeog., Palaeoclim., Palaeoecol.* **57**, 71–90.
Ellis, J. B. 1975. *Urban storm water pollution*. Middlesex Polytech. Res. Rep. no. 1.
Ellison, W. D. 1944a. Studies of raindrop erosion. *Agric. Eng.* **25**, 131–6, 181–2.
Ellison, W. D. 1944b. Two devices for measuring soil erosion. *Agric. Eng.* **25**, 53–5.
Ellison, W. D. 1945. Some effects of raindrops and surface flow on soil erosion and infiltration. *Trans.Am.Geophys.Union* **26**, 415–29.
Ellwood, J. M., P. D. Evans and I. G. Wilson 1975. Small scale aeolian bedforms. *J.Sedim.Petrol.* **45**, 554–61.
Elverhoi, A., O. Lonne and R. Seland 1983. Glaciomarine sedimentation in a modern fjord environment. *Polar Res.* **1**, 127–49.
Elvhage, C. 1983. *Terrangatergivning. En sere svenska experimentkortor*. Uppsala Naturgeog. Inst. Rapport no. **59**, Uppsala Univ. Dept Phys. Geog.
Emiliani, C. 1966. Isotopic palaeotemperatures. *Science* **154**, 851–7.
Engelhardt, H. F., W. D. Harrison and B. Kamb 1978. Basal sliding and conditions at the glacier bed as revealed by borehole photography. *J.Glaciol.* **20**, 469–508.
Engel, C. G. and R. P. Sharp 1958. Chemical data on desert varnish. *Bull.Geol.Soc.Am.* **69**, 487–518.
Engstrom, D. R. and H. E. Wright 1984. Chemical stratigraphy of lake sediments as a record of environmental change. In *Lake sediments and environmental history*, E. Y. Haworth and J. W. G. Lund (eds), 11–67. Leicester: Leicester University Press.
Epstein, S. 1978. The D/H ratio of cellulose in a New Zealand *Pinus radiata*; a reply to the criticism of A. T. Wilson and M. J. Grinstead. *Earth Planet. Sci. Lett.* **39**, 303–7.
Epstein, S. and C. Yapp 1976. Climatic implications of the D/H ratio of hydrogen in C–H groups in tree cellulose. *Earth Planet. Sci. Lett.* **30**, 252.
Epstein, S., R. Buchsbaum, H. Lowenstam and H. C. Urey 1953. Revised carbonate–water isotopic temperature scale. *Bull.Geol.Soc.Am.* **64**, 1315–26.
Eriksson, E. 1955. Air borne salts and the chemical composition of river waters. *Tellus* **7**, 243–50.
Eriksson, E. 1960. The yearly circulation of Cl^- and S^- in nature; meteorological, geochemical and pedological implications, part 2. *Tellus* **12**, 63–109.

Eriksson, E. 1966. Air and precipitation as sources of nutrients. In *Handbuch der Pflanzenernahrung und Dungung*, K. Scharrer and H. Linser (eds), 774–92. Vienna: Springer.
Eriksson, K. G. 1979. Saharan dust sedimentation in the western Mediterranean Sea. In *Saharan dust*, C. Morales (ed.), 197–210. Chichester: Wiley.
Erikstad, L. and J. L. Sollid 1986. Neoglaciation in South Norway using lichenometric methods. *Nor. Geog.Tiddskr.* **40**, 85–105.
Eronen, M. 1983. Late Weichselian and Holocene shore displacement in Finland. In *Shorelines and isostasy*, D. E. Smith and A. G. Dawson (eds), 183–207. London: Academic Press.
Escandc, L. 1953. Similitude des ondulations de sable des modélés reduits et des dunes du desert. In *Actions éoliennes*, 71–9. CNRS, Paris, Coll. Int. no. 35.
Evans, G. R. 1973. Huttan's shearwaters initiating local soil erosion in the seaward Kaikoura Range. *NZ J. Sci.* **16**, 637–42.
Evans, I. S. 1969. The geomorphology and morphometry of glacial and nival areas. In *Water, earth and man*, R. J. Chorley (ed.), 369–80. London: Methuen.
Evans, I. S. 1972. General geomorphometry, derivatives of altitude and descriptive statistics. In *Spatial analysis in geomorphology*, R. J. Chorley (ed.), 17–90. London: Methuen.
Evans, I. S. 1974. *The geomorphometry and asymmetry of glaciated mountains*. Unpubl.Ph.D. Thesis, University of Cambridge.
Evans, I. S. 1975. The effect of resolution on gradients calculated from an altitude matrix. *Statistical characterization of altitude matrices by computer*, Report 3. Department of Geography, University of Durham.
Evans, I. S. 1977a. Frequency distribution of gradient. *Statistical characterization of altitude matrices by computer*, Report 4. Department of Geography, University of Durham.
Evans, I. S. 1977b. The selection of class intervals. *Trans. Inst.Br.Geog. NS* **2**, 98–124.
Evans, I. S. 1978. An integrated system of terrain analysis and slope mapping. *Statistical characterization of altitude matrices by computer*, Report 6. Department of Geography, University of Durham.
Evans, I. S. 1984. Correlation structures and factor analysis in the investigation of data dimensionality: statistical properties of the Wessex land surface, England. In *Proc.Int. symp. on spatial data handling*, vol. 1, 98–116. Geographisches Inst., Universitat Zurich, Irchel.
Evans, I. S. 1987a. The morphometry of specific landforms. In *International geomorphology 1986*, Part **II**, V. Gardiner (ed.), 105–24. Chichester: Wiley.
Evans, I. S. 1987b. A new approach to drumlin morphometry. In *Drumlin symposium*, J. Menzies and J. Rose (eds), 119–30. Rotterdam: Balkema.
Evans, J. G. 1966. Late-glacial and post-glacial subaerial deposits at Pitstone, Bucks. *Proc.Geol. Assoc.* **77**, 347–64.
Evans, J. W. 1968. The role of *Penitella penita* (Conrad 1837) (Family Pholadidea) as eroders along the Pacific coast of North America. *Ecology* **49**, 156–9.
Everett, K. R. 1966. Slope movement and related phenomena. In *Environment of the Cape Thompson region*, 175–220. US Atomic Energy Commission, Division of Technical Information.
Evernden, J. F., D. E. Savage, G. H. Curtis and G. T. James 1964. Potassium–argon dates and the Cenozoic mammalian chronology of North America. *Am.J.Sci.* **265**, 145–98.
Eykhoff, P. 1974. *Systems identification*. New York: Wiley.
Eyles, R. J. 1966. Stream representation on Malayan maps. *J.Trop.Geog. 22*, 1–9.
Eyles, R. J. 1973. Drainage density representation on Wellington maps. *J.Hydrol. NZ* **12**, 19–31.

Faegri, K. and J. Iversen 1964. *A textbook of modern pollen analysis*. Copehangen: Munksgaard.
Faegri, K. and J. Iversen 1975. *Textbook of pollen analysis*, 3rd edn. Copenhagen: Munksgaard.
Fagerlund, G. 1973. Determination of pore-size distribution by suction porosimetry. *Matér.Constr.* **6**, 191–201.
Fahey, B. D. and T. H. Lefebure 1988. The freeze–thaw weathering regime at a section of the Niagara Escarpment on the Bunce Peninsula, Southern Ontairo, Canada. *Earth Surf. Processes Landforms* **13**, 293–304.
Fairbridge, R. W. 1961. Eustatic changes of sea-level. *Phys. Chem. Earth* **5**, 99–185.
Fairbridge, R. W. (ed.) 1981. Neotectonics. *Z.Geomorphol. Supp. NF* **40**, 226.
Fairchild, J. C. 1972. Longshore transport of suspended sediment. In *Proc. 13th conf. on coastal engineering*, Am.Soc.Civ.Eng., vol. 2, 1069–88.
Faniran, A. 1969. The index of drainage intensity: a provisional new drainage factor. *Aust. J.Sci.* **31**, 328–30.
FAO 1960. *Soil erosion by wind and measures for its control on agricultural land*. FAO Agric. Develop. Paper no. 71.
FAO 1975. *Guidelines for soil profile description*. Rome: FAO, Land and Water Development Division.

Farres, P. J. 1980. *Hardware models: simulation of rainfall*. Portsmouth Polytech. Dept Geog. Occas. Paper no. 5.

Faure, G. 1986. *Principles of isotope geology*, 2nd edn. New York: Wiley.

Federal Inter-Agency Sedimentation Project 1963. *Determination of fluvial sediment discharge*. Rep.no. 14.

Federoff, N. 1966. Contribution à la connaissance de la pédogenèse Quaternaires dans le sud-ouest du Bassin Parisien. *Bull.Assoc.Fr.Etude Quat.* **2**, 94–106.

Federoff, N. 1973. The distribution of iron oxides and hydroxides during the Pleistocene in the west Mediterranean Basin and Atlantic France. Paper presented at *Reading meeting of Quaternary Research Association*, Jan. 1973.

Federoff, N. and P. Goldberg 1982. Comparative micromorphology of two late Pleistocene paleosols (in the Paris Basin). *Catena* **9**, 227–51.

Fenwick, I. M. 1985. Paleosols: problems of recognition and interpretation. In *Soils and Quaternary landscape evolution*, J. Boardman (ed.), 3–21. Chichester: Wiley,

Fenwick, I. M. and B. J. Knapp 1981. *Soils: process and response*. London: Duckworth.

Ferguson, C. W. and D. A. Graybill 1983. Dendrochronology of bristlecone pine: a progress report. *Radiocarbon* **25**, 287–8.

Ferguson, R. I. 1973. Regular meander path models. *Water Resour. Res.* **9**, 1079–86.

Ferguson, R.I. 1975. Meander irregularity and wavelength estimation. *J.Hydrol.* **26**, 315–33.

Ferguson, R. I. 1976. Disturbed periodic model for river meanders. *Earth Surf. Processes* **1**, 337–47.

Ferguson, R. I. 1977a. Meander sinuosity and direction variance. *Bull.Geol.Soc.Am.* **88**, 212–14.

Ferguson, R. I. 1977b. On determining distances through stream networks. *Water Resour. Res.* **13**, 672–4.

Ferguson, R. I. 1979. Stream network volume: an index of channel morphometry: discussion. *Bull.Geol.Soc.Am.* **90**, 606–8.

Ferguson, R. I. 1986. Hydraulics and hydraulic geometry. *Prog. Phys. Geog.* **10**, 1–31.

Ferguson, R. I. 1987. Accuracy and precision of methods for estimating river loads. *Earth Surf. Processes Landforms* **12**, 95–104.

Feyling-Hanssen, R.W. 1964. Foraminifera in Late Quaternary deposits from the Oslofjord area. *Nor. Geol. Unders.* **225**, 1–383.

Fink, W. and H. van der Piepen 1974. Image processing methods facilitating information extraction. *S.Afr.J.Sci.* **70**, 370–4.

Finkel, H. J. 1959. The barchans of southern Peru. *J.Geol.* **67**, 614–47.

Finlayson, B. L. 1975. Measurement of the organic content of suspended sediments at low concentrations. *BGRG Tech.Bull.* **18**.

Finlayson, B. L. 1976. *Measurements of geomorphic processes in a small drainage basin*. Unpubl. Ph.D. Thesis, University of Bristol.

Finlayson, B.L. 1985. Soil creep: a formidable fossil of misconception. In *Geomorphology and soils*, K. S. Richards, R. R. Arnett and S. Ellis (eds). 141–58. London: Allen & Unwin.

Finlayson, B. L. 1986. Determination of dissolved organic matter in streamwater using visible spectrophotometry: a comment. *Earth Surf. Processes Landforms* **11**, 457–9.

Finlayson, B. L. and H. A. Osmaston 1977. An instrument system for measuring soil movement. *BGRG Tech.Bull.* **19**.

Firth, C. R. 1986. Isostatic depression during the Loch Lomond Stadial: preliminary evidence from the Great Glen, northern Scotland. *Quat. Newslett.* **48**, 1–9.

Fish, G.R. 1976a. N and P analyses of rainfall at Rotorua. *NZ J.Hydrol.* **15**, 17–26.

Fish, G.R. 1976b. The fall out of N and P compounds from the atmosphere. *NZ J.Hydrol.* **15**, 27–33.

Fisher, D. W., A. W. Gambell, G. E. Likens and F. H. Bormann 1968. Atmospheric contributions to water quality of streams in the Hubbard Brook Experimental Forest, New Hampshire. *Water Resour. Res.* **4**, 1115–26.

Fisher, J. E. 1953. Two tunnels in cold ice at 4000 m on the Breithorn. *J.Glaciol.* **2**, 513–20.

Fisher, N., R. Dolan and B. P. Hayden 1984. Variations in large scale beach amplitude along the coast. *J.Sedim.Petrol.* **54** (1), 73–85.

Fleisher, P. J. 1984. Maps in applied geomorphology. In *Developments and applications of geomorphology*, J. F. Costa and P. J. Fleisher (eds), 171–200. Berlin: Springer.

Fleming, G. 1969. Suspended solids monitoring: a comparison between three instruments. *Water & Water Eng.* 377–82.

Fleming, M. V. and A. E. DeWall 1982. *Beach Profile Analysis System (BPAS)*. Vol. 1: *System overview*. US Coastguard Oceanographic Unit, Tech. Rep. 82–1 (1).

Fleming, R. W. and A. M. Johnson 1975. Rates of seasonal creep of a silty clay soil. *Q.J.Eng. Hydrol.* **8**, 1–29.

Fleming, S. J. 1971. Thermoluminescent authenticity testing of ancient ceramics: the effects of sampling by drilling. *Archaeometry* **13**, 59–69.
Flemming, N. C. 1964. Tank experiments on the sorting of beach material during cusp formation. *J. Sedim.Petrol.* **34**, 112–22.
Flemming, N. C. 1965. Form and function of sedimentary particles. *J.Sedim.Petrol.* **35**, 381–90.
Flemming, N. C. 1969. Archaeological evidence for eustatic change of sea level and earth movements in the western Mediterranean during the last 2000 years. *Geol.Soc.Am.Spec. Paper* **109**, 1–125.
Flook, A. G. 1978. The use of dilation logic on the Quantimet to achieve fractal dimension characterisation of textural and structural profiles. *Powder Technol.* **21**, 295–8.
Folk, R. L. 1966. A review of grain size parameters. *Sedimentology* **6**, 73–93.
Folk, R. L. 1975. Glacial deposits identified by chattermark trails in detrital garnets. *Geology* **3**, 473–5.
Folk, R. L. 1976. Reddening of desert sands: Simpson Desert, N.T., Australia. *J.Sedim.Petrol.* **46**, 604–15.
Fookes, P. G. and R. Best 1969. Consolidation characteristics of some Pleistocene periglacial metastable soils of east Kent. *Q.J.Eng.Geol.* **2**, 103–28.
Ford, D. C. 1973. Development of the canyons of the South Nahanni River, NWT. *Can.J.Earth Sci.* **10**, 366–78.
Ford, D. C. 1976. Evidences of multiple glaciation in South Nahanni National Park, Mackenzie Mountains, NWT. *Can.J.Earth Sci.* **13**, 1433–45.
Ford, D. C., H. P. Schwarcz, J. J. Drake, M. Gascoyne, R. S. Harmon and A. G. Latham 1981. Estimates of the age of the existing relief within the southern Rocky Mountains of Canada. *Arctic Alpine Res.* **13** (1), 1–10.
Ford, D. C., P. Thompson and H. P. Schwarcz 1972. Dating cave calcite deposits by the uranium disequilibrium method: some preliminary results from Crowsnest Pass, Alberta. In *Research methods in Pleistocene geomorphology*, E. Yatsu and A. Falconer (eds), 247–55. Guelph, Ont.: University of Guelph.
Ford, J. P. 1984. Mapping of glacial landforms from Seasat radar images. *Quat. Res.* **22**, 314–27.
Foster, I. D. L., J. A. Dearing, A. D. Simpson, A. Carter and P. G. Appleby 1985. Estimates of contemporary and historical sediment yields in the Merevale catchment, Warks, UK. *Earth Surf. Processes Landforms* **10**, 45–68.
Foster, R. H. and J. S. Evans 1974. The measurement of soil microstructure. *Microscope* **22**, 323–39.
Fox, C. G. and D. E. Hayes 1985. Quantitative methods for analysing the roughness of the seafloor. *Rev. Geophys.* **23**, 1–48.
Franklin, J. A. and P. E. Denton 1973. The monitoring of rock slopes. *Q.J.Engl.Geol.* **6**, 259–86.
Franklin, S. E. 1987. Terrain analysis from digital patterns in geomorphometry and Landsat MSS spectral response. *Photogrammetric Eng. Remote Sensing* **53**, 59–65.
Frederiksen, R. L. 1969. A battery powered proportional stream water sampler. *Water Resour. Res.* **5**, 1410–13.
Friedkin, J. F. 1945. *A laboratory study of the meandering of alluvial rivers.* US Waterways Exp. St.
Friedman, G. M. 1961. Distinction between dune, beach and river sands from their textural characteristics. *J.Sedim.Petrol.* **31**, 514–29.
Friedman, I. and W. Long 1976. Hydration rate of obsidian. *Science* **191**, 347–52.
Friedman, I. and J. Obradovich 1981. Obsidian hydration dating of volcanic events. *Quat. Res.* **16**, 37–47.
Fries, M. and U. Hafsten 1965. Asbjornsen's peat sampler – the protoype of the Hiller sampler. *Geol.Foren.Forhandl.Stockholm* **87**, 307–13.
Frissell, S. S. 1978. Judging recreation impacts on wilderness campsites. *J.Forestry* 481–3.
Fritschen, L. J. and L. W. Gay 1980. *Environmental instrumentation.* New York: Springer.
Fritts, H. C. 1971. Dendroclimatology and dendroecology. *Quat. Res.* **1**, 419–49.
Fritts, H. C. 1976. *Tree rings and climate.* London: Academic Press.
Fromm, E. 1970. An estimation or errors in the Swedish varve chronology. In *Radiocarbon variations and absolute chronology*, I. Olsson (ed.), 163–72. Stockholm: Almqvist & Wiksell.
Fryberger, S. G. 1978. *Techniques for the evaluation of surface wind data in terms of eolian sand drift.* Open-File Report 78–405, US Geol Surv., Denver, Colo.
Fulton, K. and J. Olfe 1981. *Historical coastal erosion: a manual for researching.* Calif. Univ., Santa Cruz, Science Writing Program.

Gäggeler, H., H. R. Von Gunten and V. Nyffeler 1976. Determination of ^{210}Pb in lake sediments and air samples by direct gamma-ray measurement. *Earth Planet. Sci. Lett.* **33**, 119–21.

Gaillard, D. B. W. 1904. *Wave action*. Washington DC: Corps of Engineers, US Army.
Galehouse, J. S. 1968. Anisotropy of magnetic susceptibility as a palaeocurrent indicator: a test of the method. *Bull. Geol. Soc.Am.* **79**, 387–90.
Galehouse, J. S. 1971. Sedimentation analysis. In *Procedures in sedimentary petrology*, R. E. Carver (ed.), 69–94. New York: Wiley.
Galloway, J. N. and G. E. Likens 1976. Calibration of collection procedures for the determination of precipitation chemistry. *Water, Air Soil Pollut.* **6**, 241–58.
Galloway, J. N. and G. E. Likens 1978. The collection of precipitation for chemical analysis. *Tellus* **30**, 71–82.
Galvin, G. C. 1967. Longshore current velocity: a review of theory and data. *Rev.Geophys.* **5**, 287–304.
Gambell, A. W. and D. W. Fisher 1966. *Chemical composition of rainfall, eastern North Carolina and southeastern Virginia*. US Geol. Surv. Water Supply Paper no. 1535-K.
Gard, J. A. (ed.), 1971. *Electron-optical investigation of clays*. London: Mineralogical Society.
Gardiner, V. 1973. Univariate distributional characteristics of some morphometric variables. *Geog. Annaler* **55A**, 147–53.
Gardiner, V. 1975. Drainage basin morphometry. *BGRG Tech.Bull.* **14**, Norwich: Geo Abstracts.
Gardiner, V. 1978a. Slope maps and the traditional skills of Bacon. *Area* **10**, 205–8.
Gardiner, V. 1978b. Redundancy and spatial organization of drainage basin form indices: an empirical investigation of data from north-west Devon. *Trans. Inst.Br.Geog.NS* **3**, 416–31.
Gardiner, V. 1979. Estimation of drainage density from topological variables. *Water Resour. Res.* **15**, 909–17.
Gardiner, V. 1982. Stream networks and digital cartography. In *Perspectives in the alternative cartography*, D. P. Bickmore (ed.), 38–44. *Cartographica*, Monogr. 28, vol. 19, no. 2.
Gardiner, V. 1983a. Some recent trends in fluvial morphometry. *Natl Geog.* **18**, 1–8.
Gardiner, V. 1983b. Drainage networks and palaeohydrology. In *Background to palaeohydrology*, K. J. Gregory (ed.), 257–77. Chichester: Wiley.
Gardiner, V. 1983c. The relevance of geomorphometry to studies of Quaternary morphogenesis. In *Proc.VIth British–Polish seminar 1977*, 1–18. Norwich: Geo Abstracts.
Gardiner, V. 1987. Fluvial palaeohydrology: the morphometric contribution. *Natl Geog.* **22**.
Gardiner, V. and R. V. Dackombe 1977. A simple method for the field measurement of slope profiles. *BGRG Tech.Bull.* **18**, 9–18.
Gardiner, V. and R. Dackombe 1983. *Geomorphological field manual*. London: Allen & Unwin.
Gardiner, V. and K. J. Gregory 1982. Drainage density in rainfall–runoff modelling. In *Rainfall–runoff relationship*, V. P. Singh (ed.), 449–76. Littleton, Colo: Water Resources Publications.
Gardiner, V. and C. C. Park 1978. Drainage basin morphometry: review and assessment. *Prog.Phys.Geog.* **2**, 1–35.
Gardiner, V. and D. W. Rhind 1974. The creation of slope maps by a photo-mechanical technique. *Area* **6**, 14–21.
Gardiner, V. and D. J. Unwin 1989. *An introduction to computer-assisted cartography*. London: Croom Helm.
Gardiner, V., K. J. Gregory and D. E. Walling 1977. Further notes on the drainage density–basin area relationship. *Area* **9**, 117–21.
Gardiner, V., K. J. Gregory and D. E. Walling 1978. Drainage density–basin area relationship; discussion. *Area* **10**, 354–5.
Gardner, G. J., A. J. Mortlock, D. M. Price, M. L. Readhead and R. J. Wasson 1987. Thermoluminescence and radiocarbon dating of Australian desert dunes. *Aust. J. Earth Sci.* **34**, 343–57.
Gardner, J. S. 1969. Rockfall: a geomorphic process in high mountain terrain. *Albertan Geog.* **6**, 15–20.
Gardner, T. W. 1975. The history of part of the Colorado River and its tributaries: an experimental study. *Four Corners Geological Society handbook*, 87–95. 9th Field Conf., Canyonlands.
Gardner, T. W., D. W. Jorgensen, C. Shuman and C. R. Lemieux 1987. Geomorphic and tectonic process rates: effects of measured time interval. *Geology* **15**, 259–61.
Gartner, J. F. 1984. Some aspects of engineering geology mapping in Canada. *Bull.Assoc.Eng.Geol.* **21**, 269–93.
Gascoyne, M. 1982. Geochemistry of the actinides and their daughters. In *Uranium series disequilibrium: applications to environmental problems*, M. Ivanovich and R. S. Harmon (eds), 33–35. Oxford: Oxford University Press.
Gascoyne, M. and H. P. Schwarcz 1982. Carbonate and sulphate precipitates. In *Uraniun series disequilibrium: applications to environmental problems*, M. Ivanovich and R. S. Harmon (eds), 268–301. Oxford: Oxford University Press.
Gascoyne, M., G. J. Benjamin, H. P. Schwarcz and D. C. Ford 1979. Sea-level lowering during the Illinoian glaciation: evidence from a Bahama 'Blue Hole'. *Science* **205**, 806–8.

Gascoyne, M., A. P. Currant and T. C. Lord 1981a. Ipswichian fauna of Victoria Cave and the marine paleoclimatic record. *Nature* **294**, 652–4.

Gascoyne, M., D. C. Ford and H. P. Schwarcz 1981b. Late Pleistocene chronology and paleoclimate of Vancouver Island as determined from cave deposits. *Can.J.Earth Sci.* **18**, 1643–52.

Gascoyne, M., D. C. Ford and H. P. Schwarcz 1983. Rates of cave and landform development in the Yorkshire Dales from speleothem age data. *Earth Surf.Processes Landforms* **8**, 557–68.

Gascoyne, M., H. P. Schwarcz and D. C. Ford 1978. Uranium series dating and stable isotope studies of speleothem. Part 1: Theory and techniques. *Trans. Br.Cave Res. Assoc.* 91–111.

Gat, J. and A. Issar 1974. Desert isotope hydrology: water sources of the Sinai Desert. *Geochim. Cosmochim. Acta* **6**, 1117–31.

Geiger, R. 1965. *Climate near the ground.* Cambridge, Mass.: Harvard University Press.

Geiger, R. 1975. *The climate near the ground*, 5th edn. Cambridge, Mass.: Harvard University Press.

Geikie, A. 1880. Rock weathering as illustrated in Edinburgh church yards. *Proc.R.Soc. Edin.* 1879/80, 518–32.

Gemmell, A. M. D. 1975. The late-glacial history of the Firth of Clyde: a reply. *Trans. Inst.Br.Geog.* **64**, 132–40.

Gerlach, T. 1967. Hillslope troughs for measuring sediment movement. *Rev. Géomorphol. Dyn.* **17**, 173.

Gerrard, A. J. W. 1978. Hillslope profile analysis. *Area* **10**, 129–30.

Gerrard, A. J. W. and D. A. Robinson 1971. Variability in slope measurements. *Trans. Inst. Br.Geog.* **54**, 45–54.

Gersper, R. L. and N. Holowaychuck 1970. Effects of stemflow on a Miami soil under a beech tree. I: Morphological and physical properties. II: Chemical properties. *Proc.Soil Sci.Soc.Am.* **34**, 779–94.

Gervitz, J. L. 1976. Fourier analysis of bivalve outlines: implications on evolution and autecology. *Math. Geol.* **8**, 151–63.

Geyl, W. F. 1961. Morphometric analysis and the world-wide occurrence of stepped erosion surfaces. *J.Geol.* **69**, 388–416.

Gibbs, R. J. 1967. The factors that control the salinity and composition of the suspended solids in the Amazon River. *Bull.Geol. Soc.Am.* **78**, 1203–32.

Gibbs, R. J. 1973. Mechanisms of trace metal transport in rivers. *Science* **180**, 71–2.

Gibson, R. E. and D. J. Henkel 1954. Influence of duration of tests at constant rate of strain on measured 'drained' strength. *Geotechnique* **4**, 6–15.

Gilbert, G. K. 1914. *The transportation of debris by running water*. US Geol. Surv.Prof. Paper no. 86.

Gilbert, R. 1982. Contemporary sedimentary environments on Baffin Island, NWT, Canada: glaciomarine processes in fjords of eastern Cumberland Peninsula. *Arctic Alpine Res.* **14**, 1–12.

Gilewska, S. 1967. Different methods of showing the relief on the detailed geomorphological maps. *Z.Geomorphol. NF* **11**, 481–90.

Gilewska, S., M. Klimkowa and L. Starkel 1982. The 1:500 000 geomorphological map of Poland. *Geog. Pol.* **48**, 7–24.

Gilg, A. W. 1973. A note on slope measurement techniques. *Area* **5**, 114–17.

Gill, E. D. and J. G. Lang 1983. Micro-erosion meter measurements of rock wear on the Otway coast of southeast Australia. *Mar. Geol.* **52** (1–2), 141–56.

Gill, M. A. 1976. Analysis of one-dimensional non-Darcy vertical infiltration. *J.Hydrol.* **31**, 1–11.

Gillet, F. 1975. Steam, hot-water and electrical thermal drills for temperate glaciers. *J.Glaciol.* **14**, 171–9.

Gillette, D. A. 1978. Tests with a portable wind tunnel for determining wind erosion threshold velocities. *Atmos. Environ.* **12**, 2309–13.

Gillette, D. A. 1979. Environmental factors affecting dust emission by wind erosion. In *Saharan dust*, C. Morales (ed.), 71–91. Chichester: Wiley.

Gillette, D. A. and T. R. Walker 1977. Characteristics of airborne particles produced by wind erosion of sandy soil, High Plains of West Texas. *Soil Sci.* **123**, 97–110.

Gillot, J. E. 1970. Fabric of Leda clay investigated by optical and X-ray diffraction methods. *Eng.Geol.* **4**, 133–53.

Gilluly, J. 1964. Atlantic sediments, erosion rates and the evolution of the continental shelf: some speculations. *Bull.Geol. Soc.Am.* **75**, 483–92.

Gilman, K. and M. D. Newson 1980. *Soil pipes and pipeflow – a hydrological study in upland Wales*. BGRG Research Monograph. Norwich: Geo Books.

Giusti, E. V. and W. J. Schneider 1962. Comparison of drainage on topographic maps of the Piedmont province. *US Geol. Surv.Prof. Paper* **450E**, 118–19.

Glew, J. R. 1977. Simulation of rillenkarren. In *Proc.7th int. speleological conf.*, 218–19. Sheffield, England.

Glover, R. R. 1972. Optical brighteners: a new water tracing reagent. *Trans. Cave Res.Gp GB* **14**, 84–8.

Godfrey, G. K. 1955. A field study of the activity of the mole *(Talpa europaea)*. *Ecology* **31**, 678–85.

Gohren, H. and H. Laucht 1972. Instrument for long-term measurement of suspended matter (silt-gauge). In *Proc.13th conf. on coastal engineering*, vol. 2, 1089–104. Am.Soc.Civ.Eng.

Goldie, H. S. 1981. Morphometry of the limestone pavements of Farleton Knott (Cumbria, England). *Trans. Br. Cave Res. Assoc.* **8**, 207–24.

Goldich, S. S. 1938. A study in rock weathering. *J.Geol.* **46**, 17–58.

Golding, J. F. 1971. *Measuring oscilloscopes*. London: Iliffe.

Goldsmith, V. 1973. Internal geometry and origin of vegetated coastal sand dunes. *J.Sedim.Petrol.* **43**, 1128–42.

Goldsmith, V., J. M. Colonell and R. J. Byrne 1973a. The causes of non-uniform wave energy distribution over the shelf and along the shoreline. In *Int.symp.relations sédimentaires entre estuares et plateaux continentaux*. Inst.Géol. du Bassin d'Aquitaine, Bordeaux, France.

Goldsmith, V., W. G. Morris, R. J. Byrne and C. H. Sutton 1973b. The Virginia sea wave climate model: a basis for the understanding of shelf and coastal sedimentology. *Abstracts, Northeastern Sect., Geol. Soc. Am.*, Allentown, Pa.

Goldstein, J. I., D. E. Newbury, P. Echlin, D. C. Joy, C. Fiori and E. Lifshin 1981. *Scanning electron microscopy and X-ray microanalysis*. New York: Plenum.

Goldstein, M. N., V. A. Misumsky and L. S. Lapidus 1961. The theory of probability in relation to rheology of soils. In *Proc.5th int.conf. on soil mechanics*, vol. 1, 123–6. Paris.

Goldthwaite, R. P. 1973. Jerky glacier motion and meltwater. *Int.Assoc. Sci.Hydrol. Publ.* **85**, 183–8. Symp. on hydrology of glaciers, Cambridge, 7–13 Sept. 1969.

Golterman, H. L. and R. S. Clymo 1969. *Methods for chemical analysis of fresh waters*. IBP Handbook no. 8. Oxford: Blackwell.

Golterman, H. L., R. S. Clymo and M. A. M. Ohnstad 1978. *Methods for chemical analysis of fresh waters*. IBP Handbook, 2nd edn. Oxford: Blackwell.

Goodchild, M. F. 1980. Fractals and the accuracy of geographical measures. *Math. Geol.* **12**, 85–98.

Goodchild, M. F. and D. M. Mark 1987. The fractal nature of geographic phenomena. *Ann.Assoc.Am.Geog.* **77**, 265–78.

Goodfriend, G. A. and M. Magaritz 1988. Paleosols and late Pleistocene rainfall fluctuations in the Negev Desert. *Nature* **332**, 144–6.

Goodman, G. T., M. J. Inskip, S. Smith, G. D. R. Parry and M. A. S. Burton 1979. The use of moss-bags in aerosol monitoring. In *Saharan dust*, C. Morales (ed.), 211–32. Chichester: Wiley.

Gordon, C. M. 1975. Sediment entrainment and suspension in a turbulent tidal flow. *Mar. Geol.* **18**, 1957–64.

Gordon, D., P. L. Smart, J. N. Andrews, D. C. Ford, T. C. Atkinson, P. Rowe and N. S. J. Christopher 1989. Dating of United Kingdom Interglacials and Interstadials from speleothem growth frequency. *Quat. Res.* **31**, 27–40.

Gordon, J. E. 1977. Morphometry of cirques in the Kintail–Affric–Cannich area of north west Scotland. *Geog. Annaler* **59A**, 177–94.

Gore, A. J. P. 1968. The supply of six elements by rain to an upland peat area. *J.Ecol.* **56**, 483–95.

Gorham, E. 1961. Factors influencing supply of major ions to inland waters, with special reference to the atmosphere. *Bull.Geol. Soc.Am.* **72**, 795–840.

Gorsline, G. S. 1966. Dynamic characteristics of west Florida Gulf coast beaches. *Mar. Geol.* **4**, 187–206.

Goudie, A. S. 1969. Statistical laws and dune ridges in southern Africa. *Geog.J.* **135**, 404–6.

Goudie, A. S. 1970. Input and output considerations in estimating rates of denudation. *Earth Sci. J.* **4**, 59–65.

Goudie, A. S. 1974. Further experimental weathering by salt and other mechanical processes. *Z. Geomorphol. NF* **21**, 1–12.

Goudie, A. S. 1977. *Environmental change*. Oxford: Clarendon.

Goudie, A. S. 1983. Dust storms in space and time. *Prog.Phys.Geog.* **7**, 502–30.

Goudie, A. S. 1986. Laboratory simulation of 'the wick effect' in salt weathering of rock. *Earth Surf. Processes Landforms* **11**, 275–86.

Goudie, A. S. 1988. The geomorphological role of termites and earthworms in the tropics. In *Biogeomorphology*, H. A. Viles (ed.). Oxford: Basil Blackwell.

Goudie, A. S. and A. Watson 1984. Rock block monitoring of rapid salt weathering in southern Tunisia. *Earth Surf. Processes Landforms* **9**, 95–8.

Goudie, A. S., R. U. Cooke and J. C. Doornkamp 1979. The formation of silt from quartz dune sand by salt-weathering processes in deserts. *J.Arid Environ.* **2**, 105–12.

Gould, P. 1970. Is *statistix inferens* the geographical name for a wild goose? *Econ. Geog.* **46**, 439–48.

Graf, W. H. 1971. *Hydraulics of sediment transport*. New York: McGraw-Hill.
Graf, W. L. 1975. A cumulative stream-ordering system. *Geog.Anal.* **6**, 335–40.
Graf, W. L. 1977. Measuring stream order: a reply. *Geog.Anal.* **9**, 431–2.
Graf, W. L. 1979. Catastrophe theory as a model for change in fluvial systems. In *Adjustments in the fluvial system*, D. D. Rhodes and G. P. Williams (eds), 13–32. Dubuque, Iowa: Kendall/Hunt.
Graf, W. L. 1983. The arroyo problem – palaeohydrology and palaeohydraulics in the short term. In *Background to palaeohydrology*, K. J. Gregory (ed.), 279–302. Chichester: Wiley.
Graham, E. R. 1949. The plagioclase feldspars as an index to soil weathering. *Proc.Soil Sci.Soc.Am.* **14**, 300–2.
Granat, L. 1976. Principles in network design for precipitation chemistry measurements. *J.Great Lakes Res.* **2**, 42–55.
Grant, D. R. 1980. Quaternary sea level change in Atlantic Canada as an indication of coastal delevelling. In *Earth rheology, isostasy and eustasy*, N.-A. Mörner (ed.), 201–14. Chichester: Wiley.
Grant, K. and A. A. Finlayson 1978. The assessment and evaluation of geotechnical resources in urban or regional environments. *Eng.Geol.* **12**, 219–93.
Grant, K., T. G. Ferguson, A. A. Finlayson and B. G. Richardson 1981. *Terrain analysis, classification and evaluation for regional development purposes of the Albury–Wodonga area, NSW and Victoria.* CSIRO Aust., Div.Appl. Geomech.,Tech.Paper no. 30.
Grant, P. 1978. The role of the scanning electron microscope in cathodoluminescence petrology. In *Scanning electron microscopy in the study of sediments*, W. B. Whalley (ed.), 1–11. Norwich: Geo Abstracts.
Grant, W. H. 1969. Abrasion pH, an index of chemical weathering. *Clays Clay Min.* **17**, 151–5.
Gravenor, C. P. 1974. The Yarmouth drumlin field, Nova Scotia, Canada. *J.Glaciol.* **13**, 45–54.
Gray, J. and P. Thompson 1976. Climatic information from $^{18}O/^{16}O$ ratios of cellulose in tree rings. *Nature* **262**, 481–2.
Gray, J. M. 1972. Trends through clusters. *Area* **4**, 275–9.
Gray, J. M. 1974. The main rock platform of the Firth of Lorn, western Scotland. *Trans. Inst.Br.Geog.* **61**, 81–99.
Gray, J. M. 1975. *Measurement and analysis of Scottish raised shoreline altitudes*. Dept Geog., Queen Mary College, Univ. London, Occas.Paper no. 2, 1–40.
Gray, J. M. 1983. The measurement of relict shoreline altitudes in areas affected by glacio-isostasy, with particular reference to Scotland. In *Shorelines and isostasy*, D. E. Smith and A. G. Dawson (eds), 97–127. IBG Spec.Publ. no. 16. London: Academic Press.
Gray, M. 1981. Large-scale geomorphological field mapping: teaching the first stage. *J.Geog. Higher Educ.* **5**, 37–44.
Greeley, R. and J. D. Iversen 1985. *Wind as a geological process on Earth, Mars, Venus and Titan*. Cambridge: Cambridge University Press.
Greenwood, B. and D. J. Sherman 1984. Waves, currents, sediment flux and morphological response in a barred nearshore system. *Mar. Geol.* **60** (1–4), 31–61.
Gregory, K. J. 1966. Dry valleys and the composition of the drainage net. *J.Hydrol.* **4**, 327–40.
Gregory, K. J. 1976a. Drainage networks and climate. In *Geomorphology and climate*, E. Derbyshire (ed.), 289–315. Chichester: Wiley.
Gregory, K. J. 1976b. Lichens and the determination of river channel capacity. *Earth Surf. Processes* **1**, 273–85.
Gregory, K. J. 1977a. Stream network volume: an index of channel morphometry. *Bull.Geol. Soc.Am.* **88**, 1075–80.
Gregory, K. J. 1977b. Channel and network metamorphosis in northern New South Wales. In *River channel changes*, K. J. Gregory (ed.), 389–410. Chichester: Wiley.
Gregory, K. J. (ed.) 1977c. *River channel changes*. Chichester: Wiley.
Gregory, K. J. 1979. Drainage network power. *Water Resour. Res.* **15**, 775–7.
Gregory, K. J. 1983. *Background to palaeohydrology*. Chichester: Wiley.
Gregory, K. J. and V. Gardiner 1975. Drainage density and climate. *Z. Geomorphol. NF* **19**, 287–98.
Gregory, K. J. and C. C. Park 1974. Adjustment of river channel capacity downstream from a reservoir. *Water Resour. Res.* **10**, 870–3.
Gregory, K. J. and D. E. Walling 1968. The variation of drainage density within a catchment. *Bull.Int.Assoc. Sci.Hydrol.* **13**, 61–8.
Gregory, K. J. and D. E. Walling 1973. *Drainage basin form and process*. London: Edward Arnold.
Greig-Smith, P. 1964. *Quantitative plant ecology*, 2nd edn. London: Butterworth.
Grieve, I. C. 1985. Determination of dissolved organic matter in streamwater using visible spectrophotometry. *Earth Surf. Processes Landforms* **10**, 75–8.
Griffey, N. J. and J. A. Matthews 1978. Major Neoglacial glacier expansion episodes in southern

Norway: evidences from moraine ridge stratigraphy with ^{14}C dates on buried palaeosols and moss layers. *Geog. Annaler* **60A**, 73–90.
Griffey, N. J. and P. Worsley 1978. The pattern of Neoglacial glacier variations in the Okstindan region of northern Norway during the last three millennia. *Boreas* **7**, 1–17.
Griffin, G. M. 1971. Interpretation of X-ray diffraction data. In *Procedures in sedimentary petrology*, R. E. Carver (ed.), 541–69. New York: Wiley.
Griffiths, J. C. 1967. *Scientific method in analysis of sediments*. New York: McGraw-Hill.
Grissinger, E. H. and L. L. McDowell 1970. Sediment in relation to water quality. *Water Resour. Res.* **6**, 7–14.
Grün, R. and H. P. Schwarcz 1987. Some remarks on 'ESR dating of bone'. *Ancient TL* **5** (2), 1–9.
Guth, P. L., E. K. Ressler and T. S. Bacastow 1987. Microcomputer program for manipulating large digital terrain models. *Comput. Geosci.* **13**, 209–13.
Guy, H. P. 1964. *An analysis of some storm period variables affecting stream sediment transport*. US Geol. Surv.Prof. Paper no. 462-E.
Guy, H. P., D. B. Simons and E. V. Richardson 1966. *Summary of alluvial channel data from flume experiments, 1956–61*. US Geol. Surv.Prof. Paper no. 462-1.
Guy, M. and M. Mainguet 1975. Les courants de transport éolien au Sahara at leurs manifestations au sol. *C.R. Acad.Sci. Paris D* **281**, 103–6.

Hack, J. T. *Studies of longitudinal stream profiles in Virginia and Maryland*. US Geol. Surv. Prof. Paper no. 294B.
Hack, J. T. 1960. Interpretation of erosional topography in humid temperate regions. *Am.J.Sci.* **258A**, 80–97.
Haefeli, R. 1951. Some observations on glacier flow. *J.Glaciol.* **1**, 57–9.
Hafsten, U. 1960. Pollen analytic investigations in south Norway. In *Geology of Norway*, O. Holtedahl (ed.). *Nor. Geol. Unders.* **208**, 434–61.
Hafsten, U. 1983. Biostratigraphical evidence of Late Weichselian and Holocene sea-level changes in southern Norway. In *Shorelines and isostasy*, D. E. Smith and A. G. Dawson (eds), 161–181. London: Academic Press.
Hagen, J. O., B. Wold, O. Liestol, G. Ostrem and J. L. Sollid 1983. Subglacial processes at Bondhusbreen, Norway: preliminary results. *Ann. Glaciol.* **4**, 91–8.
Haggett, P. 1965. *Locational analysis in human geography*. London: Edward Arnold.
Haigh, M. J. 1977. The use of erosion pins in the study of slope evolution. *BGRG Tech.Bull.* **18**, 31–49.
Haigh, M. J. 1981. Micro-profile measurement using the contour gauge. *BGRG Tech. Bull.* **29**, 31–2.
Hails, J. R. 1974. A review of some current trends in nearshore research. *Earth Sci. Rev.* **10**, 171–202.
Haines-Young, R. H. 1983. Size variations of *Rhizocarpon* on moraine slopes in southern Norway. *Arctic Alpine Res.* **15**, 295–305.
Hakala, I. 1971. A new model of the Kajak bottom sampler and other improvements in the zoobenthos sampling technique. *Ann.Zool.Fenn.* **8**, 422–6.
Hakanson, L. 1974. A mathematical model for establishing numerical values of topographical roughness for lake bottoms. *Geog. Annaler* **56A**, 183–200.
Hakanson, L. 1977. On lake form, lake volume and lake hypsographic survey. *Geog. Annaler* **59A**, 1–29.
Hakanson, L. 1978. The length of closed geomorphic lines. *Math. Geol.* **10**, 141–68.
Hakanson, L. 1982. Lake bottom dynamics and morphometry: the dynamic ratio. *Water Resour. Res.* **18**, 1444–50.
Hall, C. M. and D. York 1984. The applicability of $^{40}Ar/^{39}Ar$ dating to young volcanics. In *Quaternary dating methods*, D. J.Easterbrook (ed.), 67–74. Amsterdam: Elsevier.
Hall, F.R. 1970a. Dissolved solids–discharge relationships. 1: Mixing models. *Water Resour. Res.* **6**, 845–50.
Hall, M. J. 1970b. A critique of methods of simulating rainfall. *Water Resour. Res.* **6**, 1104–14.
Hallermeier, R. J. 1973. Design considerations for a 3-D laser Doppler velocimeter for studying gravity waves in shallow water. *Appl. Opt.* **12**, 294–300.
Hallet, B. 1979. A theoretical model of glacial abrasion. *J.Glaciol.* **23**, 39–50.
Hallet, B. and R. S. Anderson 1980. Detailed glacial geomorphology of a proglacial bedrock area at Castleguard Glacier, Alberta, Canada. *Z. Gletscherk. Glazialgeol.* **16** (2), 171–84.
Hama, F.R. 1954. Boundary-layer characteristics for smooth and rough surfaces. *Soc.Naval Archit. Mar. Eng.* **62**, 333–58.
Hamblin, W. K. 1962. X-ray radiography in the study of structures in homogeneous sediments. *J.Sedim.Petrol.* **32**, 201–10.

Hamilton, R. A., C. A. Biddle and B. W. Sparks 1957. Surveying aneroids: their uses and limitations. *Geol.J.* **123**, 481–9.

Hancock, N. J. 1978. An application of scanning electron microscopy in pilot water injection studies for oilfield development. In *Scanning electron microscopy in the study of sediments*, W. B. Whalley (ed.), 61–70. Norwich: Geo Abstracts.

Hansbo, S. 1957. A new approach to the determination of the shear strength of clay by the fall-cone test. *Proc.R.Swed.Geotech.Inst.* **14**.

Harbaugh, J. W. and G. Bonham-Carter 1970. *Computer simulation in geology*. New York: Wiley.

Harbaugh, J. W. and D. F. Merriam 1968. *Computer applications in stratigraphic analysis*. New York: Wiley.

Hardcastle, P. J. 1971. A pebble measurer for laboratory use, giving a punched tape output. *J.Sedim. Petrol.* **41**, 1138–9.

Harden, J. W. 1982. A quantitative index of soil development from field descriptions: examples from a chronosequence in central California. *Geoderma* **28**, 1–28.

Hardisty, J. 1983. An assessment and calibration of formulations for Bagnold's bedload equation. *J.Sedim.Petrol.* **53**, 1007–10.

Hardisty, J. 1984. A dynamic approach to the intertidal profile. In *Coastal research: UK perspectives*, M. W. Clark (ed.), 17–36. Norwich: Geo Books.

Hardisty, J. 1986. A morphological model for beach gradients. *Earth Surf. Processes Landforms* **11**, 327–33.

Hardisty, J. 1987. Higher frequency acoustic measurements of swash processes. *Phys. Geog.* **8** (2), 169–78.

Hare, P. E. 1974. Amino acid dating: a history and evaluation. *MASCA Newslett.* **4**.

Hare, P. E., A. S. Brooks, J. E. Kokis and K. Kumin 1984. Aminostratigraphy: the use of ostrich eggshell in dating the Middle Stone Age at Gi, Botswana. *Geol.Soc.Am.Abstr.Programs* **16**, 529.

Harlin, J. M. 1978. Statistical moments of the hypsometric curve and its density function. *Math. Geol.* **10**, 59–72.

Harmon, R. S., D. C. Ford and H. P. Schwarcz 1977. Interglacial chronology of the Rocky and Mackenzie Mountains based upon $^{230}Th/^{234}U$ dating of calcite speleothems. *Can.J.Earth Sci.* **14**, 2543–52.

Harmon, R. S., H. P. Schwarcz and D. C. Ford 1978a. Late Pleistocene sea level history of Bermuda. *Quat. Res.* **9**, 205–18.

Harmon, R. S., H. P. Schwarcz, D. C. Ford and D. Koch 1979a. An isotopic palaeotemperature record of the Late Wisconsinan Stage in northeast Iowa. *Geology* **7**, 430–3.

Harmon, R. S., H. P. Schwarcz and J. R. O'Neil 1979b. D/H ratios in speleothem fluid inclusions: a guide to variations in the isotopic composition of meteoric precipitation? *Earth Planet. Sci. Lett.* **42**, 254–66.

Harmon, R. S., P. Thompson, H. P. Schwarcz and D. C. Ford 1978b. Late Pleistocene palaeoclimates of North America as inferred from stable isotope studies of speleothems. *Quat. Res.* **9**, 54–70.

Harris, C. 1974. Wind speed and sand movement in a coastal dune environment. *Area* **6**, 243–9.

Harris, C. 1987. Mechanisms of mass movement in periglacial environments. In *Slope stability: geotechnical engineering and geomorphology*, M. G. Anderson and K. S. Richards (eds), 531–59. Chichester: Wiley.

Harrison, W. 1969. Empirical equations for foreshore changes over a tidal cycle. *Mar. Geol.* **7**, 529–51.

Harrison, W. 1972. Changes in foreshore sand volume on a tidal beach: role of fluctuations in water table and ocean still-water level. In *Proc. 24th int.geology congr.*, Sect.12, 159–66. Montreal.

Harrison, W. and J. D. Boon 1972. Beach water table and beach-profile measuring equipment. *Shore Beach* **40**, 26–33.

Harrison, W. D. and B. Kamb 1973. Glacier bore-hole photography. *J.Glaciol.* **12**, 129–37.

Harrison, W. D. and B. Kamb 1976. Drilling to observe subglacial conditions and sliding motion. In *Ice-core drilling*, J. F. Splettstoesser (ed.). Lincoln, Neb: University of Nebraska Press.

Harrison, W. D., C. F. Raymond and P. MacKeith 1986. Short period motion events on Variegated Glacier as observed by automatic photography and seismic methods. *Ann. Glaciol.* **8**, 82–9.

Hartigan, J. A. 1975. *Clustering algorithms*. New York: Wiley.

Hartwell, A. D. and M. O. Hayes 1969. Hydrography of the Marionack River Estuary. In *Coastal environments: NE Massachusetts and New Hampshire* Contrib.no.1 CRG, Dept Geol.Publ.Ser., Univ.Massachusetts, Amherst.

Harvey, A. M. 1969a. Channel capacity and the adjustment of streams to hydrologic regime. *J.Hydrol.* **8**, 82–98.

Harvey, A. M. 1974. Gully erosion and sediment yield in the Howgill Fells, Westmorland. In

Fluvial processes in instrumented watersheds, K. J. Gregory and D. E. Walling (eds), 45–58. Inst.Br.Geog.Spec.Publ. no. 6. London.
Harvey, D. W. 1966. Geographical processes and point patterns: testing models of diffusion by quadrat sampling. *Trans.Inst.Br.Geog.* **40**, 81–95.
Harvey, D. W. 1968. Some methodological problems in the use of Neyman type A and negative binomial probability distributions for the analysis of spatial point pattern. *Trans.Inst.Br.Geog.* **44**, 85–95.
Harvey, D. 1969b. *Explanation in geography*. London: Edward Arnold.
Hastenrath, S. L. 1967. The barchans of the Arequipa region, southern Peru. *Z. Geomorphol. NF* **11**, 300–11.
Hawkes, I. and M. Mellor 1970. Uniaxial testing in rock mechanics laboratories. *Eng. Geol.* **4**, 177–284.
Hayes, M. O. 1969. Forms of sand accumulation in estuaries. In *Coastal environments: NE Massachusetts and New Hampshire*, Contrib.no.1 CRG, Dept Geol.Publ.Ser., Univ.Massachusetts, Amherst.
Haynes, J. M. 1973. Determination of pore properties of constructional and other materials. *Mater. Constr.* **6**, 169–74.
Haynes, R. 1983. *An introduction to dimensional analysis for geographers*. Catmog 33 Norwich: Geo Abstracts.
Head, K. H. 1986. *Manual of soil laboratory testing*. London: Pentech.
Hearty, P. J. and G. H. Miller 1987. Global trends in isoleucine epimerization: data from the circum-Atlantic, the Mediterranean and the South Pacific. *Geol.Soc.Am.Abstr. Programs* **19**, 698.
Hearty, P. J., G. F. Miller, C. E. Stearns and B. J. Szabo 1986. Aminostratigraphy of Quaternary shorelines in the Mediterranean basin. *Geol. Soc. Am. Bull.* **97**, 850–8.
Hedberg, H. D. 1976. *International stratigraphic guide*. New York: Wiley.
Hedges, R. E. M. and C. B. Moore 1978. Enrichment of ^{14}C and radiocarbon dating. *Nature* **276**, 255–7.
Heede, B. H. 1975. Mountain watershed and dynamic equilibrium. In *Watershed management symp. ASCE irrigation and drainage division*, 407–20.
Heerdegen, R. G. and M. A. Beran 1982. Quantifying source areas through land surface curvature and shape. *J. Hydro.* **57**, 359–73.
Helvey, J. D. and J. H. Patric 1965. Design criteria for interception studies. *Int.Assoc. Sci.Hydrol. Publ.* **67**, 131–7. Symp. on the design of hydrological networks.
Hem, J. D. 1970. *Study and interpretation of the chemical characteristics of natural water*. US Geol.Surv.Water Supply Paper no. 1473, 2nd edn.
Hembree, C. H., R. A. Kreiger and P. R. Jordan 1964. *Chemical quality of the surface waters and sedimentation in the Grand River drainage basin, N and S Dakota*. US Geol. Surv.Water Supply Paper no. 1769.
Hembree, C.H. and F.H. Rainwater 1961. *Chemical degradation on opposite flanks of the Wind River Range, Wyoming*. US Geol. Surv. Water Supply Paper no. 1535-E.
Hencher, S. R. 1976. A simple sliding apparatus for the measurement of rock joint friction: discussion. *Geotechnique* **26**, 641–4.
Hendrikson, E. R. 1968. Air sampling and quantity measurement. In *Air pollution*, A. C. Stern (ed.), 31–52. New York: Academic Press.
Hendy, C. H. and A. T. Wilson 1968. Palaeoclimatic data from speleothems. *Nature* **219**, 48–51.
Henin, S. and M. Robert 1964. Nouvelle observations sur le mechanisme de la désagrégation du granite. C. R. Acad.Sci.Paris **259**, 3577–80.
Henkel, D. J. 1957. Investigations of two long term failures in London Clay clay slopes at Wood Green and Northolt. In *Proc. 4th int.conf.on soil mechanics*. London.
Hennig, G. J. and R. Grün 1983. ESR dating in Quaternary geology. *Quat. Sci Rev.* **2**, 157–238.
Herbertson, J. G. 1969. A critical review of conventional bed load formuli. *J.Hydrol.* **8**, 1–26.
Herbotson, P. W., J. W. Douglas and A. Hill 1971. *River level sampling periods*. Rep.no.9., Inst.Hydrol.,Wallingford, England.
Heron, J. 1962. Determination of phosphate in water after storage in polyethylene. *Limnol. Oceanog.* **7**, 316–21.
Herschy, R. W. 1971. *The magnitude of errors at flow measurement stations*. Water Resour. Board Rep. no. TN11.
Herschy, R. W. 1976. New methods of river gauging. In *Facets of hydrology*, J. C. Rodda (ed.), 110–61. Chichester: Wiley.
Hesse, P. R. 1955. A chemical and physical study of the soils of termite mounds in East Africa. *J.Ecol.* **43**, 449–61.
Hesse, P. R. 1971. *A textbook of soil chemical analysis*. London: Murray.

Hey, R. D. 1979. Flow resistance in gravel bed rivers. *Proc. Am. Soc. Civ. Eng., J. Hydraul. Div.* **105**, **HY4**, 365–79.
Hey, R. D., J. C. Bathurst and C. R. Thorne (eds) 1982. *Gravel bed rivers*. Chichester: Wiley.
Hey, R.W., D. H. Krinsley and P. J. W. Hyde 1971. Surface textures of sand grains from the Hertfordshire pebble gravels. *Geol.Mag.* **108**, 377–82.
Hibbert, F. A. and V. R. Switzur 1976. Radiocarbon dating of Flandrian pollen zones in Wales and northern England. *New Phytol.* **77**, 793–807.
Hickin, E. J. 1978. Mean flow structure in meanders of the Squamish River, British Columbia. *Can.J.Earth Sci.* **15**, 1833–49.
Hickin, E. J. 1984. Vegetation and river channel dynamics. *Can. Geog.* **28**, 111–26.
Hickin, E. J. and G. C. Nanson 1975. The character of channel migration on the Beatton River, Northeast British Columbia, Canada. *Geol.Soc.Am.Bull.* **85**, 487–94.
Hickman, G. C. and L. N. Brown 1973. Mound-building behaviour of the southeastern pocket gopher (*Geomys pinetis*). *J. Mammalogy* **54**, 786–9, 971–4.
Higgins, C. G. 1956. Formation of small ventifacts. *J.Geol.* **64**, 506–16.
Higgitt, S. R. 1985. *A palaeoecological study of recent environmental change in the drainage basin of the Lac d'Annecy (France)*. Unpubl. Ph.D. Thesis, University of Liverpool.
High, C. and F. K. Hanna 1970. A method for the direct measurement of erosion on rock surfaces. *BGRG Tech.Bull.* **5**.
Hill, A. R. 1973. The distribution of drumlims in County Down, Ireland. *Ann.Assoc.Am.Geog.* **63**, 226–40.
Hill, J. M. 1984. *Nitrates in surface waters: observations from some rivers in the Lee drainage basin.* Water Res. Centre Tech. Rep. no. TR 203.
Hobba, L. J. and G. Robinson 1972. A suggested new approach to the analysis of drainage basin shapes. *Area* **4**, 242–51.
Hobson, R. D. 1967. FORTRAN IV programs to determine surface roughness in topography for the CDC 3400 computer. *Kansas Comput. Contrib.* **14**.
Hobson, R. D. 1972. Surface roughness in topography: quantitative approach. In *Spatial analysis in geomorphology*, R. J. Chorley (ed.), 221–45, London: Methuen.
Hodge, S. M. 1971. A new version of a steam-operated ice drill. *J.Glaciol.* **10**, 387–93.
Hodge, S.M. 1976. Direct measurement of basal water pressures: a pilot study. *J.Glacio.* **16**, 205–18.
Hodgson, J. M. 1978. *Soil sampling and soil description.* Oxford: Clarendon.
Hodgson, J. M., J. H. Rayner and J. A. Catt 1974. The geomorphological significance of the Clay-with-flints on the South Downs. *Trans.Inst.Br.Geog.* **61**, 119–29.
Hoecker, W. H. 1975. A universal procedure for deploying constant volume balloons and for deriving vertical air speeds from them. *J.Appl.Meteorol.* **14**, 1118–24.
Hoefs, J. 1987. *Stable isotope geochemistry*, 3rd edn. Berlin: Springer.
Hoek, E. and J. W. Bray 1977. *Rock slope engineering*, 2nd edn. London: Institution of Mining and Metallurgy.
Hollin, J. T. 1977. Thames interglacial sites, Ipswichian sea-levels and Antarctic ice surges. *Boreas* **6**, 33–52.
Hollingshead, A. B. 1971. Sediment transport measurements in gravel river. *Proc.Am. Soc.Civ.Eng.,J.Hydraul. Div.* **97**, **HY11**, 1817–34.
Hollingworth, S. E. 1938. The recognition and correlation of high level erosion surfaces in Britain. *Q.J.Geol.Soc.* **94**, 55–84.
Holmes, S. C. A. 1972. Geological applications of early large-scale cartography. *Proc.Geol.Assoc.* **83**, 121–38.
Hooke, J. 1980. Magnitude and distribution of rates of river bank erosion. *Earth Surface Processes* **5**, 223–56.
Hooke, J. M. and R. J. P. Kain 1982. *Historical change in the physical environment: a guide to sources and techniques.* London: Butterworth.
Hooke, R. LeB. 1974. *Shear-stress and sediment distribution in a meander bend.* UNGI Rapport no. 30, Uppsala.
Hooke, R. LeB. and W. L. Rohrer 1979. Geometry of alluvial fans: effect of discharge and sediment size. *Earth Surf. Processes* **4**, 147–66.
Hooke, R. LeB., B. Wold and J. O. Hagen 1985. Subglacial hydrology and sediment transport at Bondhusbreen, southwest Norway. *Bull.Geol.Soc.Am.* **96**, 388–97.
Hooper, J. H. D. 1958. Bat erosion as a factor in cave formation. *Nature* **182**, 1464.
Hope, R., H. Lister and R. Whitehouse 1972. The wear of sandstone by cold sliding ice. In *Polar geomorphology*, R. J. Price and D. E. Sugden (eds), 21–31. Inst.Br.Geog.Spec.Publ. no. 4.
Horam, K. 1982. A satellite altimetric geoid in the Philippine Sea. *Nature* **299**, 117–21.

Horie, S. (ed.) 1972. *Palaeolimnology of Lake Biwa and the Japanese Pleistocene*, vol. 1. Kyoto University.
Horie, S. (ed.) 1974. *Palaeolimnology of Lake Biwa and the Japanese Pleistocene*, vol. 2. Kyoto University.
Horie, S. (ed.) 1975. *Palaeolimnology of Lake Biwa and the Japanese Pleistocene*, vol. 3. Kyoto University.
Horikawa, K. and H. W. Shen 1960. *Sand movements by wind action – on the characteristics of sand traps*. Tech. Memo. no. 119, US Army Corps of Engineers, Beach Erosion Board, Washington DC.
Hormann, K. 1968. *Rechenprogramme zur morphometrischen Kartenauswertung*. Schr. Geog. Inst. Univ. Kiel no. 29.
Hormann, K. 1971. *Morphometrie der Erdoberflache*. Schr. Geog. Inst. Univ. Kiel no. 36.
Horn, H. M. and D. V. Deere 1962. Frictional characteristics of minerals. *Geotechnique* **12**, 319–35.
Horsfield, W.T. 1977. An experimental approach to basement-controlled faulting. *Geol. Mijnbouw* **56**, 363–70.
Horton, R. E. 1932. Drainage basin characteristics. *Trans. Am.Geophys. Union* **14**, 350–61.
Horton, R. E. 1945. Erosional development of streams and their drainage basins; hydrophysical approach to quantitative morphology. *Bull.Geol.Soc.Am.* **56**, 275–370.
Howard, A. D. 1977. Effect of slope on the threshold of motion and its application to orientation of wind ripples. *Bull.Geol.Soc.Am.* **88**, 853–6.
Howard, A. D., M. E. Beetch and C. L. Vincent 1970. Topological and geometrical properties of braided streams. *Water Resour. Res.* **6**, 1674–88.
Howard, A. D., J. B. Morton and M. Gad-el-Hak 1977. *Simulation model of erosion and deposition on a barchan dune*. Rep. NASA, CR-2838.
Howard, J. D. 1969. Radiographic examination of variations in barrier island facies, Sapelo Island, Georgia. *Gulf Coast Assoc.Geol.Soc.Trans.* **19**, 217–32.
Howe, S. E. and T. Webb 1983. Calibrating pollen data in climate terms: improving the methods. *Quat.Sci.Rev.* **2**, 17–51.
Hsu, S. A. 1973. Computing eolian sand transport from shear velocity measurements. *J.Geol,* **81**, 739–43.
Huang, W.H. and W. D. Keller 1972. Organic acids as agents of chemical weathering of silicate minerals. *Nature, Phys. Sci.* **239**, 149–51.
Hubbell, D. W. 1964. *Apparatus and techniques for measuring bedload.* US Geol.Surv. Water Supply Paper no. 1748.
Hucka, V. 1965. A rapid method for determining the strength of rock *in situ. Int.J.Rock Mech. Min. Sci.* **2**, 127–34.
Hudson, N. 1972. *Soil conservation*. London: Batsford.
Huh, O. L. and V. E. Noble 1977. Remote sensing of environment: achievements and prognosis for the future. *Naval Res. Rev.* July, 1–18. Office of Naval Research, Arlington, Va.
Humbert, F. L. 1968. *Selection and wear of pebbles on gravel beaches.* Rep. no. 190, Geol.Inst., Univ. Groningen, Netherlands.
Humphrey, N., C.F. Raymond and W. Harrison 1986. Discharges of turbid water during mini-surges of Variegated Glacier, Alaska, USA. *J.Glaciol.* **32**, 111, 195–207.
Hunting Technical Services 1977. *Technical Report. Sand dune movement study, south of Umm Said, 1963–76.* Govt of Qatar, Ministry of Public Works.
Huntley, B. and H. J. B. Birks 1983. *An atlas of past and present pollen maps for Europe: 0–13 000 years ago.* Cambridge: Cambridge University Press.
Huntley, D. J., D. I. Godfrey-Smith and M. L. Thewalt 1985. Optical dating of sediments. *Nature* **313**, 105–7.
Hupp, C. R. 1986. The headward extent of fluvial landforms and associated vegetation on Massanutten Mountain, Virginia. *Earth Surf. Processes Landforms* **11**, 545–55.
Hutchinson, C. S. 1974. *Laboratory handbook of petrographic techniques.* New York: Wiley.
Hutchinson, J. N. 1970. A coastal mudflow on the London Clay Cliffs at Beltinge, north Kent. *Geotechnique* **20**, 412–38.
Hutchinson, J. N. 1973. The response of London Clay Cliffs to differing rates of toe erosion. *Geol. Appl. Idrogeol.* **7**, 222–39.
Hutchinson, J. N. and R. K. Bhandari 1971. Undrained loading, a fundamental mechanism of mudflows and other mass movements. *Geotechnique* **21**, 353–8.
Hutchinson, J. N. and T. P. Gostelow 1976. The development of an abandoned cliff in London Clay at Hadleigh, Essex. *Phil.Trans.R.Soc.A* **283**, 557–604.
Hutchinson, J. N., D. B. Prior and N. Stephens 1974. Potentially dangerous surges in an Antrim mudslide. *Q.J.Eng.Geol.* **7**, 363–76.

Huttunen, P. and J. Merilainen 1978. New freezing device providing large unmixed sediment samples from lakes. *Ann.Bot.Fenn.* **15**, 128–30.

Huttunen, P., J. Merilainen and K. Tolonen 1978. The history of a small dystrophied forest lake, southern Finland. *Pol.Arch.Hydrobiol.* **25**, 189–202.

Huybrechts, W. 1985. Profile type maps of alluvial deposits: the case of the Mark R., Belgium. *Z.Geomorphol Supp. NF* **55**, 113–120.

Iken, A. and R. A. Bindschadler 1986. Combined measurements of subglacial water pressure and surface velocity of Findelengletscher, Switzerland. Conclusions about drainage system and sliding mechanism. *J.Glaciol.* **32**, 101–19.

Iken, A., H. Rothlisberger and K. Hutter 1977. Deep drilling with a hot water jet. *Z. Gletscherk. Glazialgeol.* **12**, 143–56.

Ikeya, M. 1985. Electron spin resonance. In *Dating methods of Pleistocene deposits and their problems*, N. W. Rutter (ed.), 73–87. *Geosci. Can.* Reprint Ser. 2.

Ilier, I. G. 1956. An attempt to estimate the degree of weathering of intrusive rocks from their physico-mechanical properties. In *Proc.1st int.congr.society of rock mechanics*, vol. 2, 109.

Imbrey, J. and N. G. Kipp 1971. A new micropalaeontological method of quantitative palaeoclimatology: application to a late Pleistocene Caribbean core. In *The late Cenozoic glacial ages*, J. K. Turekian (ed.). New Haven, Conn.: Yale University Press.

Imeson, A. C. 1973. Solute variations in small catchment streams. *Trans. Inst.Br.Geog.* **60**, 87–100.

Imeson, A. C. 1976. Some effects of burrowing animals on slope processes in the Luxembourg Ardennes. Part I: The excavation of animal mounds in experimental plots. *Geog. Annaler* **58A**, 115–25.

Imeson, A. C. 1977. Splash erosion, animal activity and sediment supply in a small forested Luxembourg catchment. *Earth Surf. Processes* **2**, 153–60.

Imeson, A. C. and F. J. P. M. Kwaad 1976. Some effects of burrowing animals on slope processes in the Luxembourg Ardennes. Part II: The erosion of animal mounds by splash under forest. *Geog. Annaler* **58A**, 317–28.

Imeson, A. C. and H. J. M. van Zon 1979. *Erosion processes in small forested catchments in Luxembourg*. Norwich: Geo Books.

Imhof, E. 1982. *Cartographic relief presentation.* Berlin: de Gruyter.

Ingle, J. C. 1966. The movement of beach sand: an analysis using fluorescent grains. *Developments in sedimentology*, vol. 5. Amsterdam: Elsevier.

Ingle, J. C. and D. S. Gorsline 1973. Use of fluorescent tracers in the nearshore environments. In *Tracer techniques in sediment transport*, 125–48. Vienna: International Atomic Energy Agency.

Ingram, R. L. 1971. Sieve analysis. In *Procedures in sedimentary petrology*, R. E. Carver (ed.), 49–67. New York: Wiley.

Inman, D. L. and R. A. Bagnold 1963. Littoral processes. In *The sea*, M. N. Hill (ed.), vol. III, 529–33. New York: Interscience.

Inman, D. L. and A. J. Bowen 1963. Flume experiments on sand transport by waves and currents. In *Proc.8th conf. on coastal engineering*, 137–50.

Inman, D. L. and J. D. Frautschy 1966. Littoral processes and the development of shorelines. In *Proc.coastal engineering spec.conf.*, 511–36. Am.Soc.Civ.Eng.

Inman, D. L. and N. Nasu 1956. *Orbital velocity associated with wave action near the breaker zone.* BEB Tech.Memo. no. 79.

Inman, D. L. and G. A. Rasnak 1956. *Changes in sand level on the beach and shelf at La Jolla, California.* BEB Tech.Memo. no. 82.

Inman, D. L., G. C. Ewing and J. B. Corliss 1966. Coastal sand dunes of Guerro Negro, Baja California, Mexico. *Bull.Geol. Soc.Am.* **77**, 787–802.

Inman, D. L., P. D. Komer and A. J. Bowen 1968. Longshore transport of sand. In *Proc.11th conf.coastal engineering*, vol. 1, 298–306. Am.Soc.Civ.Eng.

Innes, J. L. 1984. Lichenometric dating of moraine ridges in northern Norway: some problems of application. *Geog. Annaler* **66A**, 241–52.

Innes, J. L. 1985a. Lichenometry. *Prog.Phys.Geog.* **9**, 187–254.

Innes, J. L. 1985b. A standard *Rhizocarpon* nomenclature for lichenometry. *Boreas* **14**, 83–5.

Institute of Hydrology 1977. *Selected measurement techniques in use at Plynlimon experimental catchments.* Report no. 43.

Isaac, E. K. 1975. A multiple crest stage recorder. *BGRG Tech.Bull.* **17**, 7–12.

ISSS/INQUA Working Group 1971. Criteria for the recognition and classification of palaeosols. In *Paleopedology*, D. H. Yaalon (ed.), 153–8. Jerusalem: Int.Soc.Soil Sci. and Israel University Press.

Ivanovich, M. 1982. Uranium series disequilibria in geochronology. In *Uranium series disequili-*

brium: applications to environmental problems, M. Ivanovich and R. S. Harmon (eds), 56–78. Oxford: Oxford University Press.

Iversen, J. 1944. *Visium, Hedera* and *Ilex* as climatic indicators. *Geol. Foren. Stockholm Forhandl.* **66**, 463–83.

Jackson, R. G., II 1975. Velocity–bedform–texture patterns of meander bends in the Lower Wabash River of Illinois and Indiana. *Geol. Soc.Am.Bull.* **86**, 1511–22.

Jaenicke, R. 1979. Monitoring and critical review of the estimated source strength of mineral dust from the Sahara. In *Saharan dust*, C. Morales (ed.), 233–42. Chichester: Wiley.

Jagger, T. A. 1908. Experiments illustrating erosion and sedimentation. *Harvard Univ. Mus. Compar.Zool.Bull.* **49**, 285–304.

Jagger, T. A. 1912. Structure of esker fans experimentally studied. *Bull.Geol. Soc.Am.* **30**, 746.

Jago, C. F. and J. Hardisty 1984. Sedimentology and morphodynamics of a macrotidal beach, Pendine Sands, SW Wales. *Mar. Geol.* **60**, 123–54.

Jahn, A. 1961. Quantitative analysis of some periglacial processes in Spitzbergen. *Zesz.Nauk Univ.Wroclaw.Ser.B* **5**, 1–54.

James, P. A. 1971. The measurement of soil frost-heave in the field. *BGRG Tech.Bull.* **8**.

James, W. R. 1970. A class of probability models for littoral drift. In *Proc.12th conf.coastal engineering*, vol. 2, 813–37. Am.Soc.Civ.Eng.

Jamieson, T. F. 1865. On the history of the last geological changes in Scotland. *Q.J.Geol. Soc.Lond.* **21**, 161–203.

Janda, R. J. 1971. An evaluation of procedures used in computing chemical denudation rates. *Bull.Geol. Soc.Am.* **82**, 67–80.

Jansson, M. B. 1982. *Land erosion by water in different climates.* Uppsala Univ. Naturgeog. Inst. Rapport no. 57.

Jardine, G. W. 1976. Some problems in plotting the mean surface of the North Sea and the Irish Sea during the last 15 000 years. *Geol. Foren. Stockholm Forhandl.* **98**, 78–82.

Jarvis, R. S. 1972. New measure of the topologic structure of dendritic drainage networks. *Water Resour. Res.* **8**, 1265–71.

Jarvis, R. S. 1976a. Classification of nested tributary basins in analysis of drainage basin shape. *Water Resour. Res.* **12**, 1151–64.

Jarvis, R. S. 1976b. Link length organization and network scale dependencies in the network diameter model. *Water Resour. Res.* **12**, 1215–25.

Jarvis, R. S. 1977a. Spatial variation in ordered tributary basins: a proposed classification. *Geog. Annaler* **59A**, 241–51.

Jarvis, R. S. 1977b. Drainage network analysis. *Prog.Phys.Geog.* **1**, 271–95.

Jauhiainen, E. 1975. Morphometric analysis of drumlin fields in northern Central Europe. *Boreas* **4**, 219–30.

Jaynes, S. M. and R. U. Cooke 1987. Stone weathering in south-east England. *Atmos. Environ.* **21** (7), 1601–22.

Jeffery, P. G. 1975. *Chemical methods of rock analysis.* Oxford: Pergamon.

Jelgersma, S. 1961. *Holocene sea-level changes in the Netherlands.* Meded. Geol. Sticht. CVI 7.

Jelgersma, S. 1966. Sea level changes in the last 10 000 years. In *Int. symp. on world climate from 8000–0 BC*, 54–69. Royal Meteorological Society.

Jenkins, R. and J. L. de Vries 1970a. *Practical X-ray spectrometry.* London: Macmillan.

Jenkins, R. and J. L. de Vries 1970b. *Worked examples in X-ray spectrometry.* London: Macmillan.

Jenkinson, D. S. and J. H. Rayner 1977. The turnover of soil organic matter in some of the Rothamsted classical experiments. *Soil Sci.* **123**, 298–305.

Jennings, J. N. 1967. Topographical maps and the geomorphologist. *Cartography* **6**, 73–81.

Jennings, J. N. 1971. Karst. Introduction to *Systematic geomorphology*, vol. 7. Canberra: Australian National University Press.

Jenny, H. 1941. *Factors of soil formation.* London: McGraw-Hill.

Jochimsen, M. 1966. Ist die Grosse des Flectenthallus wirklich ein brauchbarer Masstab zur Datierung von glazialmorphologischen Relikten. *Geog. Annaler* **48A**, 157–64.

Jochimsen, M. 1973. Does the size of lichen thalli really constitute a valid measure for dating glacial deposits? *Arctic Alpine Res.* **5**, 417–24.

Johansson, C. E. 1965. Structural studies of sedimentary deposits. *Geol. Foren. Stockholm Forhandl.* **87**, 3–16.

Johnsen, S. W., W. Dansgaard, H. B. Clausen and C. C. Langway 1972. Oxygen isotope profiles through the Antarctic and Greenland ice sheets. *Nature* **235**, 429–34.

Johnson, C. M. and P. R. Needham 1966. Ionic composition of Sagehen Creek, California, following an adjacent fire. *J.Ecol.* **47**, 636–9.

Johnson, N. M., G. E. Likens, F. H. Bormann, D. W. Fisher and R. S. Pierce 1969. A working model for the variation in stream water chemistry at the Hubbard Brook Experimental Forest, New Hampshire. *Water Resour. Res.* **5**, 1353–63.

Johnson, P. L. and W. T. Swank 1973. Studies of cation budgets in the Southern Appalachians on four experimental watersheds with contrasting vegetation. *Ecology* **54**, 70–80.

Johnston, M. J. S., S. McHugh and R. D. Burford 1976. On simultaneous tilt and creep observations on the San Andreas Fault. *Nature* **260**, 691–3.

Johnstone, A. 1889. On the action of pure water and of water saturated with carbonic acid gas on the minerals of the mica family. *Q.J.Geol. Soc.Lond.* **45**, 363–8.

Joliffe, I. P. 1963. A study of sand movements on the Lowestoft sandbank using fluorescent tracers. *Geog.J.* **129**, 480–93.

Jonasson, A. 1977. New devices for sediment and water sampling. *Mar. Geol.* **24**, M13–21.

Jońca, E. 1972. Winter denudation of molehills in mountainous areas. *Acta Theriol. XVII* **31**, 407–17.

Jones, A. V. and K. C. A. Smith 1978. Image processing for scanning microscopists. *Scanning Electron Microsc.* **1**, 13–26.

Jones, J. A. A., M. Lawton, D. P. Wareing and F. G. Crane 1984. An economical logging system for field experiments. *BGRG Tech.Bull.* **31**.

Jones, P. F. and E. Derbyshire 1983. Late Pleistocene periglacial degradation of lowland Britain: implications for civil engineering. *Q.J.Eng.Geol.* **16**, 197–210.

Jones, R. L. and P. R. Cundill 1978. Introduction to pollen analysis. *BGRG Tech.Bull.* **22**.

Jordan, C. F. 1985. *Nutrient cycling in tropical forest ecosystems.* Chichester: Wiley.

Journaux, A. and J.-P. Lautridou 1976. Les études de cryoclastie au Centre de Géomorphologie du CNRS à Caen. *Geog. Glasnik* 130–9.

Jowsey, P. C. 1966. An improved peat sampler. *New Phytol.* **65**, 245–8.

Juang, F. H. T. and N. M. Johnson 1967. Cycling of chlorine through a forested watershed in New England. *J.Geophys.Res.* **72**, 5641–7.

Judson, S. 1968. Erosion rates near Rome, Italy. *Science* **160**, 1444–5.

Junge, C. E. and R. T. Werby 1958. The concentration of chloride, sodium, potassium, calcium and sulfate in rain-water of the United States. *J.Meteorol.* **15**, 417–25.

Justice, C. O., S. W. Wharton and B. N. Holben 1981. Application of digital terrain data to quantify and reduce the topographic effect on Landsat data. *Int. J. Remote Sensing* **2**, 213–30.

Kachel, N. B. and R. W. Sternberg 1971. Transport of bedload as ripples during an ebb current. *Mar. Geol.* **19**, 229–44.

Kaganov, E.I. and A. M. Yaglom 1976. Errors in wind-speed measurements by rotation anemometers. *Boundary-Layer Meteorol.* **10**, 15–34.

Kamb, W. B. 1970. Sliding motion of glaciers: theory and observation. *Rev. Geophys. Space Phys.* **8**, 673–728.

Kamb, W. B. and E. R. LaChappelle 1964. Direct observation of the mechanism of glacier sliding over bedrock. *J.Glaciol.* **5**, 159–72.

Kamb, W. B., *et al.* 1985. Glaciers surge mechanism: 1982–1983 surge of Variegated Glacier, Alaska. *Science* **227** (4686), 469–79.

Kang, B. T. and R. Lal 1981. Nutrient losses in water runoff from agricultural catchments. In *Tropical agricultural hydrology*, R. Lal and E. W. Russell (eds), Ch. 3.6, 153–61. Chichester: Wiley.

Kansky, K. J. 1963. *Structure of transport networks: relationships between network geometry and regional characteristics.* Univ. Chicago, Dept of Geography, Res. Paper no. 84.

Kanwisher, J. and K. Lawson 1975. Electromagnetic flow sensors. *Limnol. Oceanog.* **20**, 174–82.

Karcz, I. 1978. Rapid determination of lineament and joint densities. *Tectonophysics* **44**, T29–33.

Karlén, W. 1979. Glacier variations in the Svartisen area, northern Norway. *Geog.Annaler* **61A**, 11–28.

Katz, S. H., G. W. Smith, W. M. Myers, L. J. Trostel, M. Ingels and L. Greenburg 1925. *Comparative tests of instruments for determining atmospheric dusts.* US Publ.Health Bull. No. 144.

Kaufman, A. 1971. U-series dating of Dead Sea basin carbonates. *Geochim. Cosmochim. Acta* **35**, 1269–81.

Kaufman, A. 1986. The distribution of $^{230}Th/^{234}U$ ages in corals and the number of Last Interglacial high-sea stands. *Quat. Res.* **25**, 55–62.

Kaufman, A., W. Broecker, T. L. Ku and D. L. Thurber 1971. The status of U-series methods of mollusc dating. *Geochim. cosmochim. Acta* **35**, 1155–83.

Kaulagher, P. G. 1984. Steroscopic fusion of a photographic image with a directly observed view: applications to geomorphology. *Z. Geomorphol. NF* **51**, 123–40.

Kawamura, R. 1951. *Study on sand movement by wind.* Rep.Inst. Sci. Technol., Univ. of Tokyo, vol. 5, no.314.

Kawamura, R. 1953. Mouvement du sable sous l'effet du vent. In *Actions éoliennes*. CNRS, Paris, Coll. Int. no. 35, 117–51.

Kaye, B. H. 1978a. Specification of the ruggedness and/or texture of a fine particle profile by its fractal dimension. *Powder Technol.* **21**, 1–16.

Kaye, B. H. and A. G. Naylor 1972. An optical information procedure for characterising the shape of fine particle images. *Pattern Recogn.* **4**, 195–9.

Keller, W. D. 1957. *The principles of chemical weathering*. Columbia, Mo.: Lucas.

Keller, W. D. 1976. Scan electron micrographs of kaolins collected from diverse origins. III: Influence of parent material on flint clays and flint-like clays. *Clays Clay Min.* **24**, 262–4.

Keller, W. D. 1978. Classification of kaolin exemplified by their textures in scan electron micrographs. *Clays Clay Min.* **26**, 1–20.

Keller, W. D. and A. L. Reesman 1963. Glacial milks and their laboratory simulated counterparts. *Bull.Geol.Soc.Am.* **74**, 61–76.

Kellerhals, R. and D. I. Bray 1971. Sampling procedures for coarse fluvial sediments. *Proc.Soc.Civ. Eng., J. Hydraul. Div.* **97**, **HY1**, 1165–80.

Kemp, P. H. 1971a. Chemistry of natural waters. 1: Fundamental relationships. *Water Res.* **5**, 297–311.

Kemp, P. H. 1971b. Chemistry of natural waters. 6: Classification. *Water Res.* **5**, 943–56.

Kemp, R. A. 1985. The Valley Farm soil in southern East Anglia. In *Soils and Quaternary landscape evolution*, J. Boardman (ed.), 179–96. Chichester: Wiley.

Kempton, J. P. and K. Cartwright 1984. 3D geologic mapping: a basic for hydrogeologic and land-use evaluations. *Bull.Assoc. Eng. Geol.* **21**, 317–35.

Kenn, M. J. 1965. Experiments simulating the effects of mountain building initiated by the late Professor S. W. Wooldridge. *Proc.Geol.Assoc.Lond.* **76**, 21–7.

Kennedy, B. A. 1978. After Horton. *Earth Surf. Processes* **3**, 219–31.

Kennedy, G. L., K. R. Lajoie and J. F. Wehmiller 1982. Aminostratigraphy and faunal correlations of Late Quaternary marine terraces, Pacific Coast USA. *Nature* **299**, 545–7.

Kenney, T. C. 1967. The influence of mineral composition on the residual shear strength of natural soils. In *Proc.geotech.conf.Oslo*, vol. 1, 123–31.

Kerr, R. C. and J. O. Nigra 1952. Eolian sand control. *Am. Assoc. Petrol.Geol.Bull.* **36**, 1541–73.

Kershaw, K. A. 1964. *Quantitative and dynamic ecology*. London: Edward Arnold.

Kestner, F. J. T. 1970. Cyclic changes in Morecambe Bay. *Geog. J.* **136**, 85–97.

Khazi, A. and J. L. Knill 1969. The sedimentation and geotechnical properties of the Cromer Till between Happisburgh and Cromer, Norfolk. *Q.J.Eng.Geol.* **2**, 63–86.

Kheoruenromne, I., W. E. Sharp and L. R. Gardner 1976. Variation of link magnitude with map scale. *Water Resour. Res.* **12**, 919–23.

Kidson, C. and M. M. M. Manton 1973. Assessment of coastal changes with the aid of photogrammetric and computer-aided techniques. *Estuar. Coastal Mar. Sci.* **1**, 271–83.

Kidson, C., J. A. Steers and N. C. Flemming 1962. A trial of the potential value of aqualung diving to physiography on British coasts. *Geog.J.* **128**, 49–53.

Kihl, R. 1975. Physical preparation of organic matter samples for submission to a radiocarbon dating laboratory. *Quat. Newslett.* **16**, 4–6.

Kilmer, V. J., J. W. Gillian, J. W. Lutz, R.T. Joyce and C. D. Eckland 1974. Nutrient losses from fertilised grassland watersheds in western North Carolina. *J.Environ.Qual.* **3**, 214.

Kilpatrick, F. A. and H. H. Barnes 1964. *Channel geometry of Piedmont streams as related to frequency of floods*. US Geol. Surv.Prof. Paper no. 422-E.

Kimbara, K., S. Shimoda and T. Sudo 1973. An unusual chlorite as revealed by high temperature X-ray. *Clay Min.* **10**, 71–8.

Kimmins, J. P. 1973. Some statistical aspects of sampling throughfall precipitation in nutrient cycling studies in British Columbian coastal forests. *Ecology* **54**, 1008–19.

King, C. A. M. 1963. Some problems concerning marine planation and the formation of erosion surfaces. *Trans.Inst.Br.Geog.* **33**, 29–43.

King, C. A. M. 1968. Beach measurement at Gibraltar Point, Lincolnshire. *East Midl.Geog.* **4**, 295–300.

King, C. A. M. 1969a. Trend surface analysis of central Pennine erosion surfaces. *Trans.Inst. Br.Geog.* **47**, 47–59.

King, C. A. M. 1970a. Changes in the spit at Gibraltar Point, Lincolnshire, 1951–1959. *East Midl.Geog.* **5**, 19–30.

King, C. A. M. 1970b. Feedback relationships in geomorphology. *Geog. Annaler* **52A**, 147–59.
King, C. A. M. 1972. *Beaches and coasts*. London: Edward Arnold.
King, C. A. M. 1974. Morphometry in glacial geomorphology. In *Glacial geomorphology*, D. R. Coates (ed.), 147–62. Binghamton, NY: State University of New York.
King, C. A. M. 1978. Coastal geomorphology in the United Kingdom. In *Geomorphology: present problems and future prospects*, C. Embleton, D. Brunsden and D. K. C. Jones (eds), for BGRG, 224–50. Oxford: Oxford University Press.
King, C. A. M. and F. A. Barnes 1964. Changes in the configuration of the inter-tidal beach zone of part of the Lincolnshire coast since 1951. *Z. Geomorphol. NF* **8**, 105–26.
King, C. A. M. and J. Buckley 1968. Analysis of stone size and shape in Arctic environments. *J.Sedim.Petrol.* **38**, 200–14.
King, C. A. M. and M. J. McCullagh 1971. A simulation model of a complex recurved spit. *J.Geol.* **79**, 22–36.
King, L. J. 1969b. *Statistical analysis in geography*. Englewood Cliffs, NJ: Prentice-Hall.
Kirk, R. M. 1973. An instrument system for shore process studies. *BGRG Tech.Bull.* **10**.
Kirkby, A. V. T. and M. J. Kirkby 1974a. Surface wash at the semi arid break in slope. *Z. Geomorphol. Supp. NF* **21**, 151–76.
Kirkby, A. V. T. and M. J. Kirkby 1974b. The implications of geomorphic processes for archaeological reconnaissance survey of semi-arid areas. In *Geoarchaeology*, M. L. Shackley (ed.), 1976. London: Duckworth.
Kirkby, M. J. 1963. *A study of rates of erosion and mass movements on slopes with special reference to Galloway*. Unpubl. Ph.D. Thesis, University of Cambridge.
Kirkby, M. J. 1967. Measurement and theory of soil creep. *J.Geol.* **75**, 359–78.
Kirkby, M. J. 1976. Tests of the random network model and its application to basin hydrology. *Earth Surf. Processes* **1**, 197–212.
Kirkby, M. J. 1984. Modelling cliff development in south Wales: Savigear reviewed. *Z. Geomorphol. NF* **28**, 405–26.
Kirkby, M. J. 1986. Mathematical models for solutional development of landforms. In *Solute processes*, S. T. Trudgill (ed.), 439–96. Chichester: Wiley.
Kirkby, M. J. and R. J. Chorley 1967. Throughflow, overland flow and erosion. *Bull.Int.Assoc. Sci.Hydrol.* **12**, 5–21.
Kittleman, L. R., Jr 1964. Application of Rosin's distribution in size-frequency analysis of clastic rocks. *J.Sedim.Petrol.* **34**, 483–502.
Klages, M. G. and Y. P. Hsieh 1975. Suspended solids carried by the Gallatin River of southwestern Montana. II: Using mineralogy for inferring sources. *J.Environ.Qual.* **4**, 68–73.
Klappa, C. F. 1979. Lichen stromatolites: criteria for sub-aerial exposure and a mechanism for the formation of laminar calcretes (caliche). *J.Sedim.Petrol.* **49**, 387–400.
Klein, J., J. C. Lerman, P. E. Damon and E. K. Ralph 1982. Calibration of radiocarbon dates: tables based on the consensus date of the workshop on calibrating the radiocarbon time scale. *Radiocarbon* **24**, 103–50.
Klimaszewski, M. (ed.) 1963. Problems of geomorphological mapping. *Prace Geof. Warsaw* **46**.
Klug, H. P. and L. E. Alexander 1954. *X-ray diffraction procedures for polycrystalline and amorphous materials*. New York: Wiley.
Knapp, B. J. 1978. Infiltration and storage of soil water. In *Hillslope hydrology*, M. J. Kirkby (ed.), 43–72. Chichester: Wiley.
Knight, A. H., R. Boggie and H. Shepherd 1972. The effect of groundwater level on water movement in peat: a study using tritiated water. *J.Appl.Ecol.* **9**, 633–41.
Knight, S. H. 1924. Eolian abrasion of quartz grains. *Bull.Geol.Soc.Am.* **35**, 107–8.
Knighton, A. D. 1981. Asymmetry of river channel cross-sections: Part I. Quantitative indices. *Earth Surf. Processes Landforms* **6**, 581–8.
Knott, P. 1979. *The structure and pattern of dune-forming winds*. Unpubl. Ph.D. Thesis, University of London.
Knox, J. C. 1972. Valley alluviation in south-western Wisconsin. *Ann.Assoc.Am.Geog.* **62**, 401–10.
Koerner, R. M. 1970. Effect of particle characteristics on soil strength. *Proc.Am.Soc.Civ.Eng.* **SM4**, 1221–34.
Köhler, A. and E. W. Fleck 1963. *Untersuchungen zur Festlegung eines standardmessgerates fur Staunbniederschlag T.H. Darmstadt*. Inst.Meteorol.Abschlussber. no.J.(84) 2.
Kohyama, N., K. Fukushima and A. Fukami 1978. Observations of the hydrated form of tabular halloysite by an electron microscope equipped with an environmental cell. *Clays Clay Min.* **26**, 25–40.
Koide, M., A. Soutar and E. D. Goldberg 1972. Marine geochronology with ^{210}Pb. *Earth Planet Sci. Lett.* **14**, 442–6.

Komar, P. D. 1976. *Beach processes and sedimentation.* Englewood Cliffs, NJ: Prentice-Hall.
Komar, P. D. 1983. Shapes of streamlined islands on Earth and Mars: experiments and analyses of the minimum drag form. *Geology* **11**, 651–4.
Komar, P. D. 1984. The lemniscate loop – comparisons with the shape of streamlined landforms. *J.Geol.* **92**, 133–45.
Komar, P. D. 1985. Computer models of shoreline configuration: headland erosion and the graded beach revisited. In *Models in geomorphology*, M. J. Woldenburg (ed.), 155–70. Winchester, Mass.: Allen & Unwin.
Komar, P. D. 1986. Nearshore currents and sand transport on beaches. In *Physical oceanography of coastal and shelf seas,* B. Jones (ed.), 67–109. Elsevier Oceanog. Ser. no. 35. Amsterdam: Elsevier.
Komar, P. D. 1987. Selective grain entrainment by a current from a bed of mixed sizes: a reanalysis. *J. Sedim. Petrol.* **57**, 203–11.
Komar, P. D. and B. Cui 1984. The analysis of grain-size measurements by sieving and settling-tube techniques. *J.Sedim.Petrol.* **54**, 603–15.
Komar, P. D. and R. A. Holman 1986. Coastal processes and the development of shoreline erosion. *Annu. Rev. Earth Planet. Sci.* **14**, 237–65.
Komar, P. D. and D. L. Inman 1970. Longshore sand transport on beaches. *J.Geophys.Res.* **75**, 5914–27.
Komar, P. D. and M. L. Miller 1975. On the comparison between the threshold of sediment motion under water and unidirectional currents with a practical evaluation of the threshold. *J.Sedim.Petrol.* **45**, 362–7.
Komar, P. D. and C. E. Reimers 1978. Grain shape effects on settling velocity. *J.Geol.* **86**, 193–209.
Konecny, G. 1977. Software aspects of analytical plotters. *Photogramm. Eng. Remote Sensing* **43**, 1363–6.
Koontz, W. A. and D. L. Inman 1967. *A multi-purpose data acquisition system for field and laboratory instrumentation of the nearshore environment.* US Army Coastal Eng.Res. Center T.M.21.
Kopper, J. S. and K. M. Creer 1976. Palaeomagnetic dating and stratigraphic interpretation in archaeology. *MASCA Newslett.* **12**, 1–4. Applied Science Center and Archaeology, The University Museum, Pennsylvania.
Koster, E. 1964. *Granulometrische und morphometrische Messmethoden an Mineralkornen, Steinen und sonstigen Stoffen.* Stuttgart: Enke.
Kostrzewski, A. and Z. Zwolinski 1985. Chemical denudation rate in the Upper Parseta catchment, western Pomerania: research methods and preliminary results. *Quaest. Geog. (Poznan)* **1,** Special Issue 121–38.
Kraus, N. C. 1985. Field experiments on vertical mixing of sand in the surf zone. *J.Sedim.Petrol.* **55** (1), 3–14.
Krausel, R. 1961. Paleobotanical evidence of climate. In *Descriptive paleoclimatology*, A. F. M. Nairn (ed.), 227–54. New York: Interscience.
Krauskopf, K. B. 1967. *Introduction to geochemistry.* New York: McGraw-Hill.
Krcho, J. 1973. Morphometric analysis of relief on the basis of geometric aspect of field theory. *Acta Geog. Univ. Comenianae, Geog.-Phys. (Bratislava)* **1**, 7–233.
Krinitzsky, E. L. 1970. *Radiography in the earth sciences and soil mechanics.* New York: Plenum
Krinitzsky, E. L. 1972. X-ray measurement of soil densities in models. *J. Mater.* **7**, 119–30.
Krinsley, D. H. 1978. The present state and future propects of environmental discrimination by scanning electron microscopy. In *Scanning electron microscopy and the study of sediments*, W. B. Whalley (ed.), 169–79. Norwich: Geo Abstracts.
Krinsley, D. H. and J. C. Doornkamp 1973. *An atlas of quartz sand surface textures.* Cambridge: Cambridge University Press.
Krinsley, D. H. and F. W. McCoy 1977. Significance and origin of surface textures on broken sand grains in deep-sea sediments. *Sedimentology* **24**, 857–62.
Krinsley, D. H. and S. V. Margolis 1969. A study of quartz sand grain surfaces with the scanning electron microscope. *Trans. NY Acad.Sci.* **31**, 457–77.
Krinsley, D. H. and T. Takahashi 1962a. Electron microscope examination of natural and artificial glacial sand grains. *Geol.Soc.Am.Spec. Papers* **73**, 175.
Krinsley, D. H. and T. Takahashi 1962b. Surface textures of sand grains: an application of electron microscopy. *Science* **135**, 923–5.
Krinsley, D. H. and T. Takahashi 1962c. Surface textures of sand grains: an application of electron microscopy – glaciation. *Science* **138**, 1262–4.
Krinsley, D. H. and N. K. Tovey 1978. Cathodoluminescence in quartz sand grains. In *Scanning electron microscopy 1978*, O. Johari (ed.), vol.1, 887–94. Chicago: SEM.

Kroger, P. M., J. M. Davidson and E. C. Gardner 1986. Mobile very long baseline interferometry and global positioning system measurement of vertical crustal motion. *J.Geophys.Res.* **91** (B9), 9169–76.

Krumbein, W. C. 1941. Measurement and geological significance of shape and roundness of sedimentary particles. *J.Sedim.Petrol.* **11**, 64–72.

Krumbein, W. C. 1959. The sorting out of geological variables illustrated by regression analysis of factors controlling beach firmness. *J.Sedim.Petrol.* **29**, 575–87.

Krumbein, W. C. and F. A. Greybill 1965. *An introduction to statistical models in geology*. New York: McGraw-Hill.

Krumbein, W. C. and R.L. Shreve 1970. *Some statistical properties of dendritic channel networks*. Tech.Rep. no. 13, Dept Geol. Sci., Northwestern University, Evanston, Ill.

Krumbein, W. C. and L. L. Sloss 1955. *Stratigraphy and sedimentation*. San Francisco: Freeman.

Ku, T. L. 1976. The uranium-series methods of age determination. *Annu. Rev. Earth Planet. Sci.* **4**, 347–79.

Ku, T. L. 1982. Progress and perspectives. In *Uranium series disequilibrium: applications to environmental problems*, M. Ivanovich and R. S. Harmon (eds), 497–506. Oxford: Oxford University Press.

Ku, T. L. and W. Broecker 1967. Rates of sedimentation in the Arctic Ocean. In *Oceanography*, M. Sears (ed.), vol.4, 95–104. Oxford: Pergamon.

Ku, T. L. and Z. C. Liang 1984. The dating of impure carbonates with decay-series isotopes. *Nucl.Instrum.Meth.* **223**, 563–71.

Küchter, A. W. 1967. *Vegetation mapping*. New York: Ronald.

Kuenen, Ph.H 1955. Experimental abrasion of pebbles.I: Wet sand blasting. *Leidsche Geol.Meded.* **20**, 131–7.

Kuenen, Ph.H 1960a. Sand. *Sci. Am.* April.

Kuenen, Ph.H. 1960b. Experimental abrasion.4: Eolian action. *J.Geol.* **68**, 427–49.

Kuenen, Ph.H. 1964. Pivotability studies of sand by a shape sorter. In *Developments in sedimentology*. Vol.1: *Deltaic and shallow marine deposits*, L. M. J. U. van Straaten (ed.), 207–15. Amsterdam: Elsevier.

Kuenen, Ph.H. and W. G. Perdok 1962. Experimental abrasion. 5: Frosting and defrosting of quartz grains. *J.Geol.* **70**, 648–58.

Kuhn, T. S. 1962. *The structure of scientific revolutions*. Chicago: Chicago University Press.

Kuipers, H. 1957. A relief meter for soil cultivation studies. *Neth. J. Agric. Sci.* **5**, 255–62.

Kwaad, F. J. P. H. 1970. Experiments on the granular distinegration of granite by salt action. *Fys. Geog. Bodernk. Lab., Amsterdam, Publ.* **16**, 67–80.

Lally, A. E. 1982. Chemical procedures. In *Uranium series disequilibrium: applications to environmental problems*, M. Ivanovich and R. S. Harmon (eds), 79–106. Oxford: Oxford University Press.

Lam, Kin-Che 1977. Patterns and rates of slopewash on the badlands of Hong Kong. *Earth Surf. Processes* **2**, 319–32.

LaMarche, V. C. 1968. Rates of slope degradation as determined from botanical studies, White Mountains, California. *US Geol. Surv.Prof. Paper* **352–1**, 341–77.

LaMarche, V. C., Jr and H. C. Fritts 1971. Tree rings, glacial advance and climate. *Z. Gletscherk. Glazialgeol.* **7**, 125–31.

Lambe, T. W. and R. V. Whitman 1969. *Soil mechanics*. New York: Wiley.

Lancaster, N. 1981. Aspects of the morphometry of linear dunes in the Namib desert. *S. Afr. J. Sci.* **77**, 366–8.

Land, L. S. 1964. Eolian cross-bedding in the beach dune environment, Sapelo Island, Georgia. *J.Sedim.Petrol.* **34**, 389–94.

Laney, R. L. 1965. A comparison of the chemical composition of rainwater and groundwater in western North Carolina. *US Geol. Surv. Prof. Paper* **525-C**, 187–9.

Lanfredi, N. W. and M. R. Framinnan 1987. HP 67/97 calculator wave application programs. *Comput. Geosci.* **13** (4), 409–16.

Langbein, W. B. and L. B. Leopold 1966. *River meanders: theory of minimum variance*. US Geol. Surv. Prof. Paper no. 422H.

Langbein, W. B. and S. A. Schumm 1958. Yield of sediment in relation to mean annual precipitation. *Trans. Am.Geophys. Union* **39**, 1076–84.

Langmuir, D. 1971. E_h–pH determination. In *Procedures in sedimentary petrology*, R. E. Carver (ed.), 597–634. New York: Wiley.

Laronne, J. B. and M. A. Carson 1976. Inter-relationship between bed morphology and bed-material transport for a small gravel-bed channel. *Sedimentology* **23**, 67–85.

Latham, A. G., H. P. Schwarcz and D. C. Ford 1986. The paleomagnetism and U–Th dating of Mexican stalagmite, DAS2. *Earth Planet. Sci. Lett.* **79**, 195–207.

Lavalle, P. 1967. Some aspects of linear karst depression development in south central Kentucky. *Ann.Assoc. Am.Geog.* **57**, 49–71.

Lavalle, P. 1968. Karst depression morphology in south central Kentucky. *Geog. Annaler* **50A**, 94–108.

Lavelle, P. 1983. The soil fauna of tropical savannas. II. The earthworms. In *Tropical savannas*, F. Bourliere (ed.), 485–504. Amsterdam: Elsevier.

Lawler, D. M. 1978. The use of erosion pins in river banks. *Swansea Geog.* **16**, 9–17.

Laws, R. M. 1970. Elephants as agents of habitat and landscape change in East Africa. *Oikos* **21**, 51–66.

Lawson, D. E. 1979. *Sedimentological analysis of the western terminus region of the Matanuska Glacier, Alaska.* US Army CRREL Rep. no. 79–9.

Lawson, D. E. 1981. Distinguishing characteristics of diamictons at the margin of the Matanuska Glacier, Alaska. *Ann. Glaciol.* **2**, 78–84.

Leatherman, S. P. 1978. A new aeolian sand trap design. *Sedimentology* **25**, 303–6.

Lee, K. E. and T. G. Wood 1971. *Termites and soils.* London: Academic Press.

Lees, B. J. 1979. A new technique for injecting fluorescent sand tracer in sediment transport experiments in a shallow marine environment. *Mar. Geol.* **33**, M95–8.

Lees, G. 1969. Influence of boundary effects on the packing and porosity of granular materials. *Q.J.Eng.Geol.* **2**, 129–47.

Leopold, L. B. 1973. River channel changes with time: an example. *Geol. Soc.Am.Bull.* **84**, 1845–60.

Leopold, L. B. 1978. El Asunto del arroyo. In *Geomorphology: present problems and future prospects*, C. Embleton, D. Brunsden and D. K. C. Jones (eds), for BGRG. Oxford: Oxford University Press.

Leopold, L. B. and T. Dunne 1971. Field method for hillslope description. *BGRG Tech. Bull.* **7**.

Leopold, L. B. and W. W. Emmett 1976. Bedload measurements, East Fork River, Wyoming. *Proc.Nat.Acad.Sci. USA* **73**, 1000–4.

Leopold, L. B. and T. Maddock 1953. *The hydraulic geometry of stream channels and some physiographic implications.* US Geol. Surv. Prof. Paper no. 252.

Leopold, L. B., M. G. Wolman and J. P. Miller 1964. *Fluvial processes in geomorphology.* San Francisco: Freeman.

Lerman, A. (ed.) 1978. *Lakes: chemistry, geology and physics.* New York: Springer.

Lesack, L. F. W., R. E. Hecky and J. M. Melack 1984. Transport of carbon, nitrogen, phosphorus and major solutes in the Gambia River, West Africa. *Limnol. Oceanog.* **29** (4), 816–30.

Leser, H. 1975. Bemerkungen zur geomorphologische Kartierung 1:25 000 in der Bundesrepublik Deutschland am Beispeil des Blattes 7520 Mossingen (Kreis Tubingen: Baden-Wurtemburg). *Erdkunde* **29**, 166–73.

Letey, J., J. Osborn and R. E. Pelishek 1962. Measurement of liquid–solid contact angles in soils and sand. *Soil Sci.* **93**, 149–53.

Letey, J., J. Osborn and N. Valoras 1975. *Soil water repellency and the use of monionic surfactants.* Dept Soil Sci. Agric.Eng., Univ. California, Contrib. no. 154.

Lettau, K. and H. Lettau 1969. Bulk transport of sand by the barchans of the Pampa La Joya in southern Peru. *Z. Geomorphol. NF* **13**, 182–95.

Lever, A. and I. N. McCave 1983. Eolian components in Cretaceous and Tertiary North Atlantic sediments. *J.Sedim.Petrol.* **53**, 811–32.

Levin, M. J. and D. R. Jackson 1977. A comparison of *in situ* extractors for sampling soil water. *J.Soil Sci.Soc.Am.* **41**, 535–6.

Lewin, J. 1970. A note on stream ordering. *Area* **2**, 32–5.

Lewin, J. 1978. Floodplain geomorphology. *Prog.Phys.Geog.* **2**, 408–37.

Lewin, J. and M. Manton 1975. Welsh floodplain studies: the nature of floodplain geometry. *J.Hydrol.* **25**, 37–50.

Lewin, J. and P. J. Wolfenden 1978. The assessment of sediment sources: a field experiment. *Earth Surf. Processes* **3**, 171–8.

Lewis, J. R. 1964. *The ecology of rocky shores.* London: English Universities Press.

Lewis, K. 1983. A morphometric and geological study of limestone pavements in south Wales. *Trans. Br. Cave Res. Assoc.* **10**, 199–204.

Lewis, W. V. 1944. Stream trough experiments and terrace formation. *Geol.Mag.* **81**, 240–52.

Lewis, W. V. and M. M. Miller 1955. Kaolin model glaciers. *J.Glaciol.* **2**, 533–8.

Lewontin, R. C. 1966. On the measurement of relative variability. *Syst. Zool.* **15**, 141–2.

Liao, K. H. and A. E. Scheidegger 1968. A computer model for some branching-type phenomena in hydrology. *Bull.Int.Assoc. Sci.Hydrol.* **13**, 5–13.

Liedtke, H. 1984. Geomorphological mapping in the FR Germany at scales of 1:25 000 and 1:100 000. *Bochumer Geog. Abh.* **44**, 67–73. Paderborn. (Also presented to participants at *1st int.conf. on geomorphology*, Manchester 1985).

Likens, G. E. and J. S. Eaton 1970. A polyurethane stemflow collector for trees and shrubs. *Ecology* **51**, 938–9.

Likens, G. E., F. H. Bormann, N. M. Johnson and R. S. Pierce 1967. The calcium, magnesium, potassium and sodium budgets for a small forested ecosystem. *Ecology* **48**, 712–85.

Likens, G. E., F. H. Bormann, R. S. Pierce, J. S. Eaton and N. M. Johnson 1977. *Biogeochemistry of a forested ecosystem*. New York: Springer.

Likens, G. E., F. H. Bormann, R. S. Pierce and D. W. Fisher 1970. Nutrient–hydrologic cycle interaction in small forested watershed-ecosystems. In *Productivity of forest ecosystems*, P. Duvigneaud (ed.), 553–63. Paris: Unesco.

Lilberiussen, J. 1973. Till fabric analysis based on X-ray radiography. *Sedim.Geol.* **10**, 249–60.

Lin, J. and J. Hamasaki 1983. Pore geometry: a new system for quantitative analysis and 3-D display. *J.Sedim.Petrol.* **53**, 670–2.

Lindberg, S. E., R. R. Turner, N. M. Ferguson and D. Matt 1977. Walker Branch watershed element cycling studies: collection and analysis of wetfall for trace elements and sulphate. In *Watershed research in eastern North America*, D. L. Correll (ed.), vol. 1, 125–50. Edgewater: Smithsonian Institute.

Linskens, H. F. 1951. Niederschlagsmessungen unter verschiedenen Baumkronentypen im belaubten und unbelaubten Zustand. *Bericht der Deutschen Botanischen Gesellschaft*. **64**, 215–21.

Linskens, H. F. 1952. Niederschlagsverteilung unter einem Apfelbaum im Laufe einer Vegetationsperiode. *Annalen Meteorol.* **5**, 30–4.

Lisitzin, E. 1974. *Sea-level changes*. Amsterdam: Elsevier.

Lisle, L. D. and R. Dolan 1984. Coastal erosion and the Cape Hatteras lighthouse. *J. Coastal Res.* **1** (2), 129–39.

Lister, H., A. Pendlington and J. Chorlton 1968. Laboratory experiments on abrasion of sandstones by ice. *Int.Assoc. Sci. Hydrol.Publ.* **79**, 98–106. Berne Symp. 1967.

Livingstone, D. A. 1967. The use of filament tape in raising long cores from soft sediment. *Limnol.Oceanog.* **12**, 346–8.

Livingstone, D. L. and R. S. Cambray 1978. Confirmation of ^{137}Cs dating by algal stratigraphy in Rostherne Mere. *Nature* **276**, 259–61.

Livingstone, I. 1986. Geomorphological significance of wind flow patterns over a Namib linear dune. In *Aeolian geomorphology*, W. G. Nickling (ed.), 97–112. Boston: Allen & Unwin.

Lliboutry, L. A. 1968. General theory of subglacial cavitation and sliding of temperate glaciers. *J.Glaciol.* **7**, 21–58.

Locke, W. W., III, J. T. Andrews and P. J. Webber 1979. A manual for lichenometry. *BGRG Tech.Bull.* **26**, 1–47.

Lohnes, R. A. and T. Demirel 1978. SEM applications in soil mechanics. *Scanning Electron Microsc.* **1**, 643–54.

Long, J. T. and R. P. Sharp 1964. Barchan dune movement in Imperial Valley, California. *Bull.Geol. Soc.Am.* **75**, 149–56.

Longmore, M. E., B. M. O'Leary and C. W. Rose 1983a. Caesium-137 profiles in the sediments of a partial-meromictic lake on Great Sandy Island (Fraser Island), Queensland, Australia. *Hydrobiologia* **103**, 21–7.

Longmore, M. E., B. M. O'Leary, C. W. Rose and A. L. Chandica 1983b. Mapping soil erosion and accumulation with the fallout isotope caesium-137. *Aust.J.Soil.Res.* **21**, 373–85.

Longuet-Higgins, M. S. 1972. Recent progress in the study of longshore currents. In *Waves on beaches and resulting sediment transport*, R. E. Jeyer (ed.), 203–48. New York: Academic Press.

Longuet-Higgins, M. S., D. E. Cartwright and N. D. Smith 1963. Observations of the directional spectrum of sea waves using the motions of a floating buoy. *Ocean wave spectra*, 111–13. Rep. Conf. Englewood Cliffs, NJ: Prentice-Hall.

Longwell, C. R., R. F. Flint and J. E. Sanders 1969. *Physical geology*. New York: Wiley.

Lorimer, C. G. 1980. Age structure and disturbance history of a southern Appalachian virgin forest. *Ecology* **61**, 1169–84.

Loughran, R. J. and K. J. Malone 1976. *Variations in some stream solutes in a Hunter Valley catchment*. Res. Papers Geog. no. 8, Univ. Newcastle, NSW.

Loughran, R. J., B. L. Campbell and D. E. Walling 1987. Soil erosion and sedimentation indicated by caesium-137, Jackmoor Brook catchment, Devon, England. *Catena* **14**, 201–12.

Lovering, T. S. and C. Engel 1967. *Translocation of silica and other elements from rock into 'Equisetum' and three grasses*. US Geol. Surv. Prof. Paper no. 594-B.

Lowe, J. J., S. Lowe, A. J. Fowler, R. E. Hedges and T. J. F. Austin 1988. A comparison of accelerator and radiometric radiocarbon measurements obtained from Late Devensian lake sediments from Llyn Gwernan, North Wales, UK. *Boreas* 355–69.

Lowe, R. L., D. L. Inman and B. M. Brush 1972. Simultaneous data systems for instrumenting the shelf. In *Proc.13th conf.coastal engineering*, vol. 1, 95–112. Am.Soc.Civ.Eng.

Lowell, S. and J. E. Shields 1984. *Powder surface area and porosimetry*, 2nd edn. London: Chapman & Hall.

Lubowe, J. K. 1964. Stream junction angles in the dendritic drainage pattern. *Am.J.Sci.* **262**, 325–39.

Lucas, D. H. and D. J. Moore 1964. The measurement in the field of pollution by dust. *Air Water Pollut.Int.J.* **8**, 441–53.

Lucey, E. 1978. *The Mackereth corer* (8 mm film). University of Edinburgh Film Unit.

Luckman, B. H. 1971. The role of snow avalanches in the evolution of alpine talus slopes. *Inst.Br.Geog.Spec.Publ.* **3**, 93–110.

Luckman, B. H. 1976. Rockfalls and rockfall inventory data: some observations from Surprise Valley, Jasper National Park, Canada. *Earth Surf. Processes* **1**, 287–98.

Luckman, B. H. 1977. Lichenometric dating of holocene moraines at Mount Edith Cavell, Jasper, Alberta. *Can.J.Earth Sci.* **14**, 1809–22.

Ludlam, S. D. 1976. Laminated sediments in holomictic Berkshire lakes. *Limnol.Oceanog.* **21**, 743–6.

Ludwick, J. C. 1974a. Tidal currents and zig-zag sand shoals in a wide estuary entrance. *Geol.Soc.Am.Bull.* **85**, 717–26.

Ludwick, J. C. 1974b. *Variations in boundary drag coefficient in the tidal entrance to Chesapeake Bay, Virginia*. Tech. Rep. no. 19. Inst. Oceanog., Old Dominion Univ., Norfolk, Va.

Luk Shiu-Hung 1977. Rainfall erosion of some Alberta soils: a laboratory simulation study. *Catena* **3**, 295–309.

Lundquist, J. 1975. Ice recession in central Sweden and the Swedish time scale. *Boreas* **4**, 47–54.

Lutz, H. J. 1960. Movement of rocks by uprooting of forest trees. *Am.J.Sci.* **258**, 752–6.

Luz, B., Y. Kolodny and M. Horowitz 1984. Fractionation between mammalian bone and drinking water. *Geochim. Cosmochim. Acta* **48**, 1689–94.

McArthur, D. S. and R. Ehrlich 1977. An efficiency valuation of four drainage basin shape ratios. *Prof. Geog.* **29**, 290–5.

McArthur, J. L. 1977. Quaternary erosion in the upper Derwent basin and its bearing on the age of surface features in the southern Pennines. *Trans.Inst.Br.Geog.NS* **2**, 490–7.

McBride, E. F. 1971. Mathematical treatment of size distribution data. In *Procedures in sedimentary petrology*, R. E. Carver (ed.), 109–27. New York: Wiley.

McCaffrey, R. J. and J. Thompson 1974. A record of trace metal fluxes to a Connecticut salt marsh determined by ^{210}Pb dating. In *Trace elements in natural waters. Annual progress report*. K. K. Turekian (ed.). Atomic Energy Commission AEC COO-3573-8. New Haven, Conn.: Yale University.

McCall, J. G. 1954. Glacier tunnelling and related observations. *Polar Record* **7**, 120–36.

McCallan, M. E., B. M. O'Leary and C. W. Rose 1980. Redistribution of caesium-137 by erosion and deposition on an Australian soil. *Aust.J.Soil.Res.* **18**, 2.

McCammon, R. B. 1962. Efficiencies of percentile measures for describing the mean size and sorting of sedimentary particles. *J.Geol.* **70**, 453–65.

McCann, S. B. 1965. The raised beaches of north-east Islay and western Jura, Argyll. *Trans.Inst. Br.Geog.* **35**, 1–16.

McCann, S. B. 1966. The main post-glacial raised shoreline of western Scotland from the Firth of Lorne to Loch Broom. *Trans.Inst.Br.Geog.* **39**, 87–99.

McCann, S. B., P. J. Howarth and J. G. Cogley 1972. Fluvial processes in a periglacial environment, Queen Elizabeth Islands, NWT, Canada. *Trans.Inst.Br.Geog.* **55**, 69–82.

McCauley, J., D. Egbert, J. McNaughton and F. T. Ulaby 1975. Stream pattern analysis using optical processing of ERTS imagery of Kansas. *Mod. Geol.* **5**, 127–42.

McCave, I. N. 1971. Wave effectiveness at the sea bed and its relationship to bedforms and deposition of mud. *J.Sedim.Petrol.* **41**, 89–96.

McCave, I. N. and A. C. Geiser 1978. Megaripples, ridges and runnels on intertidal flats of The Wash, England. *Sedimentology* **26**, 353–69.

McCave, I. N. and J. Jarvis 1973. Use of the Model T Coulter Counter in size analysis of fine to coarse sand. *Sedimentology* **20**, 305–15.

McCave, I. N., R. J. Bryant, H. F. Cook and C. A. Coughanowr 1986. Evaluation of a laser-diffraction-size analyzer for use with natural sediments. *J.Sedim.Petrol.* **56**, 561–4.

McClean, R. 1967. Plan shape and orientation of beaches along the east coast, South Island, New Zealand, *NZ Geog.* **23**, 16–22.

McConnell, H. and J. M. Horn 1972. Probabilities of surface karst. In *Spatial analysis in geomorphology*, R. J. Chorley (ed.), 111–33, London: Methuen.
McCoy, F. W., Jr, W. J. Nokleberg and R. M. Norris 1967. Speculations on the origin of the Algodones Dunes, southern California. *Bull.Geol.Soc.Am.* **78**, 1039–44.
McCoy, R. M. 1969. Drainage networks with K-band radar imagery. *Geog.Rev.* **59**, 493–512.
McCoy, R. M. 1971. Rapid measurement of drainage density. *Bull.Geol.Soc.Am.* **82**, 757–62.
McCoy, W. D. 1987. The precision of amino acid geochronology and paleothermometry. *Quat. Sci.Rev.* **6**, 43–54.
McCready, P. B. 1965. Dynamic response characteristics of meteorological sensors. *Bull.Am. Meteorol.Soc.* **46**, 533–8.
MacCrimmon, H. R. and J. R. M. Kelso 1970. Seasonal variation in selected nutrients of a river system. *J.Fish. Res.Board Can.* **27**, 837–46.
McCullagh, M. J. and R. J. Sampson 1972. User desires and graphics capability in the academic environment. *Cartog. J.* **9**, 109–22.
McDougall, I. 1966. Precision methods of potassium–argon isotopic age determination on young rocks. In *Methods and techniques in geophysics*, vol. 2, 279–304. New York: Wiley Interscience.
McDougall, J. D. 1976. Fission-track dating. *Sci. Am* **255**, 114–22.
McElhinney, M. W. 1973. *Palaeomagnetism and plate tectonics*. Cambridge: Cambridge University Press.
McGown, A. and E. Derbyshire 1974. Technical developments in the study of particulate matter in glacial tills. *J.Geol.* **82**, 255–36.
McHenry, J. R., J. C. Ritchie and A. C. Gill 1973. Accumulation of fallout caesium-137 in soils and sediments in selected watersheds. *Water Resour. Res.* **9**, 676–86.
McKee, E. D. 1960. Laboratory experiments on form and structure of off-shore bars and beaches (abstract). *Bull.Am.Assoc. Petrol. Geol.* **44**, 1253.
McKee, E. D. 1966. Structures of dunes at White Sands National Monument, New Mexico. *Sedimentology* **7**, 1–68.
McKee, E. D. 1979. *Global sand seas*. US Geol. Surv. Prof. Paper no. 1052.
McKee, E. D. and C. S. Breed 1974. An investigation of major sand seas in desert areas throughout the world. In *Third Earth resources technology satellite-I symp.* 665–79. Washington DC 10–14 December 1973. NASA SP-351.
McKee, E. D. and J. R. Douglass 1971. Growth and movement of dunes at White Sands National Monument, New Mexico. Geol. Surv.Res. *US Geol.Surv. Prof. Paper* **750-D**, 108–14.
McKee, E. D. and G. C. Tibbitts 1964. Primary structures of a seif dune and associated deposits in Libya. *J.Sedim.Petrol.* **34**, 5–17.
McKee, E. D., J. R. Douglass and S. Rittenhouse 1971. Deformation of lee-side laminae in eolian dunes. *Bull.Geol.Soc.Am.* **82**, 359–78.
McKee, T. R. and J. L. Brown 1977. Preparation of specimens for electron microscopic examination. In *Minerals in soil environments*, J. B. Dixon and S. W. Weed (eds), 809–46. Madison, Wisc.: Soil Science Society of America.
Mackenzie, R. C. 1970. *Differential thermal analysis*, vol. I. London: Academic Press.
Mackenzie, R. C. 1972. *Differential thermal analysis*, vol. 2. London: Academic Press.
Mackereth, F. J. H. 1958. A portable core sampler for lake deposits. *Limnol. Oceanog.* **3**, 181–91.
Mackereth, F. J. H. 1965. Chemical investigations of lake sediments and their interpretation. *Proc.R.Soc.B* **161**, 295–309.
Mackereth, F. J. H. 1966. Some chemical observations on post-glacial lake sediments. *Phil. Trans.R.Soc. B* **250**, 165–213.
Mackereth, F. J. H. 1969. A short-core sampler for subaqueous deposits. *Limnol. Oceanog.* **14**, 145–51.
Mackereth, F. J. H. 1971. On the variation in direction of the horizontal component of remanent magnetization in lake sediments. *Earth Planet. Sci. Lett* **12**, 332–8.
Mackin, J. M. 1936. The capture of the Greybull River. *Am.J.Sci.* **231**, 373–85.
McLean, R. F. 1967. Measurements of beachrock erosion by some tropical marine gastropods. *Bull. Mar. Sci.* **17**, 551–61.
McNeil, D. R. 1977. *Interactive data analysis*. New York: Wiley.
Madgwicke, H.A.I. and J.D. Ovington 1959. The chemical composition of precipitation in adjacent forest and open plots. *Forestry* **32**, 14–22.
Mahanney, W. C. (ed.) 1984. *Quaternary dating methods*. Amsterdam: Elsevier.
Mainguet, M. 1972. Etude d'un erg (Fachi Bilma, Sahara central). Sédimentation sableuse et son insertion dans le paysage après les photographies prise par satellites. *C.R. Acad.Sci. D*, **274**, 1633–6.
Maizels, J. K. 1977. Experiments on the origin of kettle holes. *J.Glaciol.* **18**, 291–303.

Maizels, J. K. 1983. Proglacial channel systems: change and thresholds for change over long, intermediate and short timescales. In *Modern and ancient fluvial systems*, J. D. Collinson and J. Lewin (eds). Int.Assoc.Sedimentol. Spec. Publ. no. 6. Oxford: Blackwell.

Mandelbrot, B. 1967. How long is the coast of Britain? Statistical self-similarity and fractional dimension. *Science* **156**, 636–8.

Mandelbrot, B. B. 1977a. *Fractals: form, chance and dimension*. San Francisco: Freeman.

Mandelbrot, B. B. 1977b. *The fractal geometry of nature*. New York: Freeman.

Mandelbrot, B. 1982. *The fractal geometry of nature*. San Francisco: Freeman.

Mangerud, J. and S. Gulliksen 1975. Apparent radiocarbon ages of recent marine shells from Norway, Spitzbergen and Arctic Canada. *Quat. Res.* **5**, 263–73.

Mannion, A. M. 1980. *Pollen analysis: A technique in palaeoenvironmental reconstruction*. Geog. Papers, Dept Geog., Univ. Reading.

Marble, D. F., K. Brassel, H. W. Calkins and D. J. Peuquet 1978., *Computer software for spatial data handling*. Ottawa: IGU Commission on Geographical Data Sensing and Processing.

Marchand, D. E. 1971. Rates and modes of denudation, White Mountains, eastern California. *Am.J.Sci.* **270**, 109–35.

Mardia, K. V. 1972. *Statistics of directional data*. London: Academic Press.

Mardia, K. V. 1975. Statistics of directional data. *J. R. Stat. Soc.B* **37**, 349–93.

Margolis, S. V. and J. P. Kennett 1971. Cenozoic paleoglacial history of Antarctica recorded in subantarctic deep-sea cores. *Am.J.Sci.* **270**, 1–36.

Margolis, S. V. and D. H. Krinsley 1974. Processes of formation and environmental occurrence of microfeatures on detrital quartz grains. *Am.J.Sci.* **274**, 449–64.

Marinenko, G. and W. F. Koch 1984. A critical review of measurement practices for the determination of pH and acidity of atmospheric precipitation. *Environ. Int.* **10**, 315–19.

Mark, D. M. 1974. Line intersection method for estimating drainage density. *Geology* **2**, 235–6.

Mark, D. M. 1975a. Geomorphometric parameters: a review and evaluation. *Geog. Annaler* **57A**, 165–77.

Mark, D. M. 1975b. Computer analysis of topography: a comparison of terrain storage methods. *Geog. Annaler* **57A**, 179–88.

Mark, D. M. 1983. Relations between field-surveyed channel networks and map-based geomorphometric measures, Inez, Kentucky. *Ann. Assoc.Am.Geog.* **73**, 358–72.

Mark, D. M. and P. B. Aronson 1984. Scale-dependent fractal investigations of topographic surfaces: an empirical investigation, with applications in geomorphology and computer mapping. *Math. Geol.* **16**, 617–83.

Mark, D. M. and M. Church 1977. On the misuse of regression in earth science. *Math. Geol.* **9**, 63–75.

Marker, M. E. and M. M. Sweeting 1983. Karst development on the Alexandria limestones, East Cape Province, South Africa. *Z. Geomorphol. NF* **27**, 21–38.

Marston, R. A. 1982. The geomorphic significance of log steps in forest streams. *Ann. Assoc. Am.Geog.* **72**, 99–108.

Marthinussen, M. 1960. Coast and fjord area of Finnmark. In *Geology of Norway*, O.Holtedahl (ed.). *Nor. Geol. Unders.* **280**, 416–29

Marthinussen, M. 1962. ^{14}C datings referring to shorelines, transgressions and glacial substages in northern Norway. *Nor. Geol. Unders.* **215**, 37–67.

Martini, A. 1967. Preliminary experimental studies of frost weathering of certain rock types from the west Sudetes. *Biulteyn Peryglacjalny* **16**, 195–201.

Martin-Kaye, P. 1974. Application of side-looking radar in earth-resource surveys. In *Environmental remote sensing: applications and achievements*, E. C. Barrett and L. F. Curtis (eds). 29–48. London: Edward Arnold.

Martinson, D. G., N. G. Pisias, J. D. Hays, J. Imbrie, T. C. Moore and N. J. Shackleton 1987. Age dating and the orbital theory of the ice ages: development of a high-resolution 0 to 300 000-year chronostratigraphy. *Quat. Res.* **27**, 1–29.

Martvall, S. and G. Nilsson 1972. *Experimental studies of meandering*. UNGI Rapport no. 20. Uppsala.

Mast, R. F. and P. E. Potter 1963. Sedimentary structures, sand shape fabrics and permeability. Part II. *J.Geol.* **71**, 548–65.

Mather, P. and J. C. Doornkamp 1970. Multivariate analysis in geography, with particular reference to drainage basin morphology. *Trans. Inst. Br. Geog.* **51**, 163–87.

Matschinski, M. 1952. Sur les formations sableuses des environs de Beni-Abbes. *Soc.Géol. Fr., C.R.* **9–10**, 171–4.

Matthews, B. 1967. Automatic measurement of frost-heave results from Malham and Rodley (Yorkshire). *Geoderma* **1**, 107–15.

Matthews, E. R. 1983. Wave disturbance and texture of beaches in Palliser Bay, southern North Island, New Zealand. *NZJ. Geol. Geophys.* **26** (2), 197–212.

Matthews, J. A. 1974. Families of lichenometric dating curves from the Storbreen gletschervorfeld, Jotunheimen, Norway. *Nor. Geog. Tidsskr.* **28**, 215–35.

Matthews, J. A. 1977. Glacier and climatic variations inferred from tree-growth variations over the last 250 years, central southern Norway. *Boreas* **6**, 1–24.

Matthews, J. A. 1987. A comment on 'Neoglaciation in South Norway using lichenometric methods'. *Nor. Geog.Tidsskr.* **41**, 61–7.

Matthews, J. A. and P. Q. Dresser 1983. Intensive 14C dating of a buried palaeosol horizon. *Geol.Foren. Stockholm Forhandl.* **105**, 59–63.

Matthews, J. A., R. Cornish and R. A. Shakesby 1979. 'Saw-tooth' moraines in front of Bodalsbreen, southern Norway. *J.Glaciol.* **22**, 535–46.

Matthias, G. F. 1967. Weathering rates of Portland arkose tombstones. *J.Geol.Educ.* **15**, 140–4.

Maxwell, J. C. 1960. *Quantitative geomorphology of the San Dimas Experimental forest, California.* Off. Naval Res., Geog. Branch, Project NR 389–042, Tech. Rep. no. 19.

Mazzullo, J. and K. D. Withers 1984. Sources, distribution and mixing of late Pleistocene and Holocene sand on the south Texas continental shelf. *J.Sedim.Petrol.* **54**, 1319–34.

Meade, R. H. 1969. Errors in using modern stream-load data to estimate natural rates of denudation. *Bull.Geol.Soc.Am.* **80**, 1265–74.

Meade, R. H. 1982. Sources, sinks and storage of river sediment in the Atlantic drainage of the United States. *J.Geol.* **90**, 235–52.

Mears, R. I. 1980. Municipal avalanche zoning: constrasting policies of four western US communities. *J.Glaciol.* **26**, 355–62.

Mejdahl, Y. 1988. The plateau method for dating partially-bleached sediments by thermoluminescence. *Quaternary Science Reviews* **7**, 347–8.

Mellor, A. 1985. Soil chronosequences on Neoglacial moraine ridges, Jostedalsbreen and Jotunheimen, southern Norway: a quantitative pedogenic approach. In *Soils and geomorphology*, K. S. Richards, R. Arnett and S. Ellis (eds), 289–308. London: Allen & Unwin.

Melton, M. A. 1957. *An analysis of the relations among elements of climate, surface properties and geomorphology.* Off. Naval Res., Geog. Branch, Project NR 389–042, Tech Rep. no. 11.

Melton, M. A. 1958. Geometric properties of mature drainage systems and their representation in an E4 phase space. *J.Geol.* **66**, 25–54.

Melton, M. A. 1965. Debris-covered hillslopes of the southern Arizona desert: consideration of their stability and sediment contribution. *J.Geol.* **73**, 715–29.

Merilainen, J. and P. Huttunen 1978. The liquid nitrogen method in sampling of the uppermost lake sediment. *Pol.Arch.Hydrobiol.* **25**, 287–90.

Mesolella, K. A., R. K. Matthews, W. S. Broecker and D. L. Thurber 1969. The astronomical theory of climatic change; Barbados data. *J. Geol.* **77**, 250–74.

Meteorological Office 1968. *The measurement of upper winds by means of pilot balloons*, 4th edn. Met. Off. 804. London: HMSO.

Meyer, D. R. 1972. Geographical population data: statistical description not statistical inference. *Prof. Geog.* **24**, 26–8.

Meyer, L. D. 1982. Soil erosion research leading to development of the Universal Soil Loss Equation. In *Proc. workshop on estimating erosion and sediment yield on Rangelands*, 1–16. Tucson, Arizona, 7–9 March 1981.

Miall, A. D. 1978. *Fluvial sedimentology.* Calgary: Canada Society of Petroleum Geologists.

Michels, J. W. 1986. Obsidian hydration dating. *Endeavour NS* **10**, 97–100.

Middleton, G. V. 1976. Hydraulic interpretation of sand size distributions. *J.Geol.* **84**, 405–26.

Middleton, N. J., A. S. Goudie and G. L. Wells 1986. The frequency and source areas of dust storms. In *Aeolian geomorphology*, W. G. Nickling (ed.), 237–59. Boston: Allen & Unwin.

Midgley, D. and K. Torrance 1978. *Potentiometric water analysis.* Chichester: Wiley.

Midgley, H., G. Petts and D. Walker 1986. *Streamflow measurement.* British Standards Institute.

Miklas, J., T. L. Wu, A. Hiatt and D. L. Correll 1977. Nutrient loading of the Rhode River via land use practices and precipitation. In *Watershed research in North America*, D. L. Correll (ed.), vol. 1, 169–91. Edgewater: Smithsonian Institute.

Miller, D. S. and S. Lakatos 1983. Uplift rate of Adirondack anorthosite measured by fission-track analysis of apatite. *Geology* **11**, 284–6.

Miller, G. H. and J. Brigham-Grette 1989. Amino acid geochronology: resolution and precision in carbonate fossils. *Quaternary dating methods.* INQUA Spec. publ. Oxford: Pergamon.

Miller, G. H. and P. E. Hare 1980. Amino acid geochronology: integrity of the carbonate matrix and potential of molluscan fossils. In *Biogeochemistry of amino acids*, P. E. Hare, T. C Hoering and K. King, Jr (eds), 415–43. New York: Wiley.

Miller, G. H. and J. Mangerud 1985. Aminostratigraphy of European marine interglacials. *Quat. Sci. Rev.* **4**, 215–78.

Miller, G. H. and F. Wendorf 1986. Geochronological potential of amino acid racemization in ostrich eggshells. *Geo.Soc.Am.Abst. Programs* **18**, 696.

Miller, G. H., J. I. Andrews and S. K. Short 1977. The last interglacial–glacial cycle, Clyde Foreland,, Baffin Island, NWT: stratigraphy, biostratigraphy and chronology. *Can.J.Earth Sci.* **14**, 2824–57.

Miller, J. P. 1961. *Solutes in small streams draining single rock types, Sangre de Cristo Range, New Mexico*. US Geol.Surv. Water Supply Paper no. 1535–D.

Miller, R. L. and J. M. Zeigler 1958. A model relating dynamics and sediment pattern in equilibrium in the region of shoaling waves, breaker zone, and foreshore. *J.Geol.* **66**, 417–41.

Miller, T. K. and L. J. Onesti 1975. A multivariate analysis of the relationships between stream order and selected hydromorphic stream properties. In *The hydrological characteristics of river basins and the effects of these characteristics on better water management* 137–46. Publ. 117, Int. Assoc. Hydrol.Sci., Tokyo.

Miller, V. C. 1953. *A quantitative geomorphic study of drainage basin characteristics in the Clinch Mountain area, Virginia and Tennessee*, 1–30. Dept Geology, Columbia Univ. Contract N6 ONR 271–30. Tech.Rep. no. 3.

Milliman, J. D. and K. O. Emery 1968. Sea levels during the past 35 000 years. *Science* **162**, 1121.

Mills, H. H. 1980. An analysis of drumlin form in the north-western and north-central US: summary. *Bull.Geol.Soc.Am.* **91**, 2214–89.

Mills, H. H. and D. D. Starnes 1983. Sinkhole morphometry in a fluviokarst region: eastern Highland Rim, Tennessee, USA. *Z. Geomorphol. NF* **27**, 39–54.

Mills, H. H. and J. R. Wagner 1985. Long-term change in regime of the New River, as indicated by vertical variations in extent and weathering intensity of alluvium. *J.Geol.* **93**, 131–42.

Mills, H. M. 1984. Effect of hillslope angle and substrate on tree tilt and denudation of hillslopes by tree fall. *Phys Geog.* **5**, 253–61.

Milne, J. A. 1982. Bed forms and bend–arc spacings of some coarse-bedload channels in upland Britain. *Earth Surf. Processes Landforms* **7**, 227–40.

Minshall, N., M. S. Nichols and S. A. Witzel 1969. Plant nutrients in the baseflow of streams in southwestern Wisconsin. *Water Resour. Res.* **5**, 706–13.

Miotke, F.-D. 1974. Carbon dioxide and the soil atmosphere. *Abh. Karst-und-Höhlenk A* **9**.

Mitsuyasu, H. 1967. Shock pressures of breaking waves. In *Proc.10th Conf. on coastal engineering*, 268–83. Am.Soc.Civ.Eng.

Mitterer, R. M. 1974. Pleistocene stratigraphy in southern Florida based on amino acid diagenesis in fossil *Mercenaria*. *Geology* **2**, 425–8.

Mitterer, R. M. and P. E. Hare 1967. Diagenesis of amino acids in fossil shells as a potential geochronometer. *Geol.Soc.Am.Abst.Programs* 152.

Mizutani, S. 1963. A theoretical and experimental consideration on the accuracy of sieving analysis. *J.Earth Sci.* **11**, 1–27.

Moffat, A. J. and J. A. Catt 1986. A re-examination of the evidence for a Plio-Pleistocene marine transgression on the Chiltern Hills. III. Deposits. *Earth Surf. Processes Landforms* **11**, 233–47.

Moiola, R. J. and D. Weiser 1968. Textural parameters: an evaluation. *J.Sedim.Petrol.* **38**, 45–53.

Møller, J. J. and J. L. Sollid 1972. Deglaciation chronology of Lofoten–Vesteralen–Ofoten, north Norway. *Nor. Geog. Tidsskr.* **26**, 101–33.

Monkhouse, F. J. and H. R. Wilkinson 1971. *Maps and diagrams*, 3rd edn. London: Methuen.

Montgomery, H. A. C. and I. C. Hart 1974. The design of sampling programmes for rivers and effluents. *Water Pollut.Control* **73**, 77–101.

Moore, I. D., E.M. O'Loughlin and G. J. Burch 1988. A contour-based topographic model for hydrological and ecological applications. *Earth Surf. Process and Landforms* **13**, 305–20.

Moore, P. D. and D. J. Bellamy 1974. *Peatlands*. London: Elek.

Moore, P. D. and J. A. Webb 1978. *An illustrated guide to pollen analysis*. London: Edward Arnold.

Moore, R. F. and J. B. Thornes 1976. LEAP – a suite of FORTRAN IV programs for generating erosional potentials of land surfaces from topographic information. *Comput. Geosci.* **2**, 493–9.

Morales, C. 1979. The use of meteorological observations for studies of the mobilization, transport and deposition of Saharan soil dust. In *Saharan dust*, C. Morales (ed.), 119–31. Chichester: Wiley.

Moreira-Nordemann, L. M. 1980. Use of $^{234}U/^{238}U$ disequilibrium in measuring chemical weathering rate or rocks. *Geochim. Cosmochim. Acta* **44**, 103–8.

Morgan, R. P. C. 1970. An analysis of basin asymmetry in the Klang basin, Selangor. *Bull.Geol.Soc. Malaysia* **3**, 17–26.

Morgan, R. P. C. 1971. A morphometric study of some valley systems on the English chalkland. *Trans.Inst.Brit.Geog.* **54**, 33–44.
Morgan, R. P. C. 1976. The role of climate in the denudation system: a case study from West Malaysia. In *Geomorphology and climate*, E. Derbyshire (ed.), 319–44. Chichester: Wiley
Morgan, R. P. C. 1978. Field studies of rainsplash erosion. *Earth Surf. Processes* **3**, 295–300.
Morgenstern, N. R. and V. E. Price 1965. The analysis of stability of general slip surfaces. *Geotechnique* **15**, 79–93.
Morgenstern, N. R. and J. Tchalenko 1967. Microscopic structures in kaolin subjected to direct shear. *Geotechnique* **17**, 309–28.
Morisawa, M. E. 1957. Accuracy of determination of stream lengths from topographic maps. *Trans.Am.Geophys.Union.* **38**, 86–8.
Morisawa, M. E. 1958. Measurements of drainage basin outline form. *J.Geol.* **66**, 587–90.
Morisawa, M. E. 1959. *Relation of quantitative geomorphology to stream flow in representative watersheds of the Appalachian Plateau Province*. Columbia Univ., Off. Naval Res., Project NR 389–042. Tech.Rep.no. 20.
Mörner, N.-A. 1969. The late Quaternary history of the Kattegat Sea and the Swedish west coast. *Sver.Geol.Unders. Ser. C* **640** (63), 3.
Mörner, N.-A. 1971. Eustatic changes during the last 20 000 years and a method of separating the isostatic and eustatic factors in an uplifted area. *Palaeogeog.,Palaeoclim.Palaeoecol.* **9**, 153–81.
Mörner, N.-A. and J. P. Lanser 1974. Gothenberg magnetic 'Slip'. *Nature* **251**, 408–9.
Morrison, G. M. P., D. M. Revitt, J. B. Ellis, G. Svensson and P. Balmer 1984. Variations of dissolved and suspended solid heavy metals through an urban hydrograph. *Environ. Technol. Lett.* **7**, 313–18.
Morrison, R. B. 1967. Principles of Quaternary soil stratigraphy. In *Quaternary soils*, R. B. Morrison and H. E. Wright (eds.), vol. 9, 1–69. Proc. VII INQUA Congr., USA, 1965. Nevada: Desert Research Institute, University of Nevada.
Moses, H. and H. G. Daubek 1961. Errors in wind measurements associated with tower-mounted anemometers. *Bull.Am.Meteorol.Soc.* **42**, 190–4.
Mosley, M. P. 1973. Rainsplash and the convexity of badland divides. *Z. Geomorphol. Supp. NF* **18**, 10–25.
Mosley, M. P. 1975. A device for the accurate survey of small scale slopes. *BGRG. Tech.Bull.* **17**, 3–6.
Mosley, M. P. 1976. An experimental study of channel confluences. *J.Geol.* **84**, 535–62.
Mosley, M. P. 1978. Bedload material transport in the Tamaki River near Dannevirke, North Island, New Zealand. *NZ J. Sci.* **21**, 619–26.
Mosley, M. P. and G. L. Zimpfer 1978. Hardware models in geomorphology. *Prog.Phys.Geog.* **2**, 438–61.
Moss, A. J. 1972. Technique for assessment of particle breakage in nature and artificial environments. *J. Sedim.Petrol.* **42**, 725–8.
Moss, A. J. and P. H. Walker 1978. Particle transport by continental water flows in relation to erosion, deposition, soils and human activities. *Sedim.Geol.* **20**, 81–139.
Moss, A. J., P. H. Walker and J. Hutka 1973. Fragmentation of granite quartz in water. *Sedimentology* **21**, 637–8.
Mueller, J. E. 1968. An introduction to the hydraulic and topographic sinuosity indexes. *Ann.Assoc.Am.Geog.* **58**, 371–85.
Mueller-Dombois, D. and H. Ellenberg 1974. *Aims and methods of vegetation ecology*. New York: Wiley.
Muller, E. H. 1974. Origins of drumlins. In *Glacial geomorphology*, D. R. Coates (ed.), 187–204. Binghampton, NY: State University of New York.
Muller, F. 1976. On the thermal regime of a high Arctic valley glacier. *J.Glaciol.* **16**, 119–33.
Muller, G. 1967. *Sedimentary petrology*. Vol. 1: *Methods of sedimentary petrology*. New York: Hafner.
Mumpton, F. A. and W. C. Ormsby 1978. Morphology of zeolites in sedimentary rocks by scanning electron microscopy. In *Natural zeolites, occurrence, properties, use*, L. B. Sand and F. A. Mumpton (eds), 113–32. Oxford: Pergamon.
Munday, J. C. and T. T. Alfoldi 1979. Landsat test of diffuse reflectance models for aquatic suspended solids measurement. *Remote Sensing Environ.* **8**, 169–83.
Munday, J. C., T. T. Alfoldi and C. L. Amos 1979. Verification and application of a system for automated multidate Landsat measurement of suspended sediment. Paper presented at the *Fifth Annual W. T. Pecora Symp. on satellite hydrology*. Sioux Falls, SD, 11–14 June 1979.
Munk, W. H., G. R. Miller, F. E. Snodgrass and N. F. Barber 1963. Directional recording of swell from distant storms. *Phil.Trans.R.Soc.A* **255**, 505–84.

Murphy, C. P., P. Bullock and K. J. Biswell 1977. The measurement and characterisation of voids in soil thin sections by image analysis. Part II: Applications. *J.Soil Sci.* **28**, 509–18.
Murray, S. P. 1967. Control of grain dispersion by particle size and wave state. *J.Geol.* **75**, 612.
Musgrave, G. W. and H. N. Holtan 1964. Infiltration. In *Handbook of applied hydrology*, V. T. Chow (ed.), 12.1–12.30. New York: McGraw-Hill.
Mutchler, C. K. and L. F. Hermsmeier 1965. A review of rainfall simulators. *Trans.Am.Soc. Agric.Eng.* **8**, 67–8.

Nabholz, J. V., D. A. Crossley and G. R. Best 1984. An inexpensive weir and proportional sampler for miniature watershed ecosystems. *Water Resour. Bull.* **20** (4), 619–25.
Naeser, C. W. and N. D. Naeser 1988. Fission-track dating of Quaternary events. In *Dating Quaternary sediments*, D. J. Easterbrook (ed.), 1–12. Geol. Soc. Am. Spec. Paper no. 227.
Naeser, C. W., N. D. Briggs, J. D. Obradovich and G. A. Izett 1981. Geochronology of Quaternary tephra deposits. In *Tephra studies*. S. Self and R. S. J. Sparks (eds), 13–47. Dordrecht: Reidel.
Nakajima, T., K. Yaskawa, N. Natsuhara, N. Kawai and S. Horie 1973. Very short period geomagnetic excursion 18 000 yr BP. *Nature, Phys. Sci.* **244**, 8–10.
Natural Environmental Research Council 1975. *Flood studies report*. London: HMSO.
Neal, C. and A. G. Thomas 1985. Field and laboratory measurement of pH in low-conductivity natural waters. *J. Hydrol.* **79**, 319–22.
Nelkon, M. and P. Parker 1970. *Advanced level physics*. London: Heinemann.
Nelsen, T. A. 1976. An automatic rapid sediment analyser. *Sedimentology* **23**, 867–72.
Nelson, D. J. 1970. Measurement and sampling of outputs from watersheds. In *Analysis of temperate forest ecosystems*, D. E. Reichle (ed.), 257–67. New York: Springer.
Nesbitt, H. W. 1979. Mobility and fractionation of rare earth elements during weathering of a granodiorite. *Nature* **279**, 206–10.
Nettleton, L. L. 1934. Fluid mechanics of salt domes. *Bull. Am. Assoc. Petrol. Geol.* **18**, 1175–204.
Nevin, C. M. and D. W. Trainer 1927. Laboratory study in delta building. *Bull. Geol. Soc. Am.* **38**, 451–8.
Newell, R. E. and J. W. Kidson 1979. The tropospheric circulation over Africa and its relation to the global tropospheric circulation. In *Saharan dust*, G. Morales (ed.), 133–69. Chichester: Wiley.
Newman, L., G. E. Likens and F. H. Bormann 1975. Acidity in rainwater: has an explanation been presented? *Science* **188**, 957.
Nicholson, I. A., I. S. Paterson and F. T. Last 1980. *Methods for studying acid precipitation in forest ecosystems* ITE/NERC. Cambridge: Institute of Terrestrial Ecology.
Nickling, W. G. (ed.) 1986. *Aeolian geomorphology*. Boston: Allen & Unwin.
Niesel, K. 1973. Determination of the specific surface by measurement of permeability. *Mater. Constr.* **33**, 227–31.
Nieter, W. M. and D. H. Krinsley 1976. The production and recognition of aeolian features on sand grains by silt abrasion. *Sedimentology* **23**, 713–20.
Nieuwenhuis, J. D. and J. A. van den Berg 1971. Slope investigations in the Morvan (Haut Folin area). *Rev. Géomorphol. Dyn.* **20**, 161–76.
Nikonov, A. A. and T. Yu Shebalina 1979. Lichenometry and earthquake age determination in central Asia. *Nature* **280**, 675–7.
Nilsen, T. H., R. H. Wright, T. C. Vlasic and W E. Spangle 1979. *Relative slope stability and land-use planning in the San Francisco Bay Region, California.* US Geol. Surv. Prof. Paper no. 944.
Ninkovich, D. and N. J. Shackleton 1975. Distribution, stratigraphic position and age of ash layer 'L' in the Panama Basin region. *Earth Planet. Sci. Lett.* **27**, 20–34.
Nishiyama, T. and S. Shimoda 1974. Oblique texture electron-diffraction patterns and morphologies of mica clay minerals and their interstratified minerals. *J. Tokyo Univ., Gen. Educ. (Nat. Sci.)* **17**, 1–13.
Noble, C. A. and R. P. C. Morgan 1983. Rainfall interception and splash detachment with a Brussels sprout plant: a laboratory simulation. *Earth Surf. Processes Landforms* **8**, 569–77.
Noda, H. 1967. Suspension of sediment due to wave action. In *Proc. 14th conf. on coastal engineering*, 306–14. Japan.
Noda, H. 1971. Mechanism of bottom sediment suspension by waves. In *Proc. 14th conf. on coastal engineering*, 349–52. Japan.
Nöel, M. 1975. The palaeomagnetism of varved clays from Blekinge, southern Sweden. *Geol. Foren. Stockholm Forhand.* **97**, 357–67.
Nordin, C. F. and J. H. Algert 1966. Spectral analysis of sand waves. *Proc. Am. Soc. Civ. Eng., J. Hydraul. Div.* **92**, 95–114.
Nordin, C. F. and E. V. Richardson 1968. Statistical description of sand waves from streambed profiles. *Int. Assoc. Sci. Hydrol. Bull.* **13**, 25–32.

Norman, J. W., T. H. Leibowitz and P. G. Fookes 1975. Factors affecting the detection of slope instability with air photographs in an area near Sevenoaks, Kent. *Q. J. Eng. Geol.* **8**, 159–77.

Norris, R. M. 1966. Barchan dunes of Imperial Valley, California, *J. Geol.* **74**, 292–306.

Norton, P. E. P. 1967. Marine molluscan assemblage in the early Pleistocene of Sidestrand, Bramerton and the Royal Society Borehole at Ludham. *Phil. Trans. R. Soc. Lond. B* **253**, 161–200.

Norton, P. E. P. 1970. The crag mollusca: a conspectus. *Bull. Soc. Belge Géol,* **79**, 157–66.

Norton-Griffiths, M. 1978. *Counting animals*. Handbook no. 1. Nairobi: AWLF.

Novak, I. D. 1973. Predicting coarse sediment transport: the Hjulstrom curve revisited. In *Fluvial geomorphology*, M. Morisawa (ed.), 13–25. Binghamton, NY: State University of New York.

Nunally, N. R. 1967. Definition and identification of channel and overbank deposits and their respective roles in floodplain formation. *Prof. Geog.* **19**, 1–4.

Nye, J. F. 1970. Glacier sliding without cavitation in a linear viscous approximation. *Proc. R. Soc. Lond. A* **315**, 381–403.

Nye, J. F. 1973. Water at the bed of a glacier. *Int. Assoc. Sci. Hydrol. Publ.* **95**, 189–94. Symp. on hydrology of glaciers, Cambridge, 7–13 Sept. 1979.

Nye, J. F. and P. C. S. Martin 1968. Glacial erosion. *Int. Assoc. Sci. Hydrol. Publ.* **79**, 78–86. Berne Symp. 1967.

O'Callaghan, J. F. and D. M. Mark 1984. The extraction of drainage networks from digital elevation data. *Comput. Vision, Graphics Image Processing* **28**, 323–44.

Oeppen, B. J. and E. D. Ongley 1975. Spatial point processes applied to the distribution of river junctions. *Geog. Anal.* **7**, 153–71.

Off, T. 1963. Rhymithic linear sand bodies caused by tidal currents. *Bull. Am. Assoc. Petrol. Geol.* **47**, 324–41.

Ohmori, H. 1978. Relief structure of the Japanese mountains and their stages in geomorphic development. *Bull. Dept Geog. Univ. Tokyo* **10**, 31–85.

Oldfield, F. 1960a. Late Quaternary changes in climate, vegetation and sea-level in lowland Lonsdale. *Trans. Inst. Br. Geog.* **28**, 99–117.

Oldfield, F. 1960b. Studies in the Postglacial history of British vegetation: lowland Lonsdale. *New Phytol.* **59**, 192–217.

Oldfield, F. 1975. Pollen-analytical results. Part II. In *Late Pleistocene environments of the Southern High Plains*, F. Wendorf and J. Hester (eds). Fort Burgwin Res. Center Publ. no. 9. Taos, New Mexico.

Oldfield, F. 1977. Lakes and their drainage basins as units of sediment-based ecological study. *Prog. Phys. Geog.* **1**, 460–504.

Oldfield, F. and P. G. Appleby 1984a. Empirical testing of ^{210}Pb dating models for lake sediments. In *Lake sediments and environmental history*, E. Y. Haworth and J. W. G. Lund (eds), 93–114. Leicester: Leicester University Press.

Oldfield, F. and P. G. Appleby 1984b. A combined radiometric and mineral magnetic approach to recent geochronology in lakes affected by catchment disturbance and sediment redistribution. *Chem. Geol.* **44**, 67–83.

Oldfield, F. and B. A. Maher 1984. A mineral magnetic approach to erosion studies. In report on *Conf. on drainage basin erosion and sedimentation*, Newcastle, NSW, Australia, May 1984.

Oldfield, F. and J. Schoenwetter 1975. Discussion of the pollen-analytical evidence. In *Late Pleistocene environments of the Southern High Plains*, F. Wendorf and J. J. Hester (eds). Fort Burgwin Research Center Publ. no. 9. Taos, New Mexico.

Oldfield, F., P. G. Appleby and R. W. Battarbee 1978a. An alternative approach to ^{210}Pb dating: preliminary results from the New Guinea Highlands and from Lough Erne, N. Ireland. *Nature* **271**, 339–42.

Oldfield, F., P. G. Appleby, R. S. Cambray, J. D. Eakins, K. E. Barber, R. W. Battarbee, G. W. Pearson and J. M. Williams 1979a. ^{210}Pb, ^{137}Cs and ^{239}Pu profiles in ombrotrophic peat. *Oikos* **33**, 40–5.

Oldfield, F., P. G. Appleby and R. Thompson 1980. Palaeoecological studies of three lakes in the Highlands of Papua, New Guinea. I: The chronology of sedimentation. *J. Ecol.* **68**, 457–77.

Oldfield, F., A. F. Brown and R. Thompson 1979b. The effect of mire vegetation and microtopography on variations in particulate atmospheric influx. *Quat. Res.* **3**, 326–32.

Oldfield, F., J. A. Dearing, R. Thompson and S. E. Garret-Jones 1978b. Some magnetic properties of lake sediments and their possible links with erosion rates. *Pol. Arch. Hydrobiol,* **25**, 321–31.

Oldfield, F., S. R. Higgitt, B. A. Maher, P. G. Appleby and M. Scoullos 1985a. A new integrative methodology for desertification studies based on magnetic and short-lived radioisotope measurements. In *Desertification in Europe*, R. Fantechi and N. S. Margaris (eds), 95–109. Proc.

information symp. in the EEC Programme on Climatology, Mytilene, Greece, 15–19 April 1984. Dordrecht: Reidel.

Oldfield, F., T. A. Rummery, R. Thompson and D. E. Walling 1979c. Identification of suspended sediment source by means of magnetic measurements: some preliminary results. *Water Resour. Res.*, 211–18.

Oldfield, F., R. Thompson and K. E. Barber 1978c. The changing atmospheric fall-out of magnetic particles recorded in ombrotrophic peat sections. *Science* **199**, 679–80.

Oldfield, F. and A. T. Worsley (née O'Garra) and P. G. Appleby 1985b. Evidence from lake sediments for recent erosion rates in the Highlands of Papua New Guinea. In *Environmental change and tropical geomorphology*, I. Douglas and T. Spencer (eds), 185–96. London: Allen & Unwin.

O'Leary, M., R. H. Lippert and O. T. Spitz 1966. FORTRAN IV and map program for computation and plotting of trend surfaces for degrees 1 through 6. *Kansas Geol. Surv. Comput. Contrib.* **3**, 1–48.

Oliver, M. and R. Webster 1986. Semi-variograms for modelling the spatial pattern of landform and soil properties. *Earth Surf. Processes Landforms* **11**, 491–504.

Ollier, C. D. 1969. *Weathering*. London: Longman.

Ollier, C. D. 1977. Terrain classification: methods, applications and principles. In *Applied geomorphology*, J. R. Hails (ed.), 277–316. Amsterdam: Elsevier.

Olofin, E. A. 1974. Classification of slope angles for land planning purposes. *J. Trop. Geog.* **39**, 72–7.

O'Loughlin, E. M. 1981. A rate of rise controller for automatic water samples. *J. Hydrol.* **50**, 383–7.

Olson, J. R. 1958. Lake Michigan dune development. II: Plants as agents and tools in geomorphology. *J. Geol.* **66**, 345–51.

Olson, J. R. 1967. *Flow meters in shallow water oceanography*, 1–44. Naval Underwater Warfare Center, US Navy, San Diego, Calif., no. TP5.

Olsson, I. U. 1974. Some problems in connection with the evaluation of ^{14}C dates. *Geol. Foren. Stockholm Forhandl.* **96**, 311–20.

Omo Malaka, S. L. 1977a. A study of the chemistry and hydraulic conductivity of mound materials and soils from different habitats of some Nigerian termites. *Aust. J. Soil Res.* **15**, 87–91.

Omo Malaka, S. L. 1977b. A note on the bulk density of termite mounds. *Aust. J. Soil Res.* **15**, 93–4.

O'Neil, J. R. 1986. Theoretical and experimental aspects of isotopic fractionation. In *Stable isotopes in high-temperature processes*, J. W. Valley, H. P. Taylor, Jr, and J. R. O'Neil (eds), 1–41. Mineral. Soc. Am., Rev. Mineral. no. 16.

O'Neill, M. P. and A. D. Abrahams 1984. Objective identification of pools and riffles. *Water Resour. Res.* **20**, 921–6.

O'Neill, M. P. and A. D. Abrahams 1986. Objective identification of meanders and bends. *J. Hydrol.* **83**, 337–53.

O'Neill, M. P. and D. M. Mark 1987. On the frequency distribution of land slope. *Earth Surf. Processes Landforms* **12**, 127–36.

Onesti, L. J. and T. K. Miller 1974. Patterns of variation in a fluvial system. *Water Resour. Res.* **10**, 1178–86.

Onesti, L. J. and T. K. Miller 1978. Topological classification of drainage networks: an evaluation. *Water Resour. Res.* **14**, 144–8.

Ongley, E. D. 1970. Determination of rectilinear profile segments by automatic data processing. *Z. Geomorphol. NF* **14**, 383–91.

Opdyke, N. D. 1972. Palaeomagnetism of deep-sea cores. *Rev. Geophys. Space Phys.* **10**, 213–49.

Orford, J. D. 1975. Discrimination of particle zonation on a pebble beach. *Sedimentology* **22**, 441–63.

Orford, J. D. and W. B. Whalley 1983. The use of the fractal dimension to quantify the morphology of irregular-shaped particles. *Sedimentology* **30**, 655–68.

Orion Research 1978. *Analytical methods guide*. Cambridge, Mass.: Orion Research.

Osmond, D. A. and P. Bullock (eds) 1970. *Micromorphological techniques and applications*. Soil Surv. Tech. Monogr. no. 2. Harpenden: Soil Survey of England and Wales.

Osmond, J. K. and J. B. Cowart 1982. Groundwater. In *Uranium series disequilibrium: applications to environmental problems*, M. Ivanovich and R. S. Harmon (eds), 202–45. Oxford: Oxford University Press.

Østrem, G. 1975. Sediment transport in glacial meltwater streams. In *Glaciofluvial and glaciolacustrine sedimentation*, A. V. Jopling and B. C. McDonald (eds), 101–22. Tulsa, Okla.: Society of Economic Palaeontologists and Mineralogists.

Østrem, G. and A. Stanley 1969. *Glacier mass-balance measurements: a manual for field and office work*. Ottawa: Department of Energy, Mines and Resources.

O'Sullivan, P. E. 1983. Annually-laminated lake sediments and the study of Quaternary environmental changes – a review. *Quat. Sci. Rev.* **1**, 245–313.

O'Sullivan, P. E., M. A. Coard and D. A. Pickering 1982. The use of annually laminated lake sediments in the estimation and calibration of erosion rates. In *Recent developments in the explanation and prediction of erosion and sediment yield*, 385–96. IAHS Publ. no. 137.

O'Sullivan, P. E., F. Oldfield and R. W. Battarbee 1973. Preliminary studies of Lough Neagh sediments. 1. Stratigraphy, chronology and pollen analysis. In *Quaternary plant ecology*, H. J. B. Birks and R. G. West (Eds). Oxford: Blackwell.

Otlet, R. L., G. Fluxtable and D. C. W. Sandersun 1986. The development of practical systems for ^{14}C measurement in small samples using miniature counters. *Radiocarbon* **28**, 603–14.

Owen, A. E. B. 1952. Coast erosion in east Lincolnshire. *Lincs. Hist.* **9**, 35–48.

Owens, M., M. A. Learner and P. J. Maris 1967. Determination of the biomass of aquatic plants using an optical method. *J. Ecol.* **55**, 671–6.

Paepe, R. 1971. Dating and position of fossil soils in the Belgian Pleistocene. In *Paleopedology*, D. H. Yaalon (ed.), 261–9. Jerusalem: Israel University Press.

Paepe, R. and R. van Hoorne 1973. *The stratigraphy and palaeobotany of the late Pleistocene in Belgium*. Geol.Surv. Belg. Mem. no. 8.

Pahlke, H. 1973. Some applications using artificial radioactive tracers in Germany. In *Tracer techniques in sediment transport*, 81–96. Vienna: International Atomic Energy Agency.

Pakarinen, P. and K. Tolonen 1977. On the growth-rate and dating of surface peat (Finnish with English abstract). *Suo* **28**, 19–24.

Pandey, P. C., B. Singh and P. S. Rehill 1983. Windthrow in deodar. *Commonw. For. Rev.* **62**, 37–9.

Pannekoek, A. J. 1967. Generalized contour maps, summit level maps and streamline surface maps as geomorphological tools. *Z. Geomorphol. NF* **11**, 169–82.

Papo, H. B. and E. Gelbman 1984. Digital terrain models for slopes and curvatures. *Photogramm. Eng. Remote Sensing* **50**, 695–701.

Park, C. C. 1976. The relationship of slope and stream channel form in the River Dart, Devon. *J.Hydrol.* **29**, 139–47.

Parker, G. A. and D. F. Boltz 1971. Colorimetry. In *Modern methods of geochemical analysis*, R. E. Wainerdi and E. A. Uken (eds), 97–126. New York: Monographs in Geoscience.

Parr, J. F. and A. R. Bertrand 1960. Water infiltration into soils. *Adv. Agron.* **12**, 311–63.

Parry, J. T. and J. A. Beswick 1973. The application of two morphometric terrain classification systems using airphoto interpretation methods. *Photogrammetria (Amsterdam)* **29**, 153–86.

Parsons, A. J. 1976. A Markov model for the description and classification of hillslopes. *Math. Geol.* **8**, 597–616.

Parsons, A. J. 1977. Curvature and rectilinearity in hillslope profiles. *Area* **9**, 246–51.

Parsons, A. J. 1979. Plan form and profile form of hillslopes. *Earth Surf. Processes* **4**, 395–402.

Parsons, A. J. 1982. Slope profile variability in first-order drainage basins. *Earth Surf. Processes Landforms* **7**, 71–8.

Patel, C. Kl. N. 1978. Laser detection of pollution. *Science* **202**, 157–73.

Pathak, P. C., A. N. Pandey and J. E. Singh 1984. Overland flow, sediment output and nutrient loss from certain forested sites in the central Himalaya, India. *J.Hydrol.* **71**, 239–51.

Paton, T. R., P. B. Mitchell, D. Adamson, R. A. Buchanan, M. D. Fox and G. Boeman 1976. Speed of podzolisation. *Nature* **260**, 601–2.

Patro, B. C. and B. K. Sahu 1974. Factor analysis of sphericity and roundness data of clastic quartz grains: environmental significance. *Sedim.Geol.* **11**, 59–78.

Patterson, E. M. and D. A. Gillette 1977. Measurements of visibility vs mass-concentration for air-borne soil particles. *Atmos.Environ.* **1**, 193–6.

Patton, P. C. and V. R. Baker 1976. Morphometry and floods in small drainage basins subject to diverse hydrogeomorphic controls. *Water Resour. Res.* **12**, 941–52.

Pavoni, N. and R. Green (eds) 1975. *Recent crustal movements*. Developments in Geotectonics no. 9.

Peach, P. A. and L. A. Perrie 1975. Grain-size distribution within glacial varves. *Geology* **3**, 43–6.

Peacock, J. D., D. K. Graham, J. E. Robinson and I. Wilkinson 1977. Evolution and chronology of lateglacial marine environments at Lochgilphead, Scotland. In *Studies in the Scottish lateglacial environment*, J. M. Gray and J. J. Lowe (eds), 89–100. Oxford: Pergamon.

Pearson, F. J. and D. W. Fisher 1971. *Chemical composition of atmospheric precipitation in the northeastern United States*. US Geol. Surv. Water Supply Paper no. 1535-P.

Pearson, G. W. and M. Stuiver 1986. High-precision calibration of the radiocarbon time scale 500–2500 BC. *Radiocarbon* **28**, 839–62.

Pecsi, M. 1972. Geomorphological position and absolute age of the Lower Palaeolithic site at Vertesszollos, Hungary, *Folrajzi Kozlemenyek* **2**, 109–19.

Pedro, G. 1961. An experimental study on the geochemical weathering of cristalline rocks by water. *Clay Min. Bull.* **4**, 226–81.

Peglar, S., S. C. Fritz, T. Alapieti, M. Saarnisto and H. J. B. Birks 1984. The composition and formation of laminated lake sediments in Diss Mere, Norfolk, England. *Boreas* **13**, 13–28.

Pennington, W. (Mrs T. G. Tutin), R. S. Cambray, J. D. Eakins and D. D. Harkness 1976. Radionuclide dating of the recent sediments of Blelham Tarn. *Freshwater Biol.* **6**, 317–31.

Pennington, W., R. S. Cambray and E. M. Fisher 1973. Observations on lake sediment using fallout Cs-137 as a tracer. *Nature* **242**, 324–6.

Pennington, W., E. Y. Haworth, A. P. Bonny and J. P. Lishman 1972. Lake sediments in northern Scotland. *Phil.Trans. R.Soc.B* **264**, 191–294.

Perrin, R. M. S. 1965. The use of drainage water analysis in soil studies. In *Experimental pedology*, E. G. Hallsworth and D. V. Crawford (eds), 73–91. London: Butterworth.

Perrin, R. W., E. A. Willis and C. A. Hodge 1963. Dating of humus podzols by residual radiocarbon activity. *Nature* **202**, 165–6.

Peters, N. E. 1984. *Evaluation of environmental factors affecting yields of major dissolved ions of streams in the United States.* US Geol. Surv. Water Supply Paper no. 2228.

Pethick, J. 1975. A note on the drainage density–basin area relationship. *Area* **7**, 217–22.

Pethick, J. 1984. *An introduction to coastal geomorphology.* London: Edward Arnold.

Petrie, G. and R. J. Price 1966. Photogrammetric measurements of the ice wastage and morphological changes near the Casement Glacier, Alaska. *Can.J.Earth Sci.* **3**, 827–40.

Petrov, M. P. 1976. *Deserts of the world.* New York: Wiley.

Pettijohn, F. J. 1949. *Sedimentary rocks*, 1st edn. New York: Harper & Row.

Pettijohn, F. J., P. E. Potter and R. Siever 1972. *Sand and sandstone.* Berlin: Springer.

Peucker, T. K. 1972. *Computer cartography.* Assoc. Am.Geog. Commission on College Geog. Resource Paper no. 17.

Phillips, A. W., 1970. The use of the Woodhead seabed drifter. *BGRG Tech.Bull.* **4**.

Phillips, F. C. 1971. *The use of stereographic projection in structural geology.* London: Edward Arnold.

Phipps, R. L. 1974. The soil creep–curved tree fallacy. *J.Res. US Geol. Surv.* **2**, 371–8.

Pickup, G. and R. F. Warner 1976. Effects of hydrologic regime on magnitude and frequency of dominant discharge. *J.Hydrol.* **29**, 51–75.

Pielou, E. C. 1969. *An introduction to mathematical ecology.* New York: Wiley.

Pierce, K. L., J. D. Obradovich and I. Friedman 1976. Obsidian hydration dating and correlation of Bull Lake and Pinedale Glaciations near West Yellowstone, Montana. *Geol. Soc.Am.Bull.* **87**, 703–10.

Pieri, D. C. 1980. Martian valleys: morphology, distribution, age, and origin. *Science* **210**, 895–7.

Pike, R. J. and W. J. Rozema 1975. Spectral analysis of landforms. *Ann.Assoc. Am.Geog.* **64**, 499–514.

Pike, R. J. and S. E. Wilson 1971. Elevation–relief ratio, hypsometric integral, and geomorphic area–altitude analysis. *Geol. Soc.Am.Bull.* **82**, 1079–84.

Pilcher, J. R., M. G. L. Baillie, B. Schmidt and B. Becker 1984. A 7272-year European tree-ring chronology. *Nature* **312**, 150–2.

Pilgrim, D. H. 1976. Travel times and non-linearity of flood runoff from tracer measurements on a small watershed. *Water Resour. Res.* **12**, 487–96.

Pincus, H. J. 1956. Some vector and arithmetic operations on two-dimensional orientation variates with application to geological data. *J.Geol.* **64**, 533–57.

Pincus, H. J. 1966. Optic processing of vectorial rock fabric data. *Proc.1st cong.int.soc.rock mechanics*, vol. 1, 173–7. Lisbon.

Pinder, G. F. and J. F. Jones 1969. Determination of the groundwater component of peak discharge from the chemistry of total runoff. *Water Resour. Res.* **5**, 438–45.

Piper, D. J. 1971. The use of the D-Mac Pencil Follower in routine determinations of sedimentary parameters. In *Data processing in biology and geology*, J. L. Cutbill (ed.). Systematics Assoc. Spec. Vol. no. 3.

Pissart, A. and J. P. Lautridou 1984. Variations de longueur de cylindres de pierre de Caen (calcaire Bathonien) sous l'éffet de sécharge et d'humidification. *Z.Geomorphol. Supp. NF* **49**, 111–16.

Pittman, E. D. and R. W. Duschatko 1970. Use of pore casts and scanning electron microscope to study pore geometry. *J.Sedim.Petrol.* **40**, 1153–7.

Pitts, J. 1979. Morphological mapping in the Axemouth–Lyme Regis Undercliffs, Devon. *Q.J.Eng.Geol.* **12**, 205–17.

Pitty, A. F. 1966. Some problems in the location and delimitation of slope-profiles. *Z. Geomorphol. NF* **10**, 454–61.

Pitty, A. F. 1967. Some problems in selecting a ground surface length for slope-angle measurement. *Rev. Géomorphol. Dyn.* **17**, 66–71.

Pitty, A. F. 1968. A simple device for the field measurement of hillslopes. *J.Geol.* **76**, 717–20.
Pitty, A. F. 1969. *A scheme for hillslope analysis.* I. *Initial considerations and calculations.* Univ. Hull Occas. Papers Geog. no. 9.
Pitty, A. F. 1970. *A scheme for hillslope analysis. II: Indices and tests for differences.* Univ. Hull Occas. Papers Geog. no. 17.
Pitty, A. F. 1971. *Introduction to geomorphology*. London: Methuen.
Podmore, T. H. and L. F. Huggins 1981. An automated profile meter for surface roughness measurements. *Trans. Am. Soc. Agric. Eng.* **24**, 663–5, 669.
Poel, L. W. 1979. On the use of the katharometer for the estimation of carbon dioxide concentrations in soils. *Ann. Bot.* **43**, 285–93.
Pomeroy, D. E. 1976. Some effects of mound-building termites on soils in Uganda. *J.Soil Sci.* **27**, 377–94.
Poole, M. A. and P. N. O'Farrell 1971. The assumptions of the linear regression model. *Trans. Inst.Br.Geog.* **52**, 145–58.
Poole, R. W. and I. W. Farmer 1980. Consistency and repeatability of Schmidt Hammer rebound data during field testing. *Int.J.Rock Mech. Min. Sci.* **17**, 167–71.
Popper, K. R. 1972. *The logic of scientific discovery*. London: Hutchinson.
Porter, J. J. 1962. Electron microscopy of sand surface textures. *J.Sedim.Petrol.* **32**, 124–5.
Porter, S. C. 1972. Distribution, morphology and size-frequency of cinder cones on Mauna Kea Volcano, Hawaii. *Bull.Geol.Soc.Am.* **83**, 3607–12.
Porter, S. C. 1975. Weathering rinds as a relative-age criterion: an application to sub-division of glacial deposits in the Cascade Range. *Geology* **3**, 101–4.
Potts, A. S. 1970. Frost action in rocks: some experimental data. *Trans. Inst.Br.Geog.* **49**, 109–24.
Poulin, A. O. 1962. Movement of frost formed patterns using air photo techniques. *Photogramm. Eng.* **28**, 141–7.
Powers, M. C. 1953. A new roundness scale for sedimentary particles. *J.Sedim.Petrol.* **23**, 117–19.
Price, R. J. 1966. Eskers near the Casement Glacier, Alaska. *Geog. Annaler* **48A**, 111–25.
Price, R. J. 1969. Moraines, sandar, kames and eskers near Breidamerkurjokull, Iceland. *Trans. Inst.Br.Geog.* **46**, 17–43.
Price, W. A. 1968. Variable dispersion and its effects on the movement of tracers on beaches. In *Proc.11th conf. on coastal engineering*, vol. 1, 518–25. Am.Soc.Civ.Eng.
Price, W. A., K. W. Tomlinson and D. H. Willis 1972. Predicting changes in the plan shape of beaches. In *Proc. 13th conf. on coastal engineering*, vol. 2, 1321–9. Am.Soc.Civ.Eng.
Prior, D. B. 1977. Coastal mudslide morphology and processes on Eocene clays in Denmark. *Geog. Tiddskr.* **76**, 14–33.
Prior, D. B. and C. Ho 1971. Coastal and mountain slope instability on the islands of St Lucia and Barbados. *Eng.Geol.* **6**, 1–18.
Prior, D. B. and N. Stephens 1972. Some movement patterns of temperate mudflows: examples from north-eastern Ireland. *Bull.Geol. Soc.Am.* **83**, 2533–44.
Prior, D. B., N. Stephens and G. R. Douglas 1971. Some examples of mudflow and rockfall activity in north-eastern Ireland. *Inst.Br.Geog.Spec.Publ.* **3**, 129–40.
Pritchard, D. W. and W. V. Burt 1951. An inexpensive and rapid technique for obtaining current profiles in estuarine water. *J.Mar.Res.* **10**, 180–8.
Prospero, J. M. and T. N. Carlson 1970. Radon-222 in the north Atlantic trade winds: its relationship to dust transport from Africa. *Science* **167**, 974–7.
Prospero, J. M., D. L. Savoie, T. N. Carlson and R. T. Nees 1979. Monitoring Saharan aerosol transport by means of atmospheric turbidity measurements. In *Saharan dust*, C. Morales (ed.), 171–86. Chichester: Wiley.
Pryor, W. A. 1971. Grain shape. In *Procedures in sedimentary petrology*, R. E. Carver (ed.), ch. 7. New York: Wiley.
Przybylowicz, W. and H. P. Schwarcz 1988. Dirty calcites. II: U-series dating of artificial calcite–detritus mixtures. *Isot. Geosci.* In press.
Purdy, E. G. 1974. Reef configurations: cause and effect. *Soc. Econ.Palaeontol. Min.Spec.Publ.* **18**, 9–76.

Rabinowicz, E. 1965. *Friction and wear of materials.* New York: Wiley.
Raddke, U., R. Grün and H. P. Schwarcz 1988. Electron spin resonance dating of the Pleistocene coral reef tracts of Barbados. *Quat. Res.* **29**, 197–215.
Rahn, K. A., R. D. Borys, G. E. Shaw, L. Shutz and R. Jaenicke 1979. Long-range impact of desert aerosol on atmospheric chemistry: two examples. In *Saharan dust*, C. Morales (ed.), 243–66. Chichester: Wiley.

Rahns, P. H. 1971. The weathering of tombstones and its relationship to the topography of New England. *J.Geol.Educ.* **19**, 112–18.
Rainwater, F.H. and L. L. Thatcher 1960. *Methods for collection and analysis of water samples.* US Geol. Surv.Water Supply Paper no. 1454.
Ralph, E. K., H. N. Michael and M. C. Han 1973. Radiocarbon dates and reality. *MASCA Newslett.* **9**, 1–20.
Ralph, M. O. and C. F. Barrett 1976. *A wind-tunnel study of the efficiency of deposit gauges.* Interim Rep., Warren Springs Lab.Rep. no. LR235(AP). Stevenage, Herts.
Ramberg, H. 1964. Note on model studies of folding of moraines in piedmont glaciers. *J.Glaciol.* **5**, 207–18.
Rand, M. C., A. E. Greenberg and M. J. Jarvis 1980. *Standard methods for the examination of water and waste-water*, 15th edn. Washington, D.C.: American Public Health Association.
Rapp, A. 1960a. Recent development of mountain slopes in Karkevagge and surroundings, northern Scandinavia. *Geog. Annaler* **42**, 73–200.
Rapp, A. 1960b. Talus slopes and mountain walls at Templefjorden, Spitzbergen. *Nor. Polarinst. Skr.* **119.**
Rashidian, K. 1986. A new technique for field measurement of soilcreep displacement profiles. *Z. Geomorphog. Supp. NF* **60**, 93–103.
Rathbun, R. E. and C. F. Nordin 1971. Tracer studies of sediment transport processes. *Proc. Am.Soc.Civ.Eng., J.Hydraul. Div.* **97**, **HY9**, 1305–16.
Raymond, C. F. 1971. Flow in a transverse section of Athabasca Glacier, Alberta, Canada. *J.Glaciol.* **10**, 55–84.
Rayner, J. N. 1971. *An introduction to spectral analysis.* London: Pion.
Rayner, J. N. 1972. The application of harmonic and spectral analysis to the study of terrain. In *Spatial analysis in geomorphology*, R. J. Chorley (ed.), 283–302. London: Methuen.
Readings, C. J. and H. E. Butler 1972. The measurement of atmospheric turbulence from a captive balloon. *Meteorol.Mag.* **101**, 286–98.
Reches, Z. and O. F. Hoexter 1981. Holocene seismic and tectonic activity in the Dead Sea area. *Tectonophysics* **80**, 235–54.
Reddy, M. M. 1988. Acid rain damage to carbonate stone: a quantitative assessment based on the aqueous geochemistry of rainfall runoff from stone. *Earth Surf. Processes Landforms* **13**, 335–54.
Reed, B., C. J. Galvin, Jr and J. P. Miller 1962. Some aspects of drumlin geometry. *Am.J.Sci.* **260**, 200–10.
Rees, A. I. 1965. The use of anisotropy of magnetic susceptibility in the estimation of sedimentary fabrics. *Sedimentology* **4**, 257–83.
Reger, R. D. and T. L. Péwé 1969. Lichenometric dating in the central Alaskan Range. In *The periglacial environment: past and present*, T. L. Péwé (ed.), 223–47. Montreal: Arctic Institute of North America, McGill-Queen's University Press.
Reiche, P. 1950. *A survey of weathering processes and products.* Univ. New Mexico Publ. Geol. no. 3.
Reid, I., A. C. Brayshaw and L. E. Frostick 1984. An electromagnetic device for automatic detection of bedload motion and its field application. *Sedimentology* **31**, 269–76.
Reid, I., J. T. Laiyman and L. E. Frostick 1980. The continuous measurement of bedload discharge. *J. Hydraul. Res.* **18**, 243–9.
Reid, P. C. 1972. Dinoflagellate cyst distribution around the British Isles. *J.Mar. Biol.Assoc. UK* **52**, 939–44.
Reimnitz, E. and D. A. Ross 1971. The Sea Sled – a device for measuring bottom profiles in the surf zone. *Mar. Geol.* **11**, 1727–32.
Reiners, W. A. 1972. Nutrient content of canopy throughfall in three Minnesota forests. *Oikos* **23**, 14–22.
Renberg, I. 1978. Palaeolimnology and varve counts of the annually-laminated sediments of Lake Rudetjarn, northern Sweden. *Early Norrland* **11**, 63–92.
Renberg, I. 1981a. Improved methods for sampling, photographing and varve-counting of varved lake sediments. *Boreas* **10**, 255–8.
Renberg, I. 1981b. Formation, structure and visual appearance of iron-rich, varved lake sediments. *Verh.Int.Ver.Limnol.* **21**, 94–101.
Reniger, A. 1955. Erozja gleb na terenie podgorskim w obrebie zlewni potokn Lukawica. *Roczu.Nauk.Roln. Ser.F* **71**, 149–210.
Rescher, N. 1962. The stochastic revolution and the nature of scientific explanation. *Synthèse* **14**, 200–15.
Reynolds, B. 1981. *Methods for the collection and analysis of water samples for a geochemical cycling study.* ITE Bangor Occas. paper no.5.

Reynolds, B. 1984. Spatial variation in bulk precipitation chemistry and implications for sampling in an upland catchment. *Water Resour. Res.* **20**, 733–5.
Reynolds, R. C. 1971. Analysis of alpine waters by ion-electrode methods. *Water Resour. Res.* **7**, 1333–7.
Reybolds, R. J. and K. Aldous 1970. *Atomic absorption spectroscopy*. London: Griffin.
Reynolds, S. G. 1974. A note on the relationship between size of area and soil moisture variability. *J.Hydrol.* **22**, 71–6.
Reynolds, S. G. 1975. Soil property variability in slope studies: suggested sampling schemes and typical required samples. *Z. Geomorphol. NF* **19**, 191–208.
Rhind, D. W. 1976. Overlap: cartography. Institutes of maps. *Geog.Mag.* **48**, 244.
Rhodes, E. J. 1988. Methodological considerations in the optical dating of quartz. *Quat.Sci.Rev.* **7**, 395–400.
Rice, A. 1976. Insolation warmed over. *Geology* **4**, 61–2.
Richards, K. S. 1972. Meanders and valley slope. *Area* **4**, 288–90.
Richards, K. S. 1973. Hydraulic geometry and channel roughness: a non-linear system. *Am.J.Sci.* **273**, 877–96.
Richards, K. S. 1976. The morphology of riffle–pool sequences. *Earth Surf. Processes* **1**, 71–88.
Richards, K. S. 1977. Channel and flow geometry: a geomorphological perspective. *Prog. Phys.Geog.* **1**, 65–102.
Richards, K. S. 1978. Yet more notes on the drainage density–basin area relationship. *Area* **10**, 344–8.
Richards, K. S. 1979. Prediction of drainage density from surrogate measures. *Water Resour. Res.* **15**, 435–42.
Richards, K. S. 1980. A note on changes in channel geometry at tributary junctions. *Water Resour. Res.* **16**, 241–4.
Richards, K. S. 1982. *Rivers: form and process in alluvial channels*. London: Methuen.
Richardson, E. V., D. B. Simons and G. J. Posakony 1961. *Sonic depth sounder for laboratory and field use*. US Geol. Surv. Circ. no. 450.
Richardson, H. D. 1968. *Industrial radiography*. US Govt Printing Office. Cat. no. FS 5.284: 84036.
Ricq de Bouard, M. 1976. A method of concentrating the major impurities contained in ice by ion exchange. *J.Glaciol.* **17**, 129–33.
Rider, L. J. and M. Armendariz 1966. A comparison of tower and pibal wind measurements. *J.Appl.Meteorol.* **5**, 43–8.
Riezebos, H.Th. and E. Seyhan 1977. Essential conditions of rainfall simulation for laboratory water erosion experiments. *Earth Surf. Processes* **2**, 185–90.
Riley, J. D. and J. W. Ramster 1972. Woodhead seabed drifter recoveries and the influence of human, tidal and wind factors. *J.Cons.Int.Explor.Mar.* **34**, 389–415.
Riley, M. C. 1941. Projection sphericity. *J.Sedim.Petrol.* **11**, 94–7.
Riley, N. M. 1979. In General discussion, *J.Glaciol.* **23**, 381–400.
Riley, S. J. 1969. A simplified levelling instrument: the A-frame. *Earth Sci.J.* **3**, 51–3.
Riley, S. J. 1972. A comparison of morphometric measures of bankfull. *J.Hydrol.* **17**, 23–31.
Rim, M. 1958. Simulation, by dynamical model, of sand tract morphologies occurring in Israel. *Res.Counc.Isr. Bull.* **7-G**, 123–33.
Rittenhouse, G. 1943. A visual method of estimating two dimensional sphericity. *J.Sedim.Petrol.* **13**, 79–81.
Robbins, J. A. 1978. Geochemical and geophysical applications of radio-active lead. In *Biogeochemistry of lead in the environment*, J. O. Nriagu (ed.). Amsterdam: Elsevier.
Robbins, J. A. and D. N. Edgington 1975. Determination of recent sedimentation rates in Lake Michigan using Pb-210 and Cs-137. *Geochim.Cosmochim.Acta* **39**, 285–304.
Roberts, H. H. 1972. X-ray radiography in Recent carbonate sediments. *J.Sedim.Petrol.* **42**, 690–3.
Roberts, H. H., T. Whelan, III and W. G. Smith 1977. Holocene sedimentation at Cape Sable, south Florida. *Sedim.Geol.* **18**, 25–60.
Robertson-Rintoul, M. S. E. 1986. A quantitative soil-stratigraphic approach to the correlation and dating of postglacial river terraces in Glen Feshie, southwest Cairngorms. *Earth Surf. Processes Landforms* **11**, 605–17.
Robinson, A. H. W. 1960. Ebb-floor channel systems in sandy bays and estuaries. *Geography* **45**, 183–90.
Robinson, A. H. W. 1962. *Marine cartography in Britain*. Leicester: University of Leicester.
Robinson, A. H. W. 1964. Inshore waters, sediment supply and coastal changes of part of Lincolnshire. *East Midl.Geog.* **3**, 307–21.
Robinson, E. 1968. Effects of air pollution visibility. In *Air pollution*, A. C. Stern (ed.), 349–400. New York: Academic Press.

Robinson, G., J. A. Peterson and P. A. Anderson 1971. Trend surface analysis of corrie altitudes in Scotland. *Scot. Geog.Mag.* **87**, 142–6.
Robinson, L. A. 1976. The micro-erosion meter technique in a littoral environment. *Mar. Geol.* **22**, M51–8.
Robinson, L. A. 1977. Erosive processes on the shore platform of northeast Yorkshire, England. *Mar. Geol.* **23**, 339–61.
Rockaway, J. D. 1976. The influence of map scale on engineering geological mapping. *Bull. Assoc.Eng.Geol* **13**, 233–45.
Rockwell, T. K., D. L. Johnson, E. A. Keller and G. R. Dembroff 1985. A late Pleistocene–Holocene soil chronosequence in the Ventura basin, southern California, USA. In *Soils and geomorphology*, K. S. Richards, R. Arnett and S. Ellis (eds), 309–27. London: Allen & Unwin.
Rodda, J. C. 1970. A trend surface analysis trial for the planation surfaces of north Cardiganshire. *Trans.Inst.Br.Geog.* **50**, 107–14.
Rodriguez-Iturbe, I. 1982. The coupling of climate and geomorphology in rainfall–runoff analysis. In *Rainfall–runoff relationship*, V. P. Singh (ed.), 431–48. Littleton, Colo.: Water Resources Publication.
Rodriguez-Iturbe, I. and J. B. Valdes 1979. The geomorphologic structure of hydrologic response. *Water Resour. Res.* **15**, 1409–29.
Roels, J. M. 1985. Estimation of soil loss at a regional scale based on plot measurements – some critical considerations. *Earth Surf. Processes Landforms* **10**, 587–95.
Rogers, W. B. and R.E. Rogers 1848. On the decomposition and partial solution of minerals, rocks, etc. by pure water, and water charged with carbonic acid. *Am.J.Sci* **5**, 401–4.
Rogerson, R. J. and M. J. Batterson 1982. Contemporary push moraine formation in the Yoho Valley, B.C. In *Research in glacial, glaciofluvial and glaciolacustrine systems*, R. Davidson-Arnott, W. Nickling and B. D. Fahey (eds), 71–90. Norwich: Geo Books.
Rogerson, R. J. and N. Eyles 1979. *Berendon Glacier medial moraines, their form, sediments and terminal icemelt effects.* Res. Note no. 4, Dept Geog., Memorial Univ. Newfoundland, St Johns.
Rose, J. 1978. Glaciation and sea-level change at Bugoyfjord, south Varangerfjord, north Norway. *Nor. Geog. Tidsskr.* **32**, 121–35.
Rose, J. and P. Allen 1977. Middle Pleistocene stratigraphy in south-east Suffolk. *J.Geol. Soc.Lond.* **133**, 83–102.
Rose, J. and J. M. Letzer 1975. Drumlin measurements: a test of the reliability of data derived from 1:25 000 scale topographic maps. *Geol.Mag.* **112**, 361–71.
Rose, J. and F. M. Synge 1979. Glaciation and shoreline development between Nydal and Haukdal, south Varangerfjorden, north Norway. *Quaest. Geog.* **5**, 125–51.
Rose, J., P. Allen and R. W. Hey 1976. Middle Pleistocene stratigraphy in southern East Anglia. *Nature* **263**, 492–4.
Rose, J., P. Allen, R. A. Kemp, C. A. Whiteman and N. Owen 1985a. The early Anglian Barham soil of eastern England. In *Soils and Quaternary landscape evolution*, J. Boardman (ed.), 197–229. Chichester: Wiley.
Rose, J., J. Boardman, R. A. Kemp and C. A. Whiteman 1985b. Palaeosols and the interpretation of the British Quaternary stratigraphy. In *Soils and geomorphology*, K. S. Richards, R. Arnett and S. Ellis (eds), 348–75. London: Allen & Unwin.
Rosenfeld, A. 1976. *Digital picture analysis.* Topics in Applied Physics, Vol.II. New York: Springer.
Rosfelder, A. 1961. *Contribution à l'analyse texturel des sediments.* Bull.Serv. Carte Géol. Algérie no. 29.
Rosholt, J. N. 1980. *Uranium-trend dating of Quaternary sediments.* US Geol. Surv. Open File Rep. no. 80–1087.
Rossbacher, L. A. 1985. Ground ice models for the distribution and evolution of curvilinear landforms on Mars. In *Models in geomorphology* M. J. Woldenberg (ed.), 343–72. Binghampton Symp. Geomorphol.: Int. Ser. no. 14. Boston: Allen & Unwin.
Roth, E. S. 1965. Temperature and water content as factors in desert weathering. *J.Geol.* **73**, 454–68.
Rothlisberger, H. 1968. Erosive processes which are likely to accentuate or reduce the bottom relief of valley glaciers. *Int.Assoc.Sci.Hydrol.Publ.* **79**, 87–97. Berne Symp. 1967.
Rothlisberger, H. 1972. Water pressure in intra- and subglacial channels. *J.Glaciol.* **11**, 177–203.
Rouse, L. J. and J. M. Coleman 1976. Circulation observations in the Louisiana Bight using LANDSAT imagery. *Remote Sensing Environ.* **5**, 55–66.
Rouse, W.C. 1967. Residual shear strength of sandstone soils. *Geotechnique* **17**, 298–300.
Rouse, W. C. 1975. Engineering properties and slope form in granular soils. *Eng.Geol.* **9**, 221–35.
Rouse, W. C. and Y. I. Farhan 1976. Threshold slopes in south Wales. *Q.J.Eng.Geol.* **9**, 327–38.
Rouse, W. C. and A. J. Reading 1985. Soil mechanics and natural slope stability. In *Geomorphology and soils*, K. S. Richards, R. R. Arnett and S. Ellis (eds), 159–79. London: Allen & Unwin.

Rouse, W. C., A. J. Reading and R. P. D. Walsh 1990. Measurement and interpretation problems of tropical volcanic soil properties in Dominica, West Indies. *Eng. Geol.* **23**.

Rowe, P. W. and L. Barden 1966. A new consolidation cell. *Geotechnique* **16**, 162–70.

Roy, A. G. 1983. Optimal angular geometry models of river branching. *Geog. Anal.* **15**, 87–96.

Roy, A. G. and M. J. Woldenberg 1986. A model for changes in channel form at a river confluence. *J.Geol.* **94**, 402–11.

Royal Commission on Coastal Erosion 1911. *Reports* (3 vols). London: HMSO.

Ruhe, R. V. 1975. *Geomorphology*. Boston: Houghton Mifflin.

Ruhe, R. V., G. A. Miller and W. J. Vreeken 1971. Paleosols, loess sedimentation and soil stratigraphy. In *Paleopedology*, D. H. Yaalon (ed.), 41–60. Jerusalem: Israel University Press.

Russell, R. C. H., D. E. Newman and K. W. Tomlinson 1963. Sediment discharges measured by continuous injection of tracers from a point. In *Int.Assoc.Hydraul.Res.10th congr.*, vol.1, 69–76. London.

Rutter, N. and R. S. Edwards 1968. Deposition of airborne marine salt at different sites over the College Farm, Aberystwyth in relation to wind and weather. *Agric.Meteorol.* **5**, 235–54.

Ruxton, B. P. 1968. Measures of the degree of chemical weathering of rocks. *J.Geol.* **76**, 518–27.

Ruxton, B. P. and I. McDougall 1967. Denudation rates in north-east Papua from potassium–argon dating of lavas. *Am.J.Sci.* **265**, 545–61.

Ryley, M. D. 1969. *The use of a microwave oven for the rapid determination of the moisture content of soils*. Road Res. Lab. Rep. no. LR 280.

Saarnisto, M. 1979. Applications of annually laminated lake sediments: a review. *Acta Univ. Ouluensis* **A82,** *Geology* **3**, 97–108.

Saarnisto, M. 1986. Annually laminated lake sediments. In *Handbook of Holocene palaeoecology and palaeohydrology*, B. E. Berglund (ed.), 343–70. Chichester: Wiley.

Saarnisto, M., P. Huttunen and K. Tolonen 1977. Annual lamination of sediments of Lake Lovojarvi, southern Finland, during the past 600 years. *Ann. Bot. Fenn.* **14**, 35–45.

Sadler, P. M. 1981. Sediment accumulation rates and the completeness of the stratigraphic sections. *J. Geol.* **89**, 569–84.

Saint-Onge, D. A. 1968. Geomorphic maps. In *Encyclopedia of geomorphology*, R. W. Fairbridge (ed.), 389–403. New York: Reinhold.

Sakaguchi, Y. 1969. Development of drainage basins: an introduction to statistical geomorphology. *Bull. Dept. Geog., Univ. Tokyo* **1**, 67–74.

Sala, M. 1984. Pyrenees and Ebro basin complex. In *Geomorphology of Europe*, C. Embleton (ed.), 268–93. London: Macmillan.

Salisbury, C. R., P. J. Whiteley, C. D. Litton and J. L. Fox 1984. Flandrian courses of the River Trent at Colwick, Notingham. *Mercian Geol.* **9**, 189–207.

Sallenger, A. H., P. C. Howard, C. H. Fletcher and P. A. Howd 1983. A system for measuring bottom profile, waves and currents in the high energy nearshore environment. *Mar. Geol.* **51** (1–2), 63–76.

Salomé, A. I. and H. J. van Dorsser 1982. Examples of the 1:50 000 scale geomorphological maps of part of the Ardennes. *Z. Geomorphol. N.F.* **26**, 481–9. See also 1982. *ITC Journal* **3**, 272–4.

Salomé, A. I. and H. J. van Dorsser 1985. Some reflections on geomorphological mapping systems. *Z. Geomorphol. NF* **29**, 375–80.

Sanglerat, G. 1972. The penetrometer and soil exploration. *Developments in geotechnical engineering*, vol. 1. Amsterdam: Elsevier.

Sarin, M. M. and S. Krishnaswami 1984. Major ion chemistry of the Ganga–Brahmaputra river systems, India. *Nature* **312**, 538–41.

Sarma, S. K. 1973. Stability analysis of embankments and slopes. *Geotechnique* **23**, 423–33.

Sarnthein, M. and B. Koopmann 1980. Late Quaternary deep-sea record of northwest African dust supply and wind circulation. *Palaeoecology. Afr.* **12**, 239–53.

Sarnthein, M. and E. Walger 1974. Der aolische Sandstorm aus der W-Sahara zur Atlantikkuste. *Geol. Rundsch.* **63**, 1065–87.

Sartz, R. S. and D. N. Tolsted 1974. Effects of grazing on runoff from two small watersheds in southwestern Wisconsin. *Water Resour. Res.* **10**, 354–6.

Sauchyn, D. J. and J. S. Gardner 1983. Morphology of open rock basins, Kananaskis area, Canadian Rocky Mountains. *Can. J. Earth Sci.* **20**, 409–14.

Saunders, I. and A. Young 1983. Rates of surface processes on slopes, slope retreat and denudation. *Earth Surf. Processes Landforms* **8**, 473–501.

Sauramo, M. 1923. Studies on the Quaternary varve sediments in southern Finland. *Bull. Comm. Geol. Finl.* **60**, 1–164.

Savigear, R. A. G. 1952. Some observations on slope development in South Wales. *Transactions Institute of British Geographers* **18**, 31–2.
Savigear, R. A. G. 1965. A technique of morphological mapping. *Ann. Assoc. Am. Geog.* **55**, 514–39.
Saville, T. and J. M. Caldwell 1953. *Accuracy of hydrographic surveying in and near the surf zone*. BEB Tech. Memo. no. 32.
Sayles, R. S. and T. R. Thomas 1978. Surface topography as a non-stationary random process. *Nature* **271**, 431–4; **273**, 573.
Schalscha, E. B., H. Appelt and A. Schatz 1955. Chelation as a weathering mechanism. Part 1. *Geochim. Cosmochim. Acta* **31**, 587–96.
Scharpenseel, H. W. and H. Schiffmann 1977. Radiocarbon dating of soils: a review. *Z. Planzenernaehrung Bodenk.* **140**, 159–74.
Scheidegger, A. E. 1965. The algebra of stream-order numbers. *US Geol. Surv. Prof. Paper* **525–B**, 187–9.
Scheidegger, A. E. 1967. On the topology of river nets. *Water Resour. Res.* **3**, 103–6.
Schick, A. P. 1964. Accuracy of the 1:20 000 topographic maps of Israel for morphomentric studies. *Bull. Isr. Explor. Soc.* **28**, 43–54.
Schiffman, A. 1965. Energy measurements in the swash–surf zone. *Limnol. Oceanog.* **10**, 255–60.
Schlichting, H. 1968. *Boundary-layer theory*. New York: McGraw-Hill.
Schneider, H. E. 1970. Problems of quartz grain morphology. *Sedimentology* **14**, 325–35.
Schneider, W. J. 1961. A note on the accuracy of drainage densities computed from topographic maps. *J. Geophys. Res.* **66**, 3617–18.
Schroder, J. F. 1978. Dendrogeomorphological analysis of mass movement on Table Cliffs Plateau, Utah. *Quat. Res.* **9**, 168–85.
Schroder, L. J., R. A. Linthurst, J. E. Ellson and S. F. Vozzo 1985. Comparison of daily and weekly precipitation sampling efficiencies using automatic collectors. *Water, Air Soil Pollut.* **24**, 177–87.
Schumm, S. A. 1956. The evolution of drainage systems and slopes in badlands at Perth Amboy, New Jersey. *Geol. Soc. Am. Bull.* **67**, 597–646.
Schumm, S. A. 1960. The effect of sediment type on the shape and stratification of some modern fluvial deposits. *Am. J. Sci.* **258**, 177–84.
Schumm, S. A. 1965. Quaternary palaeohydrology. In *The Quaternary of the United States*, H. E. Wright, Jr and D. G. Frey (eds), 738–94. Princeton, NJ: Princeton University Press.
Schumm, S. A. 1970. Experimental studies on the formation of lunar surface features by fluidization. *Geol. Soc. Am. Bull.* **81**, 2539–52.
Schumm, S. A. 1973. Geomorphic thresholds and complex response of drainage systems. In *Fluvial geomorphology*, M. Morisawa (ed.), 299–310. Binghamton, NY: State University of New York.
Schumm, S. A. 1977a. *The fluvial system*. London: Wiley.
Schumm, S. A. 1977b. *Drainage basin morphology*. Benchmark Papers Geol. no. 41. Stroudsburg, Pa.: Dowden, Hutchinson & Ross.
Schumm, S. A. and R. J. Chorley 1964. The fall of Threatening Rock. *Am. J. Sci.* **262**, 1041–54.
Schumm, S. A. and R. J. Chorley 1966. Talus weathering and scarp recession in the Colorado Plateaus. *Z. Geomorphol. NF* **10**, 11–36.
Schumm, S. A. and R. F. Hadley 1961. *Progress in the application of landform analysis in the studies of semi-arid erosion* US Geol. Surv. Circ. no. 437.
Schumm, S. A. and H. R. Khan 1971. Experimental study of channel patterns. *Nature* **233**, 407–9.
Schumm, S. A. and H. R. Khan 1973. Experimental study of channel patterns. *Geol. Soc., Am. Bull.* **83**, 1755–70.
Schwarcz, H. P. 1977. Uranium series dating of travertine from Palaeolithic sites. *10th INQUA cong. (Abstracts)*, 409. Birmingham.
Schwarcz, H. P. 1980. Absolute age determination of archaeological sites by uranium dating of travertines. *Archaeometry* **22** (1), 3–24.
Schwarcz, H. P. 1982. Applications of U-series dating to archaeometry. In *Uranium series disequilibrium: applications to environmental problems*, M. Ivanovich and R. S. Harmon (eds), 302–25. Oxford: Oxford University Press.
Schwarcz, H. P. and C. Yonge 1983. Isotopic composition of paleowaters as inferred from speleothem and its fluid inclusions. In *Paleoclimates and paleowaters: a collection of environmental isotope studies*, R. Gonfiantini (ed.), 115–33. Proc. advisory group meeting. STI/PUB/621. Vienna: International Atomic Energy Agency.
Schwarzenbach, G. and H. Flaschka 1969. *Complexometric titrations*. London: Methuen.
Schwertmann, U., Murad, E. and D. G. Schulze 1982. Is there Holocene reddening (hematite formation) in soils of axeric temperate areas? *Geoderma* **27**, 209–23.
Scoggins, J. R. 1964. Aerodynamics of spherical balloon wind sensors. *J. Geophys. Res.* **69**, 591–8.

Scoggins, J. R. 1965. Spherical balloon wind sensor behaviour. *J. Appl. Meteorol.* **4**, 139–45.

Scott, C. R. 1974. *An introduction to soil mechanics and foundations*. London: Applied Science.

Seibert, P. 1974. Die Belastung der Pflanzendecke durch den Erholungsverkehr. *Forstwiss. Centralbl.* 35–43.

Selby, M. J. 1968. Morphometry of drainage basins in areas of pumice lithology. In *Proc. 5th New Zealand Geog. Soc.* 169–74.

Selby, M. J. 1970. Design of a hand-portable rainfall-simulating infiltrometer with results from the Otutira Catchment. *J. Hydrol. (NZ)* **9**, 117–32.

Self, F. S. and R. J. S. Sparks (eds) 1981. *Tephra studies*. Dordrecht: Reidel.

Seppäla, M. 1972. Stratigraphy and material of the loess layer at Mende, Hungary. *Bull. Geol. Soc. Finl.* **43**, 109–23.

Sethuraman, S. and R. M. Brown 1976. A comparison of turbulence measurements made by a hot-film probe, a bivane and a directional vane in the atmospheric surface layer. *J. Appl. Meteorol.* **15**, 138–44.

Setlow, L. W. and R. P. Karpovich 1972. 'Glacial' micro-textures on quartz and heavy mineral sand grains from the littoral environment. *J. Sedim. Petrol.* **42**, 864–75.

Shackleton, N. J. 1967. Oxygen isotope analyses and Pleistocene palaeotemperatures reassessed. *Nature* **215**, 15–17.

Shackleton, N. J. 1975. The stratigraphic record of deep-sea cores and its implications for the assessment of glacials, interglacials and interstadials in the Mid-Pleistocene. In *After the Australopithecines*, K. W. Butzer and G. L. Isaac (eds), 1–24. The Hague; Aldine.

Shackleton, N. J. 1988. Oxygen isotopes, ice volume and sea level. *Quat. Sci. Rev.* **6**, 183–90.

Shackleton, N. J. and N. D. Opdyke 1973. Oxygen isotope and palaeomagnetic stratigraphy of equatorial pacific core V28–238: oxygen isotope temperatures and ice volumes on a 10^5 and 10^6 year scale. *Quat. Res.* **3**, 39–55.

Shackleton, N. J. and N. D. Opdyke 1976. Oxygen isotope and palaeomagnetic stratigraphy of Pacific core V28–239: late Pliocene to latest Pleistocene. *Geol. Soc. Am. Mem.* **145**, 449–64.

Shakesby, R. A. and R. P. D. Walsh 1986. Micro-erosion meter measurements of erosion on limestone, Oxwich Point, Gower: some technical considerations and preliminary results. *Cambria* **13** (2), 213–34.

Shapiro, I. I. 1983. Use of space techniques for goedesy. In *Earthquakes: observation, theory and interpretation*, H. Kanamori and E. Boschi (eds), 530–68. Amsterdam: North-Holland.

Shapiro, J. 1958. The core-freezer: a new sampler for lake sediments. *Ecology* **39**, 758.

Sharp, R. P. 1949. Pleistocene ventifacts east of the Big Horn Mountains, Wyoming, *J. Geol.* **57**, 175–95.

Sharp, R. P. 1963. Wind ripples. *J. Geol.* **71**, 617–36.

Sharp, R. P. 1964. Wind-driven sand in Coachella Valley, California. *Bull. Geol. Soc. Am.* **75**, 785–804.

Sharp, R. P. 1966. Kelso dunes, Mohave Desert, California. *Bull. Geol. Soc. Am.* **77**, 1045–77.

Shaw, J. 1983. Drumlin formation related to inverted meltwater erosion marks. *J. Glaciol.* **29**, 461–79.

Sheets, R. G., R. L. Linder and R. B. Dahlgren 1971. Burrow systems of prairie dogs in South Dakota. *J. Mammalogy* **52**, 451–2.

Shennan, I. 1983. Flandrian and Late Devensian sea-level changes and crustal movements in England and Wales. In *Shorelines and isostasy*, D. E. Smith and A. G. Dawson (eds), 255–83. London: Academic Press.

Sheorei, P. R., D. Barat, M. N. Das, K. P. Mukherjee and B. Singh 1984. Schmidt Hammer rebound data for estimation of large scale *in situ* coal strength. *Int. J. Rock Mech. Min. Sci., Geomech. Abstr.* **21**, 39–42.

Shepard, F. P. 1950. *Longshore bars and longshore troughs*. BEB Tech. Memo. no. 15.

Shepard, F. P. 1963. Thirty-five thousand years of sea level. *Essays in marine geology*, 1–10. California: University of California Press.

Shepard, F. P. and D. L. Inman 1950. Nearshore circulation related to bottom topography and wave refraction. *Trans. Am. Geophys. Union* **31**, 196–212.

Shideler, G. L. 1976. A comparison of electronic particle counting and pipette techniques in routine mud analysis. *J. Sedim. Petrol.* **46**, 1017–25.

Shields, A. 1936. *Anwerdung der Aenlichkeitmechanik und Turbulenz für schung auf die Geschiebebewegang*. Mitteil. Preass. Versuchsanst. Wasser, Erd., Schiffsbau, Berlin 26.

Shotton, F. W. 1972. An example of hard-water error in radiocarbon dating of vegetable matter. *Nature* **240**, 460–1.

Shreve, R. L. 1967. Infinite topologically random channel networks. *J. Geol.* **75**, 178–86.

Shreve, R. L. 1974. Variation of mainstream length with basin area in river networks. *Water Resour. Res.* **10**, 1167–77.
Shreve, R. L. 1975. The probabilistic approach to drainage basin geomorphology. *Geology* **3**, 527–9.
Sieh, K. E. 1978. Slip along the San Andreas Fault associated with the great 1857 earthquake. *Seismol. Soc. Am. Bull.* **68**, 1421–48.
Sieh, K. E. and R. H. Jahns 1984. Holocene activity on the San Andreas Fault at Wallace Creek, California. *Geol. Soc. Am. Bull,* **95**, 883–96.
Sieveking, G. de. C. and M. B. Hart (eds) 1986. *The scientific study of flint and chert.* Cambridge: Cambridge University Press.
Sigafoos, R. S. 1964. *Botanical evidence of floods and floodplain deposition.* US Geol. Surv. Prof. Paper no. 485-A.
Silk, J. 1979. *Statistical concepts in geography.* London: Allen & Unwin.
Silverman, M. P. and E. F. Munoz 1970. Fungal attack on rock: solubilization and infrared spectra. *Science* **169**, 985–7.
Simola, H. 1977. Diatom succession in the formation of annually laminated sediment in Lovojarvi, a small eutrophicated lake. *Ann. Bot. Fenn.* **14**, 143–8.
Simola, H 1979. Micro-stratigraphy of sediment laminations deposited in a chemically-stratifying eutrophic lake during the years 1913–76. *Holarctic Ecol.* **2**, 160–8.
Simola, H., M. A. Coard and P. E. O'Sullivan 1981. Annual laminations in the sediment of Looe Pool. Cornwall, *Nature* **290**, 238–41.
Sinclair, P. C. 1969. General characteristics of dust devils. *J. Appl. Meteorol.* **8**, 43–5.
Singh, S. 1976. On the quantitative parameters for the computation of drainage density, texture and frequency: a case study of a part of Ranchi Plateau. *Natl Geog.* **11**, 21–33.
Singleton, G. A. and L. M. Lavkulich 1978. Adaptation of the Soxhlet extractor for pedologic studies. *J. Soil Sci. Soc. Am.* **42**, 984–6.
Sissons, J. B. 1962. A reinterpretation of the literature on late-glacial shorelines in Scotland with particular reference to the Forth Valley. *Trans. Edinb. Geol. Soc.* **19**, 83–8.
Sissons, J. B. 1966. Relative sea level changes between 10 300 and 8300 BP in part of the Carse of Stirling. *Trans. Inst. Br. Geog.* **39**, 19–30.
Sissons, J. B. 1967. *The evolution of Scotland's scenery.* Edinburgh: Oliver & Boyd.
Sissons, J. B. 1969. Drift stratigraphy and buried morphological features in the Grangemouth–Falkirk–Airth area, central Scotland. *Trans. Inst. Br. Geog.* **48**, 19–50.
Sissons, J. B. 1972. Dislocation and non-uniform uplift of raised shorelines in the western part of the Forth Valley. *Trans. Inst. Br. Geog.* **55**, 145–59.
Sissons, J. B. 1974. A late glacial ice cap in the central Grampians, Scotland. *Trans. Inst. Br. Geog.* **62**, 95–114.
Sissons, J. B. and D. E. Smith 1965. Raised beaches associated with the Perth Readvance in the Forth Valley and their relation to glacial isostacy. *Trans. R. Soc. Edinb.* **66**, 143–68.
Sissons, J. B., R. A. Cullingford and D. E. Smith 1966. Late-glacial and post-glacial shorelines in south-east Scotland. *Trans. Inst. Br. Geog.* **39**, 9–18.
Skärby, L. 1977. *Correlation of moss analysis to direct measurement by deposit gauge of cadmium, lead and copper.* Swed. Water Air Pollut. Res. Lab. Publ. no. B377a.
Skempton, A. W. 1953. Soil mechanics in relation to geology. *Proc. Yorks. Geol. Soc.* **29**, 33–62.
Skempton, A. W. 1964. Long-term stability of clay slopes. *Geotechnique* **14**, 77–101.
Skempton, A. W. and F. A. Delroy 1957. Stability of natural slopes in London Clay. In *Proc. 4th int. conf. soil mechanics and foundation engineering, 7th State-of-the-Art volume*, 291–340.
Skempton, A. W. and J. N. Hutchinson 1969. Stability of natural slopes and embankments. In *Proc. 7th int. conf. soil mechanics and foundation engineering*, 291–340.
Skempton, A. W. and D. J. Petley 1967. The strength along structural discontinuities in stiff clays. In *Proc. Geotech. Conf.*, vol. 2, 3–20. Oslo.
Skempton, A. W. and A. G. Weeks 1976. The Quaternary history of the Lower Greensand escarpment and Weald Clay Vale near Sevenoaks, Kent. *Phil. Trans. R. Soc. A* **283**, 493–526.
Skidmore, E. L. 1965. Assessing wind erosion forces: direction and relative magnitude. *Proc. Soil Sci. Soc. Am. 29*, 587–91.
Skidmore, E. L. and N. P. Woodruff 1968. *Wind erosion forces in the United States and their use in predicting soil loss.* US Dept Agric. Handbook no. 346.
Sklash, M. G., R. N. Farvolden and P. Fritz 1976. A conceptual model of watershed response to rainfall developed through the use of oxygen-18 as a natural tracer. *Can. J. Earth Sci.* **13**, 271–83.
Slack, K. V. and D. W. Fisher 1965. Light-dependent quality changes in stored water samples. *US Geol. Surv. Prof. Paper* **525-C**, 190–2.
Slater, C. S. 1957. Cylinder infiltrometers for determining rates of irrigation. *Proc. Soil Sci. Soc. Am.* **21**, 457–60.

Slater, J. and B. Ralph 1976. The determination of particle shape and size distributions using automatic image analysis techniques. In *Proc. 4th int. cong. on stereology*, 177–80. US Natl Bureau of Standards, Spec. Publ. no. 431. Gaithersburg.

Slaymaker, H. O. 1980. Geomorphic field experiments: inventory and prospects. *Z. Geomorphol. NF* **35**, 183–94.

Small, R. J. and G. C. Fisher 1970. The origin of the secondary escarpment of the South Downs. *Trans. Inst. Br. Geog.* **49**, 97–107.

Smalley, I. J. 1964. Representation of packing in a clastic sediment. *Am. J. Sci.* **262**, 242–8.

Smalley, I. J. 1971. 'In-situ' theories of loess formation and the significance of the calcium carbonate content of loess. *Earth Sci. Rev.* **7**, 67–85.

Smalley, I. J. 1978. Loess deposits associated with deserts. *Catena* **5**, 53–66.

Smalley, I. J. and D. J. Unwin 1968. The formation and shape of drumlins and their distribution and orientation in drumlin fields. *J. Glaciol.* **7**, 377–90.

Smalley, I. J., S. P. Bentley and C. F. Moon 1975. The St Jean Vianney quickclay. *Can. Mineralog.* **13**, 364–9.

Smart, J. S. 1970. Use of topologic information in processing data for channel networks. *Water Resourc. Res.* **6**, 932–6.

Smart, J. S. 1972. Channel networks. *Adv. Hydrosci.* **8**, 305–46.

Smart, J. S. 1978. The analysis of drainage network composition. *Earth Surf. Processes* **3**, 129–70.

Smart, J. S. and C. Werner 1976. Application of the random model of drainage basin composition. *Earth Surf. Processes* **1**, 219–33.

Smart, P. L. and I. M. S. Laidlaw 1977. An evaluation of some fluorescent dyes for water tracing. *Water Resourc. Res.* **13**, 15–33.

Smart, P. L. and C. M. Wilson 1984. Two methods for the tracing of pipeflow on hillslopes. *Catena* **11**, 158–68.

Smart, P. L., T. C. Atkinson, I. M. S. Laidlaw, M. D. Newson and S. T. Trudgill 1986. Comparison of the results of quantitative and non-quantitative tracer tests for determination of karst conduit networks: an example from the Traligill Basin, Scotland. *Earth Surf. Processes Landforms* **11**, 249–61.

Smettem, K. R. J. and S. T. Trudgill 1983. An evaluation of some fluorescent and non-fluorescent dyes in the identification of water transmission routes in soils. *J. Soil Sci.* **34**, 45–56.

Smith, A. G. and J. R. Pilcher 1973. Radiocarbon dating and vegetational history of the British Isles. *New Phytol.* **72**, 903–14.

Smith, A. G., J. R. Pilcher and G. Singh 1968. A large capacity hand-operated peat sampler. *New Phytol.* **67**, 119–24.

Smith, B. J. 1977. Rock temperature measurements from the north west Sahara and their implications for rock weathering. *Catena* **4**, 41–63.

Smith, B. N. and S. Epstein 1971. Two categories of $^{13}C/^{12}C$ ratios for higher plants. *Plant Physiol.* **47**, 380–4.

Smith, D. and W. Harrison 1970. Water current meter for mean flow measurement. In *Proc. 12th conf. coastal engineering*, 1904–15. Am. Soc. Civ. Eng.

Smith, D. B. 1973. The use of artificial radioactive tracers in the United Kingdom. In *Tracer techniques in sediment transport*, 97–104. Vienna: International Atomic Energy Agency.

Smith, D. B., R. A. Downing, R. A. Monkhouse, R. L. Otlet and F. J. Pearson 1976. The age of groundwater in chalk of the London basin. *Water Resour. Res.* **12**, 392.

Smith, D. E., R. Kolenkiewicz, P. J. Dunn and M. H. Torrence 1979. The measurement of fault motion by satellite laser ranging. *Tectonophysics* **52**, 59–67.

Smith, D. E., J. B. Sissons and R. A. Cullingford 1969. Isobases for the Main Perth raised shoreline in south-east Scotland as determined by trend surface analysis. *Trans. Inst. Br. Geog.* **46**, 45–52.

Smith, D. I. and T. C. Atkinson 1977. Underground flow in cavernous limestones with special reference to the Malham area. *Field Stud.* **4**, 597–616.

Smith, G. N. 1978. *Elements of soil mechanics for civil and mining engineers*, 4th edn. St Albans: Crossby Lockwood Staples.

Smith, J. D. 1969. Geomorphology of a sand ridge. *J. Geol.* **77**, 39–55.

Smith, J. D. and J. H. Foster 1969. Geomagnetic reversals in Brunhes normal polarity epoch. *Science* **163**, 565–6.

Smith, K. and M. E. Lavis 1969. A prefabricated flat-vee weir for experimental catchments. *J. Hydrol.* **8**, 217–26.

Smith, K. G. 1950. Standards for grading texture of erosional topography. *Am. J. Sci.* **248**, 655–68.

Smith, N. D. 1970. The braided stream depositional environment; comparison of the Platte river with some Silurian clastic rocks, N. Central Appalachians. *Bull.Geol. Soc. Am.* **81**, 2993–3014.

Smith, N. D. 1971. Transverse bars and braiding in the lower Platte River, Nebraska. *Geol. Soc. Am. Bull.* **82**, 3407–20.

Smith, N. D. 1972. Flume experiments on the durability of mud clasts. *J. Sedim. Petrol.* **42**, 378–83.

Smith, N. D., M. A. Vendle and S. K. Kennedy 1982. Comparison of sedimentation regimes in four glacier-fed lakes of western Alberta. In *Research in glacial, glaciofluvial and glaciolacustrine systems*, R. Davidson-Arnott, W. Nickling and B. D. Fahey (eds), 203–38. Norwich: Geo Books.

Smith, R. K. and L. M. Leslie 1976. Thermally driven vortices. a numerical study with applications to dust-devil dynamics. *Q. J. R. Meteorol Soc.* **102**, 791–804.

Smith, R. M. and P. C. Twiss 1965. Extensive gaging of dust deposition rates. *Trans. Kansas Acad. Sci.* **68**, 311–21.

Sneed, E. D. and R. L. Folk 1958. Pebbles in Lower Colorado River, Texas: a study in particle morphogenesis. *J. Geol.* **66**, 114–50.

Sollid, J. L., S. Andersen, N. Hamre, O. Kjeldsen, O. Salvigsen, S. Sturod, T. Tveita and A. Wilhelmsen 1973. Deglaciation of Finnmark, north Norway. *Nor. Geog. Tidsskr.* **27**, 233–325.

Soons, J. M. 1975. The geomorphological map: an aid to geographical comprehension. *NZ J. Geog.* **59**, 25–30.

Souchez, R. A., R. D. Lorrain and M. M. Lemmens 1973. Refreezing of interstitial water in a subglacial cavity of an alpine glacier as indicated by the chemical composition of ice. *J. Glaciol.* **12**, 453–9.

Soulliere, E. J. and T. J. Toy 1986. Rilling of hillslopes reclaimed before 1977 surface mining law, Dave Johnson mine, Wyoming. *Earth Surf. Processes and Landforms* **11**, 293–305.

Spalding, R. F. and J. D. Matthews 1972. Submerged stalagmites from caves in the Bahamas: indicators of low-sea-level stand. *Quat. Res.* **2**, 470–2.

Sparks, B. W. 1949, The denudation chronology of the dip slope of the South Downs. *Proc. Geol. Assoc.* **60**, 165–215.

Sparks, B. W. 1953. The effects of weather on the determination of heights by aneroid barometer in Great Britain. *Geog. J.* **119**, 73–80.

Sparny, K. R., J. P. Lodge, Jr, E. R. Frank and D. C. Sheesley 1969. Aerosol filtration by means of nucleopore filters: structural and filtration properties. *Environ. Sci. Technol.* **3**, 453–64.

Spate, A. P., J. N. Jennings, D. I. Smith and M. A. Greenaway 1985. The micro-erosion meter: use and limitations. *Earth Surf. Processes Landforms* **10**, 427–40.

Speight, J. G. 1965. Meander spectra of the Angabunga River. *J. Hydrol.* **3**, 1–15.

Speight, J. G. 1968. Parametric description of land form. In *Land evaluation*, G. A. Stewart (ed.), 239–50. Proc. CSIRO Symp. London: Macmillan.

Speight, J. G. 1971. Log normality of slope distributions. *Z. Geomorphol. NF* **15**, 290–311.

Sperling, C. H. B. and R. U. Cooke 1985. Laboratory simulation of rock weathering by salt crystallisation and hydration in hot, arid environments. *Earth Surf. Processes Landforms* **10**, 541–56.

Sperling, C. H. B., A. S. Goudie, D. R. Stoddart and G. G. Poole 1977. Dolines of the Dorset chalklands and other areas in southern Britain. *Trans. Inst. Br. Geog. N S* **2**, 205–23.

Splettstoesser, J. F. (ed.) 1976. *Ice-core drilling*. Proc. Symp., 28–30 Aug. 1974. Lincoln: University of Nebraska Press.

Splettstoesser, J. F. 1985. Note on rock striations caused by penguin feet, Falkland Islands. *Arctic Alpine Res.* **17**, 107–11.

Staff, M. D. 1978. Rill patterns derived from air photographs of the Grwyne Fechan catchment, Black Mountains. *Cambria 5*, 22–36.

Statham, I. 1973. Scree slope development under conditions of surface particle movement. *Trans. Inst. Br. Geog.* **59**, 41–53.

Statham, I. 1975. Some limitations to the application of the concept of angle of repose to natural hillslopes. *Area* **7**, 264–8.

Statham, I. 1976. Photographic filtering for extracting orientation and aspect data from maps. *Area* **8**, 209–12.

Statham, I. 1977. *Earth surface sediment transport*. Oxford: Clarendon.

Steele, T. D. 1968. Digital-computer applications in chemical-quality studies of surface water in a small watershed. *Int. Assoc. Sci. Hydrol. Publ.* **80**, 203–14.

Steen, B. 1979a. A possible method for the sampling of Saharan dust. In *Saharan dust*, C. Morales (ed.), 279–86. Chichester: Wiley.

Steen, B. 1979b. Techniques for measuring dry deposition. Summary of WMO Expert Meeting on Dry Deposition, 18–22 Apr. 1977, Gothenburg. In *Saharan dust*, C. Morales (Ed.), 287–9. Chichester: Wiley.

Steers, J. A. 1951. Notes on the erosion along the coast of Suffolk. *Geol. Mag.* **88**, 435–9.

Stein, R. 1985. Rapid grain-size analyses of clay and silt fraction by SediGraph 5000D: comparison with Coulter Counter and Atterberg methods. *J. Sedim. Petrol.* **55**, 590–3.
Sternberg,. R. W. 1968. Friction factors in tidal channels with differing bed roughness. *Mar. Geol.* **6**, 243–60.
Sternberg,. R. W. 1970. Field measurements of hydro-dynamic roughness of the deep-sea boundary. *Deep-Sea Res.* **17**, 413–20.
Sternberg, R. W. 1972. Predicting initial motion of bed-load transport of sediment particles in the shallow marine environment. In *Shelf sediment transport*, D. J. P. Smith, D. B. Daune and O. H. Pilkey (eds), ch. 3, 61–82. Stroudsburg, Pa.: Dowden, Hutchinson & Ross.
Sternberg, R. W., R. V. Johnson, D. A. Cacchione and D. E. Drake 1986. An instrument system for monitoring and sampling suspended sediment in the benthic boundary layer. *Mar. Geol.* **71** (3–4), 187–99.
Stevens, P. A. 1981a. *A bulk precipitation sampler for use in a geochemical cycling project*. ITE Bangor, Occas. Paper no. 7.
Stevens, P. A. 1981b. *Modification and operation of ceramic cup soil samplers for use in a geochemical cycling study*. ITE Bangor, Occas. Paper no. 8.
Stevenson, C. M. 1968. An analysis of the chemical composition of rainwater and air over the British Isles and Eire for the years 1959–64. *Q. J. Meteorol. Soc.* **94**, 56–70.
Stober, J. C. and R. Thompson 1977. Palaeomagnetic secular variation studies on Finnish lake sediments and the carriers of remanence. *Earth Planet. Sci. Lett.* **37**, 139–49.
Stoddart, D. R. 1965. The shape of atolls. *Mar. Geol.* **3**, 369–83.
Stoddart, D. R. 1967. Organism and ecosystem as geographical models. In *Models in geography*, R. J. Chorley and P. Haggett (eds), 511–48. London: Methuen.
Stoddart, D. R. 1986. *On Geography*. Oxford: Basil Blackwell
Stokes, M. A. and T. L. Smiley 1968. *An introduction to tree-ring dating*. Chicago: University of Chicago Press.
Stork, A. 1963. Plant immigration in front of retreating glaciers with examples from Kebnekajse area, northern Sweden. *Geog. Annaler* **45**, 1–22.
Stout, J. D., K. M. Goh and T. A. Rafter 1981. Chemistry and turnover of naturally occurring resistant organic compounds in soil. In *Soil biochemistry*, vol. 5, E. A. Paul and J. N. Ladd (eds), 1–73. New York: Marcel Dekker.
Strahler, A. N. 1950. Equilibrium theory of erosional slopes approached by frequency distribution analysis. *Am. J. Sci.* **248**, 673–96, 800–14.
Strahler, A. N. 1952. Hypsometric (area–altitude) analysis of erosional topography. *Geol. Soc. Am. Bull.* **63**, 1117–42.
Strahler, A. N. 1956. Quantitative slope analysis. *Geol. Soc. Am. Bull.* **67**, 571–96.
Strahler, A. N. 1957. Quantitative analysis of watershed geomorphology. *Trans. Am. Geophys. Union* **38**, 913–20.
Strahler, A. N. 1964. Quantitative geomorphology of drainage basins and channel networks. In *Handbook of applied hydrology*, V. T. Chow (ed.), 4.39–76. New York: McGraw-Hill.
Strahler, A. N. 1966. Tidal cycle of changes on an equilibrium beach. *J. Geol.* **74**, 247–68.
Striem, H. L. and J. Tadmor 1974. The improvement in the calculated annual wind frequency distribution at a coastal site in Israel due to observations in consecutive years. *Isr. J. Earth Sci.* **23**, 23–34.
Stromberg, B. 1983. The Swedish varve chronology. In *Glacial deposits in north-west Europe*, J. Ehlkers (ed.), 97–105. Rotterdam: Balkema.
Stuiver, M. 1978. Radiocarbon timescale tested against magnetic and other dating methods. *Nature* **273**, 271–4.
Stuvier, M. and G. W. Pearson 1986. High precision calibration of the radiocarbon time scale, AD 1950–500 BC. *Radiocarbon* **28**, 805–38.
Sturm, M. 1979. Origin and composition of clastic varves. In *Moraines and varves*, C. Schluchter (ed.), 281–5. Rotterdam: Balkema.
Sudo, T. 1974. *Clay mineralogy* (in Japanese). Tokyo: Iwanami Shoten.
Sudo, T., K. Oinuma and K. Kobayashi 1961. Mineralogical problems concerning rapid clay mineral analysis of sedimentary rocks. *Acta Univ. Carolinae, Geol. Suppl.* **1**, 189–219.
Sugden, D. E. 1969. The age and form of corries in the Cairngorms. *Scot. Geog. Mag.* **85**, 34–46.
Summerfield, M. A. 1976. Slope form and basal stream relationships: a case study in the Westend basin of the southern Pennines, England. *Earth Surf. Processes* **1**, 89–96.
Summers, H. J., H. D. Palmer and D. O. Cook 1971. Some simple devices for the study of wave-induced surges. *J. Sedim. Petrol.* **41**, 861–6.
Summers, W. K. 1972. Factors affecting the validity of chemical analyses of natural waters. *Groundwater* **10**, 12–17.

Sutherland, D. G. 1980. Problems of radiocarbon dating of deposits from newly deglaciated terrain: examples from the Scottish Late glacial. In *Studies in the Lateglacial of North-west Europe*, J. J. Lowe, J. M. Gray and J. E. Robinson (Eds), 139–49. Oxford: Pergamon.

Sutherland, D. G. 1983. The dating of former shorelines. In *Shorelines and isostasy*, D. E. Smith and A. G. Dawson (eds), 129–57. London: Academic Press.

Sutherland, D. G. 1984. Geomorphology and mineral exploration: some examples from exploration for diamondiferous placer deposits. *Z. Geomorphol. Supp. NF* **51**, 95–108.

Sutherland, D. G. 1986. A review of Scottish marine shell radiocarbon dates, their standardization and interpretation. *Scot. J. Geol.* **22**, 145–64.

Svensson, H. 1956. Method for exact characterizing of denudation surfaces, especially peneplains, as to the position in space. *Lund Stud. Geog. Ser. A* **8**, 1–5.

Swain, A. M. 1973. A history of fire and vegetation in northeastern Minnesota as recorded in lake sediments. *Quat. Res.* **3**, 383–96.

Swank, W. T. and J. E. Douglass 1977. Nutrient budgets for undisturbed and manipulated hardwood forest ecosystems in the mountains of North Carolina. In *Watershed research in eastern North America*, D. L. Correll (ed.), vol. 1, 343–62. Edgewater: Smithsonian Institute.

Swank, W. T. and G. S. Henderson 1976. Atmospheric inputs of some cations and anions to forest ecosystems in North Carolina and Tennessee. *Water Resour. Res.* **12**, 541–55.

Sweeting, M. M. 1972. *Karst landforms*. London: Macmillan.

Sweeting, M. M. and G. S. Sweeting 1969. Some aspects of the Carboniferous Limestone in relation to its landforms with particular reference to NW Yorkshire and County Clare. *Méditerranée* **7**, 201–9.

Swift, D. J. P. 1975. Tidal sand ridges and shoal-retreat massifs. *Mar. Geol.* **18**, 105–34.

Swift, J. A. 1970. *Electron microscopes*. London: Kogan Page.

Swinnerton, H. H. 1931. The post-glacial deposits of the Lincolnshire coast. *Q. J. Geol. Soc.* **87**, 360–75.

Synge, F. M. 1979. The raised shorelines and deglaciation chronology of Inari, Finland, and south Varanger, Norway. *Geog. Annaler* **51A**, 193–206.

Szabo, B. J. and J. N. Rosholt 1982. Surficial continental sediments. In *Uranium series disequilibrium: applications to environmental problems*, 246–67. Oxford: Oxford University Press.

Taber, S. 1930. The mechanics of frost heaving. *J.Geol.* **38**, 303–17.

Taira, A. and P. A. Scholle 1977. Design and calibration of a photo-extinction settling tube for grain size analysis. *J.Sedim.Petrol.* **47**, 1347–60.

Talsma, T., P. M. Hallam and R. S. Mansell 1980. Evaluation of porous cup soil water extractors: physical factors. *Aust. J.Soil Res.* **18**.

Tamm, C. O. 1974. Leaching of plant nutrients from soils as a consequence of forestry operations. *Ambio* **3**, 211–21.

Tanji, K. K. and J. W. Biggar 1972. Specific conductance model for natural waters and soil solutions of limited salinity levels. *Water Resour. Res.* **8**, 145–53.

Tanner, V. 1933. L'étude des terrasses littorales en Fenno-scandie et l'homotaxie intercontinental. *C. R. Cong.Int.Géog.,Paris 1931*, **2**, 61–76.

Tanner, W. F. 1959. Examples of departure from the Gaussian in geomorphic analysis. *Am.J.Sci.* **257**, 458–60.

Tanner, W. F. 1960. Numerical comparison of geomorphic samples. *Science* **131**, 1525–6.

Tarrant, J. R. 1970. Comments on the use of trend surface analysis in the study of erosion surfaces. *Trans. Inst.Br.Geog.* **51**, 221–2.

Task Committee on Preparation of Sedimentation Manual 1970. Sediment measurement techniques. E: Airborne sediment. *Proc.Am.Soc.Civ. Eng., J.Hydraul. Div.* **96**, **HY1**, 29–41.

Taylor-George, S., G. Palmer, J. T. Staley, D. J. Borns, B. Curtis and J. B. Adams 1983. Fungi and bacteria involved in desert varnish formation. *Microbiol. Ecol.* **9**, 227–45.

Teleki, P. G. 1966. Fluorescent sand tracers. *J.Sedim.Petrol.* **36**, 376, 468.

Teleki, P. G. 1967. Automatic analysis of tracer sand. *J.Sedim.Petrol.* **37**, 747.

Ten Cate, J. A. M. 1983. Detailed systematic geomorphological mapping in the Netherlands and its applications. *Geol. Mijnbouw* **62**, 611–20.

Ternan, J. L. and A. L. Murgatroyd 1984. The role of vegetation in baseflow, suspended sediment and specific conductance in granite catchments, SW England. In *Channel processes – water, sediment and catchment controls*, A. P. Schick (ed.). *Catena* Supplement no. 5, Braunschweig.

Ternan, J. L. and A. G. Williams 1979. Hydrological pathways and granite weathering on Dartmoor. In *Geographical approaches to fluvial processes*, A. F. Pitty (ed.), 5–30. Norwich: Geo Books.

Terzaghi, K. 1925. Principles of soil mechanics. *Eng. News Record* **95**, 742.

Terzaghi, K. 1936. The shearing resistance of saturated soils and the angle between the planes of shear. In *Proc.1st int.conf.soil mechanics and foundation engineering*.

Terzaghi, K. 1950. The mechanism of landslides. *Bull.Geol. Soc.Am.* Berkeley Vol., 83–122.

Terzaghi, K. and R. B. Peck 1967. *Soil mechanics in engineering practice*. New York: Wiley.

Thellier, E. 1981. Sur la direction du champ magnetique terrestre, en France, durant les deux derniers millenaires. *Phys.Earth Planet.Interiors* **24**, 89–132.

Thom, B. G. (ed.) 1984. *Coastal geomorphology in Australia*. New York: Academic Press.

Thomas, A. G. 1985. Specific conductance as an indicator of total dissolved solids in cold, dilute waters. *J.Hydrol.Sci.* **31** (1), 81–92.

Thomas, K. W. 1964. A new design for a peat sampler. *New Phytol.* **63**, 422–5.

Thompson, L. G. 1977a. *Microparticles, ice sheets and climate*. Inst. Polar Stud. Rep. no. 64, Ohio State Univ.

Thompson, O. R. Y. 1977b. Laboratory study of sea straits. *J. Fluid Mech.* **81**, 341–51.

Thompson, P., H. P. Schwarcz and D. C. Ford 1976. Stable isotope geochemistry, geothermometry and geochronology of speleothems from West Virginia. *Bull.Geol. Soc.Am.* **87**, 1730–8.

Thompson, R. 1973. Palaeolimnology and palaeomagnetism. *Nature, Phys. Sci.* **242**, 182–4.

Thompson, R. 1975. Long period European geomagnetic secular variation confirmed. *Geophys. J.R.Astron. Soc.* **43**, 847–59.

Thompson, R. 1977c. Stratigraphic consequences of palaeomagnetic studies of Pleistocene and Recent sediments. *J.Geol. Soc.Lond.* **133**, 51–9.

Thompson, R. 1983a. ^{14}C dating and magnetostratigraphy. *Radiocarbon* **25**, 229–38.

Thompson, R. 1983b. Global holocene magnetostratigraphy. *Hydrobiologia* **103**, 45–51.

Thompson, R. 1984. A global review of palaeomagnetic results from wet lake sediments. In *Lake sediments and environmental history*, E. Y. Haworth and J. W. G. Lund (eds), 146–64. Leicester: Leicester University Press.

Thompson, R. 1986. Palaeomagnetic dating. In *Handbook of Holocene palaeoecology and palaeohydrology*, B. E. Berglund (ed.), 313–27. Chichester: Wiley.

Thompson, R. and D. R. Barraclough 1982. Geomagnetic secular variation based on spherical harmonic and cross validation of historical and archaeomagnetic data. *J.Geomag.Geoelectr.* **34**, 245–63.

Thompson, R. and B. Berglund 1976. Late Weichselian geomagnetic 'reversal' as a possible example of the reinforcement syndrome. *Nature* **263**, 490–1.

Thompson, R. and K. J. Edwards 1982. Magnetic studies at Lough Catherine and an Irish Holocene geomagnetic master curve. *Boreas* **11**, 335–49.

Thompson, R. and F. Oldfield 1978. Evidence for recent palaeomagnetic secular variation in lake sediments from the New Guinea Highlands. *Phys. Earth Planet. Interiors* **17**, 300–6.

Thompson, R. and F. Oldfield 1986. *Environmental magnetism*. London: Allen & Unwin.

Thompson, R. and G. M. Turner 1979. British geomagnetic master curve 10 000–0 yr BP from dating European sediments. *Geophys. Res. Lett.* **6**, 249–52.

Thompson, R., R. W. Battarbee, P. E. O'Sullivan and F. Oldfield 1975. Magnetic susceptibility of lake sediments. *Limnol.Oceanog.* **20**, 687–98.

Thompson, R., J. C. Stober, G. M. Turner, F. Oldfield, J. A. Dearing, T. A. Rummery, J. Bloemendal and K. Tolonen 1980. Environmental applications of magnetic measurements. *Science* **207**, 481–6.

Thompson, W. C. and J. C. Harlett 1969. The effect of waves on the profile of a natural beach. In *Proc.11th conf.coastal engineering*, vol. 1, 353–72. Am.Soc.Civ.Eng.

Thorn, C. F. 1979. Bedrock freeze–thaw weathering regime in an alpine environment, Colorado Front Range. *Earth Surf. Processes* **4**, 211–28.

Thorne, C. R. and J. Lewin 1979. Bank processes, bed material movement and planform development in a meandering river. In *River equilibrium and channel adjustment*, D. D. Rhodes and G. P. Williams (eds). Binghamton, NY: Geomorphological Publications.

Thorne, P. D. 1986. An intercomparison between visual and acoustic detection of seabed gravel movement. *Mar. Geol.* **72** (1–2), 11–31.

Thornes, J. B. 1972. Debris slopes as series. *Arctic Alpine Res.* **4**, 337–42.

Thornes, J. B. 1973. Markov chains and slope series: the scale problem. *Geog.Anal.* **5**, 322–8.

Thornes, J. B. and D. Brunsden 1977. *Geomorphology and time*. London: Methuen.

Thornton, E. B. 1972. Distribution of sediment across the surf zone. In *Proc.13th conf.coastal engineering*, vol. 2, 1049–68. Am.Soc.Civ.Eng.

Thornton, P. R. 1968. *Scanning electron microscopy: applications to materials and device science*. London: Chapman & Hall.

Thorp, J. 1965. The nature of the pedological record in the Quaternary. *Soil Sci.* **99**, 1–8.

Thorstenson, D. C., F. T. Mackenzie and B. L. Ristvet 1972. Experimental vadose and phreatic cementation of skeletal carbonate sand. *J.Sedim.Petrol.* **42**, 162–7.
Thurber, D. L., W. S. Broecker, R. L. Blanchard and H. A. Potratz 1965. Uranium series ages of Pacific Atoll coral. *Science* **149**, 55–8.
Thurman, E. M. 1984. *Determination of aquatic humic substances in natural waters*. US Geol. Surv. Water Supply Paper no. 2262.
Tickell, F. G. 1965. The techniques of sedimentary mineralogy. *Developments in sedimentology*, vol. 4. Amsterdam: Elsevier.
Tinkler, K. J. 1971. Statistical analysis of tectonic patterns in areal volcanism: the Bunyaruguru volcanic field in west Uganda. *Math. Geol.* **3**, 335–55.
Tobler, W. R. 1969a. Geographical filters and their inverses. *Geog.Anal.* **1**, 234–53.
Tobler, W. R. 1969b. An analysis of a digitalized surface. In *A study of the land type*, C. M. Davis (ed.), 59–76, 86. Report on contract DA-31-124-ARO-D-456 for US Army Research Office, Department of Geology, University of Michigan (Ann Arbor).
Tolonen, M. 1978. Palaeoecology of annually laminated sediments in Lake Ahvenainen, S. Finland. *Ann. Bot. Fenn.* **15**, 177–240.
Toms, R. G., D. C. Hinge and J. Austin 1973. Experimental work on the continuous monitoring of water quality. *Water Pollut. Control* **72**, 20–33.
Tooley, M. J. 1974. Sea level changes during the last 9000 years in north west England. *Geog.J.* **140**, 18–42.
Tooley, M. J. 1978a. *Sea-level changes: north west England during the Flandrian stage*. Oxford: Clarendon.
Tooley, M. J. 1978b. Holocene sea-level changes: problems of interpretation. *Geol. Foren. Stockholm Forhandl.* **100**, 203–12.
Tooley, M. J. 1982. Sea-level changes in northern England. *Proc. Geol. Assoc* **93**, 43–51.
Tooley, M. J. 1985a. Climate, sea-level and coastal changes. In *The climatic scene. Essays in honour of Gordon Manley*, M. J. Tooley and G. M. Sheail (eds), 206–34. London: Allen & Unwin.
Tooley, M. J. 1985b. Sea-level changes and coastal morphology in north-west England. In *The geomorphology of north-west England*, R. H. Johnson (ed.), 94–121. Manchester: Manchester University Press.
Tooley, M. J. and I. Shennan 1987. *Sea-level changes*. Oxford: Basil Blackwell.
Topfer, F. and W. Pillewizer 1966. The principles of selection. *Cartog.J.* **3**, 10–16.
Torrenueva, A. L. 1975. *Variation in mineral flux to the forest floors of a pine and a hardwood stand in the Georgia piedmont*. Unpubl.Ph.D. Thesis, Athens, Georgia.
Tovey, N. K. 1971. *A selection of scanning electron micrographs of clays*. Univ. Cambridge, Dept Eng. Rep. no. CUED/C-SOILS/TR5a.
Tovey, N. K. and K. Y. Wong 1978. Optical techniques for analysing scanning electron micrographs. In *Scanning electron microscopy 1978*, O. Johari (ed.), vol. 1, 381–92. Chicago: SEM.
Tovey, N. K., N. Eyles and R. Turner 1978. Sand grain selection procedures for observations on the SEM. In *Scanning electron microscopy 1978*, O. Johari (ed.), vol. 1, 393–400. Chicago: SEM.
Townshend, J. R. G. (ed.) 1981. *Terrain analysis and remote sensing*. London: Allen & Unwin.
Trembour, F. and I. Friedman 1984. The present status of obsidian hydration dating. In *Quaternary Dating Methods*, W. C. Mahanney (ed.), 141–51. Amsterdam: Elsevier.
Trenhaile, A. S. 1971. Drumlins: their distribution, orientation and morphology. *Can.Geog.* **15**, 113–26.
Trenhaile, A. S. 1975a. The morphology of a drumlin field. *Ann.Assoc. Am.Geog.* **65**, 297–312.
Trenhaile, A. S. 1975b. Cirque elevation in the Canadian Cordillera. *Ann.Assoc. Am.Geog.* **65**, 517–29.
Trenhaile, A. S. 1987. *The geomorphology of rock coasts*. Oxford: Clarendon.
Tricart, J. 1956. Etude expérimentale du problème de la gélivation. *Biulteyn Peryglacjalny* **4**, 285–318.
Tricart, J. 1965. La cartographie géomorphologique detailée. In *Principes et méthodes de la géomorphologie*, 182–215. Paris: Masson.
Tricart, J. 1970. Principes de réalisation de la carte géomorphologique detaillée de la France. *Bull.Inst.Géog.Acad. Bulg. Sci.* **14**, 61–74.
Tricart, J. 1972. *Cartographie géomorphologique: travaux de la RCP77*. Mém. Doc. 1971, NS no. 12, Serv. Doc. Cartog. Géog., CNRS, Paris.
Trimble, G. R. and S. Weitzmann 1954. Effect of a hardwood forest canopy on rainfall intensities. *Trans. Am.Geophys. Union* **35**, 226–34.
Trimble, S. W. 1977. The fallacy of stream equilibrium in contemporary denudation studies. *Am.J.Sci.* **277**, 876–87.
Troeh, F. R. 1965. Landform equations fitted to contour maps. *Am.J.Sci.* **263**, 616–27.

Troels-Smith, J. 1955. *Karakterisering af lose jordater. Characterization of unconsolidated sediments*. Danm. Geol. Unders. Ser.IV no. 3.

Trollope, D. H. 1968. The mechanics of discontinua or clastic mechanics in rock problems. In *Rock mechanics in engineering practice*, K. C. Stagg and O. C. Kienkiewicz (eds), 275–320. New York: Wiley.

Trudgill, S. T. 1975. Measurement of erosional weight loss of rock tablets. *BGRG Tech.Bull.* **17**, 13–19.

Trudgill, S. T. 1977. Problems in the estimation of short-term variations in limestone erosion processes. *Earth Surf. Processes* **2**, 251–6.

Trudgill, S. T. (ed.) 1986. *Solute processes*. Chichester: Wiley.

Trudgill, S. T. and D. J. Briggs 1978. Soil and land potential. *Prog.Phys.Geog.* **2**, 321–32.

Trudgill, S. T., R. W. Crabtree, A. M. Pickles, K. R. J. Smettem and T. P. Burt 1984. Hydrology and solute uptake in hillslope soils on Magnesian Limestone: the Whitwell Wood project. In *Catchment experiments in fluvial geomorphology*, T. P. Burt and D. E. Walling (eds), 183–215. Norwich: Geo Books.

Trudgill, S. T., C. J. High and F. K. Hanna 1981. Improvements to the micro-erosion meter. *BGRG Tech.Bull.* **29**, 3–17.

Tsoar, H. 1974. Desert dunes, morphology and dynamics, El Arish, northern Sinai. *Z. Geomorphol. Supp. NF* **20**, 41–61.

Tsoar, H. 1975. Specific sampling of ripples and micro-features on desert dunes. In *Proc.9th int.cong.sedimentology*, vol. 3, 101–5. Nice.

Tsoar, H. 1978. *The dynamics of longitudinal dunes*. US Army Eur. Res. Off. Grant no. DA-ERO 76-G-072. Dept Geog., Ben Gurion Univ. Negev.

Tuddenham, W. M. and J. D. Stephens 1971. Infra-red spectroscopy. In *Modern methods of geochemical analysis*, R. E. Wainerdi and E. A. Uken (eds), 127–68. New York: Monographs in Geoscience.

Tuncer, E. R., R. A. Lohnes and T. Demirel 1977. Desiccation of soils derived from volcanic ash. *Transp. Res. Record* **641**, 44–9.

Turnbull, W. J., M. Krinitsky and F. J. Weaver 1966. Bank erosion in soils of the Lower Mississippi Valley. *Proc.Am.Soc.Civ. Eng., J. Soil Mech. Div.* **92**, **SM1**, 121–36.

Turner, C. and R. G. West 1968. The subdivision and zonation of interglacial periods. *Eiszeitalter Gegenwart* **19**, 93–101.

Turner, G. M. and R. Thompson 1979. Behaviour of the Earth's magnetic field as recorded in the sediment of Loch Lomond. *Earth Planet. Sci. Lett.* **42**, 412–26.

Turner, G. M. and R. Thompson 1981. Lake sediment record of the geomagnetic secular variation in Britain during Holocene times. *Geophys.J.R.Astron.Soc.* **65**, 703–25.

Turner, H. 1977. A comparison of some methods of slope measurement from large scale air photos. *Photogrammetria* **32**, 209–37.

Turon, J.-L. 1984. Direct land/sea correlations in the last interglacial complex. *Nature* **309**, 673–6.

Tuross, N. and P. E. Hare 1979. Collagen in fossil bone. *Carnegie Inst.Washington Year Book* **77**.

Turton, D. J. and P. J. Wigington 1984. A stage-activated triggering device for automatic pumping samplers. *Water Resour. Bull.* **20**, 443–7.

Twiss, P. C., E. Suess and R. N. Smith 1969. Morphological classification of grass phytoliths. *Proc.Soil Sci.Soc.Am.* **33**, 109–15.

United States Subcommittee on Sedimentation 1961. The single-state sampler for suspended sediment. In *A study of methods used in measurement and analysis of sediment loads in streams*. Rep. no. 13, Hydraul. Lab., St Anthony Falls.

Unwin, D. J. 1973a. The distribution and orientation of corries in northern Snowdonia, Wales. *Trans.Inst.Br.Geog.* **58**, 85–97.

Unwin, D. J. 1973b. Trials on trends. *Area* **5**, 31–3.

Unwin, D. J. 1975. An introduction to trend surface analysis. *Concepts and techniques in modern geography*, vol. 5. Norwich: Geo Abstracts, University of East Anglia.

Unwin, D. J. and J. Lewin 1971. Some problems in the trend analysis of erosion surfaces. *Area* **3**, 13–14.

USGS 1970. Methods of collection of water samples for dissolved minerals and gases. *Techniques of water resource investigations,* Book 5, ch.A1, 71–3. Washington: USGS.

Usher, M. B. 1969. The relations between mean square and block size in the analysis of similar patterns. *J.Ecol.* **57**, 505–14.

Valentin, H. 1954. Der Landralast in Holderness, Ostengland von 1852 bis 1952. *Die Erde* **6**, 296–315.

Valentine, K. W. G. and J. B. Dalrymple 1975. The identification, lateral variation, and chronology of two buried palaeocatenas at Woodhall Spa and West Runton, England. *Quat. Res.* **5**, 551–90.

Valentine, K. W. G. and J. B. Dalrymple 1976. Quaternary buried palaeosols: a critical review. *Quat. Res.* **6**, 209–22.

Valette-Silver, J. N., L. Brown, M. J. Pavich, J. Klein and R. Middleton 1986. Detection of erosion events using ^{10}Be profiles: example of the impact of agriculture on soil erosion in the Chesapeake Bay area. *Earth Planet. Sci. Lett* 80, 82–90.

Van Asch, Th. W. J. and H. Van Steijn 1973. *Computerized morphometric drainage basins from maps.* Publ.Geog. Inst. Tijksuniv. Utrecht, Ser B no. 51.

Van Burkalow, A. 1945. Angle of repose and angle of sliding friction: an experimental study. *Bull.Geol.Soc.Am.* **56**, 669–708.

Van Der Ancker, J. A. M., P. D. Jungerius and L. R. Mur 1985. the role of algae in the stabilization of coastal dune blowouts. *Earth Surf. Processes Landforms* **10**, 189–92.

Van Ghelue, P. and M. van Molle 1984. The large-scale mapping of soil erosion due to overland flow and mass movement. *Z. Geomorphol. Supp. NF* **49**, 183–94.

Van der Pers, J. N. C. 1978. Bioerosion by *Polydora* (Polychaeta, Sedentaria, Vermes) off Heligoland, Germany. *Geol.Mijnbouw* **57**, 465–78.

Van Steijn, H. 1984. Laboratory experiments of gelifluction using gréze-litée material. *Z.Geomorphol. Supp. NF* **49**, 195–201.

van Veen, J. 1935. Sand waves in the North Sea. *Hydrog.Rev.* **12**, 21–8.

van Veen, J. 1950. Eb-en vloedschaar systemen in de Nederlandse getigwarten. *Tijd-van Let. Kon.Ned.Aardr.Geol.* **67**, 303.

Van Vliet-Lanoe, B., J.-P. Coutard and A. Pissart 1984. Structures caused by repeated freezing and thawing in various loamy sediments; a comparison of active, fossil and experimental data. *Earth Surf. Processes Landforms* **9**, 553–65.

van Zon, H. J. M. 1978. *Litter transport as a geomorphic process.* Publ.Fys.Geog. Bodemk.Lab. no. 24.

van Zuidam, R. A. 1982. Considerations on systematic medium-scale geomorphological mapping. *Z.Geomorphol.NF* **26**. 473–80.

van Zuidam, R. A. 1985/6. *Aerial photo interpretation in terrain analysis and geomorphologic mapping.* The Hague: Smits.

Vaughan, P. R. and H. J. Walbancke 1973. Pore-pressure changes and the delayed failure of cutting slopes in overconsolidated clay. *Geotechnique* **23**, 531–41.

Veeh, H. H. and W. C. Burnett 1982. Carbonate and phosphate sediments. In *Uranium series disequilibrium: applications to environmental problems*, M. Ivanovich and R. S. Harmon (eds), 459–80. Oxford: Oxford University Press.

Velleman, P. F. and D. C. Hoaglin 1981. *Applications, basics and computing of exploratory data analysis.* Boston: Duxbury.

Veneman, P. L. M., P. V. Jacke and S. M. Bodine 1984. Soil formation as affected by pit and mound microrelief, in Massachusetts, USA. *Geoderma* **33**, 89–99.

Verague J. 1976. Nouveaux protocoles expérimentaux d'altération des roches. *Bull. Centre Géomorphol. Caen* **21**, 41–4.

Verein Deutscher Ingenieure 1971. *Bestimmung* dcs Partikelförmingen Niederschlags mit dem Hibernia- und Löbner-Liesgang-Geïat (Standardverfaren). VD1 2119, Blatt 3, VD1-Verlag GmbH, Düsseldorf, FRG, 4pp.

Verlaque, C. 1958. Les dunes d'In Salah. *Inst. Réch. Sahariennes Trav.* **17**, 12–58.

Versfeld, D. B. 1980. As assessment of windfall damage to *Pinus radiata* in the Bosboukloff Experimental Catchment. *S.Afr.For.J.* **112**, 15–19.

Verstappen, H.Th. 1977. *Remote sensing in geomorphology.* Amsterdam: Elsevier.

Verstappen, H.Th. 1983. *Applied geomorphology: geomorphological surveys for environmental development.* Amsterdam: Elsevier.

Veyret, P. 1971. Processus de l'érosion et de l'accumulation glaciaires en action. *Rev. Géog. Alpine* **59**, 155–70.

Vickers, B. 1978. *Laboratory work in civil engineering, soil mechanics.* St Albans: Crosby Lockwood Staples.

Vilborg, L. 1984. The cirque forms of central Sweden. *Geog. Annaler* **66A**, 41–77.

Viles, H. A. 1986. Coastal landforms. *Prog. Phys. Geog.* **10** (3), 429–36.

Viles, H. A. 1988. *Biogeomorphology.* Oxford: Basil Blackwell.

Viles, H. A. and S. T. Trudgill 1984. Long term remeasurements of micro-erosion meter rates, Aldabra Atoll, Indian Ocean. *Earth Surf. Processes Landforms* **9**, 89–94.

Vilks, G., J. M. Hall and D. J. W. Piper 1977. The natural remanent magnetism of sediment cores from the Beaufort sea. *J. Earth Sci.* **14**, 2007–12.
Vincent, P. and D. J. Poole 1978. Terrain roughness: *quo vadis? Area* **10**, 182–4.
Vincent, P. and L. Rose 1986. The Jakucs CO_2 microanalyser – problems and remedies *Earth Surf. Processes Landforms* **11**, 677–80.
Vincent, P. J. 1983. The morphology and morphometry of some arctic Trittkarren. *Z. Geomorphol. NF* **27**, 39–54.
Vink, A. P. A. and J. Sevink 1971. Soils and paleosols in the Lutterzand. *Med.Rijks.Geol.Dienst.NS* **22**, 165–85.
Vinken, R. 1986. Digital geoscientific maps: a priority program of the German Society for the Advancement of Scientific Research. *Math.Geol.* **18**, 237–46.
Visocky, A. P. 1970. Estimating the ground-water contribution to storm runoff by the electrical conductance record. *Groundwater* **8**, 5–10.
Vita-Finzi, C. 1986. *Recent earth movements – an introduction to neotectonics*. London: Academic Press.
Vivian, R. 1971. Glacier d'Argentière (Mont Blanc). *Ice* **35**, 6–7.
Vivian, R. 1975. *Les glaciers des Alpes Occidentales*. Grenoble: Imprimerie Allier.
Vivian, R. and G. Bocquet 1973. Subglacial cavitation phenomena under the Glacier d'Argentière, Mont Blanc, France. *J.Glaciol.* **12**, 439–51.
Voigt, G.K. 1960. Alteration of the composition of rainwater by trees. *Am.Midl.Nat.* **63**, 321–6.
Von Montfrans, H. M. 1971. *Palaeomagnetic dating in the North Sea Basin*. Thesis, Amsterdam.
Vukmirovic, V. 1973. Conventional methods of studying bed-load transport in rivers. In *Tracer techniques in sediment transport*, 19–30. Vienna: International Atomic Energy Agency.
Vyskočil, P., A. M. Wassef and R. Green (eds) 1983. Recent crustal movements, 1982. *Tectonophysics* **97**, 1–351.

Waddell, E. 1973. *Dynamics of swash and implications to beach response*. Tech.Rep.Coastal Stud.Inst.Louisiana State Univ. no. 139.
Wadell, H. 1932. Volume, shape and roundness of rock particles. *J.Geol.* **40**, 433–51.
Wadell, H. 1933. Sphericity and roundness of rock particles. *J.Geol.* **41**, 310–31.
Wadell, H. 1935. Volume, shape and roundness of quartz particles. *J.Geol.* **43**, 250–80.
Wagner, F. C. and A. Schatz 1967. Geo-microbiological investigations. VII: The behaviour of bacteria on the surface of rocks and minerals, and their role in weathering. *Z. Allg. Mikrobiol.* **7**, 35–52.
Wainerdi, R. E. and E. A. Uken 1971. *Modern methods of geochemical analysis*. New York: Plenum.
Walker, B. M. 1978. Chalk pore geometry using resin pore casts. In *Scanning electron microscopy in the study of sediments*, W. B. Whalley (ed.), 17–28. Norwich: Geo Abstracts.
Walker, D. 1970. Direction and rate in some British Post-glacial hydroseres. In *Studies in the vegetational history of the British Isles*, D. Walker and R. G. West (eds). Cambridge: Cambridge University Press.
Walker, M. J. C. 1975. Late glacial and Early Post-glacial environmental history of the central Grampian Highlands, Scotland. *J.Biogeog.* **2**, 265–84.
Walker, M. J. C. and J. J. Lowe 1976. The Abbey Piston Corer. *Quat. Newslett.* **19**, 1–3.
Walker, T. R. 1967. Formation of red beds in modern and ancient deserts. *Bull.Geol.Soc.Am.* **78**, 353–68.
Walling, D. E. 1974. Suspended sediment and solute yields from a small catchment prior to urbanization. In *Fluvial processes in instrumented watersheds*, K. J. Gregory and D. E. Walling (eds), 169–92. *Inst.Br.Geog.Spec.Publ.* no.6.
Walling, D. E. 1975. Solute variation in small catchment streams: some comments. *Inst.Br.Geog. Trans.* **64**, 141–7.
Walling, D. E. 1977a. Assessing the accuracy of suspended sediment rating curves for a small basin. *Water Resour. Res.* **13**, 531–8.
Walling, D. E. 1977b. Limitations of the rating curve technique for estimating suspended sediment loads, with particular reference to British rivers. *Int. Assoc. Sci. Hydrol. Publ.* **122**, 34–48.
Walling, D. E. 1978. Reliability considerations in the evaluation and analysis of river loads. *Z. Geomorphol. NF* **29**, 29–42.
Walling, D. E. 1984. Dissolved loads and their measurement. In *Erosion and sediment yield. Some methods of measurement and modelling*, R. F. Hadley and D. E. Walling (eds), 111–17. Norwich: Geo Books.
Walling, D. E. and K. J. Gregory 1970. The measurement of the effects of building construction on drainage basin dynamics. *J.Hydrol.* **11**, 129–44.

Walling, D. E. and A. Teed 1971. A simple pumping sampler for research into suspended sediment transport in small catchments. *J.Hydrol.* **13**, 325–37.
Walling, D. E. and B. W. Webb 1975. Spatial variation of river water quality: a survey of the River Exe. *Trans.Inst.Br.Geog.* **65**, 155–71.
Walling, D. E. and B. W. Webb 1978. Mapping solute loadings in an area of Devon, England. *Earth Surf. Processes* **3**, 85–99.
Walling, D. E. and B. W. Webb 1980. The spatial dimension in the interpretation of stream solute behaviour. *J. Hydrol.* **47**, 129–49.
Walling, D. E. and B. Webb 1982. The design of sampling programmes for studying catchment nutrient dynamics. In *Proc. Int. Symp. on hydrological research basins and their use in water resources planning*, vol.3, 747–58. Berne: Landeshydrologie.
Walling, D. E. and B. W. Webb 1985. Estimating the discharge of contaminants to coastal waters by rivers: some cautionary comments. *Mar. Pollut. Bull.* **16** (12), 488–92.
Wallington, C. E. 1968. A method of reducing observing and procedure bias in wind-direction frequencies. *Meteorol.Mag.* **97**, 293–302.
Walsh, F. 1972. Application of stream order numbers to the Merrimack river basin. *Water Resour. Res.* **8**, 141–4.
Walton, T. L. and R. G. Dean 1982. *Computer algorithm to calculate longshore energy flux and wave direction from a two pressure sensor array.* US Army Coastal Eng.Res.Center, Tech. Paper no. 82–2.
Ward, R. G. W., B. A. Haggart and M. C. Bridge 1987. Dendrochronological studies of bog pine from the Rannoch Moor area, western Scotland. In *Applications of tree-ring studies*, R. G. W. Ward (ed.). 215–23. BAR Int. Ser. no. 333.
Warne, S.St J. 1977. Carbonate mineral detection by variable atmosphere differential thermal analysis. *Nature* **269**, 678.
Warren, A. 1972. Observations on dunes and bimodal sands in the Ténéré Desert. *Sedimentology* **19**, 37–44.
Warren, A. 1976. Morphology and sediments of the Nebraska Sand Hills in relation to Pleistocene winds and the development of eolian bedforms. *J.Geol.* **84**, 685–700.
Washburn, A. L. 1979. *Geocryology: a survey of periglacial processes and environments.* London: Edward Arnold.
Watanabe, F. S. and S. R. Olsen 1965. Test of an ascorbic acid method for determining phosphorus in water and $NaHCO_3$ extracts from soil. *Prec.Soil Sci.Soc.Am.* **29**, 677–8.
Water Data Unit 1976a. *Punched tape river level recorders.* Tech.Mem.no.6, Dept Environ., Reading.
Water Data unit 1976b. *Autographic water level recorders.* Tech.Mem.no.13, Dept Environ., Reading.
Water Data Unit 1977. *Interrogable devices.* Tech.Mem.no.16, Dept Environ., Reading.
Water Research Association 1970. *River flow measurement by dilution gauging.* Water Res. Assoc.Tech.Paper no. TP 74.
Waters, R. S. 1958. Morphological mapping. *Geography* **43**, 10–17.
Waters, R. S. 1976. Stamp of ice on the North. *Geog.Mag.* **48**, 324–48.
Watkins, N. D. 1971. Polarity events and the problems of 'the reinforcement syndrome'. *Comments Earth Sci.: Geophys.* **2**, 36–42.
Watson, C. L. and J. Letey 1970. Indices for characterizing soil water repellency based upon contact angle–surface tension relationships. *Proc.Soil Sci.Soc.Am.* **34**, 841–4.
Watson, J. P. 1976. The composition of mounds of the termite. *Macrotermes falciger* (Gerstacker) on soil derived from granite in three rainfall zones of Rhodesia. *J.Soil Sci.* **27**, 495–503.
Watson, R. A. 1972. Limitations on substituting chemical reactions in model experiments. *Z. Geomorphol. NF* **16**, 103–8.
Watts, D. 1971. *Principles of biogeography.* London: McGraw-Hill.
Watts, S. H. 1977. Major element geochemistry of silcrete from a portion of inland Australia. *Geochim. Cosmochim. Acta* **41**, 1164–7.
Waugh, B. 1978. Diagenesis in continental red beds as revealed by scanning electron microscopy: a review. In *Scanning electron microscopy in the study of sediments*, W. B. Whalley (ed.), 329–46. Norwich: Geo Abstracts.
Waylen, M. J. 1979. Chemical weathering in a drainage basin underlain by Old Red Sandstone. *Earth Surf. Processes* **4**, 167–78.
Webb, B. W. and D. E. Walling 1974. Local variation in background water quality. *Sci. Total Environ.* **3**, 141–53.
Webb, T. 1985. Holocene palynology and climate. In *Palaeoclimate analysis and modelling*, A. D. Hecht (ed.), 163–95. Chichester: Wiley.

Webb, T. and R. A. Bryson 1972. Late- and postglacial climatic change in the northern Midwest, USA: quantitative estimates derived from fossil pollen spectra by multivariate statistical analysis. *Quat.Res.* **2**, 70–115.

Webb, T. and D. R. Clark 1977. Calibrating micropalaeontological data in climatic terms: a critical review. *Ann. NY Acad. Sci.* **288**, 93–118.

Webber, P. J. and J. T. Andrews 1973. Lichenometry: a commentary. *Arctic Alpine Res.* **5**, 295–302.

Webster, R. 1966. The measurement of soil water tension in the field. *New Phytol.* **65**, 249–58.

Weeks, E. P. 1978. *Field determination of vertical permeability to air in the unsaturated zone*. US Geol. Surv. Prof. Paper no. 1051.

Weertman, J. 1964. The theory of glacier sliding. *J.Glaciol.* **5**, 287–303.

Weertman, J. 1979. The unsolved general glacier sliding problem. *J.Glaciol.* **23**, 97–115.

Wehmiller, J. F. 1982. A review of amino acid racemization studies on Quaternary molluscs: stratigraphic and chronological applications in coastal and interglacial sites. Pacific and Atlantic coasts, United States, United Kingdom, Baffin Island and tropical islands. *Quat. Sci. Rev.* **1**, 83–120.

Wehmiller, J. F. 1984. Relative and absolute dating of Quaternary mollusks with amino acid racemization: evaluation, applications and questions. In *Quaternary dating methods*, D. J. Easterbrook (ed.), 171–94. Amsterdam: Elsevier.

Wehmiller, J. F. and P. E. Hare 1971. Racemization of amino acids in marine sediments. *Science* **173**, 907–11.

Wehmiller, J. F., D. F. Belknap, B. S. Boutin, J. E. Mirecki, S. D. Rahamin and L. L. York 1988. A review of the aminostratigraphy of quaternary mollusks from United States Atlantic Coastal Plain sites. In *Dating quaternary sediments*, D. J. Easterbrook (ed.), 69–110. Geol. Soc. Am. Spec. Paper no. 227.

Wehmiller, J. F., K. R. Lajoie, K. A. Kuenvolden, E. Peterson, D. F. Belknap, G. L. Kennedy, W. O. Addicott, J. G. Vedder and R. W. Wright 1977. *Correlation and chronology of Pacific coast marine terrace deposits of continental United States by fossil amino acid stereochemistry–techniques, evaluation, relative ages, kinetic model ages and geologic implications*. USGS Open File Rep. no. 77–680.

Wells, C. B. 1965. The formation of calcium carbonate concentrations. In *Experimental pedology*, E. G. Hallsworth and D. V. Crawford (eds), 165–75. London: Butterworth.

Wentworth, C. K. 1919. A laboratory and field study of cobble abrasion. *J.Geol.* **27**, 507–21.

Wentworth, C. K. 1922a. *The shape of beach pebbles*. US Geol. Surv. Prof. Paper no. 131C.

Wentworth, C. K. 1922b. The shape of pebbles. *Bull.US Geol. Surv.* **730C**, 91–114.

Werner, C. and J. S. Smart 1973. Some new methods of topologic classification of channel networks. *Geog.Anal.* **5**, 271–95.

Werritty, A. 1972. Accuracy of stream link lengths derived from maps. *Water Resour. Res.* **8**, 1255–64.

West, E. A. 1978. *The equilibrium of natural streams*. Norwich: Geo Abstracts.

West, P. W. 1968a. Chemical analysis of inorganic particulate pollutants. In *Air pollution*, A. C. Stern (ed.), 147–85. New York: Academic Press.

West, R. G. 1968b. *Pleistocene geology and biology*. London: Longman.

West, R. G. 1972. Relative land/sea-level changes in southeastern England during the Pleistocene. *Phil.Trans.R.Soc.Lond.A* **272**, 87–98.

West, R. G. 1977. *Pleistocene geology and biology*, 2nd edn. London: Longman.

West, R. G. 1980. *The pre-glacial Pleistocene of the Norfolk and Suffolk coasts*. Cambridge: Cambridge University Press.

West, R. G. and P. E. P. Norton 1974. The Icenian Crag of south-east Suffolk. *Phil.Trans.R. Soc.Lond.B*. **269**, 1–28.

West, R. G., B. M. Funnell and P. E. P. Norton 1980. An early Pleistocene cold marine episode in the North Sea: pollen and faunal assemblages at Covehithe, Suffolk, England. *Boreas* **9**, 1–10.

Westgate, J. A. and C. M. Gold 1974. *World bibliography and index of tephrochronology*. Edmonton: University of Alberta.

Whalley, W. B. 1972. The description and measurement of sedimentary particles and the concept of form. *J.Sedim.Petrol.* **42**, 961–5.

Whalley, W. B. 1976. *Properties of materials and geomorphological explanation*. Oxford: Oxford University Press.

Whalley, W. B. (ed.) 1978a. *Scanning electron microscopy in the study of sediments*. Norwich: Geo Abstracts.

Whalley, W. B. 1978b. An SEM examination of quartz grains from sub-glacial and associated environments and some methods for their characterization. In *Scanning electron microscopy 1978*, O. Johari (ed.), vol.1, 353–60. Chicago: SEM.

Whalley, W. B. 1979. Quartz silt production and sand grain surface textures for fluvial and glacial environments. *Scanning Electron Microsc.* **1**, 547–54.
Whalley, W. B. 1985. Scanning electron microscopy and the sedimentological characterisation of rocks. In *Geomorphology and soils*, K. S. Richards, R. R. Arnett and S. Ellis (eds), 183–201. London: Allen & Unwin.
Whalley, W. B. and D. H. Krinsley 1974. A scanning electron microscope study of surface textures of quartz grains from glacial environments. *Sedimentology* **21**, 87–105.
Whalley, W. B. and J. D. Orford 1986. Practical methods for analysing and quantifying two-dimensional images. In *The scientific study of flint and chert*, G. de C. Sieveking and M. B. Hart (eds), 235–42. Cambridge: Cambridge University Press.
Wheeler, D. A. 1979. The overall shape of longitudinal profiles of streams. In *Geographical approaches to fluvial processes*, A. F. Pitty (ed.), 241–60. Norwich: Geo Books.
Wheeler, D. A. 1984. Using parabolas to describe the cross-sections of glaciated valleys. *Earth Surf. Processes Landforms* **9**, 391–4.
Whitaker, J. H. McD. 1978. Diagenesis of the Brent sand formation: a scanning electron microscope study. In *Scanning electron microscopy in the study of sediments*, W. B. Whalley (ed.), 363–80. Norwich: Geo Abstracts.
White, E. J. and F. Turner 1970. A method of estimating income of nutrients in a catch of airborne particles by a woodland canopy. *J.Appl.Ecol.* **7**, 441–61.
White, E. J., R. S. Starkey and M. J. Saunders 1971. An assessment of the relative importance of several chemical sources to the waters of a small upland catchment. *J.Appl.Ecol.* **8**, 743–9.
Whitehead, H. C. and J. H. Feth 1964. Chemical composition of rain, dry fall out and bulk precipitation, California, 1957–59. *J.Geophys.Res.* **69**, 3319–33.
Whitney, M. I. 1978. The role of vorticity in developing lineation by wind erosion. *Bull.Geol. Soc.Am.* **89**, 1–18.
Whitney, M. I. and H. B. Brewer 1967. Discoveries in aerodynamic erosion with wind tunnel experiments. *Michigan Acad.Sci., Arts Lett. Paper* **53**, 91–104.
Whitney, M. I. and R. V. Dietrich 1973. Ventifact sculpture by wind blown dust. *Bull.Geol.Soc.Am.* **84**, 2561–2.
Whittaker, R. H., F. H. Bormann, G. E. Likens and T. G. Siccama 1974. The Hubbard Brook ecosystem study: forest biomass and production. *Ecol. Monogr.* **44**, 233–54.
Whittig, L. D. 1965. X-ray diffraction techniques for mineral identification and mineralogical composition. Part I. In *Methods of soil analysis*, C. A. Black (ed.), 671–98. Madison, Wisc.: American Society of Agronomy.
Whittington, G., K. S. O. Bevon and A. S. Mabin 1972. *Compactness of shape: review, theory and application*. Dept Geog. Environ. Stud. Univ. Witwatersrand, Johannesburg, Occas. Paper no. 7.
Wignall, B. L. 1977. A critical review of the Quantimet 720 Image Analyser in remote sensing. In *Remote sensing of the terrestrial environment*, R. F. Peel, L. F. Curtis and E. C. Barrett (eds), 71–9. London: Butterworth.
Wiklander, L. and A. Andersson 1972. The replacing efficiency of hydrogen ion in relation to base saturation and pH. *Geoderma* **7**, 159–65.
Wilcock, A. A. 1973. Drainage basin shape. *Area* **5**, 231–4.
Wilcock, D. N. 1967. Coarse bedload as a factor determining bed slope. *Int.Assoc.Sci.Hydrol.Publ.* **75**, 143–50.
Wilks, S. S. 1961. Some aspects of quantification in science. In *Quantification – a history of the meaning of measurement in the natural and social sciences*, H. Woolf (ed.), 5–12. Indianapolis: Bobbs-Merrill.
Will, G. M. 1959. Nutrient return in litter and rainfall under some exotic conifer stands in New Zealand. *NZ J.Agric.Res.* **2**, 719–34.
Willetts, B. B. and M. A. Rice 1983. Practical representation of characteristic grain shape of sands: a comparison of methods. *Sedimentology* **30**, 557–65.
Williams, C. and D. H. Yaalon 1977. An experimental investigation of reddening in dune sand. *Geoderma* **17**, 181–91.
Williams, D. F. W. 1981. Integrated land survey methods for prediction of gully erosion. In *Terrain analysis and remote sensing*, J. R. G. Townshend (ed.), ch. 7. London: Allen & Unwin, 169–83.
Williams, G. P. 1970. *Flume width and water depth effects in sediment-transport experiments*. US Geol.Surv.Prof. Paper no. 562-H.
Williams, G. P. 1978. Bankfull discharge of rivers. *Water Resour. Res.* **14**, 1141–54.
Williams, G. P. 1984a. Palaeohydrological methods of some examples from Swedish fluvial environments. II – River meanders. *Geog. Annaler* **66A**, 89–102.
Williams, G. P. 1984b. Paleohydrologic equations for rivers. In *Developments and applications of geomorphology*, J. E. Costa and P. J. Fleisher (eds), 343–67. Berlin: Springer.

Williams, J. 1969. *CERC wave gages*. Tech. Memo. no. 30, US Army Corps Eng., Coastal Eng. Res. Center, Washington DC.
Williams,, M. A. J. 1968. Termites and soil development near Brocks Creek, Northern Territory. *Aus.J.Sci.* **31**, 153–4.
Williams, P. J. 1986. *Pipelines and permafrost*, 2nd edn. Ottawa: Carleton University Press.
Williams, P. W. 1964. *Aspects of the limestone physiography of parts of Counties Clare and Galway, west Ireland*. Unpubl. Ph.D. Thesis, University of Cambridge.
Williams, P. W. 1971. Illustrating morphometric analysis of karst with examples from New Guinea. *Z. Geomorphol. NF* **15**, 40–61.
Williams, P. W. 1972a. The analysis of spatial characteristics of karst terrains. In *Spatial analysis in geomorphology*, R. J. Chorley (ed.), 135–63. London: Methuen.
Williams, P. W. 1972b. Morphometric analysis of polygonal karst in New Guinea. *Bull.Geol. Soc.Am.* **83**, 761–96.
Williams, P. W. 1982. Speleothem dates, quaternary terraces and uplift rates in New Zealand. *Nature* **298**, 257–9.
Williams, W. W. 1947. The determination of the gradient of enemy held beaches. *Geog. J.* **109**, 76–93.
Wills, T. G. and H. Beaumont 1972. *Wave direction measurement using sea surveillance radars*. R.Aircraft Establ., Tech Man., no. IR 118.
Wilson, A. 1970a. *pH meters*. London: Kogan Page.
Wilson, A. T. and M. J. Grinstead 1977. The D/H ratios of cellulose as a biochemcial thermometer: a comment on 'Climatic implications of the D/H ratio of hydrogen in C–H groups in tree cellulose' by S. Epstein and C. Yapp. *Earth Planet. Sci. Lett.* **36**, 246–8.
Wilson, I. G. 1970b. *The external morphology of wind-laid sand deposits*. Unpubl. Ph.D. Thesis, University of Reading.
Wilson, M. E. and S. H. Wood 1980. Tectonic tilt rates derived from lake-level measurements, Salton Sea, California. *Science* **207**, 183–6.
Wilson, T. 1985. Scanning optical microscopy. *Scanning* 779–87.
Wilson, T., E. M. McCabe and D. K. Hamilton 1986. A new method for displaying low-contrast optical-beam-induced contrast images in the scanning optical microscope. *J.App. Phys.* **59** (3), 702–3.
Wiman, S. 1963. A preliminary study of experimental frost weathering. *Geog. Annaler* **45**, 113–21.
Winkelmolen, A. M. 1971. Rollability, a functional shape property of sand grains. *J.Sedim.Petrol.* **41**, 703–14.
Winkler, E. M. 1975. *Stone: properties, durability in man's environment*, 2nd edn. Vienna: Springer.
Wintle, A. G. 1981. Thermoluminescence dating of late Devensian loesses in southern England. *Nature* **289**, 479–81.
Wintle, A. G. and D. J. Huntley 1982. Thermoluminescence dating of sediments. *Quaternary Science Reviews* **1**, 31–53.
Wise, S. 1977. *The use of fallout radionuclides Pb-210 and Cs-137 in estimating denudation rates and in soil erosion measurement*. Dept Geog., King's College, London, Occas. Papers no. 7.
Wise, S. 1980. Cs-137 and Pb-210: review of the techniques and some applications in geomorphology. In *Timescales in geomorphology*, J. Lewin, R. Cullingford and D. Davidson (eds), 109–27. New York: Wiley.
WMO 1974. International codes WMO-306. *Manual on codes*, vol. 1.
Wolcott, R. T. 1978. Sieving precision: sonic sifter versus Ro-Tap. *J.Sedim.Petrol.* **48**, 661–4.
Wold, B. and G. Østrem 1979. Subglacial constructions and investigations at Bondhusbreen, Norway. *J.Glaciol.* **23**, 363–79.
Woldenberg, M. J. 1967. Geography and properties of surfaces. *Harvard Papers Theor. Geog.* **1**, 95–189.
Wolman, M. G. 1955. *The natural channel of Brandywine Creek, Pennsylvania*. US Geol. Surv. Prof. Paper no. 271.
Wolman, M. G. and L. B. Leopold 1957. *River flood plains: some observations on their formation*. US Geol. Surv. Prof. Paper no. 282C, 87–107.
Wong, S. T. 1963. A multivariate statistical model for predicting mean annual flood in New England. *Ann. Assoc. Am. Geog.* **53**, 293–311.
Wong, S. T. 1979. A dimensionally homogeneous and statistically optimal model for predicting mean annual flood. *J.Hydrol.* **42**, 269–79.
Wood, C. A. 1980a. Morphometric evolution of cinder cones. *J. Volcanol. Geothermal Res.* **7**, 387–413.
Wood, C. A. 1980b. Morphometric analysis of cinder cone degradation. *J. Volcanol. Geothermal Res.* **8**, 137–60.

Wood, W. F. 1954. The dot planimeter: a new way to measure map area. *Prof. Geog.* **6**, 12–14.
Wood, W. F. and J. B. Snell 1957. The dispersion of geomorphic data around measures of central tendency (Abstract). *Ann. Assoc. Am. Geog.* **47**, 184–5.
Wood, W. F. and J. B. Snell 1960. *A quantitative system for classifying landforms.* US Dept Army, Natick, Mass., Tech. Rep. no. EP-124.
Wood, W. L. 1968. *A ducted impeller flowmeter for shallow-water measurement of internal velocities in breaking waves.* Tech. Rep. no. 1, Off.Naval Res., Contract No. N000 14–68 A0109 0002, NR 388–089.
Woodhead, P. M. J. and A. J. Lee 1960. *A new instrument for measuring residual currents near the seabed.* Int.Counc. Explor. Sea, Hydrog. Committee, Rep. no. 12.
Woodruff, J. F. and L. J. Evenden 1962. Geomorphic measurement from aerial photos. *Prof. Geog.* **14**, 23–6.
Woodruff, N. and F. H. Siddoway 1965. A wind erosion equation. *Proc.Soil Sci. Soc.Am.* **29**, 602–8.
Woodyer, K. D. 1968. Bankfull frequency in rivers. *J.Hydrol.* **6**, 114–42.
Worsley, P. 1980. Problems in radiocarcon dating the Chelford Interstadial of England. In *Timescales in geomorphology*, R. A. Cullingford, D. A. Davidson and J. Lewin (eds), 289–304. Chichester: Wiley.
Worth, C. P. and D. M. Wood 1978. The correlation of index properties with some basic engineering properties of soils. *Can.Geotech.J.* **15**, 137–45.
Wright, H. E. 1967. A square-rod piston sampler for lake sediments. *J.Sedim.Petrol.* **37**, 975–6.
Wright, H. E. 1980. Cores of soft lake sediments. *Boreas* **9**, 107–14.
Wright, H. E., D. A. Livingstone and E. J. Cushing 1965. Coring devices for lake sediments. In *Handbook of palaeontological techniques*, B. Kummel and D. M. Raup (eds), 494–520. San Franscico: Freeman.
Wright, H. E., D. H. Mann and P. H. Glaser 1984. Piston corers for peat and lake sediments. *Ecology* **65**, 657–9.
Wright, L. D. and A. D. Short 1984. Morphodynamic variability of surf zone and beaches: a synthesis. *Mar. Geol.* **56** (1–4), 93–118.
Wright, L. D., P. Nielson, A. D. Short and M. O. Green 1982. Morphodynamics of a macrotidal beach. *Mar. Geol.* **50** (1–2), 97–128.
Wright, L. D., A. D. Short, J. D. Boon, B. Hayden, S. Kimball and J. H. List 1987. The morphodynamic effects of incident wave groupiness and tide range on an energetic beach. *Mar. Geol.* **74** (1–2), 1–20.
Wright, P. 1976. A cine-camera technique for process measurement on a ridge and runnel beach. *Sedimentology* **23**, 705–12.

Yaalon, D. H. and J. Laronne 1971. Internal structures in eolianites and paleowinds, Mediterranean Coast, Israel. *J.Sedim.Petrol.* **41**, 1059–64.
Yaalon, D. H. and S. Singer 1974. Vertical variation in strength and porosity of calcrete (Nari) on chalk, Shefela, Israel and interpretation of its origin. *J.Sedim.Petrol.* **44**, 1016–23.
Yair, A. and J. Rutin 1981. Some aspects of regional variation in the amount of available sediment produced by isopods and porcupines, northern Negev, Israel. *Earth Surf. Processes Landforms* **6**, 221–34.
Yair, A., D. Sharon and H. Lavee 1978. An instrumented watershed for the study of partial area contribution of runoff in the arid zone. *Z. Geomorphol. NF* **29**, 71–82.
Yalin, M. S. 1977. *Mechanics of sediment transport*, 2nd edn. Oxford: Pergamon.
Yang, C. T. and J. B. Stall 1971. Note on the map scale effect in the study of stream morphology. *Water Resour. Res.* **7**, 709–12.
Yasso, W. E. 1964. *Geometry and development of spit-bar shorelines at Horseshoe Cove, Sandy Hook, New Jersey.* Tech.Rep. no. 5, Off. Naval Res., Geog.Branch, NR 388–057.
Yasso, W. E. 1965. Plan geometry of headland–bay beaches. *J.Geol.* **73**, 702–14.
Yasso, W. E. 1966. Formulation and use of fluorescent tracer coatings in sediment transport studies. *Sedimentology* **6**, 287–301.
Yasso, W. E. and E. M. Hartmann 1972. Rapid field technique using spray adhesive to obtain peels of unconsolidated sediment. *Sedimentology* **19**, 295–8.
Yatsu, E. 1966. *Rock control in geomorphology.* Tokyo: Sozosha.
Yatsu, E. 1967. Some problems on mass-movements. *Geog. Annaler* **49A**, 396–401.
Yatsu, E. 1988. *The nature of weathering: an introduction.* Tokyo: Sozosha.
Yoon, Y. N. 1975. Correlation of the stream morphological characteristics of the Han river basin with its mean daily and 7-day 10-year flows. In *The hydrological characteristic of river basins and the effects on these characteristics of better water managements*, 169–77. Publ.Int.Assoc. Hydrol. Sci., Tokyo, no. 117.

Index

Young, A. 1960. Soil movement by denudational processes on slopes. *Nature* **188**, 120–2.

Young, A. 1961. Characteristic and limiting slope angles. *Z. Geomorphol. Supp. NF* **5**, 126–31.

Young, A. 1971. Slope profile analysis: the system of best units. In *Slopes: form and process*, D. Brunsden (ed.), 1–13. Inst.Br.Geog.Spec.Publ. no. 3.

Young, A. 1972. *Slopes*. Edinburgh: Oliver & Boyd.

Young, A. 1978. A twelve year record of soil movement on a slope. *Z. Geomorphol. Supp. NF* **29**, 104–10.

Young, A., D. Brunsden and J. B. Thornes 1974. Slope profile survey. *BGRG Tech.Bull.* **11**. Norwich: Geo Abstracts, University of East Anglia.

Young, J. A. T. 1976. The terraces of Glen Feshie, Invernessshire. *Trans.R.Soc.Edinb.* **69**, 501–12.

Young, R. N. and B. P. Warkentin 1966. *Introduction to soil behaviour*. New York: Macmillan.

Yoxall, W. H. 1969. Discrepancies in stream mapping: the 1:250 000 series, Ghana survey. *J.Trop.Geog.* **28**, 84–6.

Zagwijn, W. H. 1975. Variations in climate as shown by pollen analysis especially in the Lower Pleistocene of Europe. In *Ice ages: ancient and modern*, A. E. Wright and F. Moseley (eds), 137–52. Liverpool: Seel House.

Zand, S. M., V. C. Kennedy, G. W. Zellweger and R. J. Avazino 1975. Solute transport and modelling of water quality in a small stream. *US Geol. Surv.J.Res.* **4**, 233–40.

Zeitler, P. K., N. H. Johnson, C. W. Naeser and R. A. K. Tahirkheli 1982. Fission track evidence for Quaternary uplift of the Nanga Parbat Region, Pakistan. *Nature* **298**, 255–7.

Zelenhasic, E. 1970. *Theoretical probability distributions for flood peaks*. Colorado Hydrol. Papers no. 42.

Zeller, E. J., P. W. Levy and P. L. Mattern 1967. Geological dating by electron spin resonance. In *Proc. symp. on radioactive dating and low level counting*, 531–40. Vienna: International Atomic Energy Agency.

Zeman, L. J. and E. O. Nyborg 1974. Collection of chemical fallout from the atmosphere. *Can.Agric.Eng.* **16**, 69–72.

Zeman, L. J. and O. Slaymaker 1978. Mass balance model for calculation of ionic input loads in atmospheric fallout and discharge from a mountainous basin. *Bull.Int.Assoc. Sci. Hydrol.* **23**, 103–17.

Zenkovitch, V. P. 1959. On the genesis of cuspate spits along lagoon shores. *J.Geol.* **67**, 269–77.

Zenkovitch, V. P. 1967. *Processes of coastal development*. Edinburgh: Oliver & Boyd.

Zernitz, E. R. 1932. Drainage patterns and their significance. *J.Geol.* **40**, 498–521.

Zevenbergen, L. W. and C. R. Thorne 1987. Quantitative analysis of land surface topography. *Earth Surf. Processes Landforms* **12**, 47–56.

Zimmerle, W. and L. C. Bonham 1962. Rapid methods for dimensional grain orientation measurements. *J.Sedim.Petrol.* **32**, 751–63.

Zingg, A. W. 1953. Wind-tunnel studies of the movement of sedimentary material. *State Univ. Iowa Stud. Eng.* **34**, 111–35. Proc.5th Hydraul. Conf.Bull.

Zingg, T. 1935. Beitzag zur Schotteranalyse. *Schweg.Mineralog. Petrog. Mitt. Bd* **15**, 39–140.

Zussman, J. 1977. *Physical methods in determinative mineralogy*. London: Academic Press.

Zvyagin, B. B. 1967. *Electron-diffraction analysis of clay mineral structures*, rev. edn, transl. S. Lyse. New York: Plenum.

Zwamborn, J. A., K. S. Russell and J. Nicholson 1972. Coastal engineering measurements. In *Proc.13th conf.coastal engineering*, 75–94. Am.Soc.Civ.Eng.